Particulate Emissions from Vehicles

Particulate Emissions from Vehicles

By

Peter Eastwood
Ford Motor Company, Ltd.

Published on behalf of:
SAE International
Warrendale, PA

John Wiley & Sons, Ltd

Copyright © 2008   John Wiley & Sons Ltd, The Atrium, Southern Gate, Chichester,
West Sussex PO19 8SQ, England

Telephone   (+44) 1243 779777

Email (for orders and customer service enquiries): cs-books@wiley.co.uk
Visit our Home Page on www.wiley.com

All Rights Reserved. No part of this publication may be reproduced, stored in a retrieval system or transmitted in any form or by any means, electronic, mechanical, photocopying, recording, scanning or otherwise, except under the terms of the Copyright, Designs and Patents Act 1988 or under the terms of a licence issued by the Copyright Licensing Agency Ltd, 90 Tottenham Court Road, London W1T 4LP, UK, without the permission in writing of the Publisher. Requests to the Publisher should be addressed to the Permissions Department, John Wiley & Sons Ltd, The Atrium, Southern Gate, Chichester, West Sussex PO19 8SQ, England, or emailed to permreq@wiley.co.uk, or faxed to (+44) 1243 770620.

This publication is designed to provide accurate and authoritative information in regard to the subject matter covered. It is sold on the understanding that the Publisher is not engaged in rendering professional services. If professional advice or other expert assistance is required, the services of a competent professional should be sought.

Other Wiley Editorial Offices

John Wiley & Sons Inc., 111 River Street, Hoboken, NJ 07030, USA

Jossey-Bass, 989 Market Street, San Francisco, CA 94103-1741, USA

Wiley-VCH Verlag GmbH, Boschstr. 12, D-69469 Weinheim, Germany

John Wiley & Sons Australia Ltd, 42 McDougall Street, Milton, Queensland 4064, Australia

John Wiley & Sons (Asia) Pte Ltd, 2 Clementi Loop #02-01, Jin Xing Distripark, Singapore 129809

John Wiley & Sons Canada Ltd, 6045 Freemont Blvd, Mississauga, ONT, L5R 4J3

Wiley also publishes its books in a variety of electronic formats. Some content that appears in print may not be available in electronic books.

ISBN 978-0-7680-2060-1

Order No. R-389

SAE International
400 Commonwealth Drive
Warrendale, PA 15096-0001 U.S.A.
Phone: (724) 776-4841
Fax: (724) 776-5760
E-mail: publications@sae.org
http://www.sae.org

British Library Cataloguing in Publication Data

A catalogue record for this book is available from the British Library

ISBN 978-0-470-72455-2

Typeset in 9/11pt Times by Integra Software Services Pvt. Ltd, Pondicherry, India
Printed and bound in Great Britain by Antony Rowe Ltd, Chippenham, Wiltshire

Contents

Preface	xi
Acronyms and Abbreviations	xiii
About the Author	xvii
1 Introduction	**1**
1.1 Air Traffic	4
1.2 Motor Vehicles	4
1.3 The Legislative Framework	6
2 Fundamentals	**9**
2.1 Introduction	9
2.2 Properties of Aerosol Particles	9
2.2.1 *Diameter and Shape*	10
2.2.2 *Size Distribution*	12
2.2.3 *Transport and Deposition*	14
2.2.4 *Transformation and Mutation*	18
2.3 Particles in the Atmosphere	22
2.3.1 *Character and Behaviour*	23
2.3.2 *Aerosols in Nature*	27
2.3.3 *Anthropogenic Aerosols*	30
2.3.4 *Environmental Implications*	37
2.4 Motor Vehicle Particulate	42
2.4.1 *Some Typical Particles Dissected*	42
2.4.2 *What Happens Within the Engine*	48
2.4.3 *What Happens Within the Exhaust*	51
2.4.4 *Number Versus Mass*	53
2.5 Closure	55
2.5.1 *Properties of Aerosol Particles*	55
2.5.2 *Particles in the Atmosphere*	57
2.5.3 *Motor Vehicle Particulate*	58

3 Formation I: Composition — 61
- 3.1 Introduction — 61
- 3.2 Carbonaceous Fraction: I. Classical Models — 62
  - 3.2.1 Empiricisms — 65
  - 3.2.2 Inception — 68
  - 3.2.3 Surface Growth — 73
  - 3.2.4 Agglomeration — 75
  - 3.2.5 Oxidation — 79
- 3.3 Carbonaceous Fraction: II. The Combusting Plume — 84
  - 3.3.1 Historical Overview — 85
  - 3.3.2 Premixed Burn — 86
  - 3.3.3 Mixing-controlled Burn — 88
  - 3.3.4 Late Burn — 90
- 3.4 Carbonaceous Fraction: III. Wall Interactions — 91
  - 3.4.1 Theoretical — 92
  - 3.4.2 Experimental — 93
- 3.5 Ash Fraction — 95
  - 3.5.1 Chemical Reactions — 96
  - 3.5.2 Gas-to-Particle Conversion — 100
- 3.6 Organic Fraction — 104
  - 3.6.1 Preparatory Chemical Reactions — 104
  - 3.6.2 Chemical Reactions in the Exhaust — 109
  - 3.6.3 Gas-to-Particle Conversion: Models — 112
  - 3.6.4 Gas-to-Particle Conversion: Measurements — 117
  - 3.6.5 White Smoke — 119
- 3.7 Sulphate Fraction — 122
  - 3.7.1 Chemical Reactions — 122
  - 3.7.2 Gas-to-Particle Conversion — 125
- 3.8 Closure — 129
  - 3.8.1 Carbonaceous Fraction I. Classical Models — 129
  - 3.8.2 Carbonaceous Fraction II. The Combusting Plume — 131
  - 3.8.3 Carbonaceous Fraction III. Wall Interactions — 132
  - 3.8.4 Ash Fraction — 132
  - 3.8.5 Organic Fraction — 133
  - 3.8.6 Sulphate Fraction — 134

4 Formation II: Location — 135
- 4.1 Introduction — 135
- 4.2 Within the Exhaust System — 136
  - 4.2.1 Storage and Release — 137
  - 4.2.2 Deposition Within Catalysts — 139
- 4.3 Within the Exhaust Plume — 141
  - 4.3.1 Long-term Ageing in the Atmosphere — 145
- 4.4 Within the Transfer Line — 145
- 4.5 Within the Dilution Tunnel — 147
- 4.6 On the Filter — 150
- 4.7 Closure — 152

		4.7.1	Within the Exhaust System	152
		4.7.2	Within the Exhaust Plume	153
		4.7.3	Within the Transfer Line	153
		4.7.4	Within the Dilution Tunnel	153
		4.7.5	On the Filter	154
		4.7.6	General Remarks	154

5 Measurement — 157

5.1 Introduction — 157
5.2 Particulate Measured Conventionally — 158
    5.2.1 Drawing a Sample of Exhaust Gas — 158
    5.2.2 Diluting the Exhaust — 160
    5.2.3 Collection onto a Filter — 165
    5.2.4 Fractionation by Gasification — 168
    5.2.5 Fractionation by Dissolution — 171
    5.2.6 Chemically Assaying the Organic Fraction — 175
    5.2.7 Biologically Assaying the Organic Fraction — 179
5.3 Particulate Measured Individually — 182
    5.3.1 Inertial Mobility — 183
    5.3.2 Electrical Mobility — 185
    5.3.3 Laser-induced Incandescence — 186
    5.3.4 Light Scattering — 189
5.4 Particulate Measured Collectively — 191
    5.4.1 Photoacousticity — 191
    5.4.2 Photoelectric and Diffusion Charging — 193
    5.4.3 Electrical Charge — 195
    5.4.4 Flame Ionisation — 197
    5.4.5 Mass — 198
    5.4.6 Smoke — 200
5.5 Closure — 204
    5.5.1 Particulate Measured Conventionally — 204
    5.5.2 Particulate Measured Individually — 206
    5.5.3 Particulate Measured Collectively — 207
    5.5.4 Further Remarks — 209

6 Characterisation — 211

6.1 Introduction — 211
6.2 Physical Characterisation — 212
    6.2.1 Microstructure — 212
    6.2.2 Morphology — 215
    6.2.3 Density — 219
    6.2.4 Surface Area — 222
    6.2.5 Electrical Charge — 224
6.3 Chemical Characterisation — 228
    6.3.1 Carbonaceous Fraction — 228
    6.3.2 Ash Fraction — 230
    6.3.3 Organic Fraction — 233
    6.3.4 Sulphate Fraction — 237

6.4	Biological Characterisation	238
6.5	Demographic Characterisation	240
6.6	Closure	244
	6.6.1 Physical Characterisation	244
	6.6.2 Chemical Characterisation	246
	6.6.3 Biological Characterisation	246
	6.6.4 Demographic Characterisation	247

7 Abatement — 249

7.1	Introduction	249
7.2	Fuel Formulation	250
	7.2.1 Sulphur	252
	7.2.2 Hydrocarbons	254
	7.2.3 Oxygenates	256
	7.2.4 Additives	261
	7.2.5 Volatility, Cetane Number and Density	262
7.3	Fuel Injection	265
	7.3.1 The Injector Nozzle	265
	7.3.2 Injection Pressure	266
	7.3.3 Injection Scheduling	269
7.4	Exhaust Gas Recirculation	271
7.5	Induction	275
	7.5.1 External to the Engine	275
	7.5.2 Internal to the Engine	277
7.6	Lubrication	279
	7.6.1 Oil in Particulate	279
	7.6.2 Particulate in Oil	285
7.7	Alternative Combustion Systems	291
7.8	Aftertreatment	293
	7.8.1 Catalytic Converters	294
	7.8.2 Particulate Filters	300
7.9	Closure	306
	7.9.1 Fuel Formulation	306
	7.9.2 Fuel Injection	308
	7.9.3 Exhaust Gas Recirculation (EGR)	309
	7.9.4 Induction	309
	7.9.5 Lubrication	310
	7.9.6 Alternative Combustion Systems	310
	7.9.7 Aftertreatment	311

8 Gasoline Engines — 313

8.1	Introduction	313
8.2	A Historical Perspective	314
	8.2.1 Organometallic Fuel Additives and Ash	314
	8.2.2 Oxidation Catalysts and Sulphates	317
8.3	Port-injection Engines	318
	8.3.1 Formation	319

	8.3.2	*Characterisation*	322
	8.3.3	*Abatement*	325
8.4	Direct-injection Engines		326
	8.4.1	*Formation*	327
	8.4.2	*Characterisation*	330
	8.4.3	*Abatement*	333
8.5	Two-stroke Engines		334
8.6	Closure		338
	8.6.1	*Port-injection Engines*	338
	8.6.2	*Direct-injection Engines*	339
	8.6.3	*Two-stroke Engines*	340

9 Disintegration — 341
- 9.1 Introduction — 341
- 9.2 Roads — 342
- 9.3 Brakes — 344
- 9.4 Tyres — 346
- 9.5 Exhausts — 348
- 9.6 Catalysts — 349
- 9.7 Closure — 352
  - 9.7.1 *Roads* — 352
  - 9.7.2 *Brakes* — 352
  - 9.7.3 *Tyres* — 353
  - 9.7.4 *Exhausts* — 353
  - 9.7.5 *Catalysts* — 353

10 Toxicology — 355
- 10.1 Introduction — 355
- 10.2 Public Exposure — 356
  - 10.2.1 *Nanoparticles* — 360
- 10.3 Public Health — 362
- 10.4 Pathogenesis — 364
  - 10.4.1 *Particle Deposition and Clearance* — 364
  - 10.4.2 *The Chemistry and Biochemistry of Particle-induced Reactions* — 367
  - 10.4.3 *Particle-induced Diseases* — 369
- 10.5 Epidemiology — 371
- 10.6 *In Vitro* — 375
- 10.7 *In Vivo* — 377
- 10.8 Humans — 381
- 10.9 Closure — 382
  - 10.9.1 *Public Exposure* — 382
  - 10.9.2 *Public Health* — 382
  - 10.9.3 *Pathogenesis* — 382
  - 10.9.4 *Epidemiology* — 383
  - 10.9.5 *In Vitro* — 383
  - 10.9.6 *In Vivo* — 384
  - 10.9.7 *Humans* — 384

		10.9.8	Which Particulate Fraction?	384
		10.9.9	Healthy Antioxidant Diet	385
	10.10	Glossary of Biomedical Terms		385

11 Closure   387
	11.1	Recommendations for Research		387
		11.1.1	Signal-to-noise Ratios	387
		11.1.2	Statutory Test Cycles and Real Emission Rates	388
		11.1.3	Inspection and Maintenance	388
		11.1.4	The Soot Sensor in Engine and Aftertreatment Management	388
		11.1.5	Surface Area Distribution	389
		11.1.6	Instrumentation for Number-based Legislation	389
		11.1.7	Nanoparticles in Real Exhaust Plumes (and the Ambient)	389
		11.1.8	How Should Primary and Secondary Particles Be Demarcated?	390
		11.1.9	Will Gas–particle Partitioning in the Wider Environment Be Affected?	390
		11.1.10	The Chemical Compositions of Individual Particles	390
		11.1.11	Toxicity as a Function of Particle Size	390
	11.2	Smaller Particles in Larger Numbers; or Larger Particles in Smaller Numbers		391
	11.3	Smaller and Smaller and Smaller		395
	11.4	Broader Questions of Policy		396

Further Reading   399

Literature Cited (Cross-referenced Against the Text)   401

Index   485

Preface

One of my immediate colleagues once claimed, jokingly (I suppose), that my interest in exhaust gas is 'obsessive and unnatural'. I do not, myself, see this as prurience, but, I suppose, a healthy interest in exhaust gas is reasonable enough for someone whose career has been associated, in one way or another, with motor vehicle pollution for nearly twenty years. Indeed, I don't see that I would have remained in this field were it otherwise. At one time I may well have thought, with the presumption that so much characterises youth, that I would one day understand everything that is to be understood about this fascinating subject. But, even after having closely studied what must, by now, be many thousands of research papers, I still come across conjectures and abstruse speculations that are new to me.

Make no mistake; there is no shortage of intellectual sustenance to be had from the subject of particulate emissions from motor vehicles. Despite having received the attention of the ablest researchers for decades, there is still very, very much that is not understood. Perhaps we should remember to turn and look back occasionally, and remind ourselves of what has been learned, even in just the last decade: this is also very much. It was the avalanche of research papers beginning in the 1990s that, in fact, convinced me of the need to supply a coherent monograph in this field. True, there are already many excellent books on internal combustion engines, on polluting particles in the atmosphere, and on aerosol science and technology. My intention has been to supply a text in the region where these three fields intersect. Yes: exhaust gas is very much an aerosol, no matter how much this term may be associated with deodorants. Exhaust gas always carries particles of one sort or another.

The gestation of the present text has, regrettably, been a long one; and I blench at the thought that my first book, *Critical Topics in Exhaust Gas Aftertreatment*, was published no less than eight years ago. Tellingly, there is no suggestion here of prolific authorship. Well, art is long and life is short. But one reason for this delay is clear: an excess of zeal to include the latest research; and there were times when it seemed that papers were being published faster than I could review them. This is a vibrant field, if ever there was one.

Hence my regular consternation (nay, exasperation) with persons who say to me, occasionally, that my field is 'specialist' – usually with an accusing frown, as if I'm expected to bow my head in shame, having been 'found out'. This reminds me of the scholastic debate about how many angels can dance on the head of a pin – used in modern times to parody esoteric and parochial academic research. Surely, the existence of many thousands of research papers in this field argues against parochialism. I look around my immediate workplace: no doubt more than half the people around me would have to seek employment elsewhere if the exhaust gas discharged into the atmosphere was of no importance in motor vehicle design and manufacture.

What distinguishes particulate emissions from other types of atmospheric pollutant is *visibility*; indeed, a high concentration of particles, emanating from some form of combustion process, is what we call 'smoke'. This visibility ensures immediacy in the eyes of the public, for smoke provides the most obvious indication that some form of pollution is in progress. All (Western) industrial cities were once plagued by smogs generated by coal-burning; this air pollution is now lodged in our cultural psyche, indissolubly linked to London of the Victorian age. For example, contemporary costume dramas featuring Sir Arthur Conan Doyle's famous sleuth would be utterly unthinkable without some 'pea-souper' as a suitable backdrop. Alas, the particles we face today are quite different, as this text shows...

It is revealing to note that pollution by the traditional 'smokestack' industries has not always been censored. In former times people were well aware of the close association between industrial production and national prosperity – I'm thinking of advertising literature in previous decades. Then there is the association between industrial production and national security, as evidenced by 'home front' posters during the world wars. The complete absence of smoking factory chimneys in today's corporate advertising betrays this shifted emphasis.

Let us bring these sermonic musings to a close. Certain persons have been kind enough to offer their advice on sections of the draft manuscript, or to elucidate specific issues that to me remained stubbornly obscure: I wish therefore to acknowledge Nicos Ladommatos, Matti Maricq, Tony Collier, Khizer Tufail and Haiwen Song. Finally, I wish to thank my doctoral supervisor Timothy Claypole for his abiding advice – namely, that one should never apologise for doing a piece of good work – which proved invaluable to me in 1995.

Peter Eastwood
Dunton, 2007

Acronyms and Abbreviations

ACS	American Cancer Society
AFR	air–fuel ratio
ATDC	after top dead centre
BC	black carbon
BHM	binary homogeneous nucleation
BMEP	brake mean effective pressure
BS	black smoke
BTDC	before top dead centre
CA	crank angle
CAD	crank angle degrees
CARB	California Air Resources Board
CNC	condensation nuclei counter
CNG	compressed natural gas
CNN	cloud condensation nuclei
CO	carbon monoxide
COPD	chronic obstructive pulmonary disease
CSHVR	city–suburban heavy vehicle route
CVS	constant-volume sampling
DEE	diethyl ether
DEP	diesel engine particulate
DI	direct injection
DIA	digital-imaging algorithms
DISI	direct-injection spark ignition
DMA	differential mobility analyser
DMC	dimethoxycarbonate
DME	dimethyl ether
DMM	dimethoxymethane
DMPS	differential mobility particle sizer
DNA	deoxyribonucleic acid
DPF	diesel particulate filter
EAA	electrical aerosol analyser
EC	elemental carbon

ELPI	electrical low-pressure impactor
EOI	end of injection
EPA	Environmental Protection Agency
ESC	European steady-state cycle
ETC	European transient cycle
EUDC	extra-urban drive cycle
FAME	fatty acid methyl ester
FBC	fuel-borne catalyst
FEV	forced expired volume
FID	flame ionisation detector
FIE	fuel injection equipment
FTP	Federal Test Procedure
FVC	forced vital capacity
GC	gas chromatograph(y)
GDI	gasoline direct injection
GEP	gasoline engine particulate
HACA	hydrogen abstraction and acetylene addition
HC	hydrocarbons (particularly those that are gases at $>190\,°C$)
HCCI	homogeneous charge compression ignition
HWFET	highway fuel economy test
IDI	indirect injection
IFN	ice-forming nuclei
IM240	Inspection and Maintenance (240)
IMEP	indicated mean effective pressure
IMN	ion-mediated nucleation
ISO	International Organization for Standardization
LES	laser elastic scattering
LII	laser-induced incandescence
LMMS	laser microprobe mass spectrometry
LPG	liquefied petroleum gas
MAF	mass air flow (sensor)
MATES	Multiple Air Toxics Exposure Study
MOUDI	micro-orifice uniform deposit impactor
MPI	multi-point injection
MS	mass spectrometry
MTBE	methyl t-butyl ether
NA	naturally aspirated
NAAQS	National Ambient Air Quality Standards (USA)
NEDC	New European Driving Cycle
NIEHS	National Institute for Environmental Health Sciences (US)
NIOSH	National Institute for Occupational Health (US)
NMMAPS	National Morbidity, Mortality and Air Pollution Study
NO_x	oxides of nitrogen
NPAH	nitro-PAH
OC	organic carbon
OFR	oil–fuel ratio
PAH	polycyclic aromatic hydrocarbon(s)
PCDD	polychlorinated dibenzo-p-dioxins

PCDF	polychlorinated dibenzofurans
PCI	premixed compression ignition
PCV	positive crankcase ventilation
PMN	polymorphonuclear neutrophil leukocytes
PPCI	partially premixed compression ignition
PGM	platinum group metals
PM	particulate matter
QCM	quartz crystal microbalance
REE	rapeseed ethyl ester
RME	rapeseed methyl ester
ROFA	residual oil fly ash
RON	research octane number
ROS	reactive oxygen species
SBR	styrene–butadiene rubber
SME	soya methyl ester
SMPS	scanning mobility particle sizer
SOA	secondary organic aerosol
SOC	secondary organic carbon (particles)
SOF	soluble organic fraction
SOI	start of injection
TC	turbocharged
TDC	top dead centre
TEL	tetraethyl lead
TEOM	tapered-element oscillating microbalance
TGA	thermal gravimetric analysis
TSP	total suspended particles
UDC	urban drive cycle
VCO	valve-covered orifice
VGT	variable-geometry turbine
VOF	volatile organic fraction
WHO	World Health Organization
WMTC	World Motor Cycle Test Cycle
WOT	wide-open throttle
XOC	XAD-2 resin organic component (i.e. vapour-phase organics)

About the Author

Dr. Eastwood graduated from University College, Swansea, UK, in 1985, with a B.Sc. in Mechanical Engineering; followed in 1992 by a Ph.D. in Mechanical and Electrical Engineering, with a thesis entitled *Exhaust Gas Sensors for Engine Management*. There followed a post as Royal Society Post-Doctoral Research Fellow at the Institute for Physical and Theoretical Chemistry, Tübingen, Germany, then under the direction of Prof. Dr. W. Göpel. Since then Dr. Eastwood has worked in industry in a variety of research and development roles, all of them associated in one way or another with the technology of motor vehicle pollution control. His monograph, *Critical Topics in Exhaust Gas Aftertreatment*, was published by Research Studies Press in 2000.

About the Author

1

Introduction

The first trade-off between *utility* and *amenity* probably arose with fire. The utility was manifold: cooking, heating and illumination; and also artwork, as facilitated by a mysterious black pigment: ancient hands, pressed against cave walls, and exposed to strongly sooting flames, have left their silhouettes for modernity. The luminosity of the fire, and to some extent the heat, stemmed from the black pigment – namely, *soot*. Yet the amenity was also degraded, as the soot particles, when inhaled, were prejudicial to health; and it was fire, perhaps, that engendered the first respiratory diseases. Such ailments still plague developing countries, where households continue to rely on a variety of stoves for cooking and heating, ventilation for which is poor or virtually nonexistent (Koshland and Fischer, 2002).[1]

Smoke became more inimical once population densities increased to those of the first civilisations, and the problem greatly accelerated once coal began to replace wood as the principal fuel (Brimblecombe, 2001). The great forests of England were swiftly dwindling before a rapidly expanding population: fuel was needed for energy, and land for agriculture. The competition, between utility and amenity, is written into the legislative record: Edward I (reigned 1272–1307) issued a royal proclamation forbidding coal-burning in London; Edward II (reigned 1307–1327) ordered the torture of persons fouling the air with coal smoke; Richard II (reigned 1377–1399), more humanely, chose to control coal-burning via taxation; Henry V (reigned 1413–1422) established a commission to regulate the entry of coal into London (Wilson R., 1996); and Elizabeth I (reigned 1558–1603) legislated against coal-burning whenever Parliament was sitting.

---

[1] Some misguided souls in the developed world express a preference for the 'naturalness' of open fires. This position is untenable even from the environmental standpoint that such people profess to adopt. Open fires are inefficient and largely uncontrolled; they consequently release into the atmosphere large quantities of soot and sundry organic compounds – with genuine toxicological risks. Contrastingly, the combustion inside today's engines is controlled to great precision; the pollution is excessive because motor vehicles are present at high density, not because they are particularly polluting on a unit basis.

Particulate Emissions from Vehicles   P. Eastwood
© 2008 John Wiley & Sons, Ltd

In 1648, coughing Londoners petitioned the government to prohibit the importation of coal from Newcastle, citing the injurious effects of the smoke; but they were unsuccessful. And in 1661, the famous diarist John Evelyn wrote:

> It is this horrid smoake, which obscures our churches and makes our palaces look old, which fouls our clothes and corrupts the waters so that the very rain and refreshing dews which fall in the several seasons precipitate this impure vapour, which with its black and tenacious quality, spots and contaminates whatever is exposed to it. (Cited by Wilson R., 1996).

What provided the greatest impetus of all was the Industrial Revolution: by 1819, the problem had become so conspicuous that Parliament appointed a committee to investigate how to make steam engines and furnaces less prejudicial to public health. It was concluded that smoke could be effectively controlled ... but no action was taken. As a question of social policy, addressing smoke was often viewed as prejudicial to the nation's economy, and public good was held more important than private need. There also seems to have been a subtle transition from regulation as 'nuisance' to regulation as 'negligence' (Farrell and Keating, 2000). A technical question was whether to prevent the formation of smoke in the first place; whether the smoke, once formed, should be collected prior to emission; or whether the smoke, once emitted, should be more effectively dispersed, under the time-honoured cynicism 'the solution to pollution is dilution'. The construction of ever-higher chimney stacks, as the Industrial Revolution advanced, bears testimony to the dispersion principle (and in the modern era, power stations send their pollution above the inversion layer).

Smoke emissions from locomotives were finally regulated by the Railway Clauses Act of 1845, and from factory furnaces by the Town Improvement Clauses Act of 1847. From 1875, English law regulated smoke from factory chimneys under clauses in the Public Health Acts; London was singled out for more severe restrictions. This legislation was subsequently modified by the Smoke Abatement Act of 1926, which invested local authorities with powers to enforce the provision of equipment in new buildings as might prevent smoke. (Domestic grates were not, however, included.) Several towns established 'smokeless zones', wherein all smoke emissions were prohibited (Fishenden, 1964).

These measures were insufficient. In the 1950s, the duration of sunshine in cities such as Leeds, Sheffield and Manchester, during the winter months (when domestic hearths saw most use), was less than half that in the outlying districts. Pea-soup smogs, or just 'pea-soupers', sometimes lasting for several days, plagued many cities, and (quite apart from respiratory ailments) were responsible for numerous road accidents. As recently as 1950 it was reported that, in Glasgow, three tonnes of soot fell to earth per acre per year (cited by Wilson R., 1996).

The incident which undoubtedly provided the greatest case for further legislation was the notorious London smog of December 4–9, 1952, which arose through a freak combination of low wind speed, temperature inversion and dry weather; local levels of atmospheric particulate peaked at $7\,\text{mg/m}^3$ (Maynard, 2001) – ten times the amount normally seen to day, even in the most heavily polluted of cities. The smog is reputed to have killed 4000 people:[2] the association between mortality, smoke and sulphur dioxide (SO_2) is pointedly illustrated by Figure 1.1 (cited by Wilson R., 1996). During this event, deaths attributed to bronchitis and pneumonia increased eightfold and threefold, respectively; there were also increases in deaths attributed to other respiratory or cardiac diseases (Higgins I.T.T.,

---

[2] This incident is wearisomely related in virtually every book on air pollution, amongst which the present one is no exception. Around 4000 deaths is the figure usually quoted. Recent research reported in the *Daily Telegraph*, Dec. 14, 2002, suggests the smog-related deaths to have been confused with those of a flu epidemic, and that the real figure could be as high as 12 000. The newspaper article credits this work to D. Davis and M. Bell, but the present author has been unable to locate any further information.

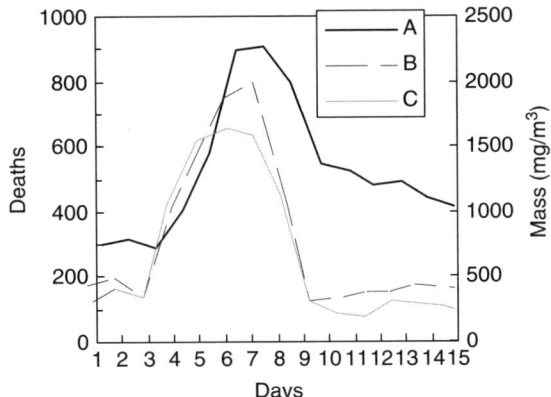

Figure 1.1 The London smog of December, 1952, and its fatal effects: A, number of deaths; B, mass concentration of SO_2; C, mass concentration of smoke. The graph is taken from Wilson R. (1996) (original reference Beaver (1953)). (The concentrations are averages over several sites).

1971). But post-mortems failed to reveal anything specifically particle-related; the common feature was that persons with cardio-respiratory diseases suffered exacerbations of their symptoms, and that some died as a result (Maynard, 2001). Morbidity rates increased also, as seen in various statistics for hospitalisations and sickness-benefit payments.

Suitably chastened, Parliament introduced the Clean Air Act of 1956, making the emission of dark smoke an offence. But mortality and morbidity in London, for certain respiratory diseases, remained until the early 1960s fairly well correlated with smoke and SO_2. After this period the proscription of smoke seems finally to have taken effect, inasmuch as mass concentrations fell from 300 to 60 $\mu g/m^3$; and there is some evidence of a concomitant improvement in public health (studies cited by Higgins I.T.T., 1971).

To be sure, the reason for the demise of pea-soup smogs was widespread conversion of domestic heating appliances to natural gas (Farrell and Keating, 2000). (The dieselisation of railways and small-scale power generation was another factor.) Smoke is a key symptom of poor air–fuel mixing: an inability to supply sufficient oxygen to the primary combustion zone is the eternal problem with solid fuels. And in the domestic arena, coal was often burned in a haphazard and uncontrolled way. Natural gas can be mixed with air automatically and far more precisely; hence the cleanliness of the combustion becomes far less dependent on the competence, or diligence, of the user.

Natural gas possesses two other advantages – it is low in sulphur and incombustible ash (principally metals). In the presence of moisture, and the catalytic effects of incombustible metals emitted as 'fly ash', fuel sulphur invariably finishes up in the atmosphere as *sulphuric acid*. The combination of soot, ash and acid was what formed the infamous smogs. This drives home forcefully the importance of *secondary* pollution: not directly emitted, but formed subsequently through ongoing atmospheric reactions. Secondary pollution greatly obfuscates the protection of ambient air quality: it can only be controlled indirectly, via the primary pollution, and a thorough knowledge of atmospheric processes.

We have followed the history of smoke, from the first beginnings to the modern era; and from primeval man, to the grimy image, etched into our cultural psyche, of the nineteenth-century industrial city. But how does this relate to particles emitted by today's internal combustion engines? This question does not, usually, concern aesthetics, as exhaust plumes are less visible, indeed approaching invisibility, just as their offending dirt and filth are evanescent. Nor is the danger nowadays a slow asphyxiation by acrid smoke, because particle concentrations in the ambient atmosphere are much lower than in the instances described – by two orders of magnitude. It should be emphasised that the pollution in the aforementioned London smog episode provided evidence of the risks to public health – evidence that was unambiguous and unmistakable. The situation today is not nearly so clear-cut; and considerably greater efforts are needed to avoid poorly focused or ineffective legislation (Hall et al., 1998).

1.1 Air Traffic

Emissions legislation, directed at aircraft, began to appear at around the same time as for motor vehicles, i.e. in the 1960s (Kittredge and NcNutt, 1971). This is not exactly a coincidence: there is to some extent a parallel between the two types of transport: similar pollutants are emitted (carbon monoxide, oxides of nitrogen, particulate and hydrocarbons), and levels of traffic increased rapidly at about the same time. But in aviation, the internal combustion engine was not the culprit. The need to control emissions, especially those of smoke, first arose towards the end of the 1950s, when the gas turbine began to replace the internal combustion engine as the preferred power plant. This was not only a problem with civilian airliners: the conspicuousness of the trails was particularly undesirable with military aircraft (Fiorello, 1968).

In fact, the earliest jets did not smoke appreciably; the problem seems to have been instigated by increases in combustion pressure and heat release rates, undertaken to improve performance (Shayeson, 1967). Smoke emitted by aircraft was particularly noticeable during take-off and on the final approach to landing. The sight and sound of jet aircraft, especially when on full thrust, further focused attention; and an aircraft's smoke trail is inherently visible against an empty sky, unlike on the ground, where neighbouring buildings render assistance in hiding the evidence (Parker, 1971).

In gas turbines, soot is formed in fuel-rich regions of the combustor; it can be burned up prior to emission, but this action is impeded by overly rapid quenching. To combat soot, therefore, the engine manufacturers were forced to redesign their combustors, as, unlike with motor vehicles, aftertreatment, that is, an exhaust gas clean-up, is impractical (Nelson, 1974). It was, apparently, possible to control the soot by directing air into the primary combustion zone, without, moreover, unacceptable increases in NO_x (Bristol, 1971). This soot–NO_x trade-off–since *both* pollutants must be reduced–is a constraint that continues up to the present time (Gupta A.K., 1997). The same trade-off is inherent in the diesel engine (to which we shall shortly turn): this is the principal reason why diesel soot has proven so intractable.

1.2 Motor Vehicles

Smoke has always been emitted by motor vehicles; and its sheer visibility, competing with the malodorous emission, ensures immediacy in the eyes, and noses, of the public. Yet in the days of Otto and Diesel, gasoline engines were denounced just as much as diesel engines. There is an interesting paper published by the newly formed US Society of Automobile Engineers [*sic*], from before the First World War, which castigates automobiles of 'the early days' [*sic*] as 'ill-smelling affairs' (Howe, 1910). It was at this time that New York City's Department of Health extended an old smoke ordinance to include motor vehicles: 'No person shall cause, suffer or allow dense smoke to be discharged from any building, vessel, stationary or locomotive engine or motor vehicle, place or premises within the City of New York.' Intriguingly, the ordinance gave no definition for a smoke density that *was* acceptable, even though it did, apparently, make a culprit liable to police arrest. (Perhaps the authorities realised just how difficult meaningful measurements of smoke are to conduct.) The author of the paper recognises the close association between smoke and tampering, and calls for manufacturers to prohibit maladjustment of carburettors such that black smoke is emitted. Apparently, it was not unknown for exhaust gas at this time to contain carbon monoxide to the tune of 12 % [*sic*].

The introduction of leaded gasoline in the 1920s gave rise to a new type of particulate emission. Tetraethyl lead was added to gasoline as an antiknock agent; and, to prevent the formation of lead deposits in the engine, ethylene dichloride and dibromide, as scavenging agents, were also added. These additives were emitted as particles of lead bromochloride, bromide and chloride. Because they were invisible, lead particles were arguably more insidious than smoke. But no changes were made until motor vehicle ownership reached modern-day proportions.

The eventual withdrawal of leaded gasoline happened for two cogent reasons. First, elevated levels of lead were discovered in the blood of persons exposed to traffic emissions. This raised a host of health issues (Russell Jones, 1987), a chief one being the suspected impairment of child development. Second, a pressing need had arisen to tackle other emissions, and catalytic converters were the only feasible technology with which to do this. Catalysts are swiftly poisoned by lead, and no aftertreatment sufficiently tolerant to this metal has ever been found.

The USA in the 1970s saw the introduction of the first catalytic converters; they were for oxidation purposes only (unlike today's three-way catalysts). Perversely, this led to another type of particulate emission: droplets of sulphuric acid. Reading the literature with the benefit of hindsight, this problem appears to have come as something of a surprise to the automotive industry; however, the appearance of sulphuric acid, after burning fuel which contains trace amounts of sulphur, and passing the exhaust gas over an oxidising catalyst, will not come as a surprise to any chemist. Again, this problem was not really solved as such; rather, technology simply moved in another direction. The introduction of three-way catalysts in the 1980s forced the adoption of stoichiometric, rather than lean air–fuel ratios; and the oxidising conditions in the exhaust necessary for the formation of sulphuric acid were lost.

Thus far we have discussed particles that are, in a sense, interlopers, inasmuch as they arise by side reactions in the combustion process. Not all particles are produced this way. Unburned fuel, if emitted in sufficiently large quantities, forms liquid droplets in the exhaust plume, and these droplets are perceived as *white* smoke. For engines that have been designed successfully, calibrated competently and maintained assiduously, this type of smoke is not an issue. Emissions legislation now places strict limits on the quantity of unburned hydrocarbons that may be emitted; and to obtain white smoke, these limits must be exceeded by an order of magnitude, so that this phenomenon is a little academic. White smoke may still be a problem, however, during cold starts, especially at low ambient temperatures. Similarly, *blue* smoke, if caused by the escape of lubricating oil, is indicative of improper or inadequate maintenance, and seldom otherwise seen.

Finally, there is the long-standing issue of *black* smoke, chiefly discharged by diesel engines, and whose characteristic blackness arises from the element carbon. Black smoke appears to be an unfortunate and ineluctable consequence of diesel combustion, if only because it has frustrated the attempts of generations of engineers to eliminate it. Solely from an engine performance perspective, black smoke reflects the efficiency, or perhaps more appropriately, the inefficiency of the combustion, because it represents lost energy. But that is not quite the point: with today's diesel engines, the carbon that is usually emitted as soot, if it were to be successfully burned, would make little difference to the overall combustion efficiency, whereas if a diesel engine at full load were to emit only 0.5 % of the fuel as black smoke, the result would be completely unacceptable from an emissions perspective. The combustion efficiency, taking into account unburned hydrocarbons, carbon monoxide and soot, is usually better than 98 %, so that, from an energy conversion perspective, the combustion is virtually complete (Heywood, 1988, p. 509). The implications for public health are a completely different issue.

1.3 The Legislative Framework

Up to now, 'smoke' and 'particulate' have been employed fairly loosely and interchangeably. From now on this practice will be insufficient, and both terms must be defined much more closely. Strictly speaking, 'smoke' denotes an *aerosol*, i.e. a suspension of particles in a gas, whereas 'particulate' describes a collection of these particles, say, on a filter.[3] But, passing from these semantic quibbles, a more important difference, of a technical (and indeed environmental) nature, arises: when there is smoke, the presence of particles may be safely inferred; but when there are particles, these are not necessarily manifested as smoke. *Smoke, then, is the visible corollary of particulate*. Thus, 'particulate' is a considerably *broader* and indeed *all-encompassing* term: it incorporates smoke, and much else besides. So, we can now state this distinction explicitly, as follows:

- *Particulate*. All material which deposits on a filter.
- *Smoke*. All material which attenuates a beam of light.

These two definitions – both of which, incidentally, exclude condensed water – should not be confused: they are often directionally consistent, but this consistency is not by any means an essential prerequisite: in fact, the mass of particulate discharged by a smoking vehicle might actually be less than for a nonsmoking vehicle (Knapp *et al.*, 2003). This is because particle masses and particle interactions with light are entirely different measures – consistent only when the particles under investigation (size and composition) are also consistent.

In the USA, smoke from heavy-duty diesels was first legislated on in 1970 as 'opacity', that is, the fraction of light successfully traversing the exhaust stream. The development of particulate-control legislation is complicated (e.g. Cucchi and Hublin, 1989; Walsh and Bradow, 1991; Walsh, 1993; Charmley, 2004), but, simplifying somewhat, the first diesel particulate standards in the world were established in the USA in 1980, and related to passenger cars and light-duty trucks; heavy-duty engines and trucks were subsequently covered in 1985. In Europe, particulate emissions from diesel engines were first controlled in 1989, via EC Directives 88/436, 91/441 and 91/542 (Hall *et al.*, 1998).

Particulate is now strictly regulated in most countries, and the mass emitted by both light-duty (g/km) and heavy-duty (g/kW h) diesels has decreased since the 1970s, by more than two orders of magnitude. It is not improper to observe that the debate between industry and legislatures about these regulations has been contentious at times (Merrion, 2003). While the enactment of this legislation has undoubtedly helped to shape, and propel forward, the technology of emission control, the diesel engine has, at times, seemed threatened with extinction (Pethers, 1998); indeed, the downward trend in statutory requirements is set to continue unabated.

Current legislative practice in most countries extends, in a belt-and-braces manner, to smoke *and* particulate. But since smoke is far more conveniently measured, it tends to be used as an in-use compliance test, i.e. for *individual* vehicles during the course of their lives – perhaps those pulled over by the authorities for roadside checks. In the UK it is an offence for any vehicle to emit smoke at levels that impair visibility for other drivers; clearly, therefore, this is not so much to protect air quality as to ensure road safety. Smoke is assessed by a roadside or garage test, in which the clutch is disengaged and the engine rapidly accelerated up to the governor run-out (on passenger cars, usually between 4000 and 5000 rpm). This is the so-called 'free acceleration' test. It is quick and convenient, but says nothing about the smoke emitted during *real* driving, i.e. with the engine operated over a

---

[3] Exact speakers of English prefer to stay with this strict definition; but in this text, 'particulate' is applied more loosely insofar as it denotes also the suspended particles *prior* to collection.

genuine duty cycle. The free acceleration test is the sum total of what a vehicle is likely to experience after it has been sold, according to the current regulations.

The regulations to be satisfied *before* sale relate to a vehicle *model*, and are known as 'type approval' or 'homologation'. These regulations are much stricter, and the manufacturer must demonstrate satisfactory levels of smoke *and* particulate. On an engine dynamometer, the smoke is assessed in two ways: first, for a progressive series of steady states along the full load curve (otherwise known as the torque limit), and second, for a free acceleration test as already described. Particulate emissions from an actual vehicle (i.e. the full power train) are assessed on a chassis dynamometer.[4] The vehicle is run over a (transient) drive cycle, during which the exhaust gas is fed into a 'dilution tunnel', mixed with air, and then filtered. This procedure is designed, among other things, to approximate the dilution of real exhaust plumes in the ambient. Current legislation mandates a dilution such that the exhaust gas temperature at the filter is less than or equal to 52°C:[5] this condition is pivotal, as it decides the transfer of material from the gas phase into the particulate phase. Once the drive cycle is completed, the filter is removed, conditioned to a certain temperature and humidity, and then, to determine the mass of particulate it has retained, simply weighed.

---

[4] This description focuses on light-duty vehicles. Legislative practice for heavy-duty vehicles is different; engine dynamometer rather than chassis dynamometer tests are preferred. Some countries also only use steady-state (cruise) tests. Test protocols vary considerably from country to country, and are continually evolving; the information here is necessarily generalised.

[5] This is the long-standing specification, the laxity of which in view of today's stringent emission control legislation is well recognised. At the time of writing, tighter specifications such as $47 \pm 5°C$ are under discussion (Wu *et al.*, 2007).

2

Fundamentals

2.1 Introduction

The present chapter is divided into three main sections. The first describes the physics, and to a lesser extent the chemistry, of particles when suspended in a gas: what has become known as 'aerosol science'. The physicochemical mechanisms in operation, from one aerosol to another, greatly depend on particle *size*. We look at the various forces that determine particle motion, or *transport*, and then turn to *mutability*, which simply means that particles are constantly changing in shape and form.

The second section, on atmospheric particles, is included in order to place motor vehicle particulate in an appropriate context. The atmosphere is not simply 79 parts nitrogen and 21 parts oxygen, with trace quantities of other gases. Such a description, although convenient, grossly overlooks a component of profound and ubiquitous influence; for the atmosphere is, in fact, thronging with innumerable particles; and this makes it not so much a gas, as a *multiphase system*. This particle soup, the 'atmospheric aerosol', is of mind-boggling complexity.

In the third section, we begin our study of the particulate actually discharged by motor vehicles. Knowledge in this area is gained, principally, by passing the exhaust gas, perhaps in raw, but more usually in diluted, form through a filter, and studying the particulate trapped thereon. The material caught this way is astonishingly diverse. The final point addressed is whether one should consider the *number* of particles or the *mass* of particles. The implications of this decision loom large in contemporary emission control.

2.2 Properties of Aerosol Particles

An immensely helpful feature of the word 'particle' is that it encompasses, without distinguishing, the liquid and solid phases of matter; indeed, this collective grouping conveniently signifies a *third* state, the 'particulate phase' – an umbrella term that will be much used in this text. But, this said, it is equally important to understand that the liquid and solid states appertain to the *macroscopic* world, i.e. when material is present in bulk. Intuitively, the physics and chemistry of the bulk cannot continue to apply no matter how finely divided the material; and this is indeed the case.

Aerosol science is a distinct scientific discipline in its own right (Spurny, 2000f), the emergence of which is fairly recent in the scientific lineage; indeed, 'aerosol' seems to have been coined in the early 1920s, although many of the important fundamental observations date from the Victorian era. A wide range of subdisciplines, and an incomprehensibly wide range of aerosols, are encompassed. To the general public, aerosols are hairsprays, deodorants and the like; but to the technologist, any collection of particles suspended in a gas is an aerosol.[1] Hence aerosols are, by their very nature, *multiphase* systems. They can be formed in numerous ways: the disintegration of liquids and solids, the resuspension of deposited particles, the break-up of agglomerated particles, and the condensation of gases into particles.

Aerosols are impermanent and highly dynamic, because their constituent particles are notoriously and capriciously protean. Interactions between particles, between particles and the gas molecules that surround them, and between particles and the solid walls that contain them are unending. Particle properties cannot, for this reason, be divorced from the conditions to which an aerosol is subjected. Heating, cooling, dilution, residence time and the like all potentially influence particle size and composition – and generally do. If not properly controlled, many of the processes outlined below give rise to variable, unpredictable or even bogus results. The paramountcy of this remark cannot be overstated.

2.2.1 Diameter and Shape

Aerosol science is fundamentally predicated on one parameter: particle *size*. But so characterising particles is problematical for various reasons. Excepting, for example, the case of liquid droplets, which offer the convenience of sphericity, irregular morphologies are the rule rather than the exception. Of course, such irregularity need not be a problem if particles of different sizes are *geometrically similar*, as one characteristic dimension may then be arbitrarily selected; but, again, in practice this condition is seldom the case. Consequently, there have arisen in aerosol science various ways by which to define particle size. This is inconvenient because the various definitions cannot always be easily interrelated. And although in aerosol science it is common to speak of a 'diameter', as of a sphere, this term requires some elucidation, as it does not refer, to any geometrical dimension that could be verified directly, as under a microscope: the suggestion of circularity should not, therefore, be taken literally.

Various definitions are based on geometry (Mark, 1998). We may, for instance, consider a spherical particle of *equivalent* surface area or volume; this is referred to as a 'fundamental' diameter. Equally, we may allow the particle to cast its shadow onto a background, and then calculate the perimeter or surface area of the shadow, referring this back to an equivalent spherical particle; this is the 'projected' diameter. The greatest possible distance between two parallel lines tangent to the particle profile is 'Feret's diameter'. Obviously, there are many possibilities.

The ponderous nature of visual or optical assessments cannot be gainsaid. And although this ponderousness is indubitably lightened by the advent of computer imaging and automated imaging processing, there seems little likelihood of real-time measurements. In any case, there is always the variability inherent in particle orientation.

Practical particle-sizing instruments do not report diameters based on, or even derived from, direct geometrical assessments. These instruments exploit the fact that particles respond to certain forces in

---

[1] Similarly, a suspension of particles in a liquid is referred to as a 'colloid'. Many of the theoretical principles for colloids are the same as for aerosols, but, with the exception of soot in oil (Section 6.3), in this work we are concerned only with the latter.

Fundamentals

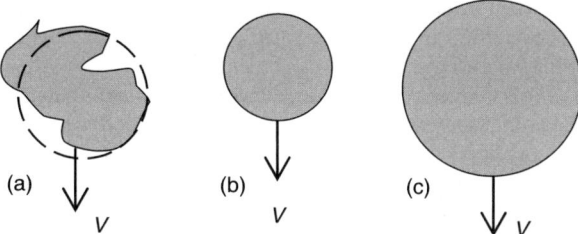

Figure 2.1 Various concepts of 'diameter': (a) an irregularly shaped particle, and, superimposed, the volume-equivalent sphere; (b) a Stokes-equivalent sphere (same density as (a)); (c) an aerodynamic-equivalent sphere (unit density). The irregular particle, Stokes-equivalent sphere and aerodynamic-equivalent sphere all have the same settling velocity (Hinds, 1999, p. 54).

a way that is size-dependent, and then report the diameter of an equivalent spherical particle that, if it were subject to the same forces, would behave in a manner identical to the real particle. Such an approach is conceptually depicted in Figure 2.1. The 'Stokes diameter' refers to a spherical particle of the same density and settling velocity; the 'aerodynamic diameter' refers to a spherical particle of unit density and the same settling velocity[2]; and the 'electrical-mobility[3] diameter' refers to the velocity of a charged particle when moving in an electrical field, as referenced to the strength of that field. There are, of course, still more definitions based on other particle behaviours.

It is not the purpose of the present section to discuss particle-sizing instruments, but, in this brief sketch, something must be said about two such units. Generally speaking the aerodynamic diameter, as measured by the electrical low pressure impactor (ELPI), is more relevant to the larger particles (inertial effects), while the electrical-mobility diameter, as measured by the scanning mobility particle sizer (SMPS), is more relevant to the smaller particles (diffusional effects). These are the two diameters, and two instruments, that have been most often used in the automotive industry. Obviously, a need regularly arises to relate the two diameters, and equations for such interconversions are available (e.g. Maricq et al., 2000b; Virtanen et al., 2002).

It will be apparent, therefore, that the measurement principle is inherent in the quoted diameter. As such, this diameter relates to the *equivalent* particle, and bears only a tenuous geometrical connection at best to the *real* particle – unless, that is, the real particle just happens to be a perfect sphere. For this reason, one should be chary of computing the volume of the particle merely by raising the quoted diameter to the third power, as this result relates to the equivalent spherical particle, and *not* to the real particle. Calculations of this sort are admittedly still undertaken, even for irregular morphologies, as they do furnish additional information; but the limitations must not be treated with careless indifference.

When one is conducting the measurement, only the *behaviour* of the real particle, when subject to a certain pattern of forces, counts, and two particles might be awarded the same diameter even though they differ greatly in surface area, mass, volume, density, shape, etc. For example, particles may appear completely different under the electron microscope, yet be wholly indistinguishable in an aerodynamic sense. This is the principal of the *derived* diameter – a flag of convenience, as it were, adopted for reasons of practicality. But one should not be too despondent, as there is a saving grace here. Capture by filters, and deposition in the respiratory tract, etc., are decided, *not* by any geometrical diameter,

---

[2] This definition also applies to cases where forces on the particle arise through inertial effects rather than a gravitational field.
[3] The mobility is the velocity per unit force.

but by particle behaviour when subject to certain forces – so, there is no need to know the shape of the particle. It makes sense, therefore, to characterise particles according to their behaviour, rather than their appearance under the microscope.

It is common (because it is convenient) to assume that density is not a function of particle diameter. Whilst this is clearly false for many types of particle, the assumption serves as a starting point from which deviations can be contemplated. For example, for soot particles, which are clusters or chains of smaller (primary) particles, the density might be supposed, because of increasing voidal fraction, to fall with increasing size, and this is generally the case. But the density functions of many aerosols – and motor vehicle particulate is no exception – are often hotly debated. Moreover, the effective or apparent density, derived from the way a particle behaves, should not be confused with the bulk density. The two are equal only for perfectly spherical, nonporous particles.

As already hinted, soot particles consist of smaller, primary particles, built into elaborate chains or anfractuous clusters; hence particle size, expressed in terms of a *single* dimension, discards extensive morphological information. This is a sop to convenience. Fortunately, structural characteristics often repeat themselves when viewed at progressively larger magnifications; that is, disordered materials are self-similar at different scales. This is where the concept of 'fractality' comes in, in which particles are described by the power law (Nyeki and Colbeck, 2000)

$$N = \varepsilon \left(\frac{d_x}{d_0}\right)^{D_f}, \qquad (2.1)$$

where N is the number of primary particles, d_x is a 'characteristic dimension' (e.g. the mobility-equivalent diameter), d_o is the primary particle diameter, ε is a constant, and D_f is the fractal dimension (a noninteger) measured in three-dimensional space. Similar fractality is often exhibited for particles of different sizes which formed under the same conditions.

It should be noted that there are different bases for fractality (e.g. light scattering, electron micrographs and electrical mobility), and these do not necessarily yield the same result. But, for fractal objects generally, D_f lies somewhere between unity and three. The closer D_f is to three, the greater the sphericity of the whole; and the close D_f is to unity, the looser the structure. The fractal dimension, therefore, captures the spectrum from open, chained morphologies to compacted, clustered morphologies, and is thus an index of the primary-particle 'packing density'. The looser the structure, the more likely it is that particle–light or particle–gas interactions are governed by a simple summation of total area, i.e. the primary particles act as if alone.

As a coda, the concept of fixed, permanent shapes is sometimes erroneous. For example, when particle surfaces acquire volatile coatings, surface tension may force some micromorphological restructuring. Then, if these coatings later evaporate, some relaxation may occur, depending on whether the deformations were plastic or elastic. There is, moreover, a 'turnover time', as particles exist in dynamic equilibrium with their environment: molecules of accreted volatile substances are continually joining and leaving, so that, eventually, all are replaced (Twomey, 1977, p. 17).

2.2.2 Size Distribution

The size range spanned by aerosol science is vast: it stretches from the minutest molecular clusters of a few nanometres to what are, by comparison, boulders of hundreds of microns. This range encompasses no less than five orders of magnitude, although, fortunately, for specific aerosols, two or three orders may be sufficient.

Should an aerosol contain more than one size of particle, and this is usually the case, it is said to be 'polydisperse'. The size distribution can be treated as continuous, although there are exceptions. If all particles in the aerosol are constructed from one primary particle, then obviously only certain sizes are allowed, in which case the size distribution is discontinuous, or discrete. Of course, in the limit, all particles are constructed from the smallest building blocks, namely molecules; but, if the smallest particles in the aerosol always contain a few dozen molecules, then the size distribution is *effectively continuous*.

The size distributions of polydisperse aerosols are statistically characterised in the same way as any other population (Hinds, 1999, pp. 81–82). The *mean* is the sum total divided by the number of particles; the *median* is the size for which half the particles are larger and half smaller; and the *mode* is the most frequently encountered size. The *geometric mean* is the nth root of the product of n values; it is widely used for characterising aerosols with lognormal size distributions.

There is more than one type of size distribution, the chief ones of interest being the number distribution $n(D_p)$, volume distribution $v(D_p)$, surface area distribution $s(D_p)$ and mass distribution $m(D_p)$. The choice of distribution has important ramifications that we shall examine shortly. For now we shall use $n(D_p)$, and follow the text by Seinfeld (1986, pp. 275–279), which provides a particularly good explanation of the mathematics that underpins size distribution functions. Figure 2.2 is derived from this work.

To elucidate $n(D_p)$: we take a unit volume of gas, and count the particles within it, as a function of their diameter. The number of particles within a size range from D_p to $D_p + dD_p$ will then be $n(D_p) dD_p$. The total number of particles per unit volume of gas, N, is then given by

$$N = \int_0^\infty n(D_p) dD_p. \qquad (2.2)$$

The units of $n(D_p)$ might be $\mu m^{-1} cm^{-3}$, and those of N, cm^{-3}. Sometimes $n(D_p)$ is normalised by dividing by N, in which case the range dD_p describes the *fraction* of the particles within it, i.e. from the total population.

The difficulty with this linear manipulation is that, because particle sizes span orders of magnitude, interesting features become obscured. Instead, therefore, we consider the function $n(\log D_p)$, i.e. the

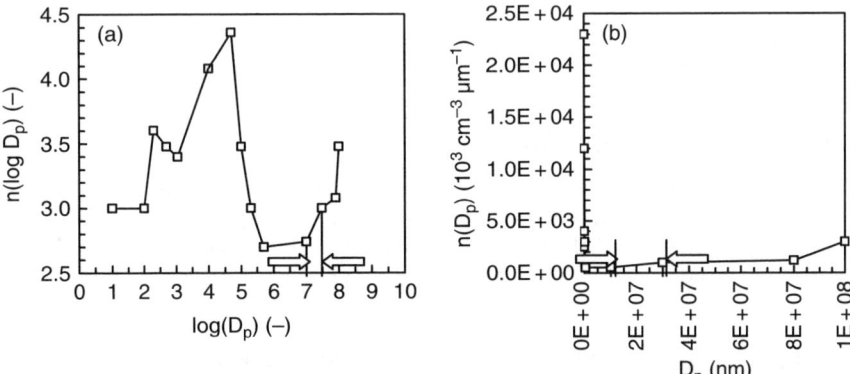

Figure 2.2 Mathematical descriptions of the number distribution for a hypothetical aerosol, where N is the total number of particles, and the number of particles per unit volume, n, is plotted as a function of size, D_p: (a) logarithmically; (b) linearly. In each case the number of particles, in the range indicated by the arrows, is the same, because the areas under the curves between these limits are also the same. (Integration of the whole curve gives N.)

number of particles as a function of the logarithm of particle diameter.[4] Note that the two functions $n(\log D_p)$ and $n(D_p)$ are *not* the same; $\log D_p$ is a *new* independent variable. Now the number of particles within the range $\log D_p$ to $\log D_p + \mathrm{d}\log D_p$ is $n(\log D_p)\mathrm{d}\log D_p$, and the total number of particles is

$$N = \int_{-\infty}^{\infty} n\left(\log D_p\right) \mathrm{d}\log D_p. \tag{2.3}$$

(Note the change in the lower bound of the integral.)

The number of particles within the range $\mathrm{d}D_p$ is equal to the number within the range $\mathrm{d}\log D_p$, since this quantity does not depend on how the number distribution is expressed, and we are considering the same aerosol. Writing this number as $\mathrm{d}N$ gives

$$\mathrm{d}N = n\left(D_p\right)\mathrm{d}D_p = n\left(\log D_p\right)\mathrm{d}\log D_p, \tag{2.4}$$

or

$$n\left(\log D_p\right) = \frac{\mathrm{d}N}{\mathrm{d}\log D_p}. \tag{2.5}$$

This derivation explains the otherwise apparently eccentric expression $\mathrm{d}N/\mathrm{d}\log D_p$, which appears on many graphs of particle size distributions. The advantage of plotting number distributions in this manner is that the number of particles in any size range is proportional to the area under the curve within that range. The same is true for the other distributions, for which the same train of analysis can be repeated, i.e. to obtain expressions for mass, volume and surface area distributions. This task is left to the reader.

Once the size distribution is known, it is possible to integrate the function and so obtain the number, volume, surface area and mass of particles per unit volume of gas. These are, variously, the number concentration N (cm^{-3}), the volume concentration V (μm^3/cm^3), the surface area concentration S (μm^2/cm^3) and the mass concentration M (μg/cm^3).[5] The choice of distribution is not just one of mathematical curiosity, as, through their differing dependencies on D_p, each distribution inherently emphasises a different range of size: the important point, and the one most pertinent to this text, is that the number distribution emphasises the smallest particles, and the mass distribution emphasises the largest particles in the distribution.

2.2.3 Transport and Deposition

Sometimes, particles are designedly caught; at other times, their capture by surfaces is inconvenient, and must be avoided. But, to be captured, particles must first be transported, and various mechanisms account for this transport, of which the six most relevant to the present topic are depicted in Figure 2.3. The simplest case, 'interception', is virtually a truism: when a gas is in motion, that is, undergoing

---

[4] Strictly speaking, it is not possible to take the logarithm of a dimensional quantity. The contrivance here is to think of dividing by a 'reference' particle of unit size, so that the quantity within the brackets becomes dimensionless (Seinfeld, 1986, p. 299). After this, the contrivance can be forgotten.

[5] These parameters are sometimes referred to as 'number density', etc. We have avoided this terminology because of the danger of confusing the word 'density' with that relating to the bulk material. Some workers use the pseudo-unit '#' for the number of particles, as the absence of a term can cause unnecessary obfuscation when manipulating units. Similarly, the units for volume concentration and area concentration are deliberately left unsimplified.

Fundamentals 15

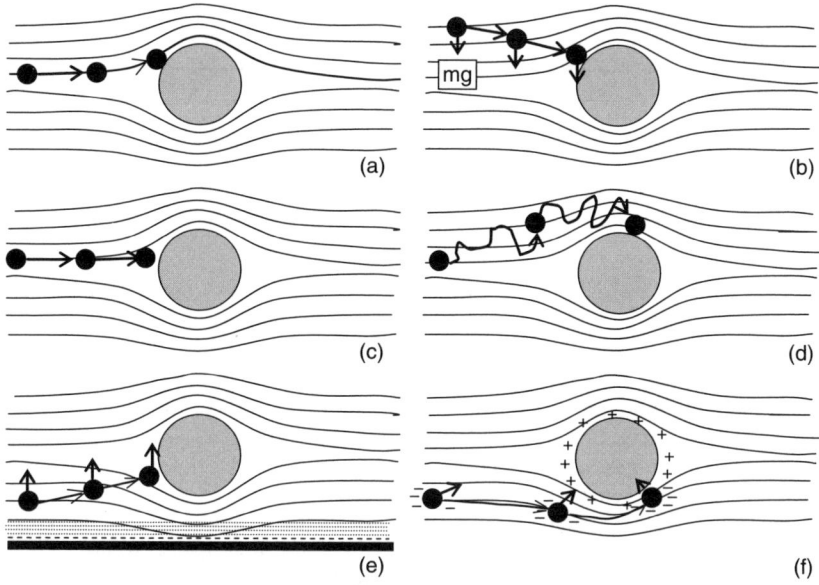

Figure 2.3 Schematic of particle transport mechanisms, and consequent deposition, onto a cylinder immersed in a gas flow: (a) interception; (b) sedimentation; (c) inertial impaction; (d) diffusion; (e) thermophoresis; (f) electrostatic attraction (adapted from Hinds, 1999, pp. 192–194).

convection, the particles suspended within it will follow, quite naturally, the streamlines, and, purely by serendipity, pass sufficiently close to a surface to be captured. For a spherical particle, for example, should its centre lie on a streamline, the critical distance for interception is the sphere's radius. With a nonspherical particle, of course, the picture is more involved, but the geometrical principle is the same. Given the fortuitous nature of the geometry required, interception is not normally a significant deposition mechanism.

Transport, as ordinarily understood, seldom concerns this simplest case, as particles are subject to a gamut of forces, the action of which ensures motion *relative* to the gas, that is, the streamlines are *not* respected. These other mechanisms of particle transport are frequently of greater importance than the simplest case; indeed, without them, particles would regularly evade capture merely by sailing triumphantly past intercepting objects.

The predominant force is, in the main, decided by particle *size*. At one end of the spectrum, particles are massive enough to experience significant inertial and even gravitational forces. At the other end of the spectrum, particles approach molecular dimensions, so that motion is influenced, and indeed dominated, by collisions with gas molecules. In this diminutive regime, the surrounding gas can no longer be considered a continuous fluid, and must, instead, be treated as rarefied with respect to the particles. The gas is, in effect, an assemblage of still smaller particles, namely *molecules*. The mean free path in exhaust gas, for example, is comparable with the size of many particles. There are thus, as outlined in Table 2.1, three regimes of behaviour; these regimes are defined nondimensionally by the Knudsen number, $2\lambda/D_p$, where λ is the mean free path of the gas molecules. This should not be confused with the mean free path of the particles, which is difficult to define, and of less use analytically (Flagan and Seinfeld, 1988, p. 294).

Table 2.1 Regimes of behaviour as determined by size of particle (D_p) in relation to the mean free path of molecules in the surrounding gas (λ).

Regime	Size	Result
Free molecular	$D_p < \lambda$	Discrete collisions between molecules and particles
Transition	$D_p \approx \lambda$	Aspects of both free and continuum
Continuum	$D_p > \lambda$	Gas acts as a continuous fluid

The drag force depends on the relative velocity between particle and gas, and this is described by the drag coefficient, which depends, in turn, on the Reynolds number Re. For $Re > 1000$, Newton's resistance law applies – derived originally for a ballistic evaluation of cannonballs, where inertial forces are much larger than viscous forces. At the opposite extreme, when $Re < 1$, Stokes's law applies, where viscous forces are more important than inertial forces. For particle sizes approaching the mean free path, Stokes's law is modified by the 'Cunningham slip correction', as the key assumption, that the relative velocity of the gas at the particle surface is zero, no longer applies; this ensures faster transport than otherwise predicted (Hinds, 1999, pp. 48–51).

When a particle is situated in a temperature gradient, so that one side is hotter than the other, the high-temperature side experiences more vigorous collisions with gas molecules than the low-temperature side. Since molecular collisions at a surface are manifested as *pressure*, a force imbalance arises; and this imbalance pushes the particle towards the lower temperature, that is, down the temperature gradient. This transport mechanism is called 'thermophoresis'. Clearly, appreciable rates of heat transfer are the controlling factor here. For $D_p < \lambda$, thermophoresis is independent of particle size; for $\lambda < D_p$, the situation is more complicated, because the particle, in acting as a thermal conduit, carries its own temperature gradient (Hinds, 1999, pp. 171–176). The practical outcome, however, of this size dependency is that, for particles of a few hundred nanometres, thermophoresis is often the only significant deposition mechanism.

Many particles are electrically charged; indeed, in the atmosphere, this situation is inevitable since air contains $\sim 10^3$ ions per cubic centimetre, courtesy of natural ionising radiation, and these ions attach themselves to particles (Hinds, 1982, p. 302). Charged particles migrate inexorably towards oppositely charged surfaces or away from similarly charged surfaces. For this reason, materials susceptible to charge build-up (such as Teflon) should be avoided in the construction of sampling lines. From an opposite standpoint, particles are sometimes deliberately charged in order to catch them: this strategy is used extensively in particle-sizing instruments. Otherwise, electrostatic attraction and repulsion are relatively unimportant for the cases we shall be studying.

Particle transport operating even in the complete absence of bulk flow,[6] that is, without any bulk-gas convection, is what chiefly distinguishes Brownian motion. This transport mechanism is notable for its stochastic rather than deterministic nature, and for its inherent homogenising action. Brownian motion arises from random buffeting by gas molecules, so that particles shudder, and trace out erratic, intoxicated trajectories. But, since particles are not generally influenced by single collisions with gas molecules – indeed, huge numbers of collisions take place per unit time – the motion is continuous and meandering, rather than, as with gas molecules, characterised by jumps and bounds (Flagan and Seinfeld, 1988, p. 294). This Brownian motion is what drives particle diffusion, as expressed by the diffusion coefficient, and this is strongly size-dependent: the diffusion of a 10 nm particle is 2000-fold

---

[6] This situation is difficult to realise, even in the most artificially favourable circumstances, but the statement nonetheless usefully emphasises the nature of Brownian motion.

faster than a 10 μm particle (Hinds, 1982, p. 137). The continual retention of particles by a surface sustains a concentration gradient: this diffusion-mediated capture mechanism is immensely significant for particles smaller than ~50 nm; indeed, it is sometimes thought, intuitively, although erroneously as it turns out, that smaller particles are harder to capture. In fact, the smallest particles are caught easily, because they diffuse rapidly towards surfaces.

As all particles possess a mass, they *ipso facto* experience forces of inertia. Consequently, whenever streamlines change direction abruptly, only the lighter particles remain faithful; the heavier ones retain their original velocity vector. In turbulent flow, particles of middling size are flung out of eddies; the heaviest particles, unable to respond to rapid velocity fluctuations, move sluggishly and phlegmatically, in accordance with the average, bulk flow. The result depends on the mass of the particle in relation to the streamline geometry, and is modelled by the dimensionless Stokes number (Hinds, 1982, pp. 113–114). As this number approaches zero, the particles follow the streamlines with ever greater assiduity. Particle deposition occurring via this means, such as at bends or contractions in the flow, is termed 'inertial impaction' or 'inertial impingement'. Inertia is also behind the concept of 'isokinetic sampling' (Hinds, 1999, pp. 206–212). This describes the way a sample of aerosol is drawn off through a probe. The ratio of flow rates and the ratio of cross-sectional areas between the sampling probe and the conduit must be equal; otherwise, the streamlines are locally distorted, and the particle size distribution of the sample is not a faithful reproduction of the population from which the sample is taken. But inertial effects are usually negligible for particles smaller than one micron.

Mass also ensures some influence from the force of gravity – this, after all, is the principle behind winnowing: separation of grain from chaff. The rate at which a particle falls merely by virtue of its own weight is termed the 'gravitational settling velocity'. If this velocity is to become significant, the size must be several microns at least, and few of the particles we shall be studying are this large. However, greatly extended horizontal sampling lines are not held up as good experimental practice. Some example calculations (cited by Lipkea *et al.*, 1978) are as follows: a 100 μm particle will fall at 30 cm/s; a one micron particle at 0.003 cm/s; and a 100 nm particle at 10^{-4} cm/s. (Such calculations are based on sphericity: nonspherical particles settle more slowly than their equivalent spheres.) The importance of gravitation, or rather the lack of it, as a deposition mechanism, is more apparent in spacecraft, where additional measures become necessary for air purification (Chu, 2004).

Particles are, therefore, disloyal to the streamlines of their carrier gas for various reasons. Conversely, using smoke or other particles to betray, visually, the streamlines (for example in wind tunnels) rests wholly on loyalty, which cannot be guaranteed without close scrutiny of the transport mechanisms just described. And, as transport is strongly size-dependent, it follows too that deposition is strongly size-dependent. The view generally entertained is that small particles (less than a few tens of nanometres) are transported by diffusion, large particles (more than many hundred nanometres) by inertia, and massive particles (more than several microns) by gravitation. Nature disdainfully fails in numerous instances to supply a strong transport mechanism for the middling particles – those of just a few hundred nanometres; and, as a result, these are destined to follow, slavishly, the streamlines of the gas, the necessary corollary of which is weak interaction with bulk matter. With filters, this weak interaction leads to a kind of 'size window', or blind spot, within which capture efficiencies are inferior (Liu Z.G. *et al.*, 2003). Other mechanisms, such as electrostatic attraction or thermophoresis, must then be promoted artificially, in order to improve the capture efficiency within this window. Equally, in cases where an aerosol passes through an open channel the width of which is large compared with the particle size, there is no reason for supposing that because no actual physical blockage, such as a filter fibre, stands in the way, then all particles will survive transit. Some particles will lodge unavoidably on the sides of the channel for the reasons explained.

Returning to the opening statement, particle capture really occurs through *two* sequential processes: transport to a surface, and retention by this surface. The first step, as it turns out, is the more difficult, because, once a particle has struck a surface, its retention is virtually assured. The question is why. Particle retention or attachment is a subject considerably less delineated mathematically than transport (Hinds, 1982, pp. 127–131). On a microscopic scale, surfaces are irregular and contact takes place at a few high points (asperities). Adhesive forces, which ultimately find their root at molecular scales (Corn, 1966), arise from three distinct sources: van der Waals interactions, electrostatic fields, and surface tension in liquid films. The first of these result from induced dipoles in the outer electron shells of the surface atoms; the last from the meniscus formed at the point of contact. Particle adhesion may involve elastic and plastic deformation in the contact *region* (rather than contact point) (Walton, 2004). These forces are decided by the nature of the particle (size, shape and composition), the wider environmental conditions (temperature and condensate), and the nature of the surface (roughness and composition). Particle hardness and the collision vector (angle and velocity) are other factors. Upon striking a surface, a particle's kinetic energy is converted into, and stored as, a localised elastic deformation. If, following the initial impact, enough of this energy of deformation is reconverted back into kinetic energy to overcome the attractive forces, then the particle rebounds. Otherwise, attractive forces quickly come to balance the deforming elastic ones, and a static condition becomes established. The plastic deformations undergone by soft particles make for stronger attachment than in the case of hard particles (Fuchs, 1964, p. 362).

2.2.4 Transformation and Mutation

Aerosols are not immutable: a host of processes modify particle composition, shape and size. These mutative processes are conveniently conceived as being of two types: gas-to-particle (or particle-to-gas) conversion, and particle-to-particle (or particle-to-surface) interaction.

Material transfer between the gas phase and the particulate phase is governed by the condition of 'saturation': this denotes the maximum amount of any volatile species that may exist in the gas phase and, mathematically expressed, this is the ratio of partial pressure to saturation vapour pressure (Hinds, 1982, p. 250).[7] Conceptually, this is identical to relative humidity, which is the more commonly employed term when water is involved, particularly in the atmosphere. When this ratio is unity, the atmosphere is said to be saturated with water; and when less than unity, it is unsaturated or subsaturated. Should conditions threaten to become supersaturated, water begins transferring from the gaseous to the liquid state via 'condensation'. As cooler air can hold less water vapour, this behaviour is triggered, most obviously, by falling temperature – as expressed by the 'dew point'.

Condensation causes gas-phase material to attach to flat surfaces, such as the bathroom mirror; yet, in principle, the surfaces of *particles* are no different. Gas-to-particle conversions of this sort are apparent in cold weather, as visible plumes formed by exhaled breath or motor vehicle exhaust. The water becomes, in effect, unstable in the gas phase, and so shifts into the particulate phase. Of course, this is unsurprising insofar as water is the chief condensable component in exhaust gas, and indeed in

---

[7] The distinction between vapour and gas is not belaboured in this text: the two terms are often enough used interchangeably. But 'vapour' often *implies* the coexistence of (indeed equilibrium between) the gas phase and the liquid phase (below the boiling point). The pressure exerted by the vapour can also be thought of as expressing the tendency of molecules to escape from the liquid. The 'saturation vapour pressure' is sometimes called the 'equilibrium vapour pressure': at a specified temperature, this is the pressure required to maintain dynamic equilibrium between gas and liquid, i.e. the mass transfer rates between the two phases are equal and opposite. Confusingly, the saturation vapour pressure is sometimes called simply the vapour pressure, for such systems are not necessarily in equilibrium. A liquid exerting a high vapour pressure at (say) room temperature is known as 'volatile'; and, at a given temperature, the higher the vapour pressure, the lower the boiling point.

exhaled breath; but identical physics applies to other condensable species. As a mutative process, the gradual accretion of material from the gas phase, via condensation, causes particle growth.

Evaporation reverses the work of condensation; indeed, through these two diametrically opposed processes, a species may exist in a state of equilibrium between the gas phase and the particulate phase. This is the direct parallel of the classical case of an enclosed chamber containing a vapour in equilibrium with a pool of its own condensed liquid. Thus, if conditions become unsaturated, the condensed component, by evaporating, returns to the gas phase, so that particles now shrink.

Particles, however, differ from pools of liquid in one critical aspect: the curvature of the surface. The partial pressure required for equilibrium is larger than for a flat liquid surface, because the attractive forces between the constituent molecules are compromised (enfeebled) by the sharpness of the curvature. The Kelvin diameter expresses the condition of stability, in which a droplet neither grows by condensation nor shrinks by evaporation; this requires an environment that is *slightly superstaturated*, rather than saturated (Hinds, 1999, pp. 281–282).

But how does growth in *existing* particles differ from the formation of entirely *new* particles (Kulmala et al., 2000)? The exact division can become difficult to define, but 'nucleation', as particle creation is styled, is believed to take place on pre-existing entities or kernels, which, being so tiny, are not commonly thought of as particles as such. Nucleation can be thought of as a special type of condensation insofar as saturation is an equally important factor. The greater the availability of condensable material, or, alternatively, the higher the saturation, the smaller the progenitor or kernel needed to commence particle creation. It turns out that two types of nucleation are possible, depending on the kernel and the saturation. Heterogeneous nucleation, or nucleated condensation, is particle formation mediated by foreign ions or condensation nuclei; and homogeneous nucleation, or self-nucleation, is particle formation mediated by molecular clusters of the nucleating substance itself. Heterogeneous nucleation is the more usual: it requires less extreme supersaturations, and, in removing material from the gas phase, restrains the saturation, and thereby suppresses its homogeneous counterpart.[8]

But either way, gas-to-particle conversion, via nucleation, produces the smallest particles of all – those of just a few nanometres: experimentally, this is inconvenient, as these particles lie below the threshold of detection with conventional particle-sizing instruments, and must be allowed to grow into the detection range. And an equally important aspect of nucleation is its switching, or 'on–off', characteristic: at some critical condition a mammoth upsurge takes place in particle creation; an example of this 'all, or nothing at all' process is shown in Figure 2.4 (Hinds, 1999, p. 284). Conversely, as the saturation falls, these extremely small volatile particles may, through the reverse process of evaporation, suddenly disappear like will-o'-the-wisps.

It should not be understood that everything in gas-to-particle conversion rests on saturation alone. If particles are hygroscopic, i.e. contain sufficient soluble material to reduce the saturation vapour pressure in their vicinity and to absorb water at relative humidities of less than 100 %, they can still grow; this is critical in atmospheric transformations such as cloud formation, and concerns the properties of 'efflorescence' and 'deliquescence' (Tang I.N., 2000). The soluble salts, of metals and ammonia, responsible for this behaviour in the atmosphere are probably not present in motor vehicle exhaust to a sufficient extent for such effects to become appreciable. And Henry's law, where the pressure exerted by a gas at an interface with a liquid is proportional to the molality of the dissolved gas, drives gas-to-particle conversion in aerosols (Jacobson, 2000); this is known to occur in the atmosphere, where water is the solvent (Tanner, 1998). The significance of this mechanism in motor vehicle exhaust has not been reported, but again, it is conceivable that, in normal circumstances, liquid-phase substances are insufficiently available for this mode of gas-to-particle conversion to become important. Then there

---

[8] The 'Aitken nucleus' refers to particle kernels in the atmosphere, as measured by forcing them to grow through the acquisition of water: this concept is used in atmospheric chemistry, and is not pursued here.

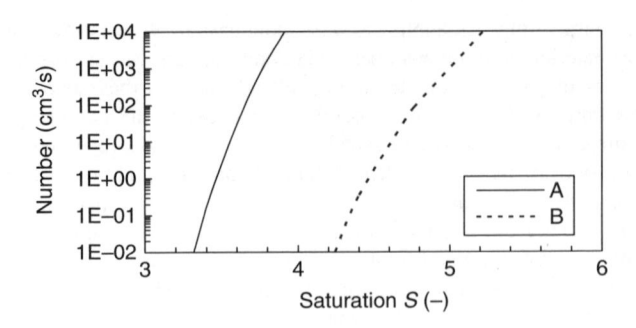

Figure 2.4 Homogeneous nucleation in pure water vapour, expressed as rate of appearance of new particles: temperature: A, 293 K; B, 273 K. The saturation S denotes the maximum amount of water that can be carried in the gas phase, i.e. the ratio of partial pressure to saturation vapour pressure (Hinds, 1999, p. 284). The Kelvin diameter is ~1.7 nm, or about 90 water molecules (Hinds, 1982, p. 255).

are physical properties, or rather geometry: capillary condensation takes place below saturation, in particle micropores or microcrevices: this is an aspect of surface tension and the contact angle between the liquid and solid surface (Kütz and Schmidt-Ott, 1992). In motor vehicle particulate this subject, although poorly reported, might be important. More generally, it can be stated that partitioning depends very much on the specific physicochemical properties of the particle; and, unlike saturation in the surrounding gas, these properties are not so readily described.

The preceding remark particularly applies to another mode of particle growth, 'adsorption', which occurs even in unsaturated conditions, and involves gas molecules attaching themselves to surfaces by means of some bond. There are two types of adsorption, classified according to whether they involve physical (van der Waals) or chemical bonds: physisorption and chemisorption. Just as with evaporation and condensation, the rates of adsorption and desorption, given sufficient opportunity, can become equal and opposite, allowing an equilibrium to exist.

The two types of adsorption are quite different. Physisorption is, usually, completely reversible; but chemisorption, since it concerns the destruction and creation of chemical bonds, often leads to desorption products quite different from the original adsorbate. This is unsurprising, since chemisorption is actually one of the key steps in catalytic processes. Consequently, the chemical specificity of the adsorption site is important, unlike in physisorption, where any gas molecule might be retained. Multiple layers can accumulate in physisorption, whereas chemisorption cannot proceed any further once the available sites have been occupied. To operate, chemisorption requires higher temperatures, or, alternatively speaking, proceeds more slowly than physisorption; this is a reflection of the greater activation energies involved.

So far, we have documented transformations that arise by gas-to-particle conversion. But at the beginning of this section, reference was made to transformations via particle–particle interactions. Particles continually collide with, and adhere to, one another; and this lascivious proclivity, known as 'coagulation' or 'agglomeration',[9] particularly distinguishes particles from molecules, whose encounters are modelled as inelastic, rebounding collisions. Strictly speaking, there is a distinction between thermal coagulation, via Brownian motion, and kinematic coagulation, via forces such as inertia (Hinds, 1982,

---

[9] These two terms are taken as synonyms in this text. Generally speaking, the choice depends on the community rather than the actual meaning: atmospheric scientists tend to speak of coagulation, whereas in combustion, agglomeration is the customary term.

p. 233); this point will not detain us. With large, loosely held agglomerations, some fragmentation may occur where there are high velocity gradients (Fuchs, 1964, p. 156); but coagulation, for all intents and purposes, is normally irreversible, causing inexorable particle growth. Small particles grow into large particles, and large particles accrete small particles. But since no new material is acquired from the gas phase, the two key signs of coagulation are, first, constant mass concentration, and second, falling number concentration.

This brings us to a well-known and much-celebrated coagulation equation derived by the Polish scientist Smoluchowski (1917). In its quintessential form, the Smoluchowski equation is often analytically intractable and hence requires numerical solutions; but, in its more commonly quoted and simplified form, the rate of coagulation, dN/dt, is expressed in terms of a coagulation coefficient, k, and the square of the number concentration, N:

$$\frac{dN}{dt} = -kN^2. \qquad (2.6)$$

Strictly speaking, this simplified version requires monodispersity, which is sufficient for the early stages, but progressively inappropriate as time passes, and also sphericity; but reasonable semiquantitative estimates are nevertheless still possible in many fields. The coagulation coefficient is derivable *a priori*, even for nonspherical particles, at least to a good approximation; and, typically, say, for 100 nm particles, $k = 10^{-9} \text{cm}^3 \text{s}^{-1}$.

Coagulation pivots on residence time and number concentration, for which the Rubicon is $kN(0)t = 1$. Taking the previous value of k, for example, this condition is fulfilled at $t = 10\,000$ s for $N(0) = 10^5 \text{cm}^{-3}$, $t = 100$ s for $N(0) = 10^7 \text{cm}^{-3}$ and $t = 1$s for $N(0) = 10^9 \text{cm}^{-3}$. The difficulty of studying aerosols at $N(0) = 10^{11} \text{cm}^{-3}$ immediately becomes clear! Figure 2.5 depicts coagulation as it unfurls according to this analysis: evidently, the number concentration cannot exceed a certain value after a given time has elapsed, regardless of the initial number concentration (Hinds, 1999, p. 266). Alternatively put, the number concentration and the elapsed time prior to measurement should be carefully assessed, to determine the amount of dilution required to arrest coagulation. A curious property of agglomeration is that, owing to competing mechanisms, broad size distributions are made narrower,

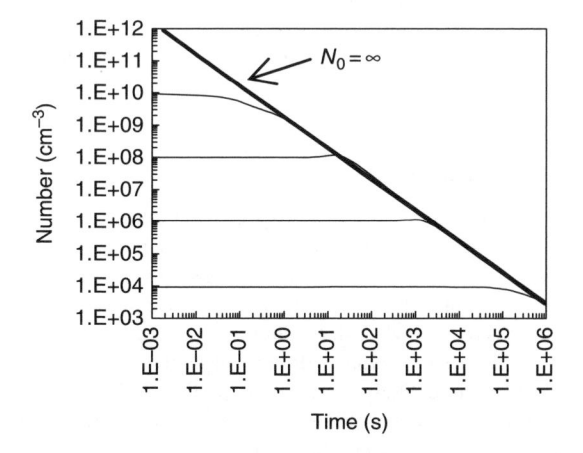

Figure 2.5 A model for the coagulation of monodisperse spherical particles, showing reduction in particle number as a function of time (coagulation coefficient k, $5 \times 10^{-10} \text{cm}^3 \text{s}^{-1}$); the asymptote represents the maximum concentration that can exist, assuming an infinite initial concentration (Hinds, 1999, p. 266).

and narrow size distributions broader, so that, given sufficient time, a stable or 'self-preserving' distribution is reached (Hinds, 1982, p. 243).

Transformations also arise through interactions with surfaces, simply because particle deposition seldom continues indefinitely. When forces of fluid shear or vibration exceed the adhesive forces, particles break off and re-enter the flow. (With a motor vehicle, pulsating flow in the exhaust system may be an important aspect, especially at locations proximate to the engine.) This phenomenon is called 're-entrainment'. For particles within the laminar layer, where viscous forces exceed inertial ones, Bernoulli's theorem does not apply (Fuchs, 1964, p. 361). In general, the adhesive forces are proportional to D_p, whereas the removal forces are, for vibration and centrifuging, proportional to D_p^3, and, for aerodynamic lift, proportional to D_p^2. This suggests that the smaller the particle, the more recalcitrant it becomes; and this is generally observed. In fact the forces of adhesion often outweigh, by orders of magnitude, the forces of transport, so that particles adhere to surfaces with remarkable tenacity. Re-entrainment has been little studied among automotive engineers, but the corresponding phenomenon of resuspension is well known in atmospheric science, in which, with increasing particle size, the drag force rises more rapidly than the adhesive force, and wind *gusts* play a significant role in raising settled particles of sand, etc. (Zufall and Davidson, 1998). The particles first begin rolling owing to the action of tangential drag, so that surface excrescences may bump them out of the laminar layer, whereafter they are taken up by turbulence (Fuchs, 1964, pp. 361, 364). Local pressures may also be depressed by vortices.

Re-entrainment is a transformation process because particles are not released in their original forms: deposits fracture along fault lines that create secondary, hybridised or agglomerated particles, of considerably greater size – several microns not being unusual. The random and largely unpredictable nature of re-entrainment makes the formulation of predictive mathematical models rather unlikely; the only real protection against factitious measurements is to avoid (especially excessive) deposit build-up, or to use preconditioning procedures.

Wall deposits retain significant volatile or semivolatile components, derived either from the particles themselves, or through subsequent take-up from the gaseous state. For example, if condensation takes place on particles, then it will presumably take place also on wall deposits. At some later stage, these wall deposits might release their volatile component to the exhaust stream. For organic compounds, this is a serious problem with low-emitting vehicles (Kado *et al.*, 2005). The ability of volatile species to modify aerosol characteristics through gas-to-particle conversion has already been described: the outgassed volatiles will briefly raise the saturation, and so encourage, for example, nucleation and condensation.

2.3 Particles in the Atmosphere

What is the size range of particles found in the atmosphere? At the small end of the spectrum, there is no clearly agreed limit, but molecular clusters smaller than one nanometre are not generally thought of as 'particles'. This includes ions to which a few neutral molecules are attached by electrostatic forces. A certain number of atoms or molecules must combine before a condition of stability is reached – below this limit, molecular clusters are short-lived and will spontaneously fragment. At the large end of the spectrum, the limit is decided by what will stay suspended for a 'reasonable' period of time, meaning seconds at least, but, more likely, hours; and particles of a few tens of microns will fall rapidly to earth. Of course, this assumes that circumstances are 'normal'. In exceptionally violent meteorological conditions, it could be argued that motor cars, and even parts of buildings, constitute 'particles'.

The variety of names accorded to atmospheric aerosols testifies to a huge diversity: dust, mist, haze, fume, smoke, fog, smog... and clouds. This creates a risk of use as synonyms, but, in their technical usage, they are all distinctly defined (Lipkea *et al.*, 1978). 'Dust' describes particles primarily of a dry, solid nature, and hence refers to processes of disintegration, e.g. erosion, abrasion, etc. 'Smoke' describes the visible component of an exhaust plume, although the word can also be applied to aerosols unconnected with combustion. A 'fume' is formed by a condensing gas, often unconnected with combustion. 'Smog', a portmanteau word constructed from 'smoke' and 'fog', is not specific to the liquid or solid phase, and relates to pollution of some form. 'Mist', 'fog' and 'clouds' betoken the condensed-water aerosols familiar in meteorology.

2.3.1 Character and Behaviour

A sample of air taken from any location in the world, and at any height, always contains a bewildering variety of particles; and, rather than ask what is found, a more appropriate question is perhaps to ask what is *not* found. In fact, virtually every element in the periodic table is airborne to some extent, even if only in trace amounts, and forty elements put in regular appearances (e.g. Herner *et al.*, 2006). To speak more concisely, however, there are six predominating particle *groups*: sulphates, organics, nitrates, minerals, carbonaceous and biological. This rich, multi-elemental profusion reflects an equally vast spectrum of sources. A broad-brush classification, into natural and anthropogenic, is convenient; and both groups, depending on the circumstances, can be described as 'pollution'. But the assumptions underpinning this classification should be considered carefully, as some sources are not easily assigned to one or the other category.

The atmospheric aerosol is a global phenomenon (Raes *et al.*, 2000): it is coupled to the biosphere, the lithosphere and the hydrosphere: for example, the weathering of the earth's crust and the formation of sediments, i.e. the rock cycle; the life cycles of many organisms; and the formation of clouds, i.e. the water cycle. The global heat balance of the earth is also affected, as some particles absorb the sun's radiation, while others scatter it back into outer space.

It is illuminating to consider just how much particulate material, each year, is included. The reader must be cautioned that such estimates are seldom anything more than educated guesses, and the errors are known to be large – although they are more likely to be factors of three or four rather than orders of magnitude. One estimate is that, on a global scale, the total mass of particulate entering the atmosphere is around $0.8-4 \times 10^{12}$ kg, with natural sources accounting for 77–94 %, and the actions of man 6–23 % (Horvath, 1998). Other estimates place the natural component at 20–50 % (Axetell, 1980). A detailed inventory is given in Table 2.2 (cited in Hinds, 1999, p. 304). According to these data, the mass emissions from natural sources far exceed the man-made ones. This evaluation, however, is on a *global* scale; in urban and industrialised areas, the anthropogenic contribution, as a percentage, is much greater. The total amount suspended at any one time is possibly $\sim 10^8$ kg (Twomey, 1977, p. 21), based on typical mass concentrations: this is small in comparison with the total flux. Thus, the rate at which material is cycled through the atmosphere and back to earth is quite rapid, making for a highly dynamic system. Most of this material resides in the lower one or two kilometres of the troposphere (Hinds, 1982, p. 275).

The atmospheric aerosol exhibits strong spatial–temporal variations, with periodicities on hourly, daily, monthly, seasonal and yearly timescales – see Figure 2.6 (cited in Salma *et al.*, 2000 and Rosinski, 2000). This reflects local variations in rates of particle input, particle removal and gas-to-particle conversion. For example, urban aerosols temporally reflect the number of vehicles on the roads or the atmospheric reactions responsible for photochemical smog, while the spatial element ensures, for example, that particles derived from combustion dominate in urban areas, whereas minerals dominate in arid areas. An example of a diurnal dispersal mechanism is provided by the height of the boundary layer, as driven by solar heating (Madhavi Latha and Highwood, 2006).

Table 2.2 Global inventory of particulate emissions (cited in Hinds, 1999, p. 304).[a]

Source	Amount (10⁹ kg/year) Range	Best estimate
Natural		
Soil dust	1000–3000	1500
Sea salt	1000–10 000	1300
Botanical debris	26–80	50
Volcanic dust	4–10 000	30
Forest fires	3–150	20
Gas-to-particle conversion[b]	100–260	180
Photochemical[c]	40–200	60
TOTAL	**2200–24 000**	**3100**
Anthropogenic		
Direct emissions	50–160	120
Gas-to-particle conversion[d]	260–460	330
Photochemical[e]	5–25	10
TOTAL	**320–640**	**460**

[a] Original references Andreae (1995) and SMIC (1971). For other global inventories see Colbeck (1995) and Raes et al. (2000).
[b] Includes sulphate from SO_2 and H_2S, ammonium salts from NH_3 and nitrate from NO_x.
[c] Primarily photochemical particle formation from isoprene and monoterpene vapours derived from trees.
[d] Includes sulphate from SO_2 and nitrate from NO_x.
[e] Primarily photochemical particle formation from anthropogenic volatile organic compounds.

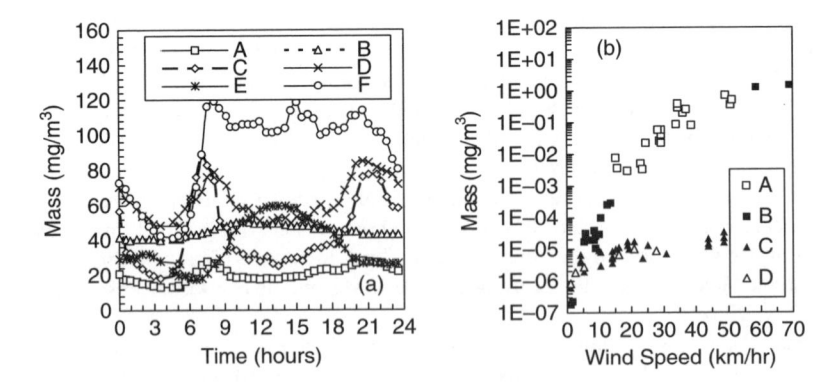

Figure 2.6 Atmospheric pollutants shifting in concentration over time: (a) diurnal variations in selected pollutants: A, CO (/10); B, SO_2; C, NO; D, NO_2; E, O_3; F, PM10; Szena Square, Budapest, averaged over 38 days from April 10 to May 17, 1996 (cited in Salma et al., 2000; original reference Salma et al., 1998); (b) particles suspended by wind: A and B, soil; C and D, sea salt (cited in Rosinski, 2000; original references Junge, 1952; Rosinski, 1996).

There are, then, numerous inputs to, outputs from and transformations within the atmospheric aerosol. Yet here we come to a remarkable and well-attested fact: the particles always fall clearly into three distinct size ranges or 'modes'. In order of increasing size, these are the *nucleation mode*, the *accumulation mode*, and the *coarse mode*. One of the modes often dominates, but this trimodal property nonetheless appears to be a universal feature – one that is always retained, no matter where or when the air sample is taken – and this property is described as 'self-preserving' (Jaenicke, 1998). Why this trimodality should show such resilience is not known for sure; perhaps it arises from the fact that a number of important physical and chemical processes (described in Sections 2.2.3 and 2.2.4) are restricted to certain ranges of size.

Mass transfer connects the three modes: this behaviour stems from the natural tendency of all particles to continue growing until they are removed from the atmosphere by processes that we shall describe shortly. Growth occurs either by agglomeration with other particles or through the acquisition of fresh material from the gas phase. The nucleation-mode particles form, as the name suggests, through the nucleation of vapours, and these then grow rapidly into accumulation-mode particles. (It is by gas-to-particle conversion that some gaseous pollutants return to the earth's surface.) However, there is little material transfer into the coarse mode; this is generated independently, such as through mechanical processes of disintegration. The two smaller modes are, to a large extent, decoupled from the coarse mode. This is also evident in the converse direction: bulk materials cannot be mechanically pulverised without limit down to ever-finer scales: processes such as grinding, for example, tend not to produce submicron particles, because the large work input causes local heating, so that re-agglomeration, through welding or melting, comes into balance with fragmentation.

Water is responsible for much of the mutability in the atmospheric aerosol; it is unique as a principal atmospheric component, inasmuch as it varies in concentration (from near zero to 3 %) and occurs in all three phases (gas, liquid and solid). There is no such thing as 'pure' water droplets as such, because pre-existing particles act as nucleation centres (Jennings, 1998), whereby they compete with one another for the available water (Ghan *et al.*, 1998). Some atmospheric particles, being highly hygroscopic, hold large amounts of water; examples are mineral acids (sulphuric and nitric), soluble inorganic salts (sodium chloride, sulphates and nitrates) and polar organic materials (Heintzenberg, 1998). The take-up and release of water, in accordance with natural fluctuations in atmospheric humidity, causes particle growth and shrinkage; when water is the dominant constituent, such particles are referred to as 'hydrometeors'. Particles act as cloud condensation nuclei (CNN) and ice formation nuclei (IFN) – both of which are central determinants in the earth's climate (Rosinski, 2000). More mundanely, this is why fogs are more frequent in cities, even though temperatures are higher and relative humidities lower than in the surrounding countryside (Seinfeld, 1986, p. 50).

Particle size, concentration and composition are, then, inherently unsettled, and impermanence in the atmospheric aerosol is guaranteed. Ongoing transformations make it a seemingly impossible task to establish provenance – there is a mixed ancestry, courtesy of persistent intermarryings between particles: part natural and part anthropogenic; part primary and part secondary; part organic and part inorganic; part liquid and part solid... leading to an almost incomprehensible anarchy. Nonetheless, air pollution abatement begins with identification of the responsible parties, and this requires *source apportionment*. Some consolation can be taken from the fact that certain particle properties are unique to a source, and so can be used to work, as it were, retroactively. We shall not dwell on source apportionment in this text, but it is worth noting that, owing to certain signatures, particles can, to some extent, be traced to motor vehicles (Schauer *et al.*, 1996).

Suspended particles ultimately return to earth at some point – this decides how far they travel, and hence the remoteness of their environmental impact. Deposition is strongly determined, as one might expect, by particle size, which is why some particles are seen regionally, and others intercontinentally. The coarse-mode particles, which are supermicron, settle quickly through gravitation, not travelling far from the source. The nucleation-mode particles are not removed as such, but grow into larger particles. Particles of 100 nm–2 μm tend to stay suspended for the longest periods: ~8 days in the lower troposphere (Mark, 1998), although those which escape from the boundary layer into the upper reaches of the troposphere may survive for several months. In fact, the accumulation mode is so named because deposition mechanisms in this size range are at their least efficient, i.e. these particles simply 'accumulate'.

The return mechanisms are of two types. Return by particle transport mechanisms alone is known as 'dry deposition', the generic principles of which were discussed in Section 2.2.3, and the atmospheric details of which are described elsewhere (Twomey, 1977; Zufall and Davidson, 1998). The counterpart, 'wet deposition', is part of *precipitation*, and refers to the scavenging of particles by water. This is usually the more dominant return mechanism, and so has a greater hand in determining atmospheric residence time. There are two types of wet deposition: 'washout', where particles are collected by falling raindrops, and 'rainout', where particles act as CNN. Falling raindrops are relatively ineffective in capturing particles of sizes lying between the diffusional and inertial regimes (Section 2.2.3); this is known as the 'Greenfield gap' (Greenfield S., 1957). Washout is thought less influential than rainout in cleansing the atmosphere of particles, but recognition that *pre-existing* particles are responsible for the formation of water droplets in clouds, fog and plumes of visible breath on cold days dates from the 1880s, and the work of Aitken and his French contemporary Coulier (Spurny, 2000f).

Figure 2.7 reveals how particle sizes in the atmospheric aerosol relate to diverse other related fields.[10] Some additional figures will be given here for reference purposes (Seinfeld, 1986, p. 276). Gas molecules such as N_2 have a diameter of 0.4 nm and a mass of 5×10^{-23} g, and are present in air at concentrations of 10^{19} cm^{-3}. Aerosol particles between, say, 10 nm and 10 μm, have masses

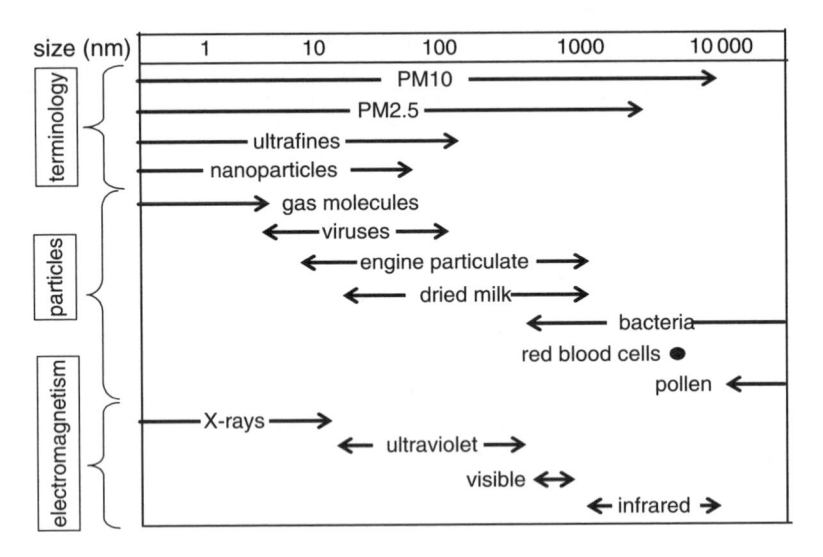

Figure 2.7 Particle size ranges for some common aerosols, compared with electromagnetic waves and other terminology used in the study of aerosols.

---

[10] A much more comprehensive version, published by the Stanford Research Institute, appears in many books on aerosols. See Hinds (1982, p. 8), and Lipkea *et al.* (1978).

of 10^{-18} g – 10^{-9} g, and are often present at concentrations of less than 10^8 cm^{-3}; in fractional terms, this is 1–100 ppb. Mass concentrations in the atmosphere range from 2 μg/m^3 in the polar regions, through 30 μg/m^3 at less remote sites, to 150 μg/m^3 in polluted urban areas. For comparison, mass concentrations in exceptional circumstances such as the London smogs of the 1950s reached 4000 μg/m^3, while sandstorms may reach 100 000 μg/m^3.

2.3.2 Aerosols in Nature

The natural atmospheric aerosol is dominated by four particle groups; these are: (a) biological (microorganisms, and necrotic cellular structures from larger organisms); (b) biogenic (nonliving chemical substances or structures released by plants); (c) oceanic (sea salt, and sulphates derived from the sea); and (d) geogenic (material of the earth's crust). *Extraterrestrial* dust, in the form of 'micrometeorites', is arguably a fifth group, although this is not discussed here.

The world's oceans are prolific contributors to the atmospheric aerosol – a natural consequence of their vastness, and relentless agitation by the wind. The particles are created predominantly via the 'bubble-bursting' mechanism, meaning the escape of air previously entrained by waves (Twomey, 1977, pp. 29–31), but also by wind shear, in which droplets are torn directly from wave crests. The concentration of sea salt particles emerging from the sea can be expressed in terms of wind speed at the water surface, and the sizes are in three modes, described as 'film drop' (100 nm), 'jet drop' (1 μm) and 'spume drop' (6 μm) (Ghan *et al.*, 1998). The sizes are, however, dependent on relative humidity (Zhang K.M. *et al.*, 2005). As these particles are soluble, indeed eminently so, they ardently act as CNN; and they are created in sufficient numbers to become major regulators of the earth's climate.

Sea salt particles are readily identifiable, as compositionally they differ clearly from terrestrial particles, with which they constantly intermingle and intermarry in coastal regions.[11] They chiefly consist, as one might expect, of sodium chloride (NaCl), although other salts, such as calcite (CaCO$_3$) and dolomite (CaMg(CO$_3$)$_2$), are also present (Hoornaert *et al.*, 1996). At their birth, sea salt particles carry the sodium–chlorine ratio of the sea, 1.8; however, as they age and proceed over land, the chlorine, through size-dependent atmospheric reactions (Yao *et al.*, 2003), especially with pollutants (Roth B. and Okada, 1998), is preferentially depleted: the characteristic composition thereafter is lost, and particle provenance thereby obscured. For example, reactions with nitric acid (HNO$_3$) and nitrogen dioxide (NO$_2$) are thought to convert particles to sodium nitrate (NaNO$_3$) (Kadowaki, 2000).

Secondary sulphate particles derived from the sea, by contrast, have a biological origin. Phytoplankton, notably during algal blooms, produces dimethyl sulphide (DMS, or CH$_3$SCH$_3$) and carbon disulphide (CS$_2$), the action of which, as particle precursors, is suspected, although the exact mechanisms are unclear. Ocean surface waters are heavily supersaturated with DMS, and this steep concentration gradient undoubtedly aids transfer across the water–air interface. Once in the atmosphere, the DMS probably reacts with the hydroxyl radical (OH), with the creation of sulphuric acid (H$_2$SO$_4$), a well-known particle precursor, or particles of ammonium hydrogen sulphate (NH$_4$HSO$_4$). Sulphate particles arising from DMS are thought to be critical in cloud formation by acting as CNN (Hewitt and Davison, 1998).

The biological aerosol concerns living or dead life forms, usually at the cellular level. Some particles are single-celled microorganisms; others are structural units from larger organisms, broken off at death or discarded during growth. Five types will be described here, with particular regard to composition and morphology (Matthias-Maser, 1998). (a) *Viruses* (5–250 nm); compositionally, they are nucleic acids (DNA or RNA) encased in protein shells, and they are morphologically symmetrical, for example

---

[11] In winter months, one should not overlook road de-icing salt (Owega *et al.*, 2004).

spherical or oval. They are noncellular, and cannot reproduce on their own, but must enter living cells to do so. (b) *Bacteria* (>200 nm): compositionally, they are DNA, RNA, proteins, lipids and cellulose, and they are morphologically varied, e.g. rods, spheres or spirals. Aside from the bacteria *per se*, some toxicity arises from structural components within the cells known as 'endotoxins'; these remains are also found in ambient particles (Osornio-Vargas *et al.*, 2003). (c) *Spores* (>500 nm): vessels of reproduction ejected by lichens, ferns, fungi, etc.; compositionally, they are protoplasm, DNA and RNA, enveloped by a chemically resistant shell of highly polymerised esters, and, morphologically, they are spherical, oval, fusiform, needle or helical. (d) *Pollen* (>5 μm): compositionally, they are proteins, carbohydrates, lipids and minerals. Morphologically, they are hugely varied; the outer surfaces are marked by numerous furrows and pores, the exact characteristics of which are unique to the progenitive plant. Most pollen particles are anemophilous;[12] some possess airbags that improve buoyancy. One plant can release a billion pollen particles per season. (e) *Debris* (>1μm): fragmented plant litter (especially leaves), parts of insects, and epithelial cells from animals. The characteristic structural materials are cellulose and starch (plants), skleroproteins (hairs) and chitin (insect fragments). Debris is not always easy to identify.

The composition of the biological aerosol varies throughout the year: spores and pollen predominate in the spring and summer; the detritus of decay processes proliferates in the autumn and winter. Distinct seasonal peaks in pollen, relating to the flowering periods of certain plants, and in debris derived from leaf fall, are detectable. But once particles have settled, they can still be disturbed. Winds refloat or re-entrain a large amount of pollen, for example, hence explaining the autumnal resurgence in hay fever. Bacteria responsible for decay processes must reach their hosts, and this is ensured by appropriate seasonal transport by the wind. An example of seasonal variation in the biological aerosol is given in Figure 2.8 (Matthias-Maser, 1998): birch and hazel pollen in March; grass pollen in June; decaying material in October; and, finally, spores in December.

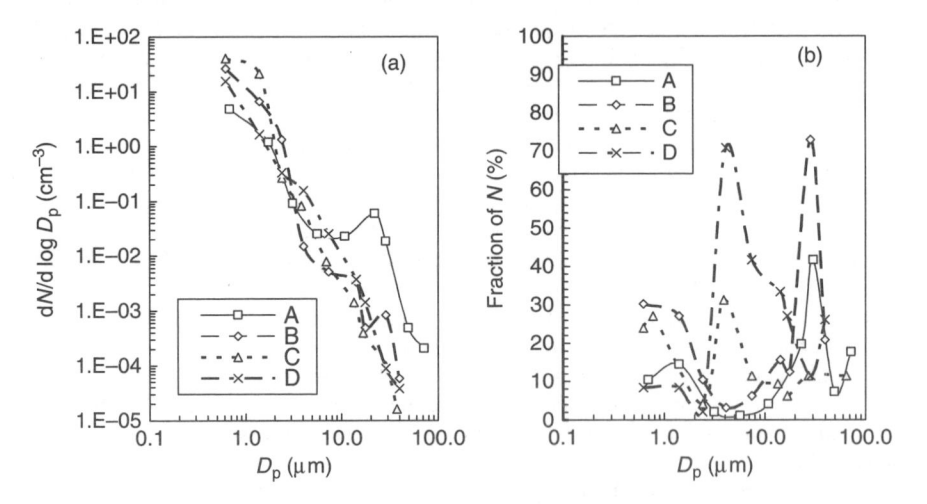

Figure 2.8 The biological aerosol monitored throughout the year: (a) number; (b) the same data as in (a), expressed as a percentage of total suspended particles. Legend: A, March; B, June; C, October; D, December. As sampled by an isokinetic two-stage impactor, and analysed under the electron microscope (standard deviations in the original have been omitted) (Matthias-Maser, 1998).

---

[12] Distributed by the wind.

The biological aerosol, in the strictest sense, is not always 'natural' in origin, as human activity is often involved. Agricultural activities, for example threshing, harrowing and ploughing, release large quantities of airborne microbes and plant debris; other strong outdoor sources of microorganisms are sewage treatment plants, livestock manure, animal husbandry, textile mills and abattoirs. Enclosed environments where airborne pathogens linger, such as in barns where livestock are kept at high density, are a hazard to workers (Rule *et al.*, 2005). Humans themselves are responsible also: coughing and sneezing release bacteria and viruses, while in the domestic arena, dressing, bed-making, shaking clothes, etc. are efficient methods of redistribution. Human bodies are prodigious sources of dead skin cells, the accumulation of which, as house dust, provides fodder for dust mites; the droppings from these mites provoke allergic reactions in humans. Lint, skin flakes, dust mites and dust-mite faeces are all part of the indoor biological aerosol. Coughing is an efficient method of spreading disease: mucous and saliva droplets of several microns size can evaporate more swiftly than they settle, allowing pathogens to survive as airborne residues, thus travelling further (Wan *et al.*, 2007). Inevitably, indoor air quality is closely linked to public health (Harrison P.T.C., 1998) via airborne allergens and microorganisms (Jones A.P., 1999).

The 'biogenic' aerosol, as distinct from the biological aerosol, is nonliving, and denotes organic compounds released by organisms (principally plants). Wax-like lipids, for example, grow on the cuticular[13] membranes of leaves, to form crystalline structures of micron and submicron dimensions (Matthias-Maser, 1998); these waxes are then released into the airstream by mechanical abrasion through leaf chafing. The principal components are *n*-alkanes (Bi *et al.*, 2002) and *n*-alkanoids. Molecules containing odd numbers of carbon atoms are more prominent in these *n*-alkanes than in crude oil (Fraser *et al.*, 1998), offering potential for source apportionment; molecules of higher carbon number are observed in warmer seasons (Gogou and Stephanou, 2000).

Plants release gas-phase organic compounds which, through ongoing atmospheric reactions, form secondary organic particles; the release, and the atmospheric reactions, depend on the temperature and light intensity. Although investigators have focused on monoterpenes and isoprene (Kroll *et al.*, 2006), plants are known to emit at least 400 different species of volatile organic compound (VOC), of which alkenes, organic acids, alcohols and C_{10}–C_{40} terpenoids are the main groupings. Why plants emit these compounds is, though, not known for certain: they might attract pollinators, act as herbicides or pesticides, or simply represent waste products. The atmospheric reactions are complex (Kavouras *et al.*, 1999), but involve particularly the reactive organics, notably isoprene (C_5H_8), the C_{10} family of monoterpenes, α- and β-pinene and *d*-limonene (Jaoui *et al.*, 2006); and also, especially in polluted environments (Kleindienst *et al.*, 2006), O_3, OH, NO_x and SO_2. Various secondary particles are created: a notable example is pinonaldehyde, the presence of which in forest air is well documented, and the precursors of which are α-pinene and ozone (Hewitt and Davison, 1998). The haze sometimes seen in deeply forested areas is ascribed to secondary aerosols derived from trees. For urban and rural settings alike, laboratory experiments and model calculations continue to suggest an abundance that is comparable with man-made primary particles; these findings have yet to be reconciled with actual field measurements, which demonstrate a considerably lower occurrence (Hewitt and Davison, 1998).

Two distinct types of geological process contribute to the natural atmospheric aerosol; the more violent is volcanism. Explosive eruptions eject extremely large quantities of debris into the atmosphere, often to high altitudes. Yet volcanoes, even when simply smouldering, are abundant sources of particles, because temperatures are sufficiently high to vaporise elements in the magma; these elements

---

[13] The cuticle is the continuous protective layer of impervious material secreted by cells of the epidermis (the outermost layer of cells).

subsequently condense or nucleate to form particles (Ammann *et al.*, 1990). Characteristic among these elements are arsenic, cadmium, selenium, zinc, lead and nickel (Claes *et al.*, 1998). Volcanism is thought to account for 40–50 % of airborne cadmium and mercury, and 20–40 % of airborne arsenic, chromium, copper, nickel and lead (Pacyna, 1998). But such estimates are problematic, because of compositional variability, and the intermittency of volcanic eruptions. Volcanoes, as prolific emitters of sulphur dioxide, also contribute significantly to secondary sulphate particles. Pyrotechnics are not, however, an essential prerequisite: following their leaching from underground rocks, geothermal energy causes the emission, at tens to hundreds of micrograms per cubic meter, of silicon, nickel, zinc, copper, molybdenum and tungsten from hot springs, fumaroles[14] and wetlands (Planer-Friedrich and Merkel, 2006).

The less violent geological process is an essential part of the geochemical rock cycle, wherein crustal material is continually broken up and re-laid in the form of sedimentary deposits that ultimately constitute new rock. This happens through freeze–thaw action, abrasion by wind-blown particles, and the incursion of plant roots. On geological timescales, ever-finer fragments are formed, until particles are sufficiently small for aeolian transport. Silicon and aluminium figure prominently, these being the two most abundant elements in the earth's crust (Kadowaki, 2000). Particle composition betrays the surface or landscape of origin, e.g. tundra, arable, savannah, etc. Clearly, arid regions, such as deserts and tundra, are the most to blame for wind-borne dust. These particles travel large distances: Saharan dust is found in northern Europe and the Caribbean, and Gobi dust in the Arctic. Besides dry dust, this aerosol concerns soil particles: these are water-insoluble clay minerals, impregnated with an organic component of decomposed vegetation (humus). Soil particles account for over half the amount of airborne chromium, manganese and vanadium, and 20–30 % of the copper, molybdenum, nickel, lead and zinc (Pacyna, 1998).

Finally, natural processes of decomposition release ammonia, the reaction of which with airborne nitric acid and sulphuric acid, both common atmospheric pollutants, creates secondary particles of ammonium nitrate and sulphate. Of course, when ammonia arises from farming, such as from decaying livestock manure (Rule *et al.*, 2005), one might argue whether these secondary particles are natural or anthropogenic. But on a global scale, the anthropogenic sources of ammonia are probably outweighed by the natural sources.

2.3.3 Anthropogenic Aerosols

As in nature, anthropogenic particles may be primary or secondary. Three prominent contributors to the primary anthropogenic aerosol are: (a) combustion (usually, but not necessarily, that of fossil fuels and biomass), (b) material disintegration (e.g. from surfaces such as roads, especially unpaved ones) and (c) sundry high-temperature industrial processes. Secondary particles are dominated by photochemical smog, London smog, nitrates and sulphates.

Agrochemical aerosols (pesticides, herbicides and fungicides), while less readily classified, are arguably the result of a kind of material disintegration. When administered from tractor booms or low-flying aircraft, these agrochemicals emerge as mists composed of coarse-mode droplets ($<100\,\mu m$), for which delivery losses are large, i.e. 60–80 % (Spurny, 2000e). Deposition onto surfaces, especially plant leaves, and subsequent drying creates powders that, in fragmenting and exfoliating, release new particles, while other particles in the atmospheric aerosol, when contaminated, act as unwitting carriers (Bester and Hühnerfuss, 2000). Being environmentally resilient, or 'biopersistent', agrochemicals travel huge distances; hence their continuing survival, in urban areas, at tens of nanograms per cubic metre,

---

[14] Holes near a volcano though which vapour emerges.

and even in the polar regions, at a few hundred picograms. There are serious health risks for farm workers, and probably for persons more distant.

Material disintegration includes explosions, such as in construction, demolition, mining and acts of war; less appreciated are the implications of propulsion explosives (e.g. potassium compounds), flame-colouring agents (e.g. copper compounds) and oxidisers (e.g. barium compounds) found in fireworks (Moreno et al., 2007). Although such festivals are short-lived, participants are nonetheless exposed to high levels of particles, and the metals released are not trivial in terms of annual airborne emissions. But, other than pyrotechnically, material disintegration concerns *surfaces*. Attrition by *weathering* is arguably a natural process, but the particles thereby created are anthropogenic insofar as man's activities expose certain vulnerable surfaces. Amongst these particle sources are spoil heaps and stockpiles (Badr and Harion, 2007), open fields, construction sites, and opencast mines. A second type of surface attrition involves relative movement, for which there are many expressions: scuffing, rubbing, grinding, friction, abrasion, erosion, fretting, scratching, chafing, rasping... and 'wear and tear'. Corrosion and thermal stresses also attack surfaces.

The disintegration of asbestos-containing materials, and the concomitant release of asbestos fibres, is a well-known, but often all too hidden health risk; and the morphology of these fibres has immediate bearing on the notorious carcinogenicity; the most menacing particles are those with widths of $<3\,\mu\mathrm{m}$, lengths of $\geq 5\,\mu\mathrm{m}$ and aspect ratios of ≥ 3 (Spurny, 2000d). The fibres have unusual aerodynamic properties: even exceptionally long ($50\,\mu\mathrm{m}$) ones are able to evade the filtering mechanisms in the upper reaches of the respiratory tract: they then lodge in the alveoli, from which the body cannot evict them.

Although the dangers to *occupational* health were recognised by the 1900s, it was, in fact, the spread of asbestosis to the general public that ultimately led to an outright ban. But, although banned, asbestos nonetheless continues to menace public health: building materials – particularly roof tiles – were at one time a significant market for asbestos; and weathering, which damages the fibrous matrix, creates a layer of weakly bound fibres. The total surface area presented to the atmosphere in the form of roofs, facades, etc., is large, and, in totality, this implies a truly staggering weight loss. In Germany alone, a total surface area of $\sim 10^9\,\mathrm{m}^2$ suggests an annual loss of several hundred tonnes (Spurny, 2000c). The fibres, moreover, because of their exposed locations on building exteriors, etc., carry other undesirable pollutants, such as polycyclic aromatic hydrocarbons (PAH) and heavy metals (Spurny, 2000c); and, because of their inertness, the fibres, once released, continue lurking in the environment, where, following deposition, they are continually disturbed, and thereby repeatedly raised.

Material disintegration, when acting within motor vehicles, encompasses rubbing, such as in brake linings and clutches, and corrosion, such as in the exhaust system. Motor vehicles generate other particles by virtue of their interaction with road surfaces – an enormous quantity of tyre material apparently 'disappears'. Road surfaces themselves disintegrate under the constant pounding; this releases asphalt and mineral dust. Prodigious quantities of soil or mineral dust are thrown up from the surfaces of unpaved roads – this problem is not limited to undeveloped countries; it occurs at all construction sites and opencast mines, and wherever large-scale landscaping is in progress. With railways, similar wear releases iron (Weckwerth, 2001) and manganese (Bukowiecki et al., 2007) particles from rails, and copper particles from overhead electrification.

Sundry high-temperature industrial processes are prolific emitters of heavy-metal particles; smelting, sintering, refining, galvanising and roasting are well-known culprits (Claes et al., 1998). The metals volatilise in the high-temperature zone, and undergo gas-to-particle conversion in the cooler conditions of the flue. Arsenic, zinc, copper and cadmium are characteristic in nonferrous-metal production; zinc in the galvanising of iron and steel products; lead, copper and zinc in pyrometallurgical processes; and chromium, manganese and graphite in the manufacture of iron and steel (Machemer, 2004). A summary of emission factors, as used in estimating the release of five metals from various industrial processes and from different types of combustion, is given in Figure 2.9 (Pacyna, 1998).

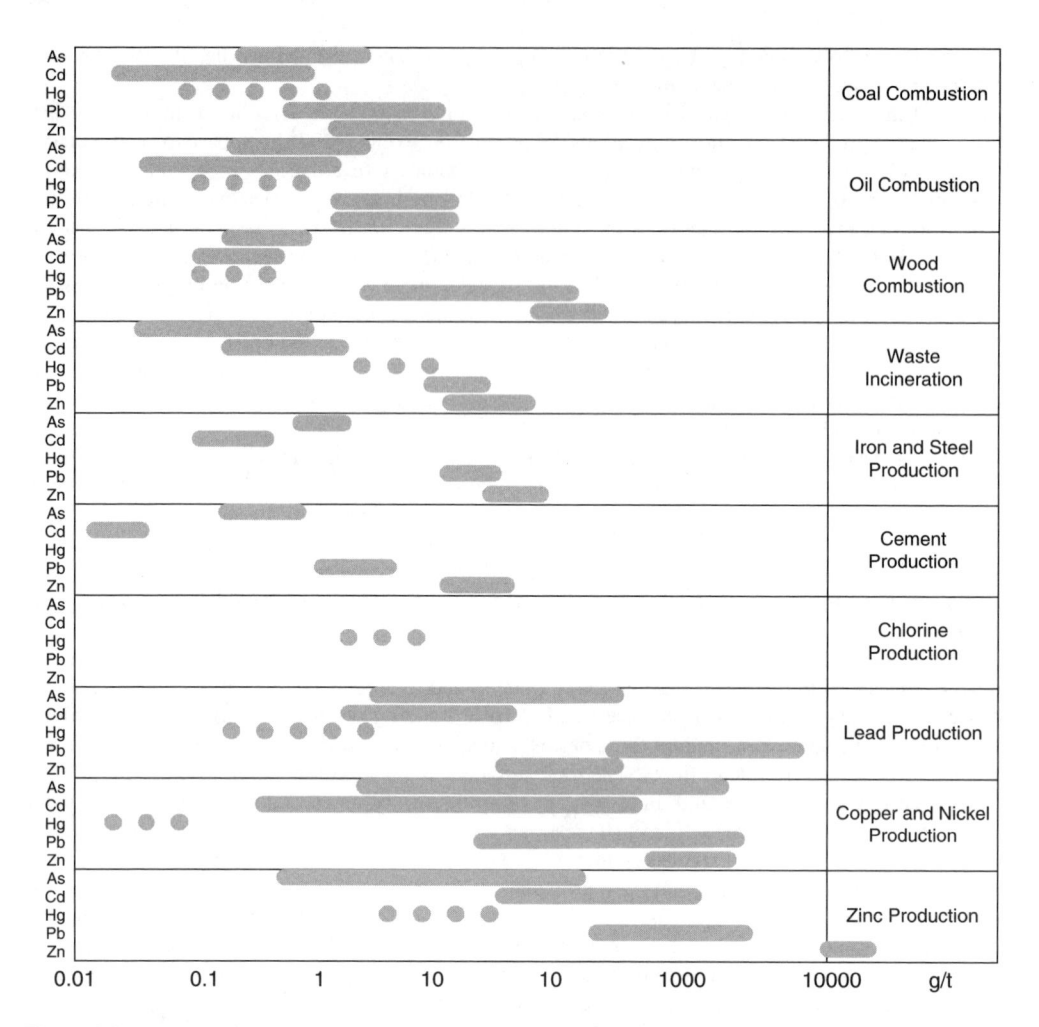

Figure 2.9 Emission factors for arsenic, cadmium, mercury, lead and zinc, as have been used in the European inventory for various combustion and industrial processes. The inference is grams of particles emitted per tonne of industrial goods produced or per tonne of raw materials consumed (cited in Pacyna, 1998) (original reference Axenfeld *et al.*, 1992).

Carbon, an element never absent from the atmospheric aerosol, is found in three distinct forms: (a) carbonate (del Monte and Rossi, 2000; Chow *et al.*, 2003); (b) myriad organic compounds, referred to collectively as organic carbon (OC); and (c) carbonaceous or quasi-graphitic material dubbed elemental carbon (EC). The first is a minor component associated with rocks and (lime)stone, as found in cement and other building materials, of which carbonates of magnesium and calcium, the mineral forms of dolomite and calcite, are common (Pooley and de Mille, 1999). The other two forms of carbon make regular appearances, and are overwhelmingly generated in the combustion of carbonaceous materials – chiefly fossil fuels and biomass. Global emissions of EC and OC, per year, are estimated at $4.3-22 \times 10^9$ kg and $17-77 \times 10^9$ kg, respectively (Bond *et al.*, 2004). Less

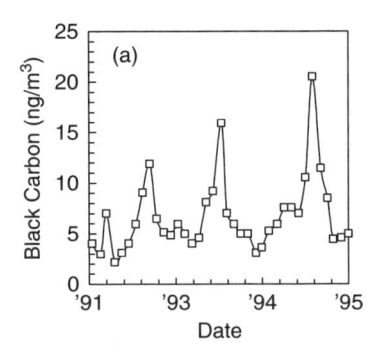

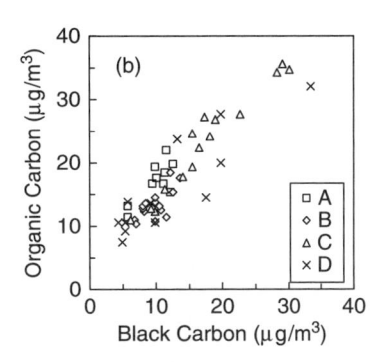

Figure 2.10 Mass of airborne carbon in the form of particles. (a) A three-year recording of elemental or black carbon at the remote site of Amsterdam Island in the Indian Ocean (77°E, 38°S), where significant incidences of biomass combustion follow each dry season in the southern hemisphere (June to September) (cited in Cachier, 1998; original reference Cachier *et al.* 1996). (b) A scatter plot showing the close association between organic carbon and black carbon (carbonaceous particulate). Location: an urban district of Nagoya, the fourth largest city in Japan. Samples were collected over 24 hours from 1984–86. Season: A, spring; B, summer; C, autumn; D, winter. The combined correlation coefficient r^2 was 0.904 (Kadowaki, 2000).

obviously, meat cooking, especially charbroiling, produces much of the OC found in urban areas, for which cholesterol and various fatty acids serve as markers (Nolte *et al.*, 1999). OC and EC are usually present in the same particle: the first, which is volatile, forms an encapsulating coating for the second, which is solid. Particles of this nature are found in the remotest areas, such as the Arctic, and far from land; Figure 2.10(a) (cited in Cachier, 1998) is an example. That the two forms of carbon are closely associated is illustrated in Figure 2.10(b) (Kadowaki, 2000); but this correlation is sometimes confounded by secondary organic carbon (SOC) or secondary organic aerosol (SOA) – not directly emitted, but created subsequently by atmospheric reactions among gases.

Aerosol carbon is produced in the combustion of gasoline and diesel in motor vehicles; heating oil in industrial and domestic boilers or furnaces; and coal in power stations. In developing or industrialising countries, there is the burning of domestic coal (Florig *et al.*, 2002) or biomass in residential boilers or cooking stoves, although in developed countries, wood-burning stoves are also known as strong sources of particles (Silva *et al.*, 1999). Biomass possesses considerable diversity; the term encompasses materials such as wood, brush, peat, straw and dung. Particle production greatly depends on whether the combustion (often poorly controlled) is smouldering or flaming. The destruction by fire of natural habitat such as grasslands and forests is a prodigious source of aerosol carbon. This destruction is sometimes prescribed, as part of sensible land management (Lee S. *et al.*, 2005), but also happens inadvertently or accidentally – sometimes with a *laissez-faire* attitude. Agricultural practices, land use and drought, contributed to the 1994 smoke haze in south-east Asia, the coverage of which was estimated at 3 million square kilometres – an area the size of western Europe (Nichol, 1997).

The carbonaceous, or solid, core is a collection, or *agglomeration*, of much smaller particles, sometimes called 'spherules', the sizes of which are ∼20–50 nm, and the microstructures of which are partly amorphous and partly graphitic or quasi-graphitic. The morphology of the whole particle, or 'agglomerate', is characteristic of the combustion process, e.g. globular for biomass, fractal for diesel, and compacted for coal (Nyeki and Colbeck, 2000). The agglomerate forms a kind of skeleton, onto which organic compounds, emitted as gases, condense as the exhaust gas cools. The relative proportions of carbonaceous and organic particulate are characteristic of the source. In biomass combustion,

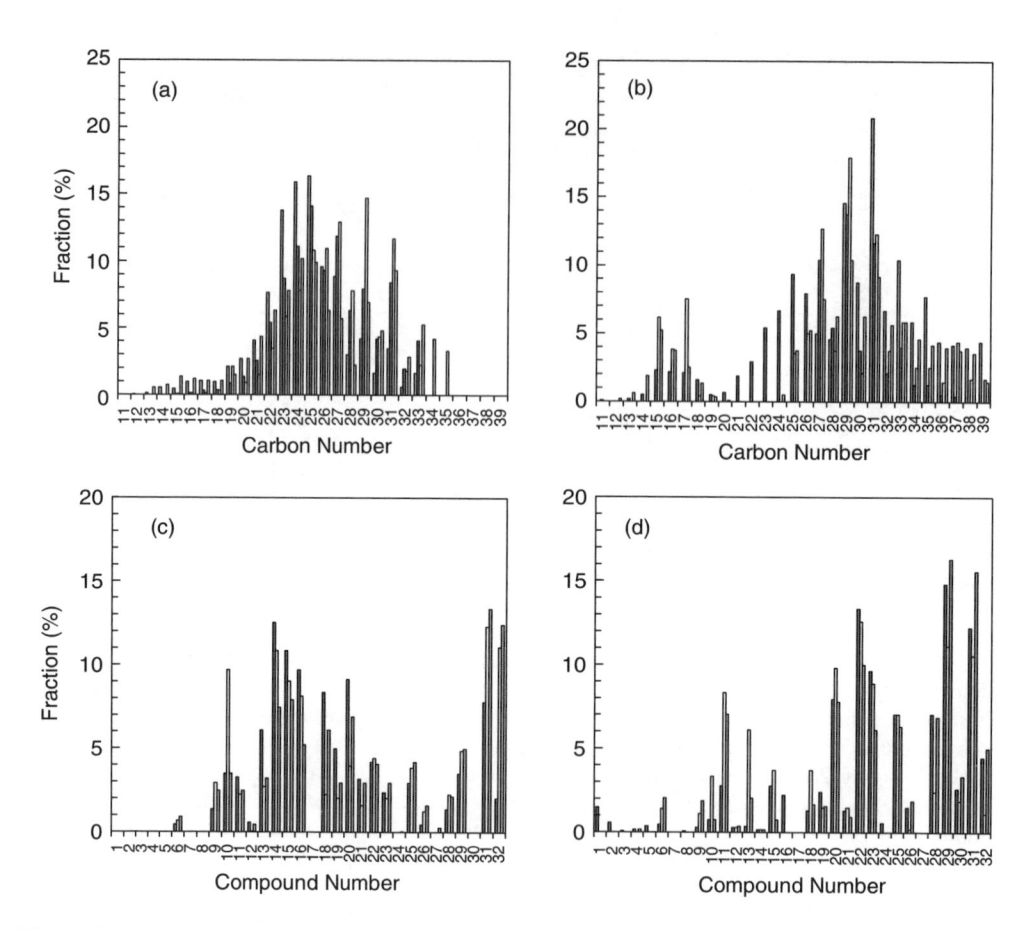

Figure 2.11 Compound distribution in the organic fraction of the atmospheric aerosol, as sampled at two locations: the rural site was an arid coastal region of Crete, and the urban site was atop a 15 m high building in the city centre of Heraklion, Crete: (a) *n*-alkanes, urban (four samples); (b) *n*-alkanes, rural (four samples); (c) PAH, urban (three samples); (d) PAH, rural (three samples). The *n*-alkanes are arranged according to the number of carbon atoms in the molecule. The PAH are arranged according to the compound number (and rising molecular weight; for their identity see original reference); the numbers of rings are as follows: 2 (1, 2); 3 (3–10, 14); 4 (11–13, 15–21); 5 (22–27); 6 (28–32) (Gogou and Stephanou, 2000).

particles are more organic; in petroleum combustion, particles are marginally more carbonaceous; and in coal combustion, particles are highly carbonaceous.

As illustrated by Figure 2.11 (Gogou and Stephanou, 2000), the chemical composition of the organic component is hugely diverse, and full speciations are normally impossible. Most of the important chemical families are represented: alkanes, alkenes, alcohols, ethers, ketones, acids, esters and aromatics. As the light C_4–C_8 compounds are also found, of the type theoretically in the gaseous state, the importance of surface interactions with the underlying carbonaceous skeleton in tying down these compounds is evident.

The organic compounds carry genuine toxicological risks (Smith D.J.T. and Harrison, 1998), of which PAH, a well-represented group, are well noted for their carcinogenicity. Residential heating, power plants, industrial boilers, municipal and commercial incineration, open burning (e.g. agriculture and forest fires), coke and carbon black production, motor vehicles... and barbecues all emit PAH; biomass and fossil fuel compounds are distinguishable by radiocarbon analysis (Kumata *et al.*, 2006). But rather than inhalation, *food* is considered the major source of human exposure, through ingestion of PAH formed during cooking. Less obvious is the failure of food-processing technology to remove organic compounds that deposit on crops prior to harvesting. Respiratory exposure is, therefore, easily confounded by dietary exposure (Larsen and Larsen, 1998). The species of PAH present, on an individual basis, vary with the type of combustion process; and this variation serves as a tracer or fingerprint (Khalili *et al.*, 1995), such that it can distinguish motor vehicle particulate from particles in the ambient generally (Currie *et al.*, 1997; Smith D.J.T. and Harrison, 1998), or even distinguish between gasoline and diesel vehicles (Nielsen, 1996). The fate and environmental impact of these 'pyrogenic' compounds, i.e. those synthesised in the combustion, differ from those of the 'petrogenic' compounds, i.e. those present in the fuel (Thorsen *et al.*, 2004). PAH are only very gradually broken down in the environment, and the oceans are thought to be important sinks (Gustafsson *et al.*, 1997), but their peregrinations are extensive, as these molecules are even found in the Arctic (Becker *et al.*, 2006).

Combustion processes emit many other forms of particle, and numerous elements other than carbon. Potassium (Silva *et al.*, 1999) and chlorine (Lee S. *et al.*, 2005) are characteristic of biomass; nickel and vanadium (Dillner *et al.*, 2005) are characteristic of heating oil. Coal is notorious for containing a wide range of unwelcome inorganic elements, and in significant but highly variable amounts[15]: some metals are associated with organic compounds (e.g. Ge, Be, Sb and B), some with inorganic compounds such as sulphides (e.g. Zn, As, Cd, Fe and Hg) and some with both groups (e.g. Al, Ti, Mo, Mg and Se) (Pacyna, 1998). Fossil-fuel power stations are estimated to account for half the global emissions of trace metals emitted by combustion processes (Pacyna, 1998). In coal combustion, these metal particles form what is known as 'fly ash'. Morphologies are variously spherical, angular, irregular or quartz-like (Malhotra *et al.*, 2002). The two commonest ash-forming metals are iron and vanadium, but there are a host of others, e.g. lithium, sodium, potassium, chromium and manganese... it is perhaps a little easier to state which metals are *not* found. Just as in industrial processes, metals vaporise in the high-temperature zone, and then, in the cooler conditions of the flue, transfer into the particulate phase. The capture of these particles with electrostatic precipitators, prior to emission, is nowadays a legal requirement.

The particles discharged from motor vehicle tailpipes, in normal circumstances, are predominantly carbon, that is, both carbonaceous and organic, the ratio of which varies significantly, depending on the duty cycle and the type of engine. Inorganic elements are, however, present, albeit at trace levels, in gasoline and diesel fuel, either naturally, from the crude; inadvertently, through contamination; or deliberately, as additives; and these elements must also be emitted. In specific instances the inorganic component is significant or even dominant: leaded gasoline is perhaps the most infamous example of this.

The incineration of municipal (Gallorini, 2000), medical (Mukerjee *et al.*, 1996) and industrial waste generates particulate as well; indeed, the presence of sundry inorganic material in the feedstock is inherent, as waste materials are so diverse, e.g. textiles, wood, paper, putrefactory substances, plastics... In some countries, incineration has overtaken landfill as the primary means of waste disposal, but the particles cannot be emitted to the atmosphere with impunity. Metals are well represented; some, such as arsenic and cadmium, are emitted more readily as particles than others,

---

[15] The environmental problems created by the release of radioactive material (radium and thorium) are comparable to those arising through radioactive waste from nuclear power plants (Wilson R., 1996).

such as cobalt and iron (Gallorini, 2000). The particles are clearly distinguishable by virtue of their enrichment in less usual elements such as zinc, phosphorus or chlorine. Waste incineration is infamous for the release of chlorine as polychlorinated dibenzo-*p*-dioxins (PCDD)[16] and polychlorinated dibenzofurans (PCDF), the potential risks of which to the public are being closely scrutinised (You *et al.*, 1996).

Two dominant anthropogenic particle groups are sulphates and nitrates, both of which are emitted, predominantly, as gas-phase particle precursors, although the secondary particles that form may actually outweigh by severalfold the directly emitted PM2.5. Interestingly, in PM2.5, the two groups are of roughly equal abundance on the east coast of the USA, whereas on the west coast, nitrates are overwhelmingly dominated by sulphates; this is a reflection of source strengths as well as climate (Spengler and Wilson, 1996).

Nitrate particles are the less abundant: some are primary, for example from the production of ammonium nitrate fertilisers (Wellburn, 1994, p. 63), but most are secondary, and arise from combustion processes, owing to the formation, through unwanted secondary reactions, of nitric oxide (NO) and nitrogen dioxide (NO_2). These gases are subsequently oxidised in the atmosphere to nitric acid (HNO_3), which does not, on its own, readily undergo gas-to-particle conversion (Kadowaki, 2000); the acid is, however, readily neutralised by ammonia, a compound of ubiquitous presence in the atmosphere, and this reaction creates particles of ammonium nitrate (NH_4NO_3). This compound is volatile, and may dissociate, or revert, into its gas-phase precursors if it becomes unstable in the particulate phase (Claes *et al.*, 1998). The fraction of secondary nitrates assigned to motor vehicles is placed at a few tens of per cent (Mensink and de Vlieger, 1998).

The primary sulphates discharged in the industrial manufacture of sulphuric acid (H_2SO_4) and from the use of sulphate minerals such as gypsum are small in comparison with secondary sulphates. Combustion processes virtually without exception discharge sulphur dioxide (SO_2) and sulphur trioxide (SO_3), simply because fossil fuels contain sulphur-bearing compounds, and coal combustion is undoubtedly the largest culprit; motor vehicles contribute only a few per cent (Mensink and de Vlieger, 1998). Other contributors to secondary sulphates are the smelting of sulphidic ores (Cu, Pb and Zn) and paper manufacture (Seinfeld, 1986, p. 104).

Oxidation of SO_2 to SO_3, in the atmosphere, via several routes – the strongest of which is reaction with OH (Claes *et al.*, 1998) – leads ultimately to sulphuric acid (H_2SO_4), which undergoes gas-to-particle conversion as such or, alternatively, neutralisation by ammonia, leading to particles of ammonium sulphate (($NH_4)_2SO_4$), ammonium hydrogen sulphate (NH_4HSO_4), letovicite (($NH_4)_2SO_4.NH_4HSO_4$) and some lesser compounds. The fate of sulphuric acid also appears to be related to that of oxalic acid, a common secondary organic compound in the atmosphere (Yu J.Z. *et al.*, 2005). Sulphuric acid resulting from coal combustion was a key factor in the infamous London smogs of the 1950s, and in the nineteenth century also; it remains a problem wherever high-sulphur coals are burned with inadequate flue gas desulphurisation. The reunification of Germany (October 3, 1990) demonstrated the logical benefit of power-plant modernisation and the displacement of coal by natural gas: sulphate particles declined (Spindler *et al.*, 2004).

London smog, which is chemically reducing, should not be confused with Los Angeles, or photochemical, smog, which is oxidising. The latter is a regular visitor to many cities during sunny periods; it consists of secondary particles generated by numerous and interwoven reactions between hydrocarbons and NO_x, in the presence of strong sunlight; for this discovery, which dates from the early 1950s, we can thank the highly esteemed scientist Haagen-Smit (1970).[17] These reactions generate

---

[16] Dioxins are chlorinated PAH. Many are highly toxic.
[17] The total quantity of hydrocarbons emitted by automobiles was by modern standards rather frightening. But not all emissions were from the tailpipe – until the 1960s, blow-by gases were simply vented to the atmosphere;

oxidants which, possessing low vapour pressures, readily undergo gas-to-particle conversion. The photochemical smog thus consists of volatile secondary organic particles (Song C. et al., 2005). The smog-forming reactions are closely linked to others which generate ozone; hence the two secondary pollutants are often experienced together. But secondary organic particles are not necessarily visible as smog: these are a widespread feature of the atmosphere, generated when alkanes, alkenes and aromatics react with OH, O_3 and NO_3 (Lim Y.B. and Ziemann, 2005).

Ammonia escaping from landfill sites, waste incineration and the industrial production of nitrogen fertilisers subsequently forms, as already related more than once, *secondary* particles of nitrate and sulphate. A fourth, far less studied source – but one increasingly demanding attention – is less obvious: the contemporary gasoline vehicle, the three-way catalyst of which manufactures ammonia. The strength of this source, though, is still uncertain, ammonia not being proscribed, and so not monitored on a routine basis. Data from remote sensing[18] reveal ammonia to be discharged predominantly during brief periods of fuel enrichment (i.e. acceleration) (Huai et al., 2005), and since its concentration in exhaust gas may actually rival that of NO_x (Baum et al., 2000), this emission should not be summarily dismissed. In totality, gasoline vehicles, as sources of ammonia, may actually reach parity with livestock manure, chemical-fertiliser production and sewage treatment plants (Fraser and Cass, 1998). Motor vehicles are likely to become still more important in this capacity, should deNO$_x$ catalysis be widely adopted: ammonia and urea are proposed as reductants (Section 7.8.1), the escape of which occurs through inadvertent overdosing, and lean-NO$_x$ traps generate ammonia in a manner akin to three-way catalysts (Hackenberg and Ranalli, 2007).

2.3.4 Environmental Implications

The environmental implications of airborne particles are arguably as multifaceted as the particles themselves. On local scales, suspended particles degrade amenity and menace public health, while on regional and global scales, the climate is undoubtedly influenced. Particles remain suspended in the atmosphere for lengthy periods; this gives them opportunity to journey over large distances, and to threaten sensitive or pristine ecosystems. The total quantity of material transported this way should not be underestimated: for example, for some pollutants, their rates of deposition into the North Sea and their rates of entry via estuaries and rivers are comparable (Claes et al., 1998).

An obvious and noticeable aspect of degraded amenity is interference with the transmission of sunlight: this 'extinction' of light happens by the combined action of 'scattering' and 'absorption'. The physics is fiendishly complex, and cannot be fully explained in merely a few words, but it depends, centrally, on the size of the particle (D_p) in relation to the wavelength of the incident light (λ): this gives rise to three different regions of particle–light interaction, covering $D_p \ll \lambda$, $D_p \approx \lambda$ and $D_p \gg \lambda$. Light extinction in the atmosphere is at its strongest in the range 100 nm – 2 µm, i.e. reductions in particle populations outside this range have little effect (Hinds, 1999, p. 357). Empirically speaking, this wavelength dependency ensures that airborne particles betray their presence by imparting various colours, tints and hues. These phenomena include the blue colour of the sky;[19] rainbows (Kerker, 1969, p. 396); reddened sunsets (Friedlander, 1977, p. 107); red skies in the aftermath of volcanic eruptions (Mark, 1998); and, following forest fires, differently hued moons (Kerker, 1969, p. 396).

---

evaporative losses from carburettors were also significant. Other notorious sources of hydrocarbons were fugitive emissions in oil refining.

[18] Roadside instrumentation designed to monitor nonintrusively the emissions from passing vehicles.

[19] The usual explanation here is 'Rayleigh scattering' by gas molecules in the earth's atmosphere. However, the same mechanism is influential with particles if these are smaller than about one-tenth of the wavelength of the light. It should be remembered that the wavelength of visible light is ~500–700 nm.

But, passing from the domain of poets and songwriters, the fact that atmospheric visibility falls far short of several hundred kilometres is adduced to the presence of airborne particles, even though the concentrations of these particles are many orders of magnitude lower than those of the air molecules (Holländer, 1995). This is the case even in ostensibly 'clean' air. But the visibility degradation caused by pollution is a glaring nuisance: this concerns local pollution, such as the offending plume emerging from a chimney stack, and regional pollution, such as the haze lying over a valley or city. In fact, the impact on visibility determines the terminology: 'haze', across distances of 1–10 km, and 'fog', for distances below one kilometre (Heintzenberg, 1998). We have seen that combustion processes emit a wide variety of particulate, but the blame for visibility degradation can, for the most part, be laid on that fraction which is carbonaceous; for, as one might expect, the blackness of this material, which, after all, is predominantly carbon, makes it an enormously effective absorber of light.

The depressing aspects of dirt and gloom are not restricted to visibility. Polluting particles lodge on any convenient surface; and this, again, is where the blackness of carbonaceous aerosols becomes immediately apparent. Deposition onto leaf surfaces compromises plant vitality by blocking sunlight (impeding photosynthesis) and by blocking stomata (impeding transpiration). Chemical reactions between particles and cells cause further loss of vitality, leaf loss and dieback. In fact, forests, by an odd perversity of nature, innocently promote their own destruction, when the α-pinene released by trees (Section 2.3.2) reacts with ozone and SO_2 to form acidic aerosols, the deposition of which onto leaves is eminently harmful to the plant (Spurny, 2000a).

Another type of pollution by deposition is readily overlooked. In places of religious devotion (Christoforou et al., 1996), wax candles, present in large numbers, and maintained continuously for many hours, reside not infrequently in close proximity to old and fragile tapestries, paintings and frescoes. Rather than merely obscuring the surface, though, these particles undergo harmful chemical reactions with paint and textiles. Sooting, or smouldering operation, as happens when candles are snuffed out, is far more polluting than normal operation. The particles are a mixture of carbonaceous and volatilised but thermally altered waxy organics (Fine et al., 1999); they are of sizes comparable with those discharged by motor vehicles. Similar problems arise with incense burning (Wang et al., 2007).

Black deposits on building facades are a conspicuous problem; in Western Europe, surface-deposited carbon in urban areas reaches concentrations of $1-3\,mg/cm^3$; various metals are also present (Nord A.G., 2000). The porosity of the stone is a factor in particle deposition and capture (Sabbioni, 2000). But the relevance of atmospheric particles to buildings does not simply concern deposition, or soiling, or the growth of encrusted surface layers – what are known as 'patinas'. Whilst such encrustations are clearly unwelcome for aesthetic reasons, particularly on ancient buildings, sculptural decorations, classical monuments of archaeological significance and the like, there are far more insidious dangers. This time, the carbonaceous particles are not to blame – at least not directly; for they act as carriers of sulphuric acid, the formation of which, from SO_2, is probably catalysed by the piggy-backed metals; iron is particularly implicated (Zappia, 2000). Exposure of stone *only* to SO_2, in environmental test chambers, does not reproduce the deleterious effects.

The sulphuric acid, in reacting with the underlying structure, causes stone sulphation (Ausset et al., 1996). Damage mechanisms differ depending on the type of stone, mortar or brick (Zappia, 2000), but the key reaction is sulphation of calcium carbonate (calcite, $CaCO_3$) to calcium sulphate dihydrate (gypsum, $CaSO_4.2H_2O$). The extent of this transformation seems more dependent on the microstructure of the stone than on the calcium carbonate content (Zappia, 2000). The gypsum is water-soluble and readily leached by rainwater; microorganisms, moreover, suffuse and colonise these sites to form slime films, with further (and circularly deleterious) reactions (Spurny, 2000a). This particle–gypsum encrustation gradually eats away at the building's facade to create a layer of embrittled material, the eventual exfoliation of which exposes a fresh surface that is in turn attacked, so that the cycle is then repeated.

The mechanisms of destruction are, then, well understood; but the human activities to declaim against are heatedly debated. The particles in the ambient are, after all, mongrels, as explained extensively in Section 2.3.1. Although heavy industry cannot be exculpated, at least not entirely, the occurrence of stone sulphation in *nonindustrial* areas implicates motor vehicles (Rodriguez-Navarro and Sebastian, 1996). Even so, the co-emission of metal-containing carbonaceous particles and SO_2 is admittedly a noted feature of coal combustion.

In ancient times, it was possible to juxtapose monuments and thoroughfares closely; in modern cities, these thoroughfares are now, most unfortunately, heavily trafficked by motor vehicles. Attempts, therefore, have been made to prevent the deposition of particles onto these monuments, through the construction of enveloping glass structures. Well-meaning but inexpert design of these structures, can, perversely, exacerbate the situation, as wind-induced turbulence, a cause of particle resuspension, is now lost, and particle sedimentation continues with impunity (Camuffo and Bernardi, 1996).

Land clearance practices involving open burning, and wanton destruction of natural habitat such as savannah and forests, are major sources of atmospheric particles in some regions (Liu X. *et al.*, 2000). It ought to be admitted, though, that there occur, courtesy of lightning, many natural wildfires, in which case the particulate emissions are not really anthropogenic. These naturally occurring fires are now widely recognised as part of a natural cycle of destruction and regrowth; but land management practices, in the developed countries, are thought (by some) to worsen needlessly the destruction when it finally does occur.

So far, we have documented the degradation of amenity; but of equal relevance, and arguably of greater significance, are the injurious effects on people. Various diseases are implicated; but pathogenesis, at a first level, is concerned with the *point of deposition* in the respiratory tract. This is where diseases of the deep lung, such as pneumoconiosis, differ from diseases of the upper airways, such as bronchitis and asthma. The point of deposition is governed by particle transport, which, as related in Section 2.2.3, depends critically on particle size. A health-related sampling convention, devised by the International Organization for Standardization (ISO), specifies four deposition curves: inhalable (through the nose or mouth); thoracic (penetration beyond the larynx); respirable (penetration beyond the ciliated airways); and 'high-risk' respirable (pertaining to children or the infirm). The distinction between 'inhalable' and 'respirable' should be noted. These deposition curves are presented, alongside agreed filter specifications for PM10 and PM2.5, in Figure 2.12(a) (Mark, 1998).

The fraction of particles depositing at various points in the respiratory system according to size has been measured extensively, and this characteristic is now mainstream biomedical science. Typical curves are shown in Figure 2.12(b) (cited by Spengler and Wilson, 1996). The exact features vary, because flow fields are determined by airway geometry and breathing rate; for example, there are differences between men and women, and between children and adults, but the dip in deposition efficiency for middling particles is a generic feature. Smaller particles are deposited via diffusion, and larger particles via inertia or sedimentation. 'Mid-wicket' particles are thus able to escape deposition to some extent, and consequently, these stand the greatest chance of being exhaled.

Even after deposition, particle size is still relevant – the smallest particles, for instance, are able to penetrate the pulmonary epithelium. But the chemical composition and morphology of the particles become much more important following deposition; this relates particularly to the nature of the particle *surface*, as this inevitably controls interactions with cells. Particles act as vehicles for the transport of other, adsorbed substances into the lung: they thus deliver more localised and concentrated doses than for inhalation of the *same* substances as gases – and in different locations. For example, SO_2 is a considerably greater respiratory irritant in the presence of particles: this suggests that, when inhaled as a gas, its great solubility in the mucous membranes allows take-up in the upper regions, whereas, when adsorbed onto particles, it hitches a ride into the deep lung (Seinfeld, 1986, p. 58).

Other routes into the human body are less obvious. Toxins gain rapid entry to the blood system if particles are readily soluble in the mucous membranes. Inhaled particles that lodge on the mucous

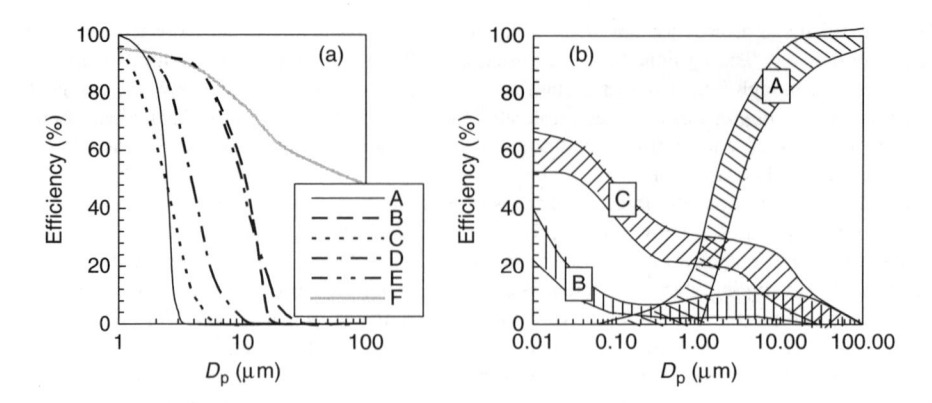

Figure 2.12 Health risks, expressed as a function of particle size. (a) Deposition or capture efficiency, in relation to ISO health-related sampling conventions: A, PM2.5; B, PM10; C, 'high-risk respirable', or pertaining to children and the infirm; D, 'respirable', or penetration beyond the ciliated airways; E, 'thoracic', or penetration beyond the larynx; F, 'inhalable', through nose or mouth (Mark, 1998). (b) Deposition efficiency in the human respiratory tract: A, nasal–pharynx; B, trachea–bronchus; C, pulmonary (cited by Spengler and Wilson, 1996; original reference International Commission on Radiological Protection 1966).

membranes are, by the action of cilia, returned to the top of the throat, where they are unconsciously swallowed. Deposition onto reservoirs and lakes, with subsequent ingestion via drinking water, is a possibility, as is deposition onto vegetation, with subsequent ingestion as food. Certain alkanes, thought to have been discharged by vehicle tailpipes, and yet not removed by food-processing technology, are present in vegetable oils (margarine) (Neukom et al., 2002). PAH, found in soils alongside roads (Johnsen et al., 2006), suggest also deposition onto crops.

The particulate pollution likely to be experienced by members of the public has changed markedly over the decades. Coarse-mode particles have declined in developed countries, along with the traditional 'smokestack' industries. But smaller particles are still being emitted; indeed, these may possibly have *increased* in abundance over the same time span (Maynard, 2000). This point is contentious, and difficult to verify since the tiniest particles were not systematically monitored until fairly recently. Intriguingly, in post-reunification Germany, the sudden modernisation of East German coal-fired power stations seems to have been marked by a corresponding shift towards less particulate mass, yet more particle abundance at smaller sizes (10–30 nm) (Kreyling et al., 2003). Such shifts have immediate bearing on the findings of epidemiology, which demonstrate, repeatedly, an association of some sort between, on the one hand, suspended particles in urban air, and on the other, morbidity and mortality in the human population. But even though numerous and undoubtedly plausible mechanisms have been posited, the true reasons for this association have not yet been unequivocally ascertained. This topic is being avidly researched at the time of writing.

Occupational diseases caused by exposure to suspended particles are well defined, because the causative agents are equally well defined, e.g. in the case of silicosis, asbestosis and pneumoconiosis. The same cannot be said for the public at large: the mishmash of suspended particles in the wider atmosphere is ill-defined, as we have seen. We shall not delve into the pathology too far, as this is covered in Chapter 10, but motor vehicle particulate emissions are certainly implicated, even if only as an accessory, in asthma and bronchitis. More disconcertingly, there are long-suspected cancer risks, especially of the pulmonary system. And, more recently, it was suggested that particles might have untoward effects on the cardiovascular system (Seaton et al., 1995); since this widely cited paper, research into cardiovascular illnesses has been swiftly gathering pace. These toxicological risks are a prime factor motivating the control of particulate emissions.

At the opposite end of the environmental scale, particles in the earth's atmosphere are of immense significance to the climate (Horvath J., 1998), but the mechanisms are hideously complex, and this is one reason why scientists are not always in agreement about climate change (Shine and de Forster, 1999). The complexity of the modelling arises from the enormous diversity, as, unlike CO_2, for example, particles possess no uniquely assignable chemical or physical properties. Particles influence, through their absorption and scattering of the sun's radiation, the planetary heat balance, as expressed by 'radiative forcing' (commonly awarded units of W/m^2). To speak generally, however, particles that scatter radiation, which include sulphates and organics, will tend to cool the atmosphere, whereas particles that absorb radiation, which concerns chiefly the carbonaceous material or soot, will tend to warm it.

Scientists are concerned about the ability of the atmosphere to tolerate modern-day levels of air traffic, i.e. to purge itself of the ensuing pollution (Machta, 1971); and there are instances where even droplets of water have had to be treated as pollutants in their own right. This water, issuing from aircraft exhausts, is manifested as contrails (condensation trails). Long atmospheric lifetimes, with concentrations creeping upwards inexorably, were feared. These contrails are still being investigated for their potential effects on climate (Gierens, 2007). In the *wider* atmosphere, supersaturations of several hundred per cent, required for homogeneous nucleation of water, do not occur; and supersaturations of ~101 %, required for heterogeneous nucleation, are uncommon. In fact, clouds form because many atmospheric particles are soluble and thus readily act as CNN (Kulmala *et al.*, 2000). The climate is, then, influenced by particles through the cloudal albedo, and the water cycle is influenced by the location and timing of rain. (In some countries, the skies are occasionally 'seeded' by discharging particles from aircraft to encourage rain.)

Sulphates, closely followed by wind-blown dust, appear to dominate the tropospheric aerosol in global climate terms. This recognises the global occurrence, and influence, of atmospheric particles – even in the absence of human activities. Yet it is quite possible that anthropogenic particles, even at current levels, have some nonnegligible effect on the global climate. The highest known particle concentrations occur over the North Pole, and generate the 'Arctic haze'. This arises when wintertime meteorological conditions favour the northward transport of particulate pollution from Europe and countries of the former Soviet Union (van Malderen *et al.*, 1996a).

The ability of atmospheric particles to bring about global climate change should not be dismissed too lightly: during the Cold War, much was said about the airborne debris thrown up by nuclear explosions, and how this might plunge the earth into a 'nuclear winter'. Cooling is observed in the Amazon basin during intense regional forest fires; and the Kuwaiti oil fires following the First Gulf War had a similar effect (cited in Cachier, 1998). Similarly, with the arrival of extraterrestrial bodies, it is not just their fragmentation in the atmosphere *per se*, but also the particulate debris thrown up at impact and also particles arising from the resulting wildfires. Anomalous particles of iridium at the Cretaceous–Tertiary boundary, on a *worldwide* basis, appear to reveal some primordial asteroid impact (Heymann *et al.*, 1998). The mass of the extraterrestrial object behind the Tunguska event in central Siberia in 1908 is still being quantified (Kolesnikov *et al.*, 1999). The fossil record is, however, punctuated with mass extinctions, wherein large percentages of life on earth oddly disappear, and asteroid impacts are blamed by some scientists for these.

But now we have strayed rather far from the point. Returning, therefore, to today's global climate concerns, the direct effects of carbonaceous particles are perhaps of subsidiary importance when placed alongside sulphate particles and greenhouse gases. Indeed, in some scenarios, anthropogenic particulate could partially reverse the global warming induced by greenhouse gases, by scattering the sun's incoming radiation back into outer space. (This is 'global dimming', but surely only the most devout optimist would expect a precise balance.) An important distinction here is that airborne particles are short-lived compared with greenhouse gases, so that their effects are regional; whether the summation of *regional* forcing from particles exceeds the *global* forcing from greenhouse gases is a moot point. And the short-lived nature of particles, relative to CO_2, might provide a faster method of combating global warming. With emissions of soot, in some regions of the world a reduction of emission from wood- and coal-burning cooking stoves, rather than from diesel engines, might be a more appropriate countermeasure (Bond and Sun, 2005).

Even so, carbonaceous particles may still exercise some disproportionate influence. The idea that particle surfaces foster reactions involved in depletion of the ozone layer is being researched (Kamm *et al.*, 1999). But better established is the disproportionate effect of blackness, as these particles are the principal light-absorbing species in the atmospheric aerosol. This effect appears to be stronger when the particles are associated with water (Markel and Shalaev, 1999). Another posited effect is on the planet's albedo, via the stability, i.e. susceptibility to melting, of icy landscapes (Fily *et al.*, 1997). If particles thus indirectly reduced the earth's albedo, and thereby diminished the amount of radiation reflected back into outer space, they might actually *reinforce* the warming influence of greenhouse gases. This would indeed be ironic, as diesel engines are regularly touted for their greenhouse friendliness, with their lower emissions of carbon dioxide compared with spark-ignition engines! This is a reminder that carbon residing in oil, coal, wood and peat is in fact *sequestered*; and returning it to the atmosphere as *particles*, rather than CO_2, is not itself without implications for global climate patterns.

2.4 Motor Vehicle Particulate

This section at last turns to motor vehicle particulate, as generated within internal combustion engines, and as emitted by motor vehicle tailpipes. Such a study equally requires, at the outset, a parallel introduction to some crucial terminology. This is rather more than just semantics: the great chemical complexity of particulate absolutely demands familiarity with these terms. One might justly add that, in the literature, many terms are insufficiently defined, blithely confused and erroneously interchanged. Inconsistency between researchers is widespread, and underlying definitions are often implicit or left unstated. The definitions, as given here, will continue to be applied throughout the remainder of this treatise.

2.4.1 Some Typical Particles Dissected

Some characteristic particles likely to be present in the exhaust stream of a motor vehicle are depicted in Figure 2.13. This diagram is schematic and conceptual, so its fidelity to the real world is inexact, although it does, nevertheless, contain many of the features that would be observed, if the exhaust gas were to be passed through a filter and the particles collected thereon were to be examined in an electron microscope. The most immediately striking aspect of Figure 2.13 is the presence of three distinct types of particle, labelled 'nucleation mode', 'accumulation mode' and 'coarse mode'. In Section 2.3.1, it was stated that particles in the atmospheric aerosol invariably exhibit this trimodality, and particulate emissions from motor vehicles are no exception.

Coarse-mode particles are of varying nature, as this mode includes, also, atypical rust and scale from the exhaust system, for example; the particles of most relevance to the present discussion, however, are not emitted directly as such, but formed from the other two modes. These predecessors lodge somewhere within the exhaust system, become attached to one another, and then re-enter the exhaust stream as much larger, composite particles. This storage–release process makes the coarse mode an inconsistent emission – one that proceeds in a random and unpredictable manner. Perhaps because

Fundamentals

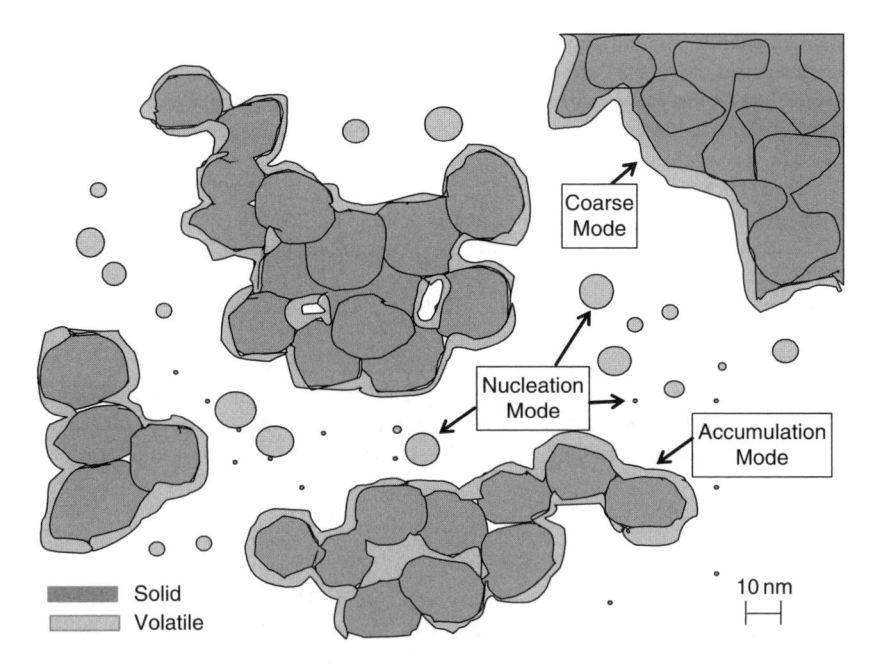

Figure 2.13 Some typical particles likely to be emitted by an internal combustion engine, depicted schematically: coarse mode (largest, shown in part), nucleation mode (smallest); accumulation mode (middling).

of this fickleness, their artifactual nature and their comparative rarity, these particles have been little studied. But, as suggested in the figure, they probably consist of a solid core (perhaps slightly more densified than in the other particles) and an outer layer of volatile material.

Nucleation-mode particles are obscure, although admittedly this situation is changing rapidly. Historically they have been even less studied, indeed, prior to the mid-1990s, hardly anything was known about them, and this is understandable, as they lie at the limit of detection for many instruments. Most, but not all, research suggests that nucleation-mode particles consist of volatile material: this is why they are depicted as spherical in Figure 2.13. Other research suggests that *some* nucleation-mode particles are in fact solid, or at least possess minute solid kernels. These questions are being avidly researched in many research laboratories at the moment.

Accumulation-mode particles have received by far and away the greatest attention. The most immediately obvious feature of accumulation-mode particles is that they consist of a collection of much smaller 'primary' particles. These primary particles are sometimes referred to as 'spherules', meaning that, while not exactly spherical, they nonetheless quite closely approximate sphericity.[20] The size range for spherules is typically ~20–50 nm. Accumulation-mode particles vary in size because they contain greater or fewer numbers of spherules, and *not* because the spherules themselves vary in size.

---

[20] Spherules should not be confused with coke particles: these are porous carbonaceous shells of far greater size (1–50 μm) (Flagan and Seinfeld, 1988, p. 372). Spherules are formed in the pyrolysis of *gaseous* fuels, or the *volatilised* components of liquid or solid fuels, whereas coke particles are formed in the *liquid-phase* pyrolysis of heavy fuel-oil droplets. In properly adjusted internal combustion engines, the coking mode of soot formation is thought unimportant, as the fuel should be vaporised prior to being burned.

The reason why spherules should be restricted to such a narrow range of size, when the assemblages they form vary so massively, is not known. The spherules are simply the 'building blocks', but, because they do vary a *little* in size, they are not so much like house bricks as like the stones in a drystone wall: some slightly smaller, some slightly larger. Their number is what decides the size of the wall, or rather, the size of the whole particle.[21]

This assemblage of spherules, which forms the skeleton, or backbone, of an accumulation-mode particle, is an 'agglomerate' or an 'aggregate'.[22] The number of spherules that may combine seems not to be subject to any natural law: it ranges from tens, through hundreds, to thousands. Nor is there any clearly defined morphology: there are highly elaborate and interlinking chains, and clusters of widely varying compactness. And, sometimes, the two ends of a chain of spherules will link up, to form a necklace-like ring. All that can really be said is that, since spherules are seldom seen in isolation, they presumably possess an astonishing proclivity for combining.

The agglomerate surface is coated by a layer of liquid or semi-liquid material: this is viscid, and penetrates into the pores and internal voids of the agglomerate. Significant compositional dissimilarity thus exists between the surface and the bulk. Agglomerates, since they rejoice in high specific surface areas, are effective at accreting and scavenging gas-phase material. The adherence of this volatile or semivolatile layer is what leads to the expression 'wet' particulate, in contradistinction to the solid core that remains after a heating process, which is the 'dry' particulate. However, as we suggest below, 'oily' is probably a better adjective than 'wet'.

Little has been said thus far about chemical composition. This can be defined according to the four-layered conceptual model depicted in Figure 2.14. At the first level, everything is included that might be captured when exhaust gas is passed through a filter, with the sole exception of condensed water; this is the legislative definition of 'particulate'. Obviously, to impose statutory limits on water *per se* would hardly be sensible, but from a health standpoint, there are no 'pure' droplets as such. Water droplets contain minute solid particles, and also absorb soluble combustion products such as aldehydes and ketones.

As already hinted, upon heating, some material evaporates, and some does not; or, alternatively, some material dissolves in certain solvents, and some does not. This divides the particulate into that which is *volatile* or *soluble*, and that which is *nonvolatile* or *insoluble*. Thereafter, there are five clear subgroups, or *fractions*:[23] sulphates, nitrates, carbonaceous, organics and ash. As we shall see presently, the chemical composition (Figure 2.14) corresponds, in a large degree, to the physical representation (Figure 2.13). But, more importantly, the five fractions are defined principally according to the analytical methods, laboratory procedures and test protocols used in their separation and quantification, rather than through any direct correspondence to features presented in Figure 2.13.

---

[21] The technical community does not universally adopt 'spherule'; but, in this work, the term is used to avoid confusing the reader; some workers carelessly use 'particle' both for the spherule and for the whole assembly of spherules. Actually, dictionaries are more likely to recognise 'spheroid' than 'spherule', and the first may be more appropriate. According to the *Shorter Oxford English Dictionary* (1993), a spheroid is 'a body resembling or approximating to a sphere in shape', and a spherule is 'a small spherical or globular body'.

[22] An 'aggregate' is a stable structure, whereas an 'agglomerate' is friable (Lipkea *et al.*, 1978), held together weakly, say, by van der Waals forces or surface tension. This subtlety is seldom observed in the literature; indeed, it is convenient to use 'agglomerate' fairly loosely, as both types of bonding are observed in motor vehicle particulate; and this is the policy adopted in this text.

[23] In one of those curious ironies unaccountably introduced by language, 'faction' is almost the same word, and might also serve as an appropriate term.

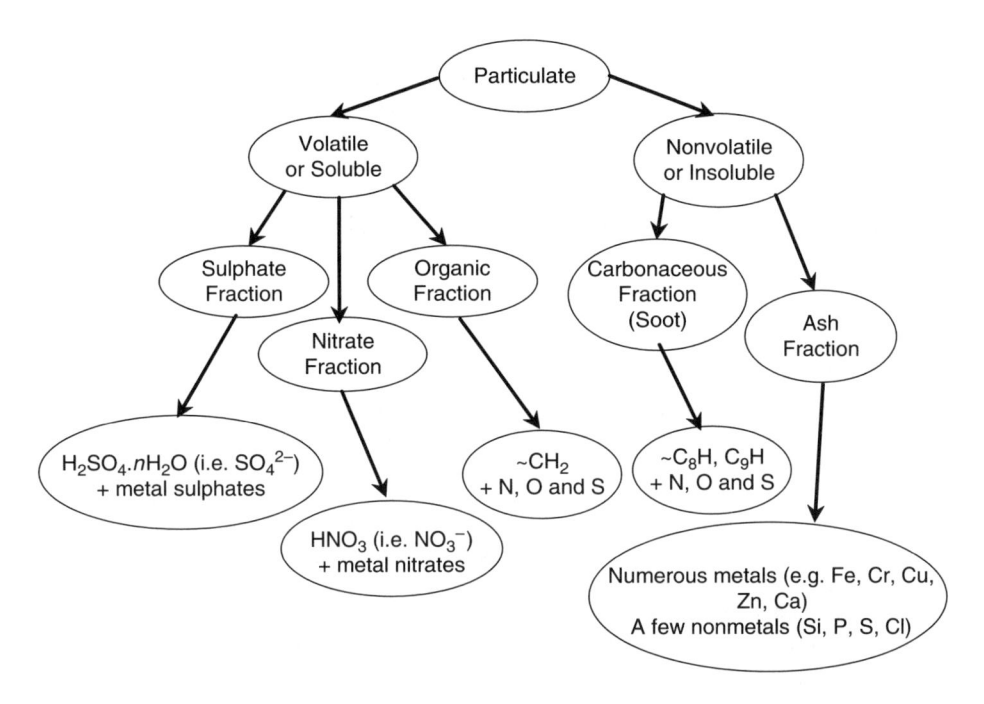

Figure 2.14 Conceptual model of particulate composition, terminating in five distinct groups or fractions: sulphates, nitrates, organics, carbonaceous and ash.

Organic Fraction

If assayed by extraction or dissolution in an organic solvent, this is the soluble organic fraction (SOF); and if assayed by heating or volatilising, this is the volatile organic fraction (VOF). Some laboratories favour one method and some the other. Fortunately, the mass of SOF is usually quite close to the mass of VOF, although, admittedly, the separation processes are based on different properties that do not necessarily generate the same end result. In any case, in this text the 'volatile' or 'soluble' qualifier is normally dropped unless of specific relevance. These quiddities aside, the organic contingent is arguably the most complex of the five fractions, as it contains several hundred, or perhaps several thousand compounds, many of which lie at the threshold of detection. Most of the major chemical families are represented, although their proportions vary widely: alkenes, alkanes, alcohols, esters, ketones, acids, esters and aromatics. Since lighter C_4 to C_8 compounds are also found – which, theoretically, should be in the gaseous state – then, presumably, surface interactions are strong.

Sulphate Fraction

This fraction concerns *water-soluble* sulphates, or the SO_4^{2-} ion, but the chief component is sulphuric acid, H_2SO_4. A distinctive property of sulphuric acid, however, is its eager association with water. (This water is *chemically* associated with the sulphuric acid; it should not be confused with condensed water, for which the association is physical.) In fact, the water content of the particulate often varies only with the sulphate content, because the particulate-bound water is found predominantly with the sulphuric acid. So, what 'sulphates' really means is the acid *and* the water. This extension has important ramifications, because the mass of water is, first, *not negligible*; and second, it is *variable*, depending on the humidity in the immediate environment of the filter. *Prior* to gravimetric assessment, therefore, filters *must* be conditioned (equilibrated) for a certain period of time in a closely defined environment – otherwise the result is meaningless. (This equilibrated state should not be confused with the hydration of the sulphate pertaining at the moment of collection on the filter.) At a relative humidity of 50 %, the mass of water is customarily taken as 130 % of the mass of sulphuric acid; this is why, to obtain the mass of sulphates, the mass of acid is simply multiplied by a factor of 2.3. This is convenient because the mass of water is not ordinarily determined in a direct fashion.

Nitrate Fraction

This term denotes *water-soluble* nitrates, or the NO_3^- ion, but the chief component is nitric acid, HNO_3. The fraction is small and, even though it is dutifully reported by many emissions laboratories, its complete elimination would be unavailing in meeting emissions legislation – this, perhaps, explains the lack of attention it receives in the literature. However, it should be said that with the latest low-sooting diesel engines, low-sulphur fuels and exhaust-mounted particulate filters, this situation may change, with nitrates set to become more prominent by proportion, if not in absolute terms.

Generally speaking, HNO_3 forms via reaction between water and NO_2; so, intuitively, the nitrate chemistry is connected to the NO_x chemistry. The detailed mechanisms pertaining in exhaust gas have not, however, been reported, and the few sundry observations or ideas to have been made do not come together to form any coherent view. The reactions

$$2NO_2 + H_2O \longrightarrow HNO_2 + HNO_3, \tag{2.7}$$

$$3HNO_2 \longrightarrow HNO_3 + 2NO + 2H_2O \tag{2.8}$$

have been mooted (Villinger *et al.*, 2002b), and the presence, in exhaust gas, of small amounts of HNO_2 and HNO_3 is indeed known (Jeguirim *et al.*, 2005). Water has some promoting role in the catalytic oxidation of carbon by NO_2 (Jeguirim *et al.*, 2005), and carbonaceous surfaces have some affinity for NO_x (Ahmad, 1980). Contrary trends exist between nitrates and sulphates, implying competition between their precursor species for exhaust gas oxygen (Lapuerta *et al.*, 2000).

Nitrates are a more prominent concern in the literature of the atmosphere, pointing to the dominance of formation *subsequent* to the emission process (Sioutas and Koutrakis, 1995), especially since this is the well-worn path to acid rain, where HNO_3 forms from NO_x (Claes *et al.*, 1998). Soot surfaces in the ambient atmosphere are thought to promote the formation of HNO_2 from NO_2 (Ammann *et al.*, 1998), and there are other relevant reactions posited (Gómez-Moreno *et al.*, 2007), for example

$$NO_2 + OH \longrightarrow HNO_3. \tag{2.9}$$

Once formed, of course, nitric acid must transfer, somehow, into the particulate phase, in order to be quantified as 'nitrates', about which there is even less information in the automotive arena. Again, this subject is better researched by atmospheric scientists, although the two fields admittedly diverge at this point, since acid–ammonia reactions (Lin Y.-C. and Cheng, 2007) are (with current technology) less important in vehicle exhaust. The nitrate fraction seems to enjoy somewhat greater volatility than the sulphate fraction, with condensation within the exhaust system less likely. (It should be noted that nitric acid has a considerably lower boiling point than sulphuric acid.)

Carbonaceous Fraction

The word 'carbonaceous' emphasises that this fraction is predominantly, but not exclusively, carbon. It is for this reason that the alternative terms 'graphitic carbon', 'elemental carbon' (EC) and 'black carbon' (BC), as used by some researchers, are misleading, although what is really meant is the absence of any coherent carbon compounds. Similarly, the ambiguous appellation 'soot' should be carefully qualified; this is a colloquialism that encompasses anything found on the inside of a chimney. True, the term is admittedly in use in the technical community as well, but this use is often inconsistent, sometimes meaning particulate, sometimes meaning the nonvolatile or insoluble particulate, and sometimes meaning just the carbonaceous fraction. *In this work, 'soot' and 'carbonaceous fraction' are synonyms.* Unsurprisingly, the inky blackness of smoke arises from this fraction, inevitably, therefore, emissions regulations framed in terms of smoke overlook the other fractions, although measures that reduce smoke may also, for example, fortuitously reduce the organic fraction.

Ash Fraction

This fraction contains sundry inorganic compounds or elements – predominantly metals, but a few nonmetals; it is a mishmash of highly variable composition, for which clearly defined chemical compounds are difficult to pin down. A well-used method of characterisation is to burn all the particulate, and then examine what remains: that is, the 'incombustible ash'. This, strictly speaking, denotes the condition immediately *after* the burning, which cannot be *quite* the same as *before*, because the burning inevitably causes partial decomposition and other chemical alterations among the ashing elements. But this distinction need not detain us for the moment. In this text, 'ash fraction' denotes, with some licence, both states – before and after – but the meaning will be apparent according to the context.

These five fractions correspond in a large measure to the features delineated in Figure 2.13: the sulphate, nitrate and organic fractions form the volatile coatings, and the carbonaceous fraction is largely represented by the spherules. In this respect, the accumulation-mode particles are sometimes called an 'aciniform' type of carbon, meaning 'clustered like grapes' (Rivin and Medalia, 1983). The location of the ash fraction is less certain: it may be atomically dispersed within spherules; present as minute nanodomains within, or attached to spherules; or, as mentioned, present as a kernel, within the nucleation-mode particles. In admittedly less usual circumstances, when ash comes to dominate the particulate emission and overwhelms the other fractions, the nucleation-mode particles are composed of ash.

The fivefold fractionation scheme allows an extremely general statement to be made about material distribution. Some figures for a typical diesel engine are: ash fraction, <1–2 %; organic fraction, 10–90 %; carbonaceous fraction, 10–90 %; sulphate fraction, <5 %;[24] nitrate fraction, <1 %. These figures are suggested with the entire operating map of the engine in mind, from light running to full load; restricted duty cycles will inevitably restrict also the compositional range, especially for the organic and carbonaceous fractions, because particulate composition is most certainly a function of engine operation.

In practice, some uncertainty arises in dividing the five fractions, as, depending on the analytical minutiae, some material may be assigned to more than one fraction, and so counted more than once, while other material may be allowed to escape quantification. These are 'boundary problems'. The significance of boundary problems has, historically, been rather small, but it is likely that, with today's

---

[24] Assuming no oxidation catalyst.

low-emitting engines, such disputes need to be taken more seriously. Boundary problems arise partly because the analytical techniques are not perfect, and partly because there are gradations or trends, rather than closely defined lines of demarcation, between fractions. For example, some organic compounds are bound so tenaciously to spherule surfaces that the fractionation is unable to remove them – the strength of this association reflects the bonding: for example, condensed compounds are removed more readily than adsorbed ones. A second example is that organic solvents are suspected of improper extraction insofar as they remove a small quantity of sulphates. A third example is that sulphuric acid reacts with metals in the ash fraction to form metal sulphates, some of which are water-soluble and others not.

Finally, an 'overlap' of sorts is apparent, between the *particulate-phase* organic compounds, as captured by a filter at $\leq 52\,°C$, and the *gas-phase* hydrocarbons, commonly designated HC, and determined with a flame ionisation detector (FID), the sampling line to which is normally operated at 190 °C. Clearly, hydrocarbons in the gas phase at $\geq 190\,°C$, yet *also* in the particulate phase at $\leq 52\,°C$, will be quantified by both methods, thus engendering a 'double determination'. In fact, the organic fraction does, on occasion, show reasonable correlations to HC (Durán *et al.*, 2002), but the occurrence of such a correlation cannot be automatically assumed, as the ranges of volatility present within these two emission groups can vary independently (Andrews *et al.*, 2000b). The vagaries of partitioning between the particulate phase and the gas phase are a huge headache in motor vehicle emissions, and will occupy much of this text.

2.4.2 What Happens Within the Engine

Of the above-mentioned five fractions, two of them, ash and carbonaceous, are generated within the engine; the others form later, in the exhaust system or, more probably, when the exhaust plume enters the surrounding air, although, of course, important precursor reactions must first take place within the engine.

Particulate emerging from the engine has arisen from four distinct sources: fuel, lubricant, air and material breakdown. The first-named one, of course, is obvious, the second one (to the uninitiated) slightly less so. Less widely appreciated are the third and fourth sources, but these are small in normal circumstances. Figure 2.15 attempts to depict, in an unabashedly subjective manner, the relative significance of each source.

A logical place to begin describing why particulate forms is the process of combustion. By combustion, we simply mean the rapid oxidation of some substance by an exothermic chemical reaction. The word 'rapid' is pertinent, as it signifies that heat energy is liberated at a faster rate than it is dissipated by the transport processes of convection, radiation and conduction.

A fuel spray entering a chamber containing air at high pressure and temperature, such as occurs in a diesel engine, undergoes two modes of combustion. If, before combustion, the fuel droplets vaporise, and the vapour has ample opportunity to mix with the air, this is a 'premixed flame'; and if fuel and air are insufficiently mixed before combustion is initiated, this is a 'diffusion flame'. The distinction between these two modes of combustion is paramount: in a premixed flame, the rate of combustion is controlled by chemical kinetics, whereas in a diffusion flame, it is the physical processes of mixing, i.e. mass transport, that are rate-determining.

Premixed and diffusion flames are both found in diesel engines: the first gives way to the second as the combustion proceeds. The premixed flame occurs because the start of injection and the start of combustion are separated by a brief interregnum, during which preparatory pre-combustion reactions occur. This short incubatory period is known as the 'ignition delay'. The consequence is that when combustion eventually *does* commence, a fraction of the charge has *already* mixed to within combustible limits. But since this premixed fraction burns much faster than the continued injection of

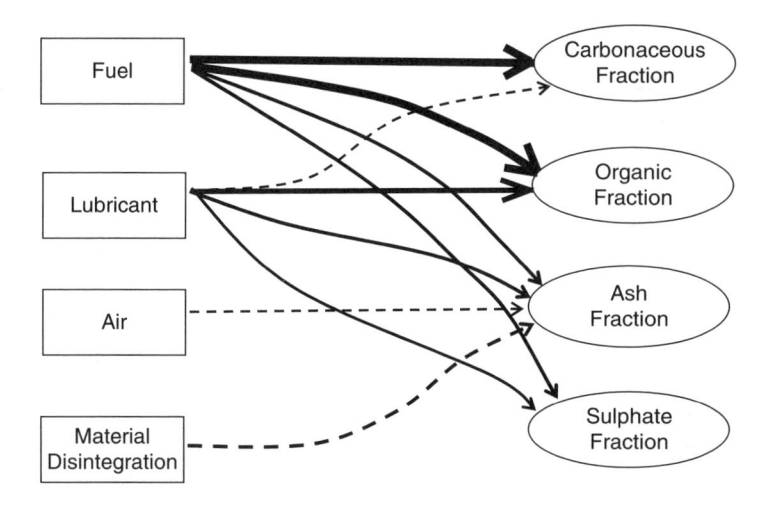

Figure 2.15 What happens within the engine: conversion of fuel, lubricant, air and engineering materials into carbonaceous, organic, ash and sulphate particulate. Purely subjectively, the line widths suggest relative importance; dashed lines suggest poorly explored possibilities. A less obvious scenario is the entry of foreign material (i.e. ash) into the fuel tank during refuelling.

fuel can replenish it, the flame, as it were, 'catches up'. After this point the combustion can proceed no faster than the rate at which combustible mixture becomes available, and so there is now a diffusion flame.

The next step is recognition that combustion involves not solely oxidation reactions. In a diesel engine, it is inevitable that *some* fuel will experience oxygen privation because, even though the global air–fuel ratio is lean, the charge is heterogeneous. When hydrocarbons are subjected to high temperature but insufficient oxygen is made available, they undergo a multiplicity of endothermic reactions known collectively as 'pyrolysis': the combustion chemistry then sets out in the direction of soot. In actual fact, the greatest sooting tendency is seen not in the complete absence, but in the presence of *some* oxygen; this is attributed to a 'sensitising' role in the pyrolytic reactions (Griffiths and Barnard, 1995, p. 117). But, more pertinently, pyrolytic and oxidation reaction pathways always compete with one another, and the lower the local oxygen concentration, the more likely it becomes that pyrolysis will prevail over oxidation. Pyrolysis is a broad term, encompassing reactions between hydrocarbons, dehydrogenation to form hydrocarbons of greater unsaturation, and cracking of hydrocarbons into fragments. Different pyrolytic reactions predominate in characteristic ranges of temperature. To gloss a complex subject, at higher temperatures, bond rupture and dehydrogenation predominate; these reactions create species of lower molecular weight; and at lower temperatures, polymerisation and condensation predominate; these reactions create species of higher molecular weight.

Pyrolysis is the process by which fuel molecules, containing a mere handful of carbon atoms, are converted into soot particles, containing tens of thousands of carbon atoms. The original fuel molecules are broken down and rebuilt as structures of ever-greater molecular weight; it is also through these reactions that the carbon–hydrogen ratio is continually increased, so that the compounds become less organic, and more carbonaceous. The fundamental building blocks in pyrolysis, from which the first recognisable soot particles are formed, as customarily understood, are the ring-structured polyaromatics, although a long sequence of interwoven pyrolytic reactions is first necessary to form them. Unsaturated hydrocarbons such as acetylene are also involved.

The organic fraction arises from fuel in two ways: direct and indirect. The direct path is when fuel escapes combustion, and simply passes, unburned, through the engine: this might happen, for example, if some fuel is overmixed with air, such that regions of the charge become too weak to support the flame. Overmixing is promoted, for example, by a long ignition delay. The indirect path is when pyrolytic reactions are for some reason interrupted in their conversion of fuel to soot: for example, if they are quenched by continued mixing. In this case the organic fraction will contain species not present in the fuel, i.e. those synthesised in the combustion. One should not, for this reason, expect the hydrocarbons in the organic fraction to mirror exactly the hydrocarbons in the fuel, although, for practical investigations, this precarious assumption is often unavoidable.

The ash fraction arises either from molecules in the fuel that are entirely inorganic, or from inorganic elements that are bonded to organic fuel molecules. These inorganic components are sometimes deliberately added to the fuel, for example to improve various aspects of the combustion; at other times they are there as stowaways or interlopers, having been inadvertently introduced through contamination in the fuel distribution network, or simply because they are natural contaminants in the crude oil (Akinlua et al., 2007). What exactly happens to these inorganic species in the combustion is poorly reported. They probably exist as vapours at typical combustion temperatures. But since their volatilities are low – much lower than any compounds in the organic fraction – their transition into the particulate phase is virtually assured before the exhaust gas is ejected from the engine. For this reason the ash fraction probably emerges slightly before or slightly after the carbonaceous fraction, depending on the elements, and the various compounds formed by these elements.

A significant nonfuel source of particulate emission is always the lubricant. Now, this fugitive oil is often enough perceived as solely contributing to the organic fraction: for example, particulate composition is occasionally depicted in pie charts, nicely divided into five sections: sulphates, ash, carbonaceous, oil organics and fuel organics. This depiction is disingenuous insofar as it conflates two aspects: particulate *composition* and particulate *origin*; for lubricant contributes to all four fractions, although admittedly its organic presence has, in the historical picture, been the most keenly felt. With the advent of new emission control technologies, this situation is now changing.

Formidable attempts are made to prevent the escape of oil, but there is no insurmountable barrier, nor is the invention of such likely for internal combustion engines as traditionally designed. For example, oil residing on the bore is nakedly exposed to hot combustion gases, and hence may evaporate; other entry routes are via the valve stem seals and piston rings, particularly when these components are worn. The amount of oil-derived particulate cannot be predicted from oil consumption, because some of the escaping oil is completely combusted. The amount combusted depends on the entry point, as this decides exposure temperature and residence time. Escaping oil burns up less readily at low load when in-cylinder temperatures are cooler, conditions which also encourage the escape of unburned fuel, so that the two sources tend to reinforce one another.

As with the fuel, no *a priori* assumption can be made that hydrocarbons in the oil will be directly reflected in the hydrocarbons of the organic fraction. Some oil components are more likely to be volatilised than others; some will burn more effectively than others; and some may more readily form partial combustion products. The fuel and oil contributions to the organic fraction are often of comparable significance. Like fuel, oil also contains ashing elements: some of them are intentionally added to improve oil properties, such as antiwear performance, and others are present naturally. A suspicion lingers that some oil is pyrolysed into soot, but research in this area is not extensive. The oil contribution to the sulphates will be described in the next section.

A poorly appreciated subtlety is that inducted air is also a source of ash fraction: what enters with the air is potentially able to leave via the exhaust. For the statutory emissions requirement, this type of ash is inconsequential, but there are other occasions when it could become important. Depending on the environment, or the time of year, or the condition of the air filter, a vehicle may ingest, for example, mineral dust or road salt. This might be nonnegligible if the other fractions are exceptionally low, for example with gasoline engines. Of course, engineers go to great lengths to *prevent* the ingestion of foreign material, but their wish here is to protect the lubrication system, and concomitantly to avoid excessive engine wear, rather than to restrict the particulate emissions *per se*.

That wear occurs in engine components suggests another potential source of ash fraction. This does not appear to have been addressed specifically, and it seems that these wear particles are more likely to enter the lubricant than find an entry into the exhaust. One might speculate whether the emission of wear debris is more of an issue for engines during their running-in period.

2.4.3 What Happens Within the Exhaust

At completion of the combustion, or opening of the exhaust valve, the particulate phase is still a very, very long way from its final form; and in the exhaust process that follows, continuing transformations take place, the repercussions of which are profound and far-reaching. Initially, exhaust temperatures are still high enough for ongoing reactions, between gases or between particles and gases. Particles grow by agglomeration, and other particles lodge on internal surfaces. The volatile components, i.e. the sulphates, organics and nitrates, adhere to existing particles or nucleate into new particles.

Gas–particle partitioning is absolutely central to the emission process. At the exhaust ports, where the temperature is high, only the carbonaceous and ash fractions are solid. But, as the exhaust temperature continues to fall, volatile components begin transferring into the particulate phase. This is the foremost problem in particulate measurement: particulate characteristics inevitably depend on how the exhaust gas has been conditioned. The volatile species are, so to speak, the 'floating voters'. Samples of particulate extracted from various points in the exhaust system demonstrate that most gas-to-particle conversion happens quite late in the emission process – often post-tailpipe.

Pivotal in the emission process is the *dilution ratio*, as this controls saturation, through cooling (heat transfer) and mixing (mass transport).[25] This is where the real-world emission process diverges from that in the laboratory; both dilutions are depicted schematically in Figure 2.16 (adapted from Frisch *et al.*, 1979). In the real world, dilution continues to an *infinite* ratio, whereas in the laboratory, dilution converges to a *finite* ratio – where it stays until the gas reaches the filter. Identicality, therefore, in particulate emissions, between the real world and the laboratory, seems rather unlikely. And, perhaps less obviously, the laboratory sampling point is also influential: gas drawn from just upstream of the tailpipe has already cooled significantly, during transit through the exhaust system, whereas gas drawn from the exhaust manifold is still hot, and denied this opportunity. Moreover, the real-world emission process is influenced by shifting parameters such as wind speed, vehicle speed and ambient humidity. But some consistent method of dilution is needed in the laboratory, just as some practical limit needs to be set on the possible permutations in these extraneous parameters.

The importance of the curve traced out by the dilution ratio, as a function of distance from the tailpipe, is currently much discussed. The upshot, however, is that, initially, cooling is the dominant factor: this decreases the saturation vapour pressures of volatile components, and encourages gas-to-particle conversion. Eventually, however, mixing catches up and becomes the dominant factor: this

---

[25] This can be confusing, insofar as 'dilution' assumes *two* meanings: one for mixing alone, the other for the parallel action of mixing and cooling. This distinction arises because it is possible to mix at constant temperature, and to cool at constant air–exhaust ratio, although these experiments are rarely done, and require specially designed apparatus. Mass transport, then, should not be confused with heat transfer, and so, for greater clarity, the *dual nature* of dilution is recognised here; and in this sense the word acts as an umbrella for the two processes.

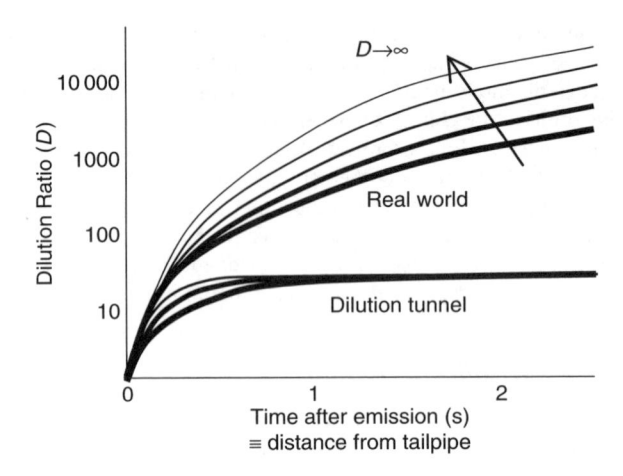

Figure 2.16 The emission process: exhaust gas as discharged into the surrounding air, depicted schematically, comparing real-world dilution with laboratory dilution (adapted from Frisch *et al.*, 1979).

decreases the partial pressures of volatile components, and encourages particle-to-gas conversion. There is a temporal dimension, since gas-to-particle conversion (and vice versa) has kinetic aspects, and there is no universal experience, since each pocket in the exhaust plume mixes and cools differently. Further aspects of this enormously complicated process can be found in Sections 3.6.3 and 4.3.

Water is the chief condensable component, and, as related in Section 2.2.4, its condensation is immediately triggered when the exhaust temperature finally falls below the *dew point*, i.e. whenever conditions of saturation are broached. Typical dew points are in the range ~45–55 °C, depending on the air–fuel ratio, and taking $\frac{1}{2}$ as a typical carbon–hydrogen ratio in the fuel (Edwards, 1977, p. 71).[26] As diesel engines always burn lean mixtures, their exhaust-gas dew points are lower than for gasoline engines burning stoichiometric mixtures. But, normally, the dew point is not transgressed within the exhaust system, except perhaps in a cold start or in outstandingly cold weather; in these instances, a dribble of water may be seen issuing from the tailpipe. In other circumstances, if dilution into the ambient air is slow in comparison with the rate of fall in temperature, then water droplets will appear in the vicinity of the tailpipe, in which case a visible exhaust plume forms.

The same physics drives gas-to-particle conversion in the other volatile components of the exhaust. The most problematical fraction to model is the organic one, because the species within it possess such remarkable diversity. Each species, of which there are thousands, is characterised by its own set of parameters, e.g. boiling point, concentration and saturation. The lightest organic compounds remain predominantly in the gas phase; the middling ones become partitioned between the gas phase and the particulate phase; and the heaviest ones reside predominantly in the particulate phase.

If unburned fuel is discharged in substantial quantities, then fuel droplets, manifested as white smoke, will form. A plume of fuel droplets is easy to confuse with a plume of water droplets, because their interactions with light are similar. Water droplets tend to be less overtly white and less opaque than fuel droplets, and are not generally seen once the power train is warm; nor are they associated with the malodorous emission that always accompanies fuel droplets. An exhaust plume of water droplets

---

[26] Water vapour, or moisture, carried into the engine with the air raises the dew point in the exhaust gas, i.e. condensation will now occur at a higher temperature, and sooner in the exhaust system, than is the case for dry air.

is shorter, suggesting swifter evaporation, and more vigorously swirling, suggesting the droplets are more faithful to the eddies.[27]

Sulphates are the other principal volatile component in exhaust gas. These are seldom much of an issue in the absence of catalytic converters, because most of the sulphur is emitted as SO_2, and, as such, this makes no difference to the amount of particulate, since it remains in the gas phase. However, any oxidation catalyst will readily convert, according to a well-understood thermodynamic law, the SO_2 to SO_3; this SO_3 is rapidly hydrolysed into sulphuric acid, H_2SO_4; and this acid, in turn, and with equal rapidity, undergoes gas-to-particle conversion. Importantly, sulphuric acid is notorious for shifting the dew point, so that condensation now begins earlier, within the exhaust system, rather than in the plume – creating a corrosion hazard for metal components.

Ash more usually exists as metal oxides or salts. It is probable that, owing to shifting equilibria, chemical compositions change as the exhaust cools, but, because this cooling occurs with such formidable rapidity, full equilibrium may not be possible, in which case these compounds remain in some quenched or nonequilibrated state. This issue has not been well researched. The ash fraction is not noted for its volatility, and is rightly viewed as effectively nonvolatile in the exhaust system; most of the gas-to-particle conversion probably happens within the engine, or close to the exhaust ports. However, one should be alert to the possibility that some ash compounds are not entirely in the particulate phase when in transit through the exhaust system.

Ash particles formed from their respective vapours should *not* be confused with others arising from processes of disintegration. *Wear* particles are still assigned to the ash fraction, on the grounds of commonality in their elemental composition, but they different distinctly in nature, being larger – sometimes much larger (microns instead of tens or hundreds of nanometres). This is the case with rust and scale, which, through the combined action of oxidation, thermal cycling, vibration and corrosion by sulphuric acid, exfoliate from the walls of the exhaust system. Various wear particles are also released from catalytic converters or other aftertreatment devices, and from mufflers.

Exhaust systems exhibit a remarkable proclivity for particle storage and release. The sudden and unexpected release of wall deposits, as triggered by vibration, temperature cycling, etc., is behind spikes, pulses or bursts of particulate, and this randomness easily leads to factitious measurements. The tailpipe emission is not, in this respect, a faithful representation of the engine-out emission; and perhaps the as-measured particulate was actually emitted by the engine in the previous hour, previous day, or previous week . . .

2.4.4 Number Versus Mass

In Section 2.2.2 it was stated that aerosol science spans five orders of magnitude in particle size. Fortunately, to study the particles emitted by internal combustion engines, three orders of magnitude are normally sufficient, while much of the important information resides in just two orders of magnitude. The size range of interest is situated between the largest viruses and the smallest bacteria, and in terms of electromagnetic radiation, this overlaps with light in the visible and ultraviolet.

If we were to take any sample of exhaust gas and examine the size distribution of particles suspended therein, the demographic information would not greatly differ from that depicted in Figure 2.17 (after Kittelson, 1998). This figure plots, using the mathematical functions developed in Section 2.2.2, the size distribution in three ways: according to *number*, *area* and *mass*. The particles fall readily into

---

[27] Since fuel droplets do not follow eddies with the same assiduity as water droplets, it seems logical to argue, using the physics described in Section 2.2.3, that they are larger or heavier. These are the personal musings of the present author.

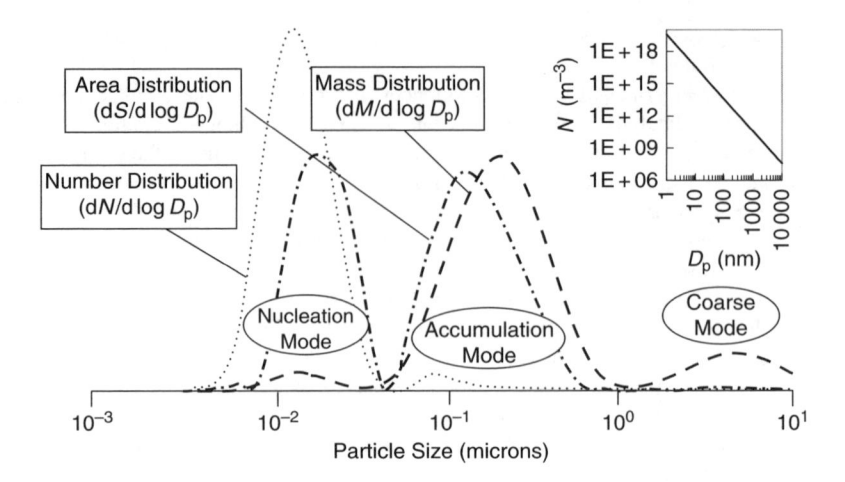

Figure 2.17 Generalised size distributions for typical particles emitted by internal combustion engines, taking spheres and constant density as a first approximation: by number N, surface area S and mass M (after Kittelson, 1998). The inset shows, as a function of size, and assuming unit density, the required N, in particles per cubic metre, in order to realise $M = 20\,\mu\text{g/m}^3$.

three groups, known as modes, and these are designated, in order of increasing size, 'nucleation', 'accumulation' and 'coarse'. The exact position of each peak on the *x*-axis varies, and the distribution is usually dominated by one mode, but the trimodality appears to be generic: this parallels the picture developed earlier for atmospheric aerosols in Section 2.3.1.

Each of the three modes, when plotted on a logarithmic *x*-axis, appears normally distributed, and this distribution is described as 'lognormal'. Actually, lognormality is not always the case, but mathematical curves generated from Gaussian distributions usually fit the data quite well.

A fundamental aspect, the profundity of which cannot be overemphasised, is the domination of different ranges by mass and number. Most of the particles, according to *number*, reside in the *nucleation* mode; while most of the particles, according to *mass*, reside in the *accumulation* mode. Each of the two functions, then, inherently emphasises different modes and different ranges of size.

This understanding can be extended to the two distributions developed in Section 2.2.2 for mass ($m(D_p)$, $\mu\text{g}\,\text{cm}^{-3}\mu\text{m}^{-1}$) and number ($n(D_p)$, $\text{cm}^{-3}\mu\text{m}^{-1}$). Since transport and transformation are determined fundamentally by particle size, then mass and number are inevitably affected by these phenomena in completely different ways. If one were solely interested in mass concentration (M, $\mu\text{g}\,\text{cm}^{-3}$), then inaccuracies in measuring the nucleation mode might not appreciably alter the end result. If, on the other hand, one just wished to investigate the number concentration (N, cm^{-3}), then inaccuracies in measuring the accumulation mode would be of lesser concern. Diffusion governs particle transport particularly strongly in the nucleation mode, and so potentially affects the number distribution, whereas inertial deposition has greater implications for the accumulation mode.

It turns out that nucleation-mode particles are vastly more vulnerable to measurement falsifications than accumulation-mode particles; indeed, measurement artefacts can completely overwhelm number concentrations. Much of the blame for this can be laid on the volatile component and the highly nonlinear nature of nucleation, as we saw in Section 2.2.4. It is, therefore, notoriously difficult to measure the nucleation mode in a meaningful, reliable or consistent way. The nucleation-mode particles are created or destroyed according to how the exhaust is diluted, whereas the accumulation-mode particles consist of a *solid* carbonaceous core, and are only affected by saturation insofar as this decides their capture or release of volatile compounds. The accumulation mode and the mass distribution are thus more likely to represent the genuine emission than are the nucleation mode and the number distribution. This mutability presents a serious problem: how can we be sure that the emission process in the laboratory adequately reproduces the particle emission that members of the public are likely to inhale? Clearly, there is no sense in enacting legislation for which unreliable measurement technology allows factitious results. This conundrum will occupy much of the text.

Fortunately, it is possible to lay down a clear principle when attempting to control the particulate emission from a motor vehicle. The overarching factor here is the proportion of volatile (sulphates and organics) to nonvolatile (carbonaceous and ash) particulate. If the latter is strongly present, it will scavenge the volatile material: this will keep down the saturation, and hence suppress nucleation, allowing the accumulation mode to dominate. Contrariwise, if the particulate emission is mainly volatile, then insufficient solid surface area, for adsorption and condensation, will promote nucleation. This insight then needs to be set alongside factors that decide the volatile/nonvolatile breakdown. For example, high engine load favours carbonaceous particulate, and low engine load favours organic particulate.

The health implications of ambient particles are gauged in terms of mass concentration, or the total weight of particulate found in a unit volume of air – historically for PM10, and now, as well, for PM2.5. But even PM2.5 is still very large in relation to motor vehicle particulate. From the foregoing, it will be appreciated that the total mass of particles, whether in motor vehicle emission control or ambient air quality monitoring, says little about the total number of particles. And in emission control, the law currently requires the capture of particulate on a filter, followed by quantification of the particulate via weighing. What Figure 2.17 implies is that such a measurement will not detect wide fluctuations in the population of nanoparticles. So, although the total mass of particulate emitted by motor vehicles has undoubtedly declined enormously in the last three decades, what has happened to the number of particles?

2.5 Closure

In this chapter we have charted three principal areas which, taken together, serve as a comprehensive backdrop to the remainder of the text. The first section looked at aerosol science: how to characterise particles, how particles are transported and how particles are altered by their experiences. The second section covered particles in the atmosphere: those that arise naturally, those present through the activities of man, how these particles behave, and the environmental implications that result. In the third section some introductory material was provided about motor vehicle particulate: chemical composition, routes of formation in the engine and exhaust, and finally a few demographic details, i.e. particle sizes, masses and numbers.

2.5.1 Properties of Aerosol Particles

Diameter and Shape

Aerosol particles span a huge range of size, from molecular clusters, at a few nanometres, to wind-blown dust, at several tens of microns. Size is often expressed in terms of a 'diameter', but this usually means a *derived* diameter, because particles are, more often than not, irregularly shaped. (Liquid droplets are an obvious exception.) A derived diameter refers to the way a particle *behaves* when subject to certain forces; it has *no direct* geometrical significance. Two commonly used derived diameters are the 'aerodynamic' and 'electrical mobility' diameters.

Size Distribution

Size *distribution* is expressed in several ways: by number, mass, volume or surface area. These distributions are handled and interrelated by well-established mathematical relationships, usually logarithmic ones. Particles are often, for expediency, assumed to have unit density, or a constant density irrespective of size; these assumptions should be adopted cautiously. Concepts of fractality are used to characterise particles of complex geometries.

Transport and Deposition

Particle transport does not necessarily come about through straightforward convection of the carrier gas, i.e. particles do not always simply follow the streamlines. Rather, transport mechanisms that cause particles to divert from the streamlines are vital. These mechanisms include thermophoresis, diffusion, inertia, sedimentation and electrostatic attraction. The largest particles are massive enough to experience significant inertial ($>1\mu m$) and even gravitational ($>10\mu m$) effects. The smallest particles (<100 nm) are of a size comparable with the mean free path of the molecules making up the carrier gas, and so themselves begin to display molecular properties, such as that of diffusion. Particle transport is often ineffective at a few hundred nanometres, since this regime falls between those of diffusion and inertia; sometimes thermophoresis is the only significant mechanism in this size range. This absence of natural transport mechanisms accounts, for example, for the mid-range dip in the capture efficiency of many filters. Artificial methods are required in these instances to improve transport mechanisms. Only in the case of interception are particles caught without them having departed from the streamlines of the carrier gas.

Transformation and Mutation

By 'mutation' is meant the notoriously protean nature of aerosols, i.e. there are continual modifications to particle size, shape, form and composition, according to the conditions. The volatile or semivolatile species are largely to blame for this mutability – these are the 'floating voters'. Because of their volatilities, these species undergo gas-to-particle and particle-to-gas conversion, and invariably become partitioned to some degree between the gas phase and the particulate phase. Consequently, characterisation difficulties multiply with the amount of volatile material. The phenomena of gas-to-particle conversion are governed by 'saturation', meaning how much of a species can exist as a gas. Above saturation, condensation into the particulate phase occurs; below saturation, adsorption is responsible for gas-to-particle conversion. Nucleation is a special type of condensation; it is responsible for the creation of entirely new particles. A critical property of nucleation is its 'on–off' nature: in the right conditions, a massive upsurge occurs in particle formation. If the volatile/nonvolatile ratio is high, then nucleation is rendered more likely; if it is low, then condensation and adsorption are more likely. Another important mutative process, which does not involve the gas phase, is coagulation: the sticking together of particles to form larger ones; the number of particles falls, but the total mass remains constant. Re-entrainment of particles deposited onto surfaces is arguably a special type of coagulation; these particles are larger than the original ones that were deposited.

2.5.2 Particles in the Atmosphere

Character and Behaviour

The atmosphere contains suspended particles of enormous diversity: most elements are found, and forty of them make regular appearances. But compositionally, these particles can be reduced to six predominating groups: sulphates, organics, carbonaceous, nitrates, minerals and biological. The sources are equally diverse, e.g. sea salt, decaying vegetation, microbes, wind-blown and volcanic dust, asbestos, and combustion. Secondary particles form through atmospheric reactions and gas-to-particle conversion; photochemical smog and sulphates are just two examples. The atmospheric aerosol forms part of the geosphere, the hydrosphere and the biosphere. Particles are eventually removed by dry deposition (transport by the wind or gravity) and wet deposition (transport by precipitation elements, principally clouds and raindrops). The composition and number of particles in the atmosphere show periodicities driven by fluctuations in the rate of input, rate of removal, and atmospheric reactions; the timescales are hourly, daily, weekly and seasonal. Atmospheric particles fall into three characteristic ranges of size called 'modes'. In order of increasing size, these are the *nucleation mode*, *the accumulation mode* and the *coarse mode*.

Aerosols in Nature

The oceans are prolific contributors to the atmospheric aerosol, through bubble bursting and the ejection of salt particles. These salt particles are central to the formation of clouds. The sea also gives rise to sulphates through the action of phytoplankton. Structures of living or dead life forms, such as spores, bacteria and pollen, and parts of insects, are frequently airborne. Decay processes release ammonia, which forms secondary nitrate and sulphate particles. Plants, especially trees, release volatile organic compounds which, through ongoing atmospheric reactions, form secondary particles. Mineral particles arise through geological forces: volcanism and the weathering of rocks.

Anthropogenic aerosols

Anthropogenic aerosols are no less diverse than natural ones. Agrochemicals are released through spraying, and asbestos particles through processes of disintegration, for example from roof tiles. The action of motor vehicles on roads, particularly unpaved ones, releases mineral particles. In many high-temperature industrial processes, particularly metal refining and extraction, metals escape when they volatilise; they subsequently cool and form particles. Carbon in the atmosphere is ubiquitous; it is derived from the combustion of fossil fuels and biomass, and takes two main forms: carbonaceous and organic; these are found together. A wide range of ash arises from combustion, courtesy of fuel components, either intentionally or unintentionally present. Sulphur is present in all fossil fuels as an interloper; it ultimately ends up in the atmosphere as secondary sulphates. A form of combustion more strongly associated with noncarbon particles is the incineration of municipal and industrial waste. Photochemical smog is an aerosol formed through the reaction of volatile organic compounds and NO_x in the presence of strong sunlight.

Environmental Implications

Atmospheric particles pose significant environmental risks: they degrade visibility, they soil and corrode building surfaces, they endanger public health, they damage vegetation, and they promote climate change. Many particles, being inert and buoyant, stay suspended for long periods – particularly in the accumulation mode, for which the residence time is one to two weeks – and so are transported to remote sites, to the detriment of pristine ecosystems. Carbonaceous particles, since they are black, are notable for being the principal light-absorbing species in the atmosphere. The amount of particles suspended in the urban environment correlates with statistics for morbidity and mortality; no firm explanations for this correlation have yet been established. But a suspicion exists that, today, there are more of the smaller particles in the atmosphere than, say, a generation or more ago. Particles in the atmosphere control the earth's climate through acting as cloud condensation nuclei and interacting with solar radiation.

2.5.3 Motor Vehicle Particulate

Some Typical Particles Dissected

As with the atmospheric aerosol, the particles emitted by motor vehicles, size-wise, fall into three characteristic ranges of size: nucleation mode, accumulation mode and coarse mode. Accumulation-mode particles are constructed from a solid core of carbonaceous building blocks called 'spherules', together forming 'agglomerates'. Spherules are fairly uniform in size, i.e. mostly 20–50 nm, and the agglomerates they form contain no defined number: it ranges from a bare handful to several thousand. There is, similarly, no unique geometry: both clusters and chains are seen. An outer layer of volatile compounds is attached to the agglomerates. The nucleation-mode particles are more arcane: most are probably formed from nucleated volatiles, but some research suggests that a few of them are solid, or contain minute solid cores. The coarse-mode particles are solid and are formed from the other two modes through a process of storage and release in the exhaust system, or through material disintegration. Composition-wise, there are five distinct 'fractions': *ash*, *carbonaceous*, *organic*, *sulphate* and *nitrate*. The first two are nonvolatile and the others volatile. This classification scheme is not perfect; dividing lines between the fractions are often difficult to trace. The organic fraction, as it contains hundreds or thousands of compounds, is arguably the most complex. Most of the major chemical families are present, although their proportions vary: esters, aromatics, alcohols, etc. The sulphate fraction is predominantly sulphuric acid and its associated water, but there can also be metal sulphates. The carbonaceous fraction is chiefly carbon. The nitrate fraction denotes nitric acid. The ash fraction consists of what is incombustible, and contains mostly metals, but also a few nonmetals.

What Happens in the Engine

The carbonaceous fraction forms in the engine via the pyrolysis of fuel molecules, i.e. when there is insufficient oxygen available for complete oxidation. The organic fraction arises either from fuel molecules that completely escape combustion or from fuel molecules that are partly modified, the reactions which form the carbonaceous fraction having been for some reason interrupted. Ash compounds present in the fuel, either deliberately or naturally, give rise to the ash fraction. Each of the principal four fractions can *also* arise from the *lubricant*, and for similar reasons; but the organic one is probably the most significant in mass terms. Other contributors to the ash fraction may be the inducted air and material disintegration, but these are small under normal circumstances.

What Happens in the Exhaust

Gas-to-particle conversion, i.e. for the sulphates and organics, happens late in the emission process, and proceeds according to the laws of saturation. In the exhaust plume (post-tailpipe), much depends on the process of dilution, and the combined effects of mixing in, and cooling by, the surrounding air. The pivotal parameter is the *dilution ratio*. The process of dilution in the laboratory inevitably differs from that in the real world; the first proceeds to a finite, and the second to an infinite, dilution ratio. Sulphuric acid, as it shifts the dew point, allows gas-to-particle conversion to occur at higher temperatures, i.e. further into the exhaust system. Shifting equilibria, as the exhaust cools, may change the composition of the ash compounds. Rust and scale from exhaust system components are another source of ash.

Number and Mass

The size distribution is lognormal. Most particulate, by mass, resides in the accumulation mode (100–900 nm), and most particulate, by number, resides in the nucleation mode (<100 nm). The coarse mode (>900 nm) is not usually significant. Vulnerability to what are known as measurement 'artefacts' varies with the metric. Mutative processes that affect the nucleation mode will modify the number distribution, and those that affect the accumulation mode will modify the mass distribution. Generally, the accumulation mode is more likely to represent what the engine has emitted; the nucleation mode, by contrast, is notoriously dependent on the way the exhaust gas has been handled, i.e. diluted. Much of the blame for this fickleness can be safely ascribed to the highly nonlinear nature of nucleation.

3

Formation I: Composition

3.1 Introduction

In this chapter and the next we chart, so to speak, the particle life story, but with two differing emphases, the present one being on the fraction, or particle *composition*. This approach makes sense when each fraction is clearly distinguishable, as origin and development, being manifestly different, are then equally distinguishable.

The carbonaceous fraction is arguably the best charted in particle formation terms, insofar as the review necessarily consumes the lion's share of the chapter. Pyrolytic reactions in oxygen-deficient regions of the charge give birth to huge numbers of spherules, which, at a certain point, stop growing, and instead begin sticking to one another. This gives rise to the particle archetype, the 'agglomerate' – those particles which form the accumulation mode. The chain of inception, growth, agglomeration and oxidation is the generalised picture of soot formation, and seems not to differ greatly, whatever the form of combustion. But these are exciting times, as laser-imaging studies of the burning fuel sprays are rapidly elucidating the peculiarities of internal combustion engines.

The ash fraction is probably the least delineated, but, given the motley nature of the elements contained, no common formation route is likely: some metals transfer into the particulate phase before the emergence of spherules, others after. (Certainly, the origins of the ash fraction and of the carbonaceous fraction are intertwined in ways still insufficiently understood.) The chemical forms taken by metals depend on the availability of nonmetals, and this, in turn, decides the moment of gas-to-particle conversion.

The organic fraction traditionally concerns those compounds which transfer into the particulate phase late in the emission process – even in the dilution tunnel. But important preparatory reactions nevertheless take place in the engine; and less characterised, in terms of particle formation, are other organic compounds found at the point of incorporation into spherules – not quite unburned fuel, but not quite carbonaceous. In this chapter we review studies of organic compounds at this highly transitory point: where pyrolytic reactions are converting fuel into spherules.

Particulate Emissions from Vehicles   P. Eastwood
© 2008 John Wiley & Sons, Ltd

The sulphate fraction is well understood insofar as the thermodynamics of the SO_2–SO_3 system are a commonplace. Difficulties ensue with the knock-on effect: formation of sulphuric acid, the gas-to-particle conversion of which is highly context-dependent. Indeed, the recently emerged picture, in studies to be reviewed here, is of sulphuric acid as a nucleating agent, around which organic compounds subsequently cluster, late in the emission process. This makes the acid the prime instigator of nucleation modes.

3.2 Carbonaceous Fraction: I. Classical Models

The formation of carbonaceous particulate, or 'soot', is not by any means restricted to diesel engines: liquid fuels are atomised and burned in a great many combustion systems, and susceptibility to the self-same problem, albeit to hugely varying degrees, is widespread – for example in gasoline engines (Chapter 8), and in gas turbines (Gupta A.K., 1997). And even the combustion of pulverised coal generates clouds of soot when outgassed organic compounds are released and pyrolysed (Fletcher *et al.*, 1997). The end product, which parallels Henry Ford's famous remark about the colour of his automobiles, is the same: '... properties of carbons formed in flames are remarkably little affected by the type of flame, the nature of the fuel being burnt and the other conditions under which they are being produced. Any complete theory of carbon formation must of course be able to account for this striking experimental finding'. (Cited by Dobbins *et al.*, 1998.)[1] This remark should not perhaps be taken too far: soots vary subtly in their *microstructures*, according to the fuel burned and the type of flame (Vander Wal and Tomasek, 2004), so that any grand unified theory must account also for diversity.

The appearance of soot in the combustion zone is harbingered by wide-band incandescence. In boilers and furnaces, this incandescence, or radiative energy loss (Ray and Wichman, 1998), is *desirable* insofar it aids heat transfer; and candles, after all, would otherwise furnish little luminosity. In diesel engines, this dissipation steals from useful work, and so is detrimental to fuel economy (Struwe and Foster, 2003). Actually, an intriguing circularity arises here, wherein soot mediates its own formation: the incandescence lowers flame temperature (Smooke *et al.*, 2005). And in diesel engines, the incoming fuel, exposed to this heat flux, burns more richly, and is more soot-producing, than otherwise (Hampson and Reitz, 1998).

Diesel engines incorporate – inadvertently, it must be said – many of the features deliberately designed into the industrial manufacture of carbon black (Lockwood F.C. and van Niekerk, 1995; Taylor R., 1997); indeed, diesel fuel is sometimes used as a feedstock, from which it seems reasonable to suppose that novel ideas may be gleaned for controlling the formation of soot: choosing the *right* conditions, for the appropriate quantity and grade of carbon black (Balthasar *et al.*, 2002a), is arguably the inverse process of avoiding the *wrong* conditions in the diesel engine. The differences arise in the immediate post-flame phase: in carbon-black production the exhaust gas is quenched rapidly, such as with water, and collected and stored immediately, whereas diesel soot is quenched by the descent of the piston principally, but also by subsequent passage to the tailpipe. Potential therefore exists for further physicochemical transformations that are denied to carbon black.

Soot appears to be inherent to diesel engine combustion: this is unfortunate inasmuch an overall surfeit of oxygen in the charge is also inherent, so that, arguably, no problem need arise. The heterogeneity of the charge, i.e. rich-burning regions of fuel, during the mixing-controlled phase is largely to blame. But the responsible mechanisms are poorly delineated compared with what is known about the other emissions (CO, HC and NO_x); indeed, many years of investigative work must pass

---

[1] Original reference: Palmer and Culliss (1965).

before the detailed soot-forming chemistry of even simple laboratory flames will be satisfactorily comprehended. The premixed and diffusion flames of laboratory burners are, actually, the fundamental tools of investigation, from the literature of which we draw thoroughly in this section.

Soot formation commences with fuel molecules containing 12 to 22 carbon atoms, and twice as many hydrogen atoms; soot formation terminates a few milliseconds later with particles, or rather spherules, containing thousands of carbon atoms, and one-tenth as many hydrogen atoms. What exactly happens between these two end-states has taxed the ingenuity of researchers for decades, and will, no doubt, continue to spawn many a PhD. But, at grass-roots level, three parameters dominate soot formation: *air–fuel ratio*, *temperature* and *pressure*.

Thermodynamic equilibria applied to diesel engine combustion suggest the formation of free carbon as solid-phase graphite (Lipkea and DeJoode, 1994), although thermodynamic end-states cannot genuinely describe what happens, as chemical kinetics rule. But the thermodynamic approach does, nevertheless, help to pinpoint the most influential factors in soot formation (Amann et al., 1980), and will at least bracket the starting and end points, even if it does not actually indicate whether the destination is reached, or reveal the path taken to that destination (Richter and Howard, 2000). The thermodynamic approach does not, however, appear to have been widely embraced by soot modellers (Mansouri et al., 1982).

Although the finer details are heatedly debated, the formation of soot, conceptually, is known to follow something like the route adumbrated in Figure 3.1. Pyrolytic reactions break down the original fuel molecules and construct soot precursors, and these precursors undergo *nucleation*, to form the first discernible particles, or nuclei, at less than 3 nm. These nuclei undergo *surface growth*, during which carbon is added, and hydrogen removed, until spherules emerge at ~20–50 nm. During this growth process the spherules themselves *agglomerate*, with surface growth occurring in parallel, until finally the particles begin to assume their familiar identities. The fourth mechanism, *oxidation*, opposes the other mechanisms, and in fact culls soot at any stage: as precursors, nuclei, spherules or agglomerates. Some workers include a *carbonisation* stage (not depicted), in which polyaromatic layers are aligned, or perhaps realigned, and amorphous carbon is transformed into graphitic carbon (Richter and Howard, 2000).

To separate and describe these mechanisms in a tidy, linear manner is convenient, but this chronology is artificial. Overlapping to hugely varying degrees, different mechanisms rise to prominence in different regions of the combustion chamber at different times, as pressure, temperature and air–fuel ratio dictate. None of these mechanisms is likely to enjoy wholehearted independence, and nucleation, surface growth and agglomeration probably compete with one another to some extent – and not just with oxidation. For example, as the amount of growth material is presumably finite, there is, perhaps, competition for this material between nucleation and surface growth.

Obviously, this chain of events presents quite a challenge for the modeller. It will be noted that two domains or levels are encompassed: the molecular system and the particle system. Models of soot formation are legion, yet they are essentially of three types (Kennedy, 1997). The most rudimentary ones claim no fundamental insights, are entirely empirical, and use correlations between soot and various experimental parameters. Next, there are models of a semi-empirical nature, for which certain terms are calibrated, or 'tuned', to obtain agreement with experimental data. Finally, there are detailed analytical models – some of them *immensely* detailed – in which chemical equations, for a large number of reactions (dozens to a few hundred), are solved simultaneously and iteratively, and the concentrations of all major species, from fuel hydrocarbons to PAH to soot, are thereby predicted – including all the radicals and short-lived intermediates. The enormous computational demands made by the detailed models have, no doubt, allowed empirical and semi-empirical approaches to persist.

This threefold classification scheme admittedly has some limitations, but it remains useful notwithstanding, insofar as it systematises a stupefying welter of disparate material – no less than 33

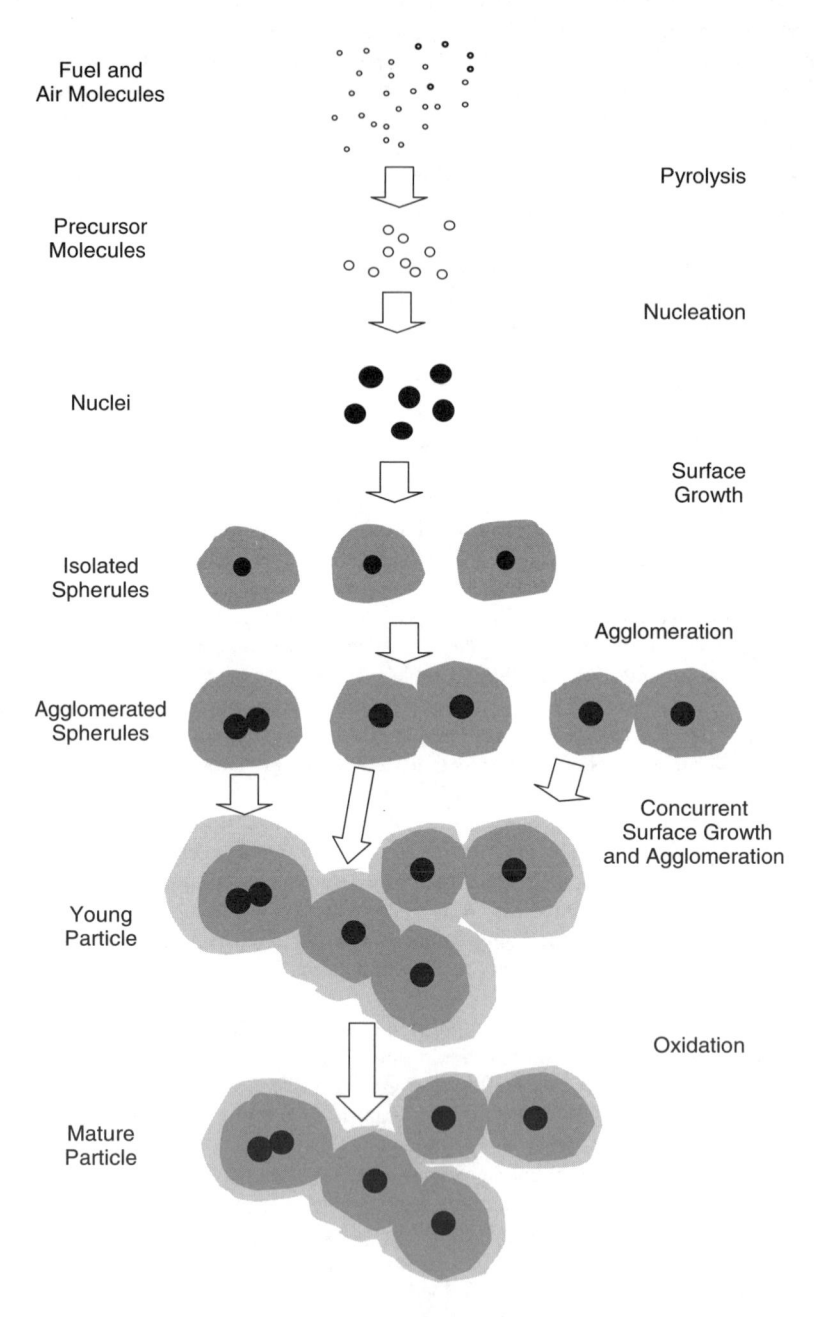

Figure 3.1 Conceptual scheme for the formation of soot (linear narrative).

models in one comprehensive review (Kennedy, 1997). Of course, these models have been devised for a host of combustion processes, not just internal combustion engines – this betrays the widespread need to control soot. To enforce some practical limit to the review, we shall focus on the models to have found acceptance amongst *engine* modellers. But, even though these models are widely applied to diesel engines, they have been derived, originally, and in the main, from studies of laboratory flames, so that their extrapolation to engines is, understandably, uncertain.

3.2.1 Empiricisms

Since soot appears at characteristic or *critical* air–fuel ratios, an obvious starting point is the global reaction scheme; for example, considering the emergence of C, H_2 and CO as intermediate products, we have (Li X. and Wallace, 1995)

$$C_\alpha H_\beta + xO_2 \rightarrow \alpha CO + \left(\tfrac{\beta}{2} - \tfrac{\alpha(1-\gamma)}{\gamma}\right)H_2 + \left(\tfrac{\alpha(1-\gamma)}{\gamma}\right)H_2O + \left(x - \tfrac{\alpha}{2\gamma}\right)O_2 \quad (3.1)$$
$$[(\alpha/2x) \leq \gamma],$$

$$C_\alpha H_\beta + xO_2 \rightarrow 2x\gamma CO + \left(\tfrac{\beta}{2} - 2x(1-\gamma)\right)H_2 + 2x(1-\gamma)H_2O + (\alpha - 2x\gamma)C(s) \quad (3.2)$$
$$[(\alpha/2x) \geq \gamma].$$

On this basis, one might suppose criticality for a C/O ratio of *unity*, where every carbon atom is at least able to find one oxygen atom, and thus form CO, and when this ratio exceeds unity, some carbon atoms appear as soot. Unfortunately, this approach has little predictive merit: critical ratios are often less than unity, reaching even 0.5 (Haynes and Wagner, 1981, p. 232), so that, evidently, not all the available oxygen atoms are used. In any case, with inhomogeneous air–fuel mixtures, the threshold will be exceeded *locally*.

In practice, critical air–fuel ratios depend on specific details, such as the nature of the fuel–air mixing, the structure of the flame, and the molecular structure of the hydrocarbon (Heywood, 1988, p. 638). Predictions of *sooting tendency*, as 'smoke point' or 'threshold soot index', through empirical indicators and apparatus such as the 'soot lamp', allow easy assessment of large numbers of hydrocarbon fuels (Ladommatos et al., 1996e), but these lack wide applicability, i.e. to other modes of combustion (Yan et al., 2005a). This should not be wholly unexpected: the chain of events depicted in Figure 3.1 is glossed over. The apparatus and methods can, however, be refined (McEnally and Pfefferle, 2007).

Other empiricisms correlate engine-out soot with various combustion parameters, or seek to supply the quantity of in-cylinder soot at each moment in the combustion, to converge on the exhaust level of soot as combustion completes. Useful predictive capability is regularly demonstrated by empiricisms, and such models are relatively undemanding computationally; hence their continuing incorporation into global combustion models is assured for the foreseeable future.

The relationship between ignition delay and smoke was noted long ago, and logically ascribed to the time available for charge preparation prior to combustion (Grigg 1976). Thus, the global air–fuel ratio at the start of combustion divided by the global air–fuel ratio at the end of combustion furnishes a logical metric, well-correlated with smoke (Bryzik and Smith, 1977). The injection timing at which maximum smoke was observed gave the smallest ignition delay, i.e. the lowest percentage of fuel injected prior to ignition, and hence the longest mixing-controlled burn. Similarly, near-linear correlations are observed between smoke and the fraction of fuel consumed in the mixing-controlled burn (Han S.B. et al., 1997). Faced with this understanding, it is apparently paradoxical that soot production in a premixed laboratory flame should correlate so well with soot production by a diesel engine (Hardenberg, 1981).

As soot forms predominantly in the mixing-controlled phase, then some correlation of sorts may be expected specifically with this phase. The index I_{soot}, in grams of soot emitted per gram of fuel burned, correlates well with calculations of the (stoichiometric, adiabatic) diffusion flame temperature T_f, via

$$I_{soot} = C_{mix} \exp\left(\frac{E}{\Re T_f}\right), \qquad (3.3)$$

where C_{mix} expresses the physical aspects of air–fuel mixing, E is an activation energy and $\Re$ is the universal gas constant (Ahmad and Plee, 1983). Figure 3.2 gives several examples of such correlations (Easley and Mellor, 2001). Data acquired from different engine operating points and from different engines sometimes collapses onto one curve; yet on other occasions, activation energies are inconsistent. The reason for this fickleness is unknown, but the shifting relationship between soot formation and soot oxidation could be a factor, as E contains contributions from both. Alternatively, the correlation, since it relates to the stoichiometric region of the flame, might fail to capture the off-stoichiometric aspects of soot formation. A related approach uses heat release; a correlation to the whole burn is not always demonstrable (Venkatesan and Abraham, 2000), but restricting the analysis to the mixing-controlled phase improves matters (Ishida et al., 1990). I_{soot} has been correlated to the inverse of the burned-gas temperature, 75 % of the way through the heat release (T_{75}) (Desantes et al., 2000):

$$I_{soot} = k \exp\left(\frac{k'}{T_{75}}\right) p_{O_2}^{-1.5} p^{-2.2}, \qquad (3.4)$$

where p is the in-cylinder pressure, p_{O_2} is the oxygen partial pressure at the end of combustion, and k and k' are constants for any particular engine operating point. Another correlation is to a 'mixing parameter', referring to that fraction of fuel which is contained in rich regions ($\lambda<0.56$) for a certain time ($t > 600\,\text{ms}$) (Dodge et al., 2002).

The Hiroyasu model (Nishida and Hiroyasu, 1989) uses an Arrhenius-type kinetic rate expression, and this has found favour with numerous engine modellers (e.g. Belardini et al., 1993; Rutland et al.,

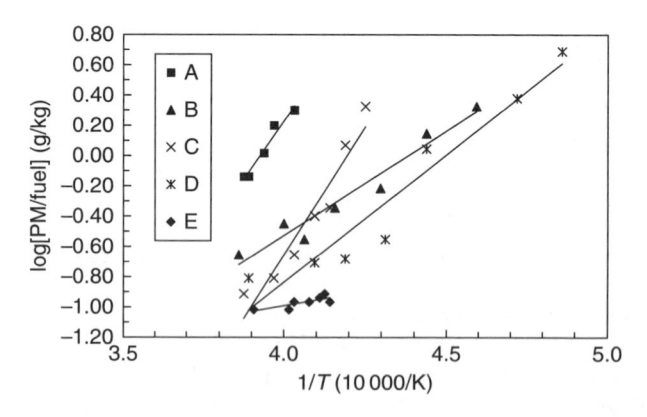

Figure 3.2 Emission of particulate matter (PM), normalised to fuel consumption, and correlated to stoichiometric flame temperature at start of combustion. Engine speed (rpm), load (kPa), injection pressure (MPa), start of injection (°BTDC), as follows: A, (2600, 880, 70, 23); B, (2000, 200, 80, 6); C, (2300, 420, 90, 12); D, (1500, 260, 60, 5); E, (1000, 10, 70, 3). Engine: 1.2-litre, four-cylinder, high-speed direct-injection, common-rail fuel injection. Fuel: aromatics, 18.9 %; sulphur, 176 ppm; cetane no., 45. Particulate was estimated from smoke measurements (Easley and Mellor, 2001).

Formation I: Composition

1995; Jung D. and Assanis, 2001; Kaario et al., 2002; Arsie et al., 2004); it considers the kinetic control of soot formation via a first-order reaction of vaporised fuel, with an exponential dependence on temperature and a square root dependence on pressure. Although the approach does not quite retain its original form (e.g. Hiroyasu and Kadota, 1976), today's workers have long settled on

$$\frac{dm_{sf}}{dt} = A_f m_{fg} p^{0.5} \exp\left(\frac{-E_{sf}}{\Re T}\right), \tag{3.5}$$

where m_{sf} is the mass of soot formed, m_{fg} is the mass of fuel vaporised, A_f is a constant, p is the pressure, T is the temperature and E_{sf} is the activation energy for soot formation. A variation on this theme is the addition of an exponent to m_{fg}, which has been both greater and smaller than unity (Lipkea and DeJoode, 1994).

Particle population, rather than total soot mass, is the focus of the Tesner model,[2] which is a two-step, chain-like kinetic process, in which 'radical nuclei', after some induction period, grow into fully fledged soot particles, although this proliferation is tempered by branching and terminating processes. The formation rate is written as

$$\frac{dn}{dt} = n_0 + (f - g)n - g_0 Nn, \tag{3.6}$$

where n is the concentration of radical nuclei, n_0 is the formation rate of radical nuclei, f is a linear branching constant, g and g_0 are linear termination constants, and N is the concentration of soot particles. n_0 is obtained via

$$n_0 = a_0 p_f \exp\left(\frac{-E_T}{\Re T}\right), \tag{3.7}$$

where a_0 is a constant, p_f is the partial pressure of fuel vapour, T is the temperature, and E_T is the activation temperature for the spontaneous production of radical nuclei. The mass of soot formed, m_{sf}, is obtained via

$$\frac{dm_{sf}}{dt} = m_p (a - bN) n, \tag{3.8}$$

where m_p is the mass of a soot particle, and a and b are model constants, the former being the reciprocal of the time needed for the smallest discernible soot particles to emerge from radical nuclei. This model, originally derived for flames, has been applied to diesel engines (e.g. Kyriakides et al., 1986; Mehta P.S. et al., 1988; Zellat et al., 1990; Nakakita et al., 1990; Taskinen et al., 1998; Taskinen, 2000). However, it requires some modification, since it insufficiently captures the effects of large variations, throughout the combustion, in pressure and temperature: for example, one must account more fully for the decline in the formation rate of nuclei as the piston descends. The coefficient g_0, being dependent on particle concentration, is therefore adjusted to the momentary in-cylinder volume at each stage in the combustion (Nakakita et al., 1990).

---

[2] Original reference: Tesner et al. (1971).

3.2.2 Inception

The sudden spawning of solid-phase carbon, or spherules, from gas-phase precursors, although bearing the hallmarks of nucleation, is customarily styled 'inception' – a term less loaded with presuppositions (Lahaye and Ehrburger-Dolle, 1994) – since it is suspected that nucleation in the *classical* sense, as described in Section 2.2.4, is not involved, and that gas-to-particle conversion occurs via other, considerably more arcane mechanisms. Inception, moreover, is an all-embracing term, as it covers, *besides* gas-to-particle conversion, the long and tortuous sequence of complex preparatory reactions: for example, a well-known model considers no less than 99 chemical species and 527 reactions (Wang H. and Frenklach, 1997).

The pre-particle gas-phase chemistry is, then, relatively well probed from a theoretical standpoint; but, as a caveat, these models, perhaps inevitably, have been developed for simple (usually one-component) fuels, and (relatively straightforward) combustion in laboratory flames; hence, understandable concerns are raised by the extrapolation of these models to multi-component fuels, as burned in internal combustion engines (McEnally *et al.*, 2006). True, the rudiments are simply stated: extensive pyrolytic reactions forge new hydrocarbons by first decomposing and then rearranging the fuel hydrocarbons. Generally speaking, pyrolytic reactions compete with oxidation reactions for the available fuel, although instances are known when oxygen, in small amounts, actually promotes pyrolysis (Arana *et al.*, 2004). Three major pyrolytic reaction groups are apparent: fragmentation (emergence of smaller molecules), polymerisation (emergence of larger molecules) and dehydrogenation (emergence of molecules of lower H/C ratio). Different pyrolytic reactions predominate in different thermal regimes. And, as will be described in Section 3.6.1, only when conditions are sufficiently hot do these reactions proceed all the way to soot; otherwise, soot precursors alone are formed, and emitted as organic particulate (Akihama *et al.*, 2001).

Precursor molecules are generally held to take two forms (Wang H. and Frenklach, 1997): unsaturated hydrocarbons, chief amongst which is acetylene (C_2H_2), a frequently found reaction intermediate in hydrocarbon combustion, and polycyclic aromatic hydrocarbons (PAH), the molecular structures of which are multitudinously built from the benzene ring. Both groups are, apparently, essential in soot formation, but the relative importance of each is debated; and, in the literature, rivalry exists between the 'aromatic model' and the 'acetylene model' (Zelepouga *et al.*, 2000; Krestinin, 2000). It is difficult to resolve the two, as the concentration of one may simply be acting as a surrogate measurement for the other (Sunderland and Faeth, 1996).

Something more clearly stated is that aliphatic hydrocarbons *must* form aromatic hydrocarbons, as proven in radiotracing experiments (Rhead *et al.*, 1990). The pyrosynthesis proceeds via low-order hydrocarbons and radicals, and leads ultimately to ring closure, or 'cyclisation', with formation of the first benzene ring. The two groups of fuel hydrocarbons, then, form soot differently: aromatic molecules by a direct path, and aliphatic molecules by an indirect path, involving acetylene and cyclisation (Martinot *et al.*, 2001). The supposition, though, that fuel aromatics sidestep cyclisation and fast-track their way directly to soot is unnecessarily sweeping, and decomposition, by ring rupture followed by recyclisation into pyrosynthesised (nonfuel) aromatics, is not ruled out, although this latter route is held to be slower (Haynes and Wagner, 1981, pp. 253–). Practical measurements do not provide simple answers: internal combustion engines, when burning purely aliphatic fuels, actually emit benzene in *greater* quantities than when burning aromatic fuels (Schulz *et al.*, 1999); while data from a drop-tube furnace demonstrate that fuel molecules do not necessarily pass through a phase of complete structural disintegration prior to their incorporation into soot (Yan *et al.*, 2005b).

Various pathways to the first aromatic rings have been mooted (Richter and Howard, 2000); these pathways are expressed, normally, with reference to benzene, the *fundamental* ring, for which there are two principal reaction groups (d'Anna *et al.*, 2000; Curran *et al.*, 2001; Wang H. and Frenklach, 1997;

Tao and Chomiak, 2002). One is the addition of acetylene to the radicals C_4H_3 and C_4H_5, followed by cyclisation into the phenyl radical (C_6H_5) or benzene (C_6H_6):[3]

$$C_4H_3 + C_2H_2 \rightarrow C_6H_5, \qquad (3.9/R\ 267)$$

$$C_4H_5 + C_2H_2 \rightarrow C_6H_6 + H. \qquad (3.10/R\ 290)$$

The other reaction group involves radicals such as propargyl (C_3H_3), allyl (C_3H_5) and cyclopentadienyl (C_5H_5): an example is the so-called propargyl recombination reaction,

$$C_3H_3 + C_3H_3 \rightarrow C_6H_6, \qquad (3.11/R\ 227)$$

$$C_3H_3 + C_3H_3 \rightarrow C_6H_5 + H. \qquad (3.12/R\ 227)$$

These one-ring aromatics then grow by acquiring *more* rings, the propensity for which varies widely amongst the progenitors. For example, the fecundity of toluene greatly exceeds that of benzene – this is consistent with the pyrolysis pathways open to the two compounds (Thijssen *et al.*, 1994). These earliest stages seem to be the bottleneck in soot formation, with the emergence of two- and three-ring aromatics perhaps being rate-limiting (d'Alessio *et al.*, 1994; Violi *et al.*, 1999).

A widely accepted mechanism for the formation of multi-ring aromatics is hydrogen abstraction and carbon addition (HACA); as the name suggests, this is a sequential two-step process. Abstraction first 'activates' the aromatic molecules, after which 'addition' propagates molecular growth and cyclisation. In the following sequence, $A_1^\bullet$ is the phenyl radical (C_6H_5):

$$A_i + H \rightarrow A_1^\bullet + H_2, \qquad (3.13)$$

$$A_1^\bullet + C_2H_2 \rightarrow A_1^\bullet C_2H_2, \qquad (3.14)$$

$$A_1^\bullet C_2H_2 + H \rightarrow A_1^\bullet C_2H + H_2, \qquad (3.15)$$

$$A_1^\bullet C_2H + C_2H_2 \rightarrow A_{i+1}. \qquad (3.16)$$

Hence, the higher-order ring A_{i+1} is constructed from its lower-order progenitor A_i, for which the subscript *i* contains the number of fused rings. This modelling sequence, owing to computational burdens, is eventually terminated, for example, at acenaphthylene (A_2R_5, $C_{12}H_{10}$), which consists of two six-membered rings (A_2) and one five-membered ring (R_5) (Rente *et al.*, 2001). The PAH molecules, then, according to this understanding, simply grow and grow. The finer details of HACA are still being probed (Xu F. and Faeth, 2000), and alternative growth routes may well exist (Hwang *et al.*, 1998). But when the same suspects are repeatedly found in the vicinity of different crime scenes,

---

[3] The following four reactions bear the same numbers as in the original paper by Wang and Frenklach (1997), in addition to the sequential numbers of this chapter.

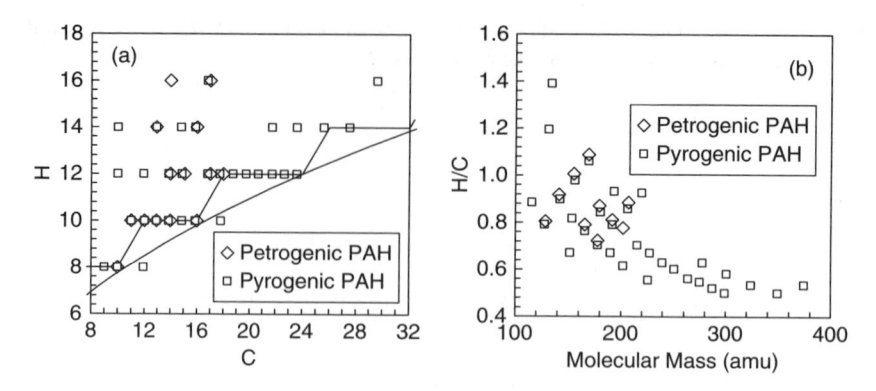

Figure 3.3 Various PAH found in diesel fuels (petrogenic) and the organic particulate discharged by diesel engines (pyrogenic). (a) Relation between carbon and hydrogen atoms in the molecule. Shown also is the 'coronene curve' for C_xH_y, where $y = (6x)^{\frac{1}{2}}$, i.e. the locus of the coronene family of compounds, and also the 'staircase curve', defined such that for even x, y is the minimum value still lying above $(6x)^{1/2}$. Isomers are not accommodated in this depiction. (b) Relation between carbon–hydrogen ratio and molecular mass. The fuel data is for three samples in the USA and UK. The particulate data is an aggregate of various engines and duty cycles. Analysis by GC-MS and LMMS (Dobbins et al., 2006).

this surely represents an invaluable clue; and, likewise, the same species of PAH repeatedly arise. This observation is compounded with a second and equally useful clue: the relatively low presence, among the usual suspects, of *odd-numbered* PAH (as identified by the number of carbon atoms in the molecule).

Evidently, the growth process favours certain routes paved with thermodynamic stepping stones, i.e. the most favoured isomers, dubbed 'stabilomers' (Dobbins et al., 1998; Dobbins, 2002) in recognition of their superior thermodynamic stabilities: those formed irrespective of the initial reactions, the fuel formulation, the fuel phase (gas, liquid or solid) or the type of flame (laminar, turbulent, premixed or diffusion). For example, three prominent stabilomers are $C_{20}H_{12}$, $C_{22}H_{12}$ and $C_{24}H_{12}$ (Dobbins et al., 1995). As shown in Figure 3.3, when the number of hydrogen atoms in each PAH molecule is plotted against the number of carbon atoms, these stabilomers appear in the form of a staircase (Dobbins et al., 2006). This distinguishes *pyrogenic* PAH – new compounds formed in the combustion – from *petrogenic* PAH – fuel compounds – which show no such staircase. A corresponding plot of hydrogen–carbon ratio against molecular mass tracks the progress of pyrogenic PAH on the pathway to soot: petrogenic PAH are again distinguishable, being restricted to the lower range.

On the basis that many isomers are potentially possible, but only a select few are theoretically stable and so become preponderant, a comprehensive network of chemical reactions can be deduced, beginning with one-ring aromatics, and extending as far as coronene ($C_{24}H_{12}$) (Richter et al., 1999); or a 'molecular growth map' can be constructed, tracking various pathways as far as 1792 amu (Lafleur et al., 1996). Starting with benzene (C_6H_6, 78 amu), larger PAH might form, say, by the addition of acetylene (C_2H_2, 24 amu) or of diacetylene (C_4H_2, 50 amu). The number of molecular structures, according to this scheme, able to accept addition of C_1 is limited, and C_3 is not present in sufficiently large concentrations; hence molecules containing even numbers of carbon atoms are favoured.

While the above-described mechanisms have a considerable following, some workers believe them too slow to account for the observed rates of soot inception, and propose, instead (or as an adjunct), an *ionic* mechanism (Calcote, 1983), wherein straight-chain hydrocarbon ions cyclise more swiftly than uncharged hydrocarbon radicals. The ions grow according to the sequence $C_3H_3^+$, $C_5H_5^+$, $C_5H_3^+$, $C_7H_5^+$,

$C_9H_7^+$, $C_{11}H_9^+$, $C_{11}H_7^+$, ... (Egsgaard, 1996). According to this theory, small, uncharged species, such as C_2H_2 and C_4H_2, attach themselves to hydrocarbon ions such as $C_xH_y^+$, with intervening stages at which H_2 is lost, leading to large gaseous polycyclic or polymeric ions such as $C_{13}H_9^+$ (three rings), $C_{17}H_{11}^+$ (four rings) and $C_{19}H_{11}^+$ (five rings) (Griffiths and Barnard, 1995, p. 120). Upon reaching ~2000 amu, these huge ions, *inter alia*, dissociate, liberating large, *uncharged* hydrocarbon molecules, which grow into spherules, and small ions such as $C_3H_3^+$, which repeat the growth curlicue (Hall-Roberts *et al.*, 2000).

Several germane observations run as follows. (a) Ions are all-pervading in hydrocarbon flames; indeed, they attain concentrations similar to spherules. (b) With the application of an external electrical field, one can manipulate the charged species in a flame (Fialkov, 1997, pp. 515–518) to reduce soot production by as much as 90% (Saito M. *et al.*, 1997); this is thought to be an aspect of the *ion flux* through the pyrolysis zone (Saito M. *et al.*, 1999), although there are numerous intriguing facets to such work (Weinberg, 1983). (c) Soot particles discharged by flames (Maricq, 2004), gas turbines (Sorokin and Arnold, 2004) and diesel engines (Section 6.2.5) are charged. (d) Reasonable grounds exist for supposing that in-flame metals react with, and deplete, hydrocarbon ions, to the detriment of nucleation, and that the rate of this ion depletion, compared with the rate of ion-mediated nucleation, is what counts (Calcote, 2001). This idea also dovetails pleasingly with the soot-suppressing action of metals (Section 3.5.1). (e) Ions detected in flames include PAH^+ containing as many as 300 carbon atoms (Weilmünster *et al.*, 1999).

Thus, for various, albeit not incontrovertible, reasons, ions cannot be summarily dismissed as instigators of nucleation. But the ion-mediated theory of nucleation remains controversial and not widely accepted: experiments where flames are purposefully seeded with easily ionised metals do not always meet with the intended soot suppression (Hall-Roberts *et al.*, 2000); the growth rates of charged and uncharged PAH might actually be on a par; and, if soot particles of ~2000 amu appear *before* the largest charged PAH of ~4500 amu (Weilmünster *et al.*, 1999), then perhaps the question is not even one of chicken and egg. Finally, ions are greatly outnumbered by radicals. Hence, the ionic theory of soot inception must be treated with caution (Fialkov, 1997, p. 500) until the ostensible interdependency with charge has been properly understood (Maricq, 2006).

These chemical transformations, from PAH to soot, are, presumably, accompanied by an even more obvious physical or geometrical transformation, as, whatever the growth mechanism, a molecule must, eventually, become so massive that its two-dimensionality, or planar structure, is no longer sustainable, and hence it acquires curvature, or *three-dimensionality* (Weilmünster *et al.*, 1999). This is where the role, if any, of fullerenes (C_{60} and C_{70}) or their structurally curved relations, in soot inception, is debated (Pope C.J. and Howard, 1994; Lahaye and Ehrburger-Dolle, 1994; Fialkov, 1997, p. 492); and fullerenic or 'fullerenic-like' structures of 2–3 nm in size are, indeed, found in electron micrographs (Hepp and Siegmann, 1998; Su D.S. *et al.*, 2004; Müller *et al.*, 2005). Fullerenes are spheroidal pure-carbon molecules, the truncated icosahedral structures of which resemble the hexagonal and pentagonal segments sewn together to make soccer balls; they are formed from curved aromatic shells, such as fluoranthene ($C_{16}H_{10}$) or corannulene ($C_{20}H_{10}$), the continued growth of which, leads, eventually, to the phenomenon of 'cage-closing'. Nevertheless, the trace levels at which atomic masses corresponding to the fullerenes C_{50}, C_{60}, C_{70} and C_{92} are found argue against any key role in soot formation (Dobbins *et al.*, 1995).

These detailed hypotheses for the emergence of massive PAH contrast markedly with the pronounced head-scratching that continues to surround the eventual crossing of the molecule – particle Rubicon, from 'macromolecule' to 'microparticle' (Fialkov, 1997, p. 503). There may be no real discontinuity as such: the PAH molecules simply grow and, through dehydrogenation and carbonisation, gradually develop the properties characteristic of soot. The view ordinarily entertained is that hexagonal geometries are somehow preserved in the basal planes of the crystallites from which the spherules are constructed (Chen H.X. and Dobbins, 2000).

Possibly the largest molecules, by virtue of their size, become unstable in the gas phase, and, as their partial pressures approach supersaturation, they nucleate into liquid microdroplets, and these microdroplets then solidify (i.e. carbonise) into soot nuclei. As noted (Prado et al., 1983), this is a circular paradox: the macromolecules would have to be so large (2000–3000 amu) that they would, in effect, already constitute particles (Haynes and Wagner, 1981, p. 263). And the emergence of soot requires relatively high activation energies inconsistent with mere physical condensation (Kennedy, 1997, p. 102). So, alternatively, nucleation may take on more of a chemical nature, with bonds forming through the simultaneous action of polymerisation, dehydrogenation and fragmentation. It was also suggested long ago that rogue airborne particles ingested into the engine might provide suitable nuclei (Edwards, 1977, p. 66); and it seems reasonable to suggest that metals derived from oil and fuel might play similar roles. Indeed, nowadays this is virtually a truism in cases where the fuel has been deliberately seeded (Section 3.5.2).

Nucleation modelling, with PAH as a soot precursor, is difficult, and acetylene is often preferred (Hong S. et al., 2005), which is, in any case, widely held to be a contributor to the emergence of PAH. Two 'graphitisation' reactions are (Tao et al., 2004)

$$A_2 \longrightarrow 4H_2 + 10C(s), \quad (3.17)$$

$$C_4H_2 \longrightarrow H_2 + 4C(s), \quad (3.18)$$

where A_2 is naphthalene, and C(s) represents 'incipient soot' – as might become fused into the geometry of a dodecahedron, representative of a soot nucleus (Rente et al., 2001). The rate of soot nucleation r is then given by

$$r = 10k_{A2}[A_2] + 4k_{C4H2}[C_4H_2], \quad (3.19)$$

where the rate coefficients k_{A2} and k_{C2H2} pertain to Equations (3.17) and (3.18).

This and analogous reactions do not fully represent nucleation, because, in reality, the carbon remains attached to some hydrogen. Also, the dimerisation of large PAH, and the coagulation of macromolecular soot precursors, is sidestepped; but the computational burden is greatly reduced. Alternatives are the coalescence of two PAH, both larger than four rings, into a solid-phase dimer, representing a soot particle of the smallest possible size class (Akihama et al., 2001), and the coalescence of pyrene into dimers (Yoshihara et al., 1994). Radically different is the polyyne model (Wen et al., 2006), in which nucleation, via polymerisation of a supersaturated polyyne vapour, leads to a polymeric globule (Krestinin, 2000).

What is the boundary between a large molecule, or 'macromolecule', and a soot nucleus? Partly, this understanding is limited by the method of measurement (Haynes and Wagner, 1981, p. 257-): for example, the smallest size visible under an electron microscope is $\sim$1.5 nm ($\sim$1600 amu) (Fialkov, 1997, p. 503); there is a gap, or blind spot, between this and chromatographic detection of PAH at 300–400 amu (Alfè et al., 2007). But the earliest stage at which carbon is thought to leave the gas-phase molecular system and to form the first identifiable particles is around 650–700 amu, for which speculative empirical formulae are $C_{54}H_{18}$ and $C_{54}H_{30}$ (Griffiths and Barnard, 1995, p. 120b). The smallest identifiable particle, or 'critical nucleus', of minimum size, is probably around one nanometre: it is modelled as a spherical ball containing 100 (Tao et al., 2004), 50 (Martinot et al., 2001) or 32 (Hong S. et al., 2005) carbon atoms – nuclei analogous to those in the model of Tesner described in the previous section. The total mass of soot, though, is still quite small, because even though these nuclei exist in colossal numbers, they are so tiny.

3.2.3 Surface Growth

Nuclei, immediately upon formation, begin accreting hydrocarbons or hydrocarbon fragments; they thus grow. This increases the mass of soot, in fact considerably so, but leaves unchanged the number of particles. Surface growth generates the *spherule*, which is the primary unit, or monomer, of the carbonaceous agglomerate; and it is responsible for the *layered* microstructure, in the outlying regions, wherein crystallites are concentrically arranged more or less parallel to the surface. This layering, or lamination, differs distinctly from the microstructure of the central region, which is disordered or amorphous, and thought to betray the progenitive nucleus.

In the strictest sense, the transition between inception and surface growth is probably not sharp, and, moreover, it is probably dependent on the method used to track the emergence of nuclei; but, in flames, the zones of nucleation and growth do appear to be distinct (Prado *et al.*, 1983). Surface growth is essentially a chemical process: acetylene is thought to serve as the predominant feedstock, but PAH are probably also agents of growth. Although freshly nucleated soot, according to the standard model, is derived from aromatics, recent evidence imputes surface growth to aliphatics (Öktem *et al.*, 2005), the chains of which shorten with advancing height above a flame, in tandem with increasing aromaticity (Santamaría *et al.*, 2007). And since soot, unlike its feedstock, is highly carbonaceous, then dehydrogenation is presumably an inherent concomitant. It has also been suggested that the spherule interior, formed from PAH, displays a higher hydrogen-to-carbon ratio than the outlying regions, which have formed through the addition of C_2 fragments (Vander Wal *et al.*, 1995).

Surface growth is not restricted to spherules in isolation: as will be described in the next section, spherules agglomerate, or adhere avidly to one another, and surface growth continues concurrently. This growth mode causes spherular fusing, in which the laminar microstructure in the outlying regions runs, in an uninterrupted fashion, from one spherule into another. When this happens, the evidence of independent formation, from an external viewpoint, is obscured by an overlying blanket, or continuous carbon bridge, of fresh material, the effect of which is to strengthen the interspherule bonding (Lahaye and Ehrburger-Dolle, 1994).

Spherular diameters are actually quite restricted (encompassing just a few tens of nanometres); why this should be so, across a wide range of operating conditions, and even across disparate forms of combustion, is not known for sure; but some common or universal termination of surface growth, transcending all particulars, is clearly implied. Perhaps, at some point, the supply of soot precursors is simply exhausted. Nor should it be forgotten that oxidation (Section 3.2.5), and consequent recession, of the spherule surface reverses the action of growth; indeed, models of surface growth require some adjustment along these lines (Sunderland *et al.*, 1995), otherwise spherule sizes are overestimated.

There is disagreement as to whether nucleation or surface growth ultimately determines the amount of soot (Kennedy, 1997, p.96); this seems to depend on how heavily or lightly sooting is the combustion. The general understanding is that surface growth, *not* inception, generates most of the soot mass, but it seems a plausible conjecture that, at some points, surface growth competes with nucleation for the available soot precursors: in laboratory flames, spherules are larger when generated in smaller numbers. An identical phenomenon is known to occur among nucleating organic compounds in the exhaust: the more abundant the nuclei, the slower the growth rates of these nuclei.

The difficulties to be faced in analytically accounting for surface growth are not, therefore, trivial, and recourse is consequently made to empiricisms. The strong association between acetylene and surface growth suggests such an empiricism; for simple hydrocarbon flames, we have (Sunderland and Faeth, 1996)

$$w_g = k_g(T) [C_2H_2]^n, \qquad (3.20)$$

where w_g is the rate of surface growth ($kg\, m^{-2} s^{-1}$), and $k_g(T)$ is an Arrhenius expression. The data indicated an approximately first-order reaction, although other hydrocarbons, as minor growth species, were also implicated. In a similar approach, w_g is approximately proportional to $[C_2H_2][H]$ (Kim C.H. et al., 2004).

Assuming each critical nucleus to reach maturity as a spherule, a theoretical expression has been derived for the mean spherular radius, r_{sph} (Smith G.W., 1982):

$$r_{sph} = \left[\frac{3M}{4\pi\, A\rho} \exp\left(\frac{4\pi r_{sph}^2 \sigma}{3kT} \right) \right]^{1/3}, \qquad (3.21)$$

where M is the molecular weight of the precursor species, A is Avogadro's number, ρ is the density of the precursor in its condensed phase, r_c is the radius of a critical nucleus, σ is the surface energy of the condensed precursor, k is Boltzmann's constant and T is the absolute temperature. It will be noted that r_{sph} is an explicit function neither of pressure, nor of the concentration of any precursor species.

Ultimately, surface growth is rooted in atomistic phenomena, or the specificity of certain sites for certain reactions. The notion that surface sites must first be 'activated', as a necessary precursive step, for surface growth might explain instances where soots exhibit different growth rates notwithstanding identicality in their surface area; this suggests variation among surface sites in their 'receptivity'. The cessation of surface growth in laboratory flames, for example, does not necessarily coincide with the disappearance of acetylene (Hanisch et al., 1994); and a transition is observed from early growth by surface reactions with aromatics to later growth by surface reactions with aliphatics (Öktem et al., 2005).

Surface growth, then, is more appropriately defined in terms of *surface sites*; and models, or posited reactions, are derived by analogy to the HACA mechanism for planar growth of PAH molecules (Hong S. et al., 2005; see also Section 3.2.2). In an 'active site' model, there are two surface states relating to carbon–hydrogen (C–H) sites and dehydrogenated, radical (C*) sites (Appel et al., 2000). The hydrogen atoms are thought to be involved in surface growth insofar as they 'activate' the surface sites, thus mediating transition between the two states. Four suggested reactions are

$$(C_{soot} - H) + H \Leftrightarrow C^*_{soot} + H_2, \qquad (3.22)$$

$$(C_{soot} - H) + OH \Leftrightarrow C^*_{soot} + H_2O, \qquad (3.23)$$

$$C^*_{soot} + H \rightarrow (C^*_{soot} - H), \qquad (3.24)$$

$$C^*_{soot} + C_2H_2 \rightarrow (C^*_{soot+2} - H) + H. \qquad (3.25)$$

Hydrogen abstraction from inchoate spherules (Reaction (3.22)) is treated as analogous to the same abstraction from benzene molecules (Section 3.2.2). Impeded surface activation, through an increasing scarcity of H atoms, offers another reason for gradually slowing surface growth rates (Xu F. et al., 1998). This scheme is supported in that hydrogen addition to a flame suppresses surface growth (Guo et al., 2006). The model includes also abstraction by OH (Reaction (3.23)). That OH serves as a mediator, or growth agent, rather than simply an oxidant (Section 3.2.5), is a significant observation, conferring on models greater predictive capability (Tao et al., 2004). Reaction (3.24), or the combination of H atoms and surface radicals, represents the formation of the C–H site; and the growth itself, that is, the addition of carbon, is contained in Reaction (3.25). It should be borne in mind that charged particles may grow faster and even bypass H abstraction (Balthasar et al., 2002b).

Formation I: Composition

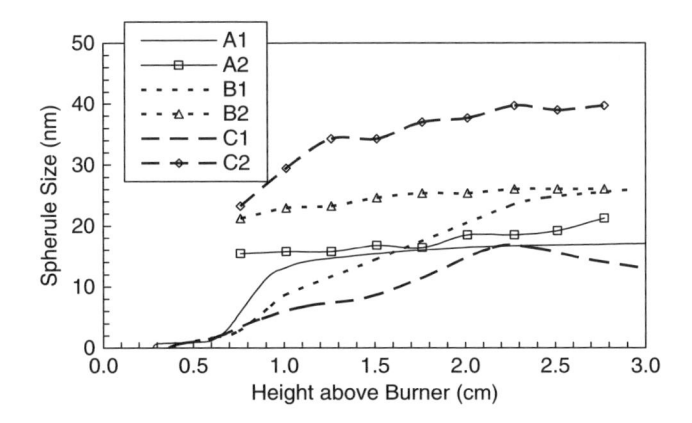

Figure 3.4 Spherule size as a function of height above a premixed flame: modelled data, 1; experimental data, 2; Carbon–oxygen (C/O) ratio: A, 0.78; B, 0.88; C, 0.98. Operating pressure, one bar; fuel, C_2H_4 (Appel et al., 2000; experimental data from Xu F. et al. (1997).

Figure 3.4, which tracks spherule size with distance along a laboratory flame, according to an active-site model, reproduces the correct order of magnitude, but close agreement is realised only in one instance (Appel et al., 2000). The model appears slightly less satisfactory as soot production increases. Reassuringly, such reaction schemes support the aforementioned empiricism relating w_g to $[C_2H_2]$ and [H]. This active-site approach, formulated originally for hydrocarbon flames, is further elucidated elsewhere (Xu F. et al., 1997; Kim C.H. et al., 2004) and has seen application to diesel engine combustion (Martinot et al., 2001).

What seems lacking is a temporal dimension to surface growth. Partly, this is a matter of residence time: by applying an electrical field to a laboratory flame, charged particles can be encouraged to leave prematurely or to dally within the pyrolysis zone, from which they emerge, respectively, as dwarves or giants (Haynes and Wagner, 1981, p. 249). But surface growth is also tempered by the fraction of surface area amenable to these reactions, and this fraction is not constant. The 'active' surface relates to graphitic, or aromatic, edges, as basal aromatic planes are relatively unreactive. A very, very young spherule is, in this understanding, a cluster of just a few PAH molecules, and its surface, consisting mostly of aromatic edges, is fully active in surface growth; but the fraction of reactive surface declines as the spherule grows, with inevitable effects on surface growth rates.

3.2.4 Agglomeration

Spherules do not, as a rule, survive as isolated individuals, because, upon collision, they stick together, to form closely bonded congregations; these congregations of spherules themselves collide, and stick together, to form still larger congregations, and so on. This is the growth process known as 'agglomeration', the action of which finally forges the morphologies characteristic of mature particles. But agglomerative growth is quite unlike surface growth: it is physical, rather than chemical, and it decreases the particle number without affecting the particle mass.

The rate of agglomeration is tightly tethered to the number concentration; hence particle demographics now assume centre stage. Perhaps understandably, owing to the complexity of the models, and also to the statutory emissions requirements as currently framed, modellers have focused, traditionally and predominantly, on soot mass rather than particle number, and, as a result, have left agglomeration comparatively untouched. (This situation is, admittedly, changing rapidly at the time of writing.) But the physics of agglomeration is well known, and extensively modelled, in aerosol science generally: for example, that large particles agglomerate more slowly than small ones and that polydisperse particles agglomerate more quickly than monodisperse ones are well-established aspects (Hinds, 1982, pp. 233–).

The bedrock for such investigations is the Smoluchowski equation, as already introduced in Section 2.2.4; this applies quite generally to aerosols, and indeed to colloids. The rate of decline in particle population is given by

$$\frac{dN}{dt} = -kN^2, \qquad (3.26)$$

where N is the number concentration, and k is the agglomeration (or coagulation) coefficient. This equation is solved to yield

$$\frac{N}{N_0} = \frac{1}{1 + (kN_0)t}, \qquad (3.27)$$

where N_0 is the initial number of particles. The strong inverse dependence of N/N_0 on time should be noted.

It should be emphasised that, in its simplest form, the Smoluchowski equation is fairly easy to solve; its precise application to the combustion chamber of an engine is quite another matter, and considerable complexity is introduced if one does not first make several simplifying but nonetheless sweeping assumptions, for example that particles remain as uniformly distributed, monodisperse spheres. In the past, solutions for polydisperse inhomogeneous particle concentrations were semi-empirical (Dolan and Kittelson, 1978). Variations in sticking probability and the propensity for nonsticking collisions during the course of combustion or as particles age raise other uncertainties (Hanisch et al., 1994). So, with advancing years, the Smoluchowski equation has been continually revised and refined; and, not unexpectedly, the analytical expressions have grown correspondingly (e.g., Tambour and Khosid, 1995; Harris and Maricq, 2001; di Stasio et al., 2002; and many references therein). The motivation for this research is a widespread need among physicists to model agglomeration in many types of aerosols and colloids, but, as the theoretical framework cannot be fully explored here, only a few remarks will be made, concerning soot particles specifically.

At the outset, one must consider the nature of particle transport (Section 2.2.3): in fact, the line of demarcation between the free molecular regime and the continuum regime shifts. At the highest pressures (>65 bar), most particles (>5 nm) are in the continuum regime, but, as the combustion unfolds, this regime continually loses ground, until, upon exhaust-valve opening, virtually all particles are in the free molecular regime (Smith G.W., 1982). This picture is supported by others (Harris and Maricq, 2001). It thus seems reasonable to conclude that many particles spend most of their in-engine lives somewhere in the *transitional* regime – a limbo, as it were, where precise analytical treatments are wanting, although, admittedly, some methods are available (Kazakov and Foster, 1998).

Age is also a factor: mature, solidified spherules exhibit a 'noncoalescent' form of agglomeration, involving point contacts; young spherules, being still of a viscid, tarry nature, are able to fuse. Hence, early-forming clusters are more closely bonded than late-forming ones (Song H. et al., 2004). In the most extreme cases, new, larger spherules are formed – an agglomeration logically dubbed 'coalescent', i.e. in which original identities have been lost – but the hidden hand of coalescence is betrayed by the presence, under the electron microscope, within one spherule, of multiple nuclei. A clue, surely, is the tendency for alkane fuels to form multiple-nucleus spherules, and aromatic fuels to form single-nucleus spherules (Song J. and Lee, 2007). Upon supposing that spherules may contain as many as thirty nuclei, $(N_c)^{1/3}$ has been added as a correction factor to Equation (3.21), where N_c is the average number of nuclei per spherule (Smith G.W., 1982). This is because Equation (3.21), as originally devised, assumed that every nucleus grows into its own uniquely identifiable spherule; and, without this correction, an underestimate in r_{sph} would result. From this consideration, agglomeration, if coalescent, is unlikely to increase spherule sizes by more than a factor of about three (Smith G.W., 1982).

The vexing question is *when*, and *why*, the transition takes place from coalescent to noncoalescent agglomeration. This question is to some extent loaded, since the particles are not necessarily fluid: agglomerated solid particles will acquire a spherical appearance if surface growth continues sufficiently. Indeed, when particles are still young and small, less crevice-filling material is required to re-establish sphericity (Balthasar and Frenklach, 2005). According to one estimate (Smith G.W., 1982), the spherules are, at this transition, just ~50 μs old – an age not significantly out of kilter with the timescales inherent in diesel engine combustion. But there is a wide range, according to recent modelling of laboratory flames, of a few hundred microseconds to tens of milliseconds (Balthasar and Frenklach, 2005). The window of opportunity for coalescent agglomeration is, apparently, narrow; but electron micrographs nonetheless regularly show spherular *fusing*, the intimacy of which depends on engine operating point (Lee K.-O. *et al.*, 2001). Rapidly increasing spherular viscosity is an obvious explanation for this transition: but dehydrogenation appears to proceed too slowly, just as the outward diffusion of hydrogen from within the spherule to the surface appears too rapid (Smith G.W., 1982). More likely, soot precursors, the adherence of which to spherule surfaces provides a gummy medium, aiding fusing, are depleted – the same depletion, incidentally, posited in determining r_{sp} (Equation (3.20)). To model all spherules above a certain size (say 25 nm) as effectively 'solid' (Smooke *et al.*, 2005) is a simplification, since the transition is continuous rather than discontinuous (Balthasar and Frenklach, 2005).

After solidification, the spherules pass to a second stage of agglomeration, where the collision *mode* has a hand in deciding the morphology (Harris and Maricq, 2001); fractal concepts have borne considerable fruit here (Lahaye and Ehrburger-Dolle, 1994), and fractality in fact betrays how an agglomerate has grown. In Section 2.2.1, the fractal dimension D_f was introduced as a factor expressing the extent of chaining or clustering, i.e. the closer this dimension is to 3, the closer or more tightly packed the cluster of spherules, and hence the more spherical the whole agglomerate. However, not only does this factor express the result of the collision, it also, for the colliding particles, expresses the type of collisions these particles are likely to experience (Harris and Maricq, 2001). To provide a full discussion would be impossible here, but collisions depend, first, on the interplay of particle cross-section and particle velocity, and second, on whether the collisions involve spherules and agglomerates, or just agglomerates. These aspects are affected in disparate ways when $D_f < 3$. A critical parameter here, again, is the mean free path: when this is *larger* than the agglomerate, growth is 'ballistic': spherules penetrate to the interior regions, leading to *clustered* morphologies. When it is *smaller* than the agglomerate, growth is 'diffusion-limited': spherules attach to outer regions, creating *chained* morphologies. These matters depend on the operating point of the engine, and the moment in the combustion (Lee K.-O. *et al.*, 2001).

These convoluted complexities aside, the relevant question is whether the mean free path, as affected, for example, by the engine operating point, accords with the particle morphologies observed under the electron microscope. Measurements do suggest a tendency for clustered agglomerates at low engine loads, and chained agglomerates at high engine loads (Lee K.-O. *et al.*, 2001; Virtanen *et al.*, 2004). There is the question of whether fractality is *itself* a function of particle size: as agglomeration proceeds, the available isolated spherules decline, so that shifting mechanisms are possible. Researchers are still probing this dense thicket of intermeshing parameters.

Morphology, once agglomerates have formed, is not, apparently, cast in stone: measurements of fractality, with distance along a laboratory flame, are intriguing in this respect – see Figure 3.5

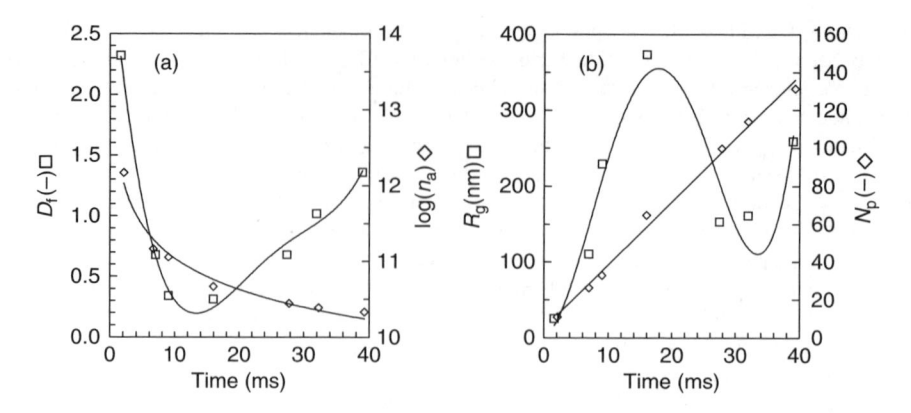

Figure 3.5 Particle morphology with increasing height above an ethylene diffusion flame, expressed as residence time: (a) fractal dimension D_f and number of agglomerates per cm^3, n_a; (b) number of spherules per agglomerate, N_P, and radius of gyration, R_g. Data obtained by light scattering and electron micrographs; standard deviations in the original have been omitted (di Stasio *et al.*, 2002).

(di Stasio *et al.*, 2002). (The significance of R_g is explained in Section 6.2.2.) The parameter D_f shows an increasing and then decreasing tendency for chained agglomeration, something implying an intermediate restructuring, or perhaps collapse, in the chained structure as the particles travelled along the flame. The significance of this observation for internal combustion engines remains to be reported.

This discussion has highlighted the precariousness of various commonly used assumptions in agglomeration studies, e.g. that particles are homogeneously distributed, monodisperse, solid, perfect spheres. Ideally, agglomeration models should be carefully coupled to the concurrent processes of surface growth and oxidation, with which there is competition. Failure to incorporate oxidation, for example, leads to the unphysical inclusion of already-annihilated particles (Xiao and Borgnakke, 1991), and hence the need to incorporate a limiting condition in surface recession (Kazakov and Foster, 1998); and agglomeration, or particle growth, proceeds in opposition to oxidation, or particle shrinkage (Tambour and Khosid, 1995). This, and accompanying fragmentation or surface growth, may explain instances where particles do not quite evolve to the asymptotic or self-preserving size distributions predicted for agglomeration-dominated aerosols, but rather to ones that are lognormal (Harris and Maricq, 2001). Interestingly, it seems that no asymptotic solution exists for aerosols which straddle the transition regime.

The flow field (i.e. turbulence), rather than simple Brownian diffusion, presumably plays some mediating role (Balthasar *et al.*, 2002a) in bringing particles together. But one aspect has been wholeheartedly ignored by combustion modellers: the role of particle charge. This is suspected to influence agglomeration in flames (di Stasio *et al.*, 2002), and particles emitted by internal combustion engines are certainly charged (Section 6.2.5). This topic particularly deserves consideration for two reasons: charged particles tend to agglomerate into chained morphologies rather than clustered ones (Zebel, 1966), and the increased rate of agglomeration between opposite charges is compensated by the decreased rate between like charges (Balthasar *et al.*, 2002b). Open morphologies and depressed rates of agglomeration will both serve to maximise surface area, and thus render particles more vulnerable to oxidants. It was once suggested (Howard and Kausch, 1980, p. 273), for example, that external electrical fields might be used to control particle charging via an electron current, and thus to control the oxidation of soot, indirectly, via the agglomerative process.

3.2.5 Oxidation

The preceding sections have described how soot comes into existence – what, collectively, might be styled soot *production*. We now progress to the final formative stage, soot *oxidation* – a converse process, since it reverses the action of soot production. The chain of events leading to soot can be terminated at any stage, and precursor molecules, nuclei, spherules and agglomerates are all culled to some extent. (With solid carbon, there are parallels in the combustion of pulverised coal particles; see Zhang M. *et al.*, 2005.) This statement belies the linear narrative of soot formation, as disingenuously portrayed in Figure 3.1; for, in reality, inception, surface growth, agglomeration and oxidation proceed concurrently (although one or another predominates at certain points).

This concurrence makes soot formation and soot oxidation difficult to distinguish experimentally. A reduction in engine-out soot might arise through inhibited production, promoted oxidation or both; sometimes both are inhibited, but unequally; and sometimes both are promoted, but unequally. In engine modelling, the instantaneous in-cylinder concentration of soot, as a function of crank angle, is computed by subtracting the oxidation model from the production model. From these studies, it is apparent that production dominates in the early stages of combustion, and oxidation in the later, with the tide turning midway.

Oxidation and formation are of *comparable importance* in determining what is emitted by the engine; and in fact, of all the soot produced, more than 90% is destroyed *by the combustion itself*. This assertion is too well attested to admit of any doubt (Section 3.3.1). Consequently, just how the remainder successfully outfoxes the oxidation and cunningly survives is paramount, as this might reveal the corrective measures necessary to eliminate soot altogether. But the prediction of engine-out soot, as a small difference between two similar numbers, magnifies the modelling difficulties. The late combustion stage, or burn-out, is probably responsible for the greater proportion of this surviving soot. At a first level, incomplete oxidation stems from temperatures that are too low, residence times that are too short and oxidants that are too scarce.

The responsible oxidant is a moot point, as there are various contenders other than the obvious O_2, e.g. O, OH, H, CO_2 and H_2O; one or another probably predominates in certain regimes. Something of a doctrinal cleft arguably exists between engine modellers, who obsessively embrace O_2, and flame modellers, who highlight the importance of OH (Kennedy *et al.*, 1996; Ezekoye and Zhang, 1997; Haudiquert *et al.*, 1997). In flames, O_2, acting alone, does not always account for the observed rates of soot oxidation (Zhu X.L. and Gore, 2005), and OH, being a more reactive oxidant, is the probable explanation (Moss *et al.*, 1995; Kim C.H. *et al.*, 2004), although uncertainty over its abundance hampers clarification of its strength of influence.

In engine models, peak concentrations of OH tend to coincide with the onset of soot depletion (Hasse *et al.*, 2000). OH, moreover, provides the only sufficiently rapid oxidation pathway (Cavaliere *et al.*, 1994); that by O_2 is slower by two orders of magnitude (Yoshihara *et al.*, 1994), or one order of magnitude (Kittelson *et al.*, 1992). As in flames, the soot emission predicted by models increases markedly if the action of OH is neglected (Yoshihara *et al.*, 1994). A simple linear summation of oxidation by OH and oxidation by O_2 appears sufficient (Xiao and Borgnakke, 1991). This attack by the two oxidants is modelled by (Tao *et al.*, 2004)

$$(C_{soot} - H) + O_2 \rightarrow (C_{soot-2} - H) + 2CO, \tag{3.28}$$

$$(C_{soot} - H) + OH \rightarrow (C_{soot-1} - H) + H + CO. \tag{3.29}$$

The predominant oxidant depends, logically enough, on spatio-temporal aspects of abundance; soot initially experiences oxidation by OH, an oxidant which is swiftly lost and tied up in H_2O and CO_2

(Smith O.I., 1981, p. 275); later oxidation is by O_2. Exposure to O_2 on the fuel side of a diffusion flame is obviously restricted, and exposure to OH is probably confined to the narrow region of the flame itself (Dec and Coy, 1996). Outside the flame, O_2, now in greater abundance, might eclipse OH (Said et al., 1997). On other occasions, OH must first consume all available CO before turning its attention to soot (Cavaliere et al., 1994).

OH and O_2, then, are proclaimed as the chief oxidants. Oxidation in internal combustion engines by CO_2 is less usually modelled (Kittelson et al., 1992), although this can predominate in laboratory flames (Xu F. et al., 1997); it involves a somewhat higher activation energy, and hence hotter conditions, than oxidation by O_2. The soot particles, on diffusional grounds, might become surrounded by CO_2 rather than O_2; the product of the reaction is CO. From the gamut of oxidants, viz. O_2, CO_2, H_2O, OH, O and H, and using Arrhenius relationships, the only one set aside *a priori* as being unimportant at representative temperatures is the last-mentioned one (Li X. and Wallace, 1995).

But whatever the oxidant, it has often been supposed that for soot particles smaller than one micron, i.e. virtually all agglomerates, and certainly all unagglomerated spherules, the inward diffusion of oxidants and the outward diffusion of oxidation products are not rate limiting, so that the oxidation is, to all intents and purposes, *kinetically* controlled (Heywood, p. 642, 1988). Yet mass transport might slow oxidation in some instances (Jung H. et al., 2004); more will be said on this score shortly. For the time being, a brief consideration of a parallel oxidation will be instructive, namely regeneration of the diesel particulate filter (DPF). Of course, the parallel should not be taken too far, as the DPF contains a contiguous deposit of soot, rather than isolated particles suspended in a gas; this is why the burn-off in a DPF can become self-sustaining.

This deposit has been modelled as an assemblage of equally sized spheres, and the regeneration as being kinetically controlled (Romero-López et al., 1996). On the other hand, when the deposit is modelled as individual spheres, burning in isolation, there is an enveloping and rate-controlling diffusion flame – apparently, kinetic considerations alone yield unrealistically high oxidation rates – notably in the later stages of regeneration, once temperatures have declined (Jørgensen and Sorenson, 1997). In this model, the carbon atoms residing at particle surfaces react with CO_2 to produce CO. The reaction of O_2 and CO, in the flame sheet, is virtually instantaneous, and the *arrival* of these species becomes rate-controlling. The modelled reaction rate does not rise rapidly with temperature, as for a purely kinetic model, and, owing to the more significant role assigned to diffusion, increases much more slowly.

But now we have digressed somewhat, so, returning, suitably edified, to the task in hand, the onset of oxidation is marked by a decline in the rate of surface growth, not necessarily because surface growth mechanisms have slowed *per se*, but because the opposing action of oxidation begins to tell. Eventually an overt transition occurs, from spherule growth to spherule shrinkage, i.e. the surface recedes (Yang B. and Koylu, 2005): for soots extracted from the combustion chamber, the surface recedes by several nanometres (Song H. et al., 2004). For spherical particles, the rate at which mass is lost, w ($g\,cm^{-2}\,s^{-1}$), is convertible to a surface recession rate via

$$\frac{dr}{dr} = -\frac{w}{\rho}, \qquad (3.30)$$

where r is the radius and ρ is the density. From this relationship, it is possible to calculate the largest particle that can be annihilated within a certain period of time or, alternatively, the period of time required to annihilate a particle of a certain size (Heywood, 1988, p. 645). For a temperature of 1500 K, this gives 100 ms at 100 nm, and 30 ms at 30 nm (Murphy et al., 1981). These figures are not dissimilar to those assembled elsewhere, following a literature survey: several milliseconds for particles of a few tens of nanometres (Smith G.W., 1982).

But an assumption of sphericity is obviously simplistic, and, strictly speaking, this limits the analysis to isolated spherules. In reality, the exposed surface of an agglomerate is unstraightforwardly related to the number of spherules within it. Morphology is an aspect of the surface – volume ratios; and, as we have learned, spherules are not only in point contact, but also partially fused. Assumptions of sphericity, therefore, lead to significant underpredictions of surface area (Ezekoye and Zhang, 1997). The implications here are bracketed by two extreme cases: a cluster of tightly packed spherules resembles a compact sphere, and a loose chain of spherules bears some similitude to a cylinder. Intuition would suggest that clusters are less readily oxidised than chains (Amann et al., 1980). There are other considerations appertaining to accessibility: oxidants could experience diffusional limitations in penetrating the pores of an agglomerate structure. It is suspected that a highly reactive oxidant (OH) attacks only the outlying perimeter: hence, a higher burn-off is necessary to destroy the agglomerate than is the case for a less vigorous oxidant (O_2), which, because it is more penetrating, attacks the internal regions (Lahaye, 1992).

Ultimately, oxidation depends on atomistic aspects of the soot *surface*, that is, the spherule microstructure; as carbon atoms, if perched on edge sites, are more susceptible to oxidative attack than when more comfortably ensconced within the lattice (Boehman et al., 2005). Hence, disordered, amorphous structures are more prone to oxidation than well-ordered crystalline ones; and lamellae are more easily pared off from rough surfaces. Spherules seem less oxidisable when their surfaces, as viewed under the electron microscope, are smoother and clearer-imaged (Song H., et al., 2004), suggesting closer and more tightly packed lamellae; and the density of reactive sites increases as oxidation strips away more and more of the surface (Yezerets et al., 2005). Finally, surface reactivity undoubtedly varies with age.

Such models of soot oxidation as exist often presume recession of the surface, or shrinkage (Jung et al., 2004); yet electron micrographs of partially oxidised soot occasionally show pits, channels and cavities, suggesting that oxidation, in some instances, proceeds from *within* (Peterson, 1987). The disordered or amorphous microstructure, with a higher hydrogen-to-carbon ratio, of the spherule centre, is perhaps more susceptible to oxidation than the well-ordered, tightly layered laminate microstructure of the outer regions. After all, spherules subjected to laser heating are not so much stripped of their surface layers as eviscerated from within (Vander Wal et al., 1995). And various metals, when entrapped, are suspected of catalytic effects (Section 3.5.2). The diffusion of oxidants through the spherule microstructure into the central region has not been subjected to any detailed study, and perhaps some oxidants are entrapped during formation. But neither scenario may be necessary: once the outer shell has been successfully penetrated (Kim C.H. et al., 2004), the inner core is rapidly oxidised, leaving a shell (Song J. et al., 2006). Should oxidation proceed quiescently, and latently, from within, the spherule density will decline in the face of unchanging external appearance – until, that is, the final structural collapse. This scenario may be difficult to distinguish from a final stage of surface oxidation, wherein spherule surfaces become porous and pock-marked (Song H. et al., 2004). Either way, a phase now begins where, because agglomerates fragment (Ishiguro et al., 1991), particle numbers may rise for a second time.

From the foregoing discussion, it will be apparent that empirical or semi-empirical models of oxidation inevitably gloss over numerous aspects; but perhaps less apparent, and more remarkable, is that such models do reproduce the salient trends rather well. The Hiroyasu[4] model has been much favoured by engine modellers for thirty years (e.g. Hiroyasu and Kadota, 1976; Kittelson et al., 1986b; Nishida and Hiroyasu 1989; Shenghua et al., 1999; Jung D. and Assanis, 2001), and is probably the

---

[4] Hiroyasu et al. (1983). Cited by Jung and Assanis (2001). (NB: the soot formation model should not be confused with the fuel spray model which bears the same name.)

second most cited. It is of an Arrhenius form, being a second-order reaction between the mass of soot, m_s, and the oxygen partial pressure, p_{O_2}:

$$\frac{dm_{so}}{dt} = A_0 m_s \frac{p_{O_2}}{p} p^{1.8} \exp\left(-\frac{E_{so}}{\Re T}\right), \qquad (3.31)$$

where m_{so} is the mass of soot oxidised, A_0 is a model constant, p is the cylinder pressure and E_{so} is an activation energy for soot oxidation. The activation energy and model constant are adjusted to bring the modelled result into line with the engine-out emission. Equation (3.31) is the parallel of, and reverses the effect of, Equation (3.5). The semi-empirical equation reported by Lee,[5] originally devised for a laminar hydrocarbon flame, is similar, but with an additional multiplicative term: a square root dependency on T as well as the exponential dependency (Kyriakides *et al.*, 1986; Song H. *et al.*, 2004).

But the most famous soot oxidation model of all is that derived by Nagle and Strickland-Constable (1962) – hereinafter referred to as the NSC model. This model, as originally formulated, concerned the oxidation of pyrographite by molecular oxygen at partial pressures of ~ 0.2 atm and temperatures of 1100–2500 K. This, of course, differs substantially from the combustion pertaining in diesel engines, although the model's predictive capabilities presumably explain why engine modellers have clung to it so tenaciously for more than a generation (e.g. Mansouri *et al.*, 1982; Kittelson *et al.*, 1986b; Rutland *et al.*, 1995; Jung D. and Assanis, 2001). That said, the NSC model admittedly has some rationale in terms of surface chemistry, unlike the fully fledged empiricisms (Song H. *et al.*, 2004): two types of site are posited, the susceptibility of which to attack by oxygen differs. The 'B' sites are less reactive, and, following their reaction with O_2, revert to 'A' sites. The equations are

$$\frac{w}{12} = \left(\frac{k_A p_{O_2}}{1 + k_C p_{O_2}}\right) x + k_B p_{O_2} (1 - x), \qquad (3.32)$$

$$x = \left(1 + \frac{k_T}{p_{O_2} k_B}\right)^{-1}, \qquad (3.33)$$

where w is the mass oxidation rate at the surface (g cm^{-2}s^{-1}), and x is the fraction of the surface occupied by type 'A' sites. The rate constant for thermal rearrangement of type 'A' into type 'B' sites is k_T. This and the other rate constants k_A, k_B and k_C are given in Table 3.1 (Song H. *et al.*, 2001).

To apply the NSC model, the initial particle diameter, which determines the initial surface area, is adjusted such that the result of the computation matches the concentration of soot discharged by the

Table 3.1 Rate constants in the Nagle – Strickland-Constable (NSC) model for the oxidation of pyrographite, relating to Equations (3.32) and (3.33) (Song H. *et al.*, 2001).

Rate constant	Value	Units
k_A	$20 \exp(-15\,100/T)$	g cm^{-2} s^{-1} atm^{-1}
k_B	$4.46 \times 10^{-3} \exp(-7640/T)$	g cm^{-2} s^{-1} atm^{-1}
k_T	$1.51 \times 10^{-5} \exp(-48800/T)$	g cm^{-2} s^{-1}
k_C	$21.3 \exp(2060/T)$	atm^{-1}

---

[5] Lee *et al.* (1962). Cited by Song H. *et al.* (2001).

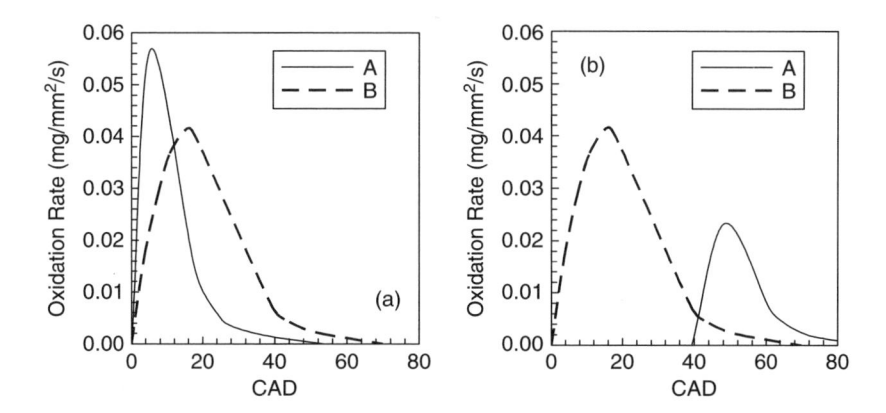

Figure 3.6 Modelling study of in-cylinder soot oxidation, based on the model of Nagle and Strickland-Constable (NSC). (a) Mixing rates: A, 'fast'; B, 'slow'. (b) Moment of burning: A, 'early' (TDC); B 'late' (−40°CA) (Amann et al., 1980).

engine. The model also needs the temperature and oxygen partial pressure as inputs. The mass of soot oxidised, per unit particle area, is obtained by integrating the curves for oxidation rate. The model is applied in opposition to a soot formation model, such as that of Hiroyasu (Equation (3.5)) (Rutland et al., 1995).

Some results obtained using the NSC model are given in Figure 3.6, starting with a pocket containing a certain amount of soot, and following this pocket with advancing crank angle (Amann et al., 1980). The oxidation rate rises rapidly at first, from zero, when air first enters a burning element, up to a maximum, after which it decays towards zero. The shape of these curves is decided by the interplay of *three* factors: the mixing rate, the pressure and the temperature.[6] Early-formed soot is burned because of a higher oxidation rate, and *not* because of a longer residence time. The initial oxidation rate is highest for the fast-mixing case, because the lower temperature is more than offset by the higher oxygen partial pressure. But, as the piston descends, the higher oxidation rate is now seen with the slower-mixing case, because the temperature has remained higher. In this second instance, it is *slower* mixing which allows the greater amount of soot to be oxidised. Oxygen needs to be introduced at an early stage, when the temperature is still high enough to support oxidation; mixing must not proceed *too* swiftly, or the oxidation will be quenched. The late-mixing soot is what escapes into the exhaust: there is now plenty of oxygen available, but the temperatures are far too low for any meaningful reactions.

Soot oxidation, it seems, can either be mixing-controlled or kinetically controlled, depending on the circumstances (Mansouri et al., 1982); and models of oxidation, if untempered by considerations of a fluid-mechanical nature, lose their predictive capability. Some workers eschew reaction kinetics entirely, focusing instead on turbulent kinetic energy (Hampson and Reitz, 1998), which tends to restrain oxidation at low temperatures (Kaario et al., 2002). This recognises the fact that soot oxidation, to some extent, is analogous to the combustion of fuel (Taskinen, 2000) and can be expressed as a 'characteristic oxidation time', defined according to local oxygen availability (Iyer and Abraham,

---

[6] It should be noted that cooling (and quenching) are caused by two factors: mixing with surrounding unreacted air, and polytropic expansion in producing useful work via the piston. The second distinguishes internal combustion engines from laboratory flames. More soot oxidation would occur without this work being done on the piston.

1998; Wadhwa and Abraham, 2000). Mediation of oxidation by mixing is what the Magnussen[7] model seeks to capture; when juxtaposed with the NSC model, the smaller of the two rates is the one adopted (Nakakita *et al.*, 1990). Temporally, there seems to be a transition from mixing-controlled oxidation near TDC to kinetically controlled oxidation with the approach of exhaust-valve opening (Hou and Abraham, 1995).

Finally, it is not possible, within an engine, to validate these expressions fully, because, as mentioned, oxidation and production contend for ascendancy; but nor do test tubes or laboratory reactors in which samples of soot are burned simulate realistic in-engine conditions. However, a novel apparatus for this purpose has been devised, wherein the soot discharged by a diesel engine is straightaway passed into a methane–air flame (Song H. *et al.*, 2001, 2004). Temperatures of 1500–1800 K were realised, although admittedly at atmospheric pressures. Experimentally determined oxidation rates, obtained from measurements of surface recession, were compared with estimates obtained from the Lee and NSC models. For low oxidation rates ($<10^{-4}\,\mathrm{g\,cm^{-2}\,s^{-1}}$), the results were closer to the Lee model; and for high oxidation rates ($>10^{-4}\,\mathrm{g\,cm^{-2}\,s^{-1}}$), the results were closer to the NSC model.

3.3 Carbonaceous Fraction: II. The Combusting Plume

The previous section was a generalised discussion of soot formation, equally applicable to many forms of combustion, and with passing references, where appropriate, to internal combustion engines. In this section we follow, within the *diesel* engine, as the combustion process unfolds, and according to various experimental techniques, the spatial–temporal distribution of soot. This presents the present reviewer with some difficulty, as, on perusing the literature, one is left with the impression that a uniform picture of soot formation in the diesel engine does not yet exist, and also that such a picture is unlikely to emerge for some time (e.g., Parker *et al.*, 1996). But this does not mean that already-published results are dubious or incorrect; rather, comparisons between workers must not be made incautiously: soot formation depends intimately on specific aspects of the fuel injection, charge temperature and pressure, fuel composition, combustion chamber geometry, etc.; and the picture, as delineated here, is necessarily generalised within this framework.

Studies of soot formation (and of the other emissions) within internal combustion engines make use of two distinct types of method: *ex situ* and *in situ*. The full details of these techniques, including how experimenters draw their conclusions, need not detain us – the reader is referred to the text by Zhao H. and Ladommatos (2001), or a paper by the same workers (Zhao H. and Ladommatos, 1998). *Ex situ* methods (Section 3.3.1) encompass cylinder sampling and cylinder dumping (Aoyagi *et al.*, 1980; Heddling *et al.*, 1981); *in situ* methods introduce light (nowadays lasers) into the combustion chamber. This second type in particular has, from the early 1990s onwards, with increasing zeal, been used to peer into the deepest recesses of diesel combustion, and the results have been fructuous to say the least. In this respect we have drawn heavily on the outstanding series of papers emanating from Sandia National Laboratories in the USA (e.g. Dec *et al.*, 1991; Dec, 1992; Espey and Dec, 1993; Espey *et al.*, 1994; Dec and Espey, 1995; Dec and Coy, 1996; Dec, 1997; Dec and Kelly-Zion, 2000; Higgins B. *et al.*, 2000; Dec and Tree, 2001; Siebers and Higgins, 2001; Tree and Dec, 2001; Musculus *et al.*, 2002a, 2002b; Siebers *et al.*, 2002; Mueller C.J. *et al.*, 2003; Pickett and Siebers, 2004).

---

[7] Magnussen and Hjertager (1977). (Cited by Wadhwa and Abraham (2000)).

3.3.1 Historical Overview

In former times the relative crudity of experimental methods forced investigators to fall back on generalisations, for example that 'soot is formed mainly in the axial position of the spray' (Hiroyasu and Kadota, 1976): a reasonable conclusion, since in this region air–fuel ratios are potently rich – effectively zero, according to the traditional understanding of diesel engine combustion. Lesser amounts of soot were held to exist at the end of the combusting plume, and concentrations were logically expected to fall sharply away from the centre line (Heywood, 1988, pp. 632–634).

Temporally, a general picture had been constructed in which the total quantity of soot inside the cylinder, through the course of a cycle, first rises swiftly to a maximum, and then, by falling at gradually diminishing rates, reaches eventually the concentration pertaining in the exhaust gas (e.g., Matsui et al., 1982; Ning et al., 1991). Some typical in-cylinder soot profiles along these lines, obtained from cylinder-dumping experiments, are given in Figure 3.7 (Kittelson et al., 1988). This rise-and-fall characteristic is generic, and arises from a struggle between two competing processes, soot formation and soot oxidation, during which the first and then the second dominates, with the tide turning at the peak (Pipho et al., 1986). But, as a cautionary remark, the *global* concentration of soot, as obtained from dumping, should not be confused with the *local* concentration, as obtained by sampling, because, owing to concurrent mixing between soot and air, the cylinder contents become progressively homogenised (Norris-Jones et al., 1984): hence, local concentrations of soot fall by mixing as much as by oxidation.

From these observations, two significant conclusions were unavoidable. First, considerably more soot forms inside the engine than is ultimately released into the exhaust: indeed, the peak concentration of soot regularly exceeds, by an order of magnitude, the concentration pertaining at exhaust-valve opening. Second, soot formation and soot oxidation are of comparable importance in determining the emission of soot.

The overall shape and position of this soot profile with respect to crank angle depend strongly on the operating conditions of the engine (e.g. speed, load, injection timing and EGR); the unpredictability,

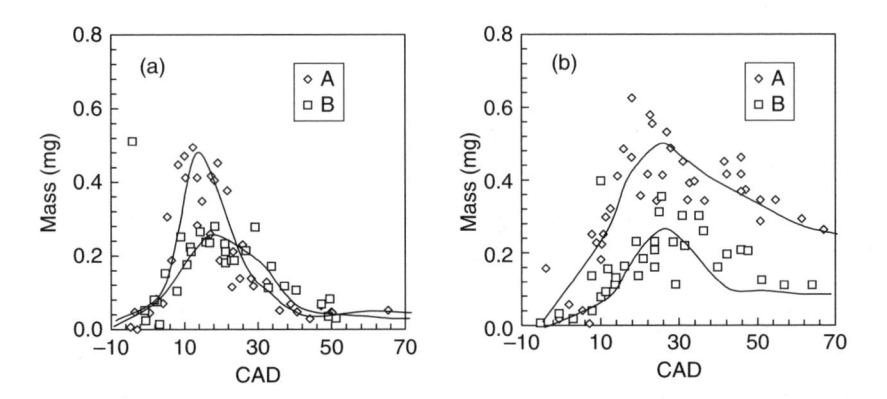

Figure 3.7 Mass of nonvolatile particulate (taken to be soot), as obtained through cylinder-dumping experiments: (a) 1000 rpm; A (−7.1, +4.1, −1.8, 0.07), B (−7.1, 4.1, 0.0, 0.03); (b) 1500 rpm; A (−8.2, 7.5, −3.5, 0.34), B (−8.2, 7.5, −1.5, 0.28). Lambda (air–fuel ratio normalised to stoichiometry): A, 2.5; B, 1.4. The figures in parentheses refer to start of injection (CAD), end of injection (CAD), start of combustion (CAD) and exhaust particulate (mg), respectively. Engine: 2.8-litre, 4-cylinder, direct-injection, high-swirl. Dilution ratios: 300–1800. CAD is given in terms of 'pressure half decay angle', i.e. the crank angle at which the cylinder pressure has fallen to half that at commencing blowdown (Kittelson et al., 1988).

or caprice, arises in part from the fact that peak soot and exhaust soot are always differently, and often contrarily, affected. For example, the observation that more soot enters the exhaust at high load is hardly startling; less obvious is a peak-to-exhaust ratio similar to that at low load (Pipho et al., 1986), while for an increase in ignition delay, the rate of soot formation may double, even though the peak concentration is cut in half (Miyamoto et at., 1990). Such trends are held to arise from disparate effects on the competing rates of soot formation and soot oxidation. Consequently, a considerably higher peak concentration of soot *does not*, on its own, correspondingly signify a higher concentration of soot in the exhaust; indeed, if the oxidation proceeds more vigorously, the exhaust gas may actually become cleaner, rather than dirtier.

It is perhaps slightly disingenuous to separate the spatial from the temporal. Soot formation studies conducted in laboratory flames invariably show, with increasing distance along the flame axis, a very similar pattern of rise and fall (Crookes et al., 1999): a point is always reached where, eventually, the rate of soot oxidation equals, and thereafter exceeds, the rate of soot formation. This is certainly because the first rises, and possibly because the second falls; the soot emitted from the flame is that fraction which avoids burn-up and successfully survives beyond the flame tip. The laboratory flame is, in this sense, the *spatial* analogue of the *temporal* sequence portrayed in Figure 3.7, although the two combustion processes are seldom advantageously contrasted in this manner (Hiroyasu et al., 1980). Of course, diesel engine combustion passes through three distinct phases (premixed burn, mixing-controlled burn and late burn), for which such a comparison is not always efficacious.

Unlike mass, measurements of particle size are vulnerable to falsifications caused by ongoing agglomeration following extraction from the combustion chamber (Xu H. et al., 1982), but, after correcting for this effect, particles do indeed grow by a hundred nanometres or so as the combustion proceeds (Luo et al., 1989). Examinations of these same agglomerates under the electron microscope show *parallel* diminutions, of a few nanometres or so, in spherule sizes (Fujiwara et al., 1984). In both of these parameters, i.e. agglomerate size and spherule size, particles simply evolve toward those dimensions observed in the exhaust. Thus, it seems, at least in these instances, that recession of spherule surfaces, as caused by oxidation, is insufficient to reverse the effects of agglomerative growth, leading ultimately to particle fragmentation.

3.3.2 Premixed Burn

In relation to soot-production tendency, the premixed burn is habitually exculpated, as if it generates no soot; but this exculpation is, in a deeper understanding, subject to certain provisos, as in-cylinder measurements during this period do indeed detect soot, albeit in small amounts (Pinson et al., 1994; Corcione et al., 2001, 2002); and these findings are corroborated by experiments of a more fundamental nature in constant-volume combustion vessels (Higgins B. et al., 2000). Of course, the oxidation of this soot, at some point prior to completion of the combustion process, seems all too probable; and, in any case, this soot is considerably overshadowed by that which forms subsequently, during the mixing-controlled phase, so that its complete extirpation would, with equal probability, be neither here nor there in meeting emissions legislation. In investigating this early-forming soot, researchers are actuated more by a desire to elucidate the secrets of diesel engine combustion than by soot suppression *per se*.

Investigations directed *specifically* at premixed-phase soot, rather than at other aspects of the combustion with tangential remarks being made *en passant*, are a fairly recent arrival in the literature. A conceptual model of ignition, premixed combustion and the emergence of soot that has rapidly gained widespread currency is depicted in Figure 3.8 (Dec, 1997); delineated therein are some admittedly idealised axial slices, through the midplane of the combusting plume, at four successive stages of development.

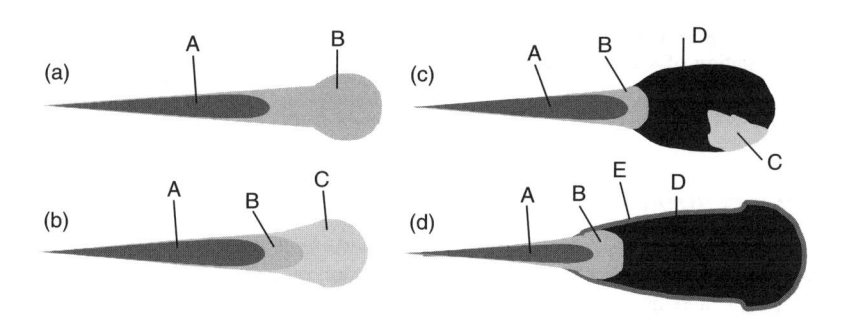

Figure 3.8 Conceptual model of plume development: (a) immediately prior to ignition; (b) just after ignition; (c) early phase of premixed burn; (d) late phase of premixed burn. A, liquid fuel; B, fuel vapour; C, PAH; D, low concentration of soot; E, diffusion-flame envelope (Dec, 1997).

Upon start of injection, the jet of liquid-phase fuel grows initially with extreme rapidity, but then gradually slackens in pace, because hot air is entrained; and this entrainment vaporises the incoming fuel, giving rise to the air–vapour segment. Prior to ignition, therefore, two clearly demarcated yet contiguous regions are observed: an *unmixed* segment of liquid-phase fuel, and a *mixed* segment of air and vaporised fuel. The air–vapour segment continues growing and, immediately prior to ignition, has mixed to a condition of $1/4 < \lambda < 1/2$.

Ignition takes place within the air–vapour segment; this is one area where the conceptual scheme under discussion diverges from previous ones, which held ignition to occur under stoichiometric conditions. The consequent breakdown and pyrolysis of fuel within this air–vapour segment gives birth to PAH – an event heralded by *chemiluminescence*. These PAH rapidly multiply and suffuse the whole of the air–vapour segment. From this understanding, it appears manifest that ignition and heat release precede the birth of PAH, as pyrolysis requires >1300 K, and these temperatures are not realised solely through charge compression. And the sudden appearance of PAH throughout this region, virtually simultaneously, strongly suggests a homogeneous reaction or multiple ignition sites, rather than a propagating flame front (Tree and Svensson, 2007).

Soot is harbingered by natural flame luminosity, or *incandescence*, which, almost immediately, supplants and overwhelms the far weaker chemiluminescence. (This moment coincides with a sudden upturn in the rate of heat release.) The first soot particles appear within pockets, the sizes, shapes and locations of which vary randomly from one cycle to another. The soot-containing pockets multiply and converge, until they cover the entire air–vapour segment. The concentration of soot increases temporally throughout the combusting plume, and spatially towards the leading edge.

A marked change in plume structure then takes place: different soot particles emerge around the *periphery* – different because they are larger, and grow much more rapidly than their siblings residing within the plume, suggesting different progenitive mechanisms. The concentration of soot at the periphery is not, however, dissimilar from that inside the plume. The zone of peripheral soot becomes thicker, perhaps because of localised turbulence-induced mixing, but the central region is not penetrated to any great degree. In fact, it is supposed, not without a considerable degree of probability, that some association exists between this peripheral soot and the diffusion flame envelope that is now becoming established around the combusting plume, in preparation for the mixing-controlled burn. In the conditions described, this peripheral soot and the enveloping diffusion flame are both established just prior to the peak in the premixed-burn heat release (Dec, 1997).

Premixed-burn soot formation is extremely rapid: for example, the interval from complete undetectability to total suffusion of the entire leading region of the plume was reported as 70 µs, or 0.5°CA, by Dec and Espey (1995). This rapid proliferation is made possible by the virtually simultaneous formation of soot at multiple points – as might be expected from premixed combustion. The first-formed soot is *not* at the edges of the combusting plume, as might be supposed for combustion proceeding solely via a diffusion flame; this peripheral soot forms slightly later.

3.3.3 Mixing-controlled Burn

Distinguishing between mixing-controlled combustion and premixed combustion is actually quite a didactic exercise. The traditional definition, using cylinder pressure traces or calculations of heat release, staked clear and unambiguous lines of demarcation, and this was where matters stood for decades. Laser imaging, on the other hand, as the combustion shifts from one mode to the other, reveals gradations, rather than abrupt transitions, in plume structure.

Various conceptual schemes have been put forward over the years to explain the *modus operandi* of mixing-controlled combustion: how fuel burns, and how pollutants are generated (Dec, 1997). But as insufficient space exists in this brief sketch to discuss lineage, we necessarily focus on the latest ideas, as derived from numerous laser-imaging experiments.

A conceptualisation of mixing-controlled combustion is depicted in Figure 3.9 (Dec, 1997, etc.). As in the previous section, this delineation represents an idealised slice, so to speak, through the midplane of the combusting plume; in reality, the edges are not so much well-defined lines as ill-defined zones. The flow fields are, after all, turbulent, and this ensures small variations in plume structure from one cycle to another, but, taken in aggregate, the overall features are consistent; and, for this reason, the plume anatomy is posited as *quasi-steady-state*. The liquid and vapour sections, in this sense, are unvarying, from which it appears logical to presuppose that reacting species are replenished and lost at roughly equal rates. The model is not, however, all-embracing, as it assumes a quiescent air charge (see Section 7.5), while interactions between the combusting plume and the wall of the surrounding combustion chamber (see Section 3.4) are not included.

This conceptual model, which is rapidly gaining widespread acceptance, is unabashedly iconoclastic insofar as it threatens to overturn all conventional notions of diesel engine combustion. This is because

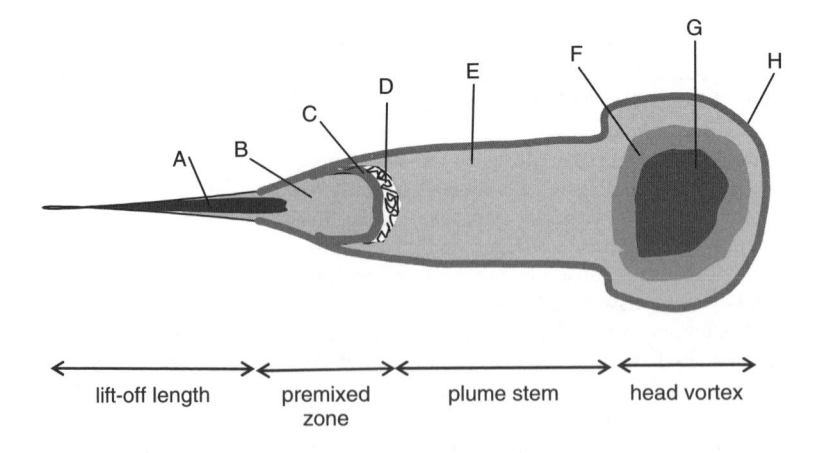

Figure 3.9 Conceptual model of combustion during the mixing-controlled phase: A, liquid fuel; B, mixture of fuel vapour and air; C, fuel-rich premixed flame; D, initial soot formation; E, F, G, in order of increasing soot concentration; H, diffusion-flame envelope (Dec, 1997, and subsequent papers).

the combustion is not conceived as completely mixing-controlled: the fuel is first *partially* burned in a *rich premixed flame*, the reaction products of which gallop, pell-mell, along the *plume stem*, to a *head vortex*, where they subsequently enter an enveloping *diffusion flame*. It is, therefore, a 'dual-mode' combustion, wherein the fuel is burned in two successive stages.

The model is, so to speak, a reductivist one, since, although many, many parameters affect the combustion (e.g. injection pressure, injector nozzle design, charge temperature, charge pressure, injection timing, engine speed and load, fuel formulation, and EGR), and more than one aspect of the plume usually changes at any one time, with often puzzling effects, it seems that *everything rests on the quantity of air entrained upstream of the premixed flame* (Pickett and Siebers, 2004) – a criterion expressed by the 'lift-off length', meaning the distance from the nozzle to the base of the diffusion flame. Very little, if any, air is entrained into the plume in regions surrounded by the diffusion flame envelope. This self-consistency undoubtedly lends the model a certain harmonious cohesion.

A key observation in laser-imaging experiments is the presence of soot *within* the combusting plume. Previous models asserted that soot formation occurs on the periphery, i.e. in the diffusion flame; but progressive inward movement of soot, from the periphery into the plume, is not observed in contemporary measurements. The logical inference, therefore, is that soot resides within the plume simply because it *formed* within the plume, and the premixed flame has been put forward as an explanation. Although the circumstantial evidence for the shadowy existence of this premixed flame is rather large, incontrovertible proof, at the time of writing, is still awaited, so that the conceptual model, as described here, is, in this respect, provisional.

Two features consistently observed in laser imaging are an increasing concentration of soot and an increasing size of soot particle along the plume stem, with the highest concentrations and the largest particles residing in the head vortex (Dec, 1992). Taken together, these findings are telling: they suggest the formation of soot precursors in the premixed flame; the consequent generation of soot particles; growth of these particles as they pass down the plume stem towards the head vortex; and, finally, oxidation of soot particles by the enveloping diffusion flame (with OH, rather than O_2, as the agent of oxidation). Thus, the conceptual model of the combusting plume described here and the conceptual model of soot formation in Figure 3.1 harmonise rather nicely, although marrying the two is a task that is only just beginning (Brahma *et al.*, 2005). The consequences for emission control rest on whether particles successfully penetrate the diffusion flame envelope; but soot is not generally found outside the plume (Tree and Svensson, 2007).

A secondary mode of soot generation seems to come into operation on the plume periphery (not shown in Figure 3.9): these particles are larger than their counterparts within the plume stem, but smaller than those within the head vortex; they are ascribed to some action of the diffusion flame, and probably arise through different progenitive mechanisms. The concentration of soot at the periphery is, however, similar to that within the plume itself. The peripheral particles can mix inwards, but seem reluctant to enter the plume to any marked extent. This thin zone of peripheral soot extends all the way back along the plume, almost to the tip of the liquid-phase segment.

It would be pleasing to confirm these ideas of soot formation via actual demographic measurements, but quantitative rather than qualitative data are sparse, and understandably lacking the fine spatial – temporal resolution required. Of course, a wealth of such data already exists for simple laboratory flames, for which optical measurements and direct sampling of soot are both regularly used, but it is unclear how these flame structures relate to the combusting plume within a diesel engine. Obviously, direct sampling from within the plume, within an engine, is impractical, but, where attempts have been made at optical measurements, particles have seldom exceeded 100 nm, with ~50 nm being the commonest size; the number concentrations are $\lesssim 10^{11} \text{cm}^{-3}$ (Pinson *et al.*, 1994). But it is unclear whether these sizes relate to spherules or to agglomerates; indeed, vigorous agglomeration is presumably in operation. Faced with the inaccessibility and transient nature of diesel engine combustion, researchers have understandably designed surrogate apparatus, such as purpose-built high-pressure chambers, into which liquid fuel is sprayed, and in which steady-state combustion is maintainable indefinitely (Crookes *et al.*, 1999). The particles extracted therefrom are agglomerates, the polydispersity of which increases with distance from the nozzle, and the spherules of which range from 20 nm to 50 nm.

3.3.4 Late Burn

And so to the *denouement*, where the injection of fuel is abruptly terminated, the consequence of which is complete collapse, and disintegration, of the plume structure. At the upstream end, the traditional picture is one where, owing to limitations in the fuel injection equipment (FIE), the last remaining fuel elements enter at low injection pressure, or simply through vaporisation from the nozzle sac, and the combustion of these tardy elements, being poor, exacerbates the emissions of hydrocarbons and soot (Section 7.3.1). More recently, in relation to the presently discussed model, it has been suggested that sudden disappearance of the premixed zone (which, after all, was maintained by the momentum of incoming fuel) allows combustion and soot formation to shift, albeit briefly, upstream, towards the nozzle (Dec and Kelly-Zion, 2000), especially if the needle closes too slowly (Tree and Svensson, 2007).

Somewhat better established are events at the opposite end, where, as sketched in Figure 3.10(a), the plume stem trundles into the head vortex, and the head vortex, after colliding with the wall of the combustion chamber, spreads out circumferentially, until, encountering head vortices from adjacent plumes that are doing the same, it doubles back, moving radially inward. In this terminating, or folding-up, process, the initially intact head vortex fragments, so that the last remaining vestiges of plume structure are lost, to be replaced by a welter of soot-containing pockets, the number and location of which vary greatly from cycle to cycle.

As suggested in Figure 3.10(b), the soot residing within these pockets is still encircled, and probably still being consumed, by diffusion flames; and, even though such flames are shortly quenched, it seems probable that, on occasion, conditions in the combustion chamber, on a global basis, remain sufficiently hot for bulk-gas soot oxidation, that is, with no discernible flame. But failure to burn

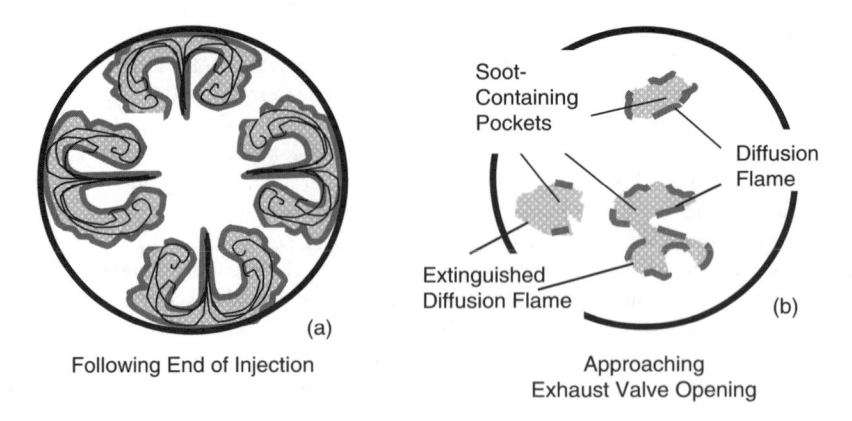

Figure 3.10 Folding up and collapse of combusting plume: (a) immediately following end of injection; (b) late burn-up of soot-containing pockets, partially surrounded by diffusion flames. Sketched from ideas developed by Dec and Kelly-Zion (2000).

this late-surviving soot, in its entirety and on a regular basis, is only too evident; and this failure is undoubtedly important, as decent correlations exist between estimates of the soot domiciled within these pockets at exhaust-valve opening, and the soot that subsequently emerges in the exhaust stream (Dec and Kelly-Zion, 2000).

Simply, the longer soot is able to survive, the greater the likelihood of its escape. But the point at which soot burn-up ceases is difficult to define; indeed, a precisely defined moment probably does not exist: the oxidation reactions just keep slowing. The soot clouds in any case continue to expand as the piston descends (Antoni and Peters, 1997). The combustion, according to customary calculations of heat release, may well be over before exhaust-valve opening, but this is not the same as the small-scale oxidation of soot, which probably has little influence on global combustion parameters; indeed, the presence of soot incandescence at exhaust-valve opening demonstrates this (Hiroyasu *et al.*, 1980), although this late-surviving soot, if cool, is not necessarily incandescent. The implications of this burn-up are disproportionate: for example, a paltry five per cent increase in the soot oxidised during the expansion stroke might *halve* the amount of soot in the exhaust (Wade *et al.*, 1986).

The vigour of this oxidation is decided, in effect, by the flow field, insofar as this controls the mixing of air and soot, i.e. where, in the combustion chamber, the soot clouds are swept by squish, swirl etc. in relation to the available oxygen: this is shown in models (Mather and Reitz, 1998) and experiments (Espey and Dec, 1993). The importance of turbulence in promoting mixing is recognised (Chikahisa and Araki, 1996): when this turbulence decays too quickly, soot oxidation is arrested (Ikegami *et al.*, 1988). As modelled, soot oxidation is proportional to the dissipation rate and inversely proportional to the kinetic energy of this turbulence (Dent, 1980).

The problem, however, in practical terms, is the control, or rather promotion, of this late-phase turbulence, as that which was imparted in the induction stroke has long since disappeared. Novel schemes for late-phase turbulence generation include high-pressure air injection into the combustion chamber, either from outside the engine (Nagano *et al.*, 1991; Kurtz *et al.*, 2000) or by installing 'microchambers' in the combustion chamber wall, from which, during the expansion stroke, the previously compressed air emerges (Rogers B.J. *et al.*, 2002). It will be noted that the aim here is not so much to furnish the oxidising reactions with more oxygen as to promote the turbulence, and thereby make better use of the already-available oxygen. But the chief difficulty with these schemes lies in inadvertent quenching of the oxidation reactions.

Just as in the earlier stages, quantitative measurements of particle demographics are scant, but, from the little work reported, the results are not wholly out of keeping with expectations. Laser-imaging studies reveal how increasing particle size through agglomeration (Pinson *et al.*, 1994) is tempered by decreasing particle size through oxidation (Pinson *et al.*, 1993). Particles, divided into three size groups, 'small', 'medium' and 'large' (admittedly according to qualitative criteria), each increase, successively, to a maximum concentration; the smallest particles rise to prominence the fastest, and to the greatest number, but also fall out of favour the fastest; and, eventually, all size classes decline towards similar concentrations (Fujimoto *et al.*, 1998).

3.4 Carbonaceous Fraction: III. Wall Interactions

One aspect of particular relevance to soot formation within internal combustion engines is the effect of constraining internal surfaces or walls. That wall effects are an aspect of soot formation more specific to internal combustion engines is evident in the scant treatment they receive in general reviews. To the contrary, however, the literature appertaining to internal combustion engines is replete with investigations, of both an experimental and a theoretical nature, on the effects of the constraining walls.

3.4.1 Theoretical

The walls of a combustion chamber are invariably dirty, i.e. they carry deposits of soot, and it seems reasonable to suppose that interactions of one sort or another between these walls and the combusting plume play some role in the emission of soot. This scenario should not be confused with another, such as occurs in small-bore engines, where the combusting plume is unable to establish itself fully, because the liquid-phase portion of the fuel spray impinges on the wall of the combustion chamber *prior* to formation of the diffusion flame, so that fuel films accumulate on the wall. These films are emitted, unburned, as white smoke (Section 3.6.5), or, if ignited, burn poorly and protractedly, in rich, late-burning, soot-forming flames (Sjöberg, 2001).

Soot particles successfully attaching to the walls of the combustion chamber are lucky, as the walls are markedly cooler ($\sim$250 °C) than the bulk gas ($\sim$2000 °C), so that the soot is, by this deposition, protected from oxidation by the thermal boundary layer. Perhaps, during expansion or blow-down, these deposits are re-entrained into the exhaust stream, whereupon they make good their escape. Clearly, the strength of this mechanism, if it exists, must be ascertained, as attempts to control soot, either by restricting its formation in the combusting plume (Section 3.3.3) or by enhancing its oxidation in the late burn (Section 3.3.4), may be misplaced.

The second observation to be made is the apparently stable thickness of these deposits, once established: there presumably arrives a point when deposition and re-entrainment occur at equal and opposite rates; or, alternatively, a point where the rate of deposition decays to zero. This condition, logically, depends on the operating point of the engine, with the deposit thickness simply readjusting according to the shifting equilibrium between deposition and re-entrainment, as driven by the shifting (average) conditions within the cylinder.

As a minor digression, wall deposits do not solely concern soot *emissions*; they also act as an insulating layer: this affects heat transfer to a degree that combustion models cannot afford to neglect (Woschni and Huber, 1991). But little is known about the cycle-to-cycle, or even instantaneous, history of these deposits: they could, for example, be regularly consumed and rapidly regenerated, or be relatively unchanging with respect to the duration of a cycle (Eiglmeier *et al.*, 2001). In a comprehensive model (Wolff *et al.*, 1997), an instantaneous deposit thickness of $\sim$15 μm is predicted, the variation in which, in the course of a cycle, does not exceed a few hundred nanometres. Yet because the soot deposits are porous, i.e. contain air pockets, and this porosity is a nonnegligible factor in their density, thermal conductivity and specific heat capacity, heat-flux models are not exactly straightforward.

Returning to the enquiry in hand, the postulate is that, when proximate to a surface, the enveloping diffusion flame extinguishes, so that a gap appears through which soot is afforded opportunity to escape. But soot particles, if they are to deposit at all, must avoid travelling *with* the gas, otherwise, following wall impingement by the plume, they flow sideways and then back, towards the bulk gas. Simply, the particle must somehow successfully traverse the boundary layer in order to strike the surface.

A consideration of several particle transport mechanisms provides sufficient evidence to impeach thermophoresis (Kittelson *et al.*, 1990; Suhre and Foster, 1992); such unwanted deposition is a recurring problem in engineering generally (Greenfield C. and Quarini, 1998). This impeachment dovetails nicely with the lesser soot emission of low-heat-rejection engines, for which one might suppose hotter walls, and hence shallower temperature gradients; we shall return to this topic shortly. Moreover, particle size becomes a factor in this deposition, unlike in the exhaust, because, inside the cylinder, the mean free path, owing to the high gas density, is shorter. Inertial forces seem unlikely to throw particles across the boundary layer, unless this layer is narrower than $\sim$30 nm; indeed soot particles are too small to travel ballistically relative to the surrounding gas (Tree and Dec, 2001). Electrostatic attraction is difficult to model because little is known about in-cylinder electrical fields, but the field strength required for significant rates of deposition is probably unrealistic. Finally, the time required to travel across the boundary layer by diffusion is at least two orders of magnitude larger than for thermophoresis.

The incorporation of thermophoresis into a combustion model is unusual (Abraham, 1996). Such modelling has suggested thermophoretic deposition to become important only for particles that are *already* in close proximity to the walls, i.e. courtesy of turbulent convection, plume orientation or soot formation. Outside the boundary layer, the characteristic time associated with turbulent convection was at least an order of magnitude shorter than the characteristic time for thermophoresis. But actual deposition was not essential to the survival of soot: these particles need only linger within a few hundred microns of the walls, where conditions are much cooler, in order to escape oxidation. For example, at $\sim100°$CA, the fraction of soot in the near-wall region was 20% of the total mass; logically, this soot is swept up during the expansion and exhaust strokes in a manner akin to the hydrocarbon-survival mechanism in gasoline engines: the 'quench layer'.

There remains to mention the break-up and re-entrainment of wall deposits, if these deposits are to escape into the exhaust. The nature of re-entrainment, which is inherently random, impedes modelling efforts, and workers have addressed this topic little, preferring, instead, to focus on the other side of the coin, deposition. But one re-entrainment model (Abu-Qudais and Kittelson, 1995) supports the notion that temporary retention of soot on internal surfaces constitutes a nonnegligible part of the soot emission.

The rate at which soot is emitted from an engine does not, however, correlate with the rate at which soot is deposited on internal surfaces, for which some astute ideas are supported by rough calculations (Tree and Dec, 2001). The soot deposit may, in acting as an insulating layer, allow the flame to approach the wall more closely; and, in supplying itself as a feedstock to the flame, the deposit may artificially protract the flame's lifespan. Moreover, a porous soot layer will contain a supply of air that a clean metal surface does not. This should not be too surprising, as porous engine deposits are known to store hydrocarbons temporarily – another hydrocarbon-survival mechanism known to occur in gasoline engines. An alternative scenario is that the deposition rate falls to zero, perhaps because thermophoresis becomes ineffective once temperature gradients have been reduced sufficiently by the build-up of an insulating layer of soot. An obvious parallel arises here with low-heat-rejection engines, the in-cylinder surfaces of which are coated with ceramics, which act as insulators or thermal barriers (Winkler and Parker, 1993); in fact, this insulation may also bring about some reduction in engine-out soot (Voss *et al.*, 1997). The application of ceramic thermal-barrier coatings, apparently, prohibits soot deposition or, alternatively permits soot oxidation once deposition has taken place. Such postulates pleasingly avoid the need to explain how re-entrained soot might be oxidised during the blow-down, when it is known that temperatures are really too low for this to occur.

3.4.2 Experimental

Experimentalists have long been teased by sundry observations suggesting that soot particles interact in some way with internal walls (Kittelson *et al.*, 1990): for example, the emission of soot is reduced by ceramic coatings, suggesting oxidation reactions; the exhaust gas carries supermicron particles (Du *et al.*, 1984), suggesting deposition and re-entrainment; and a pulse of electrostatic charge is observed at exhaust-valve opening, suggesting contact electrification. These observations are, of course, circumstantial or incidental, and do not, on their own, definitively impute walls or in-engine surfaces in the formation and emission of soot.

Direct measurements of soot deposition, that is, through short periods of operation followed by disassembly, reveal wide variations in deposition rates and variations in deposit stabilisation times between the combustion chamber walls, piston top-land, piston-top surface and cylinder head (flame face); in all of these locations, a steady-state thickness was reached in just a few minutes, following a

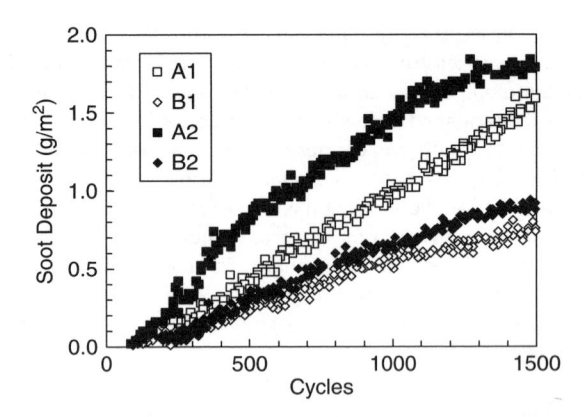

Figure 3.11 The thicknesses of soot deposits, expressed as mass per unit area, on an optical window mounted in the cylinder head, as determined by the attenuation in flame luminosity at 650 nm (measured by photodiode voltage). The window temperature in A was 100°C cooler than in B. Two test runs were performed in each case: 1 and 2. Engine: 2.33-litre, single-cylinder, injection angle 10° to 20° BTDC. Operating point, 1300 rpm; fuel–air ratio, 3. Fuel: H/C, 1.83; sulphur, 0.03%; cetane no., 48 (Suhre and Foster, 1992).

clean condition (Yamaguchi et al., 1991). The deposit thickness on a purpose-built plug, flush-mounted in the cylinder head, eventually stabilised after several hours of running (Kittelson et al., 1990). In both cases, the deposition rate, from the initial clean condition, was rapid.

Indirect measurements are possible by exploiting light transmission through an optical window or probe. Somewhat ironically, this task is the converse of engine management, in which flame luminosity measurements are annoyingly attenuated by ever-accumulating soot deposits (Nagase et al., 1985). More particularly for the present discussion, the thickness of the deposit is open to estimation via the transmission of natural flame luminosity through it: some results along these lines are reproduced in Figure 3.11 (Suhre and Foster, 1992). Here, the rate of deposition, following a clean condition, proceeded linearly for several hundreds of engine cycles, but thereafter began to decline; critically, however, the rate of deposition was markedly faster on a *cooled* optical window, thus corroborating the thermophoretic hypothesis. Soot deposition is, by a parallel or, rather, converse method, measurable by passing a laser beam *into* the engine (Tree and Dec, 2001): starting from a clean condition, the transmittance through a window mounted in the rim of the piston bowl falls sharply with each fired cycle, but has begun levelling off discernibly by fifty cycles or so.

The total quantity of soot residing on internal surfaces seems comparable with what is emitted by an engine. For example, if the deposits on the aforementioned plugs are typical of the in-engine surfaces as a whole, then, by a simple computation, the total quantity of stored soot agrees reasonably well with that expected from a thermophoretic deposition model, and represents 20–45 % of the engine-out emission (Kittelson et al., 1990). But estimates of stored soot are hampered by wide variations in deposit thickness across the various surfaces. Thicker deposits are observed where the leading edge of the combusting plume reaches the wall of the combustion chamber, and the variation in deposit thickness across a window mounted in the bowl wall is consistent with the trajectory taken by the combusting plume as it crosses the window (Tree and Dec, 2001). The reversibility of this relationship, in certain circumstances, allows, via the thickness of the deposit, an estimate of in-plume soot, i.e. how sooting the combustion is (Musculus et al., 2002a).

The sixty-four thousand dollar question is whether any relationship truly exists between this stored soot and the soot emitted by the engine. As said, this question is arguably unanswerable, and certainly problematical, given the ill-described nonuniformity in deposit thickness. But, for the sake of argument, if the deposits are assumed to be of uniform thickness, then no correlation exists with engine-out soot, and, in fact, the mass of stored soot can exceed that emitted, on a gram-per-cycle basis (Suhre and Foster, 1992). This tends to imply that stored soot makes little contribution to emitted soot. Deposition is a function of the soot concentration within the combusting plume, the time the plume is in contact with the wall, and the temperature difference between the soot cloud and the wall. The net result of these competing functions is not easily predicted; the trends in emitted soot and deposited soot, with respect to injection timing, EGR etc., are by no means predestined to agree (Tree and Dec, 2001).

The best indicator of soot deposition is, not unexpectedly, the fuel-injection quantity. Should the (single-point) deposition rate and the deposition area be proportional to the fuelling, then the soot storage would increase, as the square of the fuel quantity. Contrastingly, at a certain operating point, EGR and injection timing are relatively uninfluential on rates of soot deposition, because any increase in the (single-point) deposition rate is offset by a corresponding decrease in the deposition area (Tree and Dec, 2001). But since injection timing and EGR always wield great effects on the exhaust soot, then, presumably, *other* factors, such as the bulk burn-out of soot, are what really control the emission of soot. That said, others have reached opposite conclusions as to the importance of soot deposition in soot emission (Kittelson *et al.*, 1990), so that a universal picture is still awaited. Discouraging the deposition of this soot, oxidising the deposits *in situ*, or oxidising the deposits when re-entrained might still, therefore, represent equally laudable aims, depending on the specific circumstances.

3.5 Ash Fraction

We turn now to the smallest of the four fractions, which, in normal circumstances, is present only in trace amounts; because of this diminutive presence, its complete elimination would be unavailing in type approval procedures. But, as we have seen on various occasions, abundance need not, indeed often does not, scale with significance. The ash fraction is, for instance, a useful quality check on engine operation; and research is sometimes undertaken, for example, to assess public-health implications, to understand the undoubted catalytic effects in soot oxidation, and to discover ways of ameliorating long-term damage to aftertreatment devices.

Four sources of ash are discernible, although three of them are not, principally, the subject of this section. First, there are lubricant components (Section 7.6.1), perhaps naturally present; but those more probably present, and certainly of greater salience, are components deliberately admixed, namely phosphorus, zinc, calcium and magnesium. Second, there is airborne inorganic debris, or dust, brought in with the air, of which silicon (silica) is an example. The third category is wear metals: obviously, iron, but also, for example, magnesium (piston rings), copper and lead (bearings), and aluminium (pistons); other elements are chromium, nickel (Kimura *et al.*, 2006), manganese and molybdenum (Raux *et al.*, 2005). The high-pressure fuel pump releases ferrous particles, and the low-pressure pump releases graphite particles from motor brushes (Macián *et al.*, 2006). These being wear particles, they are of a few microns in size. Finally, the subject of the present section, there are sundry inorganic compounds or elements present in the fuel. It will be appreciated that provenance in some cases, especially trace-level cases, is obscure. And some elements are multi-sourced: for example, chlorine is found in lubricants (Mayer A. *et al.*, 1997), is a relic of catalyst manufacture (Neyestanaki *et al.*, 2004) and is ingestible as road salt (Clarke *et al.*, 1996), while silicon is also found in fuel (Owen and Coley, 1995, p. 524) and in lubricants (Caines *et al.*, 2004, p. 84) as a silicone antifoaming agent: this has also been found in particulate emissions (Tomiyasu *et al.*, 2006).

The in-engine experiences of these ashing elements, in the case of the first three sources, are flimsily documented. Inorganic compounds from the oil, unlike those legitimately entering the combustion chamber via the fuel, are fugitives; they do not necessarily experience the hottest conditions, and in fact are subjected to a range of temperatures, depending on their entry point (piston rings, valve-stem seals, etc.). Oil compounds, therefore, have potential to evade chemical breakdown, and hence to escape in their native forms. The in-engine experiences of wear and air-ingested particles are wholly unreported, but their penetration of the lubricating system is, historically, of greater interest to engineers than their emission at the tailpipe. Supermicron wear particles carried along by the fuel (if they evade retention by the fuel filter) will, of course, behave rather differently in the combustion chamber from metals that are truly dissolved in the fuel.

In the case of the fourth source, fuel carries inorganic compounds quite naturally, courtesy of the crude, and unintentionally, such as through contamination in the fuel distribution network. Given the implications here for public exposure, fuel analyses are astonishingly poorly reported: fuel-borne ash is arguably the forgotten pollutant. In a diesel fuel containing *no* additives, twenty ashing elements were present at discernible levels: for example, Si at 46 mg/litre, Zn at 5.6 mg/litre and Fe at 27 mg/litre. Tailpipe emission rates were fully accounted for by fuel concentrations; emissions of crustal elements (Al, Ca, Fe, Mg and Si) were severalfold higher than emissions of anthropogenic elements (Ag, Ba, Cd, Cr, Cu etc.). Such emission rates, scaled up to the whole vehicle population, may well exceed those from coal-fired power stations (Wang Y.-F. *et al.*, 2003). It should be noted, however, that concentrations of fuel-borne ashing elements very probably vary extensively from one region or country to another (Lim M.C.H. *et al.*, 2007).

On other occasions, inorganic, or rather organometallic, compounds are deliberately admixed in order to improve certain fuel properties, such as the combustion characteristics. Additive treat rates and rates of unintentional adulteration may be comparable: for example, an iron contamination of only ~5 ppm is already comparable to typical treat rates for ferrocene (iron dicyclopentadiene, $(C_5H_5)_2Fe$) (Du *et al.*, 1998). Of course, as actual chemical forms will differ, depending on the provenance, then so too will the manner in which these compounds burn, and form particles. Wear metals, for instance, may not be actually dissolved in the fuel as such, but present colloidally.

3.5.1 Chemical Reactions

The open literature is relatively silent on the chemical reactions between, and the chemical forms assumed by, inorganic material between combustion and emission. But, given the huge commercial significance of fuel additives, the absence of such research seems rather unlikely, and the silence, therefore, probably arises from considerations of a proprietary nature. It does not, however, require a leap of faith to envisage the breakdown, in the combustion chamber, of the original metal-containing compounds, and the consequent liberation of these metals as vapours. Then, as conditions begin to cool, these metals form new compounds, and undergo gas-to-particle conversions. True, many metals *end up* as sulphates or oxides, but this is only part of their story, as, in transit, ongoing reactions continually modify chemical compositions; and temperatures, since they plummet with such extreme rapidity, allow chemical compounds to persist in nonequilibrated states.

Despite this warning, or expectation, of nonequilibrium, workers have, nonetheless, for the sake of expediency, fallen back on thermodynamics. Equilibria suggest, for example, that ferrocene, a well-known fuel additive, decomposes in the combustion chamber to $Fe_2(OH)_2(g)$, and that, upon entering the exhaust system, the iron shifts into $Fe_2(SO_4)_3(s)$ (Du *et al.*, 1998). Others suggest the formation of FeO(g) in laboratory flames (Ritrievi *et al.*, 1987). Similar calculations for calcium, a well-known oil-derived metal, reveal an equally similar chameleon-like transition: above 2200 K, the element is present mainly as $Ca(OH)_2(g)$; from 1700 K to 1300 K, it shifts to CaO(s); and, at still lower temperatures, it is present as $CaSO_4(s)$. These modelling results are given graphically in Figure 3.12 (Abdul-Khalek *et al.*, 1998). Of course, one must bear in mind, for reasons already mentioned, that the thermal experiences of oil additives very likely differ from those of fuel additives.

Formation I: Composition

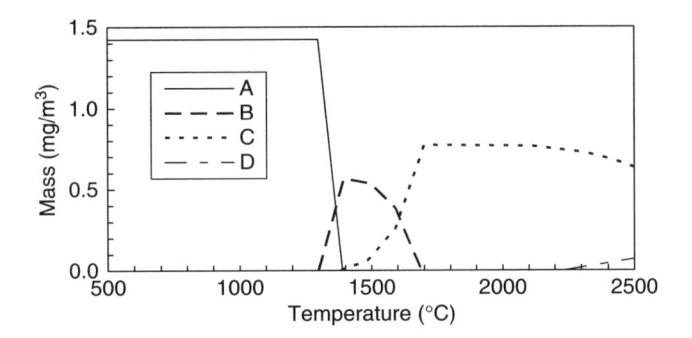

Figure 3.12 Species formed by calcium during combustion and cooling, as suggested by calculations of chemical equilibrium: A, CaSO$_4$(s); B, CaO(s); C, Ca(OH)$_2$(g); D, CaOH(g). Consumption rate of lubricant, 0.5% that of fuel; calcium content of lubricant, 3000 ppm; fuel sulphur content, 300 ppm; air–fuel ratio, 2.78; pressure, one bar (at 150 bar the solidification temperatures increase by less than 100 °C) (Abdul-Khalek et al., 1998).

That metals regularly reach their termini as sulphates is actually quite interesting in view of today's rapidly declining fuel sulphur levels; this suggests subtle shifts in the chemical identities assumed by metals over the last three decades – a topic which has received little attention in the literature. The emission of barium soot-suppressants as barium sulphate (Saito T. and Nabetani, 1973), for example, was hardly surprising in the 1970s, since fuel sulphur levels were so enormous compared with today. If sulphur is now less available or, rather, if the metal – sulphur ratio is higher, then emission simply cannot occur as sulphates. Calcium is notable, amongst this motley collection of metals, for its sulphate-forming ability (Givens et al., 2003); but, in the absence of sulphur, thermodynamic calculations suggest emission as CaCO$_3$(s) rather than as CaSO$_4$(s) (Abdul-Khalek et al., 1998). The chemical forms taken up by metals, whether these be sulphates, oxides, phosphates, etc., determine the moment of gas-to-particle conversion, and hence whether or not particles are likely to lodge in the exhaust system; and, it should be said, also determine the toxicological consequences, one particularity of which is the readiness with which these compounds dissolve in the mucus or surfactant of the respiratory tract (Section 10.3).

Faced with the aforementioned scarcity of studies directly related to internal combustion engines, it seems reasonable to mention, albeit briefly, six other forms of combustion of tangential relevance, which *have* been reported in the open literature and from which suitable information might reasonably be gleaned. In these other fields, ashing elements, either intentionally or unintentionally, are present, along with hydrocarbon fuels, to a greater extent; or, alternatively, they are the central focus of the combustion, since hydrocarbons are not always the fuel, nor indeed is O$_2$ always the oxidant.

A. In biomass combustion, inorganic compounds, not greatly dissimilar to those predicted by chemical equilibria, are generated in the flue gases (Boman et al., 2004). Various potassium-containing compounds predominate, the gas – particle partitioning of which is mediated by chlorine (Baxter et al., 1998); KCl condenses onto K$_2$SO$_4$ nuclei, in the course of which sulphur and chlorine compete for the available potassium (Jiménez and Ballester, 2005). It is believed that potassium salts are present in the flue gas as submicrometer particles, which interact with and deactivate deNO$_x$ catalysts (Larsson et al., 2007). Careful control of combustion temperature may improve the retention of certain environmentally hazardous metals, by their immobilisation or encapsulation within particles (Eldabbagh et al., 2005).

B. Diesel fuels burned in some applications, such as in maritime vessels, are significantly degraded, or adulterated, with inorganic compounds, for example Na_2SO_4, Fe_2O_3, SiO_2, CaO, V_2O_5, Al_2O3, MgO and K_2SO_4; the particle emission consists of various complex oxides or eutectic compounds (Lin C.Y. and Pan, 2001). Heavy fuel oils burned in industrial boilers contain significant amounts of vanadium and nickel (Jang H.-N. et al., 2007). These two metals remain in the vapour phase above 1400 °C and form oxides, although equilibrium calculations suggest eventual emission as sulphates, i.e. $NiSO_4$ and $VOSO_4$. Nucleation modes are prominent.

C. In the industrial-scale flame synthesis of nanoparticles (Bakrania et al., 2007), a widespread difficulty remains in predicting nucleation rates (Rosner, 2005): for example, those of TiO_2 (Kammler et al., 2002) and SiO_2 (Ehrman et al., 1998), and numerous other nanocomposites (Brezinsky, 2002). Silica and carbon black, for example, are often mixed (e.g. in the manufacture of reinforced rubber), and, as the two forms of particle are both flame-formed, their combined synthesis seems only logical (Spicer et al., 1998). The silica nuclei, once formed, may subsequently be accreted and encapsulated by carbon, while in-flame Cl, formed through decomposition of the $SiCl_4$ feedstock, may form HCl, which, in scavenging OH, a strong oxidant, enhances the production of carbon. The flame synthesis of carbon nanotubes, catalysed by iron nanoparticles, is perhaps even closer to the present concern (Vander Wal, 2002).

D. In the incineration of municipal, industrial or medical waste (Buckley et al., 2002), metals undergo gas-to-particle conversion, which is, once again, mediated by chlorine (Wendt, 1994; Lin C.-L. et al., 2005).

E. In the case of flame suppressants, retardants and inhibitors, iron pentacarbonyl ($Fe(CO)_5$) decomposes to form inhibiting gas-phase species, the effectiveness of which is thought to be impeded by an unwanted nucleation of iron particles (Linteris et al., 2004), especially when the dopant is at too high a concentration (Rumminger and Linteris, 2000).

F. Coals are notorious for containing sundry inorganic material such as aluminosilicates, pyrites and calcites, as inclusions (grains), which is emitted in the flue as 'fly ash' (Rietmeijer and Janeczek, 1997), the particles of which consist of silica (SiO_2), alumina (Al_2O_3) and iron oxide (Fe_2O_3). Other notable oxides are those of Ca, Na and Mg (Borman and Ragland, 1998, p. 520). Mineral matter is also notable for forming 'cenospheres': hollow, glass-like spheres of diameter $0.1-50\,\mu m$ (Borman and Ragland, 1993, p. 514). Some compounds decompose on release; others simply vaporise; release may also be assisted by outgassing organic compounds (Baxter et al., 1997). Fate is determined by volatility (Flagan and Seinfeld, 1988, pp. 358–372): some elements resist vaporisation, remaining behind as residual supermicron particles of unburned coal and ash; some elements vaporise and then condense onto particle surfaces; and some elements (<2 % of total ash mass) yield submicron particles, beginning with a nucleation mode (Lind et al., 1994), followed by agglomeration into an accumulation mode. The gas – particle partitioning of arsenic has been particularly studied (Senior et al., 2006).

Inorganic compounds and soot, in the combustion chamber, are interconnected via a web of chemical reactions, of which one aspect of many is the profound effect of metals on the production of soot (Flagan and Seinfeld, 1988, p. 384; Fialkov, 1997, p. 508) – either in soot promotion (e.g. sodium, potassium and caesium) or in soot suppression (e.g. barium). These effects are well documented for laboratory flames, but soot-suppressing fuel additives also have a long commercial history, not only with diesel engines (Kittelson et al., 1978b), but with many other forms of combustion (Howard and Kausch, 1980). Ferrocene, for example, as an additive for aviation fuel, was discovered in the 1950s (Shayeson, 1967), and only subsequently applied to diesel engines (Section 7.2.4); the same compound is also used as an anti-knock agent in gasoline engines (Schug et al., 1990; and Section 8.2.1).

Several mechanisms are posited for the soot-suppressing effects of metals (Howard and Kausch, 1980; Hall-Roberts et al., 2000). Ferrocene, for example, after decomposing, interacts with hydrocarbon combustion chemistry (Hirasawa *et al.*, 2004), with promoting and suppressing effects on PAH (Kasper and Siegmann, 1998), which are, after all, the feedstock for soot. But far greater attention is accorded to ions, found in *all* hydrocarbon flames in many varieties, the concentrations of which are altered drastically by adding to the fuel trace (sub-ppm) levels of *any* easily ionised species, including fuel impurities (Griffiths and Barnard, 1995, p. 112). For example, in combustion, barium undergoes chemiionisation according to

$$Ba + OH \longrightarrow BaOH^+ + e^-, \qquad (3.34)$$

$$BaO + H \longrightarrow BaOH^+ + e^-, \qquad (3.35)$$

where a treat rate of just a few ppm enormously increases the concentration of in-flame ions. Should these ions succeed in attaching to soot particles, then agglomeration will very probably be impeded by electrostatic repulsion, and, since the particles then remain small, a larger surface area is nakedly exposed to oxidants. Other additives might interfere with hydrocarbon ions involved in soot inception (Section 3.2.2), or promote the oxidation of soot by boosting the concentration of OH.

Further still, metals, if encapsulated via the inception process into spherules, will act as catalysts for carbon oxidation, for example according to

$$M_xO_y + C_{(s)} \longrightarrow CO + M_xO_{y-1}, \qquad (3.36)$$

where M is a metal (Pattas and Michelopoulou, 1992). This catalytic mechanism ought not to be too surprising, since it is thought to explain the effectiveness of fuel-borne catalysts in regenerating the diesel particulate filter. When co-deposited in the filter, the chemical form taken by the metal is inevitably a factor: the first step towards the catalytic oxidation of carbon seems not to involve CeO_2, but rather the decomposition of $Ce_2(SO_4)_3$ (Vonarb *et al.*, 2005); others point to the formation of $Ce_xO_yS_z$, a stable compound with no effect, whereas CeO_2 lowers the ignition temperature (Bianchi D. *et al.*, 2005). And activation energies for the oxidation of flame-generated soots are significantly higher than for diesel soots – implying latent catalytic activity by metals co-emitted with the diesel soots. But metals do not always oxidise carbon in a manner that is truly 'catalytic', since the activation energies are not necessarily any lower, even though oxidation rates are higher – as is the case with cerium (Kim S.H. *et al.*, 2005).

Returning, therefore, to the formative stages of the ash fraction, for an additive to be effective in soot suppression, its vaporisation and decomposition must occur at the most opportune moment – that is, with respect to the emergence of soot. Iron cyclopentadienyl, for example, has thermal and oxidative stabilities comparable to diesel fuel, and works as an effective soot suppressant, whereas iron picrate is apparently ineffective: this is adduced to unfavourable release of the iron during the combustion (Du *et al.*, 1998). Similar remarks pertain to iron acetylacetonate (Wong C., 1988). This underscores the fact that the organic portion, or 'ligand', of an organometallic compound is no disinterested carrier for the metal; it has a hand in deciding the decompositional characteristics of the molecule as a whole. Still, it should be admitted that workers do not always agree even on the soot-suppressing effects of the same compound (Ritrievi *et al.*, 1987; Du *et al.*, 1998; Wong C., 1988); and the type of combustion system is also influential, insofar as this decides the moment of gas-to-particle conversion for the metal with respect to the appearance of spherules. Seeding the fuel with nanoparticles of cerium oxide is different still; this oxide, with a melting point of 2200 °C, may not actually decompose during the combustion (Jung H. *et al.*, 2005). Finally, there is no free choice of ligand, insofar as solubility in the base fuel and decomposition chemistry are different aspects of the additive.

3.5.2 Gas-to-Particle Conversion

Ashing material, being particularly characterised by low volatility, *ipso facto* undergoes extremely early gas-to-particle conversion – before its entry into the exhaust system and, in some instances, before even the appearance of soot. This, at any rate, is the *expected* picture, unconfirmed, as yet, by direct observations of the combustion chamber. Still, this is precisely where parallel experiments with laboratory flames and *doped* fuels are profoundly revealing, and ferrocene has provided some of the best available information.

As explained in the previous section, the fates of the carbonaceous and ash fractions are closely intertwined. Ground-breaking experiments have used the photoelectric effect (Section 5.4.2) to distinguish between carbonaceous surfaces and iron oxide surfaces (Kasper *et al.*, 1999). These experiments, the results of which are reproduced in Figure 3.13, highlight distinct differences in particle formation, *throughout the flame*, between doped and undoped fuels. With increasing height above the flame, iron oxide particles appear, apparently disappear and then reappear – making three distinct zones. The accepted explanation for this artifice is the accretion and then oxidation of soot. In comparison with undoped fuels, it turns out that soot inception is accelerated; surface growth is marginally depressed and then marginally accelerated, and soot oxidation is greatly accelerated.

A. In the *nucleation* zone, particles appear earlier and in greater abundance than in undoped flames; but what is the nature of these particles? FeO(g) is, apparently, stable, and thought to be the principle vaporous iron-containing species in the flame. A steady-state, one-dimensional model of classical homogeneous nucleation is illuminating (Ritrievi *et al.*, 1987). This model predicted nucleation of FeO(g), well in advance of soot inception, rather than of Fe(g), which was, evidently, far too sluggish in terms of particle creation. FeO(g) reached enormous supersaturation levels, allowing posited nucleation rates of $10^{15} cm^{-3} s^{-1}$. Nucleation was, however, spatially restricted to a markedly narrow zone, just 3 mm in length, with nucleation rates rising and then falling across this zone through an astonishing four orders of magnitude. Particle numbers within the zone were $10^{11}-10^{12} cm^{-3}$, and the initial sizes were 1.5 nm.
B. In the *growth* zone, iron oxide nanoparticles are thought to act as nucleation centres for carbon: this is the basis for the notion that ferrocene increases the sooting propensity of flames (Hirasawa *et al.*, 2004). Sundry differences exist in spherule sizes and particle numbers between doped and undoped fuels, but the *trends* are similar. Stark behavioural differences are highlighted, however, when growth rates are normalised to surface areas. In the early growth region, doped flames exhibit decidedly lower 'specific growth rates', but in subsequent stages, these rates eventually catch up with, and even exceed, those in undoped flames. A remarkable speculative system, consisting of three mutually dependent yet superimposed factors, has been developed to account for these behavioural differences (Ritrievi *et al.*, 1987). The *first* factor is that surface growth, through the accretion of carbon, is probably reversed by iron-mediated oxidation to some extent. Logically, this might be oxidation of C to CO via the concomitant reduction of FeO to Fe. A continual reduction – oxidation cycle is thereby suggested, wherein Fe is promptly reoxidised, according to the availability of oxidants, back into FeO. The *second* factor is an expectation that Fe, in promoting the accretion of carbon, thereby encourages growth. *Both* of these factors, however, depend on a *third*, since Fe, in order to act in either capacity, must reside at the spherule surface, rather than in the interior. In order, therefore, to avoid encapsulation by continually accreting soot, Fe must, with equal rapidity, diffuse radially outward. Recapitulating, a lucid description might run thus: FeO oxidises carbon; Fe accretes carbon; and Fe diffuses outwardly. Everything depends on the rate of outward diffusion of Fe compared with the rate of accretion of carbon, as promoted by Fe, and reversed by FeO. Estimates suggested the diffusion rate to exceed the surface growth rate for lightly sooting flames, with this situation reversing for heavily sooting flames (Ritrievi *et al.*, 1987).

C. In the *burn-out* zone, accreted carbon is *oxidised*; and this, so it seems, decides predominantly the net emission of soot – although earlier stages are no less critical in deciding what is oxidis*able*. Burn-out may be *complete* or *partial*. In *complete* burn-out, iron oxide particles simply re-emerge with size distributions resembling those in the pre-accretion stage (Kasper *et al.*, 1999). The iron oxide particles may themselves agglomerate, or coalesce if liquefied by the high temperature (Yang G. *et al.*, 2001). A very faint orange-red streak, consisting of 20 nm iron oxide particles, has been observed to escape from the tip of the flame; the particles were spherical, suggesting a preceding or perhaps intermediate liquid state (the melting points of Fe, FeO and Fe_2O_3 all lay below the maximum flame temperature, >1700 °C) (Zhang J. and Megharidis, 1994a,b). In *partial* burn-out, encapsulated domains of iron, known as 'occlusions', are found within spherules; logically, these are the nucleation centres around which the carbon accreted. These occlusions, while not commonly exceeding 20 nm, evidently have no clearly defined lower limit, being detectable down to one nanometre (Zhang J. and Megharidis, 1994a,b). The trajectory of the particle through the flame decides the nature of the encapsulation: spherules passing along the flame axis contain discernible occlusions, and spherules passing close to the flame front contain atomically dispersed iron (Zhang J. and Megaridis, 1994a,b).

The relative proportions of elemental and oxidised iron are functions of the flame type and also the trajectory taken through the flame, insofar as these control the oxidant availability. The nucleation centres are iron oxide (Ritrievi *et al.*, 1987), although others argue, on the basis of oxygen paucity, for the persistence of elemental iron in the early flame, with subsequent conversion, as oxygen becomes available, to iron oxide (Zhang J. and Megharidis, 1994a,b). But it seems reasonable to suppose that such occlusions, if well buried within the carbon matrix as elemental iron, are likely to remain as such; although, following collection of these particles, the existence of oxidised iron is more strongly evidenced (Kim S.H. *et al.*, 2005). This issue is part of a larger question, namely, whether these occlusions are, in effect, Trojan horses, with the ability to oxidise, catalytically, and from within, the carbonaceous material that surrounds them. Nanoparticles generally are known for their exceptional catalytic properties; but 20 nm occlusions may, if fully encapsulated, be red herrings, with labile, atomically dispersed iron acting as the more effective form of catalyst. But the entrapped form of iron, likeliest to oxidise the surrounding carbon, is a question that does not appear to have been explicitly studied.

The present discussion has spotlighted in some detail just one metal, namely iron, as an exemplar; and from this exemplar, a generalised picture of gas-to-particle conversion for metals may be attempted (Miller *et al.*, 2007). Absolutely critical is the moment of gas-to-particle conversion *with respect to soot inception*: if earlier, metal particles serve as nucleation centres for carbon, and thus find themselves trapped as occlusions within spherules; if later, metal particles lodge on spherule surfaces, or metal vapours are perhaps taken up during surface growth. In either case, spherules subsequently enter a late oxidation stage, wherein the encapsulated metals are, to some extent, released. Of course, the details can be expected to vary extensively between metals; for example, palladium particles, although they avidly acquire carbonaceous coatings during the growth phase, are evidently unable to dislodge these coatings in the burn-out phase (Kasper *et al.*, 1999); cerium oxide, in 5–7 nm domains, decorates the surfaces of carbonaceous spherules, but also agglomerates into atypical 'spherules' in which carbon is a minor constituent (Jung H. *et al.*, 2005). But should this generalised picture be correct, then metals are not consistently emitted in any particular form: they reside in the accumulation mode and in the nucleation mode; they form taxonomically distinct particles; they are combined with carbon as occlusions; and they are combined with carbon as atomic dispersions. Obviously, this all-important information as to *dispersal*, while undoubtedly revealing about particle *formation*, is inevitably discarded whenever the ash fraction is quantified simply as whatever is 'incombustible'.

Turning back, at last, to internal combustion engines, the behaviour of ashing elements can be expected to diverge in certain respects from what is observed in laboratory flames, and this, presumably,

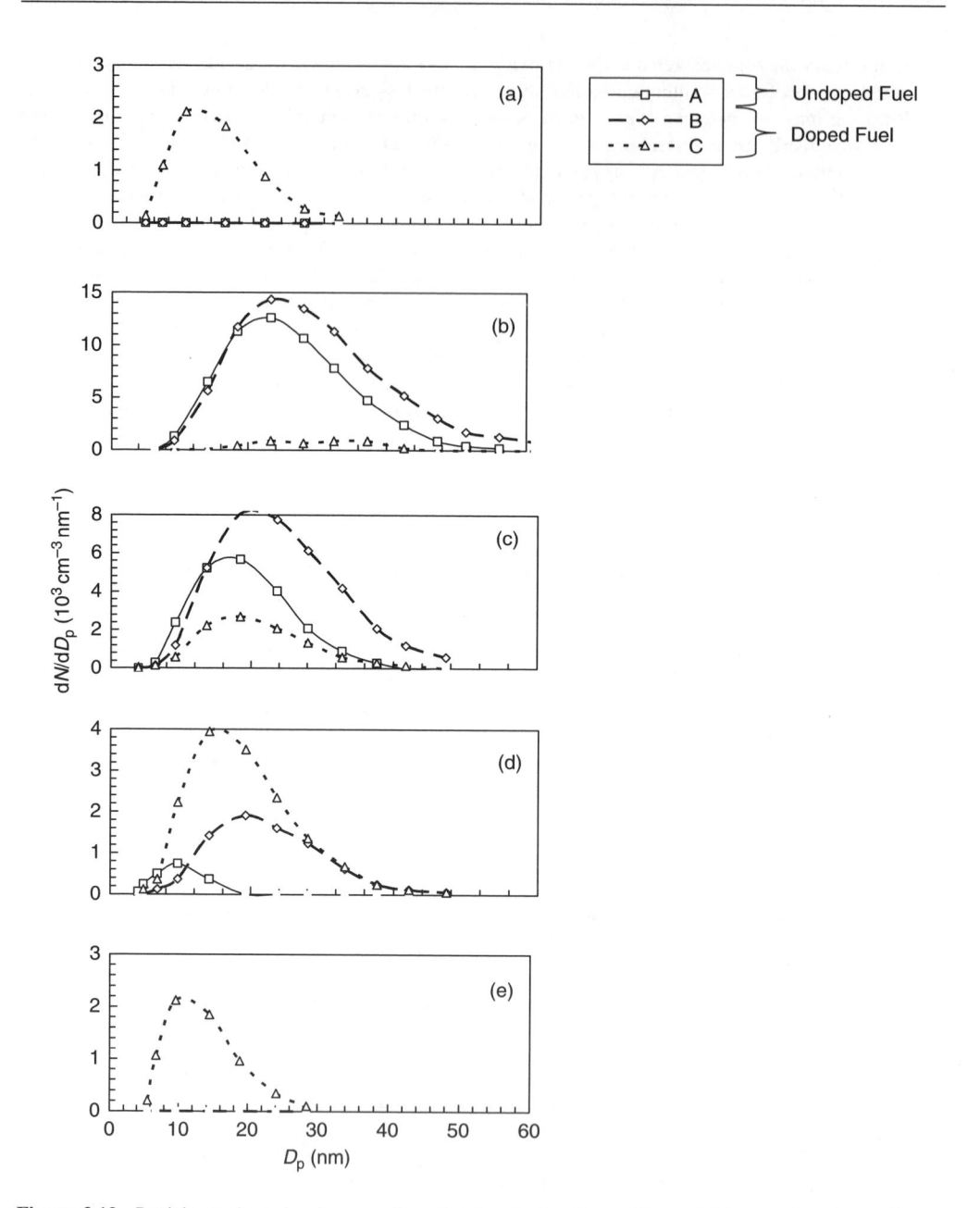

Figure 3.13 Particle numbers showing accretion of carbon by iron/iron oxide nuclei in a methane/argon flame, doped with ferrocene, as a function of height (mm): (a) 20; (b) 35; (c) 50; (d) 65; (e) 75. Particle surfaces: A, carbonaceous, undoped flame; B, carbonaceous, doped flame; C, iron/iron oxide, doped flame (Kasper *et al.*, 1999).

is why different workers do not always agree. Turbulent diffusion flames prevail in the diesel engine, whereas laminar premixed flames, for their very simplicity, are favoured in the laboratory. Nucleation is extremely nonlinear, very sensitive to local conditions of supersaturation (Ritrievi et al., 1987) and, presumably, intimately dependent on heat release rates (Du et al., 1998). The burn-out phase in an engine is often less thorough than in a laboratory flame, as it entails various additional aspects – notably the survival of soot-pockets, following end of injection (Section 3.3.4). The laboratory flame phenomena discussed in this section, though, as yet, await incorporation into the conceptual model of the combusting plume developed in Section 3.3 and delineated, in the mixing-controlled phase, in Figure 3.9. This is left to the next generation of researchers (Miller et al., 2007).

Valuable insights have, nonetheless, been gained by studying solely what is emitted by an engine. With organometallic soot-suppressing fuel additives, pronounced nucleation modes are observed (Kittelson et al., 1978b); this finding, in view of the high treat rates, is perhaps not too surprising. But the same behaviour has been observed at far lower treat rates. With ferrocene, size distributions were relatively unaffected above 30 nm; but, *below* 30 nm, a treat rate of 25 ppm Fe increased the number concentration by a factor of six, and a treat rate of 250 ppm increased the number concentration by a factor of 86 (Du et al., 1998). In gasoline engines too, the iron is emitted in particles smaller than 100 nm (Schug et al., 1990). The logical inference from these observations is the emission of metals as a taxonomically distinct nucleation mode. Of course, when naked metal/metal oxide particles are emitted from an engine in the nucleation mode, this does not on its own say whether the particles formed as such, or whether, according to the understanding developed for laboratory flames, they acquired and then lost carbonaceous coatings.

Size distributions for particles generated using cerium-doped fuels are depicted in Figure 3.14 (Burtscher and Matter, 2000). Four significant findings arise therefrom: (a) the formation of a discernible nucleation mode is guaranteed whenever treat rates exceed a certain *threshold*; (b) the magnitude of this nucleation mode depends on how far the treat rate exceeds the threshold; (c) the nucleation mode

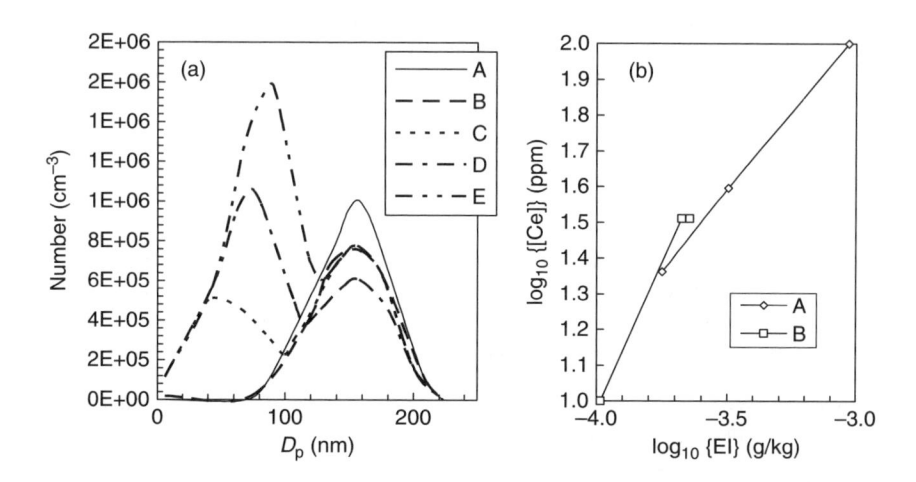

Figure 3.14 Generation of nonvolatile or ash nucleation-mode particles, with cerium fuel additive: (a) treat rates (ppm): A, 20; B, 100; C, 200; D, 500; E, 1000; (b) relation of additive dosage at which ash nucleation occurs to soot emission factor (expressed as grams of soot per kg of fuel burned). Various operating points. Engine: A, single-cylinder, direct-injection, naturally aspirated, rated at 3000 rpm, 4 kW; B, four-cylinder, direct-injection, rated at 2000 rpm, 105 kW, turbocharged. Instrumentation: SMPS (Burtscher and Matter, 2000).

is *nonvolatile*, insofar as it persists at temperatures of 400 °C; (d) a linear relationship exists between the threshold and the amount of co-emitted soot. Taken together, these findings strongly suggest the relative quantities of carbonaceous and ash fractions to be pivotal in deciding whether an ash nucleation mode appears – a remark that appears broadly applicable insofar as it encompasses more than one model of engine (Burtscher *et al.*, 1999). Others have determined, experimentally, a metal/carbon threshold, by mass, of $\sim$2 % (Haralampous *et al.*, 2004) or 1.3 % (Miller *et al.*, 2007). Logically, the take-up of metal vapour by growing spherules would tend to restrain metal saturation in the combustion chamber, and thereby act to suppress metal nucleation. Should this idea be correct, then wholesale soot reduction, without corresponding efforts to restrict the entry of metals into the combustion chamber, could be inadvisable, i.e. this might inadvertently provide the right conditions for metal nanoparticles to form. This would seem to be the case nowadays if the threshold is crossed with a fuel-borne iron concentration of 20 ppm (Miller *et al.*, 2007), as such a concentration is eminently possible if only through inadvertent contamination.

The discovery of prominent ash nucleation modes highlights an intriguing question: are metals less adept than spherules at agglomerating into the accumulation mode? (The spherules, in fact, were the original nucleation mode.) This want of conviviality remains unexplained, but one obvious conjecture is that metals nucleate much later in the combustion process, perhaps in the late burn (Section 3.3.5): they survive as nucleation modes simply because less opportunity has presented itself for agglomeration: perhaps volatilities impede gas-to-particle conversion until late in the cycle. This conjecture is, however, diametrically opposed to the previously discussed experiments which demonstrate nucleation among metals *prior* to the emergence of spherules.

Finally, gas-to-particle conversion, with many metals, is essentially complete prior to emission at the tailpipe; but this understanding should be tempered by careful consideration of exhaust temperature and also of the chemical compounds formed by these disparate elements *en route* (oxides, sulphates, carbonates, phosphates, chlorides, ...), which vary greatly in their abilities to persist as vapours. This topic is poorly reported, but there is reason to believe that, in some instances, *partitioning* occurs within the exhaust system between the gas and particulate phases, for example partitioning of boron (Takeuchi *et al.*, 2003), nickel, chlorine (Vincent *et al.*, 1999) and barium (Kittelson *et al.*, 1978b), so that some ash compounds are discharged from the tailpipe partly as vapours and partly as particulate. Some major factors in deciding chemical form are probably the relative amounts of sulphur and phosphorus available in relation to the metals, the first element having already been mentioned in this respect (Section 3.5.1).

3.6 Organic Fraction

Organic compounds in the exhaust have successfully evaded in-engine chemical reactions that would otherwise destroy them. These fortunate few, during their transit through the exhaust system, remain predominantly in the gas phase; yet temperatures remain sufficiently high for further chemical reactions, in which subtle and nuanced changes to organic speciations take place. Full-scale transfer into the particulate phase awaits significant cooling, meaning, in effect, the moment of discharge into the exhaust plume. It should moreover be noted that, by virtue of their heavier molecular masses and lower volatilities, diesel-fuel hydrocarbons more readily undergo gas-to-particle conversion than those in gasoline.

3.6.1 Preparatory Chemical Reactions

From a chemical-reaction standpoint, the organic fraction is composed of two distinct subgroups: 'petrogenic', or compounds native to the fuel, and 'pyrogenic', or newborn compounds, i.e. those

Formation I: Composition

synthesised from native compounds. The emission mechanism for native organic compounds, being comparatively straightforward, is herein mentioned only in passing. The emission mechanism for the synthesised organic compounds, links, in effect, the native organic compounds to soot; hence this subgroup might equally be styled 'organic intermediates'. This distinction matters because, if pyrosynthesis is influential in the emission of organic particulate, then control efforts directed at fuel speciation and fuel reformulation will very likely be misplaced, as the targeted compounds are *secondary*. Indeed, some organic compounds found in exhaust gas, being wholly undetectable in the fuel, *must* be generated within the combustion chamber.

That these intermediates, or partial combustion products, are found in the organic fraction is unsurprising: Section 3.2.2 detailed how, in soot formation, fuel molecules are first cracked, or fragmented, into simpler molecules, and how these simpler molecules are then built into new structures of ever-higher molecular masses – structures which act as soot precursors. The first step is irrefutable, since light hydrocarbons (e.g. methane) habitually appear in exhaust gas even when wholly absent from the fuel (Andrews *et al.*, 2000b), so that the HC emission is composed of native *and* newborn compounds (Bertoli *et al.*, 1991). When the second step, i.e. soot formation, is interrupted or forestalled, such as by insufficient temperature or quenching, soot precursors are then emitted instead as organic particulate: for instance, the petrogenic PAH, which are of molecular mass 128–206 amu, are distinguishable from the pyrogenic PAH, which are of mass 228–$\gtrsim$350 amu (Dobbins *et al.*, 2006). This understanding is soundly supported by rapidly developing models (Akihama *et al.*, 2001). For these reasons, HC, organic particulate and carbonaceous particulate are all interlinked by a web of chemical reactions, as sketched conceptually in Figure 3.15. As with soot, the organic compounds that form thus are also subsequently oxidised, especially in the late combustion phase (Section 3.3.4), and to a large degree: for example, PAH, at some point between the combustion chamber and the exhaust manifold in the expansion and exhaust strokes, are culled by an order of magnitude (Du *et al.*, 1984).

In this section we suggest the likely course of events within the engine in three ways: via pyrolytic reactions in laboratory flames, at the very moment of soot inception; via pyrolytic reactions in laboratory

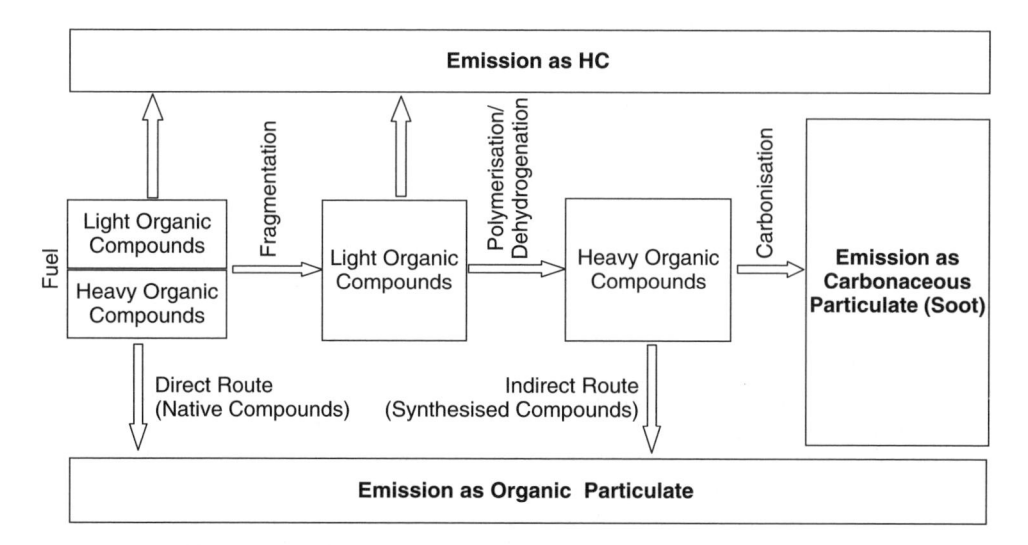

Figure 3.15 Relationship between original organic compounds in the fuel and eventual emission as HC (gaseous), organic particulate and carbonaceous particulate. (See also Figure 3.3.)

furnaces, in which fuel compositions are compared with the organic compounds that emerge; and via in-engine 'survivability' studies, or comparisons of fuel speciation with particulate composition. A final section draws on these discussions to elucidate in-engine phenomena.

Pyrolytic Reactions in Laboratory Flames

The emergence of PAH of ever-increasing structural complexity on the pathway to soot is well attested – as detailed in Sections 3.2.2 and 3.2.3. PAH, with increasing distance along a flame, become increasingly abundant, reach a peak and, at around the moment that soot emerges, decline perhaps though oxidation, perhaps through thermal decomposition, perhaps through conversion into soot, or perhaps through all three (Ciajolo *et al.*, 1994). The ratio of aromatics to aliphatics also increases until the moment of nucleation, and thereafter declines (McKinnon *et al.*, 1996). Some of this aliphatic carbon is present as separate molecules; some resides as branches or side chains on aromatic molecules. In fact, aliphatic molecules form a substantial fraction of the condensable organic contingent extracted from flames (Ciajolo *et al.*, 1994).

Organic compounds extracted from flames and found to be in the particulate phase might actually have been in the gas phase at the point of sampling. But this is not always so. Extracted particles are *already* coated with organic compounds such as PAH, the molecular weights of which increase with distance along the flame (Hepp and Siegmann, 1998). Logically, this retention is an aspect of soot formation, wherein these compounds simply represent partially used feedstock. The close relationship between the graphiticity of the underlying soot spherule and the aromaticity of the accreted organic compounds is intriguing (Yan *et al.*, 2005b), and expresses, perhaps, the readiness of these organic compounds for assimilation by the turbostratic spherular microstructure (Sections 6.3.1 and 6.3.3), i.e. these compounds have not, as it were, fully matured to the point of incorporation, but have been caught, or frozen, in the process of evolving into larger, more aromatic or, rather, more carbonaceous, structures. Moreover, evidence exists that late-adsorbing unburned fuel organic compounds are more readily lost from soot than pyrosynthesised organic compounds: hence, the two groups are not associated with spherules in the same way (Simo *et al.*, 1997). Be that as it may, from these observations it is but a short step to suggest, once the flame tip is reached, the survival of some PAH (and other organic compounds) into the flue, partly in the gas phase and partly in the particulate phase, with subsequent capture and quantification as organic particulate.

In-cylinder measurements tauntingly suggest the presence of particle entities immediately *prior* to the formation of carbonaceous spherules – entities that might not be fuel droplets (Alatas *et al.*, 1993). Weightier evidence, though, some of which is long-standing (Howard and Bittner, 1983), has been amassed from laboratory flames and *gaseous* fuels. Samples extracted from the region straddling fuel decomposition and spherule formation and examined under the electron microscope contain solitary, diffuse-appearing globules, with liquid-like or tar-like consistencies, and molecular masses of 200–300 amu (Dobbins *et al.*, 1996). *Unlike* spherules, these globules are poorly defined (although shape is probably an aspect of impaction angle), and semi-transparent or translucent under the electron beam; they possess a disordered, uncrystalline structure. They are just 2–5 nm (Vander Wal *et al.*, 1997), ~5 nm (Oh *et al.*, 2005) or 10–25 nm (Öktem *et al.*, 2005) in size, although such globules as large as ~100 nm are known (Xu F. *et al.*, 1997). Their greater polydispersity, compared with spherules, is the logical outcome of their fluid consistency, which allows wholly coalescent agglomeration. These observations are bolstered by laser-imaging experiments, in which, early in the flame, entities exist that scatter but do not adsorb light: this is consistent with the notional existence of *noncarbonised* particles (Yang B. and Koylu, 2005) connecting the regions of fluorescence (from PAH) and incandescence (from soot) (Vander Wal *et al.*, 1997), with the suggestion of an organic composition – possibly aromatic molecules of no more than two rings (d'Alessio *et al.*, 1998).

It should be admitted that by the very appellation 'precursor particle', or 'proto-particle', one adopts a teleological explanation for the formation of the spherules; but the dossier of evidence is rather more than circumstantial, i.e. it does not rest solely on the appearance of these proto-particles before spherules. Dark-field and bright-field transmission electron microscopy show increasing crystallinity (decreasing amorphousness), as if on the pathway to soot (Vander Wal, 1998); the carbon – hydrogen ratio increases with residence time in the flame from 1.75 to >2.5 (Oh et al., 2005), and some micrographs show an opaque crystalline kernel, i.e. of layered microstructure, coated with a translucent tar-like material (Oh et al., 2005), as if these proto-particles were freeze-framed in the moment of transition. Yet spatial separation by a zone of transition is not essential and, depending on the type of flame, soot and proto-particles can coexist, with the latter scavenging the former (Dobbins, 2007). Finally, the proto-particle composition passes along the same sequence of 'stabilomers', i.e. thermodynamically favoured molecules, as that described in Section 3.2.2 (Dobbins et al., 1998). Still, the evolution of proto-particles as large as 100 nm into spherules (Xu F. et al., 1997) is not easily explained by mere dehydrogenation or carbonisation, as spherules are routinely smaller. Modelling of these proto-particles is an emerging field (d'Anna and Kent, 2006). But whether these proto-particles, or organic nanoparticles, are able to escape into the exhaust of an internal combustion engine *in this form* remains to be ascertained.

Pyrolytic Reactions in Laboratory Furnaces

The spatial–temporal picture of organic intermediates in flames is closely echoed in furnace experiments – vessels in which various fuels are subjected to high temperatures in inert environments, so that pyrolysis, rather than oxidation, is promoted. Some data along these lines are provided in Figure 3.16(a) (Tosaka et al., 1989): some organic compounds are discharged in the gas phase, others in the particulate phase, but what actually counts is the filter deposit, the chemical analysis of which was conducted in the same way as for motor vehicle exhaust.

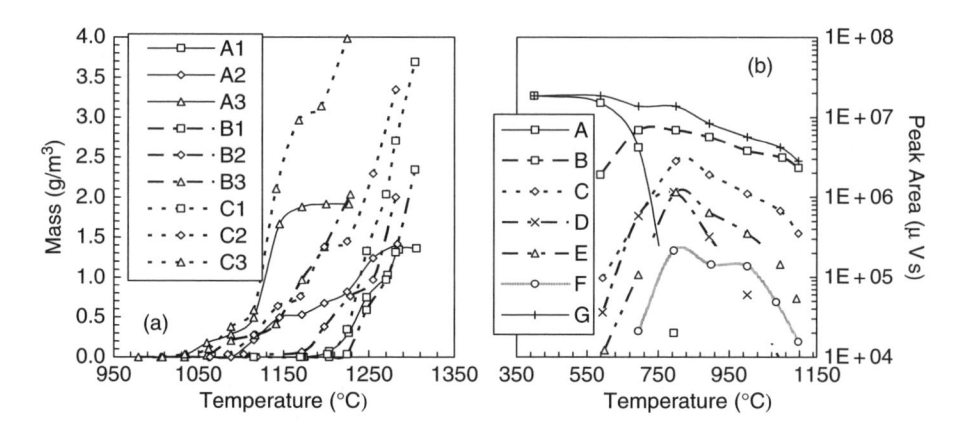

Figure 3.16 Cracking and pyrolysis of hydrocarbons in a laboratory furnace. (a) Hydrocarbon feedstock: 1, CH_4; 2, C_2H_6; 3, C_3H_8. Mass emission of particulate in effluent: A, organic; B, carbonaceous; C, total. Gas mixture: 3 % fuel in nitrogen. Space velocity, 5000/hr (Tosaka et al., 1989). (b) n-Hexane feedstock, emission quantified according to chromatogram peak area: A, *n*-hexadecane; B, low-boiling-point hydrocarbons; C, benzene; D, alkylbenzene; E, 2-ring PAH; F, 3-, 4- and 5-ring PAH; G, total. Gas mixture: 3 % fuel in nitrogen. Space velocity, 5000/hr (Tosaka and Fujiwara, 2000).

The maximal discharge of the organic fraction from such furnaces is thermally proximate to the appearance of the carbonaceous fraction; after this point, the former decreases, just as the latter increases. This transition is apparent in particulate colour also, which shifts from yellow to brown, and finally to black, and in particulate consistency, as individual particles cannot be resolved in the yellow and brown particulate, suggesting a liquid or semi-liquid nature, whereas discrete particles are observed in the black particulate, this being archetypal 'soot'. Even the simplest hydrocarbons, such as methane, ethane, acetylene and ethylene, will form organic particulate, and three-, four- and five-ring PAH are discovered in the effluent; meanwhile, triple-bonded hydrocarbons display lower soot onset temperatures than double-bonded, and double-bonded lower than single-bonded. Figure 3.15(b) reveals how three hydrocarbons (1-hexene, cyclohexane and n-hexane) are cracked into lighter hydrocarbons (acetylene, methane and ethylene) and then reformed into heavier hydrocarbons (benzene and multi-ring PAH): the abundance of each compound rises, in succession, to a peak, the magnitude of which is lower than that of the preceding, lighter compound (Tosaka and Fujiwara, 2000).

But, stepping over these details, two essential points should be emphasised: first, the emergence of *nonfuel* organic compounds in *intermediate* temperature ranges, and second, the somewhat *higher* temperatures required if soot is to emerge.

Survivability in the Engine

Speciative comparisons of fuel-related organic compounds and particulate-related organic compounds, are, perhaps, the most obvious investigative method – what are called 'survivability' studies. In effect, these are measures of the efficiency with which native compounds are combusted, but also, hopefully, suggestive of chemical reactions that forge new compounds, as survivabilities exceeding unity, that is, emission at higher rates than fuel concentrations can account for, catch pyrosynthesis red-handed, as it were. Unfortunately, evidence so conclusive is nonexistent, but survivabilities of ~0.1–0.2 % for PAH and ~0.01–0.1 % for alkanes are reasonably well established, with considerable variation from one compound to another: heavier compounds are more persistent, and reasonable evidence exists that survivabilities are independent of fuel concentrations. More pertinently, such studies conclude rather strongly that there is a dominance of the organic particulate by native or fuel-borne compounds, rather than pyrosynthesised ones (Abbass *et al.*, 1989b, 1991a, 1991b; Andrews *et al.*, 1998b). This is strengthened by operating engines on fuel that is relatively PAH-free (Abbass *et al.*, 1989c) – a similar conclusion, incidentally, is reached in fuel-doping experiments: pyrosynthesis of PAH is of little consequence in the emission of PAH (Frølund and Schramm, 1997), although it should be said that others have reached an opposite conclusion (Hori S. and Narusawa, 2001).

Unfortunately, definitive statements as to native and pyrosynthesised contributions to organic particulate, as derived from survivability studies or fuel-doping experiments, seem unlikely to emerge, because of significant confounding by engine oil (Abbass *et al.*, 1987) and exhaust-system deposits (Abbass *et al.*, 1988), the action of which as storage reservoirs for organic compounds has been well demonstrated, with the release of these stored compounds at some later stage, according to duty cycle. Organic compounds simply outgassed from the oil film on the cylinder walls, for instance, which escape direct exposure to the flame, may simply overwhelm those compounds emitted as a direct consequence of the combustion. To establish the emission mechanisms fully, a global model is required for the transport of organic compounds: one encompassing the fuel composition, the combustion chamber, the blow-by gases, the sump oil and exhaust-system deposits (Frølund and Schramm, 1997)

Inside the Engine

The synthesis and discharge of organic intermediates, described in the previous subsections, elucidates similar phenomena at work in internal combustion engines, where, as stated in the opening remarks to this section, soot-forming reactions may also be interrupted. Cylinder sampling shows, with advancing crank angle, the fragmentation of fuel hydrocarbons, the formation of new nonfuel hydrocarbons (including aromatics) and, finally, the emergence of soot (Fujiwara *et al*., 1993). Organic intermediates are able to survive because, inside internal combustion engines, charge preparation is imperfect (Ikegami *et al*., 1983) i.e. the temperature or mixing is not always ideal, leading to 'bulk quenching' – tied, not unexpectedly, to the length of the ignition delay (Kweon *et al*., 2002, 2003a), and 'wall quenching', when the combusting plume finds itself pent up within the combustion chamber (Kato S. *et al*., 1997). Naturally, these factors also help the *fuel* to survive the combustion unscathed, i.e. through mixing beyond the lean flammability limit (Tan P.Q. *et al*., 2007). The escape of such compounds into the exhaust is not guaranteed: they may be oxidised later; it is the main burn which they have evaded.

Armed with these insights, it becomes possible to explain an eminently reproducible characteristic of the particulate emission: 'dry', or predominantly carbonaceous, at high engine load, and 'wet', or predominantly organic, at low engine load, so that the two fractions exhibit, as it were, contrary trends (Ikegami *et al*., 1983; Kweon *et al*., 2002; Lombaert *et al*., 2002; Kweon *et al*., 2003a). This understanding is borne out in field measurements on heavy-duty trucks (Shah *et al*., 2004). Moreover, a minimum is seen in the particulate emission at mid-load (Shimoda *et al*., 1987; Kweon *et al*., 2003c). Recognising that engine load is, in some sense, a surrogate parameter for combustion temperature, the parallel to the aforementioned furnace experiments becomes clear: at high load, chemical reactions proceed all the way from fuel to soot, but at low load, temperatures are insufficient for soot formation; the soot-forming reactions are hence interrupted, and this leads to the emission of organic particulate.

Before laying aside in-engine phenomena, a further formative influence on the organic fraction should be mentioned: *deposition*. Wall interactions are imputed often enough in the formation of the carbonaceous fraction (Section 3.4), but, since such sooty deposits are *porous*, organic compounds are very probably taken up temporarily within these pores, and sheltered from the combustion by the thermal boundary layer, to be later released. This idea seems not too wayward, since precisely the same mechanism, as a source of HC emissions, is well recognised in spark-ignition engines: high pressures at TDC favour absorption, just as low pressures at BDC favour desorption. Organic compounds are present within these deposits to the tune of several tens of per cent (Andrews *et al*., 1992); indeed, the composition may be less rather than more carbonaceous than what is emitted by the engine, and grey and soft, rather than just hard and black (Yamaguchi *et al*., 1991). Otherwise, the emission of particle-phase PAH from a *motored* engine is hard to dismiss (Du *et al*., 1984). Actually, this topic opens up some under-appreciated subtleties: probes inserted into laboratory flames accumulate PAH of large but not small molecular weight. These stored PAH do not persist as such, but undergo *in situ* chemical reactions, driven by the temperature of the surface (Smedley *et al*., 1995). This leads to the conjecture that soot deposits arise not just through the deposition of soot, but also through the take-up of organic compounds, with subsequent conversion into soot.

3.6.2 Chemical Reactions in the Exhaust

The chemical composition of the organic fraction, while largely decided in the engine, is not thereafter constant: chemical reactions continue, albeit at rapidly diminishing rates, during ejection of the exhaust gas and *en route* towards the tailpipe. In deNO$_x$ aftertreatment (where hydrocarbons are designedly injected into the exhaust stream), ongoing homogeneous gas-phase reactions are well known:

significant differences exist between the hydrocarbon speciation prevailing at the point of injection and hydrocarbon speciation seen later at the catalyst entry (Collier and Wedekind, 1997). Presuppositions of compositional constancy in the exhaust system are therefore unwise. The temperature threshold, moreover, below which chemical reactions may safely be neglected is unclear; and this, largely, is because susceptibility to ongoing reactions depends on the species in question – of which there are many hundreds. *Some* hydrocarbons are probably affected even when conditions are not exceptionally warm. The exhaust system is not simply a conduit: it is also, in effect, a chemical reactor of immense complexity, from which the extraction of reproducible samples, particularly of trace species, is no straightforward matter. As might be expected from concerns of mutagenicity, the bulk of the work, in this field, relates to *PAH*.

The chemical reactions to which PAH fall prey are particularly quenched below 250 °C (Williams R.L. *et al.*, 1985). Above this temperature, evidence for PAH decomposition is considerable, even for residence times of a few tens of milliseconds – with corresponding underestimations of true emission rates if samples of exhaust gas are allowed to dally within sample lines. The threshold of 250 °C, however, is only a guide, as other factors, including exhaust composition, are influential: individual compounds vary widely in their 'ease of oxidation', *BaA*, *BaP* and *Per* being particularly unstable. PAH emissions with two different exhaust systems, when attached to the same engine, differed by up to an order of magnitude (Williams R.L. *et al.*, 1985). Because of in-exhaust reactions, concentrations of PAH at the exit of a sampling line did *not* reflect actual concentrations at the tailpipe. The consequences, therefore, of ignoring ongoing chemical reactions in the exhaust system are clearly demonstrated.

An extensive, but regrettably unsystematised literature concerns PAH and NPAH or, rather, the interrelationship(s) between these two compound groups (Barale *et al.*, 1992), wherein the second arises from the first via the phenomenon of 'nitration' – for example, the conversion of *Pyr* into *1-NPyr*. This is an extremely important topic that deserves to be examined in some detail, if only because NPAH are considerably more mutagenic than their forebears, PAH. But this is where the line of demarcation between particulate *characteristics* and particulate *collection* is hugely problematic (Smith D.J.T. and Harrison, 1998): was this conversion a genuine aspect of the emission process, or unwantedly provoked by the sampling and collection? Atmospheric scientists face very similar difficulties: the compound might be created on the filter, and not truly present in the atmosphere (Feilberg *et al.*, 1999).

A. *Evidence relating to genuine transformations, in the combustion process or exhaust stream.* (a) In the particulate phase, the concentration of *1-NPyr increases* by a factor of forty between the combustion chamber and the exhaust stream in the expansion and exhaust strokes, whereas the concentration of *Pyr decreases* by a factor of ten (Du *et al.*, 1984). (b) Diesel oxidation catalysts (Section 7.8.1) and diesel particulate filters (Section 7.8.2) *decrease* the mass of particulate but *increase* the mutagenicity. (c) Particulate collected by sample plugs mounted inside a combustion bomb and by a filter mounted in the exhaust gas of the same bomb contains similar amounts of *Pyr* and *1-NPyr* (Geyer *et al.*, 1987).

B. *Evidence relating to genuine transformations in the exhaust stream or artefactual transformations on the filter.* (a) NO_2, artificially added to diluted exhaust at several tens of ppm, increases by severalfold the mutagenicity of the filter deposit, and the concentration of *1-NPyr* within it (Bradow *et al.*, 1982). (b) An engine supplied with nitrogen-free oxidant rather than with air emits very little NO_2, and little *1-NPyr* can be found on the filter (Herr *et al.*, 1982).

C. *Evidence relating to artifactual transformations on the filter.* (a) NO_2, at concentrations below 150 ppm, if passed for one hour through a filter deposit at face velocities and temperatures representative of the real emission process, shows efficient conversion *on the filter*, of *Pyr* into

1-NPyr (Herr *et al.*, 1982). (b) *BaP*, on a glass filter, if exposed to just *one* ppm NO_2, for eight hours in an ambient atmosphere, is converted to 6-*NBaP* (cited by Funkenbusch *et al.*, 1979). (c) On re-exposing a fouled filter to exhaust gas, the concentration of *BaP* decreases, as the concentration of *1-NPyr* increases; the longer the residence time (up to an hour or so), the more pronounced are these trends (Saito T. *et al.*, 1982).

The transformation may thus take place either on the filter or at earlier moments in the emission process; or still earlier, in the combustion; although it should be conceded that filter deposits are artificial 'substrates', on which particle-bound PAH need only wait for nitrating species to stream past. The nitrating agent in this transformation is inconclusively identified, but plausible correlations involving NO_2 are known – plausible because of the likely impact of oxidation catalysts and exhaust-gas recirculation on the concentration of this gas. Particulate-phase NPAH does not correlate with particulate-phase PAH, but *does*, as it happens, as illustrated in Figure 3.17, correlate with a square-root product of NO_x and *total* (gas and particle) PAH (Rijkeboer and van Beckhoven, 1987). Such intercorrelations are underpinned by various reactions (Bamford *et al.*, 2003), in which the nitrating agent is not necessarily NO_2; it could, instead, be HNO_3, for which the first merely provides a surrogate measurement. Whether nitration occurs in the gas phase, with subsequent gas-to-particle conversion, or is limited by reaction kinetics, the availability of nitratable species or the availability of the nitrating agent, remains unknown, while exhaust temperature, dilution ratio and residence time seem to exert variable reinforcing or restraining effects, with unpredictable results (Bradow *et al.*, 1982). Much, therefore, in relation to nitration, remains conjectural. But, notwithstanding these mechanistic uncertainties, the simple practical consequence of especial concern is the ineffectiveness of dilution in suppressing nitration, as the hottest conditions and the highest concentrations of NO_2 are not, evidently, required for the phenomenon to arise.

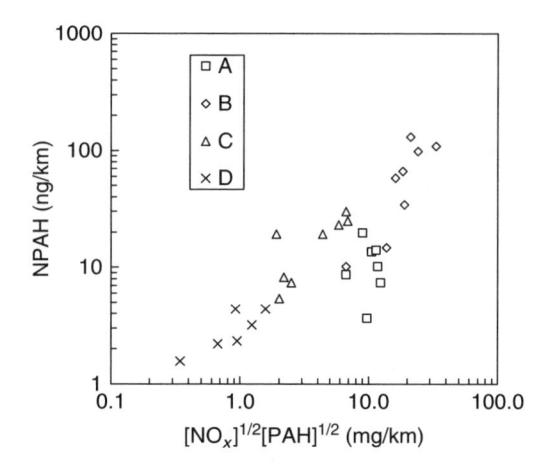

Figure 3.17 Correlation between particulate-phase NPAH and total (gas-phase and particulate-phase) PAH. Vehicles (all gasoline): A, 1.6-litre, carburettor, lean-AFR, no catalyst; B, 1.5-litre, multipoint-injection, lean-AFR, leaded fuel; C, same as B but with oxidation catalyst and unleaded fuel; D, 1.8-litre, mechanical fuel injection, stoichiometric-AFR, three-way catalyst, unleaded fuel. Duty cycles: 'mixed-mode' driving, i.e. transient and cruise. Total PAH samples collected with cryosampler at $< -20\,°C$; particulate-phase NPAH collected with CVS; dilution ratio 5–25; all extractions carried out with acetone (Rijkeboer and van Beckhoven, 1987).

The likelihood of any coherent nitration theory is frustrated by the latent role of *metals*. The nitration reactions are thought to be catalysed by iron (Enya *et al.*, 1997), a likely element in the particulate emitted by motor vehicles; increases in mutagenicity seen with a barium fuel additive were not always statistically significant (Draper *et al.*, 1988), but order-of-magnitude increases in mutagenicity were seen with a manganese fuel additive and adduced to an enrichment in NPAH (Draper *et al.*, 1987). Other cross-effects with the ash fraction are suspected: as hypothesised (Draper *et al.*, 1988), gas-phase PAH are possibly, through the presence of barium, more effectively taken up by the particulate phase, by which they evade destructive oxidising reactions, allowing their survival at higher concentrations – with *BaP*, by as much as a factor of three.

That mutagenicity is artifactually enhanced when exhaust gas is allowed to stream for long periods through a particulate deposit seems all too probable, but similar reactions also happen prior to collection, and not necessarily when particles are in transit: all exhaust systems contain dirty wall deposits. This understanding suggests that aged deposits are more inimical than freshly emitted particles. Contrarily, it might be argued that very similar transformations are predestined: *atmospheric* reactions between PAH, OH, NO_2, NO_3 and HNO_3 (Fan *et al.*, 1995; Feilberg *et al.*, 1999) might actually account for the majority of environmental exposure to NPAH. Ultimately, therefore, the question of whether NPAH emerge immediately from tailpipes or form subsequently, in the atmosphere, might prove immaterial from a public health standpoint (Matsumoto *et al.*, 1998).

3.6.3 Gas-to-Particle Conversion: Models

Models for the transfer of gas-phase organic compounds into the particulate phase, with especial relevance to the dilution experienced by the exhaust gas as it enters the air, are the subject of the present section. Except, possibly, for specific instances when their concentrations are exceptionally high (Vaaraslahti *et al.*, 2005), nucleation, according to emerging research, is not driven by organic compounds directly, although these compounds do facilitate particle growth; this topic is discussed in Section 3.7.2. Rather, the best-known models relate to adsorption and condensation – pivotal phenomena in surface chemistry: the use of activated carbon as a hydrocarbon scrubber (Rodríguez-Reinoso, 1997) presents a practical application, and an illuminating parallel to the current discussion. Hence, there is no shortage of models of a fundamental nature that may be applied and adapted to the present case – of which, here, only a small sample can be provided, to exemplify, as it were.

But before proceeding to the details, a difficulty immediately arises of a wholly expected nature, namely, great variations in gas-to-particle susceptibility among the hundreds of organic compounds. While hundreds of compounds could certainly, these days, be incorporated into models, a few archetypal ones are generally selected, and this is normally sufficient to reproduce faithfully the experimentally observed trends. In one of the earliest attempts to model the condensation of hydrocarbons within a dilution tunnel, a family of 'condensation curves' was constructed, incorporating, for example, hexadecane, nonadecane and fluorene, the saturation vapour pressures of which were calculated via standard methods or obtained from the literature (Khatri *et al.*, 1978). These saturation vapour pressures were then mapped onto conditions in the dilution tunnel such that combinations of temperature and hydrocarbon concentration likely to reach saturation could be discovered. The model successfully captured, with falling exhaust temperature and also falling air temperature, the widening range of dilution likely to promote condensation.

From Section 2.2.4, it will be recalled that condensation becomes possible whenever the saturation, i.e. the ratio of ratio of partial pressure to saturation vapour pressure, exceeds unity: hence, this is the condition which must first be established. The saturation vapour pressure of any particular species of hydrocarbon, $p_{HC(sat)}$, can be expressed as (Plee and MacDonald, 1980)[8]

$$p_{HC(sat)} = p_{atm} \exp\left[\frac{h_{fg}}{\Re}\left(\frac{1}{T_{mix}} + \frac{1}{T_{bp}}\right)\right], \tag{3.37}$$

where p_{atm} is the pressure in the dilution tunnel (slightly above atmospheric), h_{fg} is the enthalpy of vaporisation (assumed constant), T_{mix} is the temperature of the air–exhaust mixture and T_{bp} is the boiling point of the hydrocarbon. (T_{mix}, being a function of the dilution ratio D and the exhaust temperature T_{exh}, can be computed according to Equation (5.1).)

The partial pressure of the hydrocarbon p_{HC} is given by $p_{HC} = (\chi_{HC}/D)p_{atm}$, where χ_{HC} is the hydrocarbon fraction *prior* to dilution. The saturation is then simply obtained from $S = p_{HC}/p_{HC(sat)}$:

$$S = \left(\frac{\chi_{HC}}{D}\right)\exp\left[-\frac{h_{fg}}{\Re}\left(\frac{1}{T_{mix}} - \frac{1}{T_{bp}}\right)\right]. \tag{3.38}$$

Figure 3.18 depicts four likely scenarios using such an analysis (Plee and MacDonald, 1980).[9] The salient and consistent characteristic of S, as a function of D, is its rise and fall: this stems from competition between the two functions in Equation (3.38), the quotient and the exponent, which express, respectively, the effect of D on p_{HC} and $p_{HC(sat)}$. Condensation is suppressed at low D, because T_{mix} is too high; and also suppressed at high D, because p_{HC} is too low. Consequently, an intermediate range of D exists in which condensation is at its likeliest. From a detailed comparison, however, with real data, condensation was ruled out as being responsible for the experimentally observed trends,

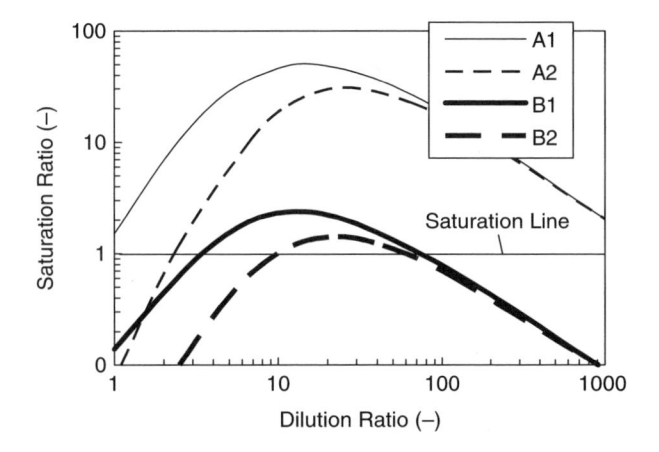

Figure 3.18 Model of saturation ratio S as a function of dilution ratio D. Temperature of dilution air: 300 K. Exhaust temperature (T_{exh}): 1, 480 K; 2, 600 K. Boiling point of hydrocarbon (T_{bp}): A, 810 K; B, 720 K. Adiabatic dilution and constant specific heats are assumed. Hydrocarbon fraction before dilution: 40 ppm (Plee and MacDonald, 1980).

---

[8] An instructive issue arises over the sign of the exponent in Equations (3.37) and (3.38), the implications of which the reader is invited to explore in the original paper.
[9] This figure is reproduced from Plee and MacDonald (1980), but is itself taken from Kittelson and Dolan (1978).

other than under relatively high hydrocarbon concentrations (Plee and MacDonald, 1980); the same conclusion was reached by others (Kraft et al., 1982).

With adsorption, too, no shortage of models is evident: for example, one paper considered no less than *seven* before finally settling on the tried-and-trusted BET model, because of its ability to handle multilayered and multicomponent adsorption – physical as well as chemical (Clerc and Johnson, 1982). (Conversely, the BET model is used in a standard laboratory procedure to determine surface areas, i.e. by allowing a certain gas to adsorb.) This model was reasonably successful in predicting the experimentally determined quantity of organic particulate, using physical adsorption, and heptadecane, a hydrocarbon of mid-range boiling point, as generally representative. Shortcomings in the model seemed to result from interference by condensation. Experimental measurements of aromatic-hydrocarbon adsorption onto soot point to the Langmuir model as the most appropriate, but, for the less volatile compounds, where evidence arises of multilayer adsorption, a BET model is deemed more suitable (Aubin and Abbatt, 2006). A Langmuir model, applied to PAH, takes the form

$$\log \frac{[\text{PAH}]_{\text{XOC}}}{[\text{PAH}]_{\text{SOF}}} = \log\left[\frac{1}{cA}\right] + \log p, \tag{3.39}$$

where c is handled as a constant, A is the (total suspended) particle surface area, and p is the saturation vapour pressure, determined as earlier described (Opris et al., 1993). The two concentrations of a certain compound, in the vapour phase (XOC) and the particulate phase (SOF), are determined experimentally, as is the surface area: taking several PAH and different test conditions, correlation coefficients >0.9 have been obtained. Another Langmuir model, also applied to PAH, and fitted to experimental measurements, yielded activation energies in accordance with those expected, if adsorption were responsible for gas-to-particle conversion (Kraft et al., 1982).

The well-known Langmuir theory of adsorption makes several customary assumptions (Plee and MacDonald, 1980): that particle surfaces contain a fixed number of adsorption sites; that the heat of adsorption is not dependent on the location of the adsorption site or the fraction of sites covered; that interactions between adsorbed molecules are negligible; and that adsorption and desorption are in equilibrium. We shall not examine the efficacy or implications of these assumptions; the reader is referred to standard texts on surface chemistry (e.g. Hayward and Trapnell, 1964). Using this understanding, the adsorption can be expressed as (Plee and MacDonald, 1980)

$$\theta = \left[1 + \left(\frac{k_\text{d}}{k_\text{a} p}\right)\right]^{-1}, \tag{3.40}$$

where k_a and k_d are Arrhenius rate coefficients for adsorption and desorption, respectively, and θ is the fraction of surface sites occupied by the adsorbing hydrocarbon. From this was derived

$$\theta = \left[1 + \left(\frac{KD}{\chi_{\text{HC}}}\right)\exp\left(-\frac{E}{\Re T_{\text{mix}}}\right)\right]^{-1}, \tag{3.41}$$

where E is an activation energy and K is a constant. This expression parallels Equation (3.38) insofar as it multiplies an exponent, containing the temperature of the air–exhaust mixture (T_{mix}), by a quotient (χ_{HC}/D), containing the partial pressure of the hydrocarbon. From this point, it is useful to make the (not unreasonable) assumption of direct proportionality between θ and the extra mass this coverage brings to the particulate. K and E, neither of which is known *a priori*, are simply chosen to suggest the likely trends, or fitted such that the model reproduces the experimental results: scenarios for two exhaust temperatures are depicted in Figure 3.19 (Amann et al., 1980). The trends resemble those in Figure 3.18, but with one important difference: the adsorption model predicts a small organic fraction,

Formation I: Composition

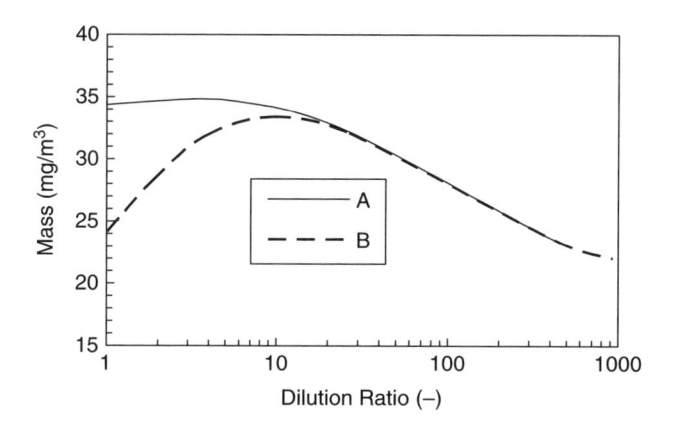

Figure 3.19 Model of hydrocarbon adsorption onto solid particles. Exhaust temperatures: A, 400 K; B, 600 K. Dilution air temperature (T_{air}): 297 K. Initial gaseous hydrocarbon mole fraction (χ_{HC}): 13 ppm (C_6). The model constants K and E in Equation (3.41) were obtained from experimental measurements at one dilution ratio, and then the modelled concentrations were calculated for other dilution ratios. Operating point: 1500 rpm, 6.2 bar IMEP. Engine: 2.1-litre, single-cylinder (Amann et al., 1980).

even at the highest dilution ratios, something observed in reality, whereas the condensation model predicts an eventual cessation of gas-to-particle conversion.

The above-described condensation (Figure 3.18) and adsorption (Figure 3.19) models are equally able, then, to follow a well-known experimentally observed phenomenon: that particulate mass, with increasing dilution ratio, increases, reaches a maximum and then decreases. There is of course no fixity or permanence in the *nature* of this maximum, and indeed considerable variations are seen in its shape; but knowledge that a certain range exists over which no appreciable change occurs to the particulate mass is useful and of immediate practical benefit, as conditions can be purposely designed to reduce sensitivity to errors in dilution ratio.

This rise-and-fall characteristic arises not only because of gas-to-particle conversion *per se*, but also because of the nature of the dilution. The two functions, of decreasing temperature (cooling) and decreasing concentration (mixing), oppose one other in their effects on particulate mass. Temperature falls rapidly at first, but, for each additional increment of air, the rate of cooling declines, until, at very high dilution, the temperature of the air–exhaust mixture asymptotically approaches the temperature of the air – an inescapable physical constraint – but since mixing continues unabated, this effect eventually catches up and ultimately outpaces the cooling effect.

Although condensation was mostly ruled out in the original study of Plee and MacDonald (1980), a recent update came to the opposite conclusion (Durán et al., 2002). One should recognise that, over the last two decades, large reductions have taken place in organic particle precursors, but *also* in solid carbonaceous surfaces, and the implications here, for gas-to-particle conversion, do not leap readily to the eye. Condensation still conceivably plays some nonnegligible role whenever hydrocarbon concentrations are exceptionally high – in the most extreme case, when there is white smoke (Section 3.6.5). Indeed, the diesel particulate filter, which, at the time of writing, is being installed on most diesel vehicles, preferentially strips the solid-phase carbon from the exhaust, while allowing much of the HC and organic particulate to pass through, so that the traditional picture must be greatly altered.

The condensation model of Figure 3.18 does not, however, address gas-to-particle conversion directly, through the available surface area, but rather, it does so indirectly, by testing the likelihood of condensation via the saturation ratio, and an instructive issue arises therein, as *both* models consider only two states or end points for the exhaust gas, namely undiluted and diluted; the *rate of transition* from one state to the other is glossed over. Figures 3.18 and 3.19 might, for example, be constructed by conducting a series of tests using different end points, or, alternatively, just one end point, and considering the events *en route*. In reality, a certain volume of raw exhaust gas does not suddenly find itself, after an infinitely fast transition, in a diluted, chilled and supersaturated state – except, perhaps, when emerging from the tailpipe of a real motor vehicle. The transition from one state to the other in a dilution tunnel is more leisurely, so that the curves depicted in Figures 3.18 and 3.19 will be traced out in *some* manner; but concurrent gas-to-particle conversion will very probably restrain the tendency for supersaturation, and this conversion might take place not only on the dilution tunnel filter but also, in transit, on suspended particles, or, for that matter, on the walls of the exhaust system or dilution tunnel. The practical outcome, however, of these musings is that, in the earlier regions of the dilution tunnel, organic compounds adsorb or condense onto particles, and then, once the 'crest' has been crossed, but before the filter is reached, these same compounds desorb or evaporate (MacDonald J.S. *et al.*, 1984).

Concentration on one 'representative' hydrocarbon overlooks a subtlety observed in the burning of agricultural waste, no less (Keshtkar and Ashbaugh, 2007). Gas-to-particle conversion by condensation occurs earlier for heavier molecules, making, for aromatics, the number of rings a factor. This might create, say, soot particles around which the PAH are layered accordingly. Condensation onto the smallest particles might, however, be restrained by the Kelvin effect, i.e. the need for slightly higher supersaturations than for larger particles. Perhaps PAH also become segregated across the size spectrum, according to the number of rings, the saturation realised at each moment in the emission process, and the surface area to which they have access.

The time spent at intermediate dilution ratios must be an aspect – this is a matter of kinetics, which is much more important when mixing and cooling are rapid, so that adsorption–desorption equilibria or condensation–evaporation equilibria are not necessarily observed at each moment in the dilution process. Examining the estimated activation energies, in the light of information gleaned from the literature, it was concluded that, in the adsorption–desorption model, the tunnel residence time exceeded the time needed for equilibration by orders of magnitude (Plee and MacDonald, 1980).[10] And, by analogy to heat transfer coefficients, a model was derived for the adsorption kinetics which suggested tunnel residence time to exceed equilibration times by a factor of two to five (Clerc and Johnson, 1982). But, taking *Pyr* as an example, PAH were believed to transfer slowly from the gas phase into the particulate phase, in comparison with residence time, so that full gas–particle equilibrium might not be achieved in all cases before the dilution tunnel filter is reached (Du *et al.*, 1984).

Laying aside kinetic considerations, several relatively untouched areas of gas-to-particle conversion are worthy of note: these concern 'microtransport', the similarity of which to surface area characterisation (Section 6.2.4), should be noted. With advancing gas-to-particle conversion, (a) sorbing compounds must diffuse ever deeper, between the spherules, into the agglomerate structure; (b) some accreted organic material may gradually solidify; (c) *ab̲sorption* may take place, into the liquid components, rather than adsorption onto surfaces; (d) the agglomerate structure could swell and become distorted; (e) with falling temperature, a 'layered' coating could develop, with more-volatile compounds being located further out from the solid surface. A pore-retarded diffusion model, in which compounds enter liquid-filled pores, has ruled out free-liquid diffusion, and pointed instead to the tortuosity of

---

[10] It should be noted that the residence time is itself a function of dilution ratio in conventional methods of dilution. The residence time in a filter matrix is much shorter than in a dilution tunnel, but the available surface area is of course much greater.

the diffusional paths winding through the agglomerate as more important constraining factors than adsorption in gas-to-particle conversion (Strommen and Kamens, 1999). Depending on the *polarity* of the gas molecule, a*d*sorption *onto* carbonaceous surfaces is perhaps of greater relevance in the vicinity of the tailpipe, with a*b*sorption *into* organic coatings becoming more important as particles age in the environment (Roth C.M. *et al.*, 2005). But evidence pertaining to adsorption versus absorption continues to point one way and then the other (Zielinska *et al.*, 2004)

Obviously, the preceding paragraph considerably opens up the number of possibilities in gas-to-particle conversion, and this brings us to a closely related but more vigorously reported area of research, namely, the partitioning of organic compounds, especially PAH, in the *atmosphere* (Jang M. *et al.*, 1997; Harner and Bidleman, 1998; Jang M.J. and Kamens, 1998; Smith D.J.T. and Harrison, 1998; Jang M.J. and Kamens, 1999) – of the greatest interest to atmospheric scientists, as it governs long-range pollutant transport (Lohmann and Lammel, 2004). A widely cited model for atmospheric partitioning of organic compounds is

$$k_p = \left[\frac{C_p}{\text{TSP}}\right] \frac{1}{C_g}, \qquad (3.42)$$

where k_p is the 'partitioning coefficient' ($m^3/\mu g$), TSP is the total suspended particles ($\mu g/m^3$), and C_p and C_g (ng/m^3) are the particle-associated and gaseous concentrations of the compound (Pankow, 1994). This model does not distinguish between absorption, adsorption and some combination thereof, but plots of k_p against the vapour pressure of a compound show respectable correlations, with gradients of ~ -1.

Redeployment of such ideas to the exhaust gas of motor vehicles could be advantageous. The partitioning of organic compounds in the atmosphere is not necessarily equilibrated (Gustafson and Dickhut, 1997); very similar questions arise in the case of tobacco smoke (Liang C. and Pankow, 1996) and of rainwater (Poster and Baker, 1996), which raises the intriguing, but wholly unanswered, question of how water-soluble and organic-soluble components in the particulate might contend with one another; and, finally, research continues into observed deviations from current theories of atmospheric partitioning, two aspects of which are retarded kinetics and irreversible adsorption (Terzi and Samara, 2004).

3.6.4 Gas-to-Particle Conversion: Measurements

Samples of exhaust gas drawn off at each point reveal the gas-to-particle conversion experienced by organic compounds as they pass along an exhaust system and into a dilution tunnel. Some evidence suggests that when organic compounds are preponderant, such as at low engine load, they will nucleate (Kati *et al.*, 2004a). But the experimental database is very much geared instead to measurements of particulate mass, and the accretion of organic compounds to soot.

To determine what is 'particulate' at each point, dilution and filtering can hardly be conducted in the usual way (Section 5.2), as this procedure, obviously, brings all samples to precisely the same end point, thus discarding the pertinent information: such measurements only show what will *eventually* become particulate, rather than suggest what is *already* particulate, at each moment in the emission process. One way around this is to filter the exhaust gas extracted from each point, at the temperature prevailing at that point, and then to quantify the organic fraction on the filter in the customary fashion. Another, indirect way - and one that is more convenient in real time (Section 5.4.4) – is to measure the *gas-phase* hydrocarbons (HC) with a flame ionisation detector (FID). Now, the exhaust-gas preconditioning for this device is normally conducted at 180–190 °C, but such a temperature is by no means inviolable (Abbass *et al.*, 1989a, 1989c). Hence, the *difference* between FID measurements conducted at any two

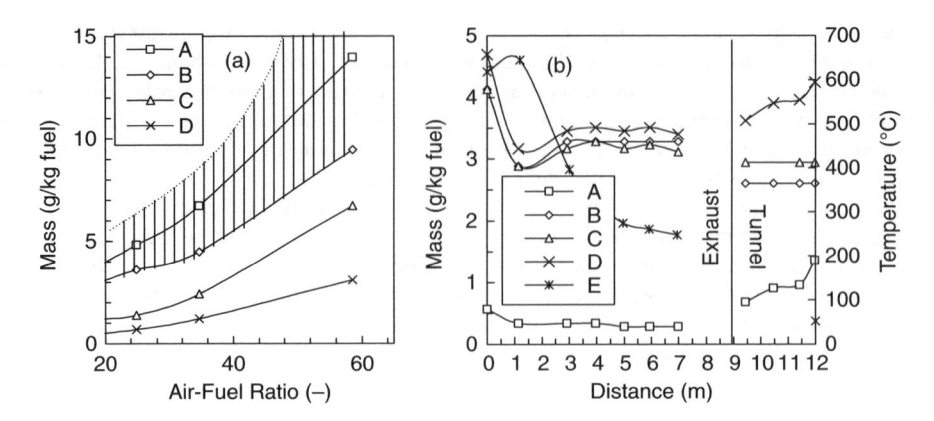

Figure 3.20 Gas-to-particle conversion in organic compounds, as suggested by measurements of the associated hydrocarbons. (a) According to preconditioning temperature of the FID. HC: A, 180 °C; B, 52 °C; C, 2 °C; D, CH$_4$ only. Engine: four-cylinder, direct-injection, naturally aspirated, converted to single-cylinder operation. Operating point: 2200 rpm, load as implied by air – fuel ratio. The shaded area suggests the total gas-phase hydrocarbons which contribute to the organic particulate (Andrews et al., 2000b). (b) According to distance from engine: A, organic particulate; B, HC (180 °C); C, carbonaceous particulate (soot); D, total particulate; E, temperature. Engine: 4.2-litre, direct-injection, converted to single-cylinder operation. Fuel: cetane number, 48. Dilution ratio, 30 (Abbass et al., 1989a).

temperatures immediately reveals the hydrocarbons susceptible to gas-to-particle conversion in that temperature range (Andrews et al., 2000b).

Such measurements, some of which are shown in Figure 3.20(a) (Andrews et al., 2000b), reveal, with falling temperature, an unbroken sequence in gas-to-particle conversion from the heaviest to the lightest hydrocarbons. (This is not too surprising, since this procedure is merely the reverse of thermogravimetry, wherein particle-to-gas conversion is measured with rising temperature; see Section 5.2.4.) Sketched in Figure 3.20(a) is an imaginary FID measurement in which *all* hydrocarbons are in the gas phase – in practical terms this is unmeasurable owing to thermal restrictions in the instrumentation. At 180 °C, the heaviest hydrocarbons of all have transferred into the particulate phase; at 52 °C, the highest temperature permitted in the dilution tunnel filter, far fewer hydrocarbons remain in the gas phase; and, finally, at 2 °C – now *below* room temperature – only the lightest hydrocarbons of all (e.g. methane) are able to escape transfer into the particulate phase.

Four pieces of information are worth emphasising here. First, of the hydrocarbons in the gas phase at 180 °C, *two-thirds* are still in the gas phase at 52 °C; and *one-third* are still in the gas phase at 2 °C; or, alternatively put, hydrocarbons still in the gas phase above 52 °C account for around one-half of the difference between 2 °C and 180 °C (Abbas et al., 1989a). Second, and admittedly to some satisfaction, the difference in HC between 180 °C and 52 °C as reported by the FID agrees quite closely with the *mass* of organic particulate as obtained in the traditional manner, via a dilution tunnel filter, also operating at 52 °C. This is quite logical. The only discrepancy appears to be caused by the heaviest hydrocarbons of all – those not in the gas phase at 180 °C, and so unquantified by the FID, but nonetheless captured by the dilution tunnel filter operating at 52 °C (Andrews et al., 2000b). The hydrocarbons thus corresponding to the organic particulate are indicated by the shaded area, of which the imaginary measurement serves as the upper bound. Third, the dangers of using a single reading of HC at 180–190 °C to suggest the amount of organic particulate are clear, for the former measurement ignores the heaviest hydrocarbons (>180 °C), just as the latter measurement ignores the lightest hydrocarbons (<2 °C). Fourth, the quantity of organic particulate captured by the dilution tunnel filter must increase substantially for excursions cooler than 52 °C – even for those only as low as room temperature.

The same method, applied at various points along the exhaust system and the dilution tunnel, also elucidates ongoing gas-to-particle conversion. The data in Figure 3.20(b) (Abbass et al., 1989a) were acquired by bleeding off samples of exhaust gas and filtering them at 52 °C i.e. *no* additional dilution was used. This distinction is important since, following this extraction, cooling alone operates on gas-to-particle conversion. Three features are apparent in Figure 3.20(b). First, the fall and brief resurgence of particulate in the first 2 m is probably the work of wall deposits, but this feature is of less interest to the current discussion. Second, along the extended exhaust system (2–9 m), no changes occur in the particulate or HC. Third – and most significantly – upon entry into the dilution tunnel, the mass of HC decreases, just as the mass of organic particulate increases. Thus, the gas-phase hydrocarbons, from the FID measurement, reappear as particulate-phase hydrocarbons on the filter. Speciation of these particulate-phase hydrocarbons into alkanes and PAH, revealed patterns broadly similar to HC, although wide variations were also observed from one compound to another (Abbass et al., 1989a). The results indicated that, upon entry into the dilution tunnel, different alkanes were taken up by the particulate to different extents, while some particulate-phase PAH increased and others decreased.

Reiterating, then, the mass of particulate collected from the dilution tunnel was higher than the mass of particulate collected from the exhaust system, even though both samples were acquired by filters operating at the *same* temperature. In this instance, therefore, mixing with air was the greater factor in controlling gas-to-particle conversion than cooling. The conclusion, that the transfer of organic compounds into the particulate phase plays little or no significant role within the exhaust system, is important in elucidating the operation of catalytic converters. For example, the mass fraction of (diesel) hydrocarbons remaining in the gas phase upon entry of the exhaust gas into the catalyst at typical operating temperatures very probably exceeds 98 % (Johnson J.E. and Kittelson, 1994), in which case *particle* transport to catalytic sites is a minor aspect, and oxidation is carried out predominantly in the gas phase. Organic compounds would not, in any case, be able to persist in the particulate phase, as their evaporation rates are rapid in comparison with residence times (Abdul-Khalek and Kittelson, 1995).

3.6.5 White Smoke

The incomprehensibly vast literature devoted to black smoke contrasts starkly with the relatively meagre attention accorded to *white smoke*. This discrepancy probably arises from the less problematical nature of white smoke, insofar as the necessary countermeasures to be undertaken, via engine calibration and engine design, are well understood and readily available, and the phenomenon is now mostly avoidable, given the evident sophistication of today's engine management systems.

Investigations into the *nature* of white smoke, as opposed to the causes, are arguably of little practical relevance, as the levels of hydrocarbons (several thousands of ppm) represent a mammoth exceedance of the statutory requirement for HC, so that engine operation, being obviously aberrant, renders the phenomenon factitious and of dubious relevance. It should also be mentioned, *en passant*, that white smoke readily accumulates as films or pools of liquid fuel within instrumentation; measurement of this emission is thus highly problematical (Section 5.4.6).

White smoke, then, as generally conceived, is the emission of *unburned fuel*, but the term is vernacular, and lacking in scientific clarity, as sulphates, when dominant, also form white smoke; hence, 'fuel smoke' represents an apter sobriquet. And, while on the subject of colour, blue smoke, by some, is associated with unburned fuel, whereas others associate it with unburned lubricant (Baumbach et al., 1995); the latter belief seems to have been confirmed by chemical analysis (Sodeman et al., 2005). The problem, though, is that colours, as perceived by the human eye, are aspects of the specific lighting conditions, and not unequivocally related to a certain emission (Fraenkle and Hardenberg, 1975). For example, an aerosol of white smoke, when collected on a dilution tunnel filter, can appear tan or brown, and not white; and oil appears yellow (Li W. et al., 2007). Colour, therefore, is to some extent a changeable ally.

Fuel smoke is, in effect, a particulate emission hugely dominated by the organic fraction. Should this fraction greatly exceed the carbonaceous one, organic molecules become deprived of pre-existing particle surfaces on which to adsorb or condense, and continued cooling forces them, instead, to nucleate into an aerosol of liquid-phase droplets. A range of colours may thus be observed, from black, through various shades of grey, to white, depending on the relative proportions of carbonaceous and organic material. And since condensation depends so strongly on temperature, via the saturation ratio, white smoke is exacerbated by cold ambients and cold starts. During these periods, particulate mass and particle numbers increase severalfold over those seen at room temperature (Aakko and Nylund, 2003), and filter deposits have a wetter consistency, being predominantly liquid-phase organics (Hara et al., 1999).

The visible manifestation at the tailpipe does not necessarily, betray the real-time engine-out emission, because catalysts (and exhaust system walls) avidly store hydrocarbons at low temperatures, such as at idle or a cold-start, and then release them at high temperatures, or at high space velocities, such as during acceleration (Tashiro et al., 1995). This particularly happens when a substantial temperature window exists between desorption/evaporation and light-off: the hydrocarbons are released before the catalyst is active. In the opposite scenario, the sudden ignition of liquid fuel films causes destructive overheating of catalysts or other aftertreatment devices.

The chain of events leading to white smoke begins with the fuel injection equipment, in which various effects are possible, a full exploration of which is impossible here, but one example is restriction of needle movement in the injector, owing to the higher viscosity of the fuel (Zahdeh and Henein, 1992). More pertinent to the current topic is the next step, namely in the combustion chamber, as, in a cold engine, the compression stroke fails to generate temperatures sufficient to vaporise the fuel, and this allows liquid films to accumulate on the cold walls of the combustion chamber. Then, when the engine begins to warm up, this stored fuel evaporates, but in an uncontrolled manner, not conducive to its combustion; hence its emission unburned. The evaporation of these films, after all, is leisurely, when compared with fuel droplets of diameters equivalent to the film thickness. The behaviour of fuel films on combustion chamber walls is an intriguing topic, for which observations may appear counter-intuitive. If the film is thick, say, 30–50 μm, heat transfer into it from the air charge is rapid; but if the thickness is only, say, 10 μm, the film becomes, thermally speaking, part of the wall, and does not boil (Pachernegg, 1975). In fact, the continual presence of these fuel films, from one cycle to the next, is akin to the film of oil residing on the cylinder walls; and there are further considerations, discussed in Section 8.4.1, relating to the Leidenfrost effect.

The proportion of injected fuel that engines are able to store (even excluding that which enters the sump) is quite large. In a *direct* measurement, a diesel engine was designed such that it could be stopped suddenly during a cold start, and the fuel within it collected and quantified: as much as 40 % of the initially injected fuel was retained. The actual amount, however, depended on combustion chamber geometry, fuel-spray orientation, engine speed, injection timing, etc., insofar as these factors control wall impingement (Tsunemoto et al., 1986). In an *indirect* measurement, using a balance computed between fuel injected (via needle lift) and fuel burned (via cylinder pressure), negative values implied fuel storage (Zahdeh and Henein, 1992).

The study of white smoke and cold starting is considerably complicated by the survival of fuel into subsequent cycles. Residual fuel left over from one cycle evaporates during the following compression stroke, generating a partially premixed charge that, at one moment, might exacerbate, and at another, ameliorate, the misfiring and intermittent combustion. Engine running is then rough: some cycles burn properly, others burn partially and others do not burn at all. Perhaps ironically, pilot injection, through its premixed character, can generate very similar effects (Osuka *et al.*, 1994). Cold-start combustion is controlled via air temperature, fuel temperature, fuel formulation, injection pressure and injection scheduling (i.e. split injections) (Lippert *et al.*, 2000).

From an engine operation standpoint, a distinction exists between cold starting and cold running. With diesel engines there is added potential to emit white smoke in cold ambients, even when an engine is warm, because the ignition delay lengthens noticeably as air-charge temperature falls. This leads to rough engine running, including misfiring, with due emission of unburned fuel. One well-known countermeasure is to advance the injection timing in order to shorten the ignition delay. It is possible that an equally well-known NO_x countermeasure, timing retardation, might have fostered susceptibility to white smoke, as this lengthens the ignition delay.

White smoke appears at high altitudes – although rarefied atmospheres *per se* are not necessarily responsible. High altitudes tend to be associated with protracted periods of downhill driving (i.e. zero fuel demand), during which the engine, since only fresh air is passing through it, has ample opportunity to cool down. Then, upon reaching an uphill section, the fuel is immediately reinstated, but the internal engine temperatures are now too low to support appropriate vaporisation of the fuel, and white smoke then reappears – often precipitously. The altitude effect is reproducible in the laboratory through air-path throttling, for which white smoke is consistently triggered at certain inlet manifold depressions; lower speeds require deeper depressions to generate the smoke (Kato M. *et al.*, 1987). Persistent low-load driving is also susceptible to white smoke, insofar as internal engine temperatures are again low, and charges of extremely lean air–fuel ratios tend to burn less reliably.

Fuel formulation is equally an aspect of white smoke. The best-correlated parameter is cetane number (Hara *et al.*, 1999), closely followed by T90 (see Section 7.2.5) – both consistent with what is known about ignition delay and fuel evaporation in the emission of white smoke (Kato M. *et al.*, 1987). A correlation between white smoke, as triggered by air-path throttling, and cetane number is shown in Figure 3.21(a). To prevent the appearance of white smoke, for example at 2000 rpm, an increase of three units in cetane number for every extra 400 m in altitude was needed, but the exact features were dependent on engine design (Kato M. *et al.*, 1987). Confounding influences by other fuel properties likely to affect atomisation and vaporisation of the fuel were not decoupled (e.g. viscosity and volatility).

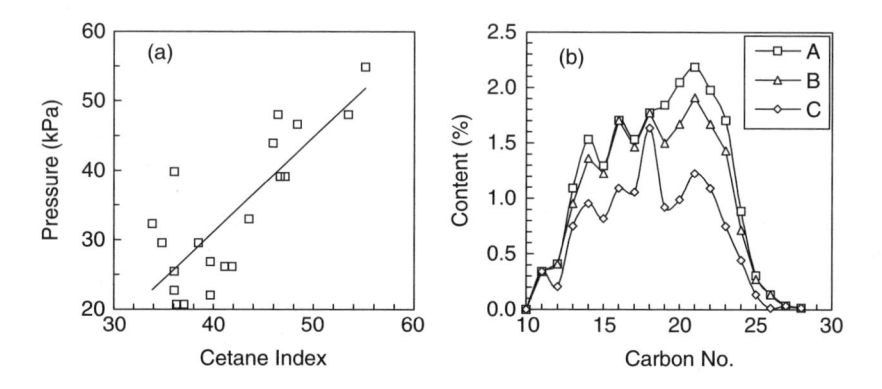

Figure 3.21 White-smoke emissions: (a) data points marking conditions for the appearance of white smoke, resulting in a correlation between inlet manifold pressure and fuel cetane number; (b) chromatograms for *n*-alkanes, as found in fuel (A), white smoke of 100% opacity (B) and white smoke of 30% opacity (C). Fuel: cetane number, 40. Vehicle: 1.8-litre, indirect-injection, naturally aspirated. Operating point: 2500 rpm (Kato M. *et al.*, 1987).

Although white smoke is equated to 'fuel droplets', strictly speaking any *a priori* assumption of identicality to fuel seems unwarranted, because partial combustion reactions and storage within the engine are selective in their effects on hydrocarbon speciation. A comparison between *n*-alkanes in the white smoke and in the fuel is given in Figure 3.21(b) (Kato M. *et al.*, 1987). The range C_{19}–C_{23}, representing one-third of the total fuel alkanes, was significantly underrepresented in the smoke. Moreover, there were indications that, when smoke was emitted more profusely, its speciation more closely resembled that of the fuel, insofar as hydrocarbons of higher boiling point became more noticeable. White smoke collected at $-8\,°C$ consisted of low-boiling-point hydrocarbons ($<200\,°C$) (Tanaka T. *et al.*, 1989). When condensed water was eliminated, the weight of the collected deposit agreed, over a range of temperature ($+10$ to $-15\,°C$), with that estimated from FID measurements of HC, but the correlation varied with the model of engine; this seemed to stem from varying hydrocarbon speciations exhausted by different engines. Chromatograms of HC and white smoke indicated the condensation of certain species, and, since this pattern varied between engines, then so did the amount of white smoke.

3.7 Sulphate Fraction

We left the sulphate fraction in Section 2.4, where it was stated that sulphur-containing compounds in fuel, and also in the oil, are broken down in the combustion, and that this sulphur is predominantly released into the exhaust as SO_2, with smaller amounts as SO_3. As such, SO_2 contributes little directly to the particulate; the more immediate problem is SO_3, which readily hydrolyses to H_2SO_4; and this acid, along with its associated water, represents the lion's share of the sulphate fraction. From a simple consideration of molecular weight, one gram of sulphur forms three grams of H_2SO_4 – quite apart from the associated molecules of H_2O. Since the conversion of SO_3 to H_2SO_4 happens virtually automatically, the preceding conversion of SO_2 to SO_3 is what must be absolutely avoided.

The emission of sulphuric acid is hardly surprising to anyone associated with industries that burn fossil fuels. But in the automotive community, this emission first came to prominence in the 1970s, when oxidation catalysts were fitted to gasoline engines (Section 8.2.2). With diesel engines, on the other hand, where particulate emissions were always of interest, the emission of sulphate particulate, even without catalysts, had always been noted, but little incentive existed to address this issue, as, for many years, statutory tailpipe standards were framed exclusively in terms of 'smoke' – effectively a measure of soot. The spotlight was only trained on sulphates upon the introduction of statutory limits for 'particulate', as this fraction contributed noticeably to the gravimetric, if not the visible emission; indeed, countermeasures became exigent when oxidation catalysts were introduced to diesel engines in the early 1990s to control HC and CO. In effect, the exact same difficulties, encountered twenty years previously with gasoline engines, had now been repeated with diesel engines, but with an added difficulty that sulphur levels in diesel fuel were much higher than in gasoline. The need to oxidise HC, CO and organic particulate, *without* oxidising SO_2 to SO_3 and thence forming H_2SO_4, proved to be quite a stumbling block; and, rather than improved catalyst formulations, it was greatly reduced fuel sulphur levels that really allowed the implementation of oxidation catalysts on diesel engines. The reduction, from 0.3–0.5 % in the 1970s to today's levels of 50 ppm, represents two orders of magnitude, while one day there may be possibly even a third order of magnitude (5 ppm).

3.7.1 Chemical Reactions

The 'sulphur system', in the combustion of fossil fuels generally, is naturally enough of widespread interest, as sulphur-containing compounds are virtually omnipresent. But the combustion chemistry of sulphur is nontrivial (Cerru *et al.*, 2006), and, the principal topic being particulate rather than the tangled skein of precursive sulphur reactions, this large subject will only be peppered with a few scattered remarks. With this proviso, then, the chain of events leading to sulphate particulate begins with the combustion and decomposition of these stowaways, amongst which organosulphur PAH, or SPAH, are characteristic: for example, dibenzothiophene ($C_{12}H_8S$) (Williams P.T. *et al.*, 1987); detailed speciations (more than a dozen compounds) are available (Liang F. *et al.*, 2006). The sulphur molecules in diesel fuel are larger and contain more rings than in gasoline (Song C., 2000). Following its liberation, the sulphur assumes the form of various compounds, of which SO_2 and SO_3 are the principal, or certainly the most pertinent, ones. Inevitably, questions of relative abundance are conflated by interconnecting reactions between these two oxides; but, thermodynamically, SO_2 is the favoured species, and, kinetically, the availabilities of H and O are also influential; in fuel-rich conditions, the conversion of SO_2 to SO_3 is suppressed, and in fuel-lean conditions, the small percentages of SO_3 are super-equilibrated (Khatri *et al.*, 1978).

Naturally enough, these remarks gloss over considerable complexity, but they do suggest the type of internal combustion engine (*inter alia*, gasoline versus diesel) to be a factor in SO_2 versus SO_3, and that the two compounds act differently towards soot (Haynes and Wagner, 1981, p. 234). In-flame sulphur compounds (e.g. H_2S, SO_2 and CS_2) are implicated in *soot suppression*: by aiding and abetting the disappearance of soot precursors, or by adsorbing onto spherule surfaces (Ni *et al.*, 1994) and forming stable carbon – sulphur complexes (Song J. *et al.*, 2002), they might inhibit surface growth (Section 3.2.3). Still more abstruse is the retardation, via the population of chemiions, of agglomeration (Sorokin and Arnold, 2004). Of course, any soot-suppressing effect, if it exists, is easily overshadowed by the litany of problems created by sulphur in emission control. And, conversely, sulphur may actually promote soot: should SO_3 remove the oxidant OH, then acetylene, a pivotal soot precursor, is able to survive (cited by Smith O.I., 1981, p. 282).

The relevance of sulphur to particulate emissions really begins in the exhaust system, and involves the interplay of thermodynamics and kinetics, as suggested in Figure 3.22 (Koltsakis and Stamatelos,

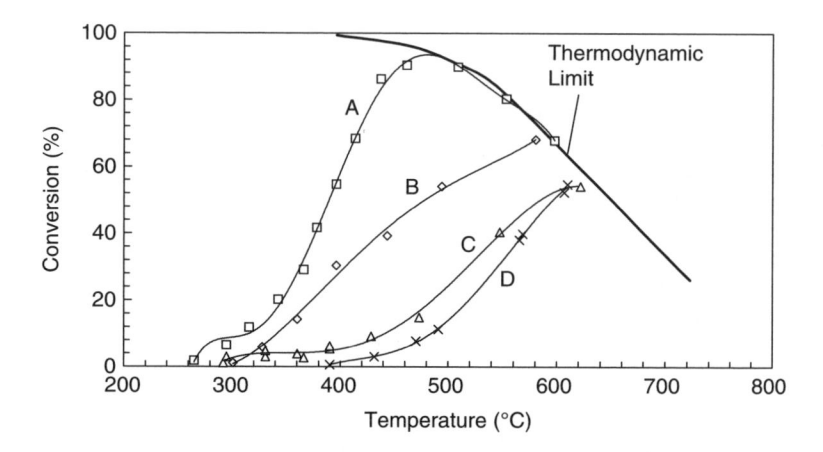

Figure 3.22 SO_2 conversion to SO_3 as a function of temperature, with 5 % O_2 (no reducing gases were present). Space velocity, $10 s^{-1}$. Catalyst formulations: A, Pt/Pd=2/1; B, Pt/Rh=10/1; C and D, Pd only (Koltsakis and Stamatelos, 1997).

1997). The thermodynamic relationship between SO_3, SO_2 and O_2 is perhaps the best-known aspect of the sulphur system:

$$SO_2 + \frac{1}{2}O_2 \Leftrightarrow SO_3. \qquad (3.43)$$

Thermodynamically, SO_3 predominates below 500 °C, and SO_2 predominates above 750 °C. This picture is, however, strongly tempered, as, *kinetically*, SO_2 is considerably more abundant. It is precisely because of this kinetic limitation that oxidation catalysts are undesirable, speeding up, as they do, the conversion of SO_2 into SO_3. Once SO_3 has appeared, the consequent appearance of H_2SO_4 is virtually assured, and, in this sense, Reaction (3.44), rather than Reaction (3.43), decides the rate of sulphate formation:[11]

$$SO_3 + H_2O \longrightarrow H_2SO_4. \qquad (3.44)$$

Molecules of H_2SO_4 have great affinity for water molecules, and readily hydrate:

$$H_2SO_4 + H_2O \longrightarrow H_2SO_4.H_2O. \qquad (3.45)$$

After n steps (Khatri *et al.*, 1978), where $n = 7$ (Janakiraman *et al.*, 2002),

$$H_2SO_4.(n-1)H_2O + H_2O \longrightarrow H_2SO_4.nH_2O. \qquad (3.46)$$

The route by which SO_2 converts to H_2SO_4 in the emission process is difficult to assign in practice. The parallel – *atmospheric* conversion – is, however, well studied (Seinfeld, 1986, pp. 196–; Pandis and Pilinis, 1995; Claes *et al.*, 1998), and involves homogeneous gas-phase reactions, mainly with OH, followed by dissolution and aqueous-phase reactions within droplets. Oxidation, heterogeneously on the surfaces of carbonaceous particles, is also known (Seinfeld, 1986, p. 226). Similarly, it is supposed, although admittedly with little direct evidence, that particles carried by the exhaust gas act as mobile catalysts (Chughtai *et al.*, 1998), facilitated, for example, by metals or metal oxides lodged on spherule surfaces. Of course, this phenomenon is well attested in the degradation of building facades (Section 2.3.4), where the particles have already deposited, and there is ample time for acid formation; the question here is whether particles act in this manner while in transit through the exhaust system. The observation that exhaust-gas dilution discourages also the formation of sulphates lends credence to the notion of particles as mobile catalysts, as dilution decreases the surface area concentration (Khatri *et al.*, 1978). The particles are not, however, necessarily mobile: for example, wall deposits of V_2O_5, formed in the combustion of vanadium-bearing heavy fuels, also have catalytic effects (Edwards, 1977, p. 75), as might the metal walls of the exhaust system. What happens on the dilution tunnel filter is different still: the presence of liquid-phase water and the continual throughput of exhaust gas have obvious implications, as SO_2, which is eminently water-soluble, might simply transition into the particulate phase via dissolution, with consequent aqueous-phase formation of H_2SO_4, just as in the atmosphere. Yet another route begins with adsorption of SO_2 onto filter fibres (Sioutas and Koutrakis, 1995).

---

[11] The general understanding through many years is that SO_2 and SO_3 contribute to particulate via H_2SO_4. It should, however, be noted that the temperature of the dilution tunnel filter is customarily ≤ 52 °C, whereas the boiling point of SO_3 is just 45 °C – this point does not seem to have been addressed in the literature. (The parallel suspicion is less tenable with SO_2, which has a boiling point of -10 °C).

Before closing this section, it is certainly worthwhile mentioning a little-known topic: sulphate particulate and nitrate particulate are suspected of synergistical or antagonistical interactions (de Lucas et al., 2001). This might relate to the concurrent gas-phase presence of SO_x and NO_x:

$$NO_2 + SO_2 \longrightarrow NO + SO_3. \tag{3.47}$$

The suggestion here is that NO_2, in serving as an oxidant for SO_2, encourages also the appearance of sulphate particulate. There seems some justification for the notion that NO_2, particularly above 200 °C, affects equilibrium levels of SO_3 (Khatri et al., 1978), while Reaction (3.47), or an alternative of similar consequence, might well occur on particle surfaces (Chughtai et al., 1998). As the conversion of SO_2 into SO_3 represents a bottleneck, so to speak, in the formation of H_2SO_4, it could be that Reaction (3.47) represents an alternative precursive step in gas-to-particle conversion; indeed, NO_2 does seem to play some mediating role in atmospheric nucleation of H_2SO_4, beginning with SO_2 (Brichard et al., 1972). Reaction (3.47), in view of its scarcity in the literature, is presumably only a bit-player in exhaust gas chemistry, although a more prominent role might become available with some of the exotic forms of aftertreatment currently being researched: the reaction has been mooted, for instance, in the frenetic search for a carbon oxidation catalyst (Liu S. et al., 2002).

3.7.2 Gas-to-Particle Conversion

In the first instance, the mass emission of sulphate particulate is decided simply by the mass throughput of sulphur, which is to say, the specific fuel consumption (SFC) and fuel sulphur content (FSC). But whereas FSC habitually steps into the limelight, SFC stands reticently off-stage: fuel formulations range more widely than engine efficiencies. This observation aside, simple linear relationships obviously exist between engine-out sulphur and SFC, and also between engine-out sulphur and FSC: what enters the engine must leave the engine. But the conversion of fuel sulphur into sulphate particulate is not quite the same thing, as the sulphur is not necessarily emitted in the particulate phase.

In the absence of a catalyst, the conversion of fuel sulphur into sulphate particulate is actually quite low: generally speaking, about 1–3 % (Springer and Baines, 1977). Data suggest that engine operating point is influential, for example low exhaust temperatures (i.e. low engine loads) are thermodynamically less stacked against SO_3 (Khatri et al., 1978), and that the duty cycle is influential, at least insofar as heavy-duty and light-duty diesels are customarily evaluated differently (den Ouden et al., 1994); but the trends are slight, and perhaps insignificant (Gomes and Yates, 1992); Figure 3.23 (Wall and Hoekman, 1984), suggests no consistent trends. Data suggest also that more sulphur appears in sulphate particulate at high FSC than linearity can account for (Opris et al., 1993), but the trends are often slight and occasionally contradictory (Frisch et al., 1979). It will be appreciated that little consistent relationship of any sort, either to SFC or FSC, seems likely, in view of the numerous reaction pathways open to sulphur, discussed in the previous section. That engines emit the same percentage of fuel sulphur as sulphate particulate, whatever the throughput, is, then, perhaps surprising.

Descending, now, into the chemistry, gas-to-particle conversion in the water – sulphate system is underpinned, mechanistically, by the profound effect of SO_3 on *acid dew point*. In the absence of SO_3, water condensation within the exhaust system seems unlikely in view of the low dew point (<50 °C). But in the presence of just 10 ppm SO_3, the dew point rises to 120 °C; while for 40 ppm SO_3, it rises still further, to 170 °C (Edwards, 1977, pp. 70–74). Sulphuric acid thus condenses earlier, or further upstream in the exhaust system, than water alone, and the location of this conversion is acutely dependent on FSC.

In extreme instances, sulphuric acid is discharged as a visible mist akin to condensed water droplets (Section 2.4.3) or condensed hydrocarbon droplets (Section 3.6.4). This mist, once collected, may appear yellowish on a filter (Kasper, 2003). However, acid mists seem rather unlikely in view of

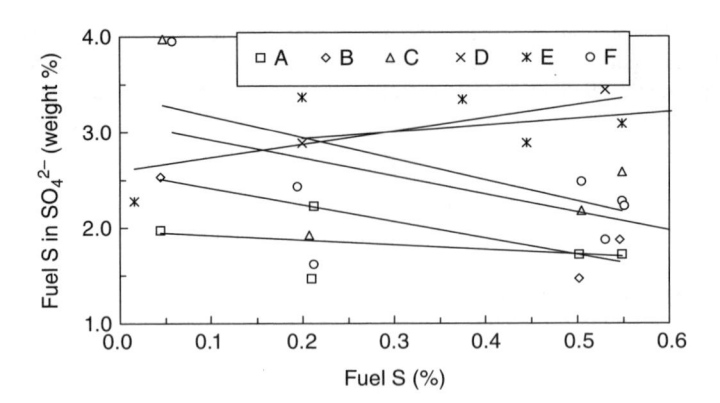

Figure 3.23 Weight percentage of fuel sulphur converted to sulphate particulate, according to a wide variety of fuel specifications and duty cycles: A, cold-start transient; B, hot-start transient; C, 1400 rpm steady-state; D, high power; E, cruise; F, idle. Also plotted are corresponding linear polynomial fits. Engine: 14-litre, six-cylinder, direct-injection, turbocharged, rated 216 kW at 2100 rpm (Wall and Hoekman, 1984).

today's low-sulphur fuels – unless sulphur storage has occurred within the exhaust system, such as in catalysts, in which case sulphate particulate can be intermittently belched. In circumstances more typical, i.e. when sulphuric acid does not dominate, one suspects adsorption and condensation onto nonvolatile particulate: thus, gas-to-particle conversion of the acid parallels that of organic molecules (Sections 3.6.3 and 3.6.4), with the proviso that freshly emitted soot is generally thought hydrophobic – a similar conclusion has been reached about aircraft contrails, in which adsorption of H_2SO_4 by soot is initially impeded, and must await surface activation (Gleitsmann and Zellner, 1998). But, in great contradistinction to organic particulate, the condensation and adsorption of sulphates onto soot in motor vehicle exhaust is ignored in the literature.

What has recently become a topic of avid interest is the role, or suspected role, of sulphuric acid in *nucleation*. That H_2SO_4 instigates nucleation seems plausible from various observations; for example, nanoparticles are strongly associated with FSC (Hall and Dickens, 2003) and oxidation catalysts (Maricq *et al.*, 2002b). This suspicion has led investigators to binary homogeneous nucleation (BHN), a general theory in which 'binary' refers to the cooperation of two molecules in particle creation, in this case H_2SO_4 and H_2O (Baumgard and Johnson, 1996). Actually, this acid – water system represents an ideal exemplar, being regularly targeted by theoreticians (Seinfeld, 1986, pp. 368–373; Noppel, 1998), of especial importance in the atmosphere (Pandis and Pilinis, 1995), and one of the few such systems to have been understood with any decency. The *classical* model, though, as devised originally for the ambient atmosphere, requires some subtle modifications if the considerably hotter exhaust-gas environment is to be suitably modelled (Vehkamäki *et al.*, 2003).

In BHN, collisions and consequent couplings between molecules of H_2O and H_2SO_4 generate *molecular clusters*. However, these clusters are not necessarily stable, and may in fact disintegrate spontaneously; to be viable as nuclei, a minimum number of molecules must gather, and this minimum number, or kernel, is in effect a threshold for particle creation, as expressed by the Gibbs surface energy G. The nucleation rate J (particles/cm³/s) is written as

$$J = C \exp\left[-\frac{\Delta G}{kT}\right], \qquad (3.48)$$

where C is a kinetic prefactor.

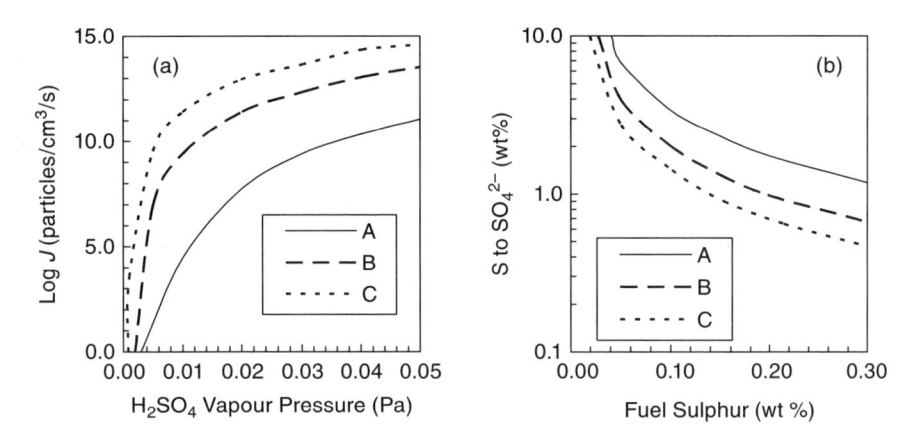

Figure 3.24 Nucleation of sulphate particulate. (a) Rate of particle formation, as a function of relative humidity and vapour pressure of sulphuric acid. Relative humidity in dilution tunnel: A, 10%; B, 20%; C, 30%. Dilution air: 46°C. The data assumes 4% conversion of fuel sulphur to sulphate. (b) Percentage conversion of fuel sulphur to sulphuric acid needed, for nucleation to take place. The data assumes 10% relative humidity in the dilution tunnel and a dilution ratio of 10. Air–fuel ratio: A, 50.0; B, 28.5; C, 20.0 (Baumgard and Johnson, 1996).

From this it becomes possible to determine the temperature, and the vapour pressures of water and acid, necessary to trigger nucleation: several scenarios are suggested in Figure 3.24 (Baumgard and Johnson, 1996). According to this analysis, sulphate nucleation is conceivable in some circumstances, even for FSC = 50 ppm. The hydrated state of the sulphuric acid, when fully incorporated into the model, leads to a predicted threshold, or molecular cluster, of ~10 molecules – corresponding to a size of ~1 nm: this is *below* the range of particle-sizing instruments. In practical terms, the particles must first grow into the countable range. J ranges widely, i.e. 10^4–10^{11} cm^{-3}s^{-1}, depending on temperature and relative humidity (Shi and Harrison, 1999). But although reasonable agreement of a *qualitative* nature is attained between theory and experiment, actual particle numbers unfortunately exceed by several orders of magnitude those expected from BHN, and the effect of FSC is weaker than predicted. This suggests the participation in particle growth of molecules besides H_2O and H_2SO_4 – seemingly confirmed by compositional analysis, of which more later.

Evidence that other species underwrite particle growth is obtained through alternative definitions of the threshold C_{crit}, for example (Khalek *et al.*, 2000)

$$C_{crit} = 0.16 \exp(0.1T - 3.5RH - 27.7), \tag{3.49}$$

where T is the temperature, and RH is the relative humidity (ranging from zero to unity). The rate of particle growth is given by

$$\frac{dD_p}{dt} = 2k \frac{(\rho_{vapour} - \rho_{sat})}{\rho_{droplet}} \left(\frac{\Re T}{2M\pi}\right)^{0.5}, \tag{3.50}$$

where M is the molecular mass, and ρ is, according to the suffix, the droplet density, the vapour density or the saturation vapour density; k accounts for concurrent take-up of water, in order to maintain equilibrium between droplet and vapour. The modelling is simplified by the fact that, in the free-molecular regime, growth proceeds independently of particle size.

Two interesting observations arise from this analysis (Khalek et al., 2000). The first is the intuitively correct expectation of interdependence between particle *creation* and particle *growth*: simply, the amount of growth material is finite, and must be shared around. For a condition which suddenly creates hyperabundant nanoparticles, competition for growth material is fiercer, and this growth material is depleted more swiftly: growth rates are therefore correspondingly slower. For this reason, growth rates in the above study varied from 2 nm/s to 24 nm/s, as defined by the expanding particle radius.

The second observation concerns organic compounds, to model which, nonadecane was taken as representative – a molecule of middling carbon number in the organic particulate (Khalek et al., 2000). The vapour pressure of this compound, even at its highest, still lies two orders of magnitude below that required for homogeneous nucleation, so that nonadecane was not, on its own, implicated in particle creation. But surface growth rates also suggested an inability of sulphuric acid droplets to grow into the detectable range (>8 nm) within the time available (<1 s). The question, then, is why particles of 20–30 nm in size appear on the same timescale, for which the answer, it seems, is the accompanying role of organic compounds as *growth* agents, evidence for which is gathering swiftly (Vouitsis et al., 2005). It should be said that not all organic compounds are growth agents, and indeed some appear to suppress or inhibit nucleation, the effects being mediated through water solubility and surface tension (Mathis et al., 2004). Studies of the atmosphere, similarly, point to sulphuric acid and water as the initial nucleating agents, with other species, such as organics and ammonia, acting as growth material (Zhang Q. et al., 2004); direct compositional analysis of nanoparticles leads to very similar conclusions about motor vehicle exhaust, where, in the absence of nucleated sulphates, organic compounds seem more willing to associate with the carbonaceous accumulation mode (Schneider et al., 2005).

This brings us to an instructive parallel, namely the exhaust plumes of jet aircraft, or airliners, where nucleation studies are far commoner than in the case of motor vehicle exhaust (Gleitsmann and Zellner, 1998; Kärcher, 1998). This reflects concern amongst atmospheric scientists about aircraft contrails, ice crystals, cirrus clouds and climate change. If >1 % of the SO_x is emitted as SO_3, BHN is dramatically enhanced; only a few per cent of the SO_3 is taken up by soot, so that levels of H_2SO_4 build up rapidly. BHN is thought to be initiated by the emergence of hydrated clusters, composed of one molecule of H_2SO_4 and one to ten molecules of H_2O: the resulting liquid-phase H_2O–H_2SO_4 particles are extremely stable (Zhao J. and Turco, 1995). But *classical* BHN theory, whether applied to jet aircraft exhaust, or to motor vehicle exhaust significantly underpredicts the population of nanoparticles, and also suggests a stronger influence of FSC than actually found (Yu F., 2001).

Investigators have thus turned to the alternative theory of chemiion-induced nucleation (CIN), or ion-mediated nucleation (IMN), in which ions work in combination with H_2SO_4 and H_2O, the idea being that *charged* molecular clusters, through electrostatic effects, grow faster than their electrically neutral counterparts. As modelled, IMN is faster in particle creation than BHN, and a nucleation-mode 'bifurcation' appears, between charged and uncharged particles. As hinted, ions are familiar territory in investigations of jet aircraft exhaust: various exotic species are detected – $HSO_4^- H_2SO_4$, $HSO_4^- SO_3$, $HSO_4^- HNO_3$, $NO_3^-(HNO_3)_m$ – which, for plume ages of a few milliseconds, reach number concentrations of $1.4 \times 10^7 cm^{-3}$ (Arnold et al., 1998), and large 'cluster ions' are created, for example $HSO_4^-(H_2SO_4)_n(H_2O)_m$, from the reaction between HSO_4^- H_2O, H_2SO_4 and HNO_3 (Yu F. et al., 1999).

The finding that nucleation-mode particles in diesel exhaust are mostly uncharged tends to argue against IMN (Jung H. and Kittelson, 2005), although the database is, admittedly, poorly stocked in this area (Section 6.2.5). It is, nevertheless, still plausible that theories originally devised for jet aircraft might be advantageously redeployed to motor vehicles. As corresponding chemiion studies for motor vehicle exhaust do not appear to exist, some reasonable assumptions are necessary for ion concentrations in the combustion (typically 10^{10}–$10^{11} cm^{-3}$ in hydrocarbon flames), and for likely rates of depletion via ion – ion recombination, wall losses, and scavenging by soot particles, in order to estimate ion abundance in motor vehicle exhaust (Yu F., 2001). The ion concentration at the moment of discharge into the atmosphere was thus estimated by Yu F. (2001) at 10^7-10^9cm^{-3} and IMN allowed more accurate predictions to be made of nanoparticle concentrations. These ideas have, with equal felicity, been applied to nucleation in the transfer line (Section 4.4), in the exhaust plume (Section 4.3), and downstream of a particulate filter (Section 7.8.2) (Yu F., 2002). In the latter case, the appearance of *more* nanoparticles in the absence of soot, the ion-scavenging action of which suppresses nucleation, is in line with IMN.

According to these ideas, nanoparticle abundance is determined *not* by the availability of H_2SO_4, as this controls growth only, but by the availability of chemiions, as this controls particle creation. This observation sheds light on the weaker influence of FSC than would otherwise be expected, i.e. when an order-of-magnitude reduction merely halves the number of nanoparticles (Yu F., 2001). When less H_2SO_4 is available, ion clusters grow more slowly, so that their period of vulnerability, to extinction by ion – ion recombination and wall losses, is protracted; they are thus denied opportunity to grow into discernible particles. The point is that once ions have grown into particles, and hence *already* passed this vulnerable stage in their development, the availability of sulphuric acid and, by inference, the FSC have no bearing on the number of particles.

As a final piece in the nucleation jigsaw, *acidic* and *organic* aerosols chemically interact in their formative stages, by which action the amount of particulate is increased. Sulphuric acid acts as an acid catalyst in converting, through heterogeneous reactions, organic compounds into the particulate phase (Jang M. *et al.*, 2005). Liquid-phase sulphuric acid reacts with short-chained unsaturated hydrocarbons, for example alkenes form alkyl hydrogen sulphates, while low-molecular-weight aldehydes (CH_2O and CH_3CHO) are very soluble in water, and this could generate charged clusters such as $CH_2OH^+(CH_2O)_n(H_2O)_m$ (Yu F. *et al.*, 1999). Organic species tend to collect around positive ions and particles, while sulphuric acid tends to condense on negative ions and particles. This leads to two variants of nanoparticle, positively charged organic and negatively charged acid, the populations of which depend on the relative abundance of precursors; for example, as FSC declines, the organic character of the aerosol is thought to be heightened. But, again, such ideas are untouched in studies of motor vehicle exhaust.

3.8 Closure

In this chapter we followed particulate formation according to the four principal fractions, commencing with the earliest moments in the combustion chamber and concluding with the final moments as the exhaust gas enters the surrounding air or the ambient atmosphere. The carbonaceous fraction, or soot, forms within the engine, this process being essentially complete by the time the exhaust valve opens; the ash fraction, depending on the component in question, is probably formed before the exhaust valve opens, but some gas-to-particle conversion within the exhaust system is not generally ruled out; the sulphate and organic fractions, being the most volatile, form the latest, perhaps at the cold end of the exhaust system or, more likely, within the exhaust plume or dilution tunnel.

3.8.1 Carbonaceous Fraction I. Classical Models

Soot formation is the process by which fuel molecules, containing 12–22 carbon atoms and twice as many hydrogen atoms, are converted, in a just few milliseconds, to 'spherules', containing thousands of carbon atoms and one-tenth as many hydrogen atoms. The straightforward *global* reaction scheme, starting with fuel and oxygen, and finishing with carbon dioxide, hydrogen and carbon, points to a critical air–fuel ratio at which soot will appear; but, in practice, this ratio is not derivable analytically, and is not universal: it varies with the nature of the combustion process, such as the fuel composition, the mixing of fuel and air, and the structure of the flame.

Empiricisms

Because of the great chemical complexity of soot formation, which involves hundreds of *intervening* reactions, researchers have understandably embraced empiricisms and global rate expressions: for example, soot has been correlated to the fraction of fuel burned in the mixing-controlled phase, the combustion temperature, and certain aspects of the heat release, while other, semi-empirical models, using Arrhenius relationships, are also well known.

Inception

Soot particles are created by the process of 'inception', the nature of which is still not understood satisfactorily, but, broadly, fuel hydrocarbons are broken up by pyrolytic reactions and rearranged into aromatic molecules; this involves the 'cyclisation' of aliphatic molecules, and the appearance of benzene, the basic molecular building block. The aromatic molecules then continue growing, in which acetylene acts as a feedstock. Carbon is incorporated, and hydrogen removed, in the 'hydrocarbon abstraction and carbon addition' (HACA) mechanism. Crossing the gas – particle boundary is the least understood aspect of inception; ideas involve macromolecules, ions, ingested airborne particles and fullerenes. The first discernible particles, or 'nuclei', are about one or two nanometres in size, and contain as few as fifty carbon atoms; they exist in massive numbers, but the total mass of soot is still very small.

Surface Growth

Additional soot is now laid down around these nuclei in a surface growth process – this creates the spherules. Surface growth greatly increases the total mass of soot, but leaves untouched the number of particles. The spherules grow into a certain range of size, say $\sim$20–50 nm, and then stop. Why spherules are of such a restricted range of size remains unknown; some common termination of the growth process seems to transcend the specific details of the combustion. Surface growth is responsible for the layered microstructure, in which crystallites are arranged concentrically around the spherule. This layering differs distinctly from the structure of the centre of the spherule, the disordered, or amorphous, structure of which is thought to betray the progenitive nucleus.

Agglomeration

Spherules avidly agglomerate, that is, they vigorously collide and adhere to one another; and, in this growth process, which is physical rather than chemical, the number of particles decreases but the mass of soot is constant. Some agglomeration occurs when the spherules are still very young, and of a tarry or fluid consistency, so that original identities are obscured; coalescence is thought to explain instances where several nuclei are found within one spherule. But even in the absence of fusing, surface growth continues concurrently, so that an overlying shared lattice structure develops, by which original boundaries are obscured. Clustered or chained morphologies are displayed, according to the nature of the collisions that generated the agglomerates.

Oxidation

More than 90 % of newly formed soot is promptly oxidised within the engine, prior to emission. There is some debate as to the responsible oxidant, i.e. not just O_2, but also O, OH, CO_2 or H_2O: this question depends on the location in the combustion chamber and the moment in the combustion process. The fluid mechanics of the combustion chamber, insofar as this controls the mixing of air and soot, and movement of the soot cloud, are also important. The oxidation of spherules is generally supposed to be kinetically limited, and to take place on the surface, as expressed by surface recession. Evidence exists, though, that some oxidation takes place *inside* spherules.

3.8.2 Carbonaceous Fraction II. The Combusting Plume

Historical Overview

Following the start of combustion, the amount of in-cylinder soot rises rapidly, reaches a peak and then, at a gradually diminishing rate, declines, to reach, eventually, and indeed asymptotically, the levels observed in the exhaust system. This pattern is held to betray the shifting balance between soot formation and soot oxidation, the first being favoured in the early stages, the second in the later; these two processes are of *equal* importance in deciding engine-out soot. Formation and oxidation are affected *differently* by engine operating parameters (e.g. EGR, injection timing and injection pressure), and hence the net effect is not always obvious: the formation of *more* soot, as indicated by a higher peak, can be offset by more vigorous oxidation, in which case there is *less* engine-out soot.

Premixed Burn

The first soot particles are thought to form during the premixed burn, in the well-mixed section of air and fuel vapour, towards the leading edge of the combusting plume. This soot appears virtually simultaneously in multiple pockets, with which the whole region is rapidly suffused. Soot formed in this phase of the combustion seems unlikely, however, to make much contribution to the total emission. During transition to the mixing-controlled phase, soot particles appear, not inside the combusting plume, but around the *periphery*. These peripheral particles are larger than those formed inside, and are thought to form in the diffusion flame that is now enveloping the combusting plume. These particles are not, however, thought predominantly responsible for the soot emission.

Mixing-Controlled Burn

The greater part of the soot emission probably arises not from the plume's periphery but from within the plume itself. The conceptual model of diesel combustion currently gaining widespread acceptance holds that during the mixing-controlled burn, soot formation occurs in a rich, premixed flame that 'stands off' from the injector nozzle. These soot particles then grow as they pass down the combusting plume along the plume stem towards the head vortex, where they are burned in the enveloping diffusion flame. Soot concentration and particle size thus increase with distance along the plume, but, at any particular cross-section, both appear to be roughly constant.

Late Burn

An inability to burn all of the in-plume soot is a prime reason for the emission of soot from the engine. Following end of injection, the combusting plume collapses, and the head vortex fragments into numerous soot-filled pockets. Diffusion flames, for a time, continue to surround these soot-filled pockets, but they are quenched prematurely, i.e. before the soot is fully consumed, and bulk temperatures are now too low to support further oxidation. This soot then survives until the exhaust valve opens.

3.8.3 Carbonaceous Fraction III. Wall Interactions

Interactions between the combusting plume and the surrounding walls are thought a significant aspect of soot formation. The walls are markedly cooler than the bulk gas; hence soot that successfully deposits will escape oxidation, with possible re-entrainment once temperatures have fallen. Thermophoresis is well chronicled as the chief deposition mechanism in fostering particle transport across the boundary layer. The trajectory and duration of the combusting plume with respect to the wall are important factors in bringing particles into the vicinity of the wall. The thickness of the wall deposit probably adjusts in line with the operating point of the engine. The soot deposit builds rapidly, perhaps linearly, from a clean condition; the rate of deposition then gradually levels off. Measurements of exhaust soot do not relate clearly to the deposited soot; in fact, the effects seem to be unpredictable: EGR and injection timing, for example, have strong effects on exhaust soot, but weak effects on deposition. The deposited soot may allow the flame to approach the wall more closely, for example if oxygen is given up from within its porous structure or if heat transfer to the wall is reduced. Rather than promoting burn-up once deposition has taken place, a better strategy might be to discourage deposition.

3.8.4 Ash Fraction

Chemical Reactions

The sources of ash particulate are lubricant, air, wear and tear, and fuel. The most studied source is the fuel. Inorganic or organometallic compounds are broken down, and their metals probably vaporised, in the combustion. The thermal and oxidative stabilities of these compounds with respect to those of the fuel hydrocarbons are paramount. The chemical compounds formed by ashing elements are poorly reported, but must, inevitably, shift as conditions cool; thermodynamic equilibria suggest oxides, hydroxides and sulphates. (The availability of sulphur must be a factor.) Given, however, the rapid cooling, nonequilibrated states are probably the rule. Some fields where these chemical transitions are better understood were highlighted, e.g. the incineration of municipal waste, and coal combustion. Inorganic species, notably metals, have soot-suppressing effects, for which various explanations are posited: improving oxidation or impeding agglomeration; not all additives operate via the same mechanism.

Gas-to-Particle Conversion

The chemical forms assumed by the inorganic material determine the moment (or temperature) of gas-to-particle conversion, and this moment, with respect to the appearance of spherules, is especially important: ashing elements are atomically dispersed within spherules; trapped within spherules as domains, or 'occlusions'; resident on spherule surfaces as domains, or present as separate nucleation-mode particles. Experiments with ferrocene in laboratory flames have been extraordinarily revealing: particles of iron oxide appear, and these acquire, and then lose, carbonaceous coatings, so that the original particles re-emerge in the burn-up zone. This burn-up may be complete or partial. A complicated relationship exists between the iron and the growing spherules, with the former diffusing outwards, keeping abreast of the depositing soot. More soot might be formed within the flame via the action of the additive; but, if the burn-up is more extensive, this leads ultimately to a lower emission of soot. Pronounced ash nucleation modes are observed from engines when treat rates for organometallic fuel additives are excessive.

3.8.5 Organic Fraction

Chemical Reactions in the Engine

Fuel compounds that successfully escape combustion give rise *directly* to organic particulate; and fuel compounds that are partially broken down, and then synthesised into new compounds, contribute *indirectly* to organic particulate. This latter route exists because the organic fraction is, in effect, the feedstock for the carbonaceous fraction, and because soot-forming reactions are prematurely interrupted; this is the concept of 'organic intermediates'. The presence of native fuel compounds *and* combustion-synthesised compounds belies the frequently expounded supposition that organic particulate emissions are a straightforward reflection of unburned fuel compounds. Three fields of research of relevance to organic intermediates were described. First, 'survivability' studies, or speciative comparisons, which, despite confounding by in-engine deposits and oil, suggest that fuel compounds, rather than pyrosynthesised compounds, dominate the organic fraction. Second, laboratory flames, which contain an intermediate or dark zone between the unburned fuel and the emergence of spherules, within which mysterious 'proto-particles' are found, the natures of which are organic and tarry. Third, furnace experiments, in which hydrocarbons are subjected to high temperatures in inert atmospheres, and, in the effluent of which, particulate emissions change with respect to temperature, from organic to carbonaceous, from fluid to solid, and from yellow to black. From consideration of these three fields, it becomes possible to explain a widely observed trend in the particulate emission from diesel engines: wet or organic at low load, and dry or carbonaceous at high load. At low load, temperatures are insufficient to drive pyrolytic reactions all the way to soot, so that organic intermediates are emitted instead. Soot-forming reactions are interrupted by wall quenching and bulk quenching, and because fuel decomposition is too slow.

Chemical Reactions in the Exhaust

Chemical reactions do not cease once the exhaust gas leaves the engine: PAH, especially, are still being modified at surprisingly low temperatures, 250 °C. This threshold is, however, difficult to define. It is, therefore, immensely difficult to estimate the tailpipe emission merely by extracting samples from the exhaust system. Ongoing chemical reactions in the exhaust gas are thought to underpin the conversion of PAH to more mutagenic NPAH, although the nitration pathway is poorly reported, and genuine reactions are poorly distinguished from artifactual ones.

Gas-to-Particle Conversion: Models

Most gas-to-particle conversion in the organic fraction happens very late in the emission process, i.e. in the exhaust plume or the dilution tunnel. Models suggest that adsorption and desorption, rather than condensation and evaporation, govern this transference. But condensation may become important when hydrocarbons are emitted at high concentrations compared with soot – a situation that occurs downstream of a DPF, for example. How closely equilibrated this gas-to-particle conversion is remains poorly researched; an admitted failing in current models is the absence of kinetic considerations.

Gas-to-Particle Conversion: Measurements

Measurements undertaken with an FID, operating at different temperatures and at different points in the exhaust system, to track the transfer of hydrocarbons into the particulate phase accord reasonably well with the quantity of organic particulate ultimately found on a dilution tunnel filter. It seems that most hydrocarbons, when in the exhaust system, stay in the gas phase, and that, consequently, their oxidation by catalysts proceeds independently of particle transport, i.e. the particles do not, at this location, act as carriers for these hydrocarbons.

White Smoke

White smoke arises whenever the organic fraction (notably unburned fuel) considerably exceeds the carbonaceous fraction; indeed, nucleation into droplets is inevitable whenever the solid surface area is overwhelmed; the hydrocarbon concentration during this time reaches thousands of ppm. Cold starts and cold ambients, as they raise the saturation, are exacerbatory factors. The root causes of white smoke, however, are long ignition delays and low in-cylinder temperatures, which permit fuel sprays to strike the walls of the combustion chamber, so that films of liquid fuel accumulate; then, when the engine begins warming up, these fuel films evaporate inappropriately, and escape combustion. The relevant fuel properties are cetane number and T90. White smoke does not reflect exactly the fuel speciation, because combustion and storage within the engine are selective. Meaningful measurements of white smoke are impossible without proper control over saturation and gas-to-particle conversion.

3.8.6 Sulphate Fraction

Chemical Reactions

Sulphur arrives in exhaust gas courtesy of both fuel and oil. It is emitted by the engine as SO_2; in the absence of a catalyst, only a few per cent exists as SO_3. Thermodynamic equilibrium favours SO_2, but this conversion is kinetically limited. Consequently, sulphates were a significant impediment to the implementation of oxidation catalysts, the nuisance action of which is to convert SO_2 into SO_3. It is also possible that, to some extent, existing particles (especially metals) catalyse the conversion of SO_2 to SO_3. The problem is really SO_3, as this promptly undergoes hydrolysis to sulphuric acid, H_2SO_4; and this acid, in turn, readily undergoes gas-to-particle conversion. Sulphuric acid shifts the dew point, so that condensation occurs at higher temperatures, i.e. earlier in the exhaust system. One gram of sulphur forms three grams of H_2SO_4, not counting the additional water of hydration. The SO_x chemistry may be coupled to the NO_x chemistry, but this aspect has not been well studied.

Gas-to-Particle Conversion

Sulphuric acid is an instigator of particle creation via nucleation; this it does by acting in concert with water, as described by the theory of binary homogeneous nucleation (BHN). Even so, the BHN model is insufficiently predictive of measurements, and other ideas, adapted from studies of jet aircraft exhaust, suggest some form of ion-mediated nucleation (IMN). Following the emergence of these sulphate kernels, organic compounds are thought to take over as the principal growth species.

4

Formation II: Location

4.1 Introduction

In this chapter particle formation will be explored still further, but with particle composition somewhat downplayed, in favour of the *location* or, alternatively put, the *moment* in the emission process. As already explained in the preamble to the previous chapter, the reason for thus blurring the compositional focus is that, in the exhaust system, particle composition is often unknown, indeed hybridised: for example, it would be fatuous to discuss separately the deposition of inorganic and carbonaceous particulate onto exhaust system walls; yet deposition is nonetheless an extremely important formative influence on the particulate emission. Examination of particle formation, in each consecutive stage of the emission process, is therefore a surer and more logical method of dissection of the subject.

As related in the previous chapter, particle formation begins in the oxygen-deficient cores of fuel sprays, with embryonic development as polyaromatic compounds, birth as carbonaceous spherules, and a coming of age as adolescent agglomerates. At around this point the exhaust valve opens, and the particles, which have now reached early adult life, pass into the exhaust system to begin their journey to the tailpipe, during which time they continue growing through the accretion of ever more gas-phase material. We leave the particles as they enter the exhaust plume and begin their airborne journey into middle age; for atmospheric ageing, senescence and their eventual fall to earth do not concern us here.

The emission process, as a whole, encompasses two clear and distinct phases: undiluted exhaust and diluted exhaust. In the first, especially, the exhaust system is a nonisothermal chemical cauldron, the contents of which abound in reactive and reacting species; and the ongoing chemical changes involve homogeneous gas-phase reactions and heterogeneous particle–gas reactions. In transitioning to the second phase, that is, in leaving the tailpipe, the exhaust gas experiences an extremely rapid dilution in the ambient, the impact of which on the particulate phase is unpredictable without detailed knowledge of exhaust-plume dynamics, that is, the duality of mixing and cooling, as the exhaust gas penetrates the surrounding air. Throughout this emission process, then, nothing is immutable, while everything is impermanent.

Particulate Emissions from Vehicles   P. Eastwood
© 2008 John Wiley & Sons, Ltd

The present chapter is divided according to the type of emission process undergone by the exhaust gas. The first two sections cover the *real-world* emission process, i.e. the exhaust system and the exhaust plume; the remaining three sections cover the *laboratory* emission process, i.e. the transfer line, the dilution tunnel and the dilution tunnel filter. It will be appreciated that a clear line is not easily traced between the last three sections in the present chapter and the measurement methods discussed in the next, but, to emphasise the distinction, the present topic is particle formation, whereas the next chapter is concerned with instrumentation.

4.2 Within the Exhaust System

Ejection of exhaust gas from the engine is characterised by two markedly distinct and, as a function of crank angle (CA), consecutive phases: 'blowdown', or expulsion by the release of cylinder pressure, and 'displacement', or expulsion by upward movement of the piston (Myers, 1983; Heywood, 1988, pp. 231–232).[1] These two phases are indelibly stamped on all emissions, particulate being no exception: this evidence has been gathered by filtering the exhaust gas, and by light-scattering measurements as a function of crank angle, although the general applicability of these findings is uncertain, as this field of research is but poorly trodden.

The picture, then, is a dual-peaked particulate emission as a function of crank angle, in which each peak relates, in turn, to blowdown and displacement. In filtering experiments, the mass emission during the blowdown phase is twice that during the displacement phase (Kittelson *et al.*, 1990), yet cycle averages are consistent with conventional measurements. Light-scattering experiments reveal a peak in the number emission at the moment of maximum valve lift (Corcione and Vaglieco, 1994); however, extensive cycle-to-cycle variations are also seen; and a second, more modest peak, observed at high load, is absent at mid-load (Klingen and Roth, 1991).

Intriguingly, light-scattering theory indicates that particles emitted during the blowdown phase are somehow 'different' from those emitted during the displacement phase (Klingen and Roth, 1991). Perhaps they are larger – as one might expect of re-entrained particles. In Section 3.4, extensive evidence was amassed pointing to in-engine deposits as a controlling factor in particulate emissions, while fragmentation and release of these deposits appears to be a central feature of CA-resolved measurements. It is only logical to ask why such deposits should break up and re-enter the flow, but mere pronouncement of 'fluid shear', with a wave of the hand, is trite and jejune. Re-entrainment could result when surfaces are struck by turbulent eddies; perhaps the pressure in the stagnation region at the wall exceeds that in the high-velocity region above the deposit, thus generating lifting forces normal to the wall, or the viscous, highly sheared flow generates drag forces. A model in which 'turbulent bursts' exfoliated the deposits by acting on the viscous sublayer gave results in qualitative agreement with CA-resolved measurements of the particulate emission (Abu-Qudais and Kittelson, 1995).

It should be admitted that the supposed 'different' nature of particles emitted during blowdown, as determined in light-scattering experiments, might be an aspect not of size but of some other optical property (e.g. refractive index) (Klingen and Roth, 1991). Again, this is intriguing: although blowdown emissions were double those in the displacement phase, the soot was responsible for only one-fifth of this difference; the rest was ascribed to *volatile* components of the particulate (Kittelson *et al.*, 1990). These blowdown particles may therefore be 'different' for several reasons: a greater volatile presence in re-entrained particles, a preferential escape of lubricant, and fuel absconding from nozzle sacs (Section 7.3.1).

---

[1] One should bear in mind a possible dilution effect caused by valve overlap.

After leaving the vicinity of the exhaust ports, particles commence their journey through the exhaust system, but, with the exception of the two subsections that follow, which address specific instances of particle interactions with walls, studies into formative processes within the exhaust system from a behavioural standpoint are infrequent. Such studies do require special sampling equipment, in order to catch particles *in situ* rather than risk confounding transformations prior to analysis; but the installation of impactors, for example, within exhaust streams (Vuk *et al.*, 1976) is an uncommon mode of investigation, so that data in this area are rare.

Logically, agglomeration continues when particle concentrations are high (Section 2.2.4) (Vuk *et al.*, 1976). Particles extracted and examined under the microscope exhibit fractalities nearly independent of distance along the exhaust system, although they are larger at high engine load, something commensurate with the higher engine-out soot emission. However, no appreciable growth occurs, even after 3 m of travel. Agglomeration is not, then, measurable or, more accurately, it proceeds slowly in comparison with the rate of passage through the exhaust system (Klingen and Roth, 1991). Why agglomerates should become more chain-like with passage along the exhaust system is unclear (Lee K.O. and Zhu, 2005). At high temperatures, it seems conceivable that particle growth through agglomeration will be reversed by oxidation; and although direct evidence of particle fragmentation is lacking, spherules do appear to become smaller in transit (Lee K.-O. *et al.*, 2001). Gas-to-particle conversion, and the progressive accretion of volatiles, is only to be expected with declining temperature: impactor plates mounted in the exhaust stream revealed yellow, gummy deposits, rather than black, carbonaceous ones, unmistakable evidence of a higher volatile contingent (Vuk *et al.*, 1976).

4.2.1 Storage and Release

Deposition, naturally, begins with particle *transport*, but, excepting two special cases, namely *catalysts* and *filters* (Sections 4.2.3 and 7.8.2, respectively), little has been published of direct relevance to exhaust systems. The obvious inference from this situation is that particle transport *within* aftertreatment devices, rather than in the exhaust system, is what really interests the automotive industry. But retention of particles by exhaust-system walls is naturally enough evidenced in blackening. Since, in the normal course of events, these deposits do not grow beyond a certain thickness, then storage and release must, presumably, come into balance. This balance depends, however, on engine operation, and shifts accordingly; in adverse circumstances, the sudden exfoliation of these deposits, in events inelegantly styled 'blow-outs' – evident in dramatic instances as smoke clouds – allows tailpipe emissions to exceed briefly the engine-out emissions (Andrews *et al.*, 2001a). Consequently, one can never be wholly sure whether the instantaneous particulate emission, as measured at the tailpipe, is really the immediate product of engine operation. Proper preconditioning procedures are thus absolutely essential.

Significant implications arise for type approval procedures, or homologation, because the driving schedule *before* the test, which decides the initial state of the exhaust system, assumes the same importance as the test itself. It should be said, however, that legally mandated cycles tend to be fairly lightly loaded, with accelerations kinder than those more probably experienced by vehicles in the field. Consequently, the true picture of storage and release might not be apparent purely from type approval. Similar remarks hold for inspection and maintenance: the first measurements of exhaust opacity are unrepresentative, so that repeats are necessary (Ragazzi *et al.*, 1986). For driving patterns encompassing extensive *urban* mileage, where high-speed/high-load purges are infrequent, particle clouds, discharged intermittently during aggressive acceleration, can be unacceptable from a consumer standpoint. Conversely, regular high-speed/high-load driving seems less prone to the phenomenon of blow-out, simply because deposition is strongly discouraged. The deposits first need time to build, so that the phenomenon has an incubation period. In other cases the storage periods are just few seconds or so, evident as a lag in the particulate emission, as compared, say, with the emission of CO_2 (Hofeldt and Chen, 1996).

The two prominent deposition mechanisms are thermophoresis, as driven by (especially rapidly shifting) temperature gradients (i.e. engine acceleration), and inertial impaction, as driven by exhaust flow rate (i.e. engine speed). The reverse process of re-entrainment is less easily summarised, but vibrations and fluid shear are obvious factors (Andrews et al., 2002). A fundamental study found little interaction with the walls of the exhaust system in the absence of thermophoresis, but gas flow rates were low (Bouris et al., 2000). More representative conditions show that thermophoretic deposition is modest for gas–wall temperature differences of 100 °C (Crane et al., 2002). The expectation that high flow rates increase inertial impaction jars with the anecdotal observation of the previous paragraph, that high-speed driving discourages particle deposition. This serves as a reminder that particles may lodge only temporarily in one location before being briefly re-entrained and then caught once more at another location, so that they bounce or cascade their way through the exhaust system. Particles released from a first muffler may simply lodge in the second muffler (Andrews et al., 2000d). As said, these ideas are conjectural, since this field has not proven attractive either to experimenters or to modellers. Lacking especially are close studies of particle size, since, according to received wisdom, deposit fragmentation releases atypical coarse-mode (supermicron) particles.

A particular case of particle storage arises at cold starts (Andrews et al., 2000c), during which the strength of the thermophoretic force, initially strong, swiftly subsides in accordance with rapidly shallowing thermal gradients. Particulate emission rates have been measured at various points along an exhaust system during this period; some results are given in Figure 4.1 (Andrews et al., 2001b). It took thirty minutes of operation from a cold start before a relatively stable emission, free of deposition and re-entrainment, was attained; this time interval corresponded, perhaps not coincidentally, to the period needed to establish a steady exhaust-system temperature throughout. The emissions, however, according to duty cycle, were not wholly intuitive. For example, extended-idle preconditioning did not

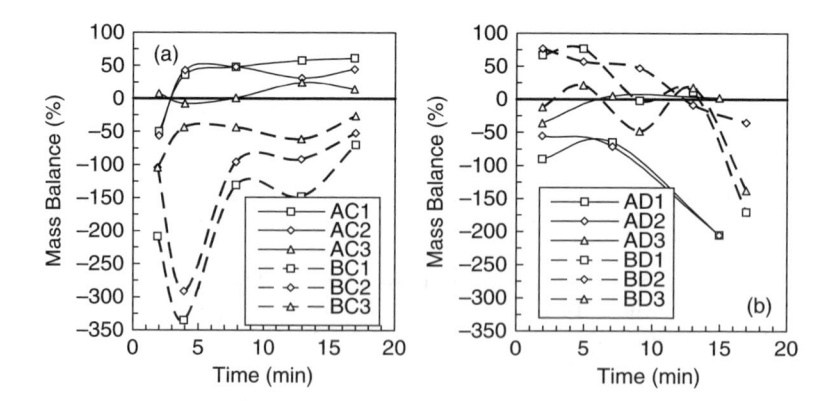

Figure 4.1 Mass balances for the particulate emission across three exhaust system components following a cold start. The legend shows the preconditioning, followed by the cold-start operating point, as follows: A, four hours idle; B, ten minutes high speed; C, 2250 rpm/35 kW; D, 3500 rpm/15 kW. Location: 1, across catalyst; 2, across first silencer; 3, across second silencer. The particulate was collected on filters over four-minute periods. Engine: 1.8-litre, four-cylinder, turbocharged, intercooled, passenger car of model year 1998 (Andrews et al., 2001b).

display high blow-out during the subsequent warm-up; yet blow-out was significant following high-speed preconditioning, when the walls were expected to be clean. As a general rule, the initially low emissions of particulate following a cold start increased to a maximum, and then fell to a steady level. This accorded with the gas-to-wall temperature differences, which displayed the same pattern – with due impeachment of thermophoresis as the dominant deposition mechanism. But, critically, deposition and re-entrainment phenomena within the exhaust system easily overwhelmed and indeed disguised the rate at which particulate left the engine.

Release events include not only solid particles re-entrained from walls but also *volatile* components outgassed from these deposits (Hall and Dickens, 2000). The take-up and release of such volatile components is much less studied. Particulate deposits residing on walls contain, to several per cent, alkanes of high molecular weight (C_{18}–C_{24}): these are stored at low temperatures (<400 °C) and returned to the exhaust gas at high temperatures (>600 °C) (Abbass *et al.*, 1989a); the effect is discernible when the engine is motored, and also can be observed by simply passing hot air through the exhaust system (Abbass *et al.*, 1988). These organic compounds, once outgassed, are not necessarily acquired by soot-particle surfaces: they also give rise to nucleation modes, suggesting brief but sufficiently large increases in saturation (Hall and Dickens, 2003). Intuitively, outgassing and re-entrainment are two distinctly differing forms of release event, distinguishable by their effects on size distributions. The countermeasure for volatile components is more readily stated: preconditioning by sustained running at high temperature will force evaporation.

4.2.2 Deposition Within Catalysts

Aftertreatment devices arguably represent special cases of particle deposition in the exhaust system. There are, though, two forms of device, the functions of which are diametrically opposed: *intentional* deposition, as occurs in the diesel particulate filter (DPF), and *unintentional* deposition, as occurs in the catalytic converter: the first is discussed in Section 7.8.2, and the second is the subject of the present section. The two fields are not, though, mutually exclusive: aftertreatment devices in which particles are encouraged to deposit onto the walls of narrow, open channels, are akin to a DPF in function, and yet akin to a catalytic converter in geometry (Rothe *et al.*, 2004). The particle transport mechanisms (Section 2.2.3) are of course common in a fundamental sense for all aftertreatment devices, and apply also to EGR coolers (Section 7.4).

But before plunging into particle transport *per se*, two sets of stage scenery should first be arranged. The first is the *sticking probability* (Section 2.2.3), for which little modelling progress has been made, and, strictly speaking, the fraction of particles that rebound and thus escape capture is unknown: an assumption that *all* particles, upon striking a wall, are captured is customary, by which an *upper limit* to deposition is implicitly calculated. The second aspect is the *flow regime*, which, within the monolith channels,[2] is laminar, although turbulence admittedly exists in the converging and diverging cones.

That catalytic converters retain particles is tritely evidenced in their blackened surfaces. This deposition compromises performance, since catalytic sites are masked or fouled, while in extreme instances, monolith channels are actually blocked. A detailed modelling study obtained expressions for six deposition mechanisms (Figure 2.3; Johnson J.E. and Kittelson, 1994), as follows.

---

[2] The convenient assumption of uniform exhaust flow distribution between the channels is normally made. In reality, however, the velocity is normally higher through the central channels.

A. *Sedimentation.* The residence time of particles within catalytic converters provides insufficient opportunity for gravitational settling, unless sizes exceed 40 μm, so this deposition mechanism is ruled out.
B. *Interception.* The fraction of particles caught providentially as streamlines pass close to the walls is negligible, including on the blunt upstream end of the monolith.
C. *Diffusion.* As modelled, (a) the particle concentration profile across a channel is never fully developed, even at the channel exit, and (b) the particle concentration at the wall is assumed to approach zero – as is commonly the case in diffusional deposition. From these assumptions, calculated deposition efficiencies were significant only for particles of size <10 nm, i.e. negligible on a mass basis but *not* on a number basis.
D. *Inertial impaction.* Some particles strike the upstream face of the monolith, rather than follow the streamlines into the channels. For particles possessing Stokes numbers of ~0.6, half will strike the surface and half will follow the streamlines. Impaction falls off rapidly with Stokes number, with no deposition below a critical value of 0.25. The calculated deposition efficiencies were significant only for particles of size >500 nm, and so this fraction could be significant on a mass basis but uninfluential on a number basis.
E. *Electrostatic attraction.* Since particle charge (Section 6.2.5) is poorly reported, estimates of capture by electrostatic attraction are conjectural. But even though particles can carry several elementary charges, with near-neutrality of the particle cloud on an overall basis, ceramic monoliths are electrical insulators. Taking, from the literature, a field strength of $200\,\text{V}\,\text{cm}^{-1}$, and assuming this strength constant across the width of the channel, an upper limit for capture by electrostatic attraction was calculated, and found significant only for particles of size <10 nm.
F. *Thermophoresis.* Constant particle concentrations were assumed across the channels: a major simplification, it must be said, but the equations thereby become tractable; the resulting equation for thermophoretic deposition was solely a function of the ratio of entry and exit temperatures, i.e. independent of channel geometry. Because of heat transfer to the surroundings, the temperature decreases significantly in the peripheral regions of the monolith, so that steep thermal gradients develop across the outlying channels; however, as these channels also experience lower flow rates, smaller numbers of particles are available for deposition. The model indicated that, for wall temperatures of 200 °C and exhaust temperatures of 300 °C, thermophoretic deposition was inefficient; *however*, at a few hundred nanometres, it exceeded by several orders of magnitude all of the other mechanisms acting in concert. The model was extended to include the effect of driving mode. For *transient* driving, where exhaust temperature and monolith temperature both fluctuate, the rate at which the second catches up with the first determines the thermophoretic deposition. (During these events, the thermophoretic force may actually reverse and hence drive particles away from the wall.) In some circumstances, such as aggressive acceleration, rates of thermophoretic deposition could exceed those observed at steady state by a considerable margin.

Some uncertainty is encountered in computing *total* deposition efficiency when several mechanisms act in concert, i.e. it is unknown whether mutual independence is shown, but a multiplicative assumption is convenient. From the above models, the particle deposition efficiency, as a function of number distribution, is shown in Figure 4.2 (Johnson J.E. and Kittelson, 1994). For steady-state, isothermal operation, and using experimentally determined size distributions, the total particle *mass* fraction captured by the catalyst was estimated not to exceed 10 %. (These results are for a single steady-state operating point.) Not unexpectedly, capture efficiencies are highest for small particles (<10 nm, diffusion) and for large particles (>500 nm, inertia), while thermophoretic deposition, which is modest in comparison, accounts for a few particles in between. Such results are reasonably reproducible experimentally, using test aerosols (Abdul-Khalek and Kittelson, 1995). In summary, soot-dominated

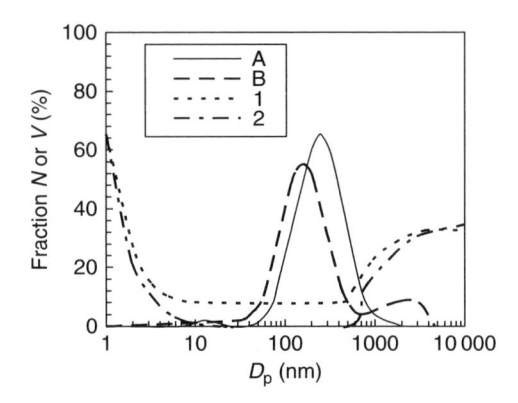

Figure 4.2 Particle interactions with a cordierite diesel oxidation catalyst during steady-state, isothermal operation. Two models are plotted for particle deposition efficiency, expressed according to *number fraction* (*N*): 1, considering diffusion, thermophoresis, electrophoresis and impaction; 2, similar but excluding thermophoresis. Taking two experimentally determined particle size distributions A and B, expressed according to *volume fraction* (*V*), the total capture efficiencies, in terms of mass, were A1, 7%; B1, 10%; A2, 1%; B2, 4%. Feed gas temperature, 300°C; monolith temperature, 200°C; wall thickness, 0.31 mm; channel width, 1.49 mm; cell density, 31 per cm^2; volume, 3.6 litres. Exhaust flow, 113 g/s (Johnson J.E. and Kittelson, 1994).

accumulation modes are deposited thermophoretically and weakly, whereas ash-dominated nucleation modes are deposited diffusionally and strongly. Indeed, a plausible corollary of ever-decreasing engine-out levels of soot is the entry of ever-smaller soot particles into the catalyst. It might, therefore, be argued, that today's catalysts are more, rather than less, likely to experience particle deposition.

4.3 Within the Exhaust Plume

Upon its discharge from the tailpipe, the exhaust gas undergoes a rapid and extremely violent retching, in the course of which fluid-mechanical interactions with the surrounding air give rise to the familiar 'exhaust plume'. This plume encompasses two equally important, concurrent and closely coupled processes: *mixing*, or mass transport, and *cooling*, or heat transfer. This *dual* aspect distinguishes the exhaust *plume* from the exhaust *system*, as, in the latter, cooling alone operates.[3] Obviously, vehicle operation (speed) has a hand in plume dynamics, but the chief distinguishing aspects of exhaust gas dilution in the ambient atmosphere are meteorological variations in wind speed, temperature and humidity, from which corresponding variations also arise in gas-to-particle conversion, from one day to the next, say; moreover, as the local dilution varies from one location to another in the plume, then so does gas-to-particle conversion. All this makes for a highly variable and inhomogeneous system.

The extreme rapidity of dilution in the ambient atmosphere suggests little opportunity for equilibration between gas phase and particulate phase, while the final dilution ratio is, effectively, infinite. Real-world dilution is thus quite different from *laboratory* dilution (Section 4.5), which proceeds in a more leisurely fashion towards a finite dilution ratio. This, inevitably, leads to differences

---

[3] One should be alert to the possibility of some small dilution slightly *upstream* of the tailpipe, as pressure pulsations in the exhaust gas might briefly reverse the flow (Abbass *et al.*, 1991c).

in the nature of the particulate phase (Frisch et al., 1979). Researchers have, therefore, been animated to study real-world, on-road, in-service exhaust plumes, and to extract particle samples therefrom, as these studies guide appropriate and representative selection of dilution ratio, dilution rate, residence time, temperature, humidity, etc. in the laboratory.

Real exhaust plumes, or perhaps those with some slight artificiality, are studied in wind tunnels (Clark N.N. et al., 2003; Gautam et al., 2003), an advantage of which is elimination of the unpredictability caused by crosswinds. Another method, suitable for real roads, is the provision of appropriate equipment, including a sampling boom, on the vehicle under test (Weaver and Balam-Alamanza, 2001), although the sampling distance from the tailpipe is admittedly quite restricted. Genuine exhaust plumes are best studied in 'chase' experiments, in which mobile aerosol laboratories pursue vehicles under test (Kolb et al., 2004): this provides the facility to extract samples at widely varying distances from, and widely varying angles with respect to, the tailpipe (Kittelson et al., 2000). Of course, chase experiments conducted on real roads must account for confounding by road dust, as stirred up by vehicle movement, although these particles are likely to be in the coarse mode (Section 9.2).

Everything, then, rests on dilution ratio, as a function of time or as a function of distance, for which the tailpipe serves as an obvious datum. The relative importance of wind-induced and exhaust-induced turbulence must be considered, as must ambient humidity and the particle background. Heat and mass transfer in the vehicle wake have traditionally been studied more with a view to reducing aerodynamic drag; and studies of pollutant dispersal, when exhaust gas is injected into this wake, are recent (Gosse et al., 2006). One should recognise the distinction between engine speed and vehicle speed, and also the distinction between flow rate and discharge velocity.[4] Others highlight the importance, in plume dispersal, of a vehicle's rear-end design (Pirjola et al., 2004). Thus, even before the entrainment of fresh air, there is admirable potential for wide variation in plume dynamics, according to vehicle model and according to vehicle operation.

Some measurements of dilution ratio pertaining to passenger cars in the open air are shown in Figure 4.3 (Dolan and Kittelson, 1979). The exhaust gas discharged from *moving* vehicles was diluted

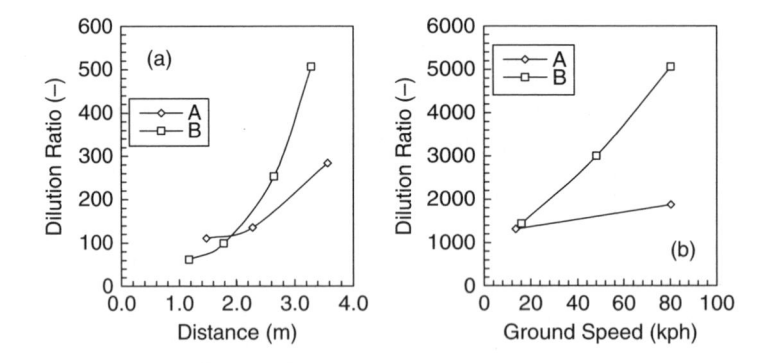

Figure 4.3 Dilution ratio based on three-minute averages of NO_x, as measured in exhaust plume: (a) idling vehicle, ~180 ms ≡ 1 m; (b) moving vehicle (standard deviations in original have been omitted). For idling, the vehicle faced a headwind of 20 kph. Vehicles: A, 5.7-litre V8; B, 1.6-litre, four-cylinder (both passenger cars, model year 1978) (Dolan and Kittelson, 1979).

---

[4] Speed *and* direction. Interestingly, some manufacturers of diesel vehicles turn tailpipes downwards, so that exhaust gas is directed towards the ground, whereas the tailpipes on gasoline vehicles are horizontal. On heavy-duty trucks, tailpipes not infrequently point upwards.

by a factor of >1000 in <1 s, and from *idling* vehicles by a factor of >100 in <2 m, but the standard deviations were large. Others have found similar dilution ratios at the same distance from the tailpipe, irrespective of vehicle speed, and a linearised relationship between dilution ratio and distance if both axes are logarithmic (Morawska *et al.*, 2007). Obviously there are numerous permutations, but some other selected measurements (not necessarily along the centre line of the plume) are as follows: (a) for a truck in a wind tunnel at 55 mph, an exhaust flow rate of $0.4 \, m^3/s$ and an exit velocity of 30 m/s, $D = 350$ at 2.5 m (Kim D.-H. *et al.*, 2002a); (b) for a truck in a wind tunnel at 55 mph, $D = \sim 75$ at 500 cm and $D = \sim 125$ at 850 cm (Clark N.N. *et al.*, 2003); (c) for a passenger car on the road at 50 kph, $D = 2500$ at 14 m, and at 120 kph, $D = 7000$ at 14 m (Ntziachristos *et al.*, 2004a). The caveat is that freely expanding plumes, as often studied, are likely to misrepresent the situation pertaining in some environments, for example in heavy, slow-moving traffic, or in underground construction sites, where dilution is probably impeded.

Traditional analyses of pollutant dispersal into the wider environment make use of Gaussian plume-dispersion models (Seinfeld, 1986, pp. 561–616), as widely applied to chimneys. One term covers horizontal spreading due to wind; another covers vertical spreading, incorporating plume 'reflection' from the ground (Spengler and Wilson, 1996). Turbulent mixing, convection, diffusion and temperature are all modelled (Kim D.-H. *et al.*, 2002b). The greater density of carbon dioxide relative to air opposes the greater buoyancy imparted by the higher temperature of the exhaust gas (Fuchs, 1964, p. 48); and, owing to hydrodynamic interactions between particles, the aerosol cloud, as a whole, is able to move faster than isolated particles (Fuchs, 1964, p. 95). Whether a crosswind blows around or through the plume is a matter of particle concentration, and the tendency to move corporately, as a coherent aerosol cloud, is gradually lost (Hinds, 1982, p. 347). Conventional plume-dispersion models, as applied to chimneys, need some modification, for example to account for the ground-level emission of vehicle tailpipes, the presence of motor vehicles at high densities, the effect of street canyons (Xie *et al.*, 2005), and the action of vehicle-induced turbulence: the vehicle wakes within about 100 m of the highway are arguably more important determining factors in mixing than natural wind-induced turbulence (Samson, 1988). On still wider scales, dispersal models incorporate population densities and the action of prevailing winds (Nigge, 1998).

The discussion has now strayed a little, insofar as formation of the *particulate phase*, rather than dispersal of the exhaust gas (Zhu and Hinds, 2005), is the primary focus of the present section. The transformations during the emission process into the ambient atmosphere are wide-ranging (Bukowiecki *et al.*, 2002); they encompass simultaneous nucleation, adsorption and agglomeration (Bessagnet and Rosset, 2001). The absence of transformations is signalled when size distributions measured at various locations within a plume are factored according to the *local* dilution ratio, whereupon all data collapse to the same curve (Gautam *et al.*, 2003).

But one transformation is eminently observable, albeit in extreme instances: visible plumes of white smoke (Section 3.6.5). According to the detailed picture developed in Section 3.6.3, progressive dilution of exhaust gas causes, *initially*, gas-to-particle conversion, during which time the temperature effect dominates, and, *subsequently*, particle-to-gas conversion, during which time the mixing effect dominates. Of course, the simple *dispersal* of white-smoke droplets cannot, by such a simple visible assessment, be divorced from *evaporation* of the same droplets, as both reduce plume visibility. The point, however, is that, with respect to distance from the tailpipe, the saturation first increases and then decreases. As said, white smoke is an extreme and atypical example, but where more typical samples of exhaust gas are suddenly diluted by factors of several hundred and bag-aged, particles shrink on timescales of a minute or so by a few nanometres – the logical outcome if volatile material evaporates or desorbs (Verrant and Kittelson, 1977).

Contrariwise, particles grow via agglomeration. A model of in-plume agglomeration in which experimentally determined size distributions at a distance of 500 cm from the tailpipe served as the initial condition highlighted how turbulence, in facilitating mixing in the ambient, also suppresses agglomeration (Kim D.-H. *et al.*, 2002a). Thus, it seems, agglomeration rates are slow in comparison with typical plume dispersal times (Vignati *et al.*, 1999); and it seems that mixing, or dispersal, into the ambient is a more important factor in reducing particle numbers than agglomeration (Jacobson and Seinfeld, 2004). Obviously, these findings must be assessed carefully against initial particle concentrations, as these also affect agglomeration rates (Section 2.2.4). In an interesting twist, modelling suggests that volatile nanoparticles may shed (evaporate) their lighter components ($<C_{24}$) rapidly, leaving a refractory kernel which, because of its shrunken size, agglomerates far more rapidly than its progenitor (Jacobson *et al.*, 2005).

The in-plume transformation currently of all-consuming interest is *nucleation* – 'all-consuming' because suspicions linger that nanoparticles are predominantly artefacts of dilution tunnels; this is justified on the basis that conditions are unrepresentative of the real world. These suspicions are undermined by the discovery of nanoparticles in real exhaust plumes (Giechaskiel *et al.*, 2005), both in wind tunnels (Gautam *et al.*, 2003) and in chase experiments (Ntziachristos *et al.*, 2004a); for example, concentrations of 10 nm particles, at distances of a few metres and vehicle speeds of 50–80 kph, reach 10^5-10^6cm^{-3} (Pirjola *et al.*, 2004). Some data are shown in Figure 4.4(a) (Kittelson *et al.*, 2000). But a (not altogether unexpected) problem immediately arises: nanoparticles are will-o′-the-wisps, appearing in real-world but not laboratory dilution (Vogt and Scheer, 2002), at one vehicle speed but not at another (Vogt and Scheer, 2002), and in one plume location but not in another (Maricq *et al.*, 2002b). Nanoparticles are, moreover, acutely sensitive to local meteorology: as illustrated in Figure 4.4(b) (Kittelson *et al.*, 2000), not unreasonable fluctuations in ambient temperature induce order-of-magnitude effects in nucleation modes.

But, while the experimental database devoted to in-plume nanoparticles is slender, the corresponding database of theoretical studies is slenderer still. Certainly, more prominent nucleation modes on *cooler* days, as revealed in Figure 4.4, are consistent with what is known about the vapour pressures of volatile particle precursors, and the implications of such for saturation and nucleation. Without modelling insights, then, the capriciousness is intuitively assignable to several factors: nucleation is acutely sensitive to saturation, and, given the right opportunity, suddenly creates nanoparticles in

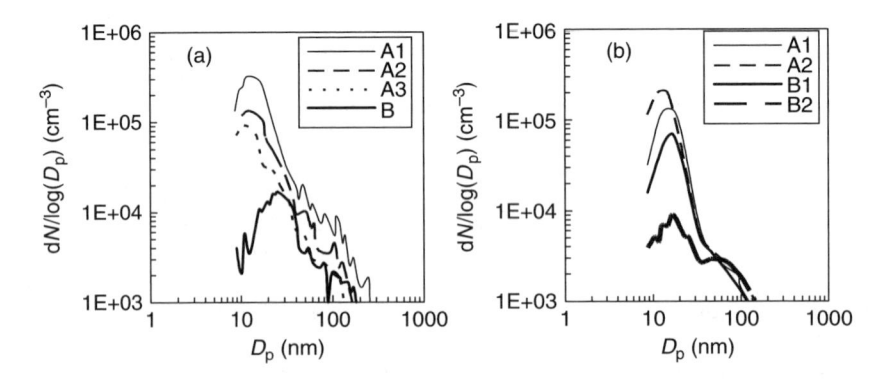

Figure 4.4 Number distributions in real exhaust plumes, as measured in chase experiments with a sample boom following at distances of 15–50 m. (a) Three different runs (A1, A2, A3) and a background average (B). Driving mode: cruise. Original data points omitted. (b) Two different ambient temperatures and driving modes: A, 11 °C; B, 21 °C; 1, acceleration (40–55 mph); 2, cruise (55 mph); data normalised to unit volume concentration. Original data points omitted. Engine: heavy-duty truck, model year 1999. Fuel: reformulated Californian diesel. Instrument: SMPS (9–300 nm) (Kittelson *et al.*, 2000).

hyperabundance, whereupon the population, perhaps with equal suddenness, declines, through either dispersal, evaporation, or agglomeration. For example, cooling (particle creation) would be offset by mixing (particle depletion), engendering different demographic trends in the nucleation mode than in accumulation mode (Maricq *et al.*, 2002b).

4.3.1 Long-term Ageing in the Atmosphere

After plume evanescence, pollutants disperse into the atmosphere on local scales (e.g. depending on the cityscape) and on regional scales (e.g. depending on atmospheric advection) (Samson, 1988), and the particulate phase, for reasons described at length in Chapter 2, is never stable. Particles have a Jekyll-and-Hyde ability to act as CNN, since the hydrophobicity of some adsorbed organic compounds is offset by the hygroscopicity of adsorbed sulphates (Dusek *et al.*, 2006). Surfaces of freshly emitted accumulation-mode particles are thought to be predominantly hydrophobic, gradually becoming, with age, hydrophilic, possibly through sulphation (Cachier, 1998), such as by the acquisition of ammonium sulphate coatings (Injuk *et al.*, 1998) or the loss of hydrophobic coatings (Simo *et al.*, 1997). This causes particle growth through the acquisition of water vapour, and increases the likelihood of action as CNN, with subsequent return to earth in the form of rain. The take-up and release of water in wet–dry cycling, in accordance with atmospheric fluctuations in humidity (Vignati *et al.*, 1999), induces stresses and strains, forcing agglomerates into more compacted morphologies (Cachier, 1998). Ongoing chemical reactions on particle surfaces, involving other pollutants such as O_3, SO_2, and NO_2 (Smith D.M. and Chugtai, 1995), modify particle composition. Ongoing dilution causes further adjustments in the organic compounds (Donahue *et al.*, 2006), the partitioning of which is not fully equilibrated, i.e. it is subject to certain kinetic constraints (Wexler and Potukuchi, 1998), and potentially affected by the 'background' of organic compounds (Shrivastava *et al.*, 2006), particle surfaces (Lipsky and Robinson, 2006), and surrounding sinks, such as vegetation (Su Y. *et al.*, 2006). Long-term ageing is, therefore, *multifaceted*, and, if the risks to public health are to be accurately assessed, a thorough understanding is required of the particle itinerary from 'tailpipe to nose'. But the extrapolation of tailpipe emissions into ambient air quality (e.g. Jensen, 1971), would deserve a chapter of its own.

4.4 Within the Transfer Line

We now turn to the laboratory analogue of the emission process, in which samples of exhaust gas are drawn off and conveyed along a 'transfer line', towards some instrument or other or, more probably, a dilution tunnel. Given the trivial and inconsequential nature of this transfer line, which is, after all, a sideshow or, rather, a brief intermission between the exhaust system and the dilution tunnel, it may reasonably be asked why a separate discussion is at all required; and, especially, why this material should appear in a chapter on particle formation, rather than particle measurement. In answer, the transfer line is far from trivial or inconsequential insofar as particulate characteristics, or rather demographics, are concerned, because in diverting the exhaust gas away from its natural path, it causes the gas to enter an unreal situation. Thus, the transfer line is the first moment in the laboratory emission process where boundary blurring occurs between particle characteristics of a *genuine* nature and other, *artificial* characteristics created by exhaust gas conditioning, prior to collection or measurement.

Although the fluid mechanics of the transfer line will not be related here, these are, in fact, analogous to the didactic picture presented to all undergraduates, of fluid flow within a circular tube, for which, in the laminar case, analytical solutions are readily available. But, in contradistinction, for example, to the parallel case of the channels within a catalyst monolith (Section 4.2.2), which are far narrower, flow within the transfer line might also, on occasion, be turbulent, in which case solutions are only approximate (Berger *et al.*, 1998). Close geometrical parallels also exist in the thermodesorber (Kitto and Colbeck, 1999; Ålander *et al.*, 2004), but these are not explored here.

The traditional concern with the transfer line is particle deposition onto walls, and the distortions such deposition imposes on size distributions. This is a particular risk, since the enclosing surface area is large in relation to the enclosed volume – a geometrical arrangement differing significantly from that in the exhaust system, and also in the dilution tunnel. Hence, all particles travel within a molecule's throw of the surrounding walls, increasing their susceptibility to capture. This deposition cannot be wholly prevented, but the likely demographic impact is discoverable by modelling; and, through these models, a correction, expressed as 'penetration', or the fraction of particles successfully surviving transit, is introduced at the transfer line exit. Moreover, by reversing the process, such models are used to design better transfer lines (Section 5.2.1), in which deposition is appropriately mitigated, if not exactly eliminated.

From these models, then, the two chief particle deposition mechanisms in the transfer line are diffusion and thermophoresis, for which the following figures are of course dependent on the design configuration. Diffusion is a particular function of residence time, and hence of transfer line length and gas flow rate (Silvis *et al.*, 2002); losses, which worsen markedly at smaller sizes, are estimated, for 10 nm particles, at 20% (Wei *et al.*, 2001a), but in less favourable circumstances at 40% (Ayala *et al.*, 2003); above 100 nm, the effects are normally negligible. Thermophoretic deposition is exacerbated by the high surface-to-volume ratio of the line, which intensifies exhaust gas cooling, and steepens radial thermal gradients: losses at 100 nm are estimated at 10–15% (Silvis *et al.*, 2002).

Figure 4.5(a) (Wei *et al.*, 2001a) shows order-of-magnitude reductions in nanoparticles if the exhaust gas is allowed to dally in the transfer line. But detailed modelling has eliminated diffusional or thermophoretic deposition, and agglomeration for that matter, as explanations for such trends. A suspicion lingers that loss of *gas-phase particle precursors* is of greater significance here than loss of particles *per se*. The key justification for this statement is the somewhat larger diffusion coefficients

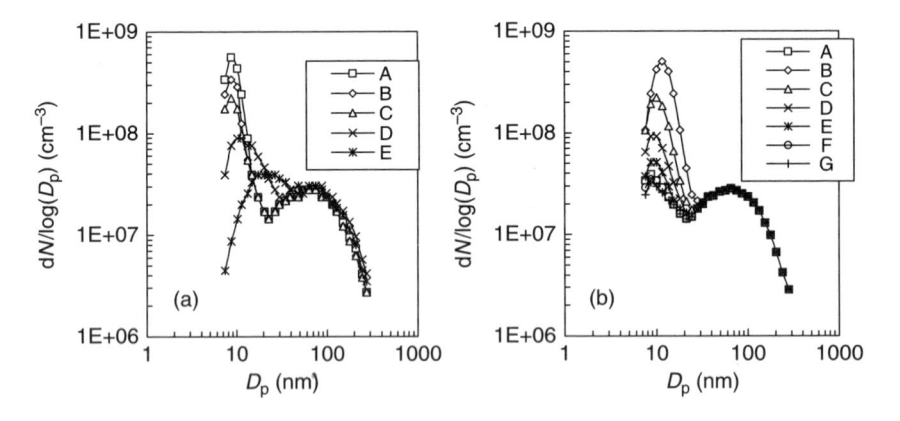

Figure 4.5 Number distributions, as affected by transit through the transfer line. (a) Residence time (s), as varied via flow rate: A, 0.096; B, 0.13; C, 0.18; D, 0.34; E, 2.1; temperature, 200 °C. (b) A step increase in temperature: A, after 2 h at 200 °C; B, immediately following a switch to 305 °C; C; after 5 min; D, after 15 min; E, after 40 min; F, after 50 min; G, after 65 mins; residence time, 0.096 s. Dilution tunnel: residence time, 1.0 s; dilution ratio, 1000±50. Engine: 4-litre, four-cylinder, direct-injection, medium-duty, model year 1995. Operating point: 1600 rpm/50% maximum load; exhaust temperature at sampling point, 305 °C. Fuel: sulphur, 0.044%. Instrumentation: SMPS (8 nm–300 nm) (Wei *et al.*, 2001a).

enjoyed by gases compared with particles: from a theoretical standpoint, the mass transport of gas-phase hydrocarbons and sulphuric acid is fast compared with typical residence times, and the adsorption of volatile material by existing particles appears to be even faster (Wei *et al.*, 2001a). The transfer line walls, accumulation-mode particles and freshly nucleated particles all compete with one another for growth material; hence, the net effect is highly context-dependent. But these investigations demonstrate that losses of gas-phase material to transfer line walls cannot be ignored as a formative mechanism: if these precursors are successfully filched, this will serve to restrain saturation, and thereby to suppress nucleation. If this is the case, the trend in Figure 4.5(a) arises not because particles are lost, but because particles are not formed.

This filching of gas-phase particle precursors by transfer line walls has a further and, as it turns out, most unfortunate corollary: whatever is taken up is, at some later stage, given back, or *outgassed*. This carries serious implications for the measurement: outgassing might push the saturation momentarily above the threshold of nucleation – a logical supposition, since the quantity of stored volatile material, while presumably the same as in any other deposit is, in the transfer line, far larger in relation to the volume of exhaust gas receiving it. Indeed, the exogenous material outgassed by silicone rubber couplings is irksomely sufficient to create nucleation modes that are wholly artificial (Maricq *et al.*, 1999b). Outgassing depends, principally, on transfer line *temperature*, thus explaining the suspiciously large nucleation modes observed at high vehicle speeds (Maricq *et al.*, 1999b), and increases and then decreases in the nanoparticle emission on shifting from one steady-state operating point to another (Maricq *et al.*, 1999b). The data of Figure 4.5(b) (Wei *et al.*, 2001a) were obtained by operating the transfer line at 200 °C for two hours and then suddenly raising its temperature, using an electrical heater, to 305 °C. In response, the nucleation mode initially increased by an order of magnitude, but then gradually, over a period of an hour or so, fell back to the original profile.

An extensive body of recently published literature, then, testifies to various falsifications that transfer lines introduce into nucleation modes – falsifications the action of which readily renders measurements misleading or downright irreproducible: one draws the somewhat exasperating conclusion that *any* nucleation mode may result, depending on the particulars. These artefacts are easily sufficient to overwhelm the genuine particle emission. But that is not all: these artefacts *mimic* the expected response, for example as vehicle speed increases, because the transfer line temperature concomitantly increases, whereupon volatile material is outgassed. Thus, the ability of these artefacts to bamboozle the most able experimenter should not be underestimated. Obviously, the need for the utmost cleanliness in the transfer line, or for measurements of sufficiently long duration to avoid transitory phases in the particle emission, is subject to practical constraints. True, the exact implications depend on the measurement objective: the nucleation mode suffers from the greatest falsification, whereas the accumulation mode is a likelier representation of the real engine-out emission. Given the historical importance of mass over number, it is easy to see why measurement artefacts occurring in the transfer line have gone unreported until recently: in all the above, the accumulation mode is virtually unaffected.

4.5 Within the Dilution Tunnel

After negotiating the transfer line, the exhaust gas issues into a dilution tunnel, where it intermingles with an air stream drawn in from the ambient atmosphere. This intermingling of air and exhaust forms an enveloping exhaust plume (Figure 5.1), inside of which dilution progresses in transit through the tunnel towards the intended fully mixed state, terminating finally in the filter. This, at any rate, is the generic picture, which, in practice, is realised in various ways: (a) the flow is filtered in its entirety; (b) a sample is drawn off and filtered; (c) a sample is drawn off, diluted again in a secondary tunnel and then filtered. From what has already been observed, slightly different transformations may be expected in each case, but, as dilution tunnel designs vary so multitudinously, the primary object of the current section is to capture the essence of these transformations in general terms; the particulars are, after all, governed throughout by common principles.

Not all particles survive transit to the filter; instead, they may deposit on the tunnel walls, an experimental nuisance for which several loss mechanisms may be cited: the reader is referred to a full-length study (Kittelson and Johnson, 1991). The implications of this deposition, as always, depend very much on the demographic focus of the measurement. Coarse-mode particles are subject to inertial deposition, but not so much in the tunnel as in the bends of auxiliary pipework; massive re-entrained particles may, therefore, bounce their way through to the filter. Nucleation-mode particles are subject, predominantly, to diffusional deposition, in which a few per cent of the mass is lost; the implications here for particle number were not addressed in the original paper of Kittelson and Johnson (1991), reflecting the concerns of the time. Large reductions in engine-out particle concentrations, say, from one test to the next, risk measurement falsifications through re-entrainment; thus, equipment for diesel and gasoline vehicles tends to be segregated.

Accumulation-mode particles are only weakly subject to diffusional and inertial deposition: the dominant mechanism in this size range is undoubtedly thermophoresis, for which mass losses were estimated at larger than 5% (Kittelson and Johnson, 1991) – this figure reflects the fact that, *unlike* the previously mentioned mechanisms, this one operates strongly on the accumulation mode – so strongly, in fact, that it easily makes the difference between 'pass' and 'fail' in statutory emissions tests. It remains to be reported whether rates of deposition, i.e. thermal gradients, eventually subside through the insulating action of the wall deposits, or whether re-entrainment is sufficient, on its own, to reverse the effects. These matters do, however, very much depend on the configuration, design and geometry of the dilution tunnel, as well as the size distribution: other workers believe the effects of thermophoretic deposition to be small (Ahlvik *et al.*, 1998). But, generically, the importance of thermophoretic deposition depends on the strength of nonadiabatic cooling (Section 5.2.2), as heat transfer ostensibly relies on intermingling of cold and hot gas, whereas energy losses to the surrounding environment, from the tunnel walls, are what drive the thermophoresis. Moreover, slow-responding thermocouples may fail to register the real strength of temperature gradients engendered by transient engine operation (Kittelson and Johnson, 1991).

The tenets of gas-to-particle conversion have already been discussed in relation to the organic fraction, in Section 3.6.3, and in relation also to the exhaust plume, in Section 4.3; the picture here is similar. Upon entering the dilution tunnel, the *cooling* effect dominates, so that the saturation initially increases; but, after a certain point, the *mixing* effect dominates, so that the saturation now decreases. Many volatile species appear to reach their peak saturations at some point *between* entry into the tunnel and capture by the filter: the initial tendency for gas-to-particle conversion thus gives way, at some intermediate moment, to a reverse tendency for particle-to-gas conversion. The interplay between *saturation ratio* and *dilution ratio* is thus paramount. Accordingly, 'the strongest driving force for nucleation, condensation and adsorption, occurs in the critical dilution range of 5 to 50, roughly the same range produced by dilution tunnels, 3 to 20' (Suresh and Johnson, 2001). This is where recent research diverges profoundly from the historical understanding – that the *final* dilution ratio, and *only* the final dilution ratio, as realised in the fully mixed condition, is what counts. This being a new vista of research, definitive conclusions are unfortunately unavailable; but well chronicled is the central importance not just of the dilution ratio *at the filter*, but also the dilution ratio taken *en route*, that is, the *dilution profile*.

This critical region of high saturation is really a thermodynamic observation, unable to elucidate any kinetic constraints. But the time spent in transitioning through this region is, intuitively at any rate, a factor in gas–particle partitioning; so, what counts is dilution rate in comparison with condensation rate, adsorption rate and nucleation rate. It should be admitted that laboratory dilution, as traditionally carried out, is fairly leisurely: this engenders a suspicion that, since the exhaust gas spends a longer period of time at more moderate dilution ratios, greater potential exists for gas-to-particle conversion than in the real-world emission process, where the exhaust gas passes through this critical region with extreme rapidity, so that super- or sub-equilibrium states seem more likely. But even so, there seems no *a priori* justification for assuming that, at each sequential moment in the dilution tunnel, the particulate phase and gas phase coexist in perfect equilibrium; and, perhaps, the particulate characteristics at the filter are governed more by kinetics than by thermodynamics.

This question is central to suggestions that nanoparticles are in fact artefacts of the laboratory emission process; hence, a welcome trend in recent times is the emergence of a new generation of dilution tunnels, able to mimic more faithfully the conditions realised in real exhaust plumes; tunnels which lend themselves particularly to studies of nucleation. Figure 4.6(a), for example, shows how decreases in dilution air temperature drive order-of-magnitude increases in nanoparticles (Wei *et al.*, 2001a). These data are directionally consistent with the expected effect of temperature on saturation. Small increases in dilution air humidity have similar effects on the nucleation mode (Shi and Harrison, 1999), and also, logically, work via saturation. Actually, considerable variations may well exist from one location to another inside the dilution tunnel, as nucleation is essentially an 'on–off' phenomenon. Hence, nanoparticles suddenly emerge in huge numbers, once precisely the right conditions are furnished: conditions which, in these measurements, lie within a remarkably narrow range of dilution ratio or, alternatively, as dilution is so rapid, a remarkably narrow interval of time. Figure 4.6(b) shows that a prominent nucleation mode has already formed just 200 ms after entry of the exhaust gas into the dilution tunnel; yet this mode does not increase much further in magnitude once the residence time exceeds 530 ms (Wei *et al.*, 2001a).

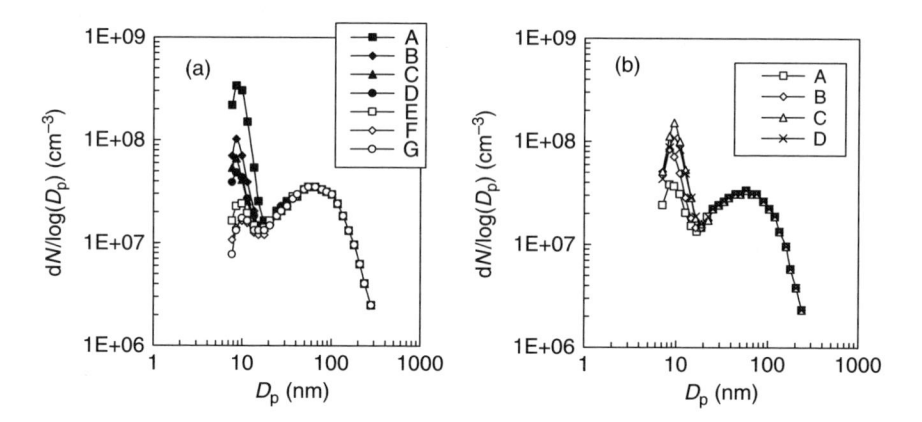

Figure 4.6 Number distributions. (a) As determined by temperature in a dilution tunnel (°C): A, 10; B, 15; C, 20; D, 25; E, 30; F, 35; G, 40. The residence time and temperature in the transfer line were held constant at 0.08 s and 305 °C. Final dilution ratio, 1000(±50) (rising linearly); residence time in tunnel, 1.0 s; exhaust and transfer line temperature, 305 °C. (b) As determined by residence time in dilution tunnel (i.e. at different points, lengthwise) (s): A, 0.2; B, 0.53; C, 1.33; D, 1.53. (Corresponding dilution ratios ranged from 240 to 1450.) The residence time in the transfer line was held constant at 0.08 s. Exhaust and transfer line temperature: 300 °C. Engine: 4-litre, four-cylinder, direct-injection, medium-duty, model year 1995. Operating point: 1600 rpm and 50 % of maximum load. Fuel: sulphur, 0.0010 %. Instrument: SMPS (8–300 nm) (Wei *et al.*, 2001a).

4.6 On the Filter

The terminus, in the laboratory emission process, is the *dilution tunnel filter*; the characteristics of which very much impose themselves on the characteristics of the particulate, so that, conversely, particulate formation, the subject of the present chapter, requires some discussion of the filter. These problems of biasing, and of 'artefactuality', are well known in the sampling of aerosols, for example in air quality monitoring (Koutrakis and Sioutas, 1996), and explain in part why sampling protocols, as used in the automotive industry, are, at the time of writing, being seriously reconsidered.

The vagaries of filtering are partly illuminated by Figure 4.7, which shows the pattern of load-up, or fouling, as a function of face velocity, for two filters mounted in series (Andersson *et al.*, 2004b). The mass of particulate retained by the *primary* filter is a *linear* function of face velocity, whereas the mass of particulate retained by the *secondary* filter is a *quadratic* function. The secondary filter, then, captures particulate at rates faster than the primary filter – something not immediately obvious if the primary filter has first call in removing the particles from the gas; the quadratic function suggests, perhaps, the dominating influence of surface area. This operating characteristic and numerous others besides are underpinned by arcane and under-appreciated phenomena which form the present topic; but, while Figure 4.7 appears to open a can of filtered worms, so to speak, some consolation can be taken from the fact that 'particulate', as defined, is whatever is captured, *regardless* of the capture mechanisms. Phrased thus, it is only necessary for sampling protocols to be *consistent*.

In an ideal world, the filter would be a perfect or disinterested arbiter, detaining assiduously the particulate phase and ignoring nonchalantly the gas phase. But the filter is not a perfect arbiter. The first problem is inconstancy in filtering characteristics as a function of time: it is normal for the filtering efficiency of, and the pressure drop across, the filter to increase as particulate accumulates. On a dilution tunnel filter, this subtlety is readily overlooked,[5] because only the mass of the final deposit counts; with the diesel particulate filter, the captured particles are well known to form a 'cake' of greater filtering efficiency than the underlying DPF (Murtagh *et al.*, 1994). Two consequences of this shift in the filter's operating characteristics are apparent. First, a rock-steady emission rate from the engine is *perceived* as a gradually increasing emission rate by the filter. Second, the pressure drop

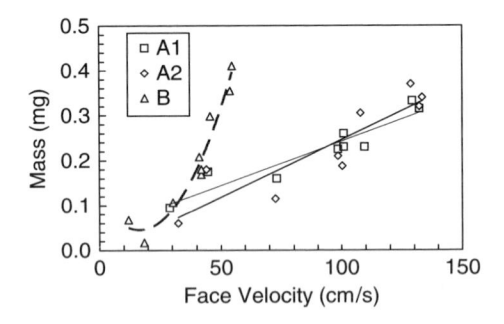

Figure 4.7 Particulate retained. A1, mass on primary filter, tested *with* secondary filter; A2, mass on primary filter, tested *without* secondary filter; B, mass on secondary filter for condition A1. Filter diameters: primary, 47 mm; secondary, 70 mm. European Transient Test Cycle. Engine: 7.8-litre, six-cylinder, rated at 259 kW, 1280 rpm (Andersson *et al.*, 2004b).

---

[5] While the effect may be small for exposures of only a few minutes, this is the justification for using a *secondary* dilution tunnel filter: the collection efficiency of the primary filter is low in the early stages.

across the filter compresses this cake, i.e. increases the packing density, and so further increases the filtering efficiency (Warner *et al.*, 2003). So, the mass of particulate, as captured by the filter, is not a simple linear function of sampling time. Moreover, it is worth noting that many particles in exhaust gas are, in fact, *charged* (Section 6.2.5), although the implications here for filtering, i.e. electrostatic effects, have not been reported.

Turning now to the vapour or volatile component, the filter temperature, obviously enough, is central in deciding what remains in the gas phase and what transitions into the particulate phase. For transient duty cycles, the exhaust temperature fluctuates, so that the filter temperature fluctuates; hence the efficiency with which volatile compounds are filtered also fluctuates. This raises questions about gas–particle partitioning of a kinetic and thermodynamic nature. But that is not all. If the same throughout of gas is to be maintained, then the depression on the obverse side of the filter must deepen, and this encourages the revolatilisation and escape of previously captured compounds (Guerrieri *et al.*, 1996). The deposited volatile compounds, as it were, are stripped from the filter, creating a *negative* artefact, that is, the particulate mass is lower than it should be.

Less well delineated capture mechanisms pertain to gas-phase compounds, and the surfaces to which these compounds attach. In this respect the filter should be acknowledged as a contrivance, insofar as it presents, to these compounds, an artificially enhanced surface area, the nature of which is a function of the loaded state (Yamane *et al.*, 1988). This raises the question of whether gas-phase compounds, on entering the filter, are encouraged to adsorb when, in the normal course of events, that is, without the filter, they would remain in the gas phase (Smith D.J.T. and Harrison, 1998; Khalek *et al.*, 2003); should this happen, the particulate emission would be exaggerated. This issue relates to the surfaces of carbonaceous or solid particles, as already captured by the filter, and also to the filter *itself*, the fibres of which, if naked, will also adsorb vapour-phase organic compounds: this is a *positive* artefact. Filter materials vary markedly in this proclivity, but, equally importantly, this makes the rate of solid-particle accumulation a factor, as this decides, contrariwise, the fraction of unfouled surface (Chase R.E. *et al.*, 2004); for example, the picture changes starkly when a DPF is fitted to a diesel vehicle, as the solid particles are preferentially removed prior to the dilution tunnel filter.

To drive home the above points, it is worth citing experiments in which exhaust gas samples are concurrently extracted both upstream and downstream of some aftertreatment device, as then the question arises of whether these two measurements truly reveal the performance of the device or reflect, instead, imperfect or nonideal arbitration by the filter. The problem is that slightly different operating environments are experienced by the two filters. Suppose this experiment is performed on a DPF (Brunner N.R., 1995). The *upstream* filter works at a higher temperature, it becomes fouled with carbonaceous particulate more rapidly, and it experiences a larger pressure drop; these three differences have some degree of influence on the measurement, *irrespective* of particulate retention by the DPF. Indeed, the particulate mass, as measured at the two locations, upstream and downstream, was found to differ by 10 %, even upon fitting a *straight pipe*, so that the point is clear.

Substantial quantities of semivolatile organic compounds, being adamantly gaseous, simply pass through the filter. (Compounds in this category particularly include PAH of four or fewer rings.) In *comprehensive* emission inventories, this 'blow-by' is quantified by installing, in series, a 'sorbent' or 'vapour-phase trap', on which these volatile compounds, after passing through the filter, are captured. The distribution, or partitioning, between the filter and the sorbent is dependent on the conditions (flow rate and temperature): for example, at one test condition, *Pyr*, *Chr* and *BaA* were retained by the filter, while at another, they were retained by the sorbent (Williams R.L. *et al.*, 1985). The distribution is also counter-intuitive: for example, the filter collected *BaP*, a five-ring compound of low vapour pressure, whereas the sorbent collected *Phe*, a three-ring compound of high vapour pressure (Hori S. and Narusawa, 2001).

Up to now, the discussion has focused exclusively on aspects of particle and vapour capture; there remains to discuss a range of phenomena *following* capture, by which the risk of measurement falsification is very great. These transformations are caused by ongoing chemical reactions on the filter itself. Some species might, courtesy of these reactions, become more toxic; others might be wholly consumed, and thereby completely disappear. Such reactions seem inevitable as the filter is, in effect, a kind of chemical reactor, in which specific surface areas are large, warm exhaust gas streams continually past, and collection periods are lengthy. Of course, chemical reactions occur in the gas phase too, either homogeneously, or heterogeneously on particle surfaces; but on the filter, reactions are deemed artefactual when, in the absence of the filter, they would not arise, or at least would be greatly attenuated. This, of course, is an idealised distinction which, in practice, is hard to prove one way or another.

Of these transformations, the conversion of PAH into NPAH requires particular study, as the latter group is markedly more mutagenic; for a discussion of this topic, see Section 3.6.2. Less obviously chemical, yet more obviously artefactual, is the mysterious improvement in capture efficiency for organic compounds (Zelenka P. *et al.*, 1990), including high-volatility PAH (Wall and Hoekman, 1984), brought about by sulphuric acid. This acid is recidivist: it repeatedly instigates on-filter transformations of one sort or another. It is suspected of destroying *primary* organic compounds such as *n*-alkanes, and of generating *secondary* organic compounds, for example via the esterification of fatty acids (Hori S. and Narusawa, 1998); the acid is also guilty as charged in the sulphation of metals or of metal oxides, and in the degradation of filter fibres (Engler *et al.*, 1993). From these observations, it seems unwise to presuppose any great stability in the mass and composition of the particulate, as captured by the filter.

4.7 Closure

The material in this chapter was divided according to the route taken by the exhaust gas in the emission process: in the real world, this is through the exhaust system and into the ambient air, and, in the laboratory, this is along a sample or transfer line and into a dilution tunnel, terminating with a dilution tunnel filter. As the particulate phase is affected in many, many ways by this journey, characterisational differences inevitably arise between the two emission processes.

4.7.1 Within the Exhaust System

Cycle-resolved measurements show that, following exhaust valve opening, particles are ejected from the engine in two pulses, the second being less pronounced. Some data suggest that the respective particles discharged in each of these two phases are of a different nature, e.g. in terms of volatile component or size. No general picture emerges of particle transformations along the exhaust system; but accretion of volatiles is observed once temperatures fall below 300 °C, with probable effects on particle morphology; coagulation seems uninfluential, at least in the accumulation mode; and oxidation at higher temperatures might cause particle fragmentation.

Storage and Release

Particles avidly lodge within the exhaust system and accumulate as wall deposits, and these deposits may, according to the duty cycle, be released in sudden bursts. Cold starts, because of the intense radial thermophoretic forces, particularly encourage deposition. High flow rates encourage deposition by inertial impaction, but also assist in the exfoliation of these deposits by fluid-shear forces. Storage in the exhaust system is understood only in this qualitative manner; the random and unpredictable nature of release, or re-entrainment, hampers modelling.

Deposition within Catalysts

Particle deposition on the channel walls of a catalyst monolith is thoroughly modelled. The flow through these channels is laminar, but particles do not necessarily remain faithful to the streamlines. Models suggest the nucleation-mode particles are caught by diffusion, and the accumulation-mode particles by thermophoresis; other deposition mechanisms are unimportant. Models should encompass transient driving, as the thermophoretic force may actually change direction, depending, for example, on the thermal inertia of the monolith.

4.7.2 Within the Exhaust Plume

In real exhaust plumes, that is, behind motor vehicles in the ambient atmosphere, the exhaust gas is diluted, in a linear or nearly linear fashion, by a factor of several hundred in less than one second. Such rapid dilution furnishes little opportunity for equilibrium between gas phase and particulate phase. The most significant finding in recent years arises from chase experiments, in which samples of diluted exhaust are extracted from exhaust plumes 'on the move'. These experiments demonstrate that nanoparticles are not, as was sometimes argued, artefacts of laboratory dilution, but are present in *real* exhaust plumes. This in-plume nucleation also follows the expected trends if saturation is the deciding factor, e.g. more nanoparticles are discovered upon dilution in colder or humid ambients.

4.7.3 Within the Transfer Line

The transfer line is the first place where exhaust gas experiences an artificial condition, and this artificiality has profound ramifications for particulate characteristics, especially demographics. Particle losses to walls, through diffusion and thermophoresis, are only too likely. The best strategy to avoid these artefacts is to minimise residence time within this line. The most widely studied aspect of the transfer line in recent years is nucleation: this mode can prove largely irreproducible, owing to the influence of test history, 'carry-over' effects, and the pattern of temperature rise and fall. These discrepancies have been traced to the transfer line walls, which take up and then release, or outgas, particle precursors (volatiles); these volatiles temporarily raise the saturation, causing nucleation. Some engineering materials will also outgas volatile material. Since these artefacts mimic expected emission behaviour, e.g. as decided by vehicle acceleration, test results for the nucleation mode can be doubly misleading.

4.7.4 Within the Dilution Tunnel

Detailed modelling of particle transport in the dilution tunnel points to thermophoresis as the predominant wall-deposition mechanism for the accumulation mode; hence mass measurements are the most affected. Diffusional losses become important below 100 nm, with due impact on the nucleation mode and measurements of particle numbers. Historically it has only been the final dilution ratio, as reached at the dilution tunnel filter, which mattered. However, an extensive body of data shows that the dilution *profile*, i.e. the pattern of mixing and cooling in transit along the tunnel, is influential – particularly the dilution *rate*. Various research groups have now designed apparatus to investigate the interplay of dilution ratio and saturation ratio. Many species appear to reach their peak saturation when the dilution ratio falls between 5 and 50; the length of time spent in transitioning through this region is therefore important. A major but unanswered question is whether full equilibrium arises between the gas phase and the particulate phase. Nucleation is observed in dilution tunnels – this might happen almost immediately the exhaust gas leaves the transfer line, but, owing to the acute sensitivity of nucleation rate to saturation, it is difficult to track the point at which the nanoparticles form.

4.7.5 On the Filter

The dilution tunnel filter is not a perfect arbiter in deciding what is 'particulate'. Its filtering efficiency inevitably rises in the course of fouling, because the retained particulate itself forms an effective filter. The pressure drop across the filter also rises, and this may cause accreted volatile species to revolatilise and abscond. Volatiles are certainly lost from the filter when the temperature rises, as happens during transient engine operation. Such real-time phenomena are nonetheless an essential part of the final cumulative result in a test cycle. Nonideal behaviour in the filter distorts measurements of the efficiency of particle capture by aftertreatment devices. As volatile species can pass through the filter, they are fully quantified by adding a subsequent adsorbent, or vapour-phase trap. Artefacts are created by the filter, which is in effect a chemical reactor, because the particulate deposit is held in hot exhaust for long periods. Chief amongst these artefacts are poorly understood reactions that generate NPAH – compounds of greater toxicity than their parent PAH. Distinguishing genuinely emitted NPAH from artefactually generated NPAH is immensely difficult. Sulphuric acid is a suspected instigator of chemical change through its reaction with organic compounds, and possibly even with the filter matrix.

4.7.6 General Remarks

A fundamental characteristic of the emission process is the constantly shifting boundary between the particulate phase and the gas phase. Gas-to-particle conversion is strongly driven by saturation, meaning (a) the temperature profile; (b) the mixing profile (when exhaust is diluted with air); and (c) the amount of volatile material available, i.e. sulphates and organics, in relation to the solid or nonvolatile material, i.e. carbonaceous material and ash. Although the general feature is one of gas-to-particle conversion, there are occasions, notably in the later stages of dilution, where particle-to-gas conversion can occur, because the saturation in the emission process first rises and then falls.

That substantial nucleation takes place in the emission process is now indisputable. The phenomenon is particularly fostered by high ratios of volatile to nonvolatile material. The implications here depend on the measurement being undertaken. If only the total mass of particulate is of interest, then nucleation can often be neglected; but if the number of particles matters, then test results are grossly distorted by failure to control nucleation appropriately. In actual fact, the recently shifted focus from mass to number is precisely why researchers have begun to scrutinise nucleation in such great detail.

Consistent measurements of particle numbers will become more difficult should any emission control strategies increase the volatile/nonvolatile ratio. When the carbonaceous or solid accumulation mode is small, volatile particle precursors are deprived of surfaces on which to condense or adsorb. Nucleation is expected to be particularly vigorous for vehicles fitted with a diesel particulate filter, as this will preferentially remove what is solid from the exhaust stream.

There are two reasons why nucleation can outfox even the most able experimenter. First, the take-up and subsequent release of volatile precursors by the surrounding walls is influential. Thus the history of use of the test apparatus becomes important. Second, nucleation is highly nonlinear: virtually an 'on–off' process. It is thus unsurprising that the number of nanoparticles can differ by orders of magnitude between two ostensibly similar test runs or laboratories. Small changes in residence time, temperature, dilution ratio, fuel sulphur concentration or humidity have profound effects on the nucleation mode. Until test conditions are fully standardised (assuming this to be possible to the extent required), measurements of particle numbers ought to be treated with greater circumspection than measurements of particulate mass.

The accumulation mode, as it contains a solid carbonaceous core, is much more stable. Consequently, measurements of this mode are undertaken with greater confidence than measurements of the nucleation mode. Many investigators have found results for the accumulation mode to be comparatively unaffected by humidity, fuel sulphur level, temperature, etc.

In summary, the danger that particulate characteristics will be purely an artefact of the test set-up in a laboratory or of the measurement protocol cannot be overstated.

5
Measurement

5.1 Introduction

The present chapter relates various instruments with which particulate emissions are studied and quantified. No bias is adopted with respect to the state of development: some of these instruments still exist as research tools; others are tried-and-trusted commercially available units. The aim is *not*, however, to compare and contrast instruments, especially commercially available instruments, according to their respective designs and specifications, or merits and demerits. It is, naturally, possible to design any number of instruments based on one operating principle alone: *here*, particulate measurement is discussed according the operating characteristics of generic instruments.

Nor is the present chapter restricted to instrumentation *per se*, as particle *characteristics* are shaped by particle *measurements*: this connection is as easy to misjudge as it is difficult to ascertain. All types of transformations are induced by handling, sampling, etc.; hence valid measurements require in-depth knowledge of numerous experimental variables. Through this connection, particles artfully resist the collection of consistent and meaningful samples; and, without considerable skill and understanding, meaningless or downright bogus results are the inevitable consequence. As the precise conditions under which measurements take place are inseparable from the particle characteristics that are measured, then 'particulate', to reverse the gist, is *ipso facto* defined *purely in terms of the method of measurement*.

Measurements must not render particulate characteristics wholly or partially unrepresentative of real human exposure, or, for that matter, unrepresentative of any other relevant environmental metric. Such a scenario would be fatuous, the central objective being the discovery of what is emitted into the atmosphere. Uninvited or unsolicited transformations created by the process of measurement are designated 'artefactual', meaning they would not otherwise have arisen in the real emission process. But, as we shall see, particulate emissions cannot be measured in a manner that does not invite falsifications of one type or another, and, often, the distinction between what is genuine and what is artefactual is difficult to define.

Particle concentrations in diluted exhaust, and sometimes also in raw exhaust, are drawing near to those in the *ambient air* – that is, the air inducted by the engine – and, by the same token,

Particulate Emissions from Vehicles P. Eastwood
© 2008 John Wiley & Sons, Ltd

many of today's instruments deliver signals that are embarrassingly close to their own noise levels. (These difficulties are at loggerheads with the practice of exhaust gas dilution.) Thus, contemporary measurements are markedly more vulnerable than in former times to sundry and extraneous influences (e.g. the cleanliness of the equipment, the calibration of the instrument, and the driving style). The absence of standardisation in many areas of sampling and measurement impedes inter-laboratory comparisons; yet, even in the same laboratory, and with the same engine or vehicle, test-to-test variability is often worse than for gaseous emissions.

The conventional measurement, as we have learned, is the mass of particulate, as collected on a dilution tunnel filter. Yet this measurement provides only an average 'mass per unit distance' for the entire test cycle, information inadequate to guide emission control efforts: the pollution emitted during steady driving, or cruising, has been far more successfully abated than that emitted during transient driving modes. Consequently, today's test cycles are disproportionately dominated by an engine's *transient* emission characteristics, for which there is a clear need for on-line, real-time instruments.

Technologists have not been slow to develop or to refine existing instruments, or to invent wholly new instruments, and there are now myriad ways in which particulate emissions can be sampled, measured, quantified and analysed. The paper by McMurry (2000b) is a *tour de force* from the related perspective (and parallel field) of atmospheric aerosols. Instruments employ a range of operating principles, some of which will be apparent from earlier chapters, e.g. interaction with light or acceleration in an electrical field. But rather than examine aerosol instruments generally, as this would enlarge the subject to impracticable proportions, the present discussion is expediently restricted to those instruments to have been reported in the automotive arena. For example, the 'diffusion battery', devised originally to measure particle diffusion coefficients, is a tool well known amongst aerosol scientists (Knutson *et al.*, 1999), but has not found any widespread application amongst engine researchers (Dolan *et al.*, 1980; Baumgard and Kittelson, 1985).

As different instruments use different operating principles to measure different particle properties, then complete agreement should never be expected. This is especially true with instruments reporting particle *size*. Indices of direct geometrical significance, i.e. from what is observed under a microscope, are impracticable; instead, some other principle of measurement is used, via which size is *inferred*. Inevitably, many assumptions underlie such inferences, the efficacy of which may be rightly disputed, and particles are normally defined in terms of the property measured, a practice which leads to a *surrogate* or *proxy* diameter. This distinction is paramount.

5.2 Particulate Measured Conventionally

To study particulate emissions, some form of collection method is first required, and, conventionally, such collection commences with a 'dilution tunnel', which, these days, is standard equipment in all emissions laboratories. In this tunnel, raw exhaust gas is diluted with ambient air, and then, after a certain period of time in transit in which to homogenise, this mixture is filtered. All tunnels are based on this concept, shown schematically in Figure 5.1 The particulate deposit captured by this filter is then fractionated and analysed using a range of techniques that are the subject of this section.

5.2.1 Drawing a Sample of Exhaust Gas

In drawing off into a probe a sample of exhaust gas, a general need arises to avoid preferential extraction of particles within certain ranges of size; otherwise, the result becomes unrepresentative of the sampled population, i.e. *biased*. The stringency of isokinetic sampling, necessary in order to avoid this bias, was mentioned in Section 2.2.3. It is, however, a long-standing perception that – for all

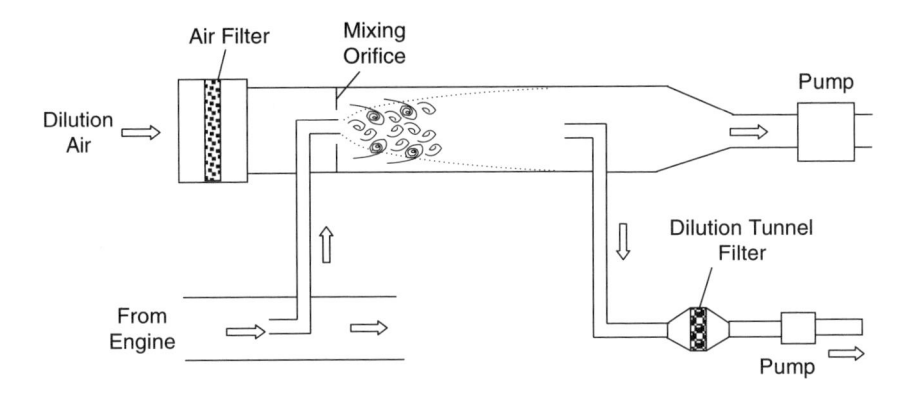

Figure 5.1 Schematic of a dilution tunnel.

experimental conditions likely to be experienced – anisokinetic sampling will lose only the massive, or supermicron, particles, of the sort re-entrained from walls, etc., i.e. those unrepresentative of the immediate engine-out emission.[1] On this basis – since secondary particles are of secondary interest – isokinetic sampling has become a custom more honoured in the breach than in the observance.

This conceit has been challenged on various grounds. One should bear in mind, for example, that estimates of anisokineticity are based on the geometry of the probe entrance – the *clean* entrance, that is – and so can be wide of the mark if particles, upon depositing in that entrance, locally divert or distort the streamlines. Nor is it implausible that *transient* engine operation, and the associated unsteady flow, exacerbates anisokinetic sampling errors, and not just for supermicron particles (Ntziachristos and Samaras, 2000). Measurements show anisokinetic mass losses of a few per cent are possible, even at 200 nm, and losses of 15 % are possible at 500 nm (Khalek *et al.*, 2002).

Particle losses through inertial effects may therefore still be an issue without proper probe design (Cunningham P.J. and Meckl, 2006): generally speaking, the problems are worse for sampling from the exhaust system than for sampling from the dilution tunnel, and worse for rear-facing probes than for forward-facing probes (Silvis *et al.*, 2002). Countermeasures involve practical compromises between exhaust pipe diameter, sample-probe diameter and flow velocity (Khalek *et al.*, 2002), and corrective models for deposition in the turbulent region of the vena contracta formed at the probe inlet during super-isokinetic sampling (Ntziachristos and Samaras, 2000)

Once the sample of exhaust gas has been safely drawn into the probe entrance, it is conveyed to a dilution tunnel (or some other apparatus) by a *transfer line*. The importance of proper transfer line design was thoroughly highlighted in Section 4.4. To avoid particle deposition by thermophoresis, one obvious strategy is to reduce, or even to eliminate, radial temperature gradients – easily accomplished electrically, by wrapping the line with heating tape (Wei *et al.*, 2001a,b); another strategy is to use a thin wall with the ability to assume rapidly the temperature of the exhaust gas. To preserve mechanical stability and to add thermal insulation, this line is sheathed by an outer one, with a dead air space between the two (Silvis *et al.*, 2002).

If diffusion losses of particles or particle precursors are significant, then pronounced concentration profiles exist across the transfer line. Should a sample of exhaust gas be bled from an orifice in the

---

[1] This distinction should also be remembered when sampling supermicron wear particles, for example from brakes, for which shallow probe entry angles, to avoid particle accumulation at bends, are equally advisable (Sanders at al, 2003).

transfer line wall, then this will be unrepresentative of the gas travelling along the central core of the line. This problem can be ameliorated by installing a mixing baffle within the transfer line, as diffusional deposition is thereby discouraged (Wei et al., 2001a).

5.2.2 Diluting the Exhaust

In the remote past, 'particulate emissions' were defined as all filterable material (with the single exception of condensed water) that resides in diluted exhaust at $\leq 52\,°C$, and this definition has remained unchanged to the present day.[2] The dilution ratio used in order to reach this condition is not directly mandated; few precepts are laid down as to the dilution *method*; and legislation does not prescribe to the finest detail every dimensional parameter of the diluting device.

In their design details, of course, there are many, many tunnels, but, in their *method* of dilution, or operating principle, there are two categories: (a) *full-flow* tunnels dilute *all* the exhaust, and *then* take a flow-proportional sample; and (b) *partial-flow* tunnels take a flow-proportional sample, and *then* dilute this sample.[3] How the data acquired in each type of tunnel relate is an obvious question, and, although not reviewed here, there have been numerous studies of this nature (e.g. Suzuki J. *et al.*, 1985; Horakouchi *et al.*, 1989; Stein and Herden, 1998; Schweizer and Stein, 2000; Odaka *et al.*, 2001; Khalek *et al.*, 2002; Silvis *et al.*, 2002; Zelenka B. *et al.*, 2004). Generally, the particulate mass reported by full-flow tunnels exceeds that for partial-flow tunnels by a few per cent, the reasons for which are various; one of them is greater thermophoretic losses to internal walls in partial-flow tunnels (see Sections 4.5 and 4.6). Others find closer agreement (Schweizer and Stein, 2000).

So, even though the mandatory condition ($\leq 52\,°C$) is studiously fulfilled, there understandably exists a multiplicity of dilution tunnel designs and configurations, even in the commercial sector; while in the research community, where purpose-built tunnels, in order to accommodate a battery of instruments not required in homologation (Ntziachristos *et al.*, 2004a), are common, considerable variation and complexity exist. And in all of these designs, due care and attention must be given to problems faced by all experimenters, for example reproducibility and repeatability (Kayes and Hochgreb, 1998). But, to avoid broadening the current section to unwieldy proportions, the focus in what follows will be the fluid-mechanical nature of the dilution, the heat transfer, and how the particulate emissions are affected thereby.

The dilution tunnel has essentially four tasks. First, dilution must be carried out in a carefully controlled and repeatable manner: the vagaries of gas-to-particle conversion demand this. Second, the dilution process must resemble the real emission process, so that the particulate emitted in the laboratory resembles the particulate emitted into the ambient atmosphere – inhalation of raw exhaust gas not being the general experience. Of course, the dilution tunnel is not an assiduously faithful reproduction of the real exhaust plume, but one must be pragmatic. Third, particle concentrations must be reduced to levels kinder to instrumentation, i.e. to avoid fouling or sooting. Fourth, water

---

[2] Sometimes the temperature in the primary mixing zone is *also* regulated, say, to 190 °C (Kittelson and Johnson, 1991). But test protocols (dilution and filtering parameters) are very much under discussion at the moment: for example, one possibility is modulation of the dilution-air temperature to restrict the range of temperature experienced by the dilution tunnel filter (Andersson *et al.*, 2004b). See also Wu *et al.* (2007).

[3] In constant-volume sampling (CVS), which is full-flow and used in certification, the dilution ratio varies with respect to engine operation, such that the combined flow (exhaust and air) remains constant (Krenn *et al.*, 2000). In partial-flow tunnels, an exhaust sample is often diluted in two successive stages, or two tunnels, for which the overall dilution ratio is simply the product of the two. This is why multistage dilution tunnels are able to reach $D = 1000$, whereas full-flow dilution tunnels are restricted to $D = 10$.

condensation must be suppressed: for example, some instruments operate at room temperature only, so that merely cooling the exhaust gas on its own is insufficient.

At the tunnel entry, an obvious need arises to *condition* the dilution air, for which there are four parameters: (a) temperature; (b) humidity, with dehumidification to a certain dew point, such as $-10\,°C$ (Chase R.E. *et al.*, 2004), or humidification by adding steam (Wei *et al.*, 2001b); (c) the organic-vapour background (Shrivastava *et al.*, 2006), removed with a carbon or charcoal filter; (d) the particle background, removed with a high-efficiency particulate air (HEPA) filter. Of concern here is not only the ingestion of foreign particulate from ambient air, but also saturation and gas-to-particle conversion. These considerations assume greater importance at high dilution ratios, where the exhaust gas becomes increasingly rarefied. Future low-emission engines, particularly those fitted with particulate filters in their exhaust systems, will pose problems: particle backgrounds sometimes *exceed* tailpipe emissions (Kasper, 2003), and assume, moreover, a significant proportion of the Air Quality Standard (Kayes *et al.*, 1999).

At the *other* end of the tunnel, the filter temperature of $T_f \leq 52\,°C$ is specified, as said, but, as the air flow required to reach this condition is *not*, this must be established – for example, to determine the sizes of the pump and ducting. By assuming adiabatic dilution and constant specific heats, a simple *ab initio* estimate is made possible (Amann *et al.*, 1980), in which the temperature of the air–exhaust mixture, T_{mix}, is given by

$$T_{mix} = T_{exh}\left(\frac{1}{D}\right) + T_{air}\left(1 - \frac{1}{D}\right), \tag{5.1}$$

where D is the dilution ratio, defined below, and T_{exh} and T_{air} are the temperatures of the exhaust gas and dilution air, respectively. (The temperatures are all absolute.) It should be noted that Equation (5.1) necessarily satisfies the two extremes of *no* dilution ($D = 1$, $T_{mix} = T_{exh}$), and *huge* dilution ($D \to \infty$, $T_{mix} \to T_{air}$). As noted in Section 3.6.3, two subtly different scenarios are covered by this analysis: dilution as *end point*, in which case T_{mix} always denotes T_f, and dilution as *spatial progression*, in which case T_{mix} approaches T_f as the air–exhaust mixture passes along the tunnel, with mixing obviously proceeding at a finite rate. Either way, this function is plotted for two different exhaust temperatures in Figure 5.2 (Amann *et al.*, 1980).

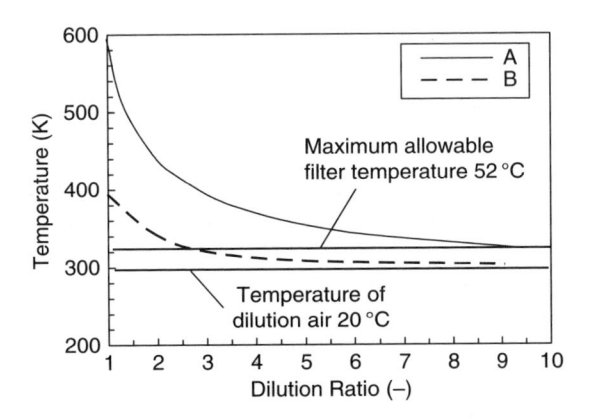

Figure 5.2 Temperature of air–exhaust mixture in dilution tunnel (T_{mix}), as modelled for one air temperature ($T_{air} = 20\,°C$) and two different exhaust temperatures (T_{exh}): A, 600 K; B, 400 K. The task is to continue diluting the exhaust gas until $T_{mix} \leq 52\,°C$, which is the temperature of the filter (T_f) (Amann *et al.*, 1980).

This analysis, although admittedly fairly rudimentary, nonetheless brackets the dilution parameters: there is, after all, a need to accommodate a *range* of T_{exh}, corresponding to the duty cycle, which might encompass all engine operation, from idle to full load. If D is insufficient, then the condition $T_f \leq 52\,°C$ will be violated. But the task is not one-sided: should one attempt to eliminate all risk of $T_f > 52\,°C$ merely by adopting an enormous value for D, the original exhaust gas would become so rarefied that little particulate would be available for collection on the filter, raising problems of sensitivity or precision in the gravimetric evaluation, or necessitating impracticably prolonged tests. And, besides this trade-off, other design constraints arise, such as practically packaging the dilution tunnel, and meeting the air demand. Actually, this compromise threatens to become a nuisance, as gas-to-particle conversion depends on D, so that more moderate values, as used in dilution tunnels, might generate particulate insufficiently representative of particulate in real exhaust plumes, for which D is large – tending to infinity, in fact.

Returning, however, to heat transfer, the adiabaticity of the preceding model is obviously idealised, and greater predictive capability is attained through due consideration of heat transfer to the surroundings, such as through the tunnel walls, and by analogy to heat exchangers (Clerc and Johnson, 1982). In fact, to obtain sufficient particulate without exceeding the temperature constraint, nonadiabatic cooling may even be essential (Kittelson and Johnson, 1991). At heart, this heat dissipation pivots on, first, the temperature difference between the tunnel and the surroundings and, second, the residence time of the gas in the tunnel. Both these parameters are, in turn, functions of dilution ratio, from which various and subtle interdependencies result (Kayes and Hochgreb, 1998), which we shall not pursue in detail, as they diverge too far from the present topic; but two examples will illustrate. First, should D decrease at a *fixed dilution-tunnel flow rate*, then the transfer-line flow rate must correspondingly increase; the transfer-line heat loss now decreases, because the transfer-line residence time decreases, and hence T_{exh} increases. Second, should D decrease at a *fixed transfer-line flow rate*, then the dilution-tunnel flow rate must correspondingly decrease; the dilution-tunnel heat loss now increases, because the dilution-tunnel residence time increases, and hence T_f becomes less dependent on D and T_{exh}. But what should not go unremarked is the unstandardised nature of dilution tunnel design, and the motley variation thereby introduced into heat transfer rates – to which one may logically ascribe inter-laboratory variation in reported emission rates, as gas-to-particle conversion is inevitably affected.

The dilution ratio is defined in more than one way (Wei *et al.*, 2001b), but is customarily based on the respective mass flow rates, m (where the suffices follow the previous convention):

$$D = \frac{m_{mix}}{m_{exh}} = \frac{[m_{air} + m_{exh}]}{m_{exh}}. \tag{5.2}$$

This equation assumes identical densities for dilution air and exhaust gas, when at the same temperature; and, at constant density, the mass flow rates can be replaced by volumetric flow rates. The concentrations of gaseous components (e.g. CO_2 and NO_x) and particles serve as alternative definitions; for example, he number concentration N can be used:

$$D = \frac{[N_{exh} - N_{air}]}{[N_{mix} - N_{air}]}. \tag{5.3}$$

A successful tunnel design faithfully reproduces, within practical constraints (some of which have been stated), the processes operating within real exhaust plumes. As we learned in Section 4.3, in real exhaust plumes, D rises linearly or nearly linearly with distance from the tailpipe; but the gradient is steep, and values of >1000 are realised in <1 s. Both the final value taken by D and the *path* by which this value is obtained – meaning the *rate* of dilution, dD/dt, or the 'dilution profile' – are equally influential in moulding particle characteristics – or, at least, recent work has strongly suggested this to be so. This is inconvenient, as conventional tunnels have little or no facility to control dD/dt – independently, that is, of other influential parameters such as flow rates, heat transfer rates and residence times.

This inflexibility, not unexpectedly, impedes efforts to discover how particle characteristics are moulded by the emission process, and so, purpose-designed tunnels, using different principles, are a recent trend in the research community. An example is shown in Figure 5.3 (Wei *et al.*, 2001a,b), for which the dilution profile, final dilution ratio, temperature and residence time are all capable of independent adjustment. To adjust the dilution profile, turbulence generators (perforated plates) and wake discs (bluff bodies) are sited where the exhaust gas enters the air stream: these features increase dD/dt in the initial stages. The dilution ratio is adjusted through the relative flow rates of exhaust gas and air. The residence time is adjusted either by flow rate or by the position at which a sample is extracted from the tunnel. The exhaust-gas flow rate is determined by the pressure difference between the tunnel and the exhaust system; the absence of a valve on the transfer line minimises particle losses. A valve and blower adjust the pressure in the exhaust system, which is maintained near ambient. It is possible to operate the tunnel over a range of pressures, from slightly above to slightly below ambient. Some mixing profiles obtained with this tunnel are shown in Figure 5.4 (Wei *et al.*, 2001a).

The fluid mechanics of this dilution tunnel, which have been modelled in some detail, allow computation of the *local* dilution ratio (Wei *et al.*, 2001b) – unlike, as is often done with traditional tunnels, where the global or *final* dilution ratio is the focus. Aside from tracking local gas-to-particle conversion, it is possible to establish whether mixing is complete, and the flow homogeneous, by the time a sample is drawn off into the probe – a necessary condition for representative samples. The answer to this question is, not unexpectedly, dependent on the size of the particles and the flow regime (Wei el al, 2001a), as follows:

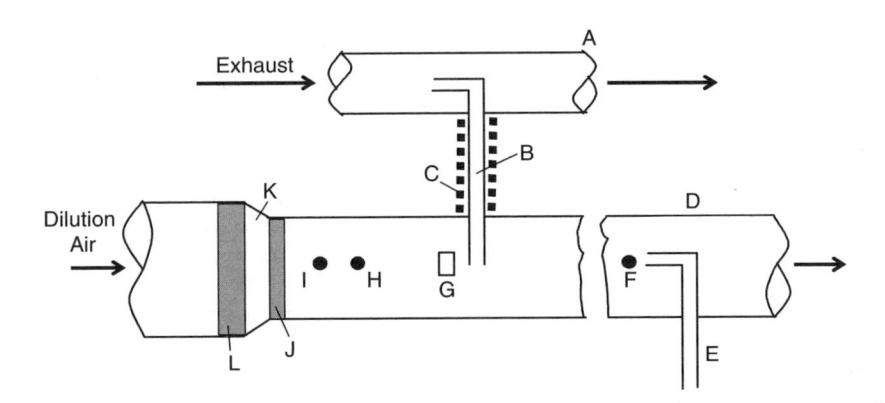

Figure 5.3 Hardware schematic of a single-stage dilution tunnel. Legend: A, exhaust pipe; B, transfer line; C, heater; D, main dilution tunnel; E, sampling probe; F, thermocouple; G, wake disc; H, humidity sensor; I, hot-wire anemometer; J, turbulence generator; K, reducer (convergent); L, filter. Specifications: dilution ratio (D), 50 to 1000; dilution rate (dD/dt), 80 to 1600s^{-1}; volumetric flow rate, 0.6 to 1.7 m^3/min; temperature, 10 to 40 °C; relative humidity, 5 to 90 %; residence time, 0.4 to 2.0 s (Wei *et al.*, 2001a,b).

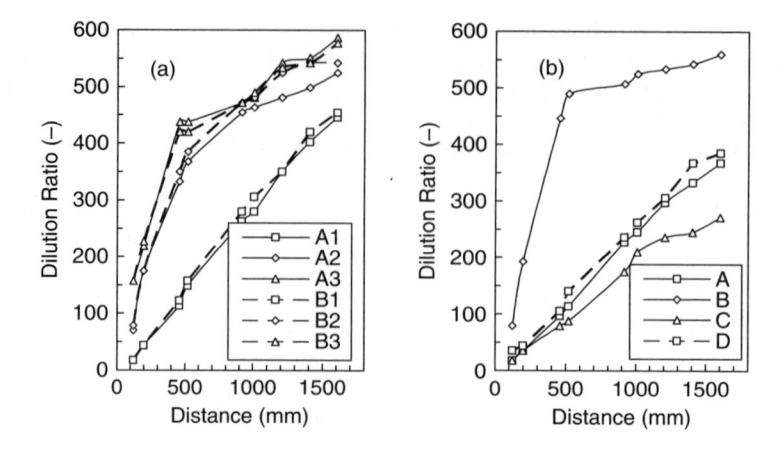

Figure 5.4 Mixing profiles along the centre line of a dilution tunnel, as determined by concentrations of NO_x and 100 nm ammonium sulphate particles. (a) Without turbulence generator. Wake disc: 1, none; 2, 34 mm; 3, 44 mm. Emission: A, particles; B, NO_x. (b) Without wake disc, NO_x only. Turbulence generator (distance from filter): A, none; B, 4 mm; C, 8 mm; D, 12 mm. Standard deviations in the original have been omitted; hardware as for Figure 5.3 (Wei et al., 2001a).

A. In *laminar* flow, and in the absence of inertial effects, mixing only proceeds by diffusion, and the mixing rate therefore depends on diffusion coefficients. Gas molecules mix faster than particles, and smaller particles mix faster than larger particles. Thus, although gases rapidly assume uniform concentrations across the tunnel cross-section, particles tend to congregate for longer periods in the central area. The essential consideration, then, is inability to predict the mixing profile for particles via localised measurements of gases.
B. In *turbulent* flow, where mixing is assisted by eddying, diffusion coefficients were assumed to be similar for all gas molecules and particles. The mixing profiles were in this case obtained with a $k-\varepsilon$ turbulence model, according to the size of the wake disc, the intensity of the turbulence, and the velocity of the exhaust gas exiting from the transfer line. Mixing was promoted by stronger vortices and heightened turbulence intensity, as one would expect. But in some cases the mixing was still very poor, leading to heterogeneity, even at the sample probe.

The present discussion has dwelled at some length on enhancements to conventional tunnels, but other, still more radical approaches have certainly been proposed – especially, those in which dilution is carried out virtually instantaneously. Miniaturisation is also helpful, as locally mounted units avoid the need for inadvisably long transfer lines. In the 'porous-tube diluter', depicted in Figure 5.5, air enters the exhaust gas via perforations in the surrounding wall of a porous tube: this arrangement helps to reduce wall deposition, and the whole assembly is cooled by an outer, enveloping jacket (Mikkanen et al., 2001). In the 'ejector diluter', the air itself entrains the exhaust gas via a venturi, so that the dilution ratio is regulated by the respective pressures: to prevent condensation, the air must be heated (Maricq et al., 1999d), which is inconvenient for cold-ambient dilution; the depression might cause some particle-to-gas conversion, and particulate deposits in the throat, as they modify the dilution ratio, must be monitored carefully (Ntziachristos et al., 2004a). In the 'rotating-disc diluter', which contains, circumferentially, a series of small chambers (Kasper et al., 2000), discrete samples of exhaust gas are successively taken, and then passed into the air stream, so that the dilution ratio is regulated by

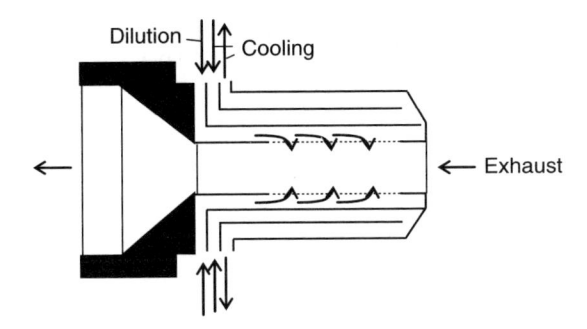

Figure 5.5 Porous-tube diluter (not to scale) (Mikkanen et al., 2001).

the rotational speed (Hueglin et al., 1996). Thus, this method is independent of exhaust-gas flow rate, and mechanically isolates upstream from downstream, preventing the transmission of exhaust-gas pulsations which might otherwise cause flow reversals (Matter U. et al., 1999). At very low rotation speeds, however, the disc will release the exhaust gas in discernible pulses.

5.2.3 Collection onto a Filter

Filters come in a wide variety of forms: the most appropriate choice depends on the intended analysis (Koutrakis and Sioutas, 1996; Mark, 1998). Conventional filters, such as those used in dilution tunnels, are made from ceramic or glass fibres, and, to guard against ongoing reactions, by which measurement artefacts are created, these fibres are protected with an inert substance; examples are borosilicate fibres coated with fluorocarbon, and glass fibres coated with Teflon. Evaporation or burn-off tests, such as are used in thermogravimetric analysis, require, to avoid adulteration of the evolved gases, uncoated fibres. Optical examinations are undertaken with membrane filters: flat surfaces of polycarbonate or PTFE, penetrated by regularly sized holes.

Filters vary greatly in their pressure drop, filtering efficiency, plugging susceptibility, artefact formation, volatile take-up, contaminating potential, friability and hygroscopicity. Coated filters contaminate the volatile or extractable particulate component (Sections 5.2.5 and 5.2.6); fibre filters are plugged by 'dry' particulate, and membrane filters by 'wet' particulate (Frisch et al., 1979); reactions between SO_2 and glass fibres create substantial positive artefacts (Claes et al., 1998); quartz fibres exaggerate particulate-phase organics through their affinity for vapour-phase organics (Chase R.E. et al., 2004); elements such as Ca, Zn and Mg, present in glass fibres, interfere with inorganic analyses (Elamir et al., 1991); and friability, or fibre loss during handling, is responsible for factitious weight losses (Akard et al., 2004). Atmospheric scientists go to considerable lengths to avoid acidity-falsifying reactions mediated by the filter (Sioutas and Koutrakis, 1995).

Significant quantities of vapour-phase organic compounds will simply pass straight through a filter: this is the 'blow-by'. To this is added 'blow-off', or low-boiling-point compounds which temporarily deposited on the filter and subsequently re-evaporated (Samara, 1995). To capture these compounds, and thus obtain a fuller emissions inventory, a sorbent or trap is added, in series. Suitable materials are activated carbon, polyurethane foam (PUF) and various types of resin. The most favoured and frequently cited sorbent is designated 'XAD-2': this is a nonionic, styrene–divinylbenzene polymeric resin of large surface area, available as beads in a 20–60 mesh size (Gautam et al., 1994), with the ability to adsorb $>C_7$ organic compounds and four-ring PAH (Dorie et al., 1987). A newer type is a

PTFE membrane impregnated with octadecyl silicon, able to collect decane – a hydrocarbon of notable volatility – with minimal blow-by (Storey *et al.*, 1999). Of course, to verify that no vapour-phase compounds successfully penetrate, subsequent traps must also be mounted, and demonstrated as having captured no further material of any significance. It should be stressed that the partitioning of organic compounds between the filter and adsorbent does not faithfully reproduce the partitioning between the particulate phase and gas phase in the ambient atmosphere.

Occasionally, larger quantities of particulate are required than the customary 47 mm diameter filters are able to accumulate – this is an important consideration when undertaking investigations into trace species, and for low-emission vehicles. The test duration can, admittedly, be prolonged, but this increases the risk of chemical transformations, i.e. measurement artefacts; and, before a trace-level compound is collected in sufficient quantity, the filter might plug. Filtering systems have therefore been devised with higher throughputs than conventionally used (Aakko and Nylund, 2003), and with the ability to capture 20–50 mg of particulate, instead of the more usual <5 mg (Rantanen *et al.*, 1996). Two different collection methods, regularly reported by researchers at Michigan Technological University, are depicted in Figure 5.6. (e.g. Waldenmaier *et al.*, 1990): naturally, the so-called 'high-volume sampler', containing a filter of size 508 mm by 508 mm, is able to capture considerably more material. One should not, however, expect complete agreement, as the two methods are not identical in filter face velocity, residence time and pressure; thus, neither are the filtering efficiencies identical – a timely reminder of nonideal behaviour in the filter, as already discussed in Section 4.6.

With today's low-emitting engines particularly, filter weighing is an increasingly nuanced and hair-raising experience, as so little material is actually captured, say, 10 μg – equivalent to a 'single latent fingerprint' (Chase R.E. and Schamp, 2007). Moreover, this must be discerned against the considerably

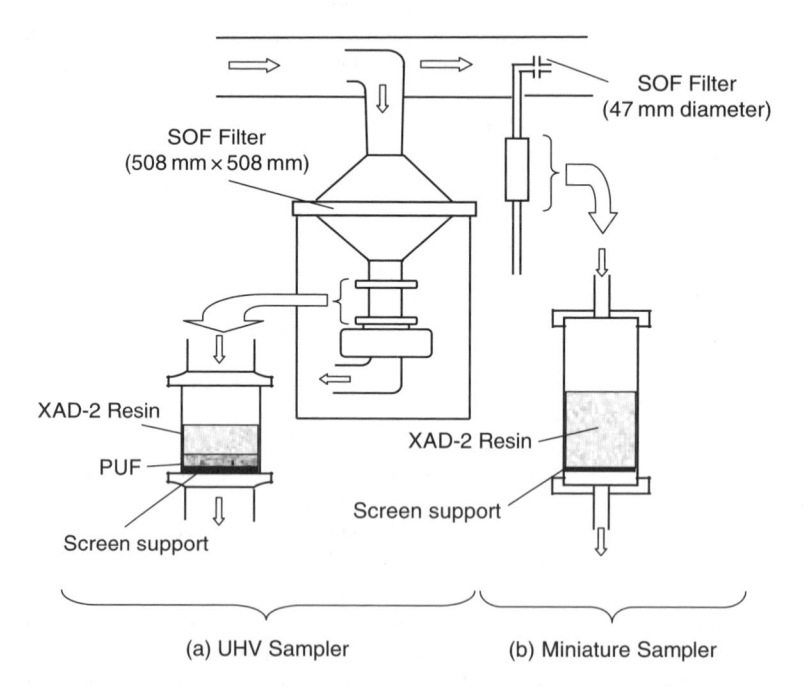

Figure 5.6 Different collection methods for SOF and XOC: (a) ultrahigh-volume (UHV) sampler, 4000 litres/min; (b) miniature sampler, 28 litres/min. (PUF, polyurethane foam.) (Waldenmaier *et al.*, 1990.)

greater mass of the filter itself. These difficulties call for considerable consistency and care in the protocol. *Before* weighing, the filter requires careful conditioning in an environmental chamber: once in the *unexposed* state (virgin or blank), and then in the *exposed* state (fouled or dirty). Aspects to observe here are the temperature, humidity and duration of the conditioning, and the possibility that volatile organic compounds will be taken up from (or lost to) laboratory air. Sulphuric acid is a notorious schemer, being hygroscopic, so that control over the take-up of water and the hydration of the acid is absolutely critical: typical conditioning is conducted at a relative humidity of $42 \pm 3\%$ and a temperature of $25 \pm 1\,°C$ (Opris *et al.*, 1993). *Buoyancy* effects arise when the air displaced by the filter influences the pre- and post-test weights if the barometric pressure has shifted between the two measurements (Chase R.E. *et al.*, 2005). *Electrostatic* effects arise through the collection of charged particles, but more probably through filter handling: the electrostatic force acting on the balance resulting from these charges becomes comparable to the gravitational force, and thus these charges must be neutralised carefully (Chase R.E. and Schamp, 2007). All in all, weight measurements these days require a precision of $0.1\,\mu g$, using a balance housed in a special purpose-built environment (Akard *et al.*, 2004).

After weighing, the filter deposit is analysed by various methods, the starting point for which is 'fractionation', according to the fivefold classification scheme of carbonaceous, organics, sulphates, nitrates and ash. As depicted in Figure 5.7, fractionation basically takes two forms: gasification (evaporation and oxidation) and dissolution (extraction by solvents), as addressed, respectively, in Sections 5.2.4 and 5.2.5. In the first method, the particulate deposit is subjected, in sequence, to two heat treatments in an inert atmosphere, and in an oxidising atmosphere, so that the first evaporates the organic and sulphate fractions, and the second oxidises the carbonaceous fraction; what remains is the ash fraction or, more correctly, the incombustible ash. (Sublimation in a vacuum is a parallel.) In the second method, the particulate deposit is subjected, in sequence, to two solvents, one organic, the other water, each removing, respectively, the organic fraction and the sulphate fraction; the ash and carbonaceous fractions remain.

How the results generated by these two fractionation schemes compare is a moot point; and, in fact, any expectation of close correspondence is questionable inasmuch as each method addresses fundamentally different properties, one being *chemical* (solubility) and the other *physical* (volatility). No comprehensive comparisons exist for the sulphate fraction, and only a few fragmentary ones exist

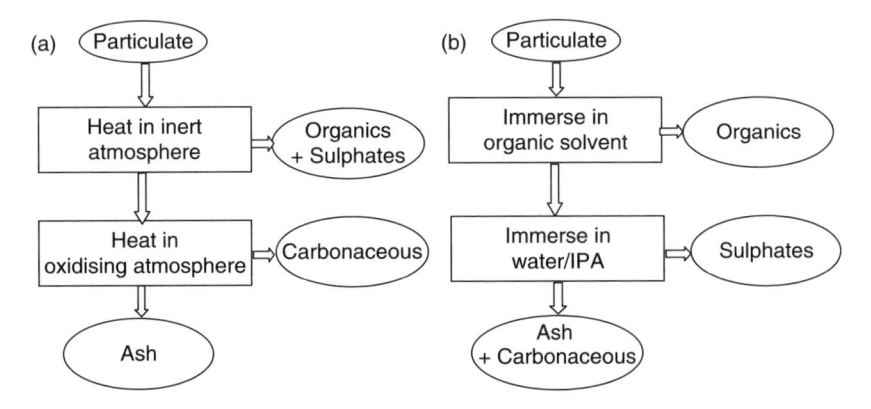

Figure 5.7 The two principal methods used in fractionating the particulate, i.e. in separating the carbonaceous, organic, sulphate and ash fractions: (b) gasification; (b) dissolution.

for the organic fraction. This question is, however, *implied* in the way these two analytical methods lend themselves to the nomenclature: namely, the soluble organic fraction (SOF)[4] and the volatile organic fraction (VOF). This topic is not reviewed here, but, purely in terms of *mass*, the two groups are at least directionally consistent, and moreover offer correlations of sorts; the VOF may be slightly larger (Postulka and Lies, 1981) or slightly smaller (Draper *et al.*, 1987; Halsall *et al.*, 1987; Lemaire and Khair, 1994) than the SOF. Others have shown that the VOF exceeds the SOF by a factor of two (Bassoli *et al.*, 1979). So, to a first approximation only, the two groups of differently fractionated organic compounds are assumed equivalent – although, as said, this equivalence is not an essential precondition (Collura *et al.*, 2005). Of course, presuppositions as to equivalent *speciations* are quite another matter.

5.2.4 Fractionation by Gasification

Thermal treatment, as a means of fractionation, has several strands. Used in air quality monitoring, but less so in the automotive industry, are 'thermooptical' methods, wherein the darkness of the deposit remaining on the filter is recorded (Cachier, 1998; Arhami *et al.*, 2006); this subject is not, however, reviewed here. In the automotive industry, the longest and best established realisation of thermal fractionation is 'thermogravimetry', wherein the filter is placed in an environmental chamber and *weighed*, as a function of gradually increasing temperature. A 'thermogram' is thereby obtained, showing the weight of particulate remaining or, alternatively, the weight of particulate escaping, as a function of time. The relationship between time and temperature is usually linear, although not necessarily so, and some measurement protocols use 'ramp-and-hold' techniques, while others incorporate intermediate periods of cooling, the reasons for which will become clear presently. But the characterisational procedure in principle is straightforward enough: different fractions are lost at different temperatures. This same information can be used to characterise the oxidative properties of the particulate, using various gases, perhaps simulated exhaust gas, and expressed in terms of Arrhenius rate expressions and thermokinetics (Stanmore *et al.*, 2001; Yezerets *et al.*, 2005; López-Fonseca *et al.*, 2006; Messerer *et al.*, 2006b); this is a large and related field, but not reviewed here.

Two key operating parameters arise, the first being the dependence of weight loss not only on temperature but also on *ramp rate*, i.e. thermograms contain a transient component, the significance of which can be difficult to define. Ramp rate is bracketed by two considerations: if too *fast*, evaporation is impeded, because volatile compounds are given insufficient opportunity to diffuse away from the immediate vicinity of the surface: this shifts thermogrammatic features such that some compounds are now released at higher temperatures (Cuthbertson *et al.*, 1979; Chan *et al.*, 1999). If too *slow*, the measurement is protracted, and uncontrolled oxidation of the sample, through penetration into the chamber of atmospheric oxygen, becomes significant: this causes suspiciously steep weight loss curves (Cuthbertson *et al.*, 1979). These factors taken together, ramp rates of 50 °C per minute are typical (Narusawa *et al.*, 1995).

The second operating parameter is the temperature at which the oxidising atmosphere replaces the inert atmosphere, dividing volatile from nonvolatile. The demarcation temperature is bracketed by two considerations: if too *low*, volatile compounds are misclassified as nonvolatile compounds; but if too *high*, the carbonaceous fraction, as it cannot oxidise, undergoes artefactual transformations which, if

---

[4] Corresponding extraction of the *sorbent* yields the *vapour-phase* organic compounds, designated XOC; these should not be confused with the *particulate-phase* organic compounds, or SOF. XOC, not being 'particulate' as such, do not figure prominently in this text. As they all consist of organic compounds (or hydrocarbons), SOF, XOC and HC are often directionally consistent (Pataky *et al.*, 1994), although this consistency is more likely with the lighter compounds in the latter two categories. (See also Section 3.6.4.)

they affect weight, are wrongly ascribed to the evaporation of volatile compounds (Lepperhoff et al., 1994). This question is to some extent an aspect of the affinity of the surface for the volatile compound: for example, the desorption temperature of *BaP* from a soot surface is 30 °C higher than from a glass-fibre filter (Spurny, 2000b).

Both operating parameters – rate of temperature increase and temperature of demarcation – are not, as yet, subject to any standardisation; the protocol that is most appropriate probably varies from one sample to another, so that some degree of compromise is necessary.

Subtler and less tractable are artefactual transformations induced in the first phase of the measurement – pyrolytic reactions being the logical outcome of exposure to high-temperature inert environments (Collura et al., 2005), as explained in Sections 3.2.2 and 3.6.1 – which falsely transfer some of the organic fraction into the carbonaceous fraction (Cuthbertson et al., 1979), such as by 'charring' (Czerwinski et al., 2007), and also synthesise new organic compounds absent in the original sample (Schulz et al., 1999). Corrections for such effects have been devised (Huffman; 1996; Turpin et al., 1990; Gertler et al., 2002). And, despite the inert environment, the evolution of CO and CO_2 betrays the oxidation of surface functional groups (Messerer et al., 2006b). Artefacts in the second phase of the measurement stem from too rapid an introduction of oxygen: local temperatures deviate substantially from those of the surrounding oven, and, owing to thermal runaway, or 'self-heating', the oxidation rate accelerates (Yezerets et al., 2005), rendering mass transport and heat transfer factitious (Neeft et al., 1996b). In extreme cases the filter fibres degrade or react with the sample, introducing further risk of falsification. These risks are contained by mixing the sample with chips of quartz, which act as heat sinks (Yezerets et al., 2002).

In attempting to negotiate or to ameliorate these problems, researchers have devised numerous thermogrammatic protocols (e.g., Abbass et al., 1991c; Fukushima et al., 2001; Cadle et al., 1999a; Warner et al., 2003); in atmospheric science particularly, no less than *seven* successive steps, of increasing temperature with intervening pauses, are used, by which four organic subfractions (inert environment) and three carbonaceous subfractions (oxygenated environment) are defined (Huffman, 1996). Since these minutiae are not standardised (Bae M.-S. et al., 2007), and yet they impact the amount of material assigned to one fraction or another, automatic agreement between different laboratories should not be expected. But a discussion of the relative merits and demerits of each protocol is beyond the scope of this text.

A thermogram obtained using a *typical* protocol for motor vehicle particulate is given in Figure 5.8 (Lepperhoff et al., 1994): organic compounds of low molecular weight evaporated below 300 °C, the sulphates and associated water evaporated between 150 °C and 350 °C, and the organic compounds of high molecular weight evaporated between 300 °C and 450 °C. With further temperature increase, the weight losses were in some measure artefactual – the result of restructuring in the carbonaceous fraction. At 650 °C, the oxygenated atmosphere was introduced, whereupon the carbonaceous fraction burned, and only the ash remained. It should be noted that ash itself can decompose to some extent, and so account for further weight loss (McGeehan et al., 2005). From this it can be seen that, since various phenomena occur concurrently, interpretation of thermograms requires some conjecture or foreknowledge of what to expect.

The mystery as to what components are being lost at any one moment is wholly answerable, since the gaseous effluent from the test chamber can simply be analysed using well-known instruments. Indeed, this additional facility is eminently commendable, because insufficient sensitivity in mass measurements is becoming an embarrassment with today's low-emitting engines, and gas analysis instruments have far greater sensitivity. Taking this a step further, the concentrations of various gases in the effluent can be used to compute, *retroactively*, a mass balance, for which the old-fashioned mass measurement serves as a useful cross-check, although this may, eventually, be discarded completely. Apparatus along these lines has received widespread coverage in the technical literature of late, in

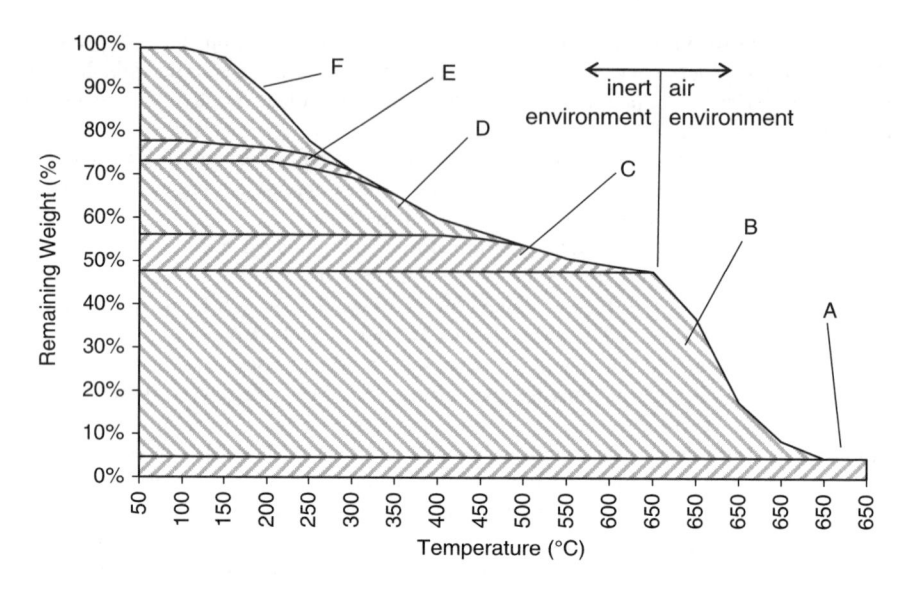

Figure 5.8 Schematic of thermogravimetric weight loss curve for particulate. Fractions: A, incombustible ash (ash fraction); B, combustion of soot (carbonaceous fraction); C, desorption, dehydrogenation and/or modification of soot structure; D, desorption of high-boiling-point hydrocarbons in organic fraction (oil); E, desorption and decomposition of sulphates, including associated water (sulphate fraction); F, desorption of low-boiling-point hydrocarbons of organic fraction (fuel). Test protocol: temperature ramp in nitrogen atmosphere at 50 °C/minute, hold for five minutes, and finally, introduction of air (after Lepperhoff *et al.*, 1994).

which the indirect, or retroactively computed, masses of the three fractions – carbonaceous, organic and sulphate – agree closely with direct measurements of mass (Fukushima *et al.*, 2001, 2003b; Akard *et al.*, 2004).

Various choices recommend themselves in gas analysis. Logically, one should opt for measurement of CO_2 and H_2O, and, to handle cases where the burn-off perhaps proceeds a little too spiritedly, CO. Measurement of H_2O might be dispensed with, if an appropriate assumption of the C:H ratio is acceptable: for example, the use of a value of 1.8 or 1.9, as in diesel fuel, to compute the mass of the organic fraction, is probably of minor influence (Akard *et al.*, 2004). One should also bear in mind that oxygen, albeit in small amounts, is also generally found in particulate, by which the mass balance may be affected. The evolving hydrocarbons might be oxidised in some subsequent step, with similar quantification as CO_2, or quantified directly, by using a flame ionisation detector (FID). The FID measurement, arguably, corresponds to the VOF; and the FID trace, in fact, depicts the first derivative of the thermogram (Cuthbertson *et al.*, 1979) – provided the sulphate fraction is small. The fate of the sulphate fraction in this process is flimsily reported, but the acid–water complex may well decompose, according to

$$H_2SO_4 \cdot nH_2O \rightarrow SO_2 + (n+1)H_2O + (1/2)O_2. \tag{5.4}$$

The sulphates are thus quantifiable in the effluent as SO_2 – provided, that is, its equilibration with SO_3 is well characterised. As before, the oxygen thus liberated may assist in consuming what is combustible – potentially, this element is a more significant confounder in the first phase, as the chamber environment is supposedly inert, i.e. unoxidising. Finally, by a similar decompositional method, this retroactive technique is able to quantify the nitrate fraction, this time via the evolved NO and NO_2 (Fukushima *et al.*, 2003a).

5.2.5 Fractionation by Dissolution

It is axiomatic that different particulate components dissolve in different solvents; fractionation thus carried out is termed 'filter extraction', or sometimes just 'extraction'. The filter is weighed, immersed in a suitable solvent and weighed again, by which the lost mass is computed *directly*; and the solute is analysed or recovered, by which the lost mass is computed *indirectly*. The direct and indirect measurements thus conveniently offer some degree of mutual corroboration, although it should be said that greater accuracy is potentially on offer from chemical analyses of the solute, rather than from gravimetric assessments of the filter. The fraction lost is generally taken to have dissolved, although there is a possibility that some components, e.g. minute metal particles, are, strictly speaking, dislodged by mechanical agitation (Shafer *et al.*, 2006).

From the beginning, one should recognise three likely sources of error. Firstly, the solvent is a source of impurities (Halsall *et al.*, 1987; Sidhu *et al.*, 2001): this is an all-pervading problem in analytical chemistry which is not addressed here. Secondly, filters are *friable*, and fibre losses are inevitably exacerbated by any agitation used to aid the process of dissolution, for example vigorous shaking or sonication. Obviously, these fibre losses will falsify mass measurements (Funkenbusch *et al.*, 1979). Thirdly, the filter *itself* is extractable to some extent – this is indisputable, since extractions of *unexposed* (virgin or blank) filters yield small quantities of unspecified material, generally supposed to be organic binders residing in the filter matrix as part of the manufacturing process, although this area is admittedly poorly documented. The mass of material released by the filter, being nonnegligible, must be taken into account; chemical analyses, e.g. chromatograms, may also be affected. An interesting question is whether unexposed and exposed filters are identical in what they release; the answer seems not to have been established. But, because of this adulteration, assiduous investigators undertake extractions not just of exposed filters but also of unexposed filters as part of their quality control methods.

Extraction of the Organic Fraction

The longest-established method of quantifying the organic fraction is 'Soxhlet extraction' (Burley and Rosebrook, 1979; Smith A.J. *et al.*, 1995), the apparatus for which is, in effect, a condenser–distiller. The exposed filter, placed in a heated vessel, is washed in distilled solvent, from which the solvent is periodically siphoned off, to be redistilled and returned to the filter, so that the solvent is gradually enriched in the extracted compounds. To avoid photodegradation, the extraction is performed under yellow light or in the dark; and, to extract effectively, the operating temperature is maintained close to the boiling point of the solvent. The extraction is terminated after a defined number of cycles or, alternatively, when the siphoned-off solvent becomes clear, signifying that no further material is extractable. The solution is then evaporated or blown to dryness, such as in a gentle stream of nitrogen, to leave a residue of organic compounds. These recovered compounds can then be used in subsequent analyses such as bioassays (Section 5.2.7).

The chief drawback of Soxhlet extraction lies in its time-consuming nature, making determination of its optimal duration doubly important, but this is easier said than done; the answer involves the law of diminishing returns, and a suitable compromise between quantitative accuracy and practical limitation. For example, extractions always yield *some* material, albeit in ever-decreasing amounts; perhaps an extraction of just one hour yields 80% of the material obtained by an extraction of 24 h

(Perez et al., 1984). One should also bear in mind the possibility of speciative biasing: for example, half the compounds of high molecular weight, such as those heavier than coronene, might survive extraction (Bassoli et al., 1979).

The possible solvents are various: for example, dichloromethane (DCM), toluene, cyclohexane, chloroform, carbon disulphide, acetone, benzene, ethanol and methanol are just some to have been investigated, of which the first mentioned is by far the commonest choice. Several obvious and not-so-obvious criteria have a hand in deciding the appropriate choice of solvent. First amongst these is extraction efficiency, but, unfortunately, no single solvent is able to extract *all* relevant organic compounds (the principal issue being polarity). Hence, extractions using more than one solvent, either sequentially or as mixtures, may prove advantageous: for example, a three-stage extraction with cyclohexane, DCM and acetonitrile extracted, respectively, nonpolar, intermediate and polar compounds (Dietzmann et al., 1980). Obviously, different protocols extract different sets of compounds: for example, the aromatic yield using benzene–methanol was twice that with DCM (Williams R.L. et al., 1985);

Other criteria in solvent choice, while not concerning the extraction of organic compounds *per se*, are not for that matter any less important. The expense of the extraction involves not just the volume of solvent used, but also, following use, disposal of the same solvent (as hazardous waste, etc.) (Storey et al., 1999). Solvents might inappropriately extract a small quantity of sulphates (Halsall et al., 1987), or oxidise the compounds they are supposed to extract (Williams R.L. et al., 1985). Extraction efficiency is related to the depth of penetration into the particulate deposit, as decided by the size of the solvent molecule in relation to the pore size (Jones C.C. et al., 2004). Other criteria in solvent choice are toxicity, volatility, purity – impurities can falsely register as particulate-derived organic compounds (Halsall et al., 1987) – and, finally, compatibility with subsequent analytical procedures (Sections 5.2.6 and 5.2.7). With these wide-ranging criteria, some degree of compromise seems inevitable in the choice of solvent.

Lest Soxhlet extraction should appear to have a monopoly, some mention is needed of other methods which, although arguably well established in other fields of analytical chemistry, have not (yet) seen wide application to motor vehicle particulate: there are also organic extraction with a sonic dismembrator and a continuous filtering system (Gautam et al., 1994), supercritical-fluid extraction (with CO_2), and subcritical-water extraction (with H_2O). Each method recovers a different set of compounds: for example, for the extraction of PAH, subcritical water is less effective at high molecular mass (Hawthorne et al., 2000), and supercritical CO_2 seems more effective than Soxhlet extraction (Jones C.C. et al., 2004).

There is an instructive distinction, though, to be made here. Historically, the aim has always been to extract the greatest quantity of organic material, in order to identify whatever is there (say, by gas chromatography); this required the highest possible recoveries, particularly as some compounds are tightly bound to the soot core, and so reluctant to leave. Establishing the long-term *environmental* implications should be recognised as a different aim, as much rests on 'bioavailability'; hence, only those compounds likely to be denuded by natural processes are of interest, in which case, these other methods may well be more suitable choices than Soxhlet extraction (Hawthorne et al., 2000; Jonker et al., 2005).

Extraction of the Sulphate Fraction

What really counts in sulphate extraction is not the sulphuric acid, in isolation, but rather the complex formed by this acid with water. This is a nuisance, because the hydrated state of this complex is acutely sensitive to water vapour, that is, the relative humidity of the air, in the vicinity of the filter deposit; and deposits in which sulphates figure prominently (e.g. Section 8.2.2) are particularly susceptible to these uncontrolled variations in mass. Thus, some laboratories, *prior* to fractionation, first expose the filter deposits to ammonia (vapour), by which process the sulphates are *stabilised*, as the resulting ammonium sulphate is less hygroscopic (Wong V.W. *et al.*, 1984). This process, known as 'ammoniation', is of little consequence to the mass, as the loss of water is offset by the acquired ammonia (Ingham and Warden, 1987), but a correction is easily enough calculated. Of course, such a deliberately induced transformation should not disturb or falsify other particulate characteristics – as seems to be the case with the mutagenicity (Bagley *et al.*, 1991), but perhaps not the mass (Gratz *et al.*, 1991), of the organic fraction.

To avoid the preparation of two separate filter deposits, one for the sulphate fraction and one for the organic fraction, with the corresponding risk of introducing inconsistencies, it is useful if both fractions are quantifiable in the *same* deposit. The question of which extraction should come first has not been comprehensively addressed, but, according to one study (Narusawa *et al.*, 1995), the sulphate fraction that results is effectively identical whether its extraction succeeds or precedes extraction of the organic fraction. As mentioned above, however, the possibility that organic extraction might improperly remove some sulphates should be considered carefully. Conversely, the implications of sulphate extraction for the organic fraction in motor vehicle emissions seem to have been ignored. However, in air quality monitoring, it is recognised that some organic compounds – predominantly aliphatic ones residing on *ambient* particles – are water-soluble (Sannigrahi *et al.*, 2006), although these compounds are, admittedly, secondary rather than primary. The secondary organic aerosol has a dual nature in that it contains hydrophobic hydrocarbon chains as well as hydrophilic polar function groups (Jenkin and Clemitshaw, 2000).

A comprehensive literature detailing methods of sulphate extraction emerged suddenly, on the introduction of oxidation catalysts to gasoline engines in the mid 1970s in the USA. Aqueous extractions used distilled, deionised water, containing a range of other compounds: ammonium carbonate, $(NH_4)_2CO_3$, and formaldehyde, HCHO (Griffing *et al.*, 1975); sodium bicarbonate, $NaHCO_3$, and sodium carbonate, Na_2CO_3 (Perez *et al.*, 1984); or, more commonly, isopropyl alcohol (IPA) (Postulka and Lies, 1981), which acted as a 'wetting agent', allowing deeper penetration of water molecules into hydrophobic samples (Cadle *et al.*, 1999a). Extraction is assisted by slight warming (50 °C) (Griffing *et al.*, 1975), vigorous shaking or sonication (Perez *et al.*, 1984).

Following extraction, the dissolved sulphates are quantified such as by the release of I_2, following reaction with barium iodate, hydrochloric acid and potassium iodide (Griffing *et al.*, 1975). Alternatively, upon reaction with barium chloranilate, the sulphates precipitate out as barium sulphate (Dietzmann, 1984), and the released chloranilate ions are then measured colorimetrically with a UV photometer; these measurements are confounded by light-absorbing hydrocarbons also in the solution, but corrections are available (Khatri *et al.*, 1978). In ion chromatography, the conductance is calibrated by solutions of potassium sulphate at known concentrations (Opris *et al.*, 1993).

The data of Figure 5.9(a) serve as a useful reminder that 'sulphate fraction' generally denotes *water-soluble* SO_4^{2-}, which, since it scales reasonably with *total* S, is obviously the dominant form of sulphur in typical particulate. But three cautionary remarks pertaining to weakly reported areas are in order. First, water-soluble SO_4^{2-} is *not* the same as S (Pierson and Brachaczek, 1976; Cadle *et al.*, 1999a), as there are also found, for example, organosulphur compounds. Second, water-soluble SO_4^{2-} is *not* the same as total SO_4^{2-}, as the acid is reactive, and some sulphated metals are, in fact, insoluble or weakly soluble. Thirdly, water-soluble SO_4^{2-} is *not* the same as total water-soluble material, particularly as this relates to nitrates, NO_3^- – often reported as a separate fraction. Other ions are also found in aqueous extracts, e.g. NH_4^-, Na^+, Mg^{2+} and Cl^-, for which the immediately preceding chemical compositions have, of course, been lost. It should be mentioned that, with low-sulphate-emitting vehicles, sufficient ammonia may exist in the exhaust gas for complete neutralisation of the acid (Cadle *et al.*, 1999a); naturally, this would upset assumptions as to the amount of associated water.

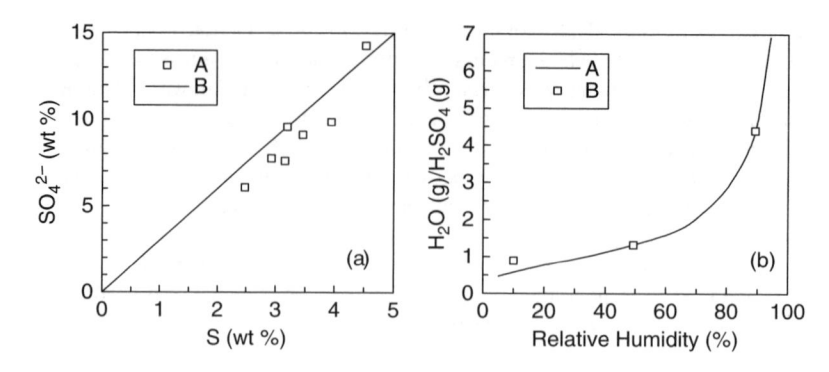

Figure 5.9 Quantification of sulphates. (a) Weight percentages of SO_4^{2-} and S in various particulate samples (A); if all the sulphur was present as sulphates, the data would lie on the solid line (B) (Wall and Hoekman, 1984). (b) Mass ratio of water to sulphuric acid, as determined by equilibration with water vapour in the surroundings. This is not condensed water; it is bound by intermolecular forces. Solid line (A) according to Stokes and Robinson (1949). The three data points (B) are from Wall and Hoekman (1984).

The associated H_2O can be directly measured by weighing the deposit before and following exposure in a desiccator (Perez *et al.*, 1984). But, customarily, only the mass of SO_4^{2-} and thereby also the mass of H_2SO_4 are known directly, as this is more convenient. Because, as aforesaid, the mass of the water–acid complex is what really counts, and this mass depends on the relative humidity, the filter deposit must be 'conditioned' *prior* to weighing, to ensure a well-defined state. The exact value chosen for this relative humidity is unimportant, so long as it is *consistent*. But, from a practical standpoint, as revealed in Figure 5.9(b), the higher end of the scale should be avoided, as small errors in relative humidity introduce large errors in mass (Wall and Hoekman, 1984).

An often-cited statistic is that, in relative humidities of 30–70%, one gram of H_2SO_4 is associated with 1.3 g of H_2O. Consequently, to account for this undetermined mass of H_2O, the mass of H_2SO_4 is simply multiplied by a factor of 2.3. This assumption seems to have served the automotive industry well for many years – but it is wise, nonetheless, at this point, to examine its efficacy, as much rests on its general validity. An assumed multiplication factor of 2.3 may not always be appropriate: the data of Figure 5.10 (Fujii *et al.*, 2002), for which water was instead directly determined, show wide divergences. For example, an assumed factor of 2.3 results in considerable overestimation of the sulphate emission from a vehicle fitted with a DPF (Section 7.8.2), the action of which ensures that a dilution tunnel filter is not overwhelmingly fouled by soot. This is interesting, since 'sulphate-make' by an oxidation catalyst creates similar conditions, and such investigations have also exposed similar shortfalls (Ketcher and Horrocks, 1990). These disparities, while as yet unexplained, are not perhaps too surprising, in view of the complexities of the filtering process (Section 4.6). Some possibilities are deeper penetration of the filter matrix by vapour, and differences between the deposited soot and the filter fibres in hydrophobicity. In this latter hypothesis, it is worth noting that Figure 5.9(b), pertaining to the acid–water equilibrium, was derived *not* from studies of particulate deposits on a filter but from *liquid surfaces* – this being for chemists a long-standing exemplar of such systems (Abel, 1946), for which the mechanisms of water transfer, from gas to liquid across the phase boundary, will not necessarily resemble adsorption of gas-phase precursors by solid surfaces, whether these be soot or filter fibres. But whatever the cause, a multiplication factor of 2.3 is really only valid within a limited range of poorly prescribed parameters.

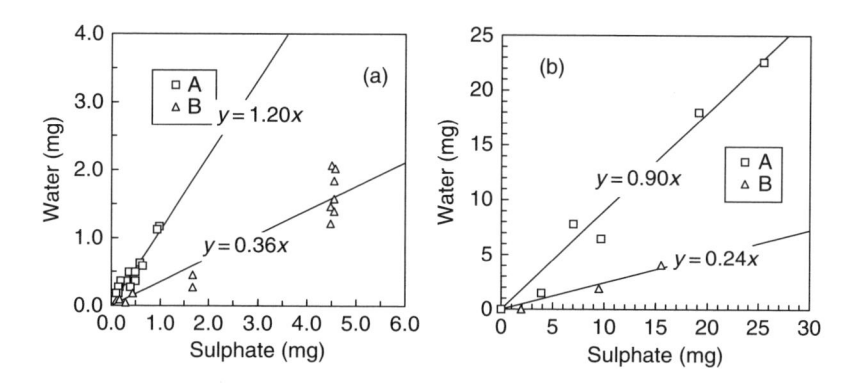

Figure 5.10 Amount of bound water as affected by DPF and type of dilution tunnel filter. (a) A, without DPF; B, with DPF. (b) Type of dilution tunnel filter: A, glass fibre; B, Teflon-coated glass fibre. Engine: 6-litre, direct-injection, model year 1999, rated at 2700 rpm/175 N m. Duty cycle: Japanese D1. Fuel: aromatics, 17.4% and 21.4%; sulphur, 443 ppm and 46 ppm; T90, 325 °C and 334 °C (Fujii et al., 2002).

5.2.6 Chemically Assaying the Organic Fraction

The isolation of organic compounds from complex mixtures, and the identification and quantification of these compounds – sometimes called 'speciation' – is the task of the analytical chemist. Legions of analytical methods are available: for example, mass spectrometry (Wood K.V. et al., 1982; Tan P.V. et al., 2002), infrared spectrometry, and nuclear magnetic resonance spectrometry (Funkenbusch et al., 1979). But the whole gamut of available apparatus cannot herein be covered; hence, emphasis is given to gas chromatography (GC) and flame ionisation detection (FID), which are, respectively, the mainstay and the mizzenmast of speciation in the automotive industry, where the first *isolates*, and the second *quantifies* (Clark N.N. et al., 1996); *identification* is then accomplished by comparison with certain standards or calibrations. Mass spectrometry (MS) is a parallel method. Even so, lengthy discussions of GC-FID or GC-MS design, calibration and operation would be inappropriate: the analytical methods are described herein only insofar as they impact on the particulate characterisations thereby obtained; and, conversely, the information sought is discussed with respect to its bearing on the chosen analytical method.

Detailed characterisation of the organic fraction is a major challenge for the analytical chemist (Samara, 1995). Unfortunately, a significant proportion of what elutes from the GC is unresolvable: 'coelution' is where individual compounds are not resolved into single-component peaks; instead, a 'hump', or undifferentiated envelope, forms in the chromatogram, on which a few identifiable peaks are superimposed – a regrettable problem also encountered in analyses of atmospheric particulate (Tanner, 1998). This obviously reflects great chemical complexity, and analytical methods always contain room for improvement (Jiao and Lafleur, 1997). It should also be stated that very similar technical hurdles are encountered in speciating a closely related and equally complex emission group, the *gas-phase* hydrocarbons or HC (Siegl et al., 1994), and, although not reviewed here, GC has been reported for C_1–C_{12} (Lepperhoff et al., 1994), C_9–C_{24} (Lanning et al., 2000) and C_1–C_{20} (Schulz et al., 1999).

Resolution of the *entire* organic fraction into clearly identifiable compounds, is thus impractical, indeed impossible; thus, this great chemical complexity is, instead, systematised according to various *subfractions*, i.e. using functional groups or molecular structures: for example, carbonyls (C=O), hydroxyls (OH), nitros (NO_2), alkanes, polycyclic aromatic hydrocarbons (PAH), nitro-PAH, oxy-PAH, hydroxy-PAH, etc. One scheme uses three subfractions, 'nonpolar', 'moderately polar' and

Table 5.1 The organic fraction systematised according to seven subfractions (cited by Heywood, 1988, p. 632).

Subgroup	Components of subgroup
Acidic	Aromatic or aliphatic Acidic functional groups Phenolic and carboxylic acids
Basic	Aromatic or aliphatic Basic functional groups Amines
Alkanes	Aliphatics, normal and branched Numerous isomers from unburned fuel and lubricant
Aromatics	From unburned fuel, partial combustion, and recombination of combustion products; from lubricant Single-ring compounds; PAH
Oxygenates	Polar functional groups but not acidic or basic Aldehydes, ketones, alcohols Aromatic phenols and quinines
Transitional[a]	Aliphatic and aromatic Carbonyl functional groups Ketones, aldehydes, esters, ethers
Insoluble	Aliphatic and aromatic Hydroxyl and carbonyl groups High-molecular-weight organic species Inorganic compounds Glass fibres from filters

[a] The chromatographic band between aromatic and oxygenates.

'highly polar' (Johnson J.H., 1988); another uses five, *n*-alkanes, alkylcyclohexanes, PAH, hopanes and steranes[5] (Kweon *et al.*, 2003b). A third, published by the US National Research Council in 1982, uses, as reproduced in Table 5.1 (cited by Heywood, 1988, p. 632), *seven* subfractions, viz., *acidics, basics, alkanes, aromatics, oxygenates, transitionals* and *insolubles*.

The analytical chain, separation procedure or 'work-up', from the organic fraction as collected to each of the subfractions, can be expressed as a *flow chart* (Waldenmaier *et al.*, 1990); that corresponding to the seven-subfraction scheme in Table 5.1 is given in Figure 5.11 (Funkenbusch *et al.*, 1979). The basic and acidic subfractions are first obtained using liquid–liquid extraction with aqueous base and aqueous acid, respectively; the neutral portion is then fractionated by column chromatography and a silica gel column, on the basis of polarity, with identification during elution by observation under UV and visible light.

Such a thorough work-up, however, is unusual, and, not infrequently, workers choose to focus on two subfractions held to be the most inimical, PAH and NPAH, as these compounds carry a very considerable proportion of the carcinogenic risk. For calibration purposes, standard reference materials

---

[5] Hopanes and steranes are sometimes used as markers for unburned lubricant (Riddle *et al.*, 2007), being neither present in fuel nor synthesised in the combustion. They are found in the atmosphere predominantly in particles of size <100 nm (Fine *et al.*, 2004). Gasoline and diesel engines are both implicated.

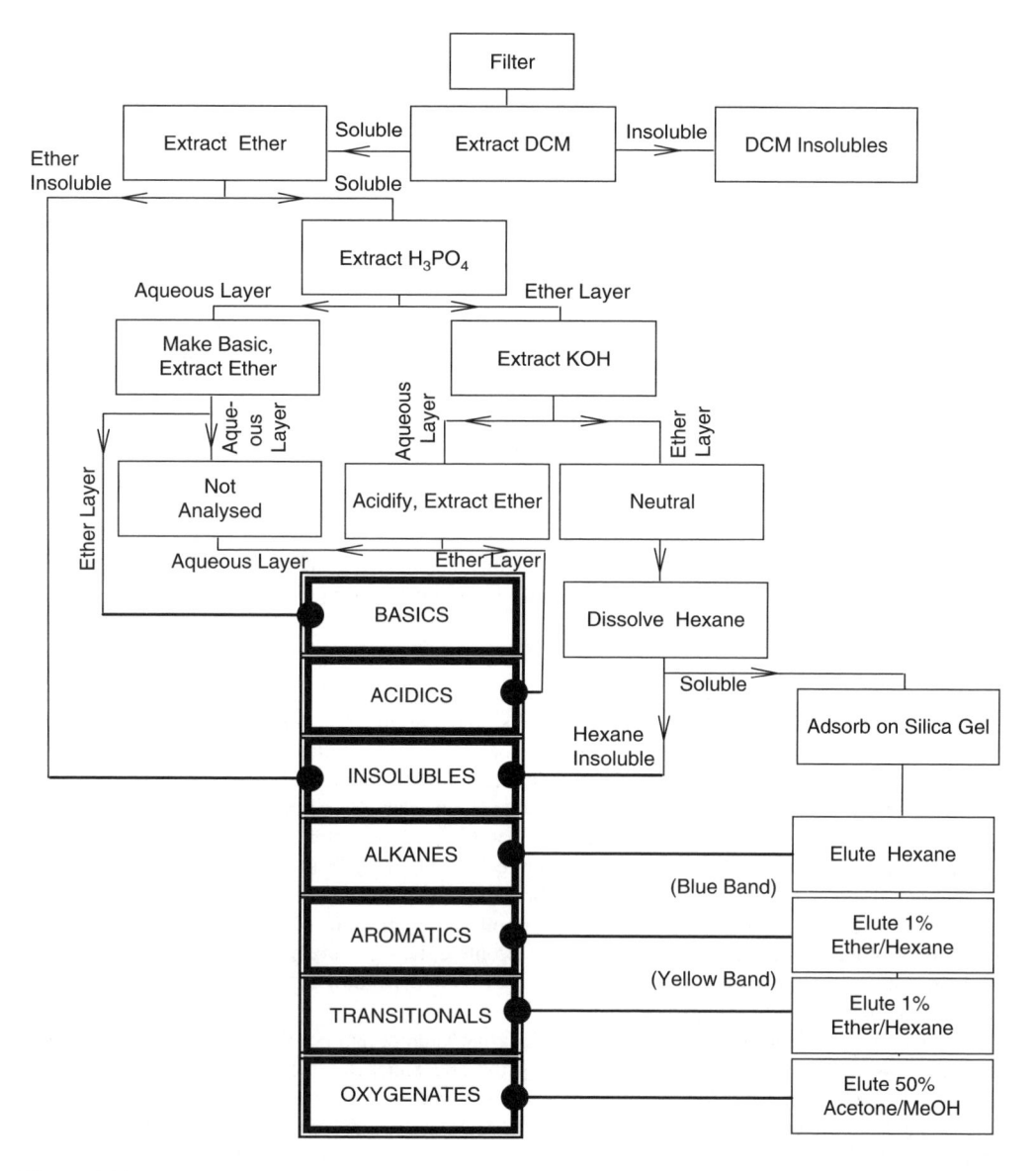

Figure 5.11 Flow chart for assaying the organic fraction of diesel particulate into seven subfractions. NB: there are eight subfractions if hexane insolubles and ether insolubles are counted separately (Funkenbusch *et al.*, 1979).

(certified mixtures) are commercially available (Bamford *et al.*, 2003). PAH are chromatographically separated and identified by *fluorescence* spectrometry (Kraft and Lies, 1981; Kraft *et al.*, 1982), by which three-dimensional chromatograms may be constructed, according to elution time and wavelength (Takada *et al.*, 2001). Quantification of NPAH is more involved, since these compounds do not fluoresce: they are first chemically reduced to fluorescent amines. Various other methods are, however, available (Johnson J.H. *et al.*, 1994). (It should be noted that, owing to their photosensitivity, PAH should only be exposed to the dark, or to yellow light.)

From what has been said, the ponderous and protracted nature of sample work-up will be apparent. This is why chemical analyses often deliver inconsistent results: each stage introduces some risk of measurement error or falsification, for example through inadvertent promotion of chemical reactions, the loss of small quantities of analyte (Swartz *et al.*, 2003) and the introduction of contaminants. Of course, these issues are common enough throughout analytical chemistry, but, even if organic fractions are analysed consistently, according to identical protocols, considerable efforts are still required to obtain inter-laboratory agreement (Eisenberg *et al.*, 1984).

To correct for losses in sampling and analysis, 'spiking' experiments are sometimes undertaken, for example using ^{14}C radiotracers (Bricklemeyer and Spindt, 1978). These compounds may be introduced at various points: to the raw exhaust, the diluted exhaust, the filter or the organic extract of the filter. These experiments show that losses *en route* can be surprisingly large: several tens of per cent. Obviously, considerable potential exists for thermal decomposition and destructive chemical reactions when tracers are injected into the hot exhaust (Section 3.6.2). But, less obviously, an assumption is required that native PAH and tracer PAH undergo *identical* losses, and this is difficult to prove. For example, the partitioning between gas phase and particulate phase may differ; this is apparently because tracer PAH, by virtue of their somewhat lower vapour pressures, are afforded greater opportunity to pass through filters. For this reason, and various other arcane ones which cannot be explored here, tracer-corrected measurements must be treated with the utmost caution (Williams R.L. *et al.*, 1985).

Separation of Fuel and Lubricant

The *separation* of fuel and lubricant is, perhaps, the commonest motivation for analysing the organic fraction (Brandenberger *et al.*, 2005).[6] This divorce is realisable by various methods. In thermogravimetry, subtle transitions are seen in the rate of weight loss because, with increasing temperature, fuel hydrocarbons tend to evaporate before lubricant hydrocarbons: a natural consequence of *boiling-point distribution*. Somewhat greater accuracy is on offer by passing the evaporating hydrocarbons to a FID (see Section 5.2.4 and Figure 5.8), the trace from which shows two envelopes or 'humps', the first for fuel, the second for lubricant. Similar features are also revealed in GC-FID (Sakurai *et al.*, 2003a), with separation according to other properties, of which one is *molecular-weight distribution*. Whatever the method, the result will resemble the signature of the fuel or the signature of the lubricant, or (more usually) take on the features of both, depending on the respective contributions these two sources make to the organic fraction.

The principal problem to be overcome is *overlap*: that is, between the 'high-end' fuel hydrocarbons and the 'low-end' lubricant hydrocarbons (Kageyama and Kinehara, 1982).[7] An intermediate range thus exists in which the two sources must be resolved, and there are various methods of doing this. One is simply to draw a line in the sand, as it were, where all compounds evaporating below a certain temperature are defined as fuel, just as all those evaporating above this temperature are defined as lubricant (Andrews *et al.*, 2000a). A similar idea can be applied to the rate of weight loss in the thermogram (Andrews *et al.*, 1998a,b). Another, more refined method is to assemble three data sets, i.e. for the organic fraction, lubricant and fuel, and, as illustrated in Figure 5.12, perform some rudimentary

---

[6] Incidentally, an identical analytical problem is encountered in determining how much fuel has entered the oil sump in the engine.
[7] Overlap is not so much of a problem with spark-ignition engines, as gasoline is more volatile than diesel fuel (Andrews and Ahamed, 1999).

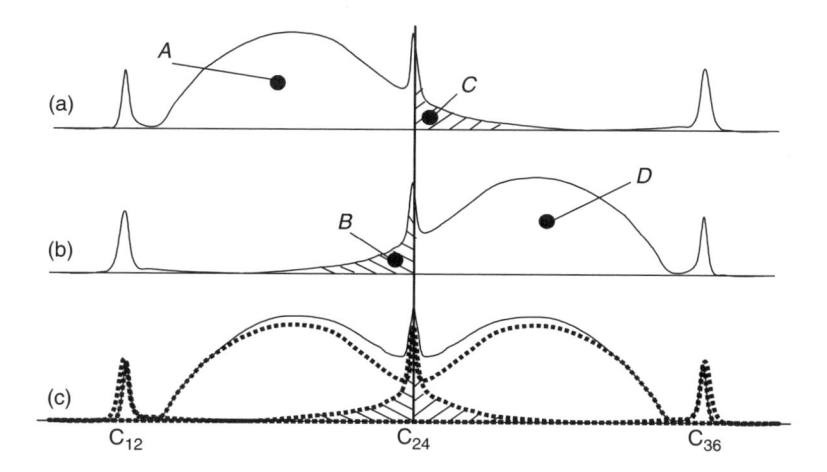

Figure 5.12 Schematic depiction of the method used to resolve lubricant and fuel contributions to the organic fraction, based on chromatograms of (a) fuel, (b) lubricant and (c) organic fraction. The spikes are tracers (markers) (Lepperhoff *et al.*, 1994).

algebra on the areas under the respective curves (Lepperhoff *et al.*, 1994). In this figure, the overlap in the chromatogram occurs at $\sim C_{20}$–C_{28}. The four 'unknowns' *A*, *B*, *C* and *D* are deduced from the four 'knowns' $A+B$, $C+D$, C/A and B/D. But whatever the chosen method, the degree of discrimination realised is easily enough established using artificially prepared mixtures of fuel and lubricant (Andrews *et al.*, 1993, 2000a). It should be emphasised, however, that Figure 5.12 is conceptual in that only the envelopes are delineated: real traces are always considerably noisier (e.g. Wall and Hoekman, 1984).

In petitioning for this divorce, two essential but nonetheless highly inconvenient assumptions are essential, but the supporting literature in both respects is remarkably threadbare. The first assumption is that heavier species in the lubricant have not been cracked into smaller molecules that resemble lighter species in the fuel – the same problem that arises in determining fuel entry into the lubricating system (Andrews *et al.*, 2001e). The second assumption is that lighter species in the fuel have not been pyrosynthesised into larger molecules that resemble heavier species in the lubricant. The consequences of these assumptions for fuel–lubricant partitioning do not seem to have been addressed, although ample evidence exists that the speciation of the organic fraction is not a straightforward superposition of the speciation of the fuel and the speciation of the lubricant.

5.2.7 *Biologically Assaying the Organic Fraction*

Many organic compounds are biologically active or 'mutagenic', in that they damage cellular DNA; and the quantification of this mutagenic potency, or mutagenicity, is the object of 'bioassays'. In the first instance, bioassays are directed at collective groups of organic compounds, that is, the particulate-phase organics and the vapour-phase organics; but, upon further refining and separation, it becomes possible to evaluate the mutagenicity of subfractions within these larger groups and, eventually, to evaluate the mutagenicity of specific compounds within these subfractions.

The Ames bioassay – easily the most popular – bears the name of the scientist who invented it (Johnson J.H. *et al.*, 1994). This was the first available *in vitro* method with which to screen chemical mutagens rapidly; it uses cells of the bacterium *Salmonella typhimurium*. The cells are a mutant strain insofar as they cannot, *unlike* the natural strain, synthesise the amino acid histidine; and mutagenicity is defined as the ability of organic compounds to revert, or 'back-mutate', the bacteria from the mutant strain to the natural strain. A spontaneous, background or natural reversion must be taken into account, and standard reference materials are used for quality control purposes (Hughes T.J. *et al.*, 1997). As what counts in mutagenicity is not necessarily the chemical substance as directly taken up, but its subsequent forms following metabolism by the body, organic extracts are sometimes also treated with 'S9', a preparation derived from rat livers, from which 'indirect-acting' and 'direct-acting' mutagens are distinguished: if the mutagenicity becomes higher in the presence of liver cells, this implies the action of some form of metabolic activation.

In the Ames bioassay, a sample of organic compounds, extracted according to procedures described in Section 5.2.5, is redissolved to known concentrations in dimethyl sulphoxide (DMSO), placed in a Petri dish along with cells of the bacterium, agar and nutrients, and incubated, for example for 48 hours at 37 °C. The emerging colonies are then counted as a function of the amount of organic extract: this yields the 'dose-response curve', which rises linearly or nearly linearly, at least in the initial portion (Ames, 1979); with increasing dosages, the response slackens and then declines, through nonrepresentative toxic rather than genuine mutagenic effects. The gradient of the curve in the linear portion, as derived by a least-squares curve fit (Campbell J. *et al.*, 1981), provides the index for mutagenicity; failure to discard the nonlinear section leads to an underestimation. Mutagenicity is then expressed as the number of revertants, for example per microgram of organic particulate, per vehicle-mile travelled, per kWh delivered by the engine, per kg of fuel consumed or per m^3 of exhaust gas.

This choice of mutagenic index should not be made incautiously, as the relative ranking, for example of vehicles according to duty cycle, manufacturer, aftertreatment, fuel formulation, mileage and numerous other influential variables, is affected quite profoundly (Gabele *et al.*, 1981): the same data, plotted according to four different indices, are given in Figure 5.13 (Hyde *et al.*, 1982). But because the preferred index varies so much among workers, according to their respective concerns or perhaps ingenuousness, meaningful comparisons are difficult. Unless otherwise stated, what 'mutagenicity' means in this text is 'revertants per unit mass of organic particulate', as this index arguably represents an intrinsic measure of the particulate, for which aspects of the power train (vehicle speed etc.) are irrelevant.

Weakly addressed, but paramount, is whether mutagenicity is itself a function of particulate mass, or of the mass of the organic fraction. Filters exposed to exhaust gas for ever-longer periods, so that they retained more mass, in fact displayed lower mutagenicities (per unit mass) (Dorie *et al.*, 1987). This may signify the loss of certain critical volatile compounds from the filter, but the cause was not fully ascertained. Conversely, one might argue that artefactual formation of more-mutagenic NPAH from less-mutagenic PAH on the filter (Section 3.6.2) was ruled out.

The Ames test cannot answer all questions: it measures microbial, *not* mammalian mutagenicity; bacterial and mammalian cells are of course quite different; it *suggests* the risk a chemical substance poses to humans. It serves as a screening test, acts as an index of mutagenicity, reveals trends and indicates where efforts should be directed. (Mutagenicity should not be confused with carcinogenicity, as a mutagen is not necessarily a carcinogen.)

Another question concerns the release of sorbed organic compounds in the preceding extraction (Section 5.2.5): mutagenicity is solvent-dependent (Seizinger *et al.*, 1979), which is logical enough since solvents are not equal in what they extract. And, more concernedly still, *bioavailability* in the lung, or the *natural* release process, into the pulmonary surfactant, is key, so that Soxhlet extraction is implausible, biochemically speaking; and surrogates for pulmonary surfactant, while perhaps more faithful to the true release process, may well generate different results (Nussear *et al.*, 1992). (See also Section 10.4.) The observation that extraction in 1-octanol reflects more appropriately the pulmonary

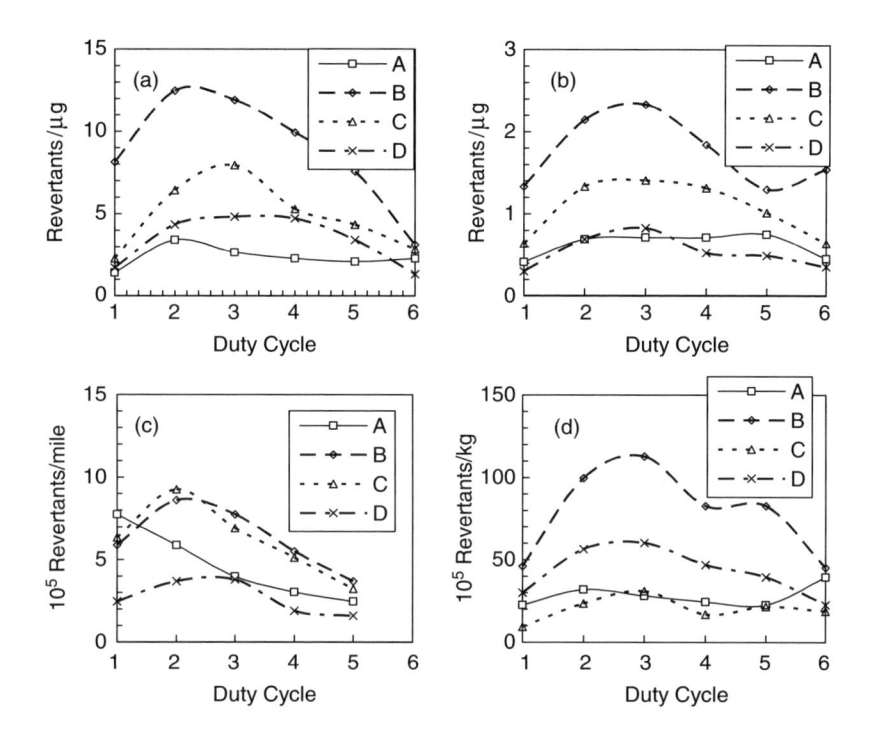

Figure 5.13 Mutagenicity expressed in four different ways: (a) revertants per unit mass of organic particulate; (b) revertants per unit mass of total particulate; (c) revertants per unit distance travelled by vehicle; (d) revertants per unit mass of fuel consumed. Duty cycles: 1, New York City Cycle; 2, Federal Test Procedure; 3, Congested Freeway Driving Schedule; 4, Highway Fuel Economy Test; 5, steady cruise, 50 mph; 6, idle. Twenty vehicles, model years 1979–80, grouped according to manufacturer, A, B, C, and D (various); see original paper for specifications (Hyde et al., 1982).

bioavailability of PAH places the widespread practice of extraction in dichloromethane in perspective (Gerde et al., 2001). A similar question arises as to the long-term impact on the environment, insofar as this rests on the rate at which the sorbed organic compounds are denuded by natural processes, once particles have been sequestered by environmental compartments (Jonker et al., 2005).

The Ames test has been criticised on the grounds of bias, namely excessive sensitivity to NPAH and insufficient sensitivity to PAH (Draper et al., 1988) – an aspect perhaps of impartial metabolism by certain enzymes in the bacterium (Rantanen et al., 1993) – so that, while the first group is impeached too readily, the second group is exculpated too readily. A second strain of the bacterium has therefore been developed in which a deficiency in the enzyme nitroreductase compromises the metabolism of NPAH, so that, when this strain is used alongside other strains without this deficiency, the mutagenicity of NPAH is resolvable from the mutagenicity of PAH (Rantanen et al., 1993). Still further strains have been developed for different strands of mutagenic research, for example to investigate the various ways in which damage is inflicted on DNA.

At the beginning of this section some remarks were made concerning the isolation and identification of certain specific chemical mutagens. But while this approach seems laudable enough in principle, such data do require cautious assessment, as interactions of a synergistical or antagonistical nature arise when mutagens are *mixed*, so that the mutagenicity of the whole is not straightforwardly computable from the mutagenicity of the component parts. Similar problems are seen, for example, with asphalt, creosote and coal tar, all of which contain mixed assortments of PAH (Jones T.D., 1995); and the need to evaluate the complete substance as a whole, rather than the component parts of the same substance in isolation, is recognised, for example, with herbicides (Oakes and Pollack, 2000).

5.3 Particulate Measured Individually

Generally speaking, aerosols encompass five orders of magnitude, from minute molecular clusters (a few nanometres or less) to boulders (tens of microns or more). The upshot is that since particle behaviour is *size-dependent*, then, by suitable contrivances, particle-sizing instruments may be devised. The downside is that identical operating principles cannot be employed throughout this vast size range, because particle behaviours are not underpinned by consistent physics. Fortunately, there is a get-out clause: the particles emitted by motor vehicles span only three orders of magnitude, and most of the useful information is contained in two orders of magnitude, so that actual instrument specifications are somewhat less demanding than for the general case. But, for comprehensive measurements, two instruments running in parallel are sometimes required, spanning different but slightly overlapping size ranges and sometimes, moreover, using different operating principles.

A broad-brush description of particle-sizing instruments might run as follows (Mark, 1998). Some instruments measure the sizes of individual particles as they pass a point; others measure the size distribution of particles contained within a volume; some instruments are *ex situ*; others are *in situ*; some instruments measure particle behaviour in certain force fields; others measure particle interactions with light; some instruments require extensive exhaust gas dilution in order to discriminate between particles; others do not. *In situ* measurements are inherently better at following transient events, and generally held to be nonperturbing and noninvasive, but they are less developed.

The instruments to be described in the present section use two basic operating principles: particle transport and interaction with light. But although they are different in operating principle, the common problematical aspect is signal conversion into particle size. Such conversions depend on particle properties, e.g. density, morphology and composition, but these are poorly prescribed, and often unknown *a priori*; hence, expediency demands the adoption of many simplifying assumptions. The possibility that instruments will register a shift in particle size simply because some confounding parameter unrelated to size has changed is ever-present. And sensitivity not infrequently declines with diminishing size: the result then becomes strongly dependent on the 'cut-off' or 'roll-off' characteristic of the instrument, should particles in this region predominate.

A salient difficulty with optical measurements is *refractive index* – a term containing the relative contributions to extinction of scattering and absorption (Heyder and Gebhart, 1986). For *in-flame* measurements, the refractive index of *soot*, at visible wavelengths, is widely taken as 1.57–0.56i; to presuppose this value in the inception zone, where proto-particles are uncarbonised (Öktem *et al.*, 2005; see Section 3.6.1), or in the exhaust gas, where volatile coatings have been acquired or agglomeration is appreciable (Zhou *et al.*, 1998), seems questionable, since then, scattering may well play a larger role. As always, assumptions considered 'customary', and used as such for very many years, eventually lose their provenance (Smyth and Shaddix, 1996); this is an interesting story we cannot pursue here. Suffice it to say, optical properties are an aspect of spherule microstructure, and also of adsorbed organics (Shaddix *et al.*, 2005), yet the fixity of either is unlikely: the first at high temperature (e.g. annealing and carbonisation); the second at low temperature (adsorption and condensation). Methods have been devised to reduce these uncertainties (de Iuliis *et al.*, 1998) or to avoid inconvenient assumptions about refractive index (Mandel and Ineichen, 1995; Lamprecht *et al.*, 1999). But it should not be assumed that optical properties are steadfast and constant, irrespective of engine operating point, or of any other factor affecting the combustion.

Sizing instruments that segregate according to particle transport mechanisms commonly require calibration aerosols (Gautam *et al.*, 2003); yet these particles differ from those under investigation. And, even with the investigated particles, biases may arise: for example, an intriguing question is raised of whether chained agglomerates acquire more charges than clustered agglomerates (Kittelson *et al.*, 1978a). There is the vexed question of interrelating results between the ELPI and the SMPS; but, since these two instruments measure different particle properties, they cannot be expected to agree entirely (Richards P. *et al.*, 1998). This topic has been studied in detail elsewhere (Maricq *et al.*, 2000b), and signal conversion into particle mass continues to be refined, particularly through more detailed assessments of density (Lehmann *et al.*, 2004).

5.3.1 Inertial Mobility

Instruments which measure inertial mobility or aerodynamic diameter are known as 'impactors'. These are of long-standing provenance in aerosol science; today's units are fully commercialised and well established (Mark, 1998), amongst which are the much-vaunted electrical low-pressure impactor (ELPI) and the micro-orifice uniform deposit impactor (MOUDI).

Impactors exploit the size dependency of the inertial force acting on particles whenever streamlines suddenly change direction: this operating principle is sketched schematically in Figure 5.14 (after Hinds, 1999, p. 129). The device consists of several stages, through each of which a number of holes are precision-drilled. The gas, forced through these perforations, emerges as a jet, and thereupon immediately encounters an obstruction, or 'impaction plate', around which the streamlines veer. What then happens depends on the inertial force in relation to the aerodynamic drag force. Small particles, since they remain faithful to the streamlines, are carried to the next stage, where the process is repeated; large particles, on the other hand, are unable to negotiate this passage, and hence they strike the surface and are captured. The size range encompassed is $\sim 30\,\text{nm} - 10\,\mu\text{m}$.

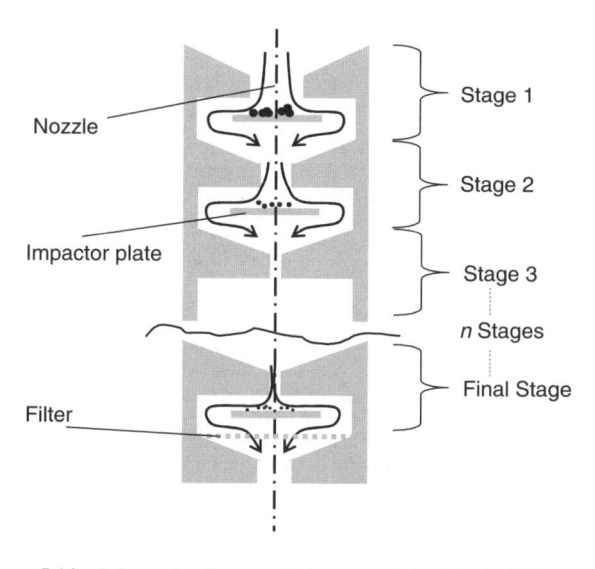

Figure 5.14 Schematic of a cascade impactor (after Hinds, 1999, p. 129).

The size segregation process works because, at each stage, hole diameters are stepped down, so that gas velocities, and consequently inertial forces, are stepped up. Each stage, therefore, with passage through the impactor, retains ever-smaller particles. The ability of each stage to pass particles *below* a certain size, and to retain particles *above* this size, is expressed as a 'cut-off' characteristic. Such characteristics must be, and usually are, vertiginously steep (Hinds, 1999, p. 125); otherwise particles are 'mis-sorted', i.e. undersized particles are falsely collected and oversized particles are falsely passed. But since theoretical predictions are complicated and lack sufficient accuracy, cut-off characteristics are instead determined experimentally with calibration aerosols (Keskinen *et al.*, 1999). As much rests upon good knowledge of, and consistency in, cut-off characteristics, any effects which may distort them, for example the accumulation of particle 'hills' immediately below the jets, must be watched carefully – incidentally, the impaction stages are rotated in the MOUDI, which facilitates even deposition.

The fact that particles are captured rather than discarded is a useful feature of the impactor, because the deposits thereby accumulated can be *weighed* – thus permitting *direct* measurement of the mass distribution. Indeed, this is how the first impactors were used; and although they are discrete (according to the number of stages), a continuous curve can be generated (McMillian *et al.*, 2002). (One should be aware that deposits might partially break up or peel off during operation, and also following disassembly and subsequent handling, thus falsifying weight measurements.) After weighing, the deposits can then be subjected to further microscopical or chemical analysis, thus allowing knowledge to be gained of morphology or composition, as a function of size. There are, however, practical, albeit not insurmountable difficulties: to obtain sufficient deposit on a lightly loaded stage, a heavily loaded stage might become overloaded or choked; and the length of time needed to obtain sufficient deposit is sometimes impractical (many hours), and easily over- or underestimated.

Two prominent and usefully juxtaposed practical issues should be mentioned (Vuk *et al.*, 1976). (a) In 'bounce', the forces of capture on the impaction surface are inadequate, so that particles return to the flow, and continue their journey to the *next* stage or attach to the underside of the *previous* stage. Hard or dry particles, being more likely to bounce than soft or wet ones (see Section 2.2.3), can be encouraged to adhere – albeit at the cost of chemical contamination – by applying grease to the impactor surface. The likelihood of bounce is also reduced by placing an *upper limit* on gas flow, that is, particle impact velocity. (b) In 'plugging', particles do not pass through the holes but, rather, adhere to the walls, to which other particles attach, leading, in extreme circumstances, to complete blockage. As gas flow through the remaining, unblocked holes necessarily increases, bounce is promoted, and cut-off characteristics are distorted. Plugging is an aspect of the near-wall flow field, for example if surfaces are rough-hewn and if particles have already deposited; but particle composition is a factor also, as wetness aids retention, just as dryness aids bounce. Moreover, should passage through the perforations induce sufficient local cooling, volatile material may condense and also plug the hole. The likelihood of plugging is reduced by placing a *lower limit* on gas flow, that is, the residence time within the hole.

Described thus, this is the *traditional* impactor, the largest drawback of which is its ponderousness, i.e. the gravimetric measurement. Nowadays this drawback has been wholly obviated with the electrical low-pressure impactor, in which indigenous particle charges are first neutralised and then, using a corona, a carefully defined unipolar distribution is imposed. The electrical current consequent upon impaction is then measured with an electrometer, for which operation is now effectively in real time (the resolution is about one second), although, owing to the time needed for traversal through the instrument, the stages may need time alignment. The electrical current, that is, the rate of charge transfer, is then related to the rate of particle impaction; hence, the ELPI measures the number distribution, unlike its predecessor, which measured only the mass distribution. But the conversion of electrical current into particle flux requires accurate knowledge of the charge distribution, or the charge acquired per particle (i.e. charging efficiency), and this is where some uncertainty is introduced. Greater description of the unit's operating principles is available elsewhere (Maricq *et al.*, 2000b).

The smallest measurable size with the ELPI, at 30 nm – obviously inconvenient in the study of nanoparticles – reflects the feebleness of inertial forces, and insufficient charging efficiencies; consequently, ways of measuring still-smaller particles are being avidly researched. From one side, the velocity of the gas jets can be increased, but there are practical limits to the manufacturability of ever-smaller perforations. Micro-orifice impactors contain large numbers of nozzles as small as 50 μm, made by chemical etching. Tackling the problem from the other side, the installation of vacuum pumps is what allowed access to 30 nm, as this concomitantly reduced the aerodynamic drag. But such low-pressure operation comes, perhaps, as a mixed blessing, as some evaporation of volatile components seems only too likely – and nanoparticles are often volatile. (Of course, volatilisation *after* charge transfer has taken place does not affect the signal.) Some investigators report access to 7 nm (Marjamäki *et al.*, 2002), but, since operation in the final stage is sonic, particular care is required to avoid particle bounce (Marjamäki *et al.*, 1999).

5.3.2 Electrical Mobility

The instruments that measure electrical mobility bear various appellations: electrical aerosol analyser (EAA), scanning mobility particle sizer (SMPS), differential mobility particle sizer (DMPS) and differential mobility analyser (DMA). These names represent slightly different variations on the main theme; and of course an acronym might, to some extent, reflect the predilections of a manufacturer. But of these, the SMPS has, of late, in the automotive industry, become the most favoured instrument, the operating principles of which are discussed in greater detail elsewhere (Maricq *et al.*, 2000b).

The *generic* operating principle is depicted schematically in Figure 5.15 (after Lüders *et al.*, 1998). The particles are first brought to a well-known bipolar charge distribution, after which the two streams of exhaust gas and particle-free air are separately introduced into one end of a tubular chamber, in which, owing to a centrally mounted electrode, particles experience radial forces: those of higher and lower mobilities are, respectively, deposited on the electrode and carried out in the excess flow

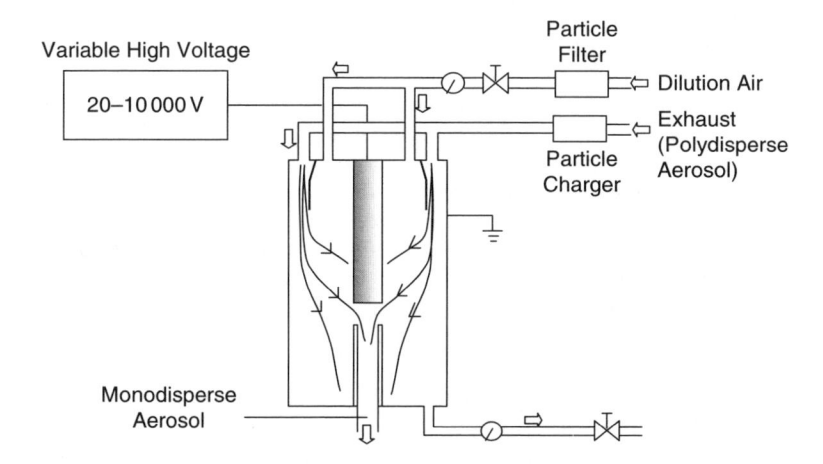

Figure 5.15 Schematic of the differential mobility analyser, or DMA (after Lüders *et al.*, 1998).

(Han H.-S. *et al.*, 2000). Thus, *only* those possessing a certain electrical mobility are able to access the central exit at the opposite end of the chamber: the device thus creates a monodisperse aerosol by selecting particles from a polydisperse aerosol, for which the segregated particles are subsequently enumerated, for example with a condensation nuclei counter (CNC) (Section 5.3.4). Thus, the number distribution is gradually constructed by sweeping or stepping the voltage of the centre electrode, so that the radial force, or the 'cut point', is also swept or stepped. Of course, for this to work, particles must carry well-defined charge distributions: when only single charges are seen, the data reduction is easy enough, since a certain mobility always corresponds uniquely to a certain size. Unfortunately, owing to the statistical nature of the charging process, particles in reality possess multiple charges, and this is where the data processing becomes rather involved.

The accessible size range is constrained by two particulars. The first is the smallest measurable size, which, for commercial instruments, is generally quoted at $\sim$7–10 nm. While not acutely embarrassing in exhaust gas measurements, this minimum nonetheless truncates many nucleation modes. The problem here is that *below* 10 nm, particles readily escape quantification, either through diffusional deposition or through incomplete charging. That said, some progress is apparent in the research community, where lower limits are reported of 3 nm (Chen D.-R. *et al.*, 1998) or even of 1–2 nm (de Juan and de la Mora, 1998). A somewhat greater nuisance is size *range*: although instruments will span 10 nm to 1 µm, such an enormous stride cannot be accomplished in one single measurement. This is why concurrent measurements are sometimes conducted with *two* instruments, operating on different but slightly overlapping ranges (Andersson *et al.*, 2001). That two instruments of identical specification fail to agree fully in their region of overlap suggests some artefact, not of the instruments *per se*, but of their calibration or data reduction.

The greatest restriction, however, in electrical-mobility measurements is the *finite duration* of the scan, being two or three minutes; indeed, the integrative nature of some instruments requires aerosols of decent stability during this time, otherwise the results are misleading, and this procedure cannot supply continuous (temporal) traces for all particle sizes during transient engine operation. The scanning period is reducible, with some instruments, to $\sim$30 s; the problem is distortion of the size distribution owing to excessively fast scan rates. A compromise is operation of the instrument in constant-size mode, i.e. with an unvarying electrical field, for which a time resolution of a few seconds is then available. To obtain a *full* size spectrum, several instruments are operated in parallel, on different settings; or, with due risk of inconsistency, the same test is repeated several times, with the same instrument, on each occasion operating on different settings. Of course, the range of the scan is also a factor in its duration, and a few tens of nanometres might be scanned in just a few seconds (Liu Z.G. *et al.*, 2004), including a scan for particles smaller than one hundred nanometres (Han H.-S. *et al.*, 2000).

This scan rate limitation, being inherent in the design of traditional instruments, has motivated research into new designs, in one of which exhaust gas is passed through a series of coaxial ring-shaped electrodes, the electrical potential of which, with distance into the chamber, increases (Collings *et al.*, 2003). This time, the particles are deflected radially outwards from the central rod, and the electrical current consequent on deposition is measured (Johnson T. *et al.*, 2004). By this technique, each size class is measurable, in effect, continually and in parallel, allowing full (albeit discretised) spectra to be obtained in real time, including during transient engine operation; three-dimensional characterisations of particle demography are thereby constructed (Goto and Kawai, 2004). Experience with this instrument is building rapidly at the moment.

5.3.3 *Laser-induced Incandescence*

Heating the particles briefly with a laser and measuring the ensuing incandescence is a rapidly evolving technique. In laser-induced incandescence (LII), two pieces of information are available: one relating to the heating period, for which the incandescence is proportional, or rather approximately proportional, to the volume concentration (ND_p^3);[8] and the other relating to the cooling period, for which the incandescence decay is an exponential function of the area concentration (ND_p^2). These two signals, taken together, thus supply, for the probe volume, the number of particles N *and* the size of the particles D_p, although what these terms denote, in this case, is the variables related to the *spherule*. The mathematical manipulation is involved, and the spherules are not truly monodisperse, but spherule sizes derived from LII are nonetheless satisfactorily in accordance with direct observations made under the electron microscope (Schraml et al., 1999; Oh and Shin, 2006). LII, moreover, seems to reveal the number of spherules in the probe volume, irrespective of their agglomerated state.

During the heating period, which lasts a few nanoseconds, particle incandescence is constant because the particle temperature is constant; this is a natural consequence of the sublimation of carbon, which serves to dissipate the incoming heat energy. The temperature of incandescence is thus the temperature of carbon sublimation, or ∼4000 K. The incandescence, being broadband, is detectable at various wavelengths; the laser wavelength, and also the wavelengths at which PAH fluoresce, are thereby avoidable. During the cooling period, which lasts a few hundred nanoseconds, the dissipation of heat is primarily conductive, as radiation and sublimation are now bit-players in the energy balance (Roth P. and Filippov, 1996), but the temperature of the surrounding gas must be taken into account. As an example, Figure 5.16 shows the normalised signal decay and the associated time constants for spherules of various sizes.

For quantitative soot measurements, LII must be calibrated, for which there are several methods (Santoro and Shaddix, 2002). Traditionally, this has required a soot aerosol of known characteristics. The method has proved somewhat sensitive to the initial temperature, size and morphology of the particles, and also to the composition of the carrier gas. Ideally, the calibration needed conducting with

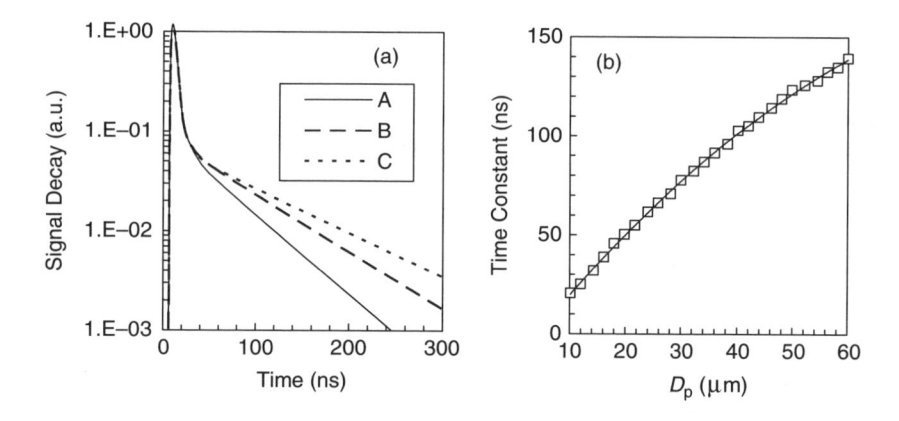

Figure 5.16 Incandescence-derived signals (modelled), as used in determining spherule size. (a) Exponential signal decay, normalised: A, 20 nm; B, 30 nm; C, 40 nm. (b) Signal decay time constant. Baseline (exhaust) temperature, 500 K (Schraml et al., 1999).

---

[8] An exponent of 3.22–3.28 is reported (Oh and Shin, 2006), although this is in fact dependent on the detection wavelength, an increase of which forces convergence on 3 (Santoro and Shaddix, 2002).

the same particles as those under investigation, and under the same conditions; but recently, the need for a known standard has been circumvented by the introduction of an *in situ* self-calibration procedure (Smallwood *et al.*, 2001), using measurements of light intensity via two- or three-colour pyrometry (Snelling *et al.*, 2000). The chief difficulty is uncertainty in the thermal state of the soot, as black-body radiation scales to the fourth power of temperature.

In the traditional configuration, LII has been widely used to study soot formation in laboratory burners and, a little more recently, soot formation in engines. The application of LII to *exhaust gas*, although comparatively speaking an even more recent development, does not run contrary to any operating principle (Snelling *et al.*, 1999). Several instructive differences between the two applications are, though, worth adumbrating. In the combustion zone, the principal need is to avoid confounding by natural flame luminosity, whereas in the cooler exhaust, *more* energy, as much as one-third of the total, is required to reach incandescence. To deliver this extra energy evidently retards the incandescence by 5–10 ns; but this delay is fortuitous, insofar as it nimbly avoids confounding by light scattering and fluorescent PAH (Gupta S. *et al.*, 2001). The longer signal decay is what permits this particle-sizing technique to be carried out on the exhaust, unlike the case for the combustion zone (Leipertz *et al.*, 2002). In the exhaust, adsorbed species are volatilised in the early stages of heating, so that the measurement corresponds principally to the spherules, or carbonaceous fraction. This is an advantage insofar as continued cooling and exhaust gas dilution, i.e. gas-to-particle conversion, does not change the end result. On the contrary, dilution is inconvenient since, in reducing the number of particles in the probe volume, it unnecessarily degrades the signal-to-noise ratio. Liquid-phase particles, however, not being particularly energy-absorbing, might not so readily evaporate. The correlation of LII to dilution tunnel measurements of particulate mass is, to a large degree, confounded by volatiles (Snelling *et al.*, 2000).

Realisation of LII as a practical measurement tool does require several important assumptions, some of which are difficult to justify completely, but all of which may be acceptable within certain limits (Axelsson and Witze, 2001). It is assumed that particles are small compared with the laser wavelength, in fact in the Rayleigh regime (Santoro and Shaddix, 2002); that particles experience identical heating rates, regardless of size; that, owing to the sublimation of carbon, the incandescence briefly becomes constant; and that particles are unchanged by the heating process. Obviously, the last two assumptions are in conflict, since the sublimation of carbon necessarily causes recession of the particle surface. Moreover, LII may induce other modifications, such as annealing, oxidation or morphological restructuring, and this could affect the optical properties of the particles in an unknown fashion; these ideas are still conjectural. But because of this risk, any blanket presupposition that LII is wholly nonperturbing seems inappropriate (Vander Wal *et al.*, 1995). Lastly, assumptions that spherules are in point contact only, and that agglomerate structures are loose enough to ignore inter-spherule bonding, are difficult to justify universally (Schraml *et al.*, 1999): the method rests on the principle that smaller spherules cool more rapidly than larger ones, but the compactness (or openness) of an agglomerate may have some effect.

LII provides what is, for many applications, effectively a real-time and continuous measurement, although in reality, repetition rate is a constraint. Theoretically, modern lasers allow operation of LII at 10–30 Hz, thus allowing one measurement per engine cycle, although practical restrictions on data processing may reduce this frequency to 3 Hz (Witze *et al.*, 2004), and crank-angle-resolved measurements are not (yet) possible. LII is, however, immensely useful for the study of soot emissions during transient engine operation (Witze *et al.*, 2004). The greatest advantage of LII is its sensitivity, which reportedly exceeds that of other instruments by an order of magnitude: the detection threshold is apparently $10\,\mu g/m^3$ (Smallwood *et al.*, 2001). Originally conceived as purely a research tool, bench-top equipment of a commercial nature may shortly become available (Gupta S. *et al.*, 2003).

5.3.4 Light Scattering

The physics and mathematics of particle–light interactions are Byzantine (e.g., see the texts by Kerker, 1969 and Bohren and Huffman, 1998) and can only be only adumbrated here. The present discussion, moreover, is restricted to *elastic* scattering, where the transmitted frequency is the same as that of the incident light. To speak generally, two well-established and mutually complementary scattering theories exist, for which the atmosphere serves admirably as an exemplar: 'Rayleigh scattering', by molecules and nanoparticles, is responsible for the blueness of the sky, and 'Mie scattering', by water droplets, is responsible for the whiteness of clouds. The explanation is the considerably greater dependency of Rayleigh scattering on the *wavelength* of the incident light, with disparate effects in the blue and in the red.

In fact, the controlling parameter throughout *all* scattering is the wavelength of the incident light in relation to the size of the *scatterer*, that is, the particle: Mie theory is quite general inasmuch as it applies from the region of classical geometrical optics down towards molecular scales, and reaches its apogee when particle sizes are comparable to the wavelength; Rayleigh scattering predominates at molecular scales, and dominates when particles are smaller than about one-tenth of the wavelength. As visible light falls between 400 nm (blue) and 700 nm (red),[9] the ascendant mechanism is clear; and, indeed, Mie theory certainly rules in all studies of motor vehicle emissions. In truth, agglomerates exhibit scattering behaviours that are not simplistic in either regime, although detailed treatments such as the Rayleigh–Debye–Gans approximation cannot be discussed here (Köylü and Faeth, 1994). Scattering by *nanoparticles*, whether in exhaust plumes (Section 4.3) or in the ambient atmosphere (Section 10.2), does not appear to have been considered in the literature.

The practical exploitation of Mie scattering highlights something of a doctrinal cleft: purists throw up their hands in exasperation, declaiming the theory's 'misapplication'; yet pragmatists point to the same theory's great utility as a practical measurement tool (Bohren and Huffman, 1998, p. 83). The problem is that practical applications demand several sweeping assumptions, and, although the detailed ramifications are still ardently debated, significant errors seem unavoidable. Controversy arises because the optical properties (refractive index), as a function of particle size, morphology and composition, are not known *a priori*. (a) Particles are assumed to be perfect spheres – surely the most contentious assumption. Real extinctions show less size dependency than predicted (Nyeki and Colbeck, 2000), and 'adjustments' are required for loosely packed, nonspherical morphologies (Scherrer *et al.*, 1981) with due consideration of fractality (Dobbins and Megaridis, 1991). (b) Particles are assumed to be of uniform composition – optical homogeneity seems an unsatisfactory assumption in cool or diluted exhaust, where viscid coatings have been acquired (Kerker, 1969, p. 189), or wherever spherules contain occlusions of metals or metal oxides, as may well be the case when fuel additives are used (Charalampopoulos *et al.*, 1992; Section 3.5.2). (c) Assumptions of monodispersity, although undoubtedly rendering the equations more tractable (Hinds, 1999, p. 353), are difficult to justify when particles have had opportunity to agglomerate appreciably. (d) Less contentiously, the aerosol is assumed to be 'dilute', thus ruling out instances of 'multiple scattering', although precise general conditions under which this criterion is satisfied cannot be easily stated (Bohren and Huffman, 1998, p. 9).

Laying, then, the theory aside with a sigh of relief, light scattering in its simplest manifestation is used not for particle sizing, but for particle *detection*, and such instruments are purely *enumerative*: this is the function of the condensation nuclei counter (CNC), or condensation particle counter (CPC). This principle of optical detection has a long history, dating back to the nineteenth century, in the

---

[9] It should be noted that investigations conducted *outside* the visible region (e.g. in the infrared) change the picture yet again. Similarly, the spectral content should be considered, since lasers are not always used as the light source.

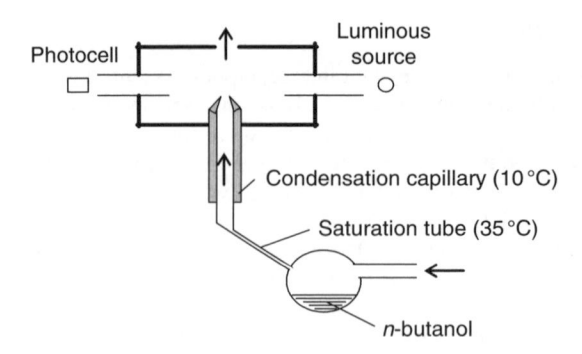

Figure 5.17 Schematic of condensation nuclei counter, or CNC (Lüders et al., 1998).

form of expansion chambers. These operated discontinuously; the continuously operating instrument dates from the 1950s (McMurry, 2000a): a schematic depiction appears in Figure 5.17. Actually, a quibble is that such devices do not detect particles as such: exhaust gas is mixed with butanol vapour; the saturation is deliberately raised; the particles thus grow into *droplets*; and these droplets are large enough to be detected by Mie scattering, on the not unreasonable expectation that one droplet corresponds to one particle.

This procedure is relatively undiscriminating in that all particles grow into droplets of the same size ($\sim 10\,\mu$m), and identical optical pulses are generated by differently sized particles. However, for reasons explained in Section 2.2.4, growth rates are retarded at small sizes (<10 nm), the upshot of which is sensitivity to the saturation profile traversed by a growing particle; and, the smaller the particle, the smaller the droplet. This problem is to some extent manageable by careful design, which explains why some instruments are able to reach 3 nm; although, viewed from the opposite standpoint, the dependence of droplet size on particle size offers some potential for a particle-sizing instrument (Weber et al., 1998).

Of greater concern is particle concentration: ideally, the exhaust gas is so dilute that only one particle at any one time is illuminated, so that only one distinct pulse of scattered light is generated; this 'single-count' mode is an absolute measurement. Corrections, though, are available for cases of 'multiple occupancy' or 'coincidence', which help extend the available range. Above particle concentrations of $\sim 10^3 \text{cm}^{-3}$, the device enters an alternative 'photometric mode', in which calibration aerosols and empirical correlations are used (McMurry, 2000a), and for which an upper limit of $\sim 10^7 \text{cm}^{-3}$ is typically cited (Liu Z.G. et al., 2004). (The quoted ranges vary between manufacturers.) When many particles are present in the probe volume, the scattering signal is some form of summation of sorts (Vojtisek-Lom and Allsop, 2001), correlatable also to mass concentration (Cheung et al., 2000), but this is a collective measurement more in keeping with opacimeters (Section 5.4.6).

Aside from detection, light-scattering signals contain certain information of a *demographic* nature (Iyer et al., 2007), from which, say, particle sizes, fractalities and numbers may be deduced; instruments that operate thus are generally called 'nephelometers'. The principles here are common to aerosols and colloids; indeed, suspensions of soot in oil provide an example of the latter application (Section 7.6.2). Nephelometers exploit the fact that light is scattered in all directions in a manner dependent on particle demography; hence, measurements are conducted over a *range* of angles to the incident light (but often at right angles). The analytical computation, though, of particle demographics from light-scattering signals, with no *a priori* knowledge of these demographics, is a daunting task, and greatly simplified by (not unreasonably) *presupposing* certain likely size distributions (Xu H. et al., 1982; Klingen and

Roth, 1991; Merola *et al.*, 2005). The measured scattering is then simply 'matched' to one of these analytical templates (Mehta D. *et al.*, 2001); poor fits are adduced to deviations from the idealities of Mie theory, for example inappropriate assumptions of refractive indices or want of sphericity, and these are rejected.

A logical extension of light scattering is to subject particles to two laser pulses of comparable energy, the first of which measures the initial scattering and also evaporates the volatile component, and the second of which again measures the scattering, so that the shift in scattering properties between the two pulses indicates that fraction which has evaporated (Witze, 2002). The technique is thus something akin to thermogravimetry (Section 5.2.4), this being laser-induced thermal desorption (Witze *et al.*, 2005). The interval between the two pulses must be long enough to ensure completion of evaporation, but short in comparison with the procession of the flow field. However, uncertainty as to refractive index precludes calculations of *real* volatile fractions; the cloud of evaporating volatiles may well have some complex effect on the scattering, and the method does not account for other possible modifications introduced into the optical properties by the laser pulse. Instead, one must content oneself with an 'apparent' volume change, although the technique is admittedly still in its infancy.

5.4 Particulate Measured Collectively

We conclude by considering several diverse instruments: to group them thus is expedient rather than strictly correct, as their operating principles are wholly unrelated; the only commonality is measurement of particulate as a whole entity, i.e. the signals represent some summation or ensemble of sorts. Expediency arguably concerns, also, what has already been discussed in the present chapter. The dilution tunnel and its filter, after all, represent a form of 'collective' measurement; and, insofar as smokemeters are the oldest form of instrument for studying particulate emissions, they arguably constitute a 'conventional' measurement.

What chiefly characterises several of the instruments to be discussed in this section is uncertainty over what their signals *actually mean* – this highlights divergence between the pragmatist and the theoretician. Instruments are available that provide useful signals, such as for monitoring and diagnostic purposes, but this information is vitiated by insufficiently developed theories, and consequent doubts as to how the data should be interpreted.

5.4.1 Photoacousticity

A particle will absorb a small proportion of the incident light it receives, and, through conservation of energy, this absorption raises the temperature of the particle, which also, in the immediate vicinity, raises the temperature of the gas. If the light intensity is *modulated* ('chopped'), the temperature fluctuates accordingly, and repeated cycles of heating and cooling in the particle drive corresponding cycles of expansion and compression in the gas – manifested, in fact, as an acoustic wave, propagating outwards from the particle. This conversion of light energy into sound energy, called the 'photoacoustic effect', was discovered in the 1880s, but remained solely a curiosity until the 1970s, when it was harnessed for measurement, because the amplitude of the acoustic wave reveals the amount of light-absorbing material in the irradiated aerosol.

Strictly speaking, an issue arises in deciding what the 'amount of light-absorbing material' really means in tangible terms. In the visible and infrared, this mysterious quantity refers to the *volume* of a submicron particle, and, in practice, the wave amplitude is held to be, in effect, directly proportional to the *mass* of irradiated soot – not implausibly, since soot is obviously the dominant light-absorbing component in the particulate. At the same time, one should remain alert, first, to the presence of other particle components (accreted volatiles) that might not be entirely nonabsorbing, and second, to the likely effect of particle morphology on light-absorbing properties. The converse step, transfer of heat energy from the particle to the gas, is, in the limit, equally open to some ambiguity, because, according to principles discussed in Section 2.2.3, heat transfer surely depends on the particle size in relation to the mean free path, although this aspect has, apparently, gone unnoticed in the literature. However, these ambiguities, being obviously difficult to quantify, are generally glossed over, and the realisation of reasonable results in photoacoustic spectroscopy seems sufficient justification for the pragmatist, if not the theoretician.

Other difficulties, of a theoretical origin, are soluble with prudent instrument design. For example, some gas molecules (NO_2 particularly), also absorb light energy: but this is either compensated, via comparisons with a second test chamber receiving *filtered* exhaust (Osada *et al.*, 1982), or avoided, through judicious choice of frequency, that is, with respect to the resonant vibration frequency of the gas molecule (Petzold and Niessner, 1993). Sensitivity is improved if the chopping frequency is tuned such that the sound resonates inside the test chamber, to create a standing wave (Burtscher, 1992, p. 560). There is a need for thermodynamic stability in the chamber, otherwise the speed of sound varies (Schindler *et al.*, 2004). In the course of time, soot accumulates on optical windows, blocking the transmission of light, the effect of which is to introduce into the signal a zero offset. Historically, however, the greatest difficulty with photoacoustic spectroscopy was not so much the eradication of these confounding phenomena as the expense and complexity of the hardware, which involved high-power lasers and microphones, and this helped to kill the technique. Contrariwise, the advent of low-power solid-state laser diodes, of electronic, rather than mechanical modulation, and of miniaturised microphones has provoked a resurgence of interest.

The design, operation and calibration of various photoacoustic spectrometers have been reported in detail (Japar and Szkarlat, 1981; Osada *et al.*, 1982; Petzold and Niessner, 1993; Krämer *et al.*, 2000; Schindler *et al.*, 2004; Arnott *et al.*, 2005). The laser wavelengths are in the green and infrared, chopping is carried out at a few thousand hertz, the acoustic wave is detected by phase-sensitive lock-in amplifiers, and test chambers are constructed from brass cylinders. One such design is shown in Figure 5.18 (Osada *et al.*, 1982).

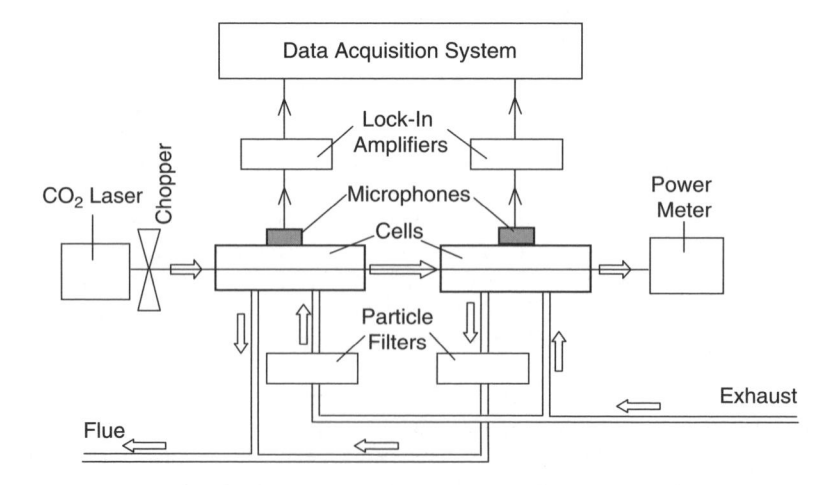

Figure 5.18 Photoacoustic spectrometer (Osada *et al.*, 1982).

It is instructive to note that, in great contradistinction to the automotive industry, photoacoustic spectroscopy has been zealously embraced by atmospheric scientists, with the object of quantifying soot in the ambient atmosphere; for this, much information has also been published on instrument design, operation and calibration (Adams *et al.*, 1989, 1990a,b; Turpin *et al.*, 1990; Arnott *et al.*, 1999). This alternative field of application, which is not reviewed here, is quite different in the technical challenges it poses to instrument designers: that soot concentrations are lower is a disadvantage, as the sensitivity becomes a larger factor; that concentrations of NO_2 and CO_2 are lower is obviously an advantage, insofar as these gases are now less confounding to the measurement.

5.4.2 Photoelectric and Diffusion Charging

The ejection of electrons from surfaces as triggered by incident light above a certain energy threshold is called the 'photoelectric effect'. This phenomenon, in the form of photoelectron spectroscopy, is widely exploited as a characterisation tool in surface chemistry, since the photoelectric threshold, or rather 'work function', is very informative about the state of a surface, especially insofar as it depends on what molecules have adsorbed. In aerosol science the principle is the same, but the outcome is different, as the particles thus irradiated become *positively* charged, and, in a potential gradient, a small (10^{-15} A) electrical current (Burtscher, 1992, p. 571) ensues which serves as a measure, or 'fingerprint', of the particle surface – of which more later.

Customarily, the photoelectron spectroscopy of surfaces is conducted *in vacuo* since gases interfere with the ejected electrons. Thus, with aerosols, the charge transfer must be deftly contrived, of which there are four aspects. (a) Indigenous ions, that is, those already present in the exhaust gas, act as confounding charge carriers, and so must be removed. (b) The irradiation must not ionise exhaust gas molecules – otherwise the ensuing ions act as confounding charge carriers. To avoid this, the energy of the incident light, besides exceeding the photoelectric thresholds for particles, must not exceed the photoionisation thresholds for gas molecules: wavelengths in the ultraviolet are preferred, for example as realised with a KrCl excimer lamp (Arnott *et al.*, 2005). (c) The ejected electrons must not return, or 'back-diffuse', to the particles from which they were ejected. This is less likely when the particles are smaller than the electron mean free path, for which one micron is an upper limit (Burtscher, 1992, p. 568); but electrons, as they possess higher mobilities than particles, can be encouraged to leave their progenitors by applying an external electrical field. (d) The ejected electrons are captured by gas molecules, resulting in negatively charged ions (chiefly of oxygen); by acquiring these ions, particles become negatively, rather than positively charged (Mohr and Burtscher, 1997). This confounding phenomenon, which is, in effect, a *second* particle-charging mechanism, is avoided by careful choice of residence time, and also through applying an electrical field, since ions have greater mobilities than particles.

From the preceding considerations, the need for various additional design features will be apparent (Burtscher, 1992, p. 572); apparatus along these lines is depicted in Figure 5.19 (Siegmann and Siegmann, 2000). The route taken by the exhaust gas is as follows: (a) an *electrical condenser*, or prefilter, where a strong electrical field removes indigenous ions; (b) an *irradiation chamber*, where exposure to UV light and photoemission take place; (c) a *precipitator*, where photoemitted electrons, and ions created through electron capture by gas molecules, are removed by a weak electrical field; and finally (d) collection of charged particles onto an electrically insulated *filter*, whereon the final charge transfer takes place, and the consequent electrical current is measured.

Precisely what photoelectric currents mean in particle characterisation is actually quite an interesting question. The significance of photoelectric charging lies in its ability to deliver information about *surfaces*, and this immediately places it apart from other instruments used to study particles. The

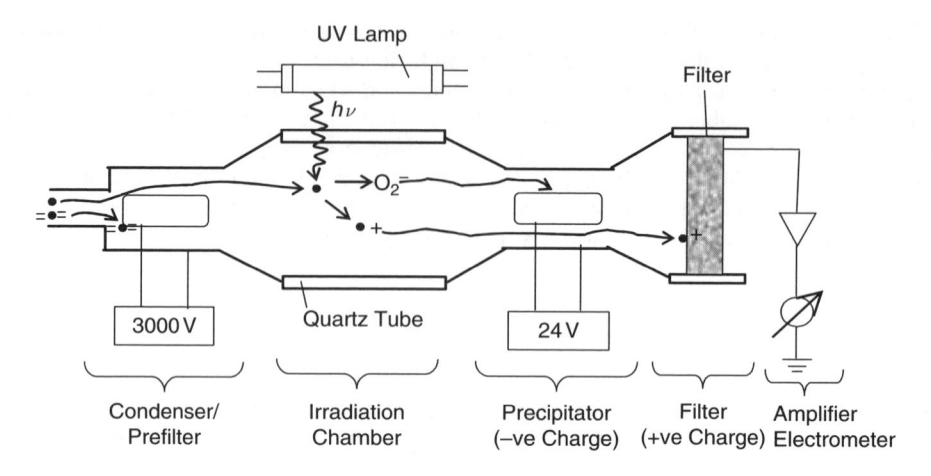

Figure 5.19 Schematic of a particle sensor using the photoelectric effect (Siegmann and Siegmann, 2000).

determinant is, in essence, the physicochemical state of the particle surface;[10] but this does *not* mean the geometrical surface, as only a fraction of this is photoelectrically active. Cavities and voids make no contribution, as electrons ejected therefrom are immediately reabsorbed upon striking opposing or proximate surfaces. This understanding leads to the concept of the 'active surface' – a fraction of the geometric surface, denoting what area is available for adsorption of gas-phase species, gas-to-particle conversion, catalytic reactions, etc. The meaning of the active surface, from a theoretical standpoint, is further elucidated elsewhere (Siegmann and Siegmann, 2000).

If an adsorbate is electronegative with respect to the particle, the photoelectric threshold is depressed, i.e. the work function is increased. (The converse occurs for an electropositive adsorbate.) In particle measurement, there are two consequences here of immediate practical relevance. First, particles of different provenance are distinguishable (Siegmann *et al.*, 1999), including those discharged by different combustion processes (e.g. diesel engines, gasoline engines, wood smoke and coal smoke), because their surface compositions also differ, although the measurement does not, of course, reveal *how* they differ (Matter D. *et al.*, 1995). Second, particles commonly give rise to far higher photoelectric currents than they do without these adsorbed substances, i.e. photoelectric thresholds are far lower than for carbonaceous agglomerates alone; and sulphates are not responsible (Burtscher, 2000), thus impeaching organic compounds or, more precisely, PAH, the influence of which on photoelectric thresholds is surprisingly strong, being clearly discernible even for coverages far below one monolayer.

The relationship between photoelectric current and adsorbed PAH is linear, and spans three decades of concentration (Siegmann and Siegmann, 2000). This is quite exciting since, as detailed extensively in Section 5.2.6, PAH are *traditionally* quantified, first, for large accumulations of particulate, and second, using long-winded analytical procedures, whereas photoelectric charging offers what is, effectively, a *real-time* indication of particulate-phase PAH – capable, that is, of tracking transient emission events (Matter U. *et al.*, 1999). But the exploitability of the relationship is weakened by four issues. First, photoelectric thresholds vary among species of PAH, so that photoelectric currents are of no use in speciative terms (Tang S. *et al.*, 2001), and, if these currents are just some summation of whatever is

---

[10] There is, admittedly, some contribution from the bulk, since the initiation step is optical absorption of photons by the particles, and the creation of electron–hole pairs (Siegmann and Siegmann, 2000).

adsorbed, the question arises of what happens to the signal when speciation alters. Second, a suspicion lingers that photoelectric currents are affected by some poorly understood aspect of the underlying carbon cores (Bukowiecki *et al.*, 2002). Third, when particles are heavily coated with PAH, that is, when coverages greatly exceed one monolayer, the outer molecules are obviously not truly adsorbed onto the carbon core. This seems to be behind cases of spurious signals – in which case, partial desorption of the outer layers, by heat treatments prior to irradiation, may be necessary (Arnott *et al.*, 2005). Fourth, vehicles fitted with a DPF emit particles in which the sulphate/soot ratio is far higher, and the sulphate is thought to quench the photoemission (Kurniawan and Schmidt-Ott, 2006).

Although not strictly speaking within the scope of this brief review, the parallel use of *photoelectric* charging, as described here, and *diffusion* charging, as mentioned here *en passant*, is certainly worth mentioning. Whereas the first charges particles by *electron emission*, the second does so by *ion attachment*; yet, although both signals are functions of surface area, these are not quite the same surface areas, as photoelectric charging incorporates aspects of surface chemistry which diffusion charging does not; this information is contained in a 'material coefficient' (Kasper *et al.*, 2000). This mutually complementary information opens up fresh vistas (Bukowiecki *et al.*, 2002), but, at the time of writing, studies of how best to exploit these vistas are still in their infancy. The practical outcome, *tout court*, is that both signals, taken together, allow even clearer differentiation between particles of different provenance (Matter U. *et al.*, 1999).

5.4.3 Electrical Charge

As mentioned in Section 3.2.2, and moreover pursued in Section 6.2.5, particles emitted by motor vehicles carry *indigenous* charges, and particle measurement might ostensibly profit thereby. But, at the same time, the fuzzy nexus between particles specifically and charges generally should be recognised. The charges reside, according to received wisdom, in the accumulation mode, that is, on the *soot* particles; but in exhaust gas, there are *other* charge carriers, obvious if only because a prodigious literature is devoted to ion sensing in gasoline engines (e.g. Witze, 1989), where, even though the combustion is, generally speaking, relatively soot-free, the so-called 'ion current' is nonetheless very informative about the progress of the combustion. Moreover, apartheid seems rather implausible, so that, logically, particles and ions will attach to one another. One cannot, therefore, guarantee exclusivity to particles in any such charge-measuring instrument, unless one takes care to segregate the charge carriers as already described in the previous section.

There are various ways of realising a charge-measuring device. For instance, particles will transfer their charges to the conducting surfaces on which they deposit, and so support a current. Of course, it should be recognised that size dependency in such deposition could engender measurement bias. Charge transfer is open to this criticism: unsurprisingly, since this is the principle of particle-sizing instruments, as already discussed in Sections 5.3.1 and 5.3.2. So, taking this deposition principle a step further, and taking also a cue from ion sensing, a potential difference might be applied to two electrodes, by which to encourage this deposition, and between which the current is measured; this, after all, is the principle of smoke alarms (Burtscher, 1992, pp. 582). Alternatively, particles might be intentionally charged by allowing them to strike a high-voltage electrode, whereafter they journey to a second electrode and are discharged (Hauser, 2006).

These *contact* methods of charge measurement are open to censure, on the grounds that soiled surfaces require regular cleaning, whereas with *noncontact* methods, particles do not so much deposit, and thereby transfer their charges, as briefly induce a redistribution of charge, i.e. an electrical current, say, in a probe tip or an encircling coil. Of course, even these measurements are still susceptible to the *unsolicited* accumulation of soot deposits, and, whatever the approach, the ensuing implications for sensor operation should be considered carefully, as these deposits are electrically conducting – a problem in ion sensing with diesel engines. Turning this problem around, the electrical conductivity of the growing deposit might *also* serve as a basis of particle measurement (Warey et al., 2004), with the conducting track intermittently incinerated by electrical heating (Hauser, 2006).

In noncontact methods the ascendancy of pragmatic or empirical investigations should be acknowledged, as sound theoretical underpinnings relating induced currents to particle charges are lacking (Collings et al., 1986): analytical solutions, even for isolated point charges moving through simple electrical coils, are nontrivial; hence, the analytical difficulties to be overcome with real, unsteady, nonuniformly distributed charge clouds will be appreciated. On the other hand, if the charge cloud, in the axial direction, is long and unvarying in comparison with the coil's dimensions, then analytical solutions are easier, because the field lines are then effectively radial. At a *practical* level, though, the instantaneous induced current, or charge flow, in the probe represents some unspecified summation of the charge carried by the passing cloud.

These theoretical questions notwithstanding, various sensor or probe designs, and the associated electronics required for signal conditioning, have been reported in the literature. The Faraday cage is a desirable design purely for signal acquisition, but undesirable for reasons of fragility in the exhaust environment (Schweimer, 1986). In one design, a probe ring, connected to a charge amplifier, encircled the exhaust duct. The output signal was the charge necessary to maintain a virtual earth; signal levels were a few picocoulombs (Collings et al., 1986). In another design, the probe was fashioned from a spark plug, on the electrode of which a small disc was welded: signal levels were <50 pC. Two probes, along with their respective outputs, are depicted in Figure 5.20 (Hong G. et al., 1987).

As typified in Figure 5.20, the charge cloud is of short duration, and virtually the entire signal occurs during *blow-down*, i.e. immediately following exhaust valve opening (EVO). Various signal characteristics have, however, been reported: one positively charged cloud (Schweimer, 1986); one positively charged cloud during steady-state operation; and two clouds, the first positive, the second negative, during transient operation (Collings et al., 1986). Offsets have been reported (Hong G. et al., 1987); certain differences reportedly exist between gasoline engines and diesel engines (Schweimer, 1986); and noticeable cycle-to-cycle variability arises. Of course, for the same total charge, bipolar but imbalanced clouds are probably indistinguishable from monopolar clouds, but these ideas cannot be fixed with any accuracy, because the experimental database is slender – a consequence, perhaps, of the fact that, historically, indigenous charges are discarded in particle-sizing instruments, because

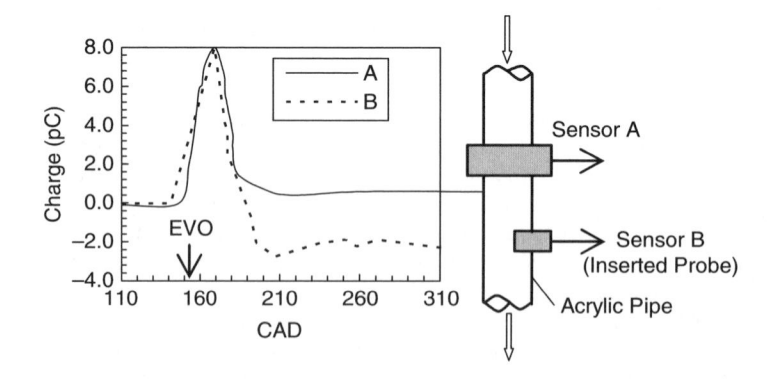

Figure 5.20 Particle sensing by induced electrical current: A, ring (encircling) probe; B: inserted probe. Engine: 1.6-litre, single-cylinder, indirect-injection. Operating point: 800 rpm/98 N m. Bosch smoke number, 1.4 (Hong G. et al., 1987).

charge distributions are not clearly defined. In this lies the technique's principal weakness: the degree of consistency to be expected in particle charge characteristics is uncertain.

But, taking an alternative tack, obvious potential exists in charge measurement for real-time closed-loop control, as used in engine management systems. For example, the signal, if integrated in the window immediately following EVO, correlates well with the smoke emitted by diesel engines; and this might be used to guide fuel metering during acceleration, so that sufficient oxygen is always available – for which operation is still mostly open-loop, and plagued by smoke puffs (Section 7.5) (Hong G. and Collings, 1989). The same sensors might also be conceivably used in the *detection* of EVO, as needed, for example, in variable valve timing; while other potential applications lie in minimising cycle-to-cycle or cylinder-to-cylinder variations (Schweimer, 1986).

5.4.4 Flame Ionisation

The flame ionisation detector (FID) is, today, fitted as standard in all emissions laboratories: this apparatus quantifies emissions of HC, an acronym which, carefully qualified, denotes *gas-phase* hydrocarbons – gas phase, that is, usually, above 190 °C. This category should not be confused with hydrocarbons which, as they undergo gas-to-particle conversion, are captured by dilution tunnel filters operating at $\leq 52\,°C$ – but we shall return to this distinction presently.

For the moment it is sufficient to observe that, in the FID, the hydrocarbons are first burned in a *hydrogen* flame, from which they emerge as ions, and that this ion flux is subsequently measured as an electrical current. Within certain limits, which need not detain us, this current is proportional to the rate at which carbon atoms emerge from the flame. This result is then expressed as an 'equivalent' – the result that would have been obtained had the HC consisted of just *one* hydrocarbon, rather than hundreds of different compounds: for example, if the equivalent is methane, then the base is C_1. The FID, therefore, does not reveal the hydrocarbon speciation; it only counts the carbon atoms as they emerge from the hydrogen flame.

In the conventional configuration, particles are filtered out prior to the flame, but an interesting question arises since, if carbon atoms from gas-phase hydrocarbons contribute to the signal, then, presumably, carbon atoms from particles will also contribute; and this might serve as a basis for particle measurement. Indeed, if dry, carbonaceous particles are allowed to burn in the flame, they do generate readily discernible signals, manifested as 'spikes' – easily confused with electrical noise (Kawai *et al.*, 1998). The area under a spike is, not implausibly, related to the size of the particle or, more correctly, the number of carbon atoms in the particle – provided, of course, the particle burns to completion. Thus, the FID is able to enumerate carbon atoms in the carbonaceous fraction as well as carbon atoms in the HC.

In principle, the organic fraction is *also* measurable by the FID, through exploiting the gas-to-particle conversion such compounds undergo, i.e. their measurement is simply governed by the sample-conditioning temperature of the FID, as already described in Section 3.6.4. Reciprocally, then, the hydrocarbons in the particulate phase at 52 °C, and so captured by a filter, are equivalent to the difference between *two* FID measurements, one at a very high temperature, when *all* hydrocarbons are gaseous, and another at 52 °C. Unfortunately, owing to thermal constraints in the apparatus, the maximum operating temperature of the FID is usually ~190 °C, so that it cannot measure the heaviest hydrocarbons, amongst which the lubricant figures prominently. There also may be some speciative sensitivity, as even the traditional FID varies slightly in its response to different hydrocarbons (Degobert, 1995, pp. 179–181), and more so for oxygenated hydrocarbons (Fialkov, 1997, p. 478). But, with these caveats, the two methods often agree reasonably well (Andrews *et al.*, 2000b).

In this dual measurement, of soot fraction and organic fraction, the flow of exhaust gas is split into two lines (Fukushima *et al.*, 2000), one of them filtered and thermally conditioned at 52 °C, and the other

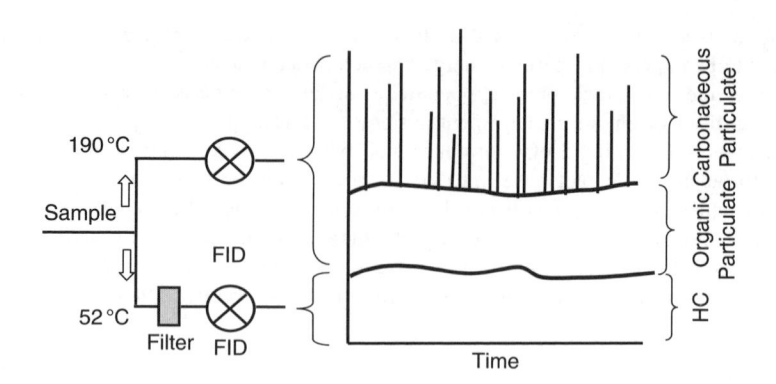

Figure 5.21 Response of flame ionisation detector (FID) to HC, carbonaceous particulate and organic particulate: in one line, the exhaust gas is filtered and conditioned at 52 °C; in the other line, the exhaust gas is unfiltered and is conditioned at 190 °C (Kawai et al., 1998).

unfiltered and thermally conditioned at 190 °C; the first, therefore, performs the customary measurement for HC, while the second supplies the additional data. This approach is depicted schematically in Figure 5.21 (Kawai *et al.*, 1998); it has much in common with the thermal fractionation methods discussed in Section 5.2.4.

5.4.5 Mass

The tapered-element oscillating microbalance (TEOM) reports the mass of particulate discharged per unit time, i.e. the *mass emission rate* (g/h).[11] This makes the instrument arguably equivalent to the dilution tunnel filter (Section 5.2), by which the mass emission rate is also obtained. But the TEOM is manifestly superior to its counterpart in two ways: the emission rate is reported continually, on a second-by-second basis, and no awkward or cumbersome assessments using weighing scales are necessary. These advantages marked a significant advance in measurement technology; and, consequently, the TEOM has quite rightly found widespread use not only in the automotive industry, but also in air quality monitoring (Ayers *et al.*, 1999).

The TEOM is a complex instrument, the inner workings of which are not herein discussed in detail; the reader is referred to other sources for ample information on design and construction, signal processing, operation, and calibration (Whitby *et al.*, 1982, 1985). For the present discussion it is sufficient to observe that the central feature, as depicted schematically in Figure 5.22 (Burtscher, 2001), is a glass tube, designed such that its cross-section changes gradually along its length: this is the *first* principal component in the four-letter acronym: the 'tapered element'. The exhaust gas is conveyed down the inside of this tube, and passed through a filter covering the narrow end. In this filtering action the TEOM bears some similitude, albeit on a different scale, to the dilution tunnel filter. Thus, some degree of agreement is expected between the TEOM and the dilution tunnel filter, of which more later.

Thus far, the particulate emissions are only captured: the next task is 'weighing' – as carried out by the *second* component in the four-letter acronym, the 'oscillating microbalance'. In this, the tapered element, the filter and the particulate together form a dynamical system forced into oscillation

---

[11] In vehicle testing, usually converted into mass per unit distance (g/km or g/mile).

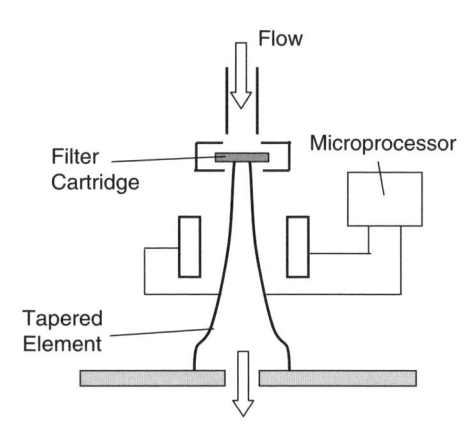

Figure 5.22 Schematic of tapered-element oscillating microbalance (TEOM) (Burtscher, 2001).

by an electrical circuit, by which the natural, or resonant, frequency is detected. By proportion, the accumulated mass must not be so large as to suppress the oscillation, yet not so small as to compromise sensitivity. Nor must the vibration dislodge particles and allow them to escape by re-entrainment.

The degree of control exercised over the surrounding environment must be of a high order (Okrent, 1998), as resonant frequencies are a function of other parameters common to all dynamical systems, and these parameters are open to extraneous influences: the forcing function (vibrations and pressure pulsations); elasticity or stiffness (temperature); damping (air density); and mass (water adsorption from the atmosphere during nonoperation). But, assuming sufficient control is exercised over these extraneous factors, the resonant frequency of this system furnishes, via long-established dynamical laws, the mass of particulate residing on the filter, a baseline calibration for which is carried out by simply attaching a known mass. These computations are repeated to obtain the mass as a function of time, and this signal is then simply differentiated to obtain the emission rate in grams per hour.

The mass emission is thus available directly – or, perhaps, one might say quasi-directly, as the particulate deposit is not actually 'weighed' in the strictest sense. This quibble notwithstanding, the TEOM stands alone in this chapter as the only true mass-measuring instrument:[12] although other devices may well *report* mass, they measure, in fact, some *other* parameter, which is then converted to mass, using empirical or analytical relationships, the precariousness of which might well be imagined: these relationships are thrown badly off course by confounding factors such as particle morphology, composition and size, whereas, for example, the TEOM is indifferent to the question of whether an agglomerate is clustered or chained. Just as on the dilution tunnel filter, whatever deposits onto the TEOM filter is honestly and ingenuously quantified as 'mass'.

Equally, the TEOM filter displays, in arbitrating what is 'particulate', precisely the same idiosyncratic biases and imperfections as the dilution tunnel filter: the filtering efficiency increases as particulate

---

[12] The quartz crystal microbalance (QCM) uses a similar operating principle (via a piezoelectric element), but little experience with this instrument exists in the automotive industry, whereas the TEOM is a regular feature in emissions laboratories. The QCM has traditionally been disadvantaged because of its smaller range, but this situation may change with modern low-emitting vehicles (Booker *et al.*, 2007).

accumulates; the temperature is a prime determinant in gas-to-particle conversion (Saito K. and Shinozaki, 1990); and the deepening depression on the downstream side of the filter encourages the evaporation, and consequent escape, of previously captured volatiles (Whitby et al., 1985). Similarly, insensitivity to face velocity is a useful characteristic insofar as more exhaust gas is filterable per unit time – advantageous with low-emitting engines. These issues are not discussed here, since they have already been fully explored in Section 4.6. But in these phenomena lies ample opportunity for divergence from dilution tunnel measurements, so that full agreement in mass emission rates between the two forms of filter is not guaranteed.

The evaporation and escape of volatile compounds from the TEOM filter courtesy, say, of brief increases in temperature consequent on engine operation, underscores the action of *two* countervailing factors in the accumulation of particulate mass. When an engine is accelerated, and briefly emits high levels of solid carbonaceous particles, the mass emission, as indicated by the TEOM, can decrease, initially at least, because volatile compounds are being stripped from the filter. Conversely, on decelerating, the sudden arrival of relatively clean and dry exhaust gas (or rather air, since during fuel-cut the engine acts simply as a pump) can also promote evaporation (Whitby et al., 1982). When the mass lost through the evaporation of volatiles exceeds the mass gained through the capture of solid particles, the TEOM registers a *negative* mass emission rate (Xu S. et al., 2005) – obviously, an unphysical result.

These artefacts seemingly place the TEOM at a disadvantage in the measurement of transient emission events. But to castigate the TEOM thus seems unduly harsh: the same phenomenon undoubtedly takes place on the dilution tunnel filter, where it goes unnoticed, simply because only the end result is measured; just as, on the TEOM, it *is* noticed, because a second-by-second trace is available. Efforts have been made to eliminate or at least to attenuate these artefacts, such as by tighter temperature control (Podsiadlik et al., 2003). But, as this loss of volatile compounds from the filter is 'real' in one sense, yet artefactual in another, its eradication seems to be more a matter of preference or policy; and a wiser tactic, perhaps, might be the pursuit of parity with the dilution tunnel filter. Indeed, reasonable correlations are attainable between total mass as obtained by integration of TEOM traces, and total mass as obtained from gravimetric measurements of dilution tunnel filters (Jarrett and Clark, 2001). The best correlations are obtained by removing transient components in the TEOM trace relating to the take-up and release of water, that is, corrections are required for moisture. Apparently, water is retained during accelerations and removed during decelerations, and these transients are responsible for higher or lower *indicated* mass rates – as distinct from real emission rates. But this is an emerging area of research at the time of writing, and the jury is still out.

5.4.6 Smoke

The attenuation of light – or, less obviously, the contrast, between attenuated and unattentuated light[13] – is what defines smoke: in this lie the optical principles used in all smokemeters. Two mechanisms are responsible for this attenuation, 'scattering' and 'absorption', which, together, constitute 'extinction'. Light extinction is not difficult to measure in principle – a comparison of incident and received light being straightforward – but if its two components must be resolved, then additional measurements become necessary (Scherrer et al., 1981). Resolution of the two components is not, however, the

---

[13] No smokemeter should respond to *gaseous* components in the exhaust, and this is generally the case. The reading from an opacimeter may, however, be falsified by NO_2, especially when this gas is unusually abundant (Rapone et al., 2006). NO_2 is strongly blue-absorbing, and will colour exhaust plumes red, brown or yellow (Seinfeld, 1986, p. 294).

business of smokemeters: what light is sent out and what fraction of that light is received are all that matter.

The expedient nature of smokemeters is undeniable: numerous aspects of particle–light interactions are glossed over: particle morphology, size distribution, refractive index, chemical composition, and the spectral distribution of the incident light ... the signal is some integrated function of all these. For this reason, identical readings might arise from two radically different aerosols. Against this, one can argue that particle characteristics likely to be encountered in engine exhaust do not range nearly as widely as those of aerosols generally. But even so, smokemeters constitute what is very much a *practical* measurement of light obscuration.

The oldest attempts to quantify smoke relate, not unexpectedly, to factory chimneys. In the last years of the nineteenth century, Ringelmann introduced a system wherein smoke-plume darkness was compared with a set of cards of various shades. Although this method appears rather archaic to the modern eye, it was once proposed in the USA as an *in-use compliance tool* for motor vehicles (Henderson, 1970) and for jet aircraft (Westfield, 1971). Obviously, a good deal of subjectivity is inherent, and results are easily confounded by natural lighting conditions, while the relative locations of emitter and observer are less readily under control with mobile sources than is the case for chimney stacks.

Recognition that greater scientific rigour was needed in the measurement of smoke provoked a flurry of activity in the late 1960s and early 1970s (Sallee, 1970). During this time many instruments came under intensive development (Hills *et al.*, 1969), but they were all variations on essentially two themes: the 'opacimeter' and the 'spotmeter' (Shaffernocker and Stanforth, 1968). The specific details of construction vary enormously, but the generic operating principles of these two forms of instrument are depicted schematically in Figure 5.23. As might well be conjectured, the two methods are related insofar as the column of gas traversed by the light in the opacimeter corresponds to the column of filtered gas in the spotmeter, although this similarity has not been widely exploited in order to develop any analytical interrelationship (Homan, 1985).

The spotmeter draws a sample of exhaust gas through a (normally white paper) filter, whereon the particles accumulate as a soiled 'spot'. (It should be noted that inadequate sealing around the circumference of the spot, as evidenced by radiating sooty streaks, falsifies measurements.) The relation between the light reflected by this *soiled* spot, R_s, and the light reflected by an *unsoiled* spot, R_u, is

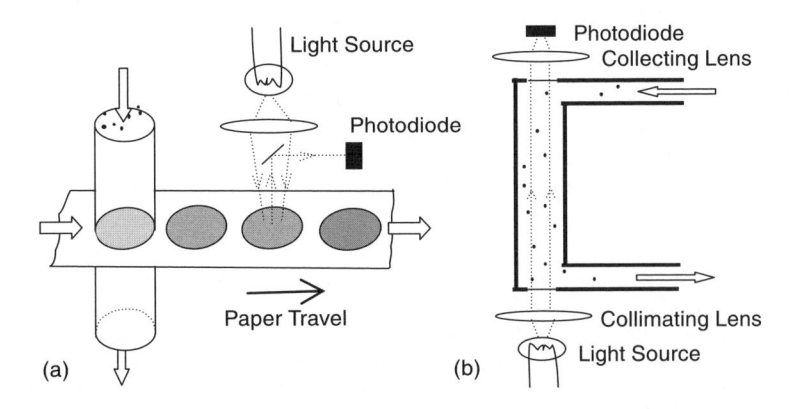

Figure 5.23 The two principal forms of smokemeter: (a) the spotmeter; (b) the opacimeter.

expressed as a quotient, or 'smoke number', S_N, on a linear scale of zero to ten or zero to one hundred, according to the chosen K (Wood A.D., 1975):

$$S_N = K\left(1 - \frac{R_s}{R_u}.\right) \quad (5.5)$$

Of course, 'smoke number' is really vernacular nomenclature, the measured parameter in scientific language being 'reflectance' – in effect, a representation of scattering and absorption, although this distinction need not detain us. In the related field of air quality monitoring, the reflectance of the soiled spot is referred to as black smoke (Madhavi Latha and Highwood, 2006), and the optical extinction *through* the spot, in the 'aethalometer', is referred to as black carbon (Kirchstetter and Novakov, 2007).[14]

The opacimeter measures the fraction of light that successfully traverses the exhaust gas, known as the 'transmittance' (Green G.L. and Wallace, 1980). This fraction, when subtracted from the total, that is, the incident light, reveals that fraction lost by absorption and scattering – what constitutes 'opacity'. The sum of the opacity and transmittance is therefore unity.[15] The attenuation of light in this manner by aerosols follows the well-known Beer–Lambert law,

$$\Theta = 1 - e^{-kL}, \quad (5.6)$$

where Θ is the opacity, L is the path length and k is the 'light extinction coefficient', the latter of which is function of soot mass (Burtscher, 1992, p. 559), and relatable to other particle properties – although, in practice, analytical solutions are not essential to the utility of smoke measurements. These measurements are generally conducted across the exhaust system or across a sampled flow taken from the exhaust. However, such measurements do not necessarily repeat the visual impact *post-tailpipe*, because of exhaust plume dynamics and mixing in the ambient air (Bascom *et al.*, 1973; see also Section 4.3).

The spotmeter appears to be the poor cousin for several reasons: one, is operating temperature. The filter paper in a spotmeter requires *cool* exhaust gas, for which the dew point must be respected – otherwise water condensation onto the spot falsifies the measurement (Shaffernocker and Stanforth, 1968) – whereas the opacimeter sends its light through hot exhaust gas. The spotmeter must draw off a *sample* of exhaust gas; an opacimeter can send its light through the whole exhaust gas. The operating principle of the opacimeter more closely resembles the experience of the human eye in observing smoke, and, if necessary, some consideration can be given to the eye's spectral response, or photo-optic characteristic (Green G.L. and Wallace, 1980). But the greatest advantage the opacimeter holds over the spotmeter is its ability to provide a continuous, real-time record: this makes it immeasurably useful in following emissions of smoke during transient engine operation. The spotmeter, by contrast, is inherently off-line and cumulative: a finite amount of particulate must first be accumulated on the filter paper, and this *ipso facto* restricts measurements to steady states. Indeed, if the smoke varies

---

[14] Yet another method is the absorption of beta rays by the soiled spot. This device is used in air quality monitoring, but has found no widespread use in the motor industry.

[15] Measurement of the successfully traversing fraction brings a commonality of sorts with light-scattering instruments (Section 5.3.4), for which line-of-sight measurements are not absolutely essential, as light is scattered not solely in the forward direction but in all directions, allowing its collection at any angle (Jones B.L. *et al.*, 1997). However, the distinction between light that has not interacted with particles and light that has interacted and yet been scattered in the forward direction, is not important with opacimeters.

during the measurement period, the result will be misleading.[16] With today's engines this restriction is irksome, because *transient* engine operation is central in the generation of smoke.

What sets smokemeters apart is their operation on *undiluted* exhaust gas – clearly, an instrument capable of measuring the bulk light-attenuating properties of exhaust gas would be unnecessarily handicapped by dilution. It need hardly be said that smokemeters generally provide some usable signal even when nothing is seen by the naked eye. But, from an opposing standpoint, smokemeters in the automotive industry are of markedly diminishing utility, because, as engines become ever cleaner, signal levels are falling into noise levels. The manufacturers of smokemeters are therefore addressing this problem (Schindler *et al.*, 2001). With opacimeters, the stability demanded in the light source is approaching a practical limit, at least for inexpensive units (Schindler *et al.*, 2004): the restriction relates also to path length, which can be increased by the use of mirrors, so that light beams repeatedly traverse the same smoke (Saito K., 1988). With spotmeters, signal strengths are increased by filtering more exhaust gas through the same size of spot.

It is not the aim of this chapter to discuss how particle-measuring instruments relate, either analytically or empirically; this is a huge subject in its own right, and more pertinent to a text on emissions measurement; detailed comparisons are available elsewhere (e.g. Moosmüller *et al.*, 2001; Mohr *et al.*, 2005). One issue, however, because of its huge importance, deserves exception – although there is space to cite only a few papers, and these are admittedly selected with bran-tub impartiality. In particulate measurement, the realisation of some credible relationship between *visibility* (smoke) and *gravimetry* (mass) has long been the Holy Grail in emission control (e.g. Matsui *et al.*, 1982; Homan, 1985; Hessami and Child, 2004). If smokemeters were able to supply, *indirectly*, the mass emission, this would represent a considerable advance, because the statutory limits are based on mass, yet, as we have seen, smoke is easier to measure than particulate mass – especially in real time. This is why measurements of smoke and filter weight have so often been conducted in parallel.

As hinted in the opening remarks to this section, certain necessary assumptions and many confounding factors are inherent in optical measurements; thus, the formulation of any universal or unequivocal relationship between smoke and mass seems rather unlikely. The light attenuation can, for example, be related to particle mass, but this relationship depends on particle size and refractive index, and these might change, say, from one engine operating point to another. The smoke-to-mass relationships as have been devised seldom possess sound theoretical underpinning: they are predominantly empirical or semi-empirical, and so of dubious validity outside the specific conditions under which they were derived. Something more widely understood, however, is the predominance of different interrelationships in different ranges of smoke (Wood A.D., 1975).

Various attempts have been made to relate S_N to the mass of retained soot in the spot, for example via particle radius, using theoretical and semi-empirical relationships that consider multiple deposition or overlaying (Muntean, 1999). Three suggested relationships are variously applied according to the range of S_N, but only one of these, quoted here, has any actual physical significance (Alkidas, 1984):

$$S_N = K \left[1 - \exp(-kM)^b \right], \tag{5.7}$$

where M is the mass concentration of soot in the exhaust, K is as above, and k and b are constants derived from experimental measurements. Widespread agreement with the results of other investigators suggested broad applicability.

Extensive confounding unquestionably arises from volatile components for measurements conducted on cooled exhaust – particularly an aspect of the spotmeter. Liquid-phase particulate, i.e. organics and

---

[16] There are instruments in which smoke is deposited onto a continuously moving filter tape. However, these are unlikely to better opacimeters.

sulphates, is generally thought to scatter light, but only slightly, in contradistinction to the carbonaceous particulate, which strongly absorbs light (Schindler et al., 2001). Equation (5.7) can be extended to incorporate the volatile component; hydrocarbons, when present on the spot, appear to have little effect on S_N (Alkidas, 1984). When the organic and carbonaceous particulate were separated, the mass of the second correlated with S_N, but not the first (Okada et al., 2003); a similar conclusion was reached about Θ (Schindler et al., 2001).

Reference to volatile particulate components raises the interesting question of how commercial smokemeters, designed with black smoke in mind, respond to white smoke. (It should be noted that in practical applications, the two smokes can be co-present.) One might well suppose that if white smoke is perceptible to the naked eye, then smokemeters should respond in some way, *mutatis mutandis*; and as already hinted, scattering is in operation rather than absorption. But the effect of volatiles on smoke readings, whether through the formation of additional droplets or condensed layers on solid particles, is poorly quantified.

As explained in Section 3.6.5, white smoke, owing to its inherent volatility, is difficult to measure meaningfully; and the vagaries of cooling and mixing are further headaches. Consistent measurements require precise control of saturation; otherwise, gas–particle partitioning is uncontrolled. Such requirements are unlikely to be satisfied in the exhaust plume; and this calls for some sort of conditioning, prior to measurement. In one apparatus (Tanaka T. et al., 1989), white smoke is regasified by heating the exhaust gas to 200 °C. A filter then removes the black smoke. Subsequent entry of chilled air regenerates the white smoke, thus allowing its independent measurement. Photography of the actual plume is another method (Tsunemoto et al., 1999).

5.5 Closure

Particulate emissions are studied using a huge variety of instruments, which, in this chapter, have been loosely classified according to their *conventionality*, their ability to discriminate between particles *individually*, and their treatment of particles *collectively*. Some of these instruments are commercially available; others remain, as yet, in a fairly rudimentary state of development, not having been pursued much beyond the research groups that originally proposed them.

5.5.1 Particulate Measured Conventionally

Drawing a Sample of Exhaust Gas

A common perception exists that isokinetic sampling is not required for most particles likely to be found in exhaust gas. This view is challenged by some workers; for example, isokinetic-sampling losses may not be entirely negligible during transient engine operation, or if the probe entrance has been fouled with deposited particles. Particle losses through thermophoretic deposition onto transfer line walls are a well-known problem; these losses are countered by, for example, thermal insulation or electrical heating.

Diluting the Exhaust

The exhaust gas enters the dilution tunnel, where it mixes with air and, after a certain period of time in which to homogenise, this air–exhaust mixture is then filtered: what then counts is the total mass of filterable material in diluted exhaust at ≤ 52 °C; this *defines* 'particulate' in the *legislative* sense. There are many designs of dilution tunnel, but all essentially fall into two categories: *full-flow* and *partial-flow*. The tunnel has several tasks; it must carry out dilution in a controlled and repeatable manner, resemble the real-world emission process, reduce particle concentrations to the requirements of the instrumentation, and avoid water condensation. The dilution air must be conditioned for humidity, particle content, temperature and organic vapours. If the dilution ratio is too small, then the necessary condition ($\leq 52\,°C$) is not fulfilled at the filter; but, if too large, the exhaust gas is too rarefied to capture sufficient particulate. Inter-laboratory comparisons are impeded by the lack of any consistent tunnel design: critical here are the tunnel's heat transfer characteristics, insofar as these control the rate at which the diluted exhaust cools (nonadiabatically). The dilution tunnel does not reproduce exactly the emission process in a real exhaust plume; there are thus pressing questions about how representative are the resulting particulate characteristics – perhaps these differ from those in the real world. Researchers have responded to these questions by devising novel ways to change, independently, the dilution rate, dilution ratio and residence time, which is to say, the *dilution profile*; and more representative methods of dilution incorporate considerations of *local* rather than global dilution ratio. Because of different mixing (different diffusion rates), gases and particles might experience different dilution ratios. Various (nontraditional) methods of imposing rapid or sudden dilution are being used.

Collection onto a Filter

There are various filters (e.g. borosilicate fibres, glass fibres and membranes): the most appropriate choice depends on the purpose of the test (thermogravimetry, chromatography or optical examination). Filter materials exhibit different propensities for plugging, artefact formation, volatile sorption, contaminating potential and material loss. *Prior* to weighing, the dirty filter must be carefully preconditioned, i.e. at a certain temperature and relative humidity, otherwise the hydrated state of the sulphate fraction is undefined. For the fullest quantification of organic compounds, the addition, in series, of a vapour-phase trap or adsorbent (such as XAD-2) is advisable, in order to catch the filter 'blow-by'. The filter deposit is separated, or 'fractionated', into the particulate components, using two principal and well-established methods: dissolving (or 'extraction') and gasification.

Fractionation by Gasification

In thermogravimetry, particulate weight is recorded in two phases as a function of increasing temperature: first in an inert atmosphere, in which the sulphate and organic fractions are evaporated, and second in an oxidising atmosphere, in which the carbonaceous fraction is burned. The features of this 'thermogram' are strongly affected by two choices: the rate at which the temperature is increased, and the temperature at which the environment is changed from inert to oxidising. Less quantifiable are, in the first phase, the transformation, by pyrolytic reactions, of organic compounds into carbonaceous material; and, in the second phase, excessively rapid burn-off rates, with their concomitant distorting effects on rates of heat and mass transfer. Numerous test protocols have been designed with the aim of ameliorating these difficulties: they are not, however, standardised. Thermogravimetry is considerably enhanced by chemical analyses of the evolved gases, by which the weight is computable retroactively.

Fractionation by Dissolving

Two phases are used. One is Soxhlet extraction, in which the organic fraction is removed through cycles of heating in an organic solvent. The appropriate length of the fractionation, or the number of cycles, is difficult to ascertain, as some material will always be extracted, albeit in ever-smaller amounts. There are many possible solvents (although the most often used is dichloromethane, DCM), but they do not all extract the same compounds; the principal issue is one of polarity. The other phase is aqueous extraction of the sulphates, by leaching by water and isopropyl alcohol, after which quantification is done by ion chromatography. Water-soluble sulphates should not be confused with sulphates generally, or with water-soluble material generally. The water associated with the sulphates is not determined directly; a value of 1.3 is customarily assumed for the mass ratio H_2SO_4/H_2O. However, recent investigations have undermined this assumption, even for filters preconditioned at the established temperature and relative humidity: the underlying chemistry here is not thoroughly understood, but one possible explanation is some aspect of the unfouled filter surface.

Chemically Assaying the Organic Fraction

The filter deposit is analysed by a battery of tests. Substantial work-up is always involved (extraction, isolation and assaying), with ample opportunity for sample contamination, formation of artefacts, and sample losses. Spiking experiments, with the aim of determining losses in the work-up, should be conducted with the utmost caution, as tracer compounds and native compounds do not necessarily undergo identical experiences, particularly in gas-to-particle partitioning. The composition of the organic fraction is particularly studied by chromatography, for which a key aim is divorcing oil-derived compounds from fuel-derived compounds: this is done by comparing chromatograms of fuel, oil and the organic fraction. The commonest taxonomy for the organic fraction is one of seven subfractions: acidics, basics, alkanes, aromatics, oxygenates, transitionals and insolubles. The PAH are quantified by fluorescence, and the NPAH by reduction to amines and then fluorescence.

Biologically Assaying the Organic Fraction

The ability of organic compounds to damage cellular DNA, known as 'mutagenicity', is measured by the Ames test, wherein an extract is incubated along with mutant cells of the bacterium *Salmonella typhimurium*. These strains cannot synthesise an amino acid, and 'mutagenicity' refers to the ability of the organic compounds to 'back-mutate' the cells into a strain that *can* synthesise this amino acid. The mutagenicity is then derived from the linear portion of the dose-response curve, expressed as revertants per µg of organics, per m^3 of exhaust gas, per kWh of delivered energy, etc. The metric should be chosen carefully, as this determines the mutagenic ranking of different technologies, fuel formulations, etc. The Ames test only *suggests* the risk to humans: it acts as screening test, reveals trends and indicates where efforts ought to be directed. It should not be confused with carcinogenicity.

5.5.2 Particulate Measured Individually

Instruments are available that report particle size. But, owing to the large size range, the same operating principle, that is, the same physics, cannot be used for all particles. Signal conversion into particle size is a highly problematical area, readily thrown off course by other aspects such as morphology and composition. Sphericity is a common and necessary assumption, but this property clearly does not apply to agglomerates. Instruments are calibrated with test aerosols that may behave in different ways from real particles discharged by motor vehicles.

Inertial Mobility

The electrical low-pressure impactor (ELPI) sorts particles according to their aerodynamic diameters (using inertial forces). Streamlines suddenly veer through right angles, whereupon small particles also do so; large particles, owing to inertial forces, retain their original velocity vectors, and thus impact on surfaces, where they are captured. Smaller particles are caught on each successive stage of the ELPI, because the velocity is stepped up; the 'cut-off' characteristic for each stage must be steep, otherwise mis-sorting occurs. The impactor relies on inertial forces, for which the lower size limit is $\sim$400 nm, but, by using lower operating pressures, this limit may be lowered to $\sim$30 nm – unfortunately, a figure still too large to measure fully the nucleation mode; moreover, such operation may cause evaporation, and loss, of some volatile material. Two important advantages of the impactor are, first, the facility to weigh its various stages, thus obtaining a direct measurement of mass distribution, and, second, the ability to measure particle sizes in (what is effectively) real time. Extensive dilution is required.

Electrical Mobility

The scanning mobility particle sizer (SMPS) sorts particles according to their electrical mobilities; it does this by gradually increasing, or scanning, the strength of an electrostatic field. The lowest measurable size is $\sim$7 nm, but one unit cannot cover, in one scan, the entire size range, up to several microns. The largest disadvantage, however, is the duration of a complete scan, two or three minutes, so that, for transient emissions, all sizes cannot be simultaneously tracked. The aerosol characteristics, moreover, must remain stable within this time frame. To circumvent this problem, several units are used in parallel, with each one operating in a different size classification. Extensive dilution is required.

Laser-induced Incandescence (LII)

Particles are heated with a laser to the carbon sublimation temperature, and the resulting incandescence is a function of volume concentration; the laser is then switched off, and, from the rate at which the incandescence decays, a single, overall size is derived for the primary particles or spherules. Sulphates and organics are not quantified, since they promptly evaporate. The laser heating may induce morphological changes in the particles that affect the result, i.e. the technique might not be completely nonperturbing. No dilution is required; indeed, this would unnecessarily degrade the signal.

Light Scattering

Light scattering is most commonly used for particle enumeration, such in the condensation nuclei counter (CNC) – this is particularly used once the particles have been sorted according to size. Application of light scattering to particle *sizing* is considerably more difficult, and application of the Mie theory requires several sweeping assumptions, not the least of which is sphericity. A nephelometer has nevertheless been devised for particle sizing, and 'best fits' obtained to the Mie theory. Light scattering, in combination with laser heating, can resolve the volatile and nonvolatile components.

5.5.3 Particulate Measured Collectively

Photoacoustic Effect

The particles become slightly warmer when irradiated by light, and some of this energy is re-emitted as heat, which, in turn, causes an increase in pressure. A modulated light beam is thus detectable as a series of acoustic waves. As the light absorbed is related to the mass of light-absorbing material, the amplitude of the signal serves as a measure of the mass of soot in the irradiated volume. Some other particulate-phase material and some gases might, though, be partly light-absorbing.

Photoelectric Effect

Particles exposed to light above the photoelectric threshold eject electrons and become ionised, such that they support a measurable electrical current. The charge carriers must be carefully managed, i.e. pre-existing ions and electrons must be removed. The current is a function of the 'active' surface area (i.e. excluding inner voids), and also of the concentration of PAH adsorbed at the surface, because photoelectric thresholds are lowered by adsorbents. No information is available, however, about *which* PAH are present, as ionisation potentials differ amongst these compounds, and the signal is some unknown summation.

Current and Charge

The natural electrical charge possessed by many particles causes charge movement within a coil through which they pass. There is no theoretical basis as to how this signal relates to the particles, i.e. what fraction are charged, how many charges per particle, etc., so that the device is indicative, and some behavioural variations remain unexplained (upgoing versus downgoing pulses, etc.). But the response is fast enough to provide crank-angle-resolved details, such as detecting exhaust valve opening.

Flame Ionisation

The flame ionisation detector (FID) has been used for many years to measure the volatile hydrocarbons, or HC, i.e. those compounds which are gaseous at $>190\,°C$; but it has recently been extended to particulate. The organic fraction is defined by what compounds condense between two temperatures, $52\,°C$ and $190\,°C$. The carbonaceous fraction is quantified by integration of the signal spikes as individual particles burn in the flame.

Mass

The tapered-element oscillating microbalance (TEOM) measures mass indirectly, by relating this to the resonant frequency of a vibrating system. The 'tapered element' refers to a specially designed hollow glass tube, one end of which holds a filter on which particles deposit. The 'oscillating microbalance' refers to the same construction, forced into resonance by an electronic circuit. Two great advantages are that particle morphology is irrelevant, and a second-by-second indication of the mass emission is available. The TEOM does, however, require good mechanical and thermal decoupling from its environment, as the vibrating system is easily perturbed by extraneous influences. The chief difficulty with the TEOM, though, is that the filter is not a perfect arbiter between the particulate phase and the gas phase, and displays the same idiosyncratic properties as the dilution tunnel filter (see Section 4.6). Shifts in engine operating point, and the attendant fluctuations introduced into flow rate and pressure, cause transient take-up and release of volatiles (including water). In extreme circumstances, volatile losses are behind negative mass emission rates – obviously an unphysical possibility. Some integrative methods are being developed to correct for these gains and losses of volatiles.

Smoke

The most long-standing instrument, the smokemeter, measures not so much particulate as the visible aspect of particulate. The smokemeter is still used widely, and there are two variants. In the opacimeter, a light beam is shone through the exhaust, and the incident light compared with the received light. In the spotmeter, the exhaust gas is passed through a filter paper, on which the particles are retained as a soiled spot, the light reflected from which is then compared with that from an unsoiled spot. The singular advantage of the opacimeter is its real-time signal; the spotmeter requires a finite time to accumulate the spot. Unlike most other particulate-measuring instruments, smokemeters are able to operate on *raw* exhaust gas. Smokemeters are of diminishing utility in the motor industry, because today's engines emit so much less particulate, and little of this material is actually visible. It is impossible to place smokemeter signals completely within the physics of the interactions of light with particles; much is empirical, and the relationship to particulate mass is application- or context-dependent (via particle size distribution, composition, morphology, etc.).

5.5.4 Further Remarks

Dilution tunnel measurements of particulate have two acute disadvantages: first, they are ponderous and dilatory, i.e. the dilution tunnel filter must be carefully conditioned and weighed, first in its blank and then in its fouled state; second, they are cumulative and off-line, i.e. the result is a summation, delivered hours after the event, containing no information whatever about *which driving modes* are most polluting. The engineer is thus left very much in the dark as to what aspects of vehicle operation need to be addressed.

It is not the purpose of the present chapter to compare and contrast instruments, or to engage in detailed discussions about test protocols; but, in view of the challenges low-emission vehicles pose to measurement technology, the following issues cannot, in the broader picture, be ignored: repeatability and reproducibility, sensitivity, sampling, and dilution (Kasper, 2003). The poor sensitivity and repeatability of the contemporary gravimetric procedure undoubtedly threatens to become an embarrassment. But although enumerative instruments (especially for nanoparticles) are far more sensitive (ranging over three orders of magnitude), this is a mixed blessing, because a significant part of the poor reproducibility relates not to imperfections in these instruments, but to instabilities in the source of particles (engine or vehicle, and dilution method): 'at low emission levels particle measurements and legal limit compliance assessment have more in common with gambling than serious technical measurement' (Kasper, 2003). Vehicles and engines, formerly regarded as reasonably stable and repeatable sources of particles, are now known to be highly variable.

In the nanorange, the need to condition the dilution air within the engine laboratory may even take on the character of so-called 'clean-air rooms', in which particle losses from human bodies (skin cells, etc.) are significant (Cooper M.G., 1986), and, in the semiconductor, pharmaceutical, health care and food/beverage industries, airborne viruses or bacteria must also be removed (Keyser and Howard, 1991), for which HEPA filters are essential. The particle background is, in contrast, less of an issue with the traditional mass emission, as a handful of coarse-mode particles are easily removed.

The ideal instrument would chemically characterise each particle as it passes – and preferably in real time, thus freeing researchers from batch-processing methods suitable only for off-line characterisations. This seems like a tall order, but instruments for single-particle analysis are certainly under rapid development (Kaufmann, 1986; Wieser and Wurster, 1986), and laser ablation and mass spectrometry already fulfil this role in ambient air quality assessments (Lazar *et al.*, 1999; Noble and Prather, 1999; Allen *et al.*, 2000; Tan P.V. *et al.*, 2002). Corresponding application to the exhausts of internal combustion engines is increasing rapidly (Ferge *et al.*, 2006; Toner *et al.*, 2006).

6
Characterisation

6.1 Introduction

Chapters 3 and 4, which described how particles are generated, form the necessary complement, and indeed backdrop, to the present chapter, which describes how particles are composed; thus, rather than what is *formed*, we now turn to what is *found*; and this, in a nutshell, is the essence of particle *characterisation*. Obviously, this information is essential in order to gauge appropriately the environmental ramifications of this nefarious emission. But, less obviously, characterisational studies provide important clues as to how particles are formed: it might be argued, for example, that, if some feature or other is observed, then one mechanism rather than another must have been responsible; and this, in turn, helps to suggest ways in which particulate emissions might be abated. Less obviously still, characterisational studies facilitate the design of laboratory apparatus able to generate 'artificial' particles (Hoard *et al.*, 2003), the characteristics of which must, naturally, reproduce those of the real emission. Such apparatus obviates the need for expensive engine test facilities; allows closer control of, and greater repeatability in, particle characteristics; and furnishes useful calibration aerosols.

The characterisational studies described herein are arranged in four sections. *Physical* characterisation encompasses the microstructure of the spherules and the morphology of the 'superstructures' constructed from these spherules, namely the agglomerates. Two common physical metrics, surface area and density, are discussed, and the section concludes with some description of the indigenous electrical charge carried by many particles. *Chemical* characterisation encompasses the compositional make-up of the four fractions: carbonaceous, sulphates, organics and ash. Sometimes these fractions have clear physical locations in the particulate; at other times they are more diffuse, and are defined in terms of the analytical methods used in their quantification. *Biological* characterisation, the third section in this chapter, is concerned solely with bacterial mutagenicity, as measured in the widely used Ames test. Of course, researchers use a considerably wider range of biological systems, including mammals, to study the biological repercussions of particles; but further discussion on these matters is reserved until Chapter 10. Finally, *demographic* characterisation is the study of particle sizes, numbers and masses, for which the database in the open literature has grown with extreme rapidity: pre-1995, for example, very little information was available.

There is a general need for critical distance in consulting papers published more than a few years ago. This is because particle characteristics cannot have remained constant during this time: the replacement of indirect injection (IDI) by direct injection (DI), the advent of low-sulphur fuels, the introduction of forced induction (turbocharging), and ever-rising injection pressures are four salient examples of the sweeping changes to have taken place. Characterisational studies nonetheless go back quite some time – particulate emissions first started coming under detailed scrutiny in the 1970s (e.g. Braddock and Gabele, 1977). This research was vigorously prosecuted in the USA in anticipation of a lucrative market for passenger-car diesels, although, as it turned out, such a market strangely never materialised. The relevance, however, of this vintage corpus of literature to *today's* large population of light-duty diesels in Europe should be considered carefully.

6.2 Physical Characterisation

Before passing to detailed studies of physical particle characteristics, brief consideration of a closely related aspect will be instructive. If the pressure drop across a dilution tunnel filter were to be straightforwardly related to the mass of particulate retained by this filter, this would provide a real-time measurement of mass and, moreover, obviate the need for cumbersome post-test weighing (MacDonald J.S. *et al.*, 1984). Indeed, an identical situation arises with the diesel particulate filter (DPF), for which engine management systems need some up-to-date knowledge of particle loading. Yet this method of mass measurement has not been adopted, in either case, with any great zeal, and the reasons are not too hard to find – pressure drop is a function of physical characteristics also, and these are not constant. First, there is particle size and shape, which determine the voidal space (Salvat *et al.*, 2000); second, there is particle dryness or wetness, i.e. powdery deposits versus oleaginous deposits; third, there is the compactness, or packing density of the deposit – logically, itself a function of the pressure drop (Mogan *et al.*, 1985) and of the distance from the wall (Koltsakis *et al.*, 2006). As, then, the physical characteristics are so variable, it seems unrealistic to expect any consistent relationship between mass *on* the filter, and pressure-drop *across* the filter – only under certain specific circumstances is this relationship of any use (Mischler and Volkwein, 2005).

6.2.1 Microstructure

As already related at numerous stages, the basic structural unit of the carbonaceous agglomerate is the *spherule* – which, up to now, has been treated as a complete, and indeed completed, entity. We now descend *into* the spherule, to examine how the carbon atoms within it are accommodated, that is, arranged. This question is not simply one of idle curiosity, since *microstructural* (or, rather, nanostructural) features are, so to speak, relics surviving from the germinal stages; hence, to reverse this logic, the delineation of these relics retrospectively elucidates spherule formation (Hays and Vander Wal, 2007). The truth of this assertion can be tested by comparing conventional notions of spherule formation, as described in Sections 3.2.2 and 3.2.3, with conventional notions of microstructure, as sketched in the present section.

The understanding of spherule microstructure has changed over the years: one source of controversy is the precise interpretation of electron micrographs or diffraction patterns, and crystallographers are not always in agreement. But since the interpretation of these images is every bit as much concerned with imaging *methods*, and a discussion of electron diffraction, X-ray or neutron scattering, Raman spectroscopy, etc., would take us too far off course, these controversies are herein avoided.

Characterisation 213

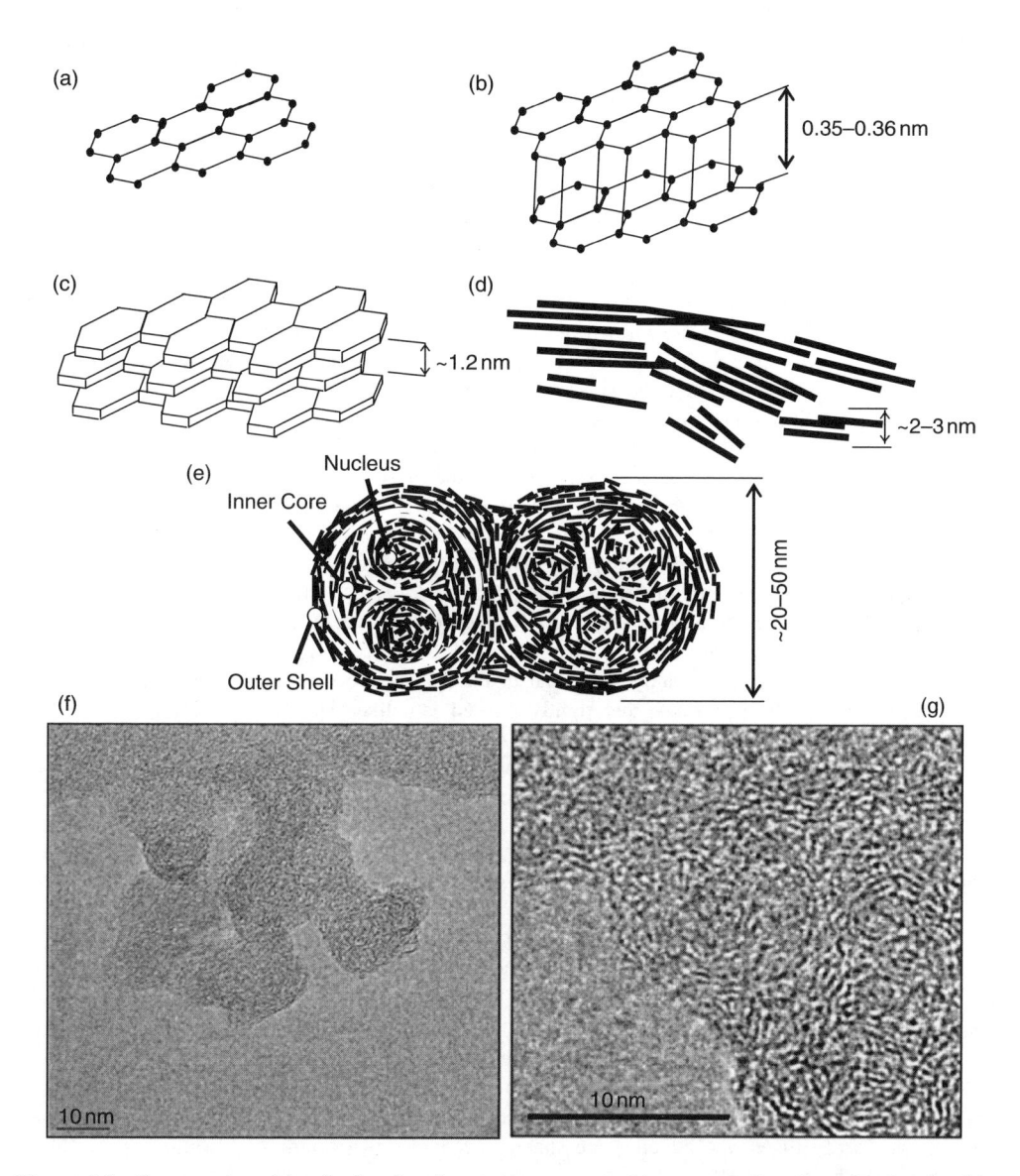

Figure 6.1 Conceptual models of spherule microstructure, arranged in an ascending dimensional scale: (a) a hexagonal array of carbon atoms; (b) two hexagonal arrays, making a platelet; (c) three platelets, making a crystallite; (d) several crystallites in a spherule (as cited in, and adapted from, Lipkea *et al.*, 1978); (e) two complete spherules, containing more than one nucleus, and bonded via a shared outer layering of crystallites (after Kawaguchi *et al.*, 1992); (f) electron micrograph of an agglomerate (direct-injection spark-ignition engine, one-cylinder, 0.56 litre, 1750 rpm, wide-open throttle); (g) close-up of (f); (courtesy of P. Price, R. Stone and K. Jurkschat, University of Oxford).

For the present moment, therefore, it is sufficient to venture a reasonably well-accepted conceptual understanding of spherule microstructure, as given in Figure 6.1 (cited in, and adapted from, Lipkea *et al.*, 1978, and after Kawaguchi *et al.*, 1992). On an ascending dimensional scale, the microstructural components are as follows: (a) carbon atoms are arranged in repeating six-membered rings, forming planar hexagonal 'arrays'; (b) hexagonal arrays are chemically bonded, forming parallel, face-centred layers or 'platelets'; (c) platelets are stacked, forming commonly oriented domains, or 'crystallites'; (d) crystallites are stacked in various orientations, forming completed spherules. By this means, nature sorts $\sim 10^5 - 10^6$ carbon atoms into $\sim 10^3 - 10^4$ crystallites. The two characteristic microstructural dimensions are the width and the stack height of the basal-plane hexagons (Chen H.X. and Dobbins, 2000).

In its distinct hexagonal ordering, this microstructure is unmistakable, being that of graphite: in carbon chemistry, these arrays are known as 'graphene sheets' (Braun *et al.*, 2004), or simply as 'graphenes' (Yan *et al.*, 2005b). The carbon atoms within layers are held together by covalent bonds, whereas the interlayer bonding is far weaker, relying on van der Waals forces. But corresponding layers of immediately neighbouring hexagons do not respect the hexagonal (ABAB...) or rhombohedral (ABCABC...) orderliness of graphite; the consequence of this, is a somewhat loosened microstructure. For example, the interlayer spacings are 0.350 nm in diesel soot, which is a 'perturbed' graphitic microstructure when compared *a fortiori* with graphite, at 0.335 nm (Fujiwara *et al.*, 1990). Crystalline structures in which basal planes appear to have slipped drunkenly relative to one another, so that the interlayer spacings have increased slightly, are called 'turbostratic'. Hence, spherule microstructures are more aptly styled 'quasi-graphitic' than graphitic (Sadezky *et al.*, 2005).

An interesting microstructural characteristic of all spherules – a relic of the germinal stages, as mentioned – is gradually shifting order as a function of radius. The *outer* regions show long-range order: crystallite orientations are consistently tangential to the surface, so that crystallites form concentric, or near-concentric, onion-like rings, and tightly packed lamellae. The *inner* regions, on the other hand, show short-range order, or rather disorder: crystallite orientations are random, haphazard or unsystematic. The electrical conductivity of the soot reveals the balance of order and disorder (Yan *et al.*, 2005b), and electron diffraction patterns show *intermediacy*, that is, a microstructure that is partly amorphous and partly crystalline (Fujiwara *et al.*, 1990) – indistinguishable, under the electron microscope, from carbon black (Clague *et al.*, 1999) – unsurprisingly, since carbon black is formed under very similar conditions. Both these forms of carbon, by virtue of their amorphousness, are easily distinguished from graphite (Rounds, 1981).

It seems plausible that these inner, amorphous domains represent progenitor nuclei, around which more carbon gathered to form the spherule; and it seems equally plausible that, when more than one amorphous domain is present within one spherule, this betrays *coalescent* agglomeration in an early stage, as described in Section 3.2.4. But, in practical observations, these amorphous domains are often indistinct and not readily counted as discrete entities. As shown in Figure 6.1(e), in cases where agglomeration is noncoalescent, crystallites accumulate in the intervening crevices to form an *inner core*, around which, in later growth stages, more crystallites accrete to form an *outer shell*. When spherules are partly fused, the lamellae are shared, i.e. laid down seamlessly (Müller *et al.*, 2006), forming part of the outer shell.

This description necessarily refers to an *archetype*; microstructures vary, in several nuanced respects, from one spherule to another: in graphiticity, or the proportions of crystallinity and amorphousness (Ning *et al.*, 1991); in concentricity of the lamellae (Lee K.-O. *et al.*, 2001); in inter-platelet spacing (Fujiwara *et al.*, 1990); and in sphericity (Lee K.-O. *et al.*, 2001). These subtleties stem, perhaps, from slight variations in the conditions of formation, and variations also in the later experiences of spherules, meaning the oxidation stage. Engine operating conditions, such as load, are then, presumably, controlling factors; but consistent relationships between microstructures and engine operation are weak and not fully established. When engines are under load, the spherules are less spherical, and the encircling crystallites trace out wavy or corrugated paths, rather than smoothly arcing ones (Lee K.-O. *et al.*, 2001), while the crystallites are larger, and the degree of crystallinity higher, that is, the microstructures are more graphitised (Braun *et al.*, 2005a). That crystallinity declines from the middle of chained agglomerates towards the ends strongly suggests the attachment of more and more immature spherules during agglomeration (Stevenson, 1982). Finally, it is suspected that, during soot inception, insufficient availability of PAH as feedstock impedes the formation of ordered microstructures (Boehman *et al.*, 2005).

These microstructural variations are of interest because possibilities for pollution abatement might thereby arise: susceptibility to oxidation is very much driven by the extent of order or disorder, insofar as this determines, concomitantly, the density of edge-site carbon atoms: first, because these atoms, being precariously located, are accessible to oxidants, and second, because these atoms have unsatisfied bonds, and so exhibit greater reactivity towards oxidants (Müller *et al.*, 2005). This makes microstructure pertinent not only to late-combustion in-cylinder burn-up (Section 3.3.4), but also to regeneration of DPFs (Section 7.8.2). For example, soot generated in the combustion of *biodiesel* burns at noticeably lower temperatures: this is ascribed to greater disorderliness in the microstructure (Boehman *et al.*, 2005).

6.2.2 Morphology

In morphological investigations, particle images are subjected to various statistical evaluations (Bérubé *et al.*, 1999) which seek to recover the three-dimensionality from the two-dimensional representation (Tian *et al.*, 2006). But first, particles must be captured on a suitable substrate, and the accumulation mode does, after all, interact reluctantly with bulk matter; capture methods include filtration (Armas *et al.*, 2001), and the use of a suitable probe inserted into the exhaust stream, using inertial deposition (Carpenter and Johnson, 1979), electrostatic deposition (Mathis *et al.*, 2005) or thermophoretic deposition (Dobbins and Megaridis, 1987). Obviously, the sampled particles must be representative, that is, of the general population, so one must bear in mind the possibility of biased selection, as particle transport is, after all, size-dependent. Other potential forms of bias are subtle: in electrostatic capture, particles approach the substrate with their longest dimension oriented parallel to the electric field, so that photomicrographs depict them 'end-on', as it were (Dolan *et al.*, 1980).

One must consider also whether a particle has actually deposited as such, or whether two or more particles merely appear as one, by virtue of having deposited at the same site. Instructive, for example, are experiments in which particles are first size-selected, such as according to electrical mobility (Section 5.3.2), and then captured for examination under the electron microscope (Dolan *et al.*, 1980): obviously, such data are misleading if deposits are allowed to become too cluttered (Armas *et al.*, 2001). This uncertainty is lessened, although not, admittedly, eliminated, by temperately acquiring just a few particles – naturally, as controlled by probe residence time in relation to particle concentration. But in cases where rare or atypical particles are required, this constraint is obviously a nuisance. A less appreciated subtlety is particle deformation or fragmentation when impacting on the surface – a risk contained by maintaining low impact velocities (Wittmaack and Strigl, 2005).

Particle morphology operates distinctly on *two* levels, the *spherule* and the *agglomerate*, the second of which is discussed immediately below. At the spherular level, the result of the investigation is, to some small extent, affected by nonsphericity – although many spherules appear spherical, as verified by tilting specimens under the electron microscope (Lee K.-O. and Zhu, 2005). The result is also affected to a larger extent by interconnectedness as, when agglomerated, spherules display widely varying degrees of intimacy, sometimes touching coquettishly, at other times fusing like Siamese twins,

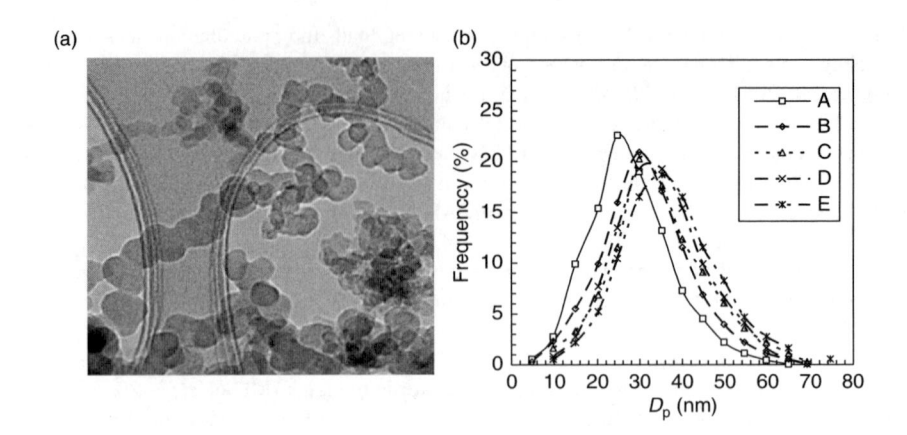

Figure 6.2 Morphology on two levels: (a) electron micrograph of some agglomerates, showing chained morphology (centre) and clustered morphology (lower right) (direct-injection spark-ignition engine, one-cylinder, 0.56 litre, 1750 rpm, wide-open throttle; courtesy P. Price, R. Stone and K. Jurkschat, University of Oxford); (b) spherule size distributions (by number), as assessed under the electron microscope by viewing 2000–3000 agglomerates; the legend shows increasing distance from engine, A to E (indirect-injection engine, single-cylinder, 900 rpm/full load) (Fujiwara and Fukazawa, 1980).

so that a spherule perimeter can be difficult to define. A third difficulty is overlaying, even within single agglomerates, so that low-lying spherules are obscured.

These caveats aside, observations of hundreds or even thousands of spherules show normal or close-to-normal size distributions, ranging, say, from a few nanometres to several tens of nanometres, with the most favoured range being ∼20–50 nm. Several such distributions are presented in Figure 6.2 (Fujiwara and Fukazawa, 1980). There is a hint of asymmetry: the drawn-out tail-off for larger spherules is more prominent in the work of others (Smekens *et al.*, 2005). And, clearly, assumptions of monodispersity – often the practice in soot formation models – are appropriate only to a first approximation (Lee K.O. *et al.*, 2003). Worthy of note is the fact that spherules, if extracted from precisely the same point in a laboratory flame, tend to be of the same size (Xu F. *et al.*, 1997), from which one might suppose that the width of the size distribution merely reflects diversity in the formative stages, i.e. differently sized spherules merely took different routes through the flame.

Such distributions are remarkably and oddly consistent, in the face of various parameters known to affect the combustion strongly. In a bench-top furnace, variations in the percentage of oxygen in the formation zone caused spherule sizes to vary by 20–30 nm (Lahaye *et al.*, 1996). The conditions inside internal combustion engines, however, are generally seen to wield far lesser influence; the trends with operating point are so small as to be uncertain. For example, spherules were smaller by just 2–3 nm at idle, compared with off-idle conditions (Braun *et al.*, 2005a), and larger by ∼5 nm at high load compared with low load (Neer and Koylu, 2006). Fuel injection pressure, injection timing and exhaust gas recirculation, if widely varied, modify spherule sizes by 10–15 nm or so (Mathis *et al.*, 2005). These trends are not, as yet, soundly anchored in any in-cylinder combustion parameter (Lee K.O. *et al.*, 2003); and, although heat-release analysis implies some form of relationship between spherule size and flame temperature (Mathis *et al.*, 2005), surface growth and oxidation, as already said, are inevitably conflated.

Whether *spherules* bear any kind of morphological relationship to *agglomerates* is an intriguing question. This topic has not been addressed as vigorously as it might, but, from the investigations to have been published, agglomerates, it would seem, are larger or smaller simply because they contain greater or fewer numbers of spherules. The sizes of spherules are not seemingly related to the sizes of agglomerates (Vuk et al., 1976). It seems plausible therefore that, other factors being equal, today's engines emit less particulate mass than in former times because agglomerates are smaller, and that agglomerates are smaller not because spherules are any smaller, but because spherules are fewer in number.

The number of combined spherules per agglomerate ranges from a mere handful to many hundreds. (Isolated individuals are seemingly rare, although the literature is stonily mute on this point.) In a laboratory flame, spherule number followed a lognormal distribution, although the smallest agglomerates were uneasily accommodated by such a curve (Tian et al., 2004). The relationship between the *size* of an agglomerate and the *number* of spherules within the agglomerate is worth establishing since, upon examining the former, one could estimate indirectly the latter, i.e. without resorting to laborious counting; and, moreover, an estimate of the 'effective' density, or closeness of the packing, can be derived – one that includes the voids between spherules (Section 6.2.3). Two diameters have been derived from agglomerate images: one, the 'smallest sphere', corresponding to the minor side of a circumscribing rectangle, and the other, the 'largest sphere', corresponding to a circumscribing circle (Vuk et al., 1976). If the number of spherules was then plotted against each of these two functions, an agglomerate of size, say, one micron, according to the 'smallest-sphere' diameter, contained forty spherules, and according to the 'largest-sphere' diameter, one hundred spherules. This method thus encompasses, say, low-density agglomerates with loosely-packed chains of spherules, and high-density agglomerates with tightly packed clusters of spherules.

This underscores two morphological aspects, *complexity* and *diversity*: every particle is anfractuous and curlicued, just as every particle is unique and inimitable. Clusters and chains represent two extreme ends of a broad spectrum: other descriptive terms used are 'feathery', 'tenuous', 'cauliflower', 'spongy', etc., reflecting, perhaps, the fecund imagination of the investigator. Laudable attempts are being made to capture this morphology within certain expressions, using characteristic (measurable) parameters. Of course, this reductionism is a large subject by no means restricted to motor vehicle emissions, and various methods have been devised in many other fields to characterise diverse and complex shapes (e.g. Dunnu et al., 2006); yet the techniques can only be sketched here. Two fortuitous trends, however, are worthy of note. First, where early investigations were cumbersome, i.e. hampered by laborious examinations of photomicrographs, the advent of digital image processing has been an invaluable fillip, as large numbers of particles are now rapidly examined; moreover, the subjectivity inherent in human assessments (Carpenter and Johnson, 1979) is avoided. Second, computerised image processing has greatly facilitated the application and development of *fractal* concepts.

Fractality is computed according to the expression

$$n = k_f \left(\frac{R_g}{d_s}\right)^{D_f}, \qquad (6.1)$$

where n is the number of spherules per agglomerate; d_s is the diameter of a spherule; k_f is a 'prefactor'; R_g is the radius of gyration, which refers to the square of the distance from the centre of each spherule to the geometric centre of the agglomerate (Lee K.O. and Zhu, 2005); and, finally, D_f is the fractal dimension. In a necessary twist, n is found from

$$n = \left(\frac{A_a}{A_s}\right)^{\alpha}, \qquad (6.2)$$

where A_a is the area of the agglomerate and A_s is the area of a spherule: this equation corrects for the fact that, in reality, agglomerates are three-dimensional objects viewed in two-dimensional space, and spherules, as they overlie one another, are sometimes obscured. The means used to obtain the

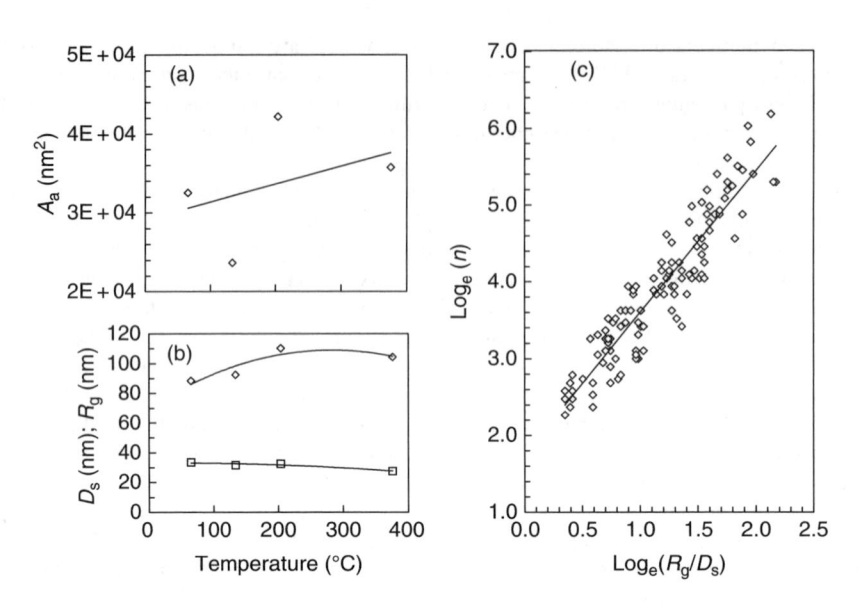

Figure 6.3 Fractal analysis of agglomerate morphology: (a) agglomerate area (A_a); (b) spherule diameter (D_s) and radius of gyration (R_g) (four operating points, as follows (rpm/% load): 675/0, 1400/0, 900/15, 1400/50); (c) from the number of spherules in an agglomerate (n), fractal analysis gave $D_f = 1.8$ at one operating point (1400 rpm/50 %). Engine: 2.5-litre, single-cylinder, direct-injection, supercharged, rated at 1800 rpm/75 hp. Fuel: no. 2 diesel. The particle analysis used high-resolution transmission electron microscopy and image processing (Lee K.-O. et al., 2001).

exponent α need not detain us; thus, the only two terms left unknown, k_f and D_f, are obtained simply by plotting n against (R_g/D_s), as shown in Figure 6.3.

From this analysis, the particles emitted at four different engine operating points held fractal dimensions ranging consistently between 1.5 and 1.9 (Lee K.-O. et al., 2001, 2003), similar to values obtained elsewhere (Chandler et al., 2007) – these should not be confused with values obtained indirectly, via measurements of electrical and aerodynamic mobility, which are ~2.2 (Maricq and Xu, 2004). In fact there are various methods of computing fractality, but how they relate need not concern us: even the most appropriate method, using electron micrographs, is still being discussed (Lapuerta et al., 2006). More germanely, fractality is a term containing certain formative aspects of agglomeration (Section 3.2.4), but it is, at best, a weak function of engine operating point (Neer and Koylu, 2006; Chandler et al., 2007), and not greatly affected by passage through the exhaust system (Klingen and Roth, 1991); others find more open morphologies with increasing engine load (Virtanen et al., 2004). If D_f is truly a function of size – with smaller agglomerates tending to be more compact, and larger agglomerates tending to be more open (Virtanen et al., 2004) – then an additional layer of complexity is created.

The morphological picture thus far delineated concerns solely the archetypal 'dry' carbonaceous agglomerates, or soot – a picture to which some retouching is required when vapour-phase compounds have transferred appreciably into the particulate phase. Uninitiated opinion on this point is that *volatile particulate components are rapidly lost under the electron microscope, through the action of the vacuum and the electron beam*; but the availability of special greases and oils for vacuum pumps testifies, presumably, to the stubbornness of certain hydrocarbons, and photomicrographs do show that *some* volatile material – plausibly, high-molecular-weight compounds – survives, although not, it should be said, indefinitely.

Particularly when emissions of organic compounds or of unburned hydrocarbons are high, spherules appear slightly larger (Vuk *et al.*, 1976) or surrounded by liquid films (Burtscher, 1992, p. 561), with crevices filled (Park K. *et al.*, 2003); and, in extreme circumstances, the underlying agglomerate morphologies are entirely obscured. It seems plausible that volatile compounds, by filling voids in the agglomerates and evening-out excrescences, disguise the underlying carbonaceous skeleton (Ntziachristos and Samaras, 2000), and thereby enhance the sphericity (Slowik *et al.*, 2004) – in extreme cases, perhaps resulting in droplets (Virtanen *et al.*, 2002). Equally, surface tension in the enveloping liquid films will, in striving to minimise the total surface area, force spherules to slide with respect to one another, resulting in compacted morphologies (Kütz and Schmidt-Ott, 1992). It remains to be ascertained whether such deformations, if they truly occur, are plastic or elastic. But the importance of agglomerate elasticity in such transformations is already recognised in powder technology (Friedlander *et al.*, 2001).

It should be said that photomicrographs also contain atypical features, or nonarchetypal particles, from time to time, and that sufficient discrimination needs to be designed into automated image-processing algorithms, otherwise the aforementioned statistical analyses will be distorted. Liquid 'pools', at a size of a hundred nanometres or so, may well consist of heavy hydrocarbons or other organic compounds, while minute particles, at a size of tens of nanometres, appear be to 'droplets', and strongly suggest a nucleation mode consisting of volatile components (Carpenter and Johnson 1979); metallic or ash nanoparticles have not, on the other hand, been reported. Particles of size ~300 nm, thought to be sulphates, retain a remarkable sphericity under the electron microscope (Lanni *et al.*, 2001), suggesting a want of fluidity.

6.2.3 Density

For particles, two types of density – often confusedly conflated – are apparent: *bulk* and *effective* (DeCarlo *et al.*, 2004). The first applies solely to the volume of solid or liquid substance; the second incorporates particle porosity, i.e. the volume enclosed by the particle periphery, including the gas-filled micropores or voids, or, alternatively, relates to an equivalent, nonvoidal homogeneous sphere that would behave similarly. The bulk density is therefore an upper limit, so to speak, approached by the effective density as the voidal space decreases or, alternatively, as the compactedness increases: for a solid sphere or liquid droplet, for example, the effective density and bulk density are one and the same.

Rough estimates of *bulk* density are relatively straightforward, and obviously vary for each of the four fractions. Generally speaking, since the carbonaceous and organic fractions usually dominate, these control also the particle mass. But in less usual cases, for example when fuels contain significant quantities of metal-containing fuel additives or when high-sulphur fuels are burned, and the exhaust gas has passed over an oxidation catalyst, the ash or sulphate fraction might dominate. But this is straying a little from the point: the mass could, in certain circumstances, be dominated by a small-volume, high-density component; for example, the density of metal-containing compounds, in relation to other components, should be considered carefully.

A. *Carbonaceous fraction:* carbon is the principal element, the atoms of which are arranged, as discussed in Section 6.2.1, in a quasi-graphitic microstructure. The particle density, therefore, might be that of graphite, i.e. 2.0–2.5 g/cm^3 (Abdul-Khalek and Kittelson, 1995). However, since the spherule microstructure is looser and more disordered than that of graphite, the real density is probably somewhat lower.

B. *Organic fraction:* effectively, this is the liquid remnants of unburned fuel and unburned oil, the first (diesel) being quoted at 0.85–0.88 g/cm^3, and the second at 0.9–1.0 g/cm^3 (Schneider et al., 2005). The effect of pyrosynthesised components is not reported, but probably small; an alternative value, however, is the density of C$_{22}$–C$_{32}$ n-alkanes, quoted at 0.8 g/cm^3 (Baumgard and Johnson, 1996).
C. *Sulphate fraction:* this is the acid–water complex H$_2$SO$_4$.nH$_2$O, the density of which depends on the hydrated state. Taking n as (say) 7, one arrives at 1.5 g/cm^3 (Warner et al., 2002); others have used 1.83 g/cm^3 (Schneider et al., 2005). The hydrated state, though, of particles when suspended and in transit should not be confused with that pertaining in the filter deposit, following conditioning in an environmental chamber.
D. *Ash fraction*: great diversity (see Section 6.3.2) militates against generalisations, but estimates of density might be based, for example, on metal sulphates, phosphates and oxides. The bulk density of Fe$_2$O$_3$, for example, is 5.18 g/cm^3 (Kim S.H. et al., 2005), and that of BaSO$_4$, 4.5 g/cm^3 (Kittelson et al., 1978b).

From these figures, and likely particle compositions, structures and sizes, some intuitive expectations of *effective* density can be constructed. In the *accumulation* mode, dry carbonaceous agglomerates, especially those consisting of arcing chains of spherules, enclose substantial gas-filled voids; while at the contrary extreme, so much volatile material may be taken up that agglomerates assume the character of droplets, and become spherical. A plausible conjecture is that, with decreasing particle size, agglomerate morphologies are progressively able to accommodate, by proportion, lesser voidal space, so that the effective density will gradually approach the bulk density. In the *nucleation* mode, if particles are spherical droplets of volatile material, then the effective density is the bulk density; and the same equality is probably true of ash (Kittelson et al., 1978b)

Obvious practical difficulties are encountered in handling light, feathery, friable, flocculent, lacy and porous structures. This plasticity is a particular problem with the *direct*, sledgehammer method: collecting sufficiently large quantities of particulate for which volume and weight can be measured. Quite apart from the obvious difficulty of defining a volume boundary, samples are inherently vulnerable to uncontrolled or ill-defined compaction. Actually, a parallel uncertainty arises with the DPF, wherein the thickness of the particulate 'cake' is difficult to define.

Some measurements of effective density are as follows: (a) from the thickness of a particulate cake in a DPF: 0.05–0.06 g/cm^3 (Howitt and Montierth, 1981); (b) from a sample of particulate cake shaken from a DPF, subjected to vigorous tapping: 0.056 g/cm^3 (Otto et al., 1980); (c) from particulate removed from an electrostatic precipitator, subjected to vigorous tapping: 0.02 g/cm^3 (Otto et al., 1980); (d) from weighing filled capsules, 0.38–0.51 g/cm^3 (Stanmore et al., 1999); and (e) from particulate removed from an EGR valve: 0.4 g/cm^3 (Otto et al., 1980). Obviously, the last two estimates are different: in (e), the volatile fraction was twice that in (b) and (c), suggesting that voids within carbonaceous agglomerates had been filled by liquid material more than in the other cases. Interestingly, upon the addition and then evaporation of acetone, the other samples also assumed 0.4 g/cm^3 (Otto et al., 1980), indicating considerable collapse in the flocculent structure. However, what stands out is that all of these measurements are noticeably lower than for the above-listed estimates of bulk density.

The second, more elegant and *indirect* method of measuring effective density is via particle-sizing instruments. The 'diameter' reported by these instruments is – assuming sphericity – converted to volume, and this volume is then referenced to direct gravimetric measurements. Now, this practice raises a host of issues which, although they undoubtedly loom very large, cannot be examined here in any detail. First, one should scrutinise extremely carefully the *operating principles* of the particle-sizing instruments (Maricq et al., 2000b; Andrews et al., 2001d; Maricq and Xu, 2004); and second, the computation of volume, simply by raising a reported diameter to its third power seems reasonable

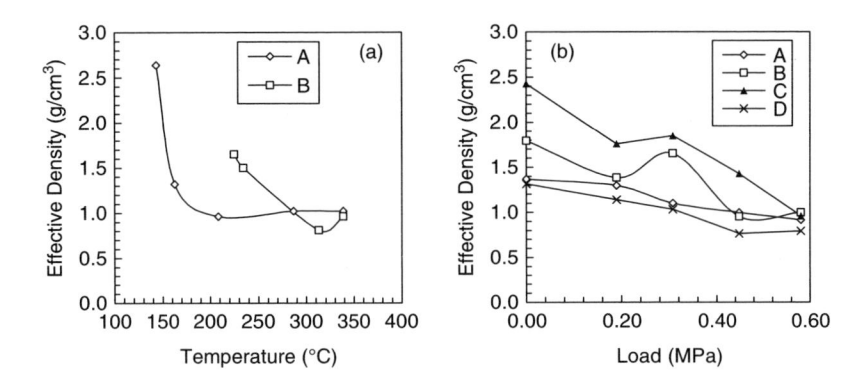

Figure 6.4 Effective particle density expressed as an ensemble, calculated from volume (via an electrical aerosol analyser, 10 nm–10 μm) and mass (via a dilution tunnel filter). (a) As an indirect function of engine load, with the exhaust temperature at the entry point in the dilution tunnel acting as a surrogate. Operating point: A, 1800 rpm; B, 3000 rpm; engine, 575 in^3, V8, medium-duty; fuel, sulphur 0.23–0.35 %; dilution ratio, 8–18 (Khatri et al., 1978). (b) As a direct function of engine load. Engine speed, 1800 rpm; engine, 492 cm^3, single-cylinder; four fuels, A, B, C, D, formulations unspecified (Kittelson et al., 1978b).

enough for the nucleation mode, but precarious for agglomerates, and hence, the limitations should be properly inculcated: for example, density is inextricably linked to fractality (Section 6.6.6 and Maricq and Xu, 2004). To avoid losing sight of this second caveat it might be appropriate to adopt the phrase 'mass-to-volume ratio' in place of density, as a reminder.

Using these methods, effective densities are quoted in two ways: of particles as a function of size, dealt with below; and of particles as an ensemble, i.e. total mass divided by total volume (Khatri et al., 1978; Kittelson et al., 1978b): it should be noted that the dilution tunnel filter (mass) experiences different dilution conditions from a particle-sizing instrument (volume), so that the gas-to-particle conversion is very probably dissimilar. This point notwithstanding, Figure 6.4 displays ensemble data; these densities are supported by other workers (Laymac et al., 1991). From such analyses, effective density consistently declines with increasing engine load (or exhaust temperature). This trend is revealing because, from the above-suggested values for *bulk* densities, one might expect an *opposite* trend, i.e. because the organic fraction dominates at low load, and the carbonaceous fraction dominates at high load (Section 3.6.1). That this expectation is not borne out suggests a converse and more dominant factor, namely, the preponderance at high load of dry, open-structured agglomerates enclosing significant voidal (empty) space, and the preponderance at low load of wet agglomerates sodden with organic compounds. This supposition is lent further credence by electron micrographs (Park K. et al., 2003).

Particle-sizing instruments really come into their own when effective densities are determined *as a function of size*, that is, the 'density function', for which the motivation is not so much the determination of particle densities *per se* as the derivation of a reliable indirect measurement of mass distribution via number distribution. The investigative methods used here lie outside the scope of the present discussion; but they involve weighing the individual stages of an impactor and obtaining agreement, iteratively, with a conventional filter or between different types of particle-sizing instrument, insofar as the relationship between aerodynamic and electrical-mobility diameters offers a definition of effective density (Ahlvik et al., 1998; Andrews et al., 2001d; Virtanen et al., 2002; Zervas et al., 2005). Intuitively and as aforesaid, one might suppose greater voidal space by proportion in larger particles, and hence lower effective densities: a conjecture that many experimental observations support. Indeed,

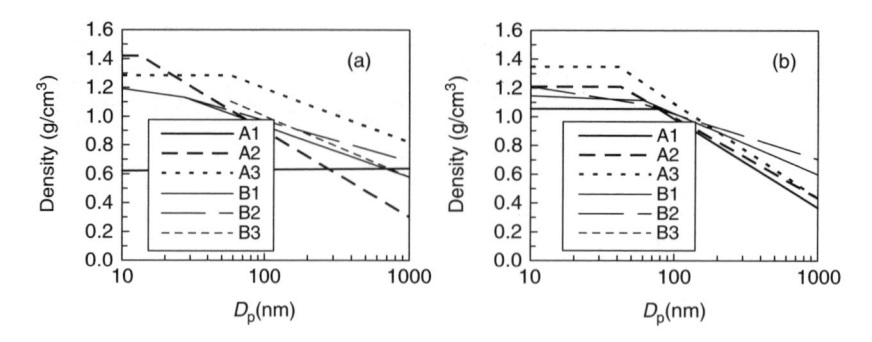

Figure 6.5 Density distribution as determined by matching ELPI and SMPS. (a) Using tunnel dilution (A) and ejector dilution (B); fuel, sulphur 430 ppm. (b) Using ejector dilution; fuel A, sulphur 23 ppm, aromatics 20.6%; fuel B, sulphur 430 ppm, aromatics 35.8%. Vehicle speed (kph): 1, 55; 2, 80; 3, 120. Ejector pump: dilution ratio of 64 within one second. Dilution tunnel: dilution ratio 7–13, within three seconds. Vehicle: 1.9-litre, turbocharged, direct-injection, oxidation catalyst, passenger car of model year 1995 (Virtanen et al., 2002).

an assumption of constant unit density, irrespective of size, leads to orders-of-magnitude overestimation in the mass of supermicron particles (Andrews et al., 2001d). And at the opposite end of the size spectrum, the densities of volatile nucleation-mode particles might be expected to approach those of bulk sulphuric acid and organic compounds (Schneider et al., 2005). Some density functions are shown in Figure 6.5 (Virtanen et al., 2002): they depend on the method of dilution and the composition of the fuel – logically, since the first drives gas-to-particle conversion, just as the second determines the quantity of volatile particle precursors.

It seems apposite to end this section with a few remarks about incombustible ash, as collected in a DPF, for which the ascertainment of density is of the greatest interest, as this information is required in order to estimate the working life of the device. The particle size, of course, is hardly the same as for the ash *as emitted*, as repeated cycles of heating cause annealing and sintering; ash, moreover, varies greatly in composition, depending on the additive packages in the fuel and lubricant. Following collection (over tens of thousands of miles), this ash is removed using, say, compressed air, vigorous shaking or simply pouring; the choice is not simply personal preference, as ash varies in its recalcitrance.[1] This ash, if placed in a graduated vessel, gives densities of 0.4–1.0 g/cm^3 (Konstandopoulos et al., 2000), 0.6 g/cm^3 (Caprotti et al., 2003), 0.45 g/cm^3 (Blanchard et al., 2003) and 0.43 g/cm^3 (Young et al., 2004).

6.2.4 Surface Area

The characterisation of surface area takes two forms, although one of them, namely diffusion charging of suspended ('in-flight') particles (Ntziachristos et al., 2004b), is not the subject of the present section. Rather, our current focus is the familiar 'test-tube' measurement often carried out on powders using an adsorbing gas (e.g. CO_2 or N_2) and usually, although not necessarily, through consideration of the well-known Brunauer–Emmett–Teller (BET) adsorption isotherm (Brunauer et al., 1938).

It should be stressed that the surface areas thus derived are not absolutes but, rather, very much dependent on the context: for example, the mean free path of the gas molecules, insofar as this relates to the sizes of the pores; whether adsorption is subject to kinetic or diffusional limitations, or only

---

[1] Ash removed from a DPF should be treated as industrial waste.

partly reversible; whether gas molecules become trapped; and whether microstructures are distorted by adsorption-induced stresses (e.g. swelling) (Durán et al., 2002). For this and various esoteric reasons which need not detain us, characterisations of surface area do not transcend the specific conditions under which they are conducted; moreover, generalised guidelines for surface-area measurement appear to require some revision for porous carbonaceous materials (Zerda et al., 1999).

Simple geometrical constraints provide a useful sanity check, and these calculations are straightforward enough for an *idealised* geometry, i.e. if the spherules are considered as isolated, monodisperse spheres (Otto et al., 1980). Taking a carbon density of $2\,\text{g/cm}^3$, the specific surface is then given by $(3 \times 10^{-6})/d$, where d is the spherule diameter, and the units are m^2/g. For typical spherules, say, where d is 20–50 nm, this expression suggests specific surfaces of 60–150 m^2/g. This obviously rudimentary calculation agrees quite reasonably with experimental measurements: for example, 105–140 m^2/g (Peterson, 1987) or 86–104 m^2/g (Otto et al., 1980). There is some evidence of a dependence on engine operating point, in that light loads tend to generate denser particulate (31 m^2/g) than medium loads (90 m^2/g), and medium loads denser particulate than high loads (165 m^2/g) (Stanmore et al., 1999).

Such close correspondence between experimental measurements and geometrical expectations tends to suggest that spherules possess, in effect, only point contacts with their neighbours; or alternatively, that the specific surfaces are dominated by effectually isolated spherules. This implication, however, sits uncomfortably with the close interspherule bonding commonly observed within agglomerates (Song H. et al., 2004), which ought to produce specific surfaces smaller than the simple geometrical estimate. There are, however, other, more elaborate ways to estimate surface area, including from electron micrographs (Smekens et al., 2005). It should also be remembered that the above calculation was for ensembles, i.e. collections of particulate, rather than individual agglomerates, in which lightly touching neighbours may be the dominant factor. That said, the assumption of separateness makes the geometrical calculation an upper limit. The implications of the parallel assumption, namely spherule monodispersity, are less easy to judge.

As emitted, agglomerates are highly flocculent, i.e. capable of considerable compaction; soot recovered from engine oil by centrifuging is denser: 44–84 m^2/g (Rounds, 1981). The distinction between 'dry' and 'wet' particulate is also important, since adsorbed components act to reduce specific areas in two ways: by blocking off pore entrances, and by smoothing out excrescences. This understanding accords with the aforementioned remarks about operating point, since particulate deposits are wetter at light load, and density measurements (Section 6.2.3) strongly suggest open morphologies at high load. Surface areas in an experiment, before and after extraction (Section 5.2.5), were 12 m^2/g and 200 m^2/g, respectively; the same experiment, when carried out on carbon black, showed little change, something easily rationalised in terms of the lower organic fraction carried by these particles, i.e. their drier nature (Dasenbrock et al., 1996). Measurements as a function of agglomerate size have not been reported, save for one study, in which organic-extracted coarse-mode particles gave specific surfaces of 25 m^2/g and 72 m^2/g (Durán et al., 2002).

There is the natural question of where this surface area predominantly resides. Interspherule pores can be estimated by returning to simple geometrical considerations, for which the smallest pore, i.e. the cavity enclosed by three closely packed 20 nm-diameter spheres, is 3 nm (Otto et al., 1980). Strictly speaking, though, a distinction arises in degree of scale between macropores (>50 nm), mesopores (2–50 nm) and micropores (<2 nm) (Rodríguez-Reinoso, 1997). So, the surface area does not relate solely to the obvious voids lying between spherules; whether spherules *themselves*, in contradistinction to the spaces between them, are truly microporous remains an open question for the time being (Braun et al., 2004); but in-engine carbonaceous deposits do seem to hold most of their area between graphene layers in the form of slit-type micropores of size 0.5 nm (Zerda et al., 1999). The total pore volume has been measured at 0.23 cm^3/g and 0.36 cm^3/g (Durán et al., 2002).

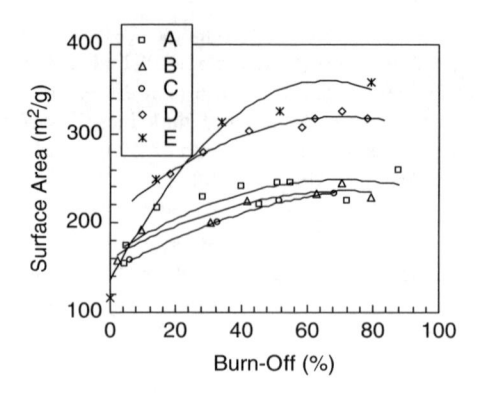

Figure 6.6 Specific surface areas (by use of the BET isotherm) for diesel soot, as a function of gasification (weight loss or burn-off). Vehicle: 2.1 litre. Duty cycle: mixed. Filter: resin-bonded borosilicate glass fibres. Particulate samples were removed by an acetone washing followed by evaporation at 120 °C. Gasification temperature (°C): A, 410; B, 425; C, 495; D, 700; E, 800 (Otto *et al.*, 1980).

In a commonly used extension of surface-area characterisation, measurements are made between stages of burn-off (Stanmore *et al.*, 2001); this process of burning off the surface, or 'gasification', is used more generally to determine some intrinsic reactivity index for carbonaceous materials (Salatino *et al.*, 1994), and to verify soot oxidation models (Peterson, 1987). It also has the effect of opening up the pore structure (Ishiguro *et al.*, 1991): in Figure 6.6 are depicted surface-area measurements for gasification in air (Otto *et al.*, 1980).[2] Carbon dioxide was a stronger gasification agent than water and oxygen; this difference became pronounced as gasification proceeded. This is an interesting, and not inconsequential finding: over and above surface-area characterisation *per se*, burn-off characteristics are extremely important in regenerating DPFs. But these measurements are, with great regularity, conducted on bench-top equipment using simple gas mixtures unrepresentative of the real exhaust, so that *these* specific surfaces, presumably, are quite different from those of most relevance to regeneration in real exhaust environments (Yezerets *et al.*, 2005). This topic has much in common, also, with the oxidation of particulate as a method of fractionation (Section 5.2.4).

6.2.5 Electrical Charge

Charge with an *indigenous* origin, or that conferred naturally by the combustion process, is, in comparison with numerous other characterisational aspects, poorly documented. This neglect is ascribable, but only in part, to the removal, or electrostatic neutralisation, of this charge prior to measurement, since, if left in place, it causes sampling losses and measurement errors. An understandable attitude has thus developed that indigenous charge is irrelevant and bothersome; but it should not be viewed too disdainfully, for three reasons. First, it elucidates the processes of particle formation, as we shall see shortly. Second, it facilitates particle capture where this *is* required: for example, in electrostatic precipitation, the power demand of the corona discharge is lessened when particles are *already* charged (Kittelson *et al.*, 1986a). Thirdly, it furnishes a convenient means of particle measurement via charge probes (Section 5.4.3).

---

[2] Gasification being a surface process, it makes more sense to express it in terms of unit surface area rather than unit weight.

Charge probes, which respond to the whole charge cloud, cannot of course elucidate the demography of particle charges. This is studied, instead, by conducting two measurements of electrical mobility: one in which the indigenous charge is left in place, and the other in the customary fashion (Section 5.3.2), where the indigenous charges are supplanted by a new, *known* distribution (Kittelson et al., 1986a; Kittelson and Collings, 1987). For example, the first unit can be used to select particles according to a certain electrical mobility, after which a second unit measures the range of charges carried by particles in this size class. This practice creates an additional layer of demography – one overlying, so to speak, the demography of particle size, i.e. charge characteristics are superimposed on particle characteristics.

The three salient aspects of charge demography are as follows. (A) The *fraction* of charge-carrying particles, variously reported as 80% (Kittelson et al., 1991) or two-thirds (Collings et al., 1986). The fraction shows a clear size dependency (Kittelson and Collings, 1987); detailed measurements are given in Figure 6.7(a) (Kittelson et al., 1986a). (B) The overall charge *balance*: as reported by charge probes (Section 5.4.3), this is neutral (Kittelson et al., 1991) or positively biased (Schweimer, 1986); or, for light smoke, negatively biased, and, for heavy smoke, positively biased (Collings et al., 1986). Electrical-mobility measurements, some of which are shown in Figure 6.7(b), show a size dependency (negative bias at several hundred nanometres) and temporal dependency (stronger negative bias during blowdown) (Kittelson and Collings, 1987). (C) The *number* of charges carried per particle: this exhibits a bipolar, symmetrical distribution, encompassing five elementary charges (plus and minus) (Figure 6.7(c)) (Kittelson et al., 1986a). An intriguing feature is the systematic, rather than random distribution of average charge, as a function of particle size, according to a power index of $\sim\frac{3}{4}$ (Figure 6.7(d)), ranging from one charge at 40 nm to four charges at 300 nm (Kittelson et al., 1986a).

The possibility that some charge is, in fact, a relic of ion-mediated soot inception (Section 3.2.2) is, apparently, discordant with measurements in hydrocarbon flames (Maricq, 2005). Another possibility is particle charging via 'thermionic emission' (Calcote, 1983), when electron energies exceed the work function (Kittelson et al., 1986a). Such ionisation of a thermal nature is known in hydrocarbon flames, chiefly above 2500 K, the threshold for which falls with increasing particle size; but the same phenomenon is probably depressed by the high pressures inside a combustion chamber (Balthasar et al., 2002b). Modelled charge levels via this charging mechanism, for two temperatures and three particle concentrations, are depicted in Figure 6.7(d) (Kittelson et al., 1986a): these are not altogether unrealistic. Although thermionic emission must occur in the combustion process, it seems not to establish any sustained significance, being restricted to the highest temperatures and particle concentrations; but the chief difficulty with this mechanism lies in its unipolarity, and hence it does not explain the bipolarity observed. However, thermionic emission may still prevail in the early (hottest) stages of the combustion, and merely give way to other mechanisms as the combustion unfolds.

A somewhat stronger contender is bipolar charging, in which particles act as sinks for ions: the formation of ions in hydrocarbon flames is, after all, comfily established (Calcote, 1957), as are the electrical aspects these ions lend to combustion (Lawton and Weinberg, 1969). In premixed-charge gasoline engines, ions are generated in two phases: in the first phase, chemiionisationally in the immediate aftermath of the spark, when chemical reactions between neutral species during growth of the flame kernel generate sufficient energy to ionise the reaction products; and in the second phase, thermally, when the energy generated at the outwardly propagating flame front, and probably through the added action of ongoing compression, heats the burned gas lying behind the flame (Reinmann et al., 1997). In a diesel engine, where combustion is inhomogeneous and occurs in two phases (premixed and mixing-controlled), this demarcation between chemiionisation and thermal ionisation is less clear-cut (Kubach et al., 2004), but ions are nonetheless still generated.

Ionisation of gas molecules creates mainly positive ions (e.g. H_3O^+, CHO^+ and a large range of hydrocarbon polymerics $C_nH_n^+$, including especially $C_3H_3^+$), although, when other molecules capture the ejected electrons, negative ions are also formed (e.g. CH_3O^-, OH^-, O^- and O_2^-) (Griffiths and

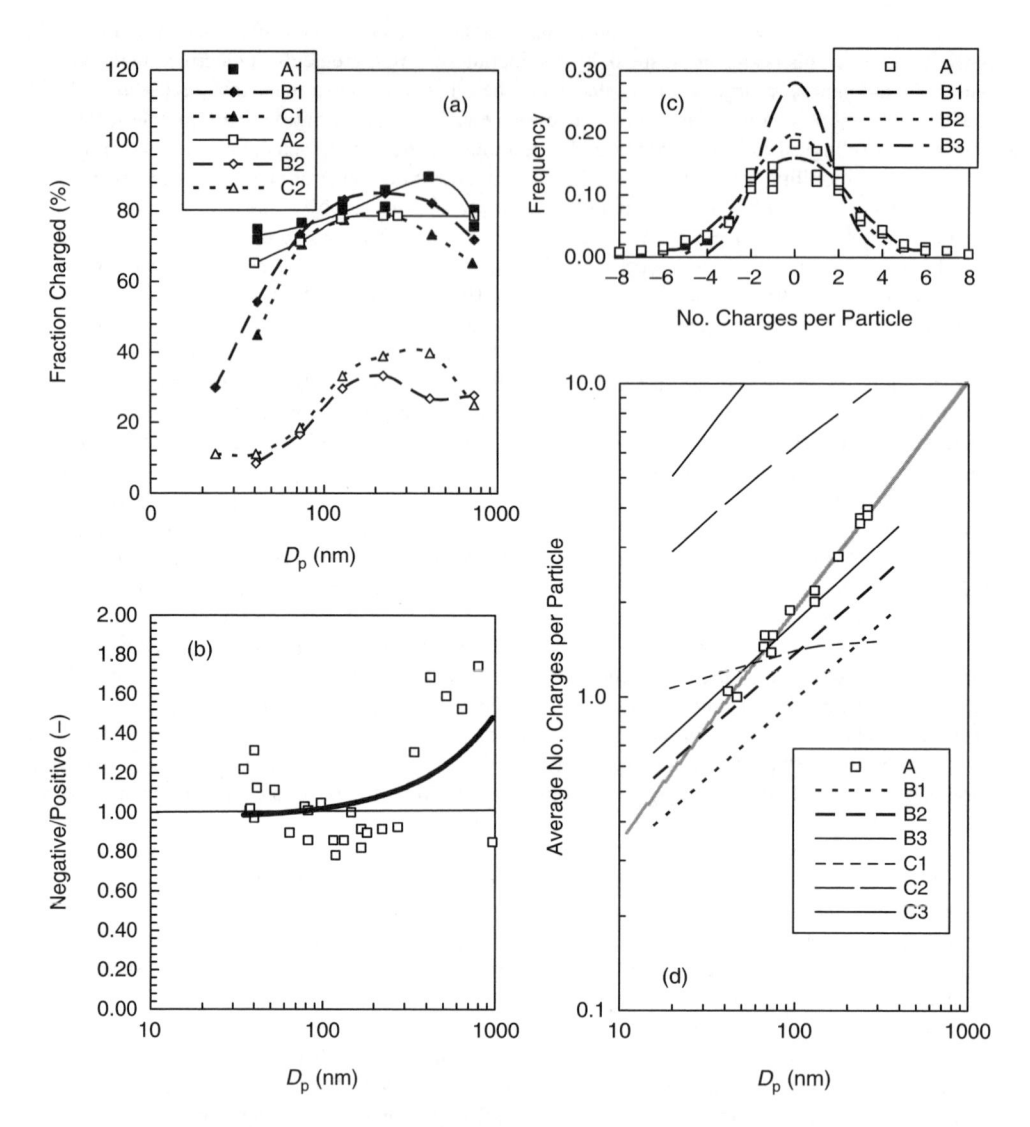

Figure 6.7 Indigenous electrical charge. (a) Fraction of charge-carrying particles. Engines: A, 1.5-litre, 4-cylinder, indirect-injection, naturally aspirated, light-duty; B, 3.6-litre, 4-cylinder, direct-injection, medium-duty; C, 5.1-litre, 6-cylinder, direct-injection, turbocharged, medium-duty. Duty cycle: 1, heavy load; 2, light load (Kittelson et al., 1986a). (b) Overall charge balance. Engine: 2.5-litre, direct-injection, naturally aspirated. Duty cycle: 1050 rpm/full load (Kittelson and Collings, 1987). (c) Charge distribution at 133 nm: A, experimental data; B, three Boltzmann charge equilibria (temperatures: 1, 500 K; 2, 1000 K; 3, 1500 K) (Kittelson et al., 1986a). (d) Average number of charges per particle, as compared with two models: A, experimental data; B, according to Boltzmann charge equilibria (temperatures: 1, 500 K; 2, 1000 K; 3, 1500 K); C, according to thermionic emission (particle concentrations and temperatures: 1, $N = 10^{10}\,\text{cm}^{-3}$, 2100 K; 2, $10^{10}\,\text{cm}^{-3}$, 2300 K; 3, $10^{9}\,\text{cm}^{-3}$, 2300 K). Instruments: CNC, DMA, EAA (Kittelson et al., 1986a).

Barnard, 1995, pp. 112–116). Subsequent events depend on ion lifetime, as ions are rapidly destroyed through recombination reactions; and this is where mobility comes in, as opposite charges must find one another. For example, hydrocarbon flames will bend towards a negatively charged electrode: free electrons (the major negative charge carrier), since they rejoice in much greater mobilities than ions, are rapidly collected by the positive electrode, thus leaving behind a positively biased flame. Similarly, the small, swift-footed ions are neutralised more quickly than the sluggish, heavy hydrocarbon ions. So, even if ion clouds are originally neutral, this neutrality is not necessarily conferred on particle clouds. This question does not seem to have been addressed.

In *smoke detection*, the equilibration of particle charge depends very much on the type of fire: sub-equilibrium (slow-smouldering fires), equilibrium (low-smoke fires) and super-equilibrium (rapid-sooting fires) (Burtscher, 1992, p. 582). This is why, for example, open fires can be distinguished from cigarette smoke. Bell-shaped models of the Boltzmann charge distribution – a bipolar, symmetrical charge equilibrium – are depicted for one particle size and three temperatures in Figure 6.7(c). It should be admitted that the theory assumes particle *sphericity*, and chained agglomerates would probably be able to accommodate higher charge levels. Nevertheless, experimentally observed results correspond reasonably to the theoretical distribution pertaining at 1500 K, even though the tail-ends are somewhat further out – this is reflected in Figure 6.7(d), where the average charge shows a somewhat stronger dependency on particle size than predicted. The average charge observed per particle corresponds, in fact, to lower and lower temperatures in the Bolzmann equilibrium with diminishing size. On the basis of such analyses, it was suggested (Kittelson *et al.*, 1986a) that this charge equilibrium continually shifts as the combustion proceeds, keeping step with falling temperatures – *until* the point where the ion concentrations are too low, such as at 1500 K (the average temperature obtaining shortly after completion of combustion), at which point the charge distribution becomes frozen.

Two features in Figure 6.7 are inconsistent with the notion that bipolar charging by flame ions establishes a Boltzmann charge equilibrium: the charge *distribution* is not symmetrical but biased, and the *average* charge shows a stronger dependency on particle size than expected. These two shortcomings are at their most pronounced between 300 nm and 500 nm, where negative exceeds positive by 3:2; and a temperature of 2600 K would seemingly be required to generate the average charge, as measured – an unlikely scenario, since this is the peak combustion temperature (Kittelson and Collings, 1987). So, although bipolar charging operates, some other mechanism is implicated, at least for the larger particles.

This leads to an alternative conjecture, namely, charging via *surface interactions*, of which there are two strands: (a) contact electrification, or charge transfer when two objects touch, as might happen when particles strike a surface and then rebound; and (b) the thermoelectric effect, or charge carrier diffusion within a material subject to a temperature gradient, as might happen in soot deposits prior to re-entrainment. Regrettably, the implications in either case are not easily predictable. In contact electrification, electrons are transferred to the material with the higher work function; yet iron and carbon are on a par in this respect, and impaction sites may, in any case, be already covered with sooty deposits. In the case of the thermoelectric effect, the charge carriers are not immediately obvious, because the particles are not pure carbon, and ill-described trace elements may be influential, in the same way that impurities (dopants) control the properties of semiconductors.

Nevertheless, such suppositions are intriguing because surface interactions are more likely to be an aspect of the *largest* particles (inertial impaction, and deposition re-entrainment) – those unfaithful to the trends expected in bipolar charging. During blowdown particularly, the sonic gas jet issuing from the gap at the exhaust valve might eject particles from the cylinder at sufficient velocities to strike the walls of the exhaust ports; and the forces of fluid shear, being at their apogee, are more likely to dislodge soot deposits. Evidence exists that the particles ejected during the blowdown phase are in some way different from those ejected during the displacement phase (Section 4.2). Indeed, the negative bias is more pronounced in particles ejected during a 10 ms window centred on blowdown than at later moments in the exhaust stroke (Kittelson and Collings, 1987). Hence, tentative but plausible explanations exist for the particle size dependency in the charge bias, and also for temporal variations in this bias. But why this bias should suddenly 'switch', as a function of particle concentration or smoke level, remains unknown.

Upon entry into the exhaust system, the charge distribution can, potentially at any rate, continue to change, but only as a result of particle *agglomeration*, increasing further the number of charges per particle, but also decreasing, through mutual annihilation of opposites, the total number of charges. Such effects appear to be comparatively small (Kittelson *et al.*, 1986a), so that the charge distribution within the exhaust gas is effectively that which prevails once the particles leave the engine. Later in the emission process, volatile compounds begin nucleating, and, on the basis that these freshly created particles are not exposed to flame-generated ions, the nucleation mode is notionally neutral. Whether sufficient ions still exist in the exhaust to impart charges to nucleation-mode particles, and indeed to promote nucleation in the exhaust stream, is currently under research (Section 3.7.2).

This charge is carried predominantly, according to conventional understanding (Kittelson *et al.*, 1991), by accumulation-mode *soot* particles, rather than nucleation-mode particles (whether volatile or ash) (Jung H. and Kittelson, 2005). Thus, significant numbers of *uncharged* nucleation-mode particles, forming later in the emission process from volatile particle precursors, will dilute this charged fraction at the smallest particle sizes, for reasons unrelated to charges *per se*, as implied in Figure 6.7(a). Low *charged* fractions are therefore seen in parallel with high *volatile* fractions. But the size dependency of the charged fraction seems, though, to relate specifically to soot, notwithstanding this charge dilution effect.

6.3 Chemical Characterisation

An elemental breakdown of the cake is not particularly informative to the baker, since this discards the real items of interest, namely the ingredients, as listed on the recipe; hence, the preferred approach – the focus of this section – is the characterisation of chemical *forms* or *compounds*. As such, chemical compositions accord quite closely with physical appearance, as depicted in Figure 2.13 – but not *too* closely. In the broader picture, the spherules obviously correspond to the carbonaceous fraction, just as the volatile coatings attached to these spherules correspond to the organic and sulphate fractions. In actuality, however, some molecules or functional groups are bound extremely tenaciously to the surface; other elements are encapsulated within spherules; and some material is divided between the nucleation mode and the accumulation mode. Characterisation in the compositional sense is, therefore, very much defined by the analytical tools employed, especially in fractionating the particulate, i.e. the lines of demarcation between the four fractions are imperfectly drawn.

6.3.1 Carbonaceous Fraction

This fraction is aptly named: 'carbonaceous' drives home the illuminating distinction that membership, while dominated by carbon, is in fact open to a minority of dissenting elements. This dissension is problematical, since membership depends just as much on practical limitations in fractionation methods as it does on identifying chemical entities that are genuinely carbonaceous as such. Thus, the question of what is genuinely and what is disingenuously carbonaceous is, in the final analysis, unanswerable – and drawing the line between carbonaceous and organic is a particularly contentious and messy business. Expediently, though, the carbonaceous fraction can be regarded as a system of *five* elements: CHONS. Of these five elements, hydrogen is certainly the second most cited: typically quoted formulae for

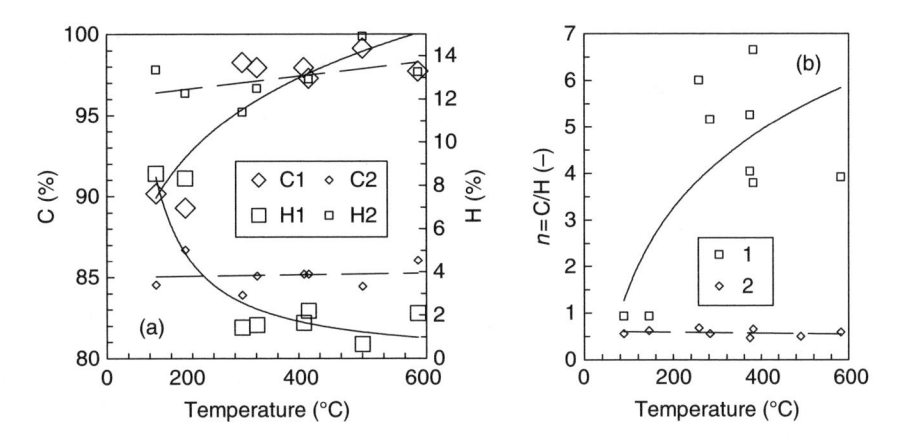

Figure 6.8 Carbon (C) and hydrogen (H) present in particulate: (a) as a mass percentage; (b) as an atomic ratio. Filters: 1, primary (solid line); 2, secondary (dashed line). The primary filter operated at the exhaust temperature; the secondary filter operated at 65 °C; the x-axis indicates the exhaust temperature. Elemental composition as determined during combustion of the particulate in a thermal-conductivity analyser. The particle sizes were <400 nm (Vuk et al., 1976).

the carbonaceous fraction are C_8H, C_9H and $C_{10}H$. These formulae, which, incidentally, are purely empirical, and should not be thought of as anything else, are generalisable as C_xH_y, or rather as C_nH, where $n = x/y$: this expresses the atomic ratio of carbon to hydrogen, and thus serves also as an index of 'carbonaceousness'. Simply, the higher the value of n, the greater is the degree carbonaceousness.

The concept of n is equally applicable to *unfractionated* particulate, that is, it suggests the relative abundance of carbonaceous fraction and organic fraction. For example, samples of particulate were obtained by passing, sequentially, exhaust gas through *two* filters: the temperature of the primary filter varied with that of the exhaust gas, while the temperature of the secondary remained at 65 °C (Vuk et al., 1976). The hydrogen and carbon residing within the particulate deposits were then quantified, as presented in Figure 6.8. The point is that, with increasing temperature, organic compounds undergo less gas-to-particle conversion; so that on the *primary* filter, n increased, i.e. the particulate composition became more carbonaceous and less organic, whereas on the *secondary* filter, n was constant, i.e. the relative proportions of organic and carbonaceous remained similar. It should moreover be noted that, on the primary filter, n increased not just because the percentage of C increased, but also because the percentage of H decreased: this is a logical consequence of the fact that, in the organic fraction, $n \approx 0.5$, as will be explained in Section 6.3.3. Generally speaking, the higher the value of n, the harder, drier or blacker will be the particulate.

Quantitatively speaking, the relative presences of the other elements are poorly reported, but one study places oxygen and nitrogen, respectively, at 5–10 % and ~0.5 % by weight, or at atomic ratios of, say, 200 and 50 with respect to carbon; the sulphur presence is <0.2 % by weight (Collura et al., 2005). Empirical formulae do not, of course, reveal how H, O, N and S are chemically accommodated within the carbonaceous fraction: yet although the microstructure of the soot spherule is now well established as a quasi-graphitic form of carbon (Section 6.2.1), the chemical natures of these dissenting elements are poorly reported. In carbon black, sulphur is chemically combined within the graphitic layers (Taylor R., 1997). But it is possible, indeed more than probable, that H, O, N and S do not actually reside in spherule interiors as such, but rather on spherule *surfaces* (Collura et al., 2005). Indeed, *some* organic compounds are bound extremely tenaciously to spherule surfaces, resisting extraction (Albers et al.,

2002) and evaporation (Hepp and Siegmann, 1998). The fractionation methods are not, after all, perfect: recalcitrant organic compounds are consigned *ipso facto* to the carbonaceous fraction; this is believed to account for anomalous increases in soot across oxidation catalysts (Janakiraman et al., 2002).

That said, evidence of hydrogen lurking *within* spherules, in some undisclosed form, does exist. The index n decreases with depth into the spherule, suggesting two reciprocal trends: decreasing aromaticity and increasing graphiticity (Albers et al., 2002). Subjection of exhaust-system deposits to temperatures of >600 °C in inert atmospheres causes hydrogen to be lost from within the soot (Andrews et al., 1992), perhaps, as suggested in Figure 5.8 (Lepperhoff et al., 1994), through continued dehydrogenation; such deposits are thus aptly labelled a kind of 'petroleum coke'. For particulate partially organic-extracted, n was initially 37; and, following a surface burn-off, where 70 % of the mass was lost, n became 167. But only *part* of this trend towards carbonaceousness was associated with the combustion of adsorbed organic compounds. Quoting from the paper, 'a small portion of hydrogen is bonded to soot carbon and not merely adsorbed as a component of organic molecules on the carbon surface. These hydrogen atoms are presumably left with carbon atoms from the fuel, which have been incompletely stripped during soot formation.'

The co-presence of hydrogen and oxygen in soot, rather than just in adsorbed organic compounds, raises pertinent questions about particulate *oxidation* mechanisms – oxidation is a common characterisational tool in the laboratory, but it also, under realistic but less controlled conditions, occurs, on-vehicle, in a DPF. Hydrogen and oxygen presumably *assist*, to some extent, in the ignition or combustion of this deposit; and there is evidence that greater carbonaceousness impedes oxidation. This subject, however, is virtually untouched, as so-called 'burn-off' studies have focused, instead, on the ash fraction as a catalyst, and the organic fraction as a 'lighter-fluid' (Otto et al., 1980; Müller et al., 2005). Nevertheless, when carbon black is used as a substitute, or surrogate, for diesel soot, the data need careful interpretation, as the former is, relatively speaking, free of oxygen and hydrogen, and this paucity arguably impedes oxidation.

6.3.2 Ash Fraction

The chemistry of ash is extraordinarily diverse; membership lists for this fraction are, perhaps, slightly disingenuous: this is a motley fraction with no consistent or clear-cut following. Some salient elements are phosphorus, silicon, chromium, calcium, copper, iron, zinc and sulphur; some less commonly cited elements are nickel, cadmium, potassium, palladium, platinum, rhodium, cerium, chlorine, sodium, magnesium, aluminium, arsenic and mercury. And, equally logically, generalisations as to relative abundance within this fraction are rendered precarious by the *trace* levels: for example, small variations in fuel contamination by metals from one batch to another or one refiner to another, while of no consequence in terms of engine performance, cause, relatively speaking, extremely large variations in emission rates. There may be neither rhyme nor reason to emission rates: for example, magnesium (a common oil component) may be prominent, and also comparable to silicon (of no unique provenance); whereas zinc (another common oil component) may be barely detectable, and yet comparable to sodium (of no unique provenance) (Sakamoto et al., 1997). On a surer footing, though, when metal-containing fuel additives are used at, say, hundreds of ppm, these metals easily and overwhelmingly dominate the ash fraction.

As already fully explained (Section 3.5.2), metals are found *combined* with spherules, where they confer marked microheterogeneity; but they are accommodated (dispersed) in a variety of ways: it should be stressed that these differences reflect the specific details of formation, e.g. the concentration and decomposition behaviour of the precursor compound, and not just the predilections of the metal. Simplifying somewhat, (a) copper and chromium (Otto et al., 1980) are sequestered in large domains, yet apparently absent elsewhere; (b) iron (Otto et al., 1980) and lead (Peterson et al., 1987) are found in 'nanodomains' or 'nanocrystals'; (c) barium (Otto *et al.*, 1980) is unresolvable, and apparently atomised; (d) cerium is dispersed (Lahaye *et al.*, 1996), yet also present in nanodomains (Campenon *et al.*, 2004).

These nanodomains, or, when encapsulated, 'occlusions', range in size from domination of the host spherule, at ∼20 nm or so, to just a few nanometres – this probably being more a resolvable than a genuine limit. Clearly, these metals are microstructurally accommodated within spherules in *some* way, although a comprehensive study, along the lines of Section 6.2.1, does not seem to have appeared: a useful cue might be taken from materials science, where so-called 'nanoencapsulates' are widely researched for their novel properties (Zhang H. *et al.*, 1998). When diesel fuel is doped with ferrocene, the resulting soot displays a far less graphitic-like microstructure, and the iron is found as maghemite, Fe_2O_3 (Braun *et al.*, 2006).

Metals are also emitted as separate, distinct, noncarbonaceous ash particles, for example supermicron flakes of iron, presumably released from exhaust system walls (Vuk *et al.*, 1976). Especially when organometallic fuel additives are used in high concentrations, ash forms pronounced nucleation-mode particles, as is notably the case with cerium (Burtscher and Matter, 2000) and barium (Kittelson *et al.*, 1978b). These metals subsequently spread themselves throughout the size spectrum, into the accumulation mode, either through agglomeration amongst themselves, or through agglomeration with the carbonaceous particles. Metal nanoparticles, emitted as such and subsequently adhering to the sides of spherules, are less integrated into the carbonaceous matrix than the aforesaid metal nanodomains, co-formed with spherules at the earliest moments. Conversely, when particles contain trace levels of carbon, mingling with a more dominant metal, this suggests late-phase burn-out, as described in Section 3.5.2.

The characterisation thus far described is incomplete, in that ash elements are seldom found in elemental form, but rather are combined; and obtaining this compositional information is necessary if only because the oxidation state determines the environmental repercussions – especially the toxicological repercussions. For example, whether iron is present as Fe(II) or Fe(III) is important, since the former tends to be more soluble and thus more bioavailable. The iron in gasoline particulate seemed to be more soluble than that in gasoline (Majestic *et al.*, 2006).

The compounds are diverse: they include, saliently, oxides, sulphates and phosphates, and, occasionally, chlorides, carbonates and silicates. Combinations of these compound groups (e.g. oxysulphates) are also found, rendering the characterisation still more difficult. The chemical form is an aspect of the aforementioned divided state: for example, cerium, as a *sulphate*, is finely dispersed within spherules, yet, as an *oxide*, is segregated into nanodomains (Vonarb *et al.*, 2005). A boundary problem moreover exists, as ashing elements are occasionally found to reside not just in purely inorganic compounds, but also in organic compounds, for example as organometallics or organohalides. The tendency for the first will of course be obvious if the metal was introduced into the fuel as an organometallic fuel additive, and not fully combusted. Of the *nonmetals*, sulphur and phosphorus tend, since they predominate, to grab the limelight; their poor relation, chlorine, has a fate that, while obscure and poorly reported, is worth highlighting nonetheless, because of the potential for invidious compounds: polychlorinated dibenzo-*p*-dioxins (PCDD) and polychlorinated dibenzofurans (PCDF) (Miyabara *et al.*, 1999). The emission of these compounds is, unfortunately, not predictable from the rate at which chlorine enters the combustion chamber with the fuel and oil (Dyke *et al.*, 2007).

A significant impediment to chemical characterisation, then, is undoubtedly the trace-level complexity of this detritus, with possible compounds multiplying virtually without limit as the fraction is analysed to finer and finer resolutions. One such attempt to systematise this detritus is given in Table 6.1 (Huggins *et al.*, 2000). Obviously, the identification of several compounds is uncertain, as signified by question marks. In summary, the findings are: (a) *sulphur*, most (60–90 %) as bisulphate,

Table 6.1 An analysis of the ash fraction within urban PM10 (collected over 12 months in St Louis, MO, USA) and diesel particulate (collected from heat exchangers of several heavy-duty engines), with some suggestions as to chemical forms (Huggins *et al.*, 2000).

Element	Urban particulate	Diesel particulate
Sulphur	Mostly sulphate; <5 % as organosulphur	Mostly bisulphate (HSO$_4^-$); some sulphate, some organosulphur (thiophene?)
Chlorine	Mixture of organic and inorganic	Mostly organochlorine
Vanadium	Mostly V(IV)	Not detected
Chromium	Cr(III) in spinel, [(Fe,Mg) (Al,Fe,Cr)$_2$O$_4$]	Cr(III) sulphate (Cr$_2$(SO$_4$)$_3$.xH$_2$O)
Manganese	Unknown mixture of Mn(II), Mn(III)? and Mn(IV)?	Mostly Mn(II)?
Iron	>80 % Fe(III), <20 %Fe(II)	No data
Copper	Cu(II) sulphate, CuSO$_4$.xH$_2$O, minor other forms	Cu(II) sulphate, CuSO$_4$.xH$_2$O
Zinc	Mostly sulphate, ZnSO$_4$.xH$_2$O	Sulphate, ZnSO$_4$.xH$_2$O
Arsenic	90 % As(V) (AsO$_4^{3-}$), 10 % As(III)	90 % As(V) (AsO$_4^{3-}$), 10 % As(III)
Bromine	Similar to chlorine (?)	Similar to chlorine (?)
Cadmium	Sulphate or silicate?	Sulphate or silicate??
Lead	Pb(II) in oxygen coordination?	Pb(II) in oxygen coordination?

some (0–30 %) as sulphate, a minor fraction (10 %) as organosulphide (thiophene, C$_4$H$_4$S); (b) *halogens* (Cl, Br), a major fraction as organohalides; (c) *chromium*, almost entirely as Cr(III), probably sulphate; (d) *manganese*, inconclusively identified as Mn(II); (e) *copper*, neither metallic nor sulphidic, but probably as a hydrated sulphate; (f) *zinc*, as for copper; (g) *cadmium*, possibly as sulphate or silicate, but no evidence of Cd(VI); (h) *arsenic*, as As(V) in arsenate, a small fraction (10%) As(III). This last observation is significant, since As(III) is many times more toxic than As(V). These chemical identities differed demonstrably from those in the PM10 generally found in an urban environment, and similarly detailed in Table 6.1.

Studies of Incombustible Ash

One should distinguish very carefully between compositional studies of *as-emitted* ash, and other, far more numerous compositional studies of *incombustible ash*. The two are horses of very different colours: incombustible ash denotes whatever is left after the particulate deposit has burned to completion. Obviously, the *initial* inorganic chemistry is, by this burning procedure, discarded or destroyed, just as is allied and no less important physical information – whether the inorganic material resided within spherules or was emitted separately.

These comments extend to the ash retained, on-vehicle, in a DPF (Sappok and Wong, 2007) – doubly so, as, in service, this ash is subjected to repeated cycles of collection and burning, thus introducing additional factors in chemical composition, these being ideal conditions for solid-state reactions between ash components or between these components and the DPF itself. The DPF is, nevertheless, advantageous in one crucial respect: namely, chemical analyses are not restricted to trace amounts of ash, as the period of collection can be greatly prolonged, say, to tens of thousands of miles. The chemical composition of this ash is dominated, obviously, by the fuel additive, or fuel-borne catalyst (FBC), used to regenerate the DPF, as such additives invariably contain metals. In other respects, the composition is highly variable for reasons already mentioned.

Ash residing in a DPF assumes various forms: early on (no FBC), discrete shiny dots; with increasing mileage, these dots merge into continuous, dense layers (Hardenberg et al., 1987a); white and powdery (no FBC), easily distinguishable from black carbonaceous deposits (Sherwood et al., 1991); uniform coatings (Cu FBC), penetrating the porous walls (Ludecke and Bly, 1984); globular deposits (Pb FBC), on inlet face, housing and walls, without penetrating the walls (Ludecke and Bly, 1984); loose and fluffy powders (Mn and Cu FBC), resistant to compaction (Montierth, 1984).

This ash is composed of various compounds: $PbSO_4$ (Peterson, 1987); $BaSO_4$ (Kittelson et al., 1978b); $CaSO_4$ (Sherwood et al., 1991); CeO_2 (Pattas et al., 1992); $Mg_3(PO_4)_2$, $Zn_2Mg(PO_4)_2$ and Fe_2O_3 (Yumlu, 1988); $Zn(PO_3)_2$, $ZnSO_4$, ZnO and $CaCO_3$ (Nemoto et al., 2004); Fe_3O_4 (Givens et al., 2003); $Zn_2P_2O_7$ (McGeehan et al., 2005); and Na_2SO_4 (Krutzsch and Wenninger, 1992). Mixed forms are also possible, i.e. $Ca_xFe_yO_z$ and $Ca_xCu_y(PO_4)_z$ (Givens et al., 2003). Generally speaking, calcium forms hydrated and nonhydrated sulphates (Manni et al., 2006); zinc and magnesium form phosphates; and iron forms oxides. Again, whenever an organometallic FBC is used, its ensuing metal will always dominate the composition of the ash (e.g. Pb, Ba, Ce and Na in the aforementioned cases); otherwise, what counts are the relative availabilities of *other* ashing elements (i.e. Mg, Fe, Ca, S and P), in which case $CaSO_4$ usually dominates.

6.3.3 Organic Fraction

To describe the organic fraction as a 'complex cocktail' sounds trite nowadays, but any supposed similitude to the alcoholic beverage rests not so much on the bottled ingredients as might be selected by the cocktail waiter, as on the chemical compounds *within* the ingredients, the number of which is obviously unconstrained by any clearly defined upper limit; anecdotal estimates range widely, from a few hundred (Gertler et al., 2002) to a few thousand (Figler et al., 1996), and even to ten thousand (Walsh, 1987). Naturally, only a general guide can be given here to characterisation studies of this mind-boggling melange.

Insofar as the organic fraction consists of particulate-phase *hydrocarbons* – as often pictured – the chief elements are hydrogen and carbon, contained in aliphatics (e.g. alkanes) and aromatics (i.e. PAH). But the broader description 'organic', as used in the present text, expresses the fact that other organically bound elements, besides hydrogen and carbon, are present: these are the 'heteroatoms': as in the case of the afore-described carbonaceous fraction (Section 6.3.1), a common view of the organic fraction is that it consists of five elements, CHONS (Hare et al., 1976): for example, *oxygen* as hydroxyl (R–OH), *nitrogen* as nitro (R–NO$_2$), and *sulphur* as thiophene (C_4H_4S) and sulphonates (SO_2O^-). Some organosulphur compounds are formed in the combustion process, rather than just being unburned fuel remnants (Liang F. et al., 2006). The organosilicon compounds found are possibly antifoaming agents from the lubricant (Collura et al., 2005). Yet another contender, *chlorine* – probably the rarest – might reside in halogenated organic compounds such as dioxins (Lohman and Seigneur, 2001).

Carbon and hydrogen, the two most commonly cited elements, are conveniently assigned an empirical formula of C_nH: it should be noted that this *summation* of the whole fraction says very little about speciation, or the identities of the individual compounds composing this fraction – of which more presently. The suffix n, which expresses the extent of saturation (or of unsaturation), takes a value of ~ 0.5 (Hare et al., 1976): not startling, as this ratio in both diesel fuel and lubricant is very similar: hence, organic fractions, if dominated by unburned fuel or lubricant compounds, have their n-values constrained accordingly. In fact, as already related (Section 6.3.1), the value taken by n in *unfractionated* particulate inevitably suggests the relative abundance of two fractions, insofar as the composition is bracketed carbonaceously on the *upper* side ($n \approx 10$) and organically on the *lower* side ($n \approx 0.5$). Unfractionated particulate will therefore sit somewhere between the two extremes, as decided, say, by

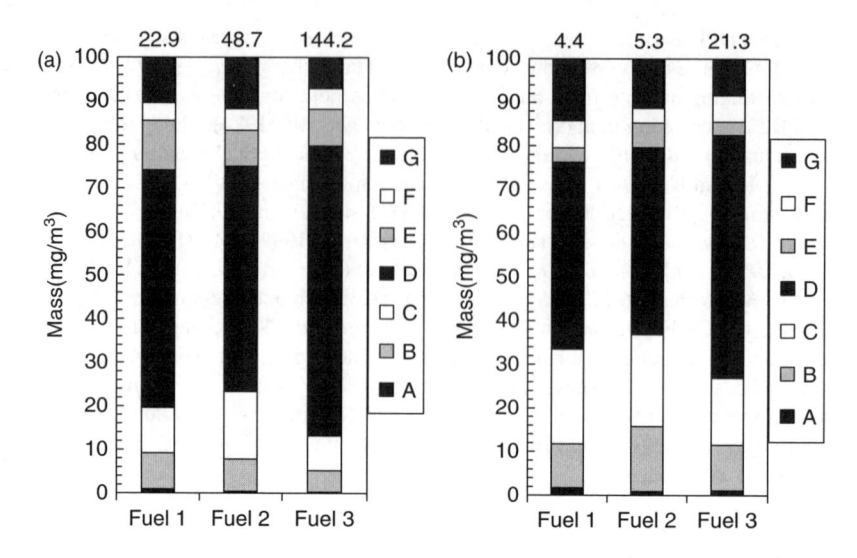

Figure 6.9 A detailed compositional breakdown (by mass) of the organic fraction, according to the seven subfractions outlined in Section 5.2.6: A, basics; B, acidics; C, insolubles; D, alkanes; E, aromatics; F, transitionals; G, oxygenates. The total mass is also given in mg. Engine: V8, naturally aspirated, medium-duty. Duty cycle: EPA '13-mode'. Operating point: (a) Mode 3 (1700 rpm, 25 % maximum load); (b) Mode 5 (1700 rpm, 75 % maximum load). Three fuels (for full specifications, see original paper): T90, fuel 1, 241 °C; fuel 2, 310 °C; fuel 3, 371 °C. Dilution ratio: 8 (Funkenbusch et al., 1979).

collection temperature, insofar as cooler conditions invite more gas-to-particle conversion in organic compounds: for example, at >300 °C, $n \approx 5$, and at <100 °C, $n \approx 1.0$ (Vuk et al., 1976). As a particulate deposit is burned, the trend from oxidation of the organic fraction to oxidation of the carbonaceous fraction is similarly evident in the evolution of CO_2 and H_2O, i.e. progressively lesser amounts of hydrogen are given off, relative to carbon (Müller et al., 2006).

An extensive characterisation, according to the seven-subfraction scheme described in Section 5.2.5, appears in Figure 6.9 (Funkenbusch et al., 1979), from which several interesting observations are worth noting. The alkanes, aromatics, transitionals and oxygenates appeared viscous and oily; the other subfractions were solid. The alkanes were, not unexpectedly, saturated ($n \approx 2.0$); the other subfractions were considerably unsaturated – the ether-insolubles especially so ($n \approx 1.0$). *By proportion*, the ether-insolubles carried most nitrogen, although this element was poorly represented (<2 % by mass) in all fractions; the oxygenates and acidics carried most oxygen (10–20 % by mass); the aromatics and transitionals intermediate levels; and alkanes the lowest levels (<2 % by mass). The presence of greater amounts of oxygen in the organic fraction than in the fuel underscores the prominence of *partial-reaction products*. The presence of such large proportions of alkanes – up to two-thirds – should be noted, since aromatics, because of their mutagenicity, always receive most attention, i.e. the greater percentage of organic compounds are not thought to be particularly bioactive, and so act as diluents for those that are.

The next stage is identification and quantification of actual, coherent compounds, of which Table 6.2 is a representative list of commonly encountered ones; lengthier lists are available elsewhere (e.g. Johnson J.H., 1988) and, ultimately, dozens of compounds are known, and hundreds remain unidentified. Because of their cumbersome names, PAH and NPAH attract foreshortened nomenclature:

Table 6.2 Commonly found or frequently investigated compounds in the organic fraction: polycyclic aromatic hydrocarbons (PAH), nitrated or nitro-PAH (NPAH), and alkanes. Names are often abbreviated, as indicated; these abbreviations are used in this text. Asterisks signify the sixteen 'priority' PAH, as specified by the US EPA.

PAH	NPAH	Alkanes
Acenaphthylene (*Acy*)	**Mononitro**	Cyclohexane
Acenaphthene* (*Ace*)	9-Nitroanthracene (*9-NAnt*)	Docosane
Anthracene* (*Ant*)	7-Nitrobenz(a)anthracene (*7-NBaA*)	Eicosane
Benzo(a)anthracene* (*BaA*)	6-Nitrobenzo(a)pyrene (*6-NBaP*)	Hexacosane
Benzo(a)pyrene* (*BaP*)	1-Nitrofluoranthene (*1-NFlu*)	Hexadecane
Benzo(b)fluoranthene* (*BbF*)	3-Nitrofluoranthene (*3-NFlu*)	Hexane
Benzo(e)pyrene* (*BeP*)	2-Nitrofluorene (*2-NFlr*)	Hexetriacontane
Benzo(ghi)perylene* (*BgP*)	6-Nitrochrysene (*6-NCry*)	Methylcyclohexane
Benzo(k)fluoranthene* (*BkF*)	3-Nitrophenanthrene (*3-NPhe*)	Octacosane
Chrysene* (*Chr*)	1-Nitropyrene (*1-NPyr*)	Octadecane
Dibenzo(ah)anthracene* (*DbA*)		Octane
Fluoranthene* (*Flu*)	**Dinitro**	Pentane
Fluorene* (*Flr*)	1,3-Dinitropyrene (*1,3-DNP*)	Nonadecane
Indeno(1,2,3-cd)pyrene* (*InP*)	1,6-Dinitropyrene (*1,6-DNP*)	Tetracosane
Naphthalene* (*Nap*)	1,8-Dinitropyrene (*1,8-DNP*)	Tetradecane
Perylene (*Per*)		Triacontane
Phenanthrene* (*Phe*)		
Pyrene* (*Pyr*)		

these are not fully standardised, but Table 6.2 provides some italicised versions, as used in the present text. More importantly, 16 so-called 'priority' PAH, as singled out by the US EPA, are marked by asterisks (Smith D.J.T. and Harrison, 1998). *BaP* is frequently viewed as representative of PAH, when more detailed analyses are unavailable; *1-NP* enjoys a similar status among NPAH. (It should be noted that collection methods frequently fail to gather sufficient quantities of NPAH other than *1-NP*). In contrast to PAH, characterisations of which are nowadays too voluminous to review in detail, their straight-chain brethren, the alkanes, are poorly reported – a reflection, no doubt, of the lesser health concerns raised by this compound group.

Insufficient space exists to review the reported emission rates of dozens of organic compounds, as determined by different vehicle technologies, various fuel formulations, disparate duty cycles, etc. Such rates are in any case extremely variable for any of these test conditions. Generalising and necessarily selecting, then, several analyses indicate prominent compounds as follows: among PAH, *Nap*, *Pyr*, *Phe* and *Flu* (Kado *et al.*, 2005; Shah *et al.*, 2005); among NPAH, *9-NAnt*, *3-NPhe* and *1-NPyr* (mononitro) and *1,3-DNP* and *1,6-DNP* (dinitro) (Bamford *et al.*, 2003); and among alkanes, octadecane, nonadecane, eicosane and docosane (Shah *et al.*, 2005). More detailed inventories are available from these and other papers (e.g. Cadle *et al.*, 1999a,b).

Characterisations of the organic fraction according to distributions of molecular weight – that is, according to the number of carbon atoms in the molecule, or 'carbon number' – must take cognisance of *two* shifting boundaries: partitioning between the particulate phase and vapour phase (Section 5.2.3), and partitioning between fuel and lubricant (Section 5.2.6). This dual-domain boundary system is extremely interesting, although one does tend to blench at the *embarras de richesses*: Figure 6.10 is an example (Black and High, 1979). This figure suggests that (a) $<C_{14}$, all organic compounds are fuel-derived, and all are in the vapour phase; (b) $>C_{23}$, all organic compounds are lubricant-derived, and a

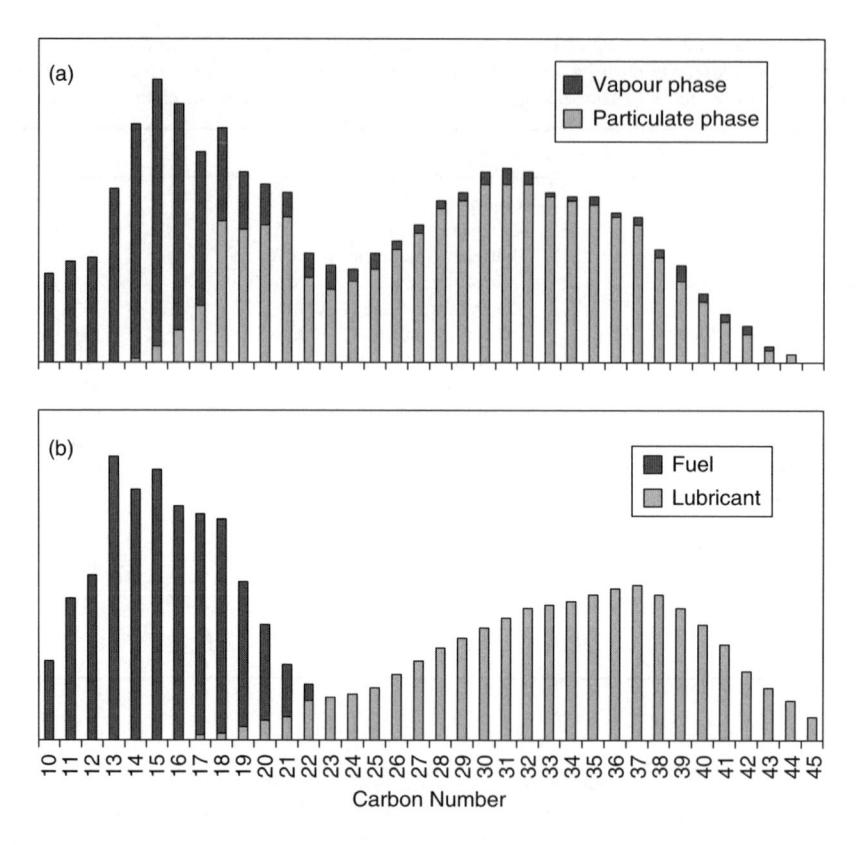

Figure 6.10 Composition of organic fraction: molecular-weight distribution according to (a) vapour phase and particulate phase; (b) fuel and lubricant. The y-axis shows the relative abundance. Collection by polymer traps (vapour) and Teflon-coated glass fibres (particulate); extraction with DCM. Vehicle: 5.5-litre diesel passenger car. Fuel: aromatics, 32.5 %; sulphur, 0.16 % (Black and High, 1979).

small proportion are in the vapour phase; (c) ~C_{13} is the fuel peak; (d) ~C_{15} is the vapour-phase peak; (e) C_{37} is the lubricant peak; (f) C_{31} is the particulate-phase peak; (f) ~C_{23} is the vapour–particulate boundary; (g) ~C_{24} is the fuel–lubricant boundary. The generality of these findings is unknown, since such thorough characterisations are extremely rare in the literature; but data such as exist are supportive (Johnson J.E. and Kittelson, 1994).

When the partitioning of *individual* compounds is examined, the multilayered complexity of organic-fraction characterisation becomes immediately apparent: obviously, this is very much a matter for the specialist and should not be attempted at home. Again, greater attention has been accorded to PAH. Generally speaking, the vapour phase carries compounds of three or four rings, whereas the particulate phase contains those of five or more; but this categorisation is not mutually exclusive, and there is some overlap (Waldenmaier *et al.*, 1990). More specifically, compounds of high molecular weight are predominantly in the particulate phase (e.g. *DbA*, *BgP* and *InP*); compounds of intermediate molecular weight are equally represented in both phases (e.g. *BaA* and *Chr*); and compounds of low molecular weight are predominantly in the vapour phase (e.g. *NaP*, *Ace* and *Flr*) (Pataky *et al.*, 1994; Collier *et al.*, 1998). Corresponding data for NPAH are scant, but compounds are similarly partitioned between the particulate phase (*1-NP*, *3-NFlu*, *6-NC* and *7-NBaA*) and the vapour phase (*3-NFlu* and *6-NC*) (Laymac et al., 1991).

Lastly, several characterisations of the organic fraction, as a function of particle *size* (Phuleria et al., 2006), are worth relating, although such studies are admittedly rather scant. This scarcity bears witness to the fact that plentiful quantities of particulate must first be collected in order to assay for trace species, so that researchers have tended to work on whole or bulk samples only. Ultimately, a generalisation is disingenuous since gas-to-particle conversion is dependent on the interplay between available surface area and ring size, making it context-specific (Section 3.3.3). PAH, especially at low engine loads, tend to congregate on the smaller stages of an impactor (Zielinska et al., 2004), just as smaller agglomerates do seem to be composed of larger fractions of PAH (Leonardi et al., 1992), especially below 150 nm (Westerholm et al., 1999), with *molecular* size possibly being related in some subtle way to *particle* size (Collin et al., 2001). The picture is different still in the case of nanoparticles, i.e. *below* the sizes of spherules, where the composition cannot be carbonaceous, and seems predominantly organic (Kerminen et al., 1997), although these compounds are not, it seems, PAH (Collin et al., 2001), but rather, in their volatilities, they resemble – in fact might possibly be – lubricant compounds (Sakurai et al., 2003a). These various observations point to several size-fractionating mechanisms in organic compounds.

6.3.4 Sulphate Fraction

Compositional characterisation of the sulphates is, at a first level, wholly uncomplicated, because the fraction encompasses not, as in the other fractions, a melange of poorly defined chemical substances in trace amounts, but two prominent and closely defined chemical compounds, i.e. the 'acid–water' complex $H_2SO_4.nH_2O$, for which n is obtained via relationships described in Section 5.2.5. At the next level, however, complications ensue, because, in reality, this acid is hardly inert with respect to other particulate components, or, for that matter, with respect to other exhaust gas constituents.

Metal sulphates are also formed – engendering a disjunctive boundary problem, since these compounds are also ascribable to the ash fraction (Section 6.3.2). The fraction to which sulphated metals are ultimately assigned depends, obviously, on the fractionation *method*, and also on the nature of the sulphate. But, in the face of inadequate experimental evidence, some conjecture is necessary: the sulphate might decompose on heating, releasing sulphur as, say, SO_2 or SO_3, and leaving behind the metal as ash; or it might be insoluble or only weakly soluble, engendering a subtle distinction between 'sulphates' and 'water-soluble sulphates'. In either case, some part of the original acid is able to escape quantification as 'sulphate', via its reaction with the metal: for example, perhaps just half of the sulphates are actually volatile (Johnson J.E. and Kittelson, 1994).

One such compound – appearing regularly in the literature – is calcium sulphate, $CaSO_4$, which may be the predominant nonvolatile sulphate (Johnson J.E. and Kittelson, 1994). This should not be too astonishing, inasfar as limestone is used for flue gas desulphurisation in coal combustion (Li F. et al., 2006). The acid is not, however, involved: $CaCO_3$ decomposes to CaO, which then reacts directly with SO_2 to form $CaSO_4(s)$. The same route goes unreported in motor vehicle exhaust gas. $CaSO_4$ appears on dilution tunnel filters whenever high exhaust temperatures and high engine loads act in combination: a clear sign that loss of lubricating oil (the supply of calcium) and catalytic oxidation (conversion of SO_2 to H_2SO_4) are both essential prerequisites; the sulphate is not formed, for example, at the same operating point if the catalytic converter is removed. Similarly, oil consumption, if excessive, and occurring in parallel with the combustion of high-sulphur fuels, leads to $CaSO_4$-mediated blockage of a DPF (Mayer A. et al., 1993). This in-catalyst deposition is quite different from mere condensation of the precursor, namely sulphuric acid, as CaSO$_4$ is less readily driven off by high temperature.

The size range within which sulphates are emitted is not well documented. Take-up of sulphuric acid by carbonaceous surfaces is, perhaps, unlikely, if freshly emitted particles are hydrophobic, as is thought. The general picture of gas-to-particle conversion emerging at the time of writing, and detailed in Section 3.7.2, is that sulphuric acid acts as the dominant nucleating agent, and, as such, is responsible for pronounced late-forming nucleation modes. Sulphur and calcium, for example, reside together in the nucleation mode (Gertler et al., 2002, p. 29). If present, however, in exceptionally high concentrations, sulphuric acid emerges as an aerosol of liquid droplets – well into the accumulation mode. Some spherical particles of a few hundred nanometres in size that have been observed under the electron microscope are thought to be sulphates (Chatterjee et al., 2002).

6.4 Biological Characterisation

The discovery that organic extracts of diesel particulate, i.e. compounds within the organic fraction, cause bacterial mutations was significant (e.g. Griffin et al., 1981, p. 17), and, since then, this result has been repeated by workers far too numerous to credit in full (e.g. Mogan et al., 1985; Draper et al., 1988, Opris et al., 1993; McDonald J.F. et al., 1995). The bacterial mutagenicity, therefore, of the organic fraction is mainstream biomedical science: indisputable, comfortably established, and widely accepted as such. The touchstone in such investigations is the bacterial mutagenic bioassay known as the *Ames test*, as detailed in Section 5.2.7, to which the present section is restricted, as other bioassays, and more detailed discussions of mutagenicity, are reserved until Chapter 10.

Nor does the present section seek to address bacterial mutagenicity as affected by the technology of emission control: these matters are discussed separately when they arise (principally in Chapter 7). But the implications of *aftertreatment* are certainly worth emphasising at this point, because vapour-phase organic compounds *also* induce bacterial mutations – this effect is comparable with, and sometimes more than, that of particulate-phase organic compounds (Bagley et al., 1991). Hence, the DPF (Section 7.8.2.), which preferentially captures that which is solid in the exhaust stream and thereby alters gas-to-particle conversion significantly, correspondingly alters also the partitioning of mutagenicity between vapour and particle. For this reason, studies of particulate-phase mutagenicity should, ideally, go hand in hand with studies of vapour-phase mutagenicity, otherwise the results may well be misleading.

That *increases* in mutagenicity regularly go hand in hand with *decreases* in mass (Hyde et al., 1982; Draper et al., 1988; Bagley et al., 1991) is certainly intriguing. Why this perversely oppositional trend should reappear across numerous forms of emission control technology is unknown, although in specific instances, such as when oxidation catalysts are fitted, the conversion of less mutagenic PAH into more mutagenic NPAH is justly suspected (Sections 3.6.2 and 7.8.1). Whether or not, environmentally speaking, the smaller mass truly compensates for the larger mutagenicity, remains a moot point.

Little investigated is mutagenicity as a function of particle size: samples collected from an impactor above 100 nm were an order of magnitude more mutagenic than those below 100 nm (McMillian et al., 2002). This difference, which seems sufficiently large to be unequivocal, remains unexplained: it would be convenient to ascribe it to more mutagenic carbonaceous agglomerates, coated with organic compounds, and less mutagenic organic droplets, lacking a carbonaceous core, but further research is required.

It is interesting, and indeed necessary, to explore how *vehicle operation* affects mutagenicity. This question is difficult to answer because of the fragmentary nature of the data. For example, mutagenicity at light load was less than half that at full load (Barale et al., 1992). The particulate emissions from 21 production vehicles, acquired from a range of manufacturers, were evaluated mutagenically (Hyde et al., 1982). Mutagenicity was not consistently affected by mileages spanning tens of thousands of miles; in *Phase 1* of the Federal Test Procedure (FTP), that is, the cold start, the mutagenicity was markedly worse than for the hot-start counterpart of that phase, *Phase 3*, and the type of test cycle exerted a strong influence: for example, the New York City Cycle (NYCC) generated particulate of noticeably lesser mutagenicity than others. Another, more recent study highlighted greater mutagenicity for intermediate engine speeds and high loads (McMillian et al., 2002). It will be apparent that diverse aspects of vehicle operation (duty cycle particularly) could induce considerable variations in mutagenicity, for which little data exist at the present time.

The roots of these trends lie in subtle speciative alterations to the organic fraction, and, in fact, only a small component of this fraction is actually thought responsible for the mutagenicity: that is, the other compounds act as *diluents*. Quantifying the mutagenicity of the fraction in its entirety is not, though, quite the same as identifying the principal mutagenic agents: obviously, this issue is greatly complicated by the number of organic compounds, of which there are probably many thousands, and also the mutagenic synergisms between these compounds. Evaluated according to *polarity*, three-quarters of the mutagenicity was carried by 'moderately polar' compounds, and the remainder by highly polar compounds (Johnson J.H., 1988); evaluated according to the seven-subfraction scheme (Section 5.2.5), and as depicted in Figure 6.11, transitionals, acidics, ether-insolubles, oxygenates and basics were the most mutagenic, while hexane-insolubles, aromatics and alkanes showed little activity (Campbell J. et al., 1981). Hence, Figure 6.11 points not so much to aromatic compounds lacking nonpolar functional groups as to aromatic or aliphatic compounds containing ether groups, ketones and aldehydes (Funkenbusch et al., 1979).

The mutagenic finger of blame, though, is repeatedly pointed at two chief compound groups: PAH, and their derivatives NPAH; and, within these groups, certain compounds are most certainly more mutagenic than others; indeed, lists of relative ranking have even been drawn up (Degobert, 1995, p. 72–73). The quantity of certain specific compounds is a more useful metric than the total quantity

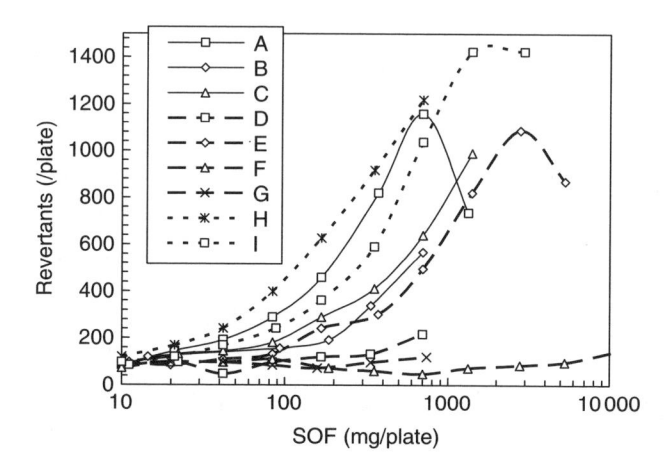

Figure 6.11 Dose-response curves for the SOF components. Ames tester strain TA100. Engine: 11.0 litre, six-cylinder, direct-injection, turbocharged, Fuel: saturates, 72.3%; alkenes, 2.9%; aromatics, 24.8%; cetane, 52.0; sulphur, 0.26%. Components: A, acidic; B, basic; C, ether-insoluble; D, hexane-insoluble; E, oxygenated; F, alkanes; G, aromatics; H, transitional; I, SOF (Campbell J. et al., 1981).

of PAH. Mutagenicity (revertants per km) did not correlate with the summed emission of 14 PAH, but did correlate well with four compounds individually (*Chr, BbF, BaA* and *BbF*) (Rantanen *et al.*, 1996). NPAH are widely held to be more potent mutagenically than PAH. The 'moderately polar' compounds, owing to the presence of NPAH, were the most mutagenic (Dorie *et al.*, 1987). For practical reasons, *inter alia* that very little material exists on which to conduct characterisations, *BaP* is often used alone to suggest the mutagenicity of PAH as a whole: this practice is questionable insofar as other compounds exhibit order-of-magnitude greater carcinogenic or mutagenic potency (Okona-Mensah *et al.*, 2005); a relatively recently identified NPAH of exceptionally high mutagenicity is *3-nitrobenzanthrone*, which forms readily from *benzanthrone* (Enya *et al.*, 1997). Finally, it should be noted that *gas*-phase PAH or NPAH may actually dominate the particulate phase in mutagenicity (Rijkeboer and van Beckhoven, 1987).

6.5 Demographic Characterisation

Demography, or the study of population statistics, might conceivably be applied to various particle properties; but the present discussion is devoted solely to *mass* and *number*, properties for which undoubtedly the greatest corpus of information exists at this time. Indeed, since the mid 1990s the database has rapidly become stupefyingly large, in both the light-duty and heavy-duty sectors. It is utterly impossible, in a just few hundred words, to cover comprehensively these *embarras de richesses*, and there is a desperate need to avoid cluttering the narrative with jungles of technical detail: the problem is that direct quantitative statements are extremely context-dependent (e.g. Holmén and Qu, 2004), and it would be tedious to state, at each and every turn, that so-and-so measured so many particles of a certain size, for this or that duty cycle, fuel formulation or power train design, and at a certain certification level or vehicle age. But, on the other hand, some mention of particle masses and numbers likely to be discharged from motor vehicles should be made, and, if this accompanying context is not provided, such statements risk misleading the reader.

To enforce some concision, and to avoid drawing up lists and tables of great prolixity, only the general trends in particle demography, according to operating point, duty cycle or driving mode, are discussed. Moreover, power trains currently on the market, or at any rate near-market models or prototypes, are preferred, as extrapolation to public exposure is then more certain. It should also be said that groups of vehicles selected from production, as part of an audit or borrowed from members of the public for in-use compliance tests, show statistical variations that are overlaid, in turn, on particle demography; and insufficient space exists to discuss these variations. To quote a simple average emission or a range for a sample of vehicles can be grossly misleading, since ensembles are often grossly distorted by ill-behaved individuals. Similarly, there are always trends according to vehicle age, not only because statutory limits have gradually decreased over the years, but also because of mileage ('wear and tear'). Last but not least, there are variations in hardware: the only additional factor covered here is the DPF, the reason being the profound demographic impact of this device. (The underlying principles of the DPF are, however, reserved until Section 7.8.2.) Similarly, other aspects of the power train of relevance to demography (e.g. fuel formulation and fuel injection) are also reserved until Chapter 7.

Another difficulty occurs with the choice of demographic metric: to name the principal ones, emissions per unit distance travelled, emissions per unit volume of exhaust gas, emissions per unit mass of fuel burned, emissions per unit of work done by the engine, emissions per unit time, and emissions as a function of air–fuel ratio (Färnlund *et al.*, 2001). Unsurprisingly, these metrics produce graphs of slightly or rather markedly different forms (Lapuerta *et al.*, 2000). The chosen metric, naturally, reflects the preferences or specific concerns of researchers; and comparisons are difficult, as papers often fail to provide sufficient information for easy interconversions. But one metric alone cannot contain all relevant information: for example, if the particulate emission increases with vehicle speed, this metric does not, on its own, say whether this increase arises from some intrinsic aspect of the combustion, or whether the output of exhaust gas has simply increased.

At the same time, it seems unfair to castigate researchers thus, since most particle characteristics are not subject to statutory constraints of any sort, by which some commonly agreed-upon metric might become established as 'standard'. The choice, often, takes its cue from homologation: particulate *mass* is totalled for the whole of the test cycle, and, for light-duty diesels, divided by the distance travelled, to give an average emission of mg/km or mg/mile; or, for heavy-duty vehicles, the equivalent of mg/kW h. Enumerative regulations might logically be expressed similarly: number of particles per km, or number of particles per kW h. These metrics are, however, whole-cycle *averages*: nothing is said about the instantaneous or second-by-second nature of the emission as the cycle unfolds. Moreover, the number of particles per unit volume (cm^3) of exhaust gas seems a more *intrinsic* measure of the emission, i.e. irrespective of engine size or vehicle inertia class.

The mass emission is a trite topic insofar as statutory limits are *already* in place, so that 'characterisations' would simply be restatements of these limits. But characterisations of mass are still pertinent, because statutory limits are obviously not universals: they apply only to certain closely defined statutory (homologation) duty cycles, e.g. the Federal Test Procedure (FTP) (in the USA) and the New European Driving Cycle (NEDC) (in the EU); and thus they offer no direct verification of emissions performance in the field, where *customer* duty cycles and driving patterns vary so widely, although, naturally, the intention is for legislated cycles to act as surrogates for the real world. With this caveat in mind, the statutory test cycle evaluations to have been reported for light-duty vehicles, with model year or homologation level specified in parentheses, run as follows. (A) NEDC *without* DPF: (a) one vehicle (Stage III) and four fuels, 15–25 mg/km (Andersson *et al.*, 2001); (b) six vehicles (Stage III) and four fuels, 20–40 mg/km (Ntziachristos *et al.*, 2004b). (B) NEDC *with* DPF: (a) one vehicle (Stage III) and four fuels, <5 mg/km (Andersson *et al.*, 2001); (b) five vehicles (Stage III) and four fuels, 0.2–8 mg/km (Ntziachristos *et al.*, 2004b).

In these studies, all vehicles stayed below their prescribed certification limit, namely Stage III (50 mg/km); indeed, even some non-DPF cases showed reasonable margins. So, all in all, mass emissions from non-DPF light-duty vehicles were on the order of a few tens of milligrams per mile. The installation of a DPF provides considerably greater margin, since emission rates then decline to just a few milligrams per kilometre – on a par with gasoline engines (Sections 8.3.2 and 8.4.2), and paving the way for Stage IV and Stage V. Field performance, or off-cycle driving, may, however, be quite different. The same non-DPF vehicles, driven on a 'motorway' driving schedule, emitted at 20–150 mg/km (Ntziachristos *et al.*, 2004b): this range is quite intriguing, since some vehicles actually fared better than on the NEDC, and some worse. This shows that the mass of particulate discharged per unit distance is not a simple function of exhaust flow rate. With the DPF vehicles, the picture is more complicated still: these were also evaluated on a 'motorway' schedule, and emitted at rates of 20–300 mg/km (Ntziachristos *et al.*, 2004b) – showing considerable exceedances of the nominal standard in some instances (although not all). High rates are transient, caused by the release of stored volatile components in the DPF, although it should be stressed that exhaust and sampling-line walls may also have the same effect. But it certainly seems that although particulate mass is not infrequently curtailed considerably by the DPF, the increased variability brought by the take-up and subsequent release of volatile particle precursors does not always lead to the intended effect.

The *distribution* of this mass as a function of particle size is certainly of interest: this is *not* subject to any statutory constraint, it receives far less attention than the number distribution, and an interesting question is how efforts to restrict *total* mass have affected the distribution of this mass. Prominent accumulation-mode particles, accompanied by somewhat less prominent coarse-mode particles are the common observation. Evaluations of light-duty engines (Sirman *et al.*, 2000), light-duty vehicles

(Aakko and Nylund, 2003) and heavy-duty engines (Fanick *et al.*, 2001), with no DPF, on both transient and steady-state duty cycles, show that this mass resides predominantly between several tens of nanometres and one or two hundred nanometres. Mass distributions for two heavy-duty diesel engines, one homologated to Stage II, the other more stringently to Stage III, are given in Figure 6.12(a) (Hall *et al.*, 2000). Interestingly, the engine satisfying the more stringent legislation appears to do so

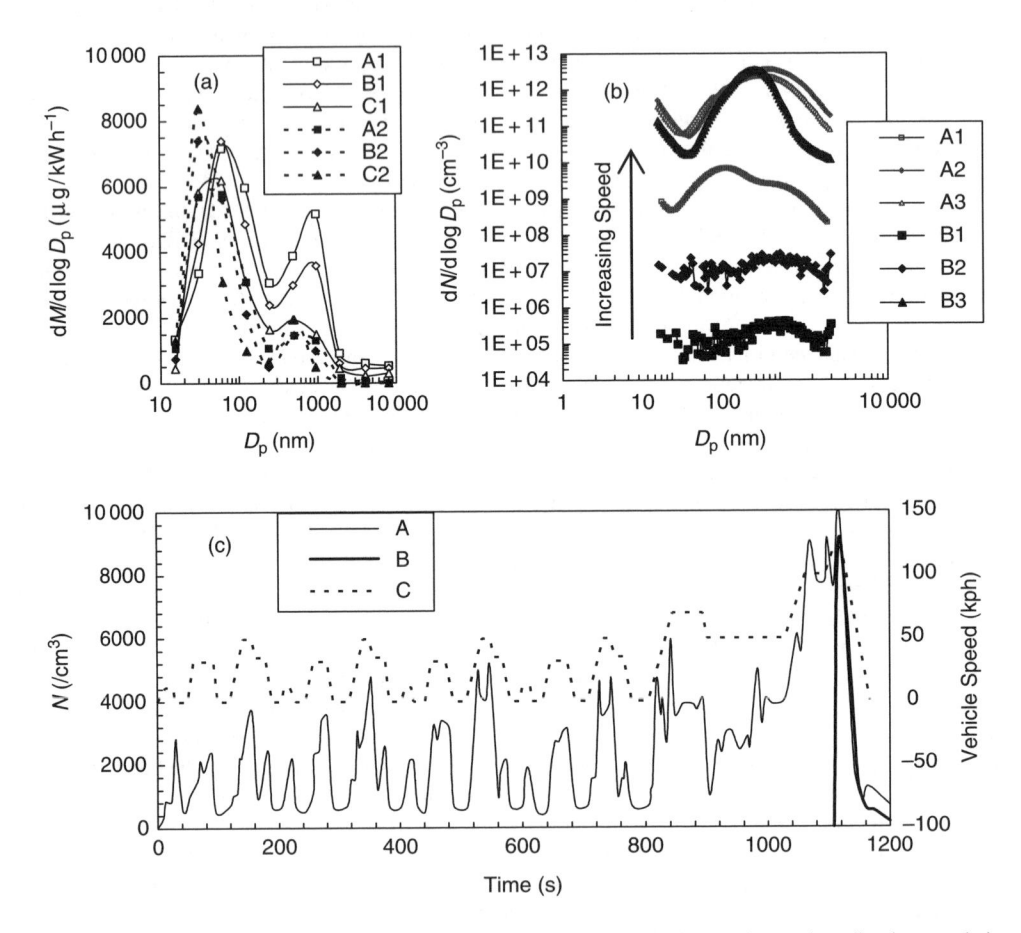

Figure 6.12 Particle demography for diesel engines. (a) Mass distribution for two heavy-duty diesels; cumulative for the whole drive cycle. 1, Stage II (test cycle, ECE), 7.3-litre, 6-cylinder, in-line, no EGR; 2, Stage III (test cycle, ESC), 9.2-litre, 6-cylinder, in-line, with EGR. Fuels (cetane number, T95 (%), sulphur (ppm), aromatics (%)): A, 50.5, 367, 498, 33.8; B, 52.2, 338, 418, 21.0; C, 59.4, 283, <1, 6.7. Instrument: Berner low-pressure impactor, >16 nm (Hall *et al.*, 2000). (b) Number distribution for steady-state operation. 1, idle; 2, 50 kph; 3, 120 kph. Vehicles: A, Stage III, 2-litre, passenger car (with catalyst); B, Stage III, 2.2-litre (with catalyst, DPF and fuel-borne catalyst). Fuel: aromatics, 21.2 %; cetane number, 52; T95, 340 °C; sulphur, 300 ppm. Instrument: SMPS, 7–320 nm (Andersson *et al.*, 2001; data courtesy of J. Andersson, Ricardo Consulting Engineers). (c) Transient or *instantaneous* number emissions at 70 nm, as measured on the NEDC. A, without DPF; B, with DPF; C, vehicle speed. (DPF silicon carbide, aged to 80 000 km; cerium fuel-borne catalyst). Vehicle: nonproduction, 2.2-litre, four-cylinder. Fuel: aromatics, 24 %; cetane number, 52; T70, 294 °C; sulphur, 289 ppm. Instrument: SMPS (Jeuland *et al.*, 2004).

predominantly by emitting a less pronounced coarse mode, although there are also, perhaps, indications of a more sharply peaked accumulation mode. Characterisations conducted with a DPF may be affected by the aforementioned releases of particle precursors. But, generally, a DPF eliminates the coarse mode and, although significantly attenuating the accumulation mode, shifts it to below 100 nm (Andersson et al., 2001).

With enumerative characterisations, reliable and consistent measurements of the smallest particles are for many reasons fraught with difficulty: these measurements are strongly dependent on the specific sampling conditions, as we saw in Chapters 3 and 4. Furthermore, because nucleation-mode particles emerge below and then grow rapidly into the measurement range, the *minimum size threshold* of the instrument must be borne in mind in the reported particle count: particles below this range are obviously overlooked. Particle numbers, therefore, since they are frequently and overwhelmingly dominated by these smallest particles, vary widely, and should be assessed with the utmost caution – this is the case, not just for nucleation-mode particles, as they appear in number *distributions*, but also for total particle *counts*.

With this caveat in mind, the statutory test cycle evaluations to have been reported for light-duty vehicles, with model year or homologation level specified in parentheses, run as follows. (**A**) NEDC *without* DPF: (a) two vehicles (Stages II and III) and three fuels, $2 \times 10^{14} - 4 \times 10^{14}$/km (Andersson et al., 2001); (b) two vehicles (model years 1996 and 1999), $\sim 1.5 \times 10^{14}$/km and $\sim 9 \times 10^{14}$/km (Aakko and Nylund, 2003); (c) six vehicles (Stage III) and four fuels, 5×10^{13}/km-3×10^{14}/km (Ntziachristos et al., 2004b); (d) two vehicles, 8×10^{13}/km and 3×10^{14}/km (Mohr et al., 2003a); two vehicles (c. Stage IV), 5×10^{13}/km and 6×10^{13}/km (Bosteels et al., 2006). (**B**) NEDC *with* DPF: (a) one vehicle (Stage III) and three fuels, $2 \times 10^{13} - 9 \times 10^{13}$/km (Andersson et al., 2001); (b) five vehicles (Stage III) and four fuels, 2×10^{10}/km-5×10^{12}/km (Ntziachristos et al., 2004b); (c) one vehicle (c. Stage IV), 1×10^{10}/km (Bosteels et al., 2006).

From these studies, it would appear that non-DPF light-duty vehicles emit in a range of about 5×10^{13}/km to 5×10^{14}/km. As with mass, it is instructive to compare performance on statutory test cycles with other forms of driving: the same vehicles running on a 'motorway' schedule emitted at $1 \times 10^{14} - 2 \times 10^{15}$/km (Ntziachristos et al., 2004b), indicating still greater particle numbers. As before, the installation of a DPF is, on occasion, highly obfuscating: the aforementioned results for the NEDC show wide ranges; this is the result of intermittent storage and release of particle precursors, a problem which worsens as vehicle speed increases. The same vehicles on a 'motorway' driving schedule, for example, emitted at $9 \times 10^8 - 2 \times 10^{15}$/km (Ntziachristos et al., 2004b) – a very large range indeed. So, a DPF vehicle, certified to a more stringent mass-based standard, may, on occasion, emit particles in greater numbers than a non-DPF vehicle, certified to a less stringent mass-based standard.

Distributions of particle number for two vehicles operated at three steady-state test conditions are shown in Figure 6.12(b) (Andersson et al., 2001). The general effect of increasing vehicle speed is to raise the whole distribution in a manner not obviously size-dependent, although there is, admittedly, evidence of increasing intensification in the nucleation mode. At idle, approximately four orders of magnitude separate the DPF vehicle from the non-DPF vehicle, in both the nucleation mode and the accumulation mode; but this gap closes with increasing speed, until, at 120 kph, the two vehicles are approximately on a par, emitting particles within a band of $10^{11} - 10^{12}$cm^{-3}, irrespective of size: as already related more than once, this behaviour is adduced to hotter conditions, which release volatile particle precursors. At 120 kph, only for particles of size >80 nm does the DPF vehicle retain some advantage.

An example of instantaneous or second-by-second emissions for particles of size 70 nm, as measured on the NEDC, is given in Figure 6.12(c): for the same vehicle, the cases of a DPF and no DPF are compared (Jeuland et al., 2004). The number emission is strongly dependent on vehicle speed, although close examination of the trace reveals broad peaks, on which spikes are superimposed. The broad peaks are associated with cruises, and the spikes are associated with accelerations and decelerations either side of the cruises – a pattern broadly also observable in the number of particles per second (Aakko and Nylund, 2003). Baseline emissions at engine idle are $\sim$600 cm^{-3}, while peak emissions reach 4000 cm^{-3} in the urban drive cycle (UDC), and 10 000 cm^{-3} in the extra-urban drive cycle (EUDC). For the DPF vehicle, the linear y-scale obscures information for most of the trace, but, towards the close of the EUDC, a spike of 10 000 cm^{-3} is emitted – a figure on a par with the non-DPF vehicle. Again, this spike is adduced to an increasingly hot exhaust system, which releases volatile particle precursors that subsequently nucleate. The release of volatile particle precursors in the first minute or so during system warm-up is also known (Andersson *et al.*, 2001). Clearly, transient releases considerably inflate particle numbers from what might be expected purely by examination of steady-state performance.

6.6 Closure

The characterisational studies in this chapter were dissected according to their four principal fields. In the first section after the introduction we looked at microstructure, beginning at the atomic level, and concluding with the basic building block, or spherule; this then progressed to how the spherules are built into agglomerates, and the various morphological features that result. In the next section the chemical composition was described according to the four principal fractions: carbonaceous or soot, organics, ash, and sulphates. The subsequent section (Section 6.4) examined the bioactivity of particulate, as defined by its effects on bacteria, and expressed as 'mutagenicity'; this serves as a prelude to the biomedical material reserved until Chapter 10. The final section (Section 6.5) covered demography, and described size distributions in terms of particle numbers and masses.

6.6.1 Physical Characterisation

Microstructure

The carbon atoms are arranged in face-centred hexagonal arrays called 'platelets'. Several of these platelets are laid one on top of another to form 'crystallites'. A few thousand of these crystallites form a 'spherule'. The degree of order in the microstructure depends on position in the spherule. In the outer regions, the crystallites run tangentially to the surface; in the inner regions, the orientation is random and disordered. The structure thus ranges from graphite-like to amorphous. Microstructural investigations suggest that the inner cores are of a different nature, perhaps betraying the progenitor nuclei.

Morphology

The spherule sizes are normally distributed, or nearly so; they are weakly dependent on engine operation. The number of spherules in an agglomerate ranges from a handful to a few thousand. The degree of interconnectedness exhibited by neighbouring spherules is thought to reveal physical states (tarry or solid) at the moment of collision. Spherule size does not seem to bear any relationship to agglomerate size. Various data-processing methods are used to characterise the complex morphology; today's research employs the latest ideas about fractality. Fractal dimensions in electron micrographs are below 2, usually around 1.8. Fractality is not an obvious function of engine operating point. The underlying agglomerate morphology is disguised by the take-up of volatiles, especially when this take-up is extensive, shifting morphologies toward sphericity. Some volatile material, thought to be organic in nature, is able to survive at least for a short time under the electron microscope: this appears as liquid-phase droplets.

Density

Bulk density should not be confused with effective density – the latter includes gas-filled pores and voids; the two are the same for a perfect sphere. The ash fraction (metal compounds) can easily dominate the density. Nucleation-mode particles are more likely to be governed by the bulk density. Ensemble estimates range from 0.8 to 2.4 g/cm^3, depending on engine operating point; the density tends to decline with increasing engine load, suggesting a trend towards dry, open structures. This finding is not immediately obvious, since the carbonaceous fraction, which dominates at high load, has a higher bulk density than the organic fraction, which dominates at low load. Considerable controversy rages about particle density as a function of size, because particle-sizing instruments are purely enumerative, and convert number distributions into mass distributions via several sweeping assumptions, thereby introducing large errors, e.g. because of nonsphericity and variable effective density. Something widely established, however, is that density falls with increasing particle size; this has been adduced to the larger voidal spaces enclosed by agglomerates. The entry of volatile material into these voids tends to reverse this effect.

Surface Area

Estimates of surface area based on simple geometrical considerations, i.e. derived from idealised, isolated spherules, suggest values of 100–150 m^2/g; some experimentally determined values are within this range, but others are lower. But it seems that almost the entire spherule surface is available to the gas phase – this does not accord with the close bonding that is commonly observed under the electron microscope. Interspherule pores are not thought small enough to distort surface area measurements by restricting the rate of inward diffusion of adsorbing gas. Some surface area may reside within porous spherules, rather than between spherules.

Electrical Charge

Indigenous charge is poorly studied in comparison with other aspects of the particulate emission. The general understanding is that these charges are associated with soot. Particles may carry anything up to several elemental charges; the distribution is bipolar; some are also neutral. The charging mechanism is not known conclusively, but could be chemiionisation, thermionic emission, diffusion charging by flame ions, some remnant of nucleation, contact electrification or the thermoelectric effect. Calculations suggest that this charge distribution becomes frozen when combustion temperatures fall below 1500 K; this happens when the concentration of gaseous ions is insufficient to maintain Boltzmann equilibrium. The number of charges is a function of particle size, and may follow some power index. Some studies show overall neutrality of the particle cloud, others a small imbalance. Nucleation-mode particles that form later in the exhaust system are reportedly uncharged, but the database here is slender – the charged fraction seems lower when volatiles have nucleated.

6.6.2 Chemical Characterisation

Carbonaceous Fraction

Soot is what corresponds most closely of all to the *spherules*. Carbon predominates, but the fraction is conveniently conceived of as a system of *five* elements: CHONS. The atomic ratio of carbon to hydrogen is ~9. The chemical form taken by the other four elements is not reported; they may be present as organic compounds that have resisted the separation process, but some residency *within* the spherules is not ruled out.

Ash Fraction

This fraction is primarily metals, e.g. Cr, Ca, Cu, Fe, Zn and Na; some nonmetals are also found, e.g. P, S and Cl. These elements confer considerable microheterogeneity on spherules, but, given their great diversity, the ash fraction is not emitted in any consistent form: it may exist as occlusions within spherules, on the surfaces of spherules, and dispersed within spherules on an atomic scale; all are possible, depending on the circumstances, as is emission as separate particles. Similarly, the compounds actually formed are not well reported: these are mainly oxides and sulphates, but there are also phosphates, and possibly carbonates, silicates and chlorides.

Organic Fraction

This fraction contains hundreds or even thousands of compounds; it can also be conceived as a *five-element system*, CHONS. The carbonaceous fraction and the fuel tend to bracket the atomic ratio of carbon to hydrogen for the entire particulate; the quantity of organic compounds to have been taken up decides where the actual ratio resides, between these limits. According to the seven-fold classification scheme, some subfractions are viscous and oily; others are solids. The amount of oxygen in the organic fraction can exceed that found in the fuel, suggesting the existence of many partial-reaction products, and the synthesis of oxygenated hydrocarbons in the combustion. The demarcations between vapour-phase and particulate-phase compounds, and between fuel and lubricant compounds, are at ~C_{23}–C_{24}. Particulate-phase PAH tend to contain five or more rings. Size-fractionated measurements of the organic fraction are not well reported, but organic compounds may be better represented in the nucleation-mode particles.

Sulphate Fraction

Sulphuric acid reacts with the metals of the ash fraction, but where this happens is uncertain; the dilution tunnel filter is, however, the likely place. Some sulphates are nonvolatile or insoluble in water. Calcium sulphate makes regular appearances. Electron microscopy reveals what are thought to be high-sulphur particles of a few hundred nanometres in size.

6.6.3 Biological Characterisation

Studies of mutagenicity focus exclusively on the organic fraction, for which the Ames bioassay regularly yields positive results. Generally, only the total mutagenicity of this fraction is reported, but attempts have been made to discover which subfractions are mostly responsible; most studies point to PAH and NPAH, particularly the latter. Poorly reported topics are the mutagenicity as a function of vehicle operation, and as a function of particle size. Many compounds are not mutagenic, and act as diluents. Emission control strategies that reduce the mass of organic particulate often increase also the mutagenicity of this particulate (per mg).

6.6.4 Demographic Characterisation

Measurements of particulate demography easily become bogged down in the numerous metrics (e.g. particles per unit volume and particles per km) and the vast array of extraneous factors such as duty cycle, fuel formulation and hardware (e.g. EGR, aftertreatment and FIE). Storage and release by the exhaust system is a constant nuisance – this is greatly exacerbated by the diesel particulate filter (DPF). Because of the release of volatile particle precursors, vehicles fitted with a DPF on occasion (transiently) emit more particulate mass, or greater numbers of particles, than vehicles without. The mass distribution is dominated by the accumulation mode, but the coarse mode is nonnegligible. Most of the mass resides within a range from few tens of nanometres to a few hundred nanometres. Measurements of number distribution, particularly, should be treated with caution because of the vagaries of nucleation. Particle numbers are extremely variable for vehicles fitted with a DPF, but they can reach parity for vehicles with no DPF at high speed. Emissions from light-duty vehicles on the NEDC (no DPF) are on the order of 10^{13}–10^{14}/km.

7
Abatement

7.1 Introduction

The present chapter relates several disparate technologies to have been pursued in particulate abatement. First amongst these are the so-called 'internal engine' methods of emission control, or those relating specifically to the *combustion*. Now, the literature in this area is, of course, formidable, and, inevitably, combustion threatens to oust particulate as the central topic, for which the salient aspect is the well-known trade-off in the diesel engine between emissions of particulate and emissions of NO_x; this text is not the place to discuss NO_x. The coverage of combustion that follows, therefore, of necessity, remains tightly focused on particulate, although this bias should certainly not be taken to suggest that other aspects of combustion, not herein related, are unimportant in the general picture.

As we have learned, engine-out soot is the ultimate product of two competing processes, *formation* and *oxidation*. Which dominates is often unclear, but the *rate* of one can be modified *relative* to the other, and this is why combustion offers some facility to control emissions of soot. Absolutely critical are the characteristics of the fuel spray, as these control the mixing of fuel and air, which, in turn, decides the heterogeneity of the charge, and hence the amount of soot. Of relevance here are injector nozzle design, injection pressure and, with today's technology, the increasingly elaborate injection schedules, in which the incoming fuel spray is intermittently cut and then reinstated. Again, this area has a vast literature, but, as already explained, it is only practical to discuss that which is of direct relevance to soot.

A disquisition on pollution abatement would be incomplete without some consideration of fuel formulation. Exhaust gas composition is very much a function of what is burned, as is true for any combustion, and, to a first level of understanding, the mixing of fuel and air is driven by the physical properties of the fuel, just as the cleanliness of the burn is driven by the chemical composition of the fuel. In short, some fuels will form soot more readily than others, and this in itself opens up a fruitful area of investigation.

Another section relates the great interconnectedness between the lubricating system and the combustion system, the two hardly being separated by a hermetic seal. Some oil contributes to the particulate emission by escaping into the exhaust; and some particulate escapes into the oil, which affects, and indeed deteriorates, engine lubrication.

The remainder of the chapter is given over to post-combustion strategies, i.e. the field of 'aftertreatment'. With this technology, one accepts that everything feasible in the engine has been done, and the focus now shifts to some form of exhaust gas clean-up, prior to emission at the tailpipe. The literature in this area is equally formidable and relates not just to particulate but to *all* emissions; however, a comprehensive review of aftertreatment by the present author is already available.

In all of the above-mentioned sections, a consideration solely of total particulate mass, when placed against the backdrop of current concerns, would clearly be inadequate. Mass is what concerns the automotive industry inasmuch as legislation is currently framed this way; and, most certainly, this is what most literature focuses on. But no policy maker, or member of the public for that matter, is likely to be content with an emission control strategy that, in reducing the mass of particulate discharged into the environment, worsens concomitantly some other aspect of the same emission (Ulfvarson, 2000). More recently, therefore, research has burgeoned into the impact emission control strategies have on wider aspects, e.g. PAH, mutagenicity, size distribution and number concentration. This material is also reviewed in the present chapter.

7.2 Fuel Formulation

As the composition of the exhaust reflects, in some measure, the composition of the fuel, then fuel *reformulation* is rather self-evident as an emissions abatement strategy – not just for particulate, but also for HC, NO_x and CO – indeed, research along these lines is not, in itself, a recent event (e.g. Shamah and Wagner, 1973). But fuel reformulation offers one particular reward certainly worth fighting for: where new vehicles are purchased relatively slowly, new fuels are purchased relatively rapidly; hence, rather than waiting a decade or more for a complete turnover in vehicle technology to occur in the marketplace, air quality improvements through fuel reformulation are realisable *immediately*, courtesy of the filling station or garage forecourt – provided, of course, no concomitant changes are demanded in vehicle specifications. Moreover, two beneficial side effects are greater rates of exhaust gas recirculation (EGR) for NO_x control if emissions of engine-out soot are lower (Section 7.4), and greater conversion efficiencies in oxidation catalysts if particulate composition shifts from soot to organic (Section 7.8.1; Last *et al.*, 1995). Still, the automotive industry does not exactly have a free hand: although the details need not concern us here, reformulation is equally a matter for the oil industry; fuels must still be economic and practical to manufacture, from the refiner's commercial standpoint.

The benefits, though, even for quite drastically reformulated fuels, are not easily attained, either because reductions in particulate mass are rather modest, compared with those attainable through new vehicle technologies (Kwon *et al.*, 2001), or because the aforementioned trade-off between NO_x and soot in EGR is insufficiently eased (Kenney *et al.*, 2001), or because particle demography is little affected (Mayer A. *et al.*, 1998). Finally, findings are frequently inconsistent, contradictory or lacking in wide applicability: for example, ostensibly identical vehicles respond *differently*, some even exhibiting opposing, rather than supporting, trends (Kwon *et al.*, 2001); hence, corroborative evidence on several levels remains evasive. On the other hand, it might be argued that contemporary emissions standards are now so stringent that even slight modifications to fuel formulation easily make the difference between 'pass' and 'fail' (Bielaczyc *et al.*, 2003).

One problem in particular pervades *all* fuel formulation studies: compositional parameters are frequently, indeed incestuously interrelated, and, because of inability in numerous instances to uncouple, or decouple, these cross-influences, the fuel-related parameter truly responsible for any perceived emissions reduction is seldom conclusively identified. This coupling advises against the practice of treating any fuel property as a true independent variable (Wall and Hoekman, 1984). Of course, this failure is not necessarily a disadvantage to the pragmatist, if a suitable and eminently commercial formulation is ultimately identified merely through trial and error. Such studies should not be dismissed simply because they do not establish any fundamental insights.

The previous remark highlights a frustrating imbalance in the vast literature on fuel formulation and emission control: for this is chock-full with so-called 'black-box' studies of fuel (engine-in) and emissions (engine-out). Ranged on the other side of the fence are just a few scattered instances of *in-engine* studies, the aim of which is elucidation of *how* reformulated fuels burn, and hence *how* reformulated fuels generate a different melange of pollutants. The 'how' question is unanswerable by the black-box approach. But, in black-box studies particularly, the possibility admittedly exists that some beneficial change will be speciously adduced to a certain fuel property, i.e. this property might simply be acting as a *surrogate* for the unknown property that really counts.

Simplifying somewhat, fuel properties are held to influence particulate emissions in two distinct ways (Rink *et al.*, 1987). *Physical* properties, such as volatility and density, tend to control spray characteristics, such as the cone angle and the sizes of droplets, and hence also the game of musical chairs in which fuel molecules must mix with and find oxygen molecules. *Chemical* composition, i.e. molecular structure, insofar as this determines the reaction pathways, controls also the pyrolytic alleyway to soot and the oxidation corridor to CO_2 and H_2O. These two aspects of fuel formulation are unlikely to be mutually exclusive; yet they are not easily distinguished.

An essential caveat in all fuel formulation studies concerns engine *calibration*. This is the process in which the manufacturer, in the development stage, painstakingly optimises EGR, timing, etc. to obtain the best compromise between economy, emissions, driveability, etc.; and this is, of necessity, carried out using a fuel of a certain specification, that is, one possessing certain closely defined properties. Subsequent operation of the same engine with the optimised calibration on some fuel of a *different* specification, assuredly, then, renders the calibration *nonoptimal* – leading to potentially deceptive emissions. For example, if the power developed by the engine is modified, the driver compensates by adjusting the pedal, and the engine management system, which uses the pedal position in 'look-up' tables (for timing and EGR), shifts the engine operation accordingly. In practice it is difficult to unravel this incestuousness, and a dichotomy also arises: in one sense, the intention could simply be to furnish filling stations with a low-emission formulation involving no retuning; but, if fundamental insights are required, engine calibrations must be *reoptimised* – otherwise the perceived 'benefits' can be quite artificial (Kwon *et al.*, 2001).

The nexus between fuel formulation and emissions is a vast field; and the nexus between fuel formulation and *particulate* emissions, while obviously a subset, forms a database that is still formidable. The present review cannot focus on all possible permutations; indeed, the list lengthens in accordance with the exoticism of the approach. Five areas not considered explicitly here are as follows.

(**A**) *Fischer–Tropsch (FT) fuels*, which possess three admirable properties, namely zero sulphur (Section 7.2.1), negligible aromatics (Section 7.2.2) and narrow distillation range (Section 7.2.5) (Atkinson C.M. *et al.*, 1999, Nishiumi *et al.*, 2004). The particulate mass is not consistently reduced, nor is the mass distribution or mutagenicity greatly affected (McMillian *et al.*, 2002). (**B**) *Water – diesel emulsions* (Nazha *et al.*, 1998; Song K.H. *et al.*, 2000; Matheaus *et al.*, 2002; Farfaletti *et al.*, 2005), which, through 'microexplosions' (Sheng *et al.*, 1994), or increased air entrainment into the combusting plume (Musculus *et al.*, 2002b), improve fuel–air mixing and thereby reduce, also, engine-out soot, for which demographic (Warner *et al.*, 2002; Bertola *et al.*, 2003; Mathis *et al.*, 2005) and speciative (Lombaert *et al.*, 2002) characterisations are available. The soot, moreover, is for some reason, probably microstructural, more amenable to burn-off in a diesel particulate filter (DPF) (Schmelzle and Chandes, 2004). (**C**) *Natural gas*, which brings about substantial reductions in particulate (Wong W.Y. *et al.*, 1991; Hountalas and Papagiannakis, 2002), and which may, in some countries, entirely displace diesel fuel (Kathuria, 2004). **(D)** *Hydrogen* (Tsolakis and Megaritis, 2005), for which demographic characterisations are available (Tsolakis *et al.*, 2005). **(E)** *Hydrogen peroxide*, for which demographic characterisations are also available (Trapel *et al.*, 2004).

7.2.1 Sulphur

The longed-for desire in the automotive industry is complete eradication of fuel sulphur: this is primarily because of the havoc this element wreaks in emission control, for example its inhibition of essential reactions in the catalytic converter, and its emission in the form of sulphate particulate; and secondarily, because sulphuric acid is a corrosive agent involved in engine wear (Takakura *et al.*, 2005), requiring neutralisation by certain compounds in the lubricant (McGeehan *et al.*, 1993). From these and many other standpoints, sulphur is present strictly on *sufferance*.

But cogent reasons advise against the removal of fuel sulphur willy-nilly: the depth of fuel desulphurisation is, obviously, constrained by practical limitations (Takatsuka *et al.*, 1997; Ho, 2004), while conventional methods of hydrodesulphurisation, at low ppm, become prohibitively costly, and new methods are required to eradicate the more refractory compounds (Kim J.H. *et al.*, 2006). Diesel fuel possesses natural self-lubricating (Anastopoulos *et al.*, 2005) and oxidation-resisting properties which are compromised, or even destroyed, in the desulphurisation process, thus rendering powertrains vulnerable to injector wear, fuel pump seizure and fuel instability; and, to compensate, suitable fuel additives or special protective coatings (Erdemir *et al.*, 2000) must be found.

The stringency of legal restrictions on fuel sulphur has rapidly tightened in recent years – and continues to tighten still (although, from one country to another, there may be field issues for some time yet). The motivation for these restrictions, in particulate terms, does *not* concern the engine-out emission, because, at this point, as explained in Section 3.7.2, only two or three per cent of the sulphur forms sulphate particulate, and this small quantity is overshadowed easily enough by the carbonaceous and organic particulate (Kwon *et al.*, 2001). Rather, the original reason for these restrictions lay in the *catalytic converter* – which oxidises organic particulate, HC and CO, but, through unwanted side reactions, oxidises also SO_2 into sulphate particulate (see Section 7.8.1), which then, relatively easily, turns the tables completely on the other two fractions. This effect turns strongly on catalyst temperature, as driven by duty cycle (Bielaczyc *et al.*, 2002). Added to this original constraint there is, latterly, a new one, namely an inability to introduce lean NO_x traps, since SO_x squats tenaciously on sites intended for NO_x – this is now a major headache for the automotive industry.

Besides fuel sulphur's *overt* contribution to sulphate particulate, some *covert* influence is suspected on organic particulate (Wall and Hoekman, 1984; Tanaka S. and Shimizu, 1999; Hori S. and Narusawa 2001; Oyama and Kakegawa, 2003), emissions of which are, actually, lower when fuel sulphur content is reduced. Several remarks were already made on this score in Section 4.6. The question is whether this cross-influence is genuine or artefactual: genuine, if the engine really does emit less organic particulate; artefactual, if gas-to-particle conversion has been mysteriously modified. This question is not as easy to answer as one might think: direct back-to-back tests require fuels desulphurised to different levels, but for which the base stock remains identical; yet the desulphurisation process modifies the hydrocarbon speciation (especially that of the aromatics), so that the burning characteristics are also, albeit unintentionally, affected (Section 7.2.2). And since natural sulphur-bearing compounds tend to be found towards the higher end of the distillation curve, there is some correlation between sulphur content and volatility (Wall and Hoekman, 1984). Hence, differences are even measurable in engine performance (Ariga *et al.*, 1992). On the other hand, the same fuel, if artificially doped with sulphur-containing compounds such as tertiary butyl disulphide (TBD, $C_8H_{18}S_2$) or thiophene (C_4H_4S), should be proven to burn in a similar fashion to undoped fuels, and to reproduce, if not exactly to imitate, the molecular structure of natural sulphur-bearing compounds (Wall and Hoekman, 1984).

Nevertheless, it is difficult to see how one or two dopants can satisfactorily reproduce the behaviour of more than a dozen naturally occurring organosulphur fuel compounds (Liang F. *et al.*, 2006).

It nonetheless seems logical to suspect sulphuric acid as the instigator, or perhaps mediator, of this covert cross-influence. For example, the acid might well, especially on the dilution tunnel filter, instigate the formation of secondary organic compounds; or, in the exhaust stream, act as a nucleating agent, and so encourage gas-to-particle conversion through the acquisition of vapour-phase organic compounds (Section 3.7.2). It could well be that sulphuric acid residing on the dilution tunnel filter acts as a kind of scrubber towards vapour-phase organic compounds; indeed, the same effect may even occur in catalysts (Khair and Bykowski, 1992). Even the *lubricant*-derived organic particulate, is, in some way, and in part, determined by *fuel*-derived sulphur: as much as one-third of the organic fraction is ascribable to the acid-mediated scrubbing effect (Zelenka P. *et al.*, 1990). Such a scrubbing action would skew the speciation of the organic fraction towards a more volatile distribution (Wall and Hoekman, 1984).

A corresponding shift in the partitioning of mass between the *filter* (particulate-phase organics) and the subsequent *sorbent* (vapour-phase organics) was not observed, but the partitioning of individual PAH was undoubtedly affected: where *Chr* and *BbF* were found solely in the particulate phase with high-sulphur fuel, they appeared in the vapour phase with low-sulphur fuel; moreover, the quantities of particulate-phase *Pyr*, *Chr* and *BaP* were *greater* with the low-sulphur fuel (Opris *et al.*, 1993). A comprehensive picture has not emerged, but, when one is faced with these parallel modifications to the amount and to the speciation of organic particulate, a further corollary is wholly unsurprising: *greater* mutagenicity of the vapour- and particulate-phase organic compounds is, somewhat perversely, observed with the *low*-sulphur fuel (Opris *et al.*, 1993). (It should be noted that the sulphur compound used as a dopant here was not, on its own, mutagenic.) The reason for this contrary trend remains unknown.

In contrast, the demographic implications of low-sulphur fuels are clear-cut and satisfactorily demonstrated: accumulation modes are stolidly consistent; the loss of acid coatings, which may aid agglomeration, is perhaps an aspect (Chatterjee *et al.*, 2001), but this effect is of subsidiary importance, and difficult to resolve from other, adjunctive modifications to fuel formulation. But, if less sulphuric acid is in the exhaust, then presumably nanoparticles are now less abundant: this acid is the prime instigator of nucleation, according to the ideas presented in Section 3.7.2. The satisfactory side to this argument is that particle enumerations with low-sulphur fuels are correspondingly less sensitive to the interplay of nucleation and dilution, and to the confusion thereby introduced. The unsatisfactory concomitant is that nanoparticle emissions for the newest vehicles have become more sensitive to variations in fuel sulphur content, because the accumulation-mode soot, which provides solid surface by which saturation is suppressed, has declined so significantly (Ristovski *et al.*, 2006).

This two-stranded understanding is borne out in Figure 7.1 (Wei *et al.*, 2001a): an order-of-magnitude difference is manifested in the nanoparticle count between fuel sulphur contents of 440 ppm and 10 ppm. Others have obtained very similar results (Opris *et al.*, 1993; Hall and Dickens, 2003). The question is at what fuel sulphur content might these nucleation modes *disappear*: a diesel vehicle fitted with a diesel particulate filter (DPF), and fuelled at only 15 ppm sulphur, *still* emits nanoparticles (Herner *et al.*, 2007). Any 'threshold', such as two parts per million (Kati *et al.*, 2004b), is, logically, context-specific, i.e. dependent on the vagaries of dilution. In any case, the *lubricant* sulphur will begin to make a nonnegligible contribution prior to this point (Hall and Dickens, 2003; Vaaraslahti *et al.*, 2005).

It is interesting to ask whether the advent of low-sulphur fuels has made a real difference in field terms. Reassuringly, demographic shifts are not solely artefacts of laboratory dilution: very similar conclusions have been reached in chase experiments (Schneider *et al.*, 2005), and nanoparticles in a street canyon diminished significantly from 1999 to 2000 – a trend consistent with a concurrent

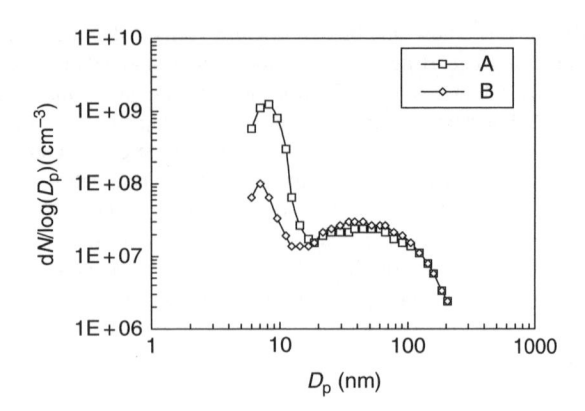

Figure 7.1 Number distribution for two different fuel sulphur contents: A, low sulphur, 440 ppm; B, ultra-low sulphur, 10 ppm. The residence time and temperature in the transfer line were held constant at 0.08 s and 300 °C, respectively. Dilution tunnel: dilution ratio 1000; residence time 1.0 s; air temperature 15 °C; exhaust temperature 300 °C. Engine: 4-litre, four-cylinder, direct-injection, medium-duty, model year 1995. Operating point: 1600 rpm, 50 % maximum load. Instrument: SMPS, 8 nm to 300 nm (Wei *et al.*, 2001a).

reduction in diesel fuel sulphur from 500 ppm to 50 ppm (Wåhlin *et al.*, 2001). Thus, in order to restrict or even to eradicate the nucleation of volatile particle precursors, fuel sulphur content *must* be reduced: this is very clear. The unfortunate irony here is that soot particles will become less sulphated in the emission process, rendering them less likely to act as cloud condensation nuclei, thus impeding their removal from the atmosphere via rainout (Huang X.-F. *et al.*, 2006).

7.2.2 Hydrocarbons

Soot-forming proclivity varies considerably from one hydrocarbon to another, so that the quantity of carbonaceous particulate reflects, inherently, the hydrocarbon *speciation* of the fuel. Soot formation, in this sense, is not so much a matter of fuel molecules finding oxygen molecules as of *bonding* within hydrocarbon molecules, i.e. molecular structure. Emissions are not, of course, the be all and end all: oil refiners labour under myriad technical and commercial constraints, so that hydrocarbon speciations, whether of low soot-forming proclivity or not, must still be economic and practical to manufacture (Cooper B.H. and Donnis, 1996), although these broader aspects are not covered here.

Much understanding relates to empirical observations of simple hydrocarbon flames: for example, with aliphatics, sooting tendency increases with the length of the molecule, the number of side chains, from single bonds to double bonds, and from double bonds to triple bonds (Ladommatos *et al.*, 1996e). Saturated rings soot more readily than saturated chains (e.g. cyclohexane versus hexane). Sooting tendency is also indexed to the hydrogen-carbon ratio (Ogawa, 2005) – a term which expresses the degree of unsaturation. Among the chief hydrocarbon groups, sooting tendency in laminar diffusion flames increases in the sequence alkanes, alkenes, alkynes and aromatics (Griffiths and Barnard, 1995, p. 117): on this basis, 'aromaticity', or the fraction of aromatic compounds in the fuel, is often taken as the key sooting index, so that low aromaticity is viewed as superior for internal combustion engines (Otani *et al.*, 1988; Tsunemoto *et al.*, 2001).

Experiments with radioactive tracers show that carbon atoms within aromatic molecules are far more likely to appear in soot than carbon atoms within alkane molecules (Burley and Rosebrook, 1979).

Section 3.2.2 related how aromatics are widely held to be the immediate soot precursors, although the exact mechanisms are still imperfectly understood. But whether soot forms directly from fuel aromatics, or indirectly from aromatics pyrosynthesised from the broken remnants of fuel aromatics, is not answerable from black-box studies – of which, as stated in Section 7.2, there are dozens in the literature: the headlong pursuit of low-aromatic fuels in the automotive industry rests almost entirely on observations of an empirical nature, lent further impetus by the toxicological risks carried by these compounds.

Controversy, however, surrounds the aromaticity–particulate relationship, and the trends, or supposed trends, do not always stand out: one such study is illustrated in Figure 7.2 (den Ouden et al., 1994). In extenuation, one should recognise that several issues are confusedly mixed. First, there is some form of dependency on the actual engine, notably involving the combustion system (e.g. IDI versus DI) and calibration (e.g. EGR, timing and injection pressure) (van Beckhoven, 1991); instances may, for example, exist where air–fuel mixing is a greater factor in soot formation than hydrocarbon chemistry. Secondly, fuel structures play a role in the sooting proclivity of diffusion flames, but not of premixed flames, as all hydrocarbons break down to similar species (Glassman, 1996, p.407), so that the relative proportions of premixed and diffusion burning, and the ignition delay, should be taken into account, whereafter, the particulate emissions seem insensitive to aromaticity (Ladommatos et al., 1997c). Thirdly, aromaticity is not easily decoupled from volatility, cetane number, density (Tritthart et al., 1993) and C/H ratio. Fourthly, the effect is dual-faceted: the organic fraction *and* the carbonaceous fraction are, in some way, both affected; thus measurements of total particulate, on their own, are unrevealing or misleading. Fifthly, *total* aromatics content is a red herring insofar as subsets of aromaticity, that is, according to the number of rings (monoaromaticity, diaromaticity, triaromaticity, ...) are also important. Sixthly, there may be some contrary effect via flame temperature (Borman and Ragland, 1998, p.419).

Recognition that both the organic and the carbonaceous fraction are affected by fuel aromatics chimes well with the ideas advanced in Section 3.6.1, in which the first acts as a feedstock for the second, and sometimes these soot-forming reactions are prematurely interrupted. Hence, low-aromatic fuels are sometimes able to reduce the particulate emission not through the carbonaceous fraction, which can, in fact, increase marginally, but instead through the organic fraction (Andrews et al., 1998b). Cases also

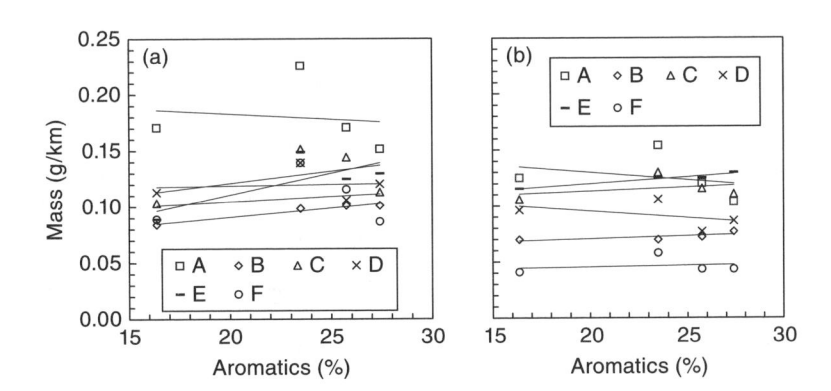

Figure 7.2 Fuel aromatics, effect on mass emission of particulate: (a) without oxidation catalyst; (b) with oxidation catalyst. All light-duty vehicles: A, direct injection, turbocharged; B, indirect injection, turbocharged; C, indirect injection, naturally aspirated; D, indirect injection, turbocharged; E, indirect injection, naturally aspirated; F, indirect injection, turbocharged. Duty cycle: NEDC (den Ouden et al., 1994).

exist where the organic fraction is unaffected (Otani et al., 1998; Tsunemoto et al., 2001), but these need considering in the light of other phenomena discussed in Section 3.6.1, such as the well-known predominance of organic particulate at low engine load and of carbonaceous particulate at high engine load, so that low-aromatic fuels might simply bestow their particulate reduction one way or the other according to engine operating point (Kweon et al., 2003c). These remarks are conjectural.

One area rather less conjectural is the inculpation of molecular size, that is, the number of rings. Some findings are as follows: polyaromaticity is more important than total aromaticity (Sienicki et al., 1990; Lee R. et al., 1998); diaromaticity and polyaromaticity are more important than total aromaticity (Li X. et al., 1996); polyaromaticity is twice as influential as monoaromaticity (Kwon et al., 2001); and diaromaticity and triaromaticity are more important than total aromaticity or monoaromaticity (den Ouden et al., 1994). These remarks apply to particulate, taken as a whole. Fundamental studies are not wholly supportive, however: in a rapid-compression machine, the ring number exerted no consistent influence (Takahashi et al., 2001); and in a high-pressure continuous-flow (diffusion-flame) combustor, low monoaromaticity could be just as sooting as high diaromaticity, depending on the pressure or air – fuel ratio (Rink et al., 1987).

Not unexpectedly, fuel aromaticity is closely linked to the emission not just of particulate-phase organic compounds *per se*, but also of particulate-phase PAH (Tsujimura et al., 2007). Again, polyaromaticity seems more important than monoaromaticity (Wall and Hoekman, 1984). Various attempts to establish a firm link between exhaust PAH and fuel PAH, along with several important caveats, were described at length in Section 3.6.1. The understanding here is that PAH emitted by the engine are predominantly fugitive fuel compounds, rather than pyrosynthesised interlopers, in which case the benefits conferred by fuel reformulation are direct, rather than indirect: clearly, if less aromatics are in the fuel, then less aromatics are emitted (Johansen et al., 1997). Research into PAH *speciation* is more sketchy: for example, aromaticity correlated with the release of *Phe*, *Pyr* and *Flu* (Johansen et al., 1997), and triaromaticity correlated with the release of *BkF*, *BbF* and *Chr* (Rantanen et al., 1996).

Reductions in mutagenicity, equally expectantly, go hand in hand with reductions in aromaticity – especially multi-ringed aromaticity. The mutagenicity (revertants per mg) of particulate extracts correlated closely with triaromaticity (Rantanen et al., 1993); and reductions of 30–90 % (revertants per km) were achieved *solely* via fuel aromaticity (Rantanen et al., 1996). An all-encompassing, definitive trend is lacking, but, in general, organic particulate generated by burning fuels of the lowest aromaticity, especially the lowest triaromaticity, tends to exhibit also the lowest mutagenicity (revertants per kg of fuel) (Johansen et al., 1997).

Studies of aliphatics, by contrast, are infrequent: the potential for soot abatement via alkanes and alkenes may have not received the attention it deserves (Ng et al., 2005). When low-boiling-point alkanes such as heptane were burned, emissions of organic particulate were lower than for regular diesel fuel, and by an order of magnitude (Tsunemoto et al., 2001). Soot emissions increased in the order heptane, decane and dodecane – in the same order as the boiling point. Generally, alkenes yielded more particulate, both carbonaceous and organic, than did alkanes of identical carbon number. On this basis, the authors of those studies suggest that double-bonded molecular structures facilitate the formation of PAH, leading to more carbonaceous particulate at high load, and more organic particulate at low load.

7.2.3 Oxygenates

Oxygenates are simply organic compounds, the functional groups of which contain oxygen in some form: several commonly encountered groups are alcohols (R–OH), esters (R–CO$_2$), ethers (R–O) and carbonates (R–CO$_3$). Some compounds, such as methanol and ethanol, are almost vernacularly labelled;

others, being more complex molecules, are known for convenience via their acronyms, for example dimethoxymethane (DMM), dimethyl carbonate (DMC), dimethyl ether (DME), diethyl ether (DEE) and methyl *t*-butyl ether (MTBE). A very important oxygenated fuel is 'biodiesel' – really an umbrella term for various mixtures of esters which, because they derive, via the process of transesterification (Demirbas, 2005), from *vegetable oils* (e.g. cottonseed, rapeseed, soybean and peanut), are much promoted as 'renewable' fuels (Patterson *et al.*, 2006): examples are fatty acid methyl ester (FAME), soya methyl ester (SME), rapeseed ethyl ester (REE) and rapeseed methyl ester (RME).

These are just a few scattered examples taken from a colossal multitude (e.g. Tree and Svensson, 2007), as the possible molecular structures, even within closely defined functional groupings, reflect, inevitably, the great prolixity of organic chemistry. This prolixity is a factor in perusing the literature: one must consider carefully the number, nature and variety of oxygenates investigated by any one researcher (Natarajan *et al.*, 2001), and the fact that considerable variations arise in combustion characteristics even within one functional group; conclusions, after all, are unlikely to transcend the range of compounds investigated. Because of this prolixity, a discussion of each oxygenate individually is impossible; hence only the general characteristics of oxygenates are covered here.

Oxygenates are burned neat or, more typically, as oxygenate–diesel blends, and the particulate-suppressing properties of such are well documented; indeed, the literature is threatening to become somewhat overstocked in this area. Of course, before any particular oxygenate is deemed technologically viable, a wide range of criteria, *other* than reliable clean-burning combustion, must first be demonstrated: examples are ignitability, long-term stability, corrosivity, elastomer compatibility, toxicity, lubricity (Geller D.P. and Goodrum, 2004), solubility (in diesel) (Zhao X. *et al.*, 2005), viscosity, volatility (Vertin *et al.*, 1999) and biodegradability; there is also susceptibility to injector nozzle coking (Labeckas and Slavinskas, 2006); but this is not the place to review these wider aspects. An initial screening encompassed no less than 71 oxygenates, but many were deemed unsuitable for reasons other than emissions and, consequently, this sizeable group was whittled down to a subset of just eight compounds (Natarajan *et al.*, 2001).

In their particulate-suppressing action, oxygenates parallel organometallic fuel additives (Section 7.2.4), but with two important differences: there is no incombustible ash, and blends are at percentage rather than ppm levels. Because of this second difference, oxygenates are more aptly designated fuel *components* rather than fuel *additives*, as modifications to energy release rates are only to be expected (Rakopoulos *et al.*, 2004). Indeed, if bulk fuel properties such as density and viscosity are altered, then certain effects might arise not through the combustion *per se*, but through the geometry of the fuel spray, the atomisation of the fuel, or even through the flow characteristics of the fuel within the injector nozzle.

The particulate fraction targeted with oxygenates is *soot*. The absence of incombustible ash, insofar as 'oxygenate' denotes purely organic compounds, means no contribution to the ash fraction, and sulphur levels are low or negligible, which restricts contributions to the sulphate fraction; these are welcome corollaries, arising straightforwardly through displacement of conventional diesel fuel. Variable effects are observed on the organic fraction: for example, the particulate mass is lower, but the carbonaceous – organic ratio remains unchanged, suggesting that both fractions are proportionally decreased (Yeh *et al.*, 2001); or there is no consistent trend (Litzinger *et al.*, 2000). More commonly, filter deposits are 'wetter', and lighter in colour (Suppes *et al.*, 1999): both observations suggest a greater organic fraction (Sidhu *et al.*, 2001), i.e. preferential reduction in soot (Lapuerta *et al.*, 2005) or *also*, perhaps, of greater gas-to-particle conversion among organic compounds than with diesel fuel. Filter deposits may also become lighter simply because the reduction in carbonaceous particulate allows *lubricant*-derived organic particulate to become more prominent by proportion (Bechtold *et al.*, 1991).

Contrary to expectations, the total mass of particulate sometimes actually increases (Cole *et al.*, 2001), or the increase in organic particulate outweighs the decrease in carbonaceous particulate

(Rantanen *et al.*, 1993). Smoke readings, responding to blackness, register only the decrease in carbonaceous particulate (Grimaldi *et al.*, 2002); this 'improvement' is largely cosmetic if organic compounds are emitted in greater quantities, undetected. But lest oxygenates should be denigrated too readily, a note of caution should be sounded. Contrary trends between HC and organic particulate (Romig and Spataru, 1996) suggest subtle differences in gas-to-particle conversion and in volatilities between oxygenates and diesel fuel, or perhaps differences even in the response of the FID. For example, with biodiesel, the organic fraction has a larger contingent of high- molecular-weight compounds than with diesel (Trapel *et al.*, 2005). Contemporary exhaust gas analysis is, after all, very much designed with conventional diesel fuel in mind, and factitious conclusions may arise through blind adherence to these same protocols with oxygenated fuels.

Which oxygenate is superior, purely in particulate mass-emission terms, really requires closer scrutiny than possible here, as the benefits, if any, are specific to the conditions and depend, principally, on the combustion system, fuel injection system, engine calibration and duty cycle. A study of eight oxygenates indicated an ether as providing the greatest particulate reduction potential (González *et al.*, 2001), and a study of fourteen oxygenates indicated an alcohol (Yeh *et al.*, 2001).

Something more decently agreed upon is *proportionality* between the mass of soot in the particulate and the mass of oxygen in the fuel (González *et al.*, 2001). (Comparison of fuels by their oxygen content, rather than oxygenate content, seems appropriate.) The seemingly inevitable consequence of this proportionality is complete disappearance of soot at a certain diesel – oxygenate blend; and, while the exact strength of the proportionality is still debated, this change in fortune is frequently cited to occur at oxygen contents of $\sim$ 30–40 % (Miyamoto *et al.*, 1998).

This proportionality, one might suppose, supplies the most obvious clue as to *how* oxygenates reduce soot. The soot suppression seems not to arise simply through displacing diesel fuel; and it might equally proceed through discouraging soot formation or encouraging soot oxidation, as both routes, after all, accomplish the same aim. Intuitively, it is tempting to argue that the combustion process, particularly if fuel-rich, is simply furnished with more oxygen (Rakopoulos *et al.*, 2006), and that oxidising, rather than pyrolytic reactions are thereby fostered. In this understanding the mixture is, in effect, leaned – through the provision of 'internal oxygen' (via the fuel), in contradistinction to 'external oxygen' (via the air). In fact, the same justification is used for supplying ethanol – gasoline blends in the wintertime: to provide the combustion process with more fuel-borne oxygen, and thereby to restrict emissions of CO. Should the same or a parallel mechanism apply to diesel engines in controlling emissions of soot, then fuel oxygen content, rather than the molecular structure or functional group of the oxygenate, would presumably become the pivotal factor. As yet, some studies point one way, and some the other (Miyamoto *et al.*, 1998; Litzinger *et al.*, 2000; González *et al.*, 2001; Yeh at al, 2001); and some are less polarised, suggesting, on the one hand, that molecular structure is an aspect of soot formation, and, on the other, that oxygen content is an aspect of soot oxidation (Rakopoulos *et al.*, 2004).

Investigations into the soot-suppressing *mechanisms* of oxygenates divide conveniently into the universal and the particular, that is, studies of combustion chemistry generally and studies of diesel engine combustion specifically, akin to Section 3.2 (Curran *et al.*, 2001) and Section 3.3 (Rakopoulos *et al.*, 2004), respectively; and, in the latter case, especially according to the picture of the combusting plume (Musculus *et al.*, 2002a) delineated in Figure 3.9. In both areas of research, the fuel oxygen content of 30–40 % as determined, empirically, for the complete disappearance of soot is felicitously supported (Curran *et al.*, 2001; Musculus *et al.*, 2002a; Kitamura *et al.*, 2003): the soot suppression seems to arise through chemical *and* physical mechanisms which, fortuitously, lend one another mutual support.

As regards the *combustion chemistry*, soot precursors are either formed less readily or oxidised more readily: for example, smaller amounts of unsaturated hydrocarbons, benzene and PAH emerge from a thermal cracker (Kitagawa *et al.*, 2001); this is supported by modelling studies (Kitamura *et al.*, 2003). A reasonable consensus exists that carbon atoms, when bound to oxygen atoms, are less able to participate in soot-forming reactions, because this type of bond is broken less readily. This understanding is corroborated by experiments with radioactive tracers (Buchholz *et al.*, 2002). Since carbon is now effectively 'sequestered', the feedstock of soot precursors is diminished (Kitamura *et al.*, 2001). The relevant index of performance, then, is the 'effective' hydrogen-carbon ratio in the oxygenate molecule, as derived by excluding carbon atoms bonded to oxygen atoms (González *et al.*, 2001). Single or double bonding is also important: carbonates and esters, for example, have oxygen atoms bound to the same carbon atom: they are less effective in soot suppression than alcohols, which tend to contain oxygen atoms bound to different carbon atoms (Yeh *et al.*, 2001); this understanding is again supported by experiments with radioactive tracers (Buchholz *et al.*, 2004). A secondary and possibly allied explanation is the improved oxidation of these soot precursors, should oxygenates boost concentrations of CO and CO_2 (Litzinger *et al.*, 2000; Nag *et al.*, 2001) or of OH (Kocis *et al.*, 2000).

As regards the *diesel engine*, various ideas cover the gamut of combustion from start of injection to late-phase burn-up. (a) In the *charge preparation*, air – fuel mixing is improved, either because droplet size is a function of surface tension and viscosity (Kajitani *et al.*, 1994), or because vaporisation (Yeh *et al.*, 2001) is a function of boiling point (Stoner and Litzinger, 1999), or because air entrainment is a function of heat of vaporisation (Curran *et al.*, 2001). (b) In the *premixed burn*, a greater percentage of fuel is burned in this relatively soot-free phase (Kocis *et al.*, 2000) because the ignition delay is longer (Huang Z. *et al.*, 2005), or alternatively because the cetane number is higher (Lü *et al.*, 2005). (c) In the *mixing-controlled burn*, the soot-producing phase is completed more swiftly (Gui *et al.*, 2004; Rakopoulos *et al.*, 2004), in-plume soot rises to a lower peak (Musculus *et al.*, 2002a); shorter lift-off lengths, and consequently poorer air entrainment prior to the posited premixed flame, appear to be outweighed by the need for less air entrainment (Kitamura *et al.*, 2001). (d) In the *late burn*, prolongation of (Gui *et al.*, 2004) and greater vigour in (Kajitani *et al.*, 1994) the soot burn-up compensate for a higher soot peak (Rakopoulos *et al.*, 2004).

From the foregoing, it will be apparent that the *formation* of particulate via oxygenated fuels is well reported; this situation contrasts oddly with *characterisations* of the same particulate, which are flimsily reported; hence, the following remarks are somewhat tentative. Morphologically, agglomerates were observed to be clustered for an alcohol fuel and, contrarily, chained for a diesel fuel (Nord K. *et al.*, 2004); whether this difference is sufficient to impute subtle alterations in the process of agglomeration (Section 3.2.4) awaits further research. Microstructurally, the spherules were not discernibly different with an alcohol fuel (Nord K. *et al.*, 2004); but, with biodiesel, greater amorphousness was apparent, with prominent wrinkling evident in the turbostratic microstructure (Boehman *et al.*, 2005). According to ideas sketched in Section 6.2.1, wrinkling is an aspect of temperature of exposure, and possibly of partial oxidation; but, these temperatures not having been greatly changed by biodiesel, this not unreasonably implicates the fuel's composition and decomposition chemistry in such differences. Chemically, the functional groups attached to the soot surface differ in subtle ways (Braun *et al.*, 2005b), and the oxygen-containing ones facilitate oxidation (Song J. *et al.*, 2006). This greater amorphousness or oxygen presence seems to render particulate in a DPF more combustible (Williams A. *et al.*, 2006), and provide still further reasons why engine-out soot is reduced (Vander Wal and Mueller, 2006). Yet another interesting facet is potassium, co-present with soot generated in biomass combustion, which may provide some catalytic assistance (Zabetta *et al.*, 2006).

The literature devoted to oxygenates and PAH recapitulates Section 3.6.1: that is, whether fuel compounds or pyrosynthesised compounds dominate. The starting point is simple displacement of diesel fuel, as oxygenates are not noted for their PAH content. Then, as mentioned above, either PAH are formed less readily in the combustion or, once formed, they are oxidised more readily (Kitagawa *et al.*, 2001). The only remotely corroborative experimental data here relate to biodiesel, which does, indeed, produce a smaller amount of PAH in the particulate phase (Ziejewski *et al.*, 1991) and also in the gas phase (Krahl *et al.*, 2003). Among 15 PAH well represented with diesel fuel, only one, *Phe*, was significantly present with biodiesel; and, summed together, PAH were reduced, as a group, by an order of magnitude (Hansen and Jensen, 1997). For biodiesel – diesel blends, 15 PAH were reduced; but, probably owing to their greater presence in the biodiesel compared with diesel, three (*Flu*, *Pyr* and *BaP*) increased (Masjuki *et al.*, 2000); other workers report significantly lower PAH (*Chr* and *BaA*), marginally lower PAH (*BkF*, *BaP* and *Phe*), and lower NPAH (*1-NP*) (Chase C.L. *et al.*, 2000). It should be noted that, with oxygenates, increases in organic particulate not infrequently accompany decreases in PAH, so that other organic compounds besides aromatic ones (perhaps simply unburned fuel) are worthy of consideration (Rantanen *et al.*, 1993).

Similarly, few papers examine mutagenicity. Alcohols either lessen (Lestz *et al.*, 1984) or worsen (Heisey and Lestz, 1981) the mutagenicity. The mutagenicity for biodiesel – diesel blends compared with diesel was similar or slightly lower (Rantanen *et al.*, 1993), or severalfold lower (Chase C.L. *et al.*, 2000); and, for neat biodiesel, lower by a factor of two or three (Krahl *et al.*, 2003). These findings presumably capture the lower PAH emission, expressed as revertants per unit mass of organic particulate. The caveat here is the higher emission of organic particulate, as, when this is expressed as revertants per unit distance, the trend reverses (Bünger *et al.*, 1998).

Even though the mass emission of particulate is reduced, indeed drastically so in some instances, this reduction may prove insufficient to meet present-day statutory requirements, in which case other, adjunctive types of emission control, e.g. a DPF, will still be necessary (Haupt *et al.*, 2004); neither can the demographic implications of oxygenates be overlooked (Oguma and Goto, 2007). As described above, the mass reduction occurs predominantly in the carbonaceous accumulation mode (Krahl *et al.*, 2003; Nord K.E. and Haupt, 2005), and this has obvious implications for nucleation; for example, while neat alcohol reduces the mass emission by an order of magnitude, the total number of particles (especially of nanoparticles) is not greatly altered (Nord K. *et al.*, 2004); nor is it too surprising when oxygenated fuel blends tend to produce more particles (Bertola *et al.*, 2001), since the absence of carbonaceous surfaces pushes saturation upwards, thus aiding nucleation. That said, the nature, or rather volatility, of these nanoparticles seems different. Figure 7.3 (Bertola *et al.*, 2001) displays

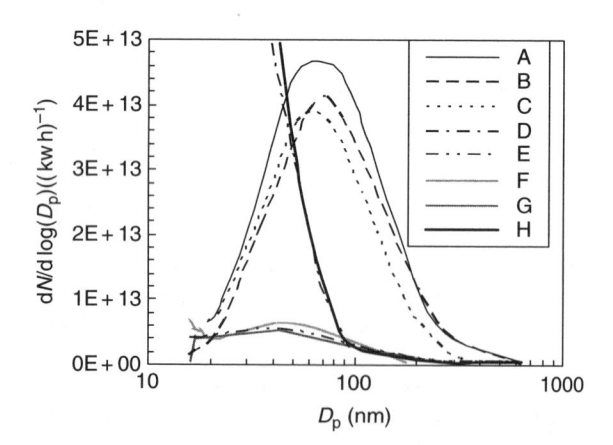

Figure 7.3 Effect of various oxygenates on number distribution: A, 15 % butylal, 24 % EGR; B, diesel, sulphur 50 ppm, 22 % EGR; C, alkane-only, 22 % EGR; D, RME, 14 % EGR; E, diesel, sulphur 50 ppm, no EGR; F, 15 % butylal, no EGR; G, alkane-only, no EGR; H, RME, no EGR. Engine: 2.1-litre, single-cylinder, research engine. Operating point: 1460 rpm, 50 % maximum load, start of injection 10 °BTDC. Instrumentation: SMPS, with pretreatment by thermodesorber, 350 °C. Original data points omitted (Bertola at al, 2001).

some demographic data for a range of oxygenates; RME was noticeable for its propensity to form nanoparticles, but these were not apparently eradicated by exposure to 350 °C, suggesting a less volatile nature than is generally the case for regular diesel.

7.2.4 Additives

Fuel additives act on particulate emissions *directly* and *indirectly*. Detergents, for example, act indirectly: they discourage the build-up of gums, varnishes and hard, tenacious deposits which, if allowed to accumulate on injector nozzles, disrupt fuel atomisation and upset spray geometry, leading, inevitably, to poor air – fuel mixing, mediocre combustion, and larger emissions, such as of carbonaceous and organic particulate (Yoshida *et al.*, 1986; Reading *et al.*, 1991; Caprotti *et al.*, 1993). These indirect-acting additives are not, however, the focus of the present section.

Of the direct-acting type of fuel additive, two distinctly different aims (and periods of use) should not be confused. The *current* application concerns *aftertreatment*, i.e. regeneration of the DPF, as described below and also in Section 7.8.2, while the *earlier* application, on which the groundwork was done, concerned the *engine*, i.e. smoke suppression. A smoke suppressant or, alternatively, soot suppressant works in the combustion chamber either by impeding the formation or by promoting the oxidation of soot, for which the posited mechanisms have already been detailed in Section 3.5.1. Generally this is an aspect of the mixing-controlled burn: but if the ignition delay is prolonged, some soot suppression is accomplished in a manner akin to that of the cetane number (Section 7.2.5). Effects on the organic fraction are usually equivocal (Draper *et al.*, 1987), although emissions of PAH may be reduced slightly (Farfaletti *et al.*, 2005). The suspicion that metals increase mutagenicity, perhaps through mediating the conversion of PAH into NPAH, has already been mentioned (Section 3.6.2), but should, perhaps, be reiterated here.

Smoke-suppressing organometallic compounds once stood high in respectability and popularity; but to the wiser, contemporary eye, this seems to have been something of a Faustian pact. *Smoke* represents that particulate component which interacts most strongly with light, namely *soot*; other components interact weakly or negligibly with light, while still contributing to the mass emission; and this includes the ash fraction. This creates a risk that although the visibility of the exhaust plume improves, the same exhaust plume is, in fact, discharging more particulate mass (Hare *et al.*, 1976). The benefits, then, conferred by these additives are arguably aesthetic, or even cosmetic.

The greater part of the literature on smoke-suppression undoubtedly relates to *barium*. This metal was not without applications issues: the formation of in-engine deposits, particularly on valves, betrayed *premature* gas-to-particle conversion, that is, the metal had insufficient opportunity to enter the exhaust. The deposits were hard and tenacious, and white or grey – readily distinguishable from soot (Brandes, 1970). A mass balance revealed that 85–95 % of the barium entered the exhaust as *insoluble* barium sulphate, $BaSO_4$ (Saito T. and Nabetani, 1973); this figure should be considered in the light of fuel sulphur levels at the time of the study (0.54 %); the remainder was present as *soluble* carbonate, oxide and nitrate, $BaCO_3$, BaO and $Ba(NO_3)_2$, respectively. This compositional breakdown is of especial importance in that the carbonate possesses notable toxicity.

Part of a comprehensive investigation into the demographic implications of barium appears in Figure 7.4 (Kittelson *et al.*, 1978b). The trends are entirely logical: the accumulation mode (soot), through the action of the additive, declined, in tandem with the decline in smoke; but, conversely, a nucleation mode emerged, the magnitude of which clearly depended on the barium treat rate. This observation follows ideas already sketched at length in Section 3.5.2, wherein metals are emitted in the nucleation mode – especially when present in large concentrations compared with the carbonaceous agglomerates. Ash nucleation modes are another reason why this method of soot suppression is, arguably, a cosmetic measure – one unlikely to find favour with present-day authorities. Clearly, treat

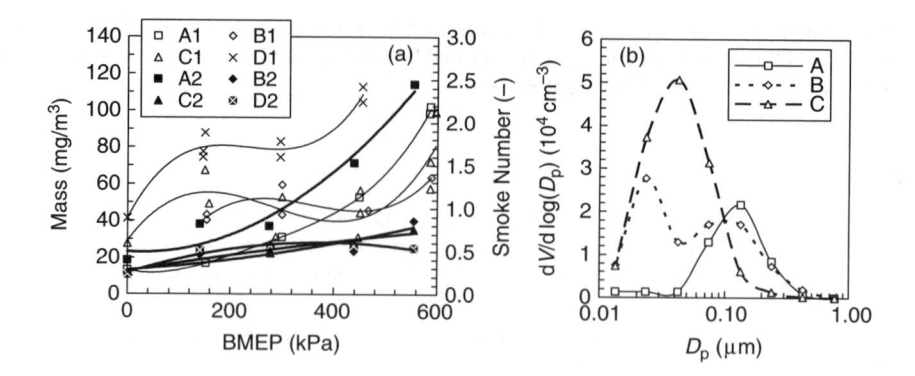

Figure 7.4 Effect of barium fuel additive on particulate emissions: (a) as a function of engine load: 1, mass concentration; 2, smoke (bold line); (b) as a function of particle size. Additive concentrations in fuel (by volume): A, none; B, 0.125 %; C, 0.250 %; D, 0.500 %. (The additive was 22.5 % barium by mass). Engine: 0.5-litre diesel, single-cylinder. Duty cycle: 1800 rpm, 150 kPa BMEP. Fuel: cetane no. 50. Particle-sizing instrument: EAA, 10 nm to 1 µm. Dilution: two stages, primary ratio 20, secondary ratio, 40–100 (Kittelson *et al.*, 1978b).

rates should not exceed those justified by the corresponding elimination of accumulation-mode soot, although this condition would be difficult, no doubt, to guarantee in practice.

The risk of fuel additives engendering emissions problems of their own, worse, even, than those they are supposed to solve, has been mostly obviated by the second-mentioned application, the DPF, in which the resulting ash is now, hopefully, captured prior to emission. The fuel treat rates, though, for the two applications, are worlds apart: a few ppm for regenerating the DPF, versus 0.1–1 % for smoke suppression, i.e. a difference of one or two orders of magnitude. But, as commonality exists between the two applications in the way these additives act towards soot, then it is not altogether unexpected, or even uncoincidental, that soot-suppressing activity in the engine should go hand in hand with the promotion of regeneration in the DPF (Valentine *et al.*, 2000). Indeed, even when additives are present in fuels at trace levels, engine-out reductions in soot emissions can be significant: this is a bonus, by which regeneration intervals are pleasingly widened, as the DPF fills with soot more slowly, there now being less soot in the feed gas. To impart regeneration successfully, the metal compounds and the carbonaceous spherules inside the DPF should be in intimate contact: the conditions for this were discussed in Section 3.5. Concomitant reductions in engine-out organic particulate are less common, but not ruled out (Burtscher *et al.*, 1999).

Nonmetal soot-suppressing fuel additives are far less reported. A purely organic (i.e. ashless) additive reduced the mass of engine-out particulate by 75 %; the volatile fraction, presumably organic, was slightly more affected than the soot (Andrews and Charalambous, 1991). The benefit arose not through the ignition delay, i.e. the cetane number, but via some other, albeit unknown, chemistry. In another paper, a nonmetallic additive reduced particle numbers, including in the nucleation mode, by 45–80 %, with no concomitant changes in size distribution (Ahmed *et al.*, 2000). The PAH were also reduced: on this basis, the authors of that paper suggested some beneficial interference with soot inception and surface growth.

7.2.5 *Volatility, Cetane Number and Density*

Diesel fuels, as they contain a welter of hydrocarbons, evaporate over *ranges* of temperature, and these evaporation ranges are referred to as 'distillation curves', specific points on which are expressed as T_x, meaning the temperature at which $x\%$ of the fuel has evaporated. The high- and low-temperature points, referred to as 'front-end' and 'back-end' volatility, are T10 ($\sim 180\,°C$) and T90 ($\sim 320\,°C$) (Owen and Coley, 1995, pp. 397–). Inevitably, fuel volatility is critical in the central process of air – fuel mixing, and thus, also, in the formation of particulate – meaning the organic and carbonaceous fractions. In the extreme case, for example the combustion of a mixture of low-volatility and high-volatility fuel, the evaporation range results in the emission of less soot, because 'flash boiling' of the first component promotes atomisation and fuel – air mixing (Fujimoto et al., 2004).

Empirical correlations between fuel volatility and particulate emissions are open to suggestion and counter-suggestion. *Positive* correlations have been observed between back-end volatility and (a) organic particulate (Bouffard and Beltzer, 1981; Rantanen et al., 1993; Tanaka S. and Shimizu, 1999), (b) carbonaceous particulate (Bouffard and Beltzer, 1981; Ullman and Human, 1991; Tanaka S. and Shimizu, 1999) and (c) total particulate (Burley and Rosebrook, 1979; Bouffard and Beltzer, 1981; Sienicki et al., 1990; Ullman and Human, 1991; Tamanouchi et al., 1997). *No* correlations have been observed between back-end volatility and (d) organic particulate (Burley and Rosebrook, 1979) and (e) total particulate (Tamanouchi et al., 1997; Bertoli et al., 1991). *Negative* correlations have been observed (f) between back-end volatility and total particulate (Desai and Watson, 1997), and (g) between T50 and organic particulate (Korn, 2001).

To dismantle this unprepossessing assemblage of black-box studies, it is worth stepping back to consider, qualitatively, the likely impact of fuel volatility on air – fuel mixing, for which there are surely two opposite extremes: *undermixing* and *overmixing*, to which can be added the two formation routes in the organic fraction (Section 3.6.1). *Direct* organic particulate denotes fuel that, in escaping the combustion, passes through the engine, and *indirect* organic particulate arises through the premature interruption of soot-forming reactions. Contrast two fuels, of high and low volatility. The high-volatility fuel will overmix: this restricts the rich, soot-generating pockets, but leads to organic particulate of the direct type. The low-volatility fuel will undermix: this promotes rich, soot-generating pockets, leading to indirect organic particulate, and, if fuel droplets survive for longer, perhaps also more direct organic particulate too. This *aperçu*, while speculative, does nonetheless help to suggest why presuppositions about volatility are sometimes wide of the mark, and also why back-end volatility has received the greater attention in controlling soot.

Similar remarks appertain to cetane number, usually in the range 45–50 (Owen and Coley, 1995, pp. 393–), which is a measure of ignition delay, and hence also of the relative proportions of premixed and mixing-controlled combustion. As mixing-controlled combustion is disproportionately responsible for the formation of soot, then a low cetane number, i.e. a long ignition delay, is expectedly beneficial. That particulate emissions accord with the duration of the mixing-controlled phase, as represented by cetane number, is thus intuitive (Li X. et al., 1996), as remarked long ago in Section 3.2.1. On the other hand, a long ignition delay furnishes the fuel with opportunity for overmixing, wall impingement, and absorption by engine deposits – factors expected to promote organic particulate, since they temporarily shelter hydrocarbons from the combustion. Taken to the extreme, such an understanding explains the strong association between low cetane number and white smoke (Wade and Jones, 1984), as explained in Section 3.6.5. The final piece in the jigsaw is duty cycle, i.e. premixed combustion assumes an ever-smaller proportion of the whole burn as engine load increases.

The empiricisms run as follows. *Positive* correlations have been observed between cetane number and (a) total particulate (Desai and Watson, 1997; Bielaczyc et al., 2003) and (b) carbonaceous particulate (Tanaka S. and Shimizu, 1999). *No* correlations have been observed between cetane number and (c) organic particulate (Tanaka S. and Shimizu, 1999) and (d) total particulate (Rantanen et al., 1993). *Negative* correlations have been observed between cetane number and (e) organic particulate (Arai, 1992; Tamanouchi et al., 1997) and (f) total particulate (den Ouden et al., 1994; Schulz et al., 1999); (g) a minimum in carbonaceous particulate is reported for cetane numbers of ~ 46 (Wade and Jones, 1984). This widespread disagreement can possibly be elucidated by similar ideas to those advanced for volatility.

But an obvious problem here is strong interdependency between cetane number and aromaticity. This confounding is avoidable if fuel *additives* are used to adjust the cetane number, rather than modifying the hydrocarbon speciation. A positive correlation between cetane number and total particulate results (Burley and Rosebrook, 1979; Li X. *et al.*, 1996), but emissions for different fuels can just as well be dissimilarly affected (Cunningham L.J. *et al.*, 1990) or unaffected (Sienicki *et al.*, 1990); and a negative correlation has also been reported (Yoshida *et al.*, 1986). Illustrative data are given in Figure 7.5(a) (Kwon *et al.*, 2001). The increase in carbonaceous particulate at high cetane number evidently outweighed, by a considerable margin, any decrease in organic particulate (Schulz *et al.*, 1999); yet a contrary trend is also known (Schwab *et al.*, 1999).

Data are scant on the implications of cetane number for the composition of the organic fraction. Low cetane number has been associated with emissions of *Flu*, *Pyr* and *Chr* (Arai, 1992), and a fivefold increase in the mass emission of PAH (Lepperhoff and Houben, 1990). Demographic characterisations are scant too; a low cetane number appears to reduce particle numbers (Tanaka S. and Shimizu, 1999), which, perhaps, relates to the accumulation-mode soot; indeed, lower agglomeration rates may serve to restrict particle growth (Kittelson *et al.*, 1978b). Unfortunately, what is gained on the swings is lost on the roundabouts, as the higher organic – carbonaceous ratio for fuels of low cetane number appears to encourage the formation of a nucleation mode (Kittelson *et al.*, 1978b).

Turning now to the third parameter to be examined in this section, density, this is $\sim 0.84-0.86\,\text{g/cm}^3$ (Owen and Coley, 1995, pp. 404–). It has been concluded that high-density fuels generate more particulate, mostly in the carbonaceous fraction. Illustrative data are given in Figure 7.5(b) (Fløysand *et al.*, 1993). In explanation of these data, one might, for example, posit some effect on spray geometry, for example greater penetration if droplets are heavier; but there are almost certainly various interwoven effects.

The observed influence of fuel density is, however, according to other workers, demonstrably quite artificial, and not simply because of confounding by interrelationships among various fuel properties.

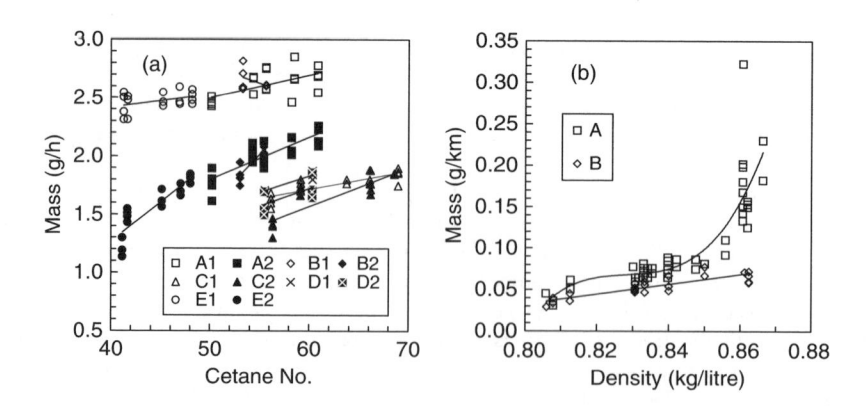

Figure 7.5 (a) Effect on particulate emissions of cetane number: 1, particulate; 2, soot; for specifications of the five fuels A – E, including cetane improvers, see the original paper. Duty cycle: four steady-state points selected from NEDC. Engine: 2-litre diesel, four-cylinder, direct-injection (Kwon *et al.*, 2001). (b) Effect of fuel density; A, with oxidation catalyst; B, without oxidation catalyst. Vehicle: 2.1-litre, indirect-injection, with EGR. Duty cycle: NEDC. For fuel properties, see the original paper (Fløysand *et al.*, 1993).

Fuel metering is volumetrically based, and engine management systems are obviously designed for a fuel density of a certain specification (Desai and Watson, 1997). Introduction of a low-density fuel, if uncompensated, will modify the emissions, if only because the engine now develops a lower load for the same pedal position. At part load, the driver compensates by further depression of the pedal. Once this effect is taken into account, around half the emissions benefit disappears: this part, then, is only 'apparent' (Rantanen *et al.*, 1993); others declare all the emissions benefit factitious (Heinze *et al.*, 1996); and similar warnings are given by others (Kwon *et al.*, 2001). A large proportion of the 'apparent' density effect, therefore, is traceable to incestuous interactions between the fuel formulation and the engine management system, through timing, EGR, etc.; and there is, here, a problem in making apple-with-orange comparisons. Finally, that this dependency declines for catalyst-fitted vehicles suggests a less investigated effect via the organic fraction (den Ouden *et al.*, 1994); see again Figure 7.5(b) (Fløysand *et al.*, 1993).

7.3 Fuel Injection

The characteristics of the fuel spray – chiefly *divergent angle* and *penetration depth* (Desantes *et al.*, 2006) – insofar as they control fuel atomisation, vaporisation and dispersion into the surrounding air, determine also the quality and thoroughness of the combustion. This is an extremely large subject; and the present section, for brevity, explores only cursorily *how* the fuel injection is linked, via the fuel spray, to the particulate emission. Another topic which cannot be addressed here is optimisation of combustion chamber geometry with respect to the fuel-spray characteristics (Bergstrand and Denbratt, 2003). But, to a first level of understanding, the spray must mix, optimally, the fuel with air; for, if the fuel is *under*mixed, rich pockets experience pyrolytic soot-forming reactions, and, if *over*mixed, lean pockets fail to burn, generating organic particulate.

Equally, it is not germane here to compare and contrast the numerous types of contemporary fuel injection equipment (FIE), that is, the inner workings of the hardware, including the fuel injector and the fuel pump. But a point worth emphasising is the significance of FIE as a major enabler in meeting today's stringent emissions legislation; and chief amongst these advances is the introduction of 'common rail' technology, by which the range of choices in injection parameters (pressure, timing and multiple events) is greatly widened (Yamaki *et al.*, 1994; Guerrassi and Dupraz, 1998; Schommers *et al.*, 2000).

7.3.1 The Injector Nozzle

In the injector nozzle, fuel is forced through a narrow gap or orifice of between one hundred and two hundred microns in size. The distortions introduced into the streamlines in negotiating this bottleneck engender localised pressure variations, flow separation, and the formation of a vena contracta wherein the flow cavitates (Payri R. *et al.*, 2005). This flow field, in turn, determines to a substantial degree the ensuing characteristics of the fuel spray upon emerging from the nozzle, making the *design* of the orifice a factor in soot formation, i.e. the area, the length-to-diameter ratio, the orientation with respect to the angle of entry, whether the ends are round-edged or sharp-edged (Su T.F. *et al.*, 1995), and the 'conicity', i.e. whether divergent, convergent (Bergstrand, 2004) or cylindrical (Petkar *et al.*, 2004).

An undoubted trend in recent years is a gradual diminution in orifice *diameter*, the rationale for which is concomitant reduction in droplet *size* – thus giving a leg-up, so to speak, to fuel atomisation and evaporation, avoiding unacceptably rich-burning, soot-producing pockets. This tidy, and intuitively reasonable, explanation cannot be a universal truth, as *greater* rather than lesser emissions of soot are sometimes reported (Nikolic *et al.*, 2001), the directional trends depending, to some degree, on engine operating point (Bergstrand and Denbratt, 2001). Explanations for this ambivalence are multilayered, but one all-too-obvious and countervailing supposition arises from precisely the opposite observation, that larger droplets are *also* associated with less, rather than more soot (Montgomery *et al.*, 1996): simply, greater momentum ensures a more deeply penetrating spray, and hence improved air utilisation (El-Hannouny *et al.*, 2003). Small droplets are not from this standpoint necessarily advantageous. Other reported facets of smaller-droplet sprays are denser clouds of soot, less susceptible to oxidation (Koyanagi *et al.*, 1999); foreshortened liquid-phase segments (Zhang L. *et al.*, 1997); lesser propensity for impingement on combustion chamber walls (Cartellieri and Wachter, 1987); and a shorter ignition delay (Bergstrand and Denbratt, 2003). Ultimately, this multifaceted subject may boil down to the model of the combusting plume described in Section 3.3.3, in that the premixed flame, in the mixing-controlled phase, is leaner and so less soot-producing.

Another well-worn aspect of nozzle geometry, of relevance to particulate emissions, concerns events immediately following the end of injection, and the small reservoir of fuel lodged within the nozzle vestibule, known as the 'sac'. Here lies the purported advantage of the 'mini-sac', or of the valve-covered orifice (VCO), in which the sac is eliminated (Bae C. *et al.*, 2002), and which is held to be less soot-producing (Leipertz *et al.*, 2002), although a drawback is greater variation in spray characteristics between orifices. The problem is that, in the traditional injector, following end of injection, the smidgen of fuel in the sac, in the late-combustion phase, evaporates or dribbles into the combustion chamber in an uncontrolled manner – in effect, constituting an uncontrolled form of metering. The significance of this uncontrolled drippage is easily underestimated: the volume of the sac might, conceivably, represent several per cent of the total fuel quantity, particularly at low load, i.e. low fuelling, where it assumes a greater fraction of the whole. This escape is not, however, a foregone conclusion, as cooler nozzle temperatures at low loads may restrict evaporation (Myers, 1983).

More pertinently, this late-escaping but nonburning fuel contributes, inevitably, to emissions of HC, which, in crank-angle-resolved traces, appear as late-emitted spikes (Myers, 1983). But this fuel does not emerge solely as HC, as many of these hydrocarbons readily undergo gas-to-particle conversion in the exhaust stream. Figure 7.6(a) (Andrews *et al.*, 2000b) shows, as measured by methods already described (Sections 3.6.4 and 5.5.4), unburned hydrocarbons prevailing at two FID temperatures, and organic particulate at 50 °C, as a function of sac volume. Figure 7.6(b) suggests a correlation of sorts between several PAH in the organic fraction, and the sac volume; but since this correlation is evidently weak, other factors are perhaps at work (Farrar-Khan *et al.*, 1992); the sac volume is obviously not the only emission mechanism.

But this fuel is not simply emitted as such: because of the poor atomisation and low temperature, these hydrocarbons burn poorly in the combustion chamber, generating partial-reaction products, and these also enter the organic fraction (Saito T. *et al.*, 1982). Speciation of the organic fraction shows that some compounds are more concentrated and others less concentrated than accountable for by mere fuel survival; this suggests unequal evaporation from the sac, or unequal combustion among the hydrocarbons after they emerge; indeed, it seems reasonable to suspect that pyrolytic PAH-forming reactions act, to some extent, on this late-escaping fuel (Andrews *et al.*, 1998b).

7.3.2 Injection Pressure

The sooting propensity of the diesel engine undoubtedly hangs, to an extremely large degree, on the pressure at which fuel is injected into the cylinder: this is apparent if only because, over the last three decades, soot emissions have fallen as inexorably as injection pressures have risen; for example, where high-speed direct-injection diesels once used less than 1000 bar, they are now fast encroaching on 2000 bar; and, in the research community, 2600 bar (Wickman *et al.*, 2000) or even 3000 bar (Dodge

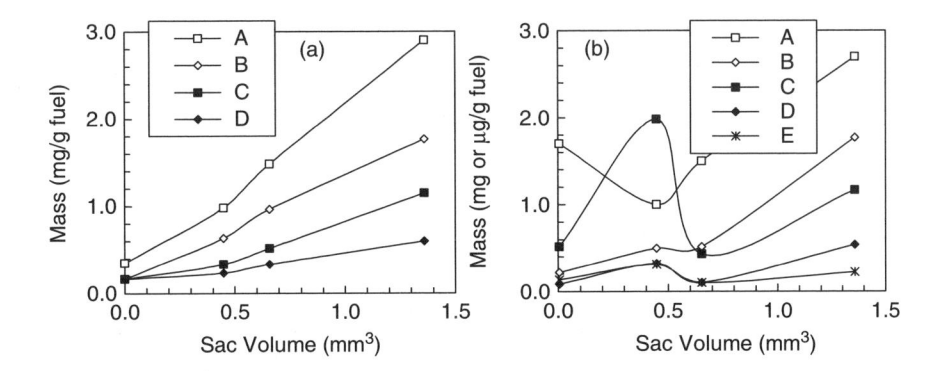

Figure 7.6 Emission of HC and organic particulate, as influenced by the injector-nozzle sac volume. (a) A, unburned hydrocarbons, 180 °C; B, unburned hydrocarbons, 2 °C; C, unburned hydrocarbons, between 180 °C and 2 °C; D, organic fraction collected at 50 °C; the hydrocarbons were measured using a FID and two different sample-conditioning temperatures. Engine: single-cylinder diesel, direct-injection, naturally aspirated. Operating point: 1500 rpm, 60 N m (Andrews *et al.*, 2000b). (b) A, total particulate (mg/g); B, organic fraction (mg/g); C, *Phe* (μg/g); D, *Flu* (μg/g); E, *Pyr* (μg/g). Engine: 4.2-litre, reduced to single-cylinder. Injectors: four holes per nozzle, 20° offset angle, 150° spray cone angle, opening pressure 215 bar. (Farrar-Khan *et al.*, 1992).

et al., 2002) is known. This inexorable increase will presumably be arrested eventually, if only through practical limitations in FIE, relating, say, to stresses and strains undergone by various components in the high-pressure fuel supply, and the ensuing design and manufacturing costs.

The benefit of high injection pressure in reducing emissions of soot is readily demonstrable: but accounting for this benefit mechanistically, i.e. in terms of what is known about soot formation mechanisms, is no simple matter, and researchers have invoked a surprisingly wide gamut of explanations. Simplistically, one might argue that the higher the injection pressure, the more thoroughly atomised is the fuel: a full discussion of droplet fragmentation mechanisms would be impossible here (Gelfand, 1996), but the expectation is one of concomitantly improved contact, or mixing, between fuel vapour and air, so that rich-burning, soot-producing pockets are now less prevalent. Another explanation is that deeper penetration (El-Hannouny *et al.*, 2003) and greater air entrainment (Kamimoto *et al.*, 1987) enhance air utilisation, which serves not only to discourage the formation but also to promote the oxidation of soot (Wadhwa *et al.*, 2001), especially in the later stages (Kobayashi *et al.*, 1992; Wakisaka and Azetsu, 2002), perhaps facilitated by more vigorous turbulence (Ikegami *et al.*, 1988). More mundanely, the (relatively soot-free) premixed burn may also account for a greater fraction of the fuel (Kamimoto *et al.*, 1987). Finally, one particular explanation dovetails plausibly with contemporary notions of the combusting plume during its mixing-controlled phase (Section 3.3.3): greater amounts of air entrainment upstream of the premixed flame (Tree and Svensson, 2007), so that subsequent soot-forming reactions along the plume stem are arrested.

At the same time, it should be admitted that the intended benefit does not always emerge, for which several explanations may also be advanced. A small-droplet spray emerging at high velocity decelerates more rapidly than a large-droplet spray emerging at low velocity (Binder and Hilburger, 1981). If the fuel droplets are too small, their momentum is insufficient to carry them far into the surrounding air, or their evaporation is completed more swiftly (Zhang L. *et al.*, 1997); the fuel spray is then *less*, rather than more penetrating, and the rich-burning pockets are not, therefore, alleviated. On the other hand, if the fuel spray penetrates *too* deeply, droplets impinge on the wall of the combustion chamber, where they bounce or splash (Lu *et al.*, 2001), to form a fuel film adhering to the wall (Sjöberg, 2001), which burns poorly, with a soot-producing flame. Perhaps, on occasion, high injection pressures promote soot formation more than they promote soot oxidation (Kamimoto *et al.*, 1987). Finally, if the fuel is overmixed, the charge becomes too lean to burn, and unconsumed hydrocarbons are released from the engine: the increase in organic fraction then outweighs the decrease in carbonaceous fraction (Otani *et al.*, 1988).

Passing now to the *demographic* implications, these are particularly apposite, as the automotive industry has relied so extensively on injection pressure with which to meet contemporary emissions legislation. Although spherules reportedly become slightly smaller as injection pressures increase (Schraml *et al.*, 2000), the reason why the *mass* of soot declines is largely because fewer spherules are formed; hence, agglomerates are now smaller and less abundant. (See Figure 7.9.) Moreover, a considerable controversy hangs over a possible side effect: whether these great reductions in the mass emission of soot – and also, of course, in smoke – accomplished through high injection pressure have inadvertently promoted, *contrariwise*, the emission of more, smaller particles – that is, nanoparticles (Raatz and Mueller, 2001). Some data showing, with increasing injection pressure, a decline in the accumulation mode and a concurrent growth in the nucleation mode are given in Figure 7.7 (Bertola *et al.*, 2001). The nucleation mode consisted of volatile material, in that thermal conditioning at 350 °C eradicated it. Such behaviour is hardly arcane: the wholesale reduction of soot, if carried out *preferentially*, or perhaps one might say injudiciously, removes solid surfaces on which gas-to-particle conversion might take place – with obvious implications for the saturation experienced by *other* particulate components. This concerns not just sulphates and organics; but whether in-cylinder nucleation in ashing elements is encouraged also, remains to be ascertained.

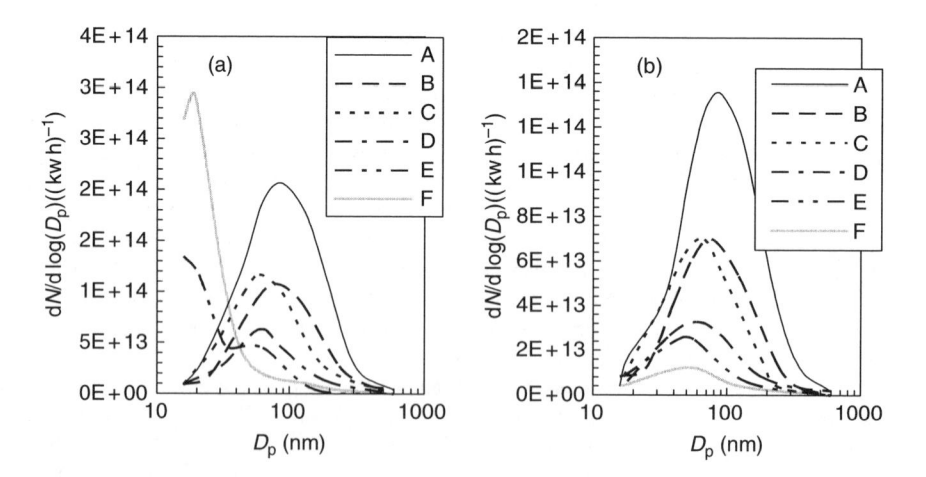

Figure 7.7 Effect of injection pressure on number distribution: (a) without thermodesorber; (b) with thermodesorber, 350 °C. Pressure (bar) and start of injection (BTDC): A, 300/8; B, 300/16; C, 500/6; D, 500/13; E, 900/5; F, 900/12. Engine: 2.1-litre diesel, single-cylinder. Operating point: 1180 rpm, 25 % maximum load. Fuel: commercial grade, sulphur 50 ppm. Instrumentation: SMPS (Bertola at al, 2001).

7.3.3 Injection Scheduling

The decisions to be made in injection *scheduling* are essentially twofold: how much fuel to inject (*quantity*), and when exactly to inject this fuel (*timing*). Naturally, these choices are subject to numerous constraints not germane to the present section, the *fundamental* one of which concerns, inevitably, the whole purpose of the engine, namely, delivery of the desired load, i.e. commensurate with driver demand.

The pertinent issue here is controlling emissions of *particulate*. Pared to its basic elements, scheduling alters the net production of soot simply because formation and oxidation are affected differently. Perhaps, therefore, a timing which encourages formation but which encourages oxidation far more introduces some benefit in the *net* production of soot. This principle has already been examined at various points in the present text. On a different tack, if the ignition delay, which is a function of timing, is prolonged, the premixed phase, which is not particularly soot-producing, will assume a greater proportion of the whole burn; but, on the other hand, greater opportunity is furnished for overmixing, in which case unburned fuel is discharged as organic particulate. The implications for the other two fractions are subsidiary; but worth mentioning is a possible impact, via combustion temperature, on sulphate particulate, via the SO_2–SO_3 chemistry (Section 3.7.1). Taken together, these controlling mechanisms may, at times, reinforce one another, and, at other times, oppose one another, thus precluding all-embracing statements as to the net effect of injection scheduling on particulate emissions: various trends, dependent on operating point, are observed among the carbonaceous, organic and sulphate fractions (Kweon *et al.*, 2003b).

Prior to the mid 1990s, the scheduling task was simple enough, since diesel engines universally used just one, single injection event: needle open, needle shut. Unfortunately, with such a limited choice it proved impossible to keep abreast of rapidly tightening emissions legislation, and hence manufacturers of FIE began providing additional features. For a time, 'boot-shaped' injections were very much in vogue, in which injection rates, mid-injection, were suddenly increased (Desantes *et al.*, 2004). But the sea change came with the introduction of *multiple injections* – opening up numerous vistas down which researchers are only just venturing. Multiple injections are proving enormously advantageous in negotiating the many and interwoven trade-offs between emissions, fuel economy and driveability. At the time of writing, interest is burgeoning in *pilot* injection and *post*-injection, the first preceding and the second succeeding the main injection; and some pundits believe that, eventually, *two* pilot, *two* main and *two* post-injections will ultimately be necessary – at least at some operating points.

The burning question is how best to exploit this technology, it being something of a double-edged sword: for example, four injections, engendering a choice of four timings and four quantities, present the engineer with a formidable optimisation problem, the complexity of which will be appreciated since, with one injection alone, once the timing is selected the quantity then follows automatically, as the engine must deliver the required load. With multiple injections this clear-cut relationship is lost since, obviously, there exists any number of possible permutations. However, this is not the place to discuss the optimisation of injection schedules; the interested reader is referred to the prodigious literature (e.g. Nehmer and Reitz, 1994; Kong *et al.*, 1995; Chen S.K., 2000; Bianchi G.M. *et al.*, 2001; Montgomery and Reitz, 2001; Badami *et al.*, 2002; Greeves *et al.*, 2003; Lee T. and Reitz, 2003; Zhang Y. and Nishida, 2003; Park C. *et al.*, 2004).

Our present concern is how these split injections affect emissions of particulate (principally of soot). If brief interruption renders the fuel spray less penetrating, then this diminishes the likelihood of soot formation through spray – wall interactions (Bianchi G.M. *et al.*, 2001). The splitting of soot clouds, under the principle of 'divide and rule', is to be welcomed inasmuch as this increases vulnerability to oxidants (Wickman *et al.*, 2000). But the pilot and post-injections operate on soot dissimilarly in certain other respects. The principle use of pilot injection is to control the ignition delay of the main injection, often enough with the aim of reducing combustion noise. Any exacerbation of soot is then

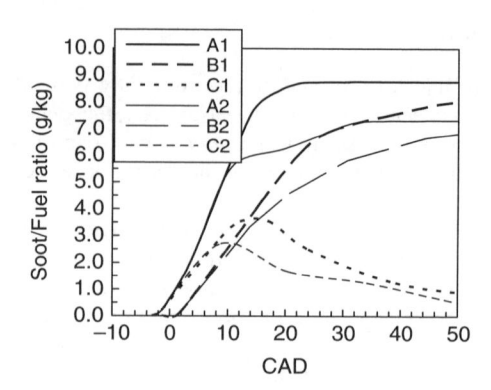

Figure 7.8 Modelled soot formation (A), oxidation (B), and net production (C), for two different injection schedules: single injection (1) and double injection (2), for which the first:second ratio of the fuel quantity was 75%:25%, with 8°CA dwell between; start of injection 10°ATDC. Engine: 2.4-litre, single-cylinder. Operating point: 1600 rpm. Fuel: tetradecane, $C_{14}H_{30}$. Injection pressure: 60 Mpa (Han Z. et al., 1996).

a side effect: the main injection either has a longer mixing-controlled burn (Campbell P.H. et al., 1995) or follows the pilot *too* closely, entering an already rich, high-temperature region (Chi et al., 2002). The period of cessation between the two injections is thus critical. The soot emission is reduced by post-injection, in some cases significantly, although a weaker consensus exists as to the cause. Turbulence is imparted to the remnants of the main injection, promoting either air – fuel mixing, leading to less soot formation (Zhang Y. and Nishida, 2003), or soot – air mixing, leading to more soot oxidation (Kong et al., 1995). It could also be that combustion of this post-injected fuel allows high temperatures to persist, furthering late-phase soot oxidation (Payri F. et al., 2002). The flip side is that post-injections may have a cooling effect, and thus impede the burn-up; or arrive too late, when the temperature is already insufficient (Badami et al., 2002). Again, the period of cessation is critical.

Modelling indicates that, with multiple injections, the instantaneous concentration of soot differs quite significantly from the case of single injection. Figure 7.8 compares a single injection and a split injection (Han Z. et al., 1996). Oxidation and formation are both affected, but *not* equally: engine-out soot for the split injection was lower by a factor of four; the predominant reason was impeded formation. With a single injection, a rich mixture is continually replenished by the arrival of more fuel, whereas with a split injection, this rich zone is no longer sustained, and the brief suspension ensures a leaner reception for the following injection. Obviously, the length of the cessation is central: it must be long enough to avoid replenishing the rich zone, but not too long, otherwise the second injection fails to benefit from the high-temperature environment left by the first injection. This explanation glosses over various interesting intricacies of multiple injections that are further explored elsewhere (Hampson and Reitz, 1998; Zhang Y. and Nishida, 2003). It should be said that the relative quantities of fuel distributed between the various injections are also critical.

Pilot and post-injections are not without their implications for particle demography, although, in comparison with measurements of particulate mass, these aspects have received scant treatment. Some demographic characterisations, comparing the two conditions of pilot and no pilot, are shown in Figure 7.9 (Jaeger et al., 1999): in this case the pilot injection brought about a diminution in the accumulation mode. Others register a growth (Bertola et al., 2001), while very similar effects are observed for post-injection (Bertola et al., 2001). Taken together, these demographic characterisations simply reflect the mass measurement insofar as this is dominated by accumulation-mode soot. The

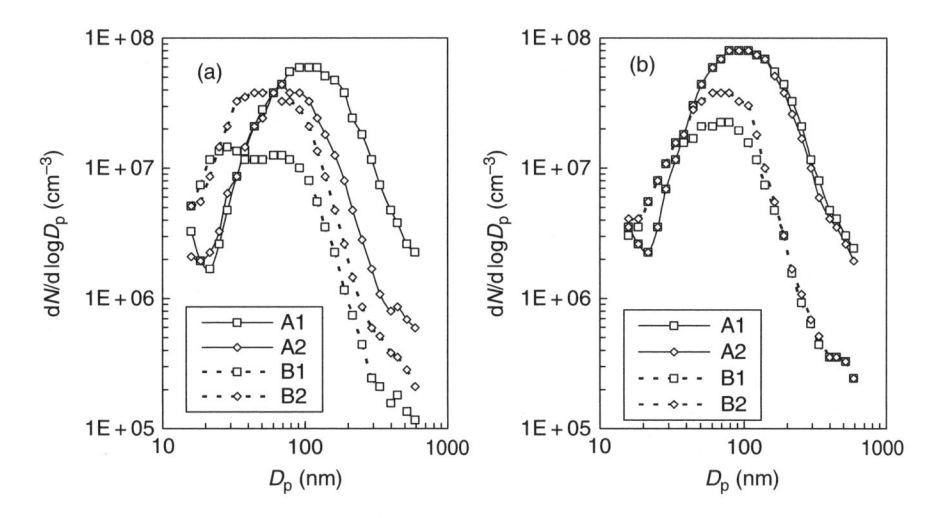

Figure 7.9 Number distribution, as affected by injection pressure and injection scheduling: 1, with pilot; 2, without pilot; A, injection pressure 40 MPa; B, injection pressure 120 MPa. Injection timing: (a) SOI 6°BTDC; (b) SOI at TDC. Engine: 2-litre, four-cylinder, common-rail. Operating point: 2000 rpm, 2 bar BMEP. Fuel: sulphur <50 ppm. Instrument: SMPS, 16 to 630 nm (Jaeger *et al.*, 1999).

corresponding formation of a nucleation mode (Raatz and Mueller, 2001) is easily explained if less surface area is available on which volatile particle precursors may condense: again, this follows naturally from injection schedules which preferentially reduce the soot emission (Kweon *et al.*, 2003b).

7.4 Exhaust Gas Recirculation

A fraction of the exhaust gas, if bled off, conveyed to the engine intake, and mixed with the inducted air[1] – an ostensibly rather strange practice, to the outsider – induces extremely wide-ranging effects on the combustion. But although the effects are wide-ranging, exhaust gas recirculation (EGR) has one object, and *only* one object: to decrease engine-out emissions of NO_x. Thus, this is the only emission control strategy discussed in the present chapter which does not actually seek to address emissions of particulate; *conversely*, this technology is included by virtue of its principal side effect: a vexing and ever-present corollary of EGR is the leg-up given to soot. This increase in soot mass is not generally linear, but curves upwards, accelerating as more EGR is scheduled. Effects on other particulate fractions are equivocal and, by comparison, normally rather small: the sulphur equilibrium, it is argued, shifts in favour of SO_3, bringing a concomitant, but slight, increase in sulphate particulate (Kreso *et al.*, 1998); and, courtesy of higher-temperature exhaust, the late-phase burn-up may improve, thus explaining instances where a decrease in organic particulate partially offsets, and occasionally outweighs (Kawano *et al.*, 2004b), the increase in carbonaceous particulate.

---

[1] The discharge of exhaust gas into the airstream entering the engine is a very similar situation to that pertaining in a dilution tunnel – although this parallel has not, apparently, been pursued in the literature.

The impossibility of meeting contemporary limits for NO_x by any other means makes EGR essential in many applications.[2] The aim, therefore, is to ameliorate or to contain, as far as practical, the deleterious side effects: no more EGR should be scheduled than absolutely necessary. Moreover, the air – exhaust mixture must be properly homogenised in the inlet manifold, so that all cylinders operate at identical points on the trade-off (William *et al.*, 2003); otherwise, say, one cylinder emits too much NO_x (too little EGR), and another too much soot (too much EGR), and this inhomogeneity thus compromises the whole.

The soot–NO_x trade-off depends, centrally, on engine operating point; hence, careful scheduling, or metering, of EGR according to duty cycle is mandatory. At a first level, this scheduling depends on the available oxygen, a surfeit of which exists at low load and at idle, where, as a result, EGR is used more freely; conversely, smoke emissions prohibit EGR at high load. EGR is thus scheduled automatically, via the position of the EGR valve, according to engine operation and preprogrammed maps calibrated by the vehicle manufacturer. Some systems are open-loop; others are closed-loop, using the signal from the MAF sensor. Actually, steady-state control is not usually an issue: the greater control problem is engendered by *transient* operation, during which spikes in emissions of soot or of smoke reveal brief instances of excessive EGR. These shortcomings might be addressed, for example, using exhaust gas oxygen (Chen S.K. and Yanakiev, 2005) or carbon dioxide (Sutela *et al.*, 2000; Green R.M., 2000) as feedback. These issues are not restricted to EGR, but form part of induction system design and air path management (Section 7.5).

A mammoth literature testifies, inferentially, to the huge importance of EGR; but only brief descriptions, *en passant*, of its wide-ranging effects on combustion are possible here, and these effects are necessarily restricted to particulate – and, especially, to soot. As revealed already at numerous points (see particularly Section 3.3.1), instantaneous, in-cylinder soot – the net result of shifting competition between soot formation and soot oxidation – is quite different from engine-out soot, which, as aforesaid, increases. Engine-out soot is largely unrevealing as to the exact mechanisms through which EGR acts, whereas observations and models of in-cylinder soot reveal various subtleties.

Figure 7.10(a) depicts a model of in-cylinder soot concentration, for various rates of EGR. Soot emerges consistently at around the same moment, yet, as EGR continues to increase, peak concentrations become slightly retarded, a feature attributable to slower rates of combustion (Kyriakides *et al.*, 1986). But the *gradients* reveal that, *with* EGR, soot concentrations rise more rapidly, yet also fall more rapidly; evidently, the second factor insufficiently compensates for the first. The figure, then, suggests that more soot is emitted from the engine simply because more soot forms. On the other hand, reasonable correlations of engine-out soot with late or end-of-combustion parameters such as oxygen and heat release have been used to suggest that oxidation is the more affected (Desantes *et al.*, 2000); and this is corroborated by direct measurements (Hentschel and Richter, 1995). Instances are also known where EGR appears to *depress* in-cylinder soot (Arcoumanis *et al.*, 1995; Bertoli *et al.*, 1998; Bruneaux *et al.*, 1999). It should be admitted that such observations may be an artefact of the experimental methods: nevertheless, this is a reminder that peak in-cylinder soot and exhaust soot are horses of very different colours, and not necessarily affected in a directionally consistent manner; perhaps *less* rather than more soot forms, in which case the higher exhaust concentrations must be the result of an effete burn-up.

It should first be said that the relative proportions of premixed and mixing-controlled burn are modified, as the ignition delay first shortens slightly, and, for very high EGR rates, lengthens dramatically (Pitsch *et al.*, 1996). These high EGR rates are currently under intensive research for use in alternative combustion systems (Section 7.7); the present section focuses only on what is 'traditional'. So, although alterations in ignition delay account for *some* of the effect of EGR on soot emissions,

---

[2] This situation will change if a breakthrough eventually occurs in the technology of $deNO_x$ aftertreatment, which is the principal impediment in NO_x control at the moment.

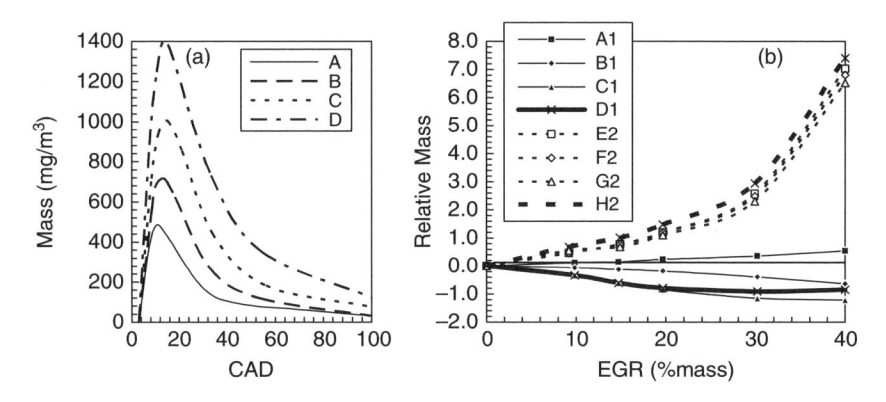

Figure 7.10 Effect of exhaust gas recirculation on particulate emissions. (a) As modelled, in-cylinder soot concentration. EGR rates (%): A, 0; B, 20; C, 37; D, 53. Engine: 1.9-litre diesel, direct-injection. Operating point: 2000 rpm/8 mg fuel quantity, start of injection, 8°BTDC (Pitsch *et al.*, 1996). (b) With the various formative mechanisms isolated. Mass is expressed as relative to no EGR (g/kg of fuel). Organic fraction (1): A, dilution mechanism (CO_2 and H_2O) only; B, with thermal mechanism (CO_2 and H_2O) subtracted; C, with chemical mechanism (CO_2) subtracted; D, with chemical mechanism (H_2O) added (final result shown as bold line). Carbonaceous fraction (2): E, dilution mechanism (CO_2 and H_2O) only; F, thermal mechanism (CO_2 and H_2O) subtracted; G, chemical mechanism (CO_2) subtracted; H, chemical mechanism (H_2O) added (final result shown as bold line). Engine: 2.5-litre, diesel, four-cylinder, direct-injection, naturally aspirated. Operating point: 2000 rpm, 40% maximum load, start of injection, 10°BTDC (Ladommatos *et al.*, 1997b).

this is minor. The greater and less readily accountable effect is restricted to the mixing-controlled burn. Obviously, on a simplistic level, the displacement of some oxygen in the air intake promotes soot-forming reactions, and, equally arguably, impedes soot-oxidising reactions. But, as mentioned, EGR operates on many levels, and, therefore, this simplistic explanation does not account for all observations.

This brings us to a series of landmark papers (Ladommatos *et al.*, 1996a,b,c,d, 1997a,b, 2000) which examined *three* distinct EGR mechanisms: *dilution*, or replacing part of the oxygen in the air; *thermal*, or raising the heat capacity of the charge; and *chemical*, or supplying reactive and dissociated species with the exhaust gas; each of them may operate through recirculated combustion products (CO_2 and H_2O). This *ménage à trois* has proven extraordinarily illuminating: although the experimental details need not detain us, it is possible to *isolate*, by the judicious selection of certain diluents, each mechanism in turn: that is, supplying the engine *not* with recirculated exhaust gas, but with synthesised mixtures of gases (e.g. carbon dioxide, nitrogen, argon, helium, . . .).

By these methods, the particulate emission, as depicted in Figure 7.10(b), has been divided according to the three mechanisms (Ladommatos *et al.*, 1997a). This figure shows the soot fraction to be controlled principally by the dilution effect (H_2O and CO_2); and the organic fraction to be controlled by all three effects. It should, moreover, be noted that some effects are *negative*, i.e. they reverse the action of others, and so decrease the emitted mass. Curiously, the linear summation of these three mechanisms, acting via H_2O in *isolation*, fails to account for the emission of organic particulate observed when the same three mechanisms act *together* (Ladommatos *et al.*, 1997a): this suggests an unknown, *nonadditive* effect – perhaps relating to the ignition delay. The *ménage à trois* also explains, perhaps, any lack of consensus that may exist as to how EGR affects the combustion process: the specifics are likely to be dependent on operating point and calibratables (timing, etc.): others suggest a dominant thermal effect, although their test gases were not as extensive (Mitchell *et al.*, 1993).

The return of exhaust gas to the intake and its passage through the engine, to be returned yet again, etc., engenders a merry-go-round, for which the intriguing question is posed of whether soot particles in the exhaust gas formed in the immediately preceding cycle, or whether they were *already*, at some earlier stage, in the exhaust, and returned to the intake, after which they survived transit through the engine, to emerge a second time. Of course, this question of *déjà vu* applies equally to other recirculated emissions: upon adjusting to a new EGR rate, the system comes into balance swiftly, so that engine-out soot does not continue to rise with advancing cycles.

Particles entering the cylinder with the air are located in a different place on the merry-go-round, *outside* the combusting plume, although this distinction has not, apparently, been addressed. Experiments in which the engine intake was seeded with particles of carbon black suggested that most (>80%) of the recirculated particles are, in fact, destroyed, so that those emerging from the engine are, predominantly, freshly created (Kittelson *et al.*, 1992). The increased emissions of soot with EGR, according to this understanding, are the result of more soot production in the engine, rather than of soot being returned to the intake and passed again into the exhaust. The two-micron particles of carbon black used in this study were, admittedly, of an unrepresentative size, but, according to what is understood about the size dependency of soot oxidation, still smaller particles would be annihilated on shorter timescales, and so more rather than less likely to disappear. Experiments in which recirculated exhaust gas was *filtered* again suggested that recirculated particles make little contribution to the engine-out emission: most particles are newly formed (McTaggart-Cowan *et al.*, 2006). Thus, purely in terms of soot formation, little is to be gained by filtering the recirculated gas; indeed, the recirculated particles did not seem to affect any other aspect of the combustion process. The motivation for filtering the recirculated exhaust gas arises instead from considerations of engine wear. The flip side is that exhaust gas returned in a cooler, low-pressure circuit is far more likely to transgress the dew point, the corollaries of which are corrosion from the condensation of sulphuric acid (McKinley, 1997), and damage to compressor blades caused by the impact of hydrocarbon droplets, not to mention other fouling problems in the air path (Mueller V. *et al.*, 2005).

The prodigious literature devoted to EGR and particulate mass or soot mass highlights, by contrast, the slender database of demographic characterisations. EGR, in increasing emissions of soot (see Figure 7.3), logically increases also the magnitude of the accumulation mode (Tsolakis *et al.*, 2005). This is not startling. Spherules are a few nanometres smaller (Smallwood *et al.*, 2002), with a narrower size distribution (Kawano *et al.*, 2004b). Not so obvious is what happens to the nucleation mode: this appears to be suppressed, or even eradicated (Kreso *et al.*, 1998; Jaeger *et al.*, 1999; Bertola *et al.*, 2001; Kawano *et al.*, 2004b); the likely reasons for this are not arcane: first, the larger surface area, in the accumulation mode, scavenges particle precursors, and thereby suppresses saturation; second, higher exhaust temperatures also suppress saturation; and third, organic particle precursors are better oxidised by the higher late-combustion temperatures.

Lastly, EGR has ramifications over and above those relating to emissions *per se*. One is the entry of soot into lubricant (Section 7.6.2); another is the encouragement given to carbonaceous deposits in the combustion chamber (Singh *et al.*, 2006). The subject addressed here is fouling of internal surfaces, such as within the EGR valve, the seating of which is compromised, leading possibly to sticking or leakage, and within the EGR cooler, the pressure drop along which is raised, and heat transfer through the wall of which is attenuated (Majewski and Pietrasz, 1992). This situation is unfortunate for systems using *cooled* EGR, which, *ipso facto*, create excellent conditions for thermophoretic deposition. Sulphuric acid, the corrosive action of which is well known, is also encouraged to condense onto and corrode metal surfaces (Usui *et al.*, 2004; Takakura *et al.*, 2005).

The physics of particle deposition onto (and particle removal from) these walls is actually quite nuanced (Lepperhoff and Houben, 1993): deposit growth rates depend on the radial temperature gradient (thermophoresis), the speed at which particles traverse the duct (flow velocity), and the consistency of the particles (oily or dry). Eventually, a stable thickness may be reached, say, of 100–200 nm (Stolz et al., 2001), because growth rates are asymptotic: the deposit attenuates heat transfer, and this reduces temperature gradients, thus arresting thermophoretic deposition. Because EGR is scheduled predominantly at low load/low speed, exhaust temperatures are also low, making them unfavourable for purging; and the particulate phase is oilier, i.e. more inclined to adhere to the surface, perhaps as a gummy deposit. The EGR valve, if downstream of the EGR cooler, receives exhaust gas in which more gas-to-particle conversion has taken place. Particles, however, can be persuaded not to deposit, by advantageous manipulation of the flow fields, for example as accomplished with riblets (Usui et al., 2004). Other information of a more general nature is available in the literature, since the fouling of heat exchanger surfaces by particle deposition is of widespread interest in engineering (Grillot and Icart, 1997).

7.5 Induction

Here we examine the preparation of the air charge, which involves factors both internal and external to the engine. The salient and universal issue is *external*: cramming sufficient air into the cylinder; this is a regular constraint insofar as oxygen paucity engenders *soot*. Internally, the optimisation of pressure, temperature and flow field for the lowest soot rests on making the best use of this available oxygen, once inducted.

7.5.1 External to the Engine

Various aspects of the induction system are responsible for the historical picture of the 'dirty' diesel', i.e. emissions of black smoke – appearing especially as 'puffs'. Soot or smoke emissions, monitored in real time and on transient duty cycles, invariably display ugly spikes briefly intercalated by gear changes – see Figure 7.11 (Hofeldt and Chen, 1996). This smoke relates, as already said, to instances where the cylinder contains insufficient oxygen to support the combustion. Not all of these instances are the responsibility of the manufacturer: a dirty air filter, for instance, will restrict

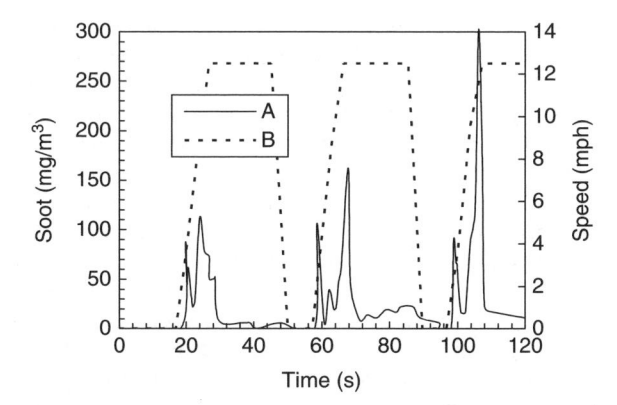

Figure 7.11 Transient nature of soot emission. Engine: 1.5-litre diesel, 6-cylinder, two-stroke, heavy-duty diesel, turbocharged and aftercooled, model year 1994. Instrument: light extinction at 710 nm, converted to soot mass (Hofeldt and Chen, 1996).

airflow; this is obviously an aspect of vehicle maintenance. Another example is vehicle operation at high altitude, where the air is rarefied (Chaffin and Ullman, 1994); this is an aspect of vehicle environment.

It should be said that many advances have been made, and continue to be made, in air path management (Mueller V. *et al.*, 2005): two salient areas are the transition from natural aspiration to forced induction (Human *et al.*, 1990), and the advent of electronic control systems (Chernich *et al.*, 1991). And, currently, much is expected of the technology of model-based control. These and other advances also facilitate defter control of soot. Air path management, though, is a subject too large to cover comprehensively, and moreover is advancing rapidly at the time of writing, while there are many other operational constraints besides that of soot emissions (e.g. those of various pressures and temperatures), which cannot be discussed here. But the implications of one particularly prominent component for soot, namely the turbocharger, do deserve to be considered in some detail.

The issues here concern steady-state and transient operation. In a steady state, some difficulty is encountered at full load/low engine speed: in this region of the operating map, the turbocharger fails to develop sufficient boost pressure, i.e. to supply sufficient oxygen: hence, emissions of black smoke are a particular aspect of maximum torque. The problem is that, in turbocharger design, boost pressure at full load/low speed has to be traded against boost pressure at full load/high speed.

The greater difficulty, however, is encountered in *transient* operation: today's turbochargers have a generic shortfall known as 'turbo-lag', a colloquialism expressing what is, in effect, a vicious cycle. Sudden driver demand requires the injection of more fuel, which, to be burned properly, requires more oxygen; which to be supplied, requires the turbocharger to be accelerated; which requires acceleration of the engine... which requires the injection of more fuel. The role of acceleration, and of the concomitant yet momentary enrichment of air – fuel ratio, in the emission of soot is quite disproportionate: for example, 80 % of the soot was generated during periods when less than half of the fuel was consumed; this is pointedly illustrated in Figure 7.11 (Hofeldt and Chen, 1996). Generally speaking, the more rapid the acceleration, the higher and narrower is the soot spike – making the integrated area under this spike the key index (Hagena *et al.*, 2006).

This engenders a conflict of interest between *emissions* and *driveability*. If the fuel is introduced too slowly, the response of the vehicle is sluggish and uncompetitive; yet if this is done too rapidly, a spike of soot is generated which, in extreme instances, manifests itself as a visible puff of smoke. Of course, even naturally aspirated engines suffer from this trade-off to some extent, yet turbocharged engines can be unfavourably disadvantaged in emissions terms during these brief transients (Sams and Tieber, 1997). On forced-induction engines, the trade-off is ameliorated through turbocharger design (Matsura *et al.* 1992) and defter fuelling management (Bazari, 1990). Some diesel engines are fitted with 'smoke puff limiters': hardware that restricts the rate at which fuelling may be increased in response to driver demand; modern software versions of this strategy take into account ambient air density as well, i.e. vehicle altitude. Regrettably, the concomitant restriction on vehicle acceleration creates a strong tampering incentive – a study of California trucks, for example, discovered widespread evidence of unscrupulous maintenance practices (Weaver and Klausmeier, 1986). Fortunately, modern software control systems are less vulnerable to such unscrupulousness.

Matching the fuel quantity to the available oxygen to obtain the best trade-off between driveability and emissions during transient manoeuvres is actually quite a complex control problem. Several aspects are relevant here. Firstly, today's engine management approximates transient operation to a continuous sequence of steady states (Stemler and Lawless, 1997; Wijetunge *et al.*, 1999). Clearly, the earlier remarks about turbo-lag belie this approach: passing through an operating point as part of a transient manoeuvre, and remaining at that point in a steady-state or cruise cannot be the same; the harsher the acceleration, the greater the disparity. Secondly, today's fuel metering is open-loop, relying on predetermined calibrations, programmed into look-up tables or software maps. The control system thus lacks self-compensating ability, i.e. to handle variations in production, variations with age, variations in environment, etc. Thirdly, a tendency exists for the variable-geometry turbine (VGT) and the EGR to antagonise one another (Brace et al., 1999). Actually, transient engine control is being frenetically researched at the moment: examples are estimating the oxygen supplied to the engine (Nakayama et al., 2003), model-based cooperative control between VGT and EGR (Shirakawa et al., 2001), and UEGO-sensor feedback (Chen S.K. and Yanakiev, 2005). This list is not exhaustive.

7.5.2 Internal to the Engine

We turn now to the *cylinder*, in which air motion, or the flow field, is influential on soot in a preparatory sense (mixing of fuel and air), and also in a secondary sense (mixing of soot and air). That quiescent flow fields are detrimental to air utilisation and the mixing process was recognised long ago; and *swirl* was introduced originally as an anti-smoke measure (Brandl et al., 1979). This is where inlet ports are purposely designed to impart to the incoming air a swirling movement (Uchida et al., 1992); today's more sophisticated systems use an adjustable inlet-mounted flap or throttle, such that the swirl intensity can be varied (Koyanagi et al., 1999). The benefit of swirl is derived from injecting fuel into a 'crosswind', so to speak: the reduction realised in engine-out soot is correlated to the tangentially swirling air momentum, and inversely correlated to the radially outward fuel momentum (Dani et al., 1990).

With swirl, a descending vortex pursues the piston downward on the induction stroke, and then ascends, preceding the piston upward, on the subsequent compression stroke. During combustion, the hot, burned gases are pushed by this swirling motion circumferentially around the cylinder, so that incoming fuel is injected into continually replenished fresh air (Binder and Hilburger, 1981). As a bonus, the same action tends to restrict fuel impingement onto combustion chamber walls (Timoney, 1985), and thus avoid soot-producing flames thereon (Lu et al., 2001). Depending on the interaction between swirl and squish, the flame might, more satisfactorily, be confined to the combustion chamber, thus steering clear of oil residing on the cylinder wall (Cartellieri and Herzog, 1988).

A key behavioural characteristic is minimal engine-out soot at some *intermediate* level of swirl – an intermediate level which is, itself, a function of engine operating point – hence the need for an adjustable flap on the inlet port. A solid theoretical explanation for the existence of an optimal swirl seems to be lacking. For *underswirl*, intuition suggests that locally rich regions are being replenished by incoming fresh fuel, exacerbating the formation of soot. For *overswirl*, either fuel-spray growth is stunted, hampering air utilisation in the outlying regions of the combustion chamber and again restricting combustion to rich, soot-forming pockets (Ikegami et al., 1983), or the combustion products are confined to the bowl, with detriment to soot oxidation (Liu Y. and Reitz, 2005). Overswirl is also likely to give rise to organic particulate, if the mixture is pushed beyond the lean limit (Ikegami et al., 1983). Actually, swirl is a deeper subject than commonly perceived: a converse swirl – soot characteristic has been observed (Zhu Y. et al., 2003); this may be an aspect of injection pressure, and interference between sprays emerging from neighbouring holes in the injector nozzle.

The mixing process is also assisted by *turbulence*. This is well known to promote soot burn-up in atmospheric flames (Said et al., 1997), and its in-cylinder use for the same purpose, in the late combustion phase, was mentioned in Sections 3.2.5 and 3.3.4. The same phenomenon, at earlier crank angles, is of benefit in fuel – air mixing, and some authors emphasise its importance, positing beneficial interactions with swirl (Ikegami et al., 1988). Turbulence is promoted during the induction stroke by, for example, shrouds mounted on the inlet valves. However, this strategy does not always meet with the intended effects (Timoney et al., 1997). The problem is that, unlike swirl, turbulence dissipates rapidly, and tends not persist into the compression stroke, i.e. little remains at the start of injection – hence the need for novel schemes to promote turbulence in the late combustion phase.

Charge temperature, although not usually a calibratable as such, is influenced by heat exchangers mounted in the air path and EGR path. Intuitively, hotter conditions will shorten the ignition delay, and hence lengthen the soot-producing mixing-controlled burn (Nikolic *et al.*, 2001). When this effect is compensated by a cetane improver, the action of charge temperature seems, in this respect, inconsequential (Ladommatos *et al.*, 1996b), and similar conclusions were reached by others by a different route of reasoning (Pinson *et al.*, 1994). Presumably, therefore, charge temperature acts on soot formation via other mechanisms than simply the relative durations of the premixed and mixing-controlled phases.

As usual, laser-imaging studies of the combusting plume have greatly helped to explain the impact of charge conditions on soot emissions (Stetter *et al.*, 2005). Figure 7.12(a) shows, for the mixing-controlled phase, the concentration of soot, averaged across the combusting plume, as a function of distance along the plume (Pickett and Siebers, 2004). (To interpret this figure, the reader may care to re-peruse Section 3.3.3.) The general feature of a soot peak somewhere along the plume suggests competition between formation and oxidation – such that, at the peak, the two are in balance. The peak is greater, and closer to the nozzle, for a higher charge temperature. In this situation, the lift-off length has shrunk; there is less air entrainment into the (admittedly still hypothetical) premixed flame; and a richer premixed burn is in progress. Since the curves tend to converge at the end of the plume, then, presumably, more extensive oxidation compensates for the larger amount of soot initially formed. Intriguingly, at 850 K (not shown), *no* in-plume soot could be detected, for which λ, at the premixed flame, was estimated at 0.7.

The final parameter to be discussed here is air density. One might suspect any increase in this density to foreshorten the ignition delay, and hence also to lengthen the soot-producing mixing-controlled burn (Zhang L. *et al.*, 1997); yet conversely, more oxygen is now available to the combustion, so that the end result is not so immediately obvious. High charge density is beneficial only up to a certain point: there are cases where emissions of soot actually *worsen* (Tanin *et al.*, 1999); and it seems probable that, eventually, penetration of the fuel spray into the surrounding air is impeded, so that poorer air

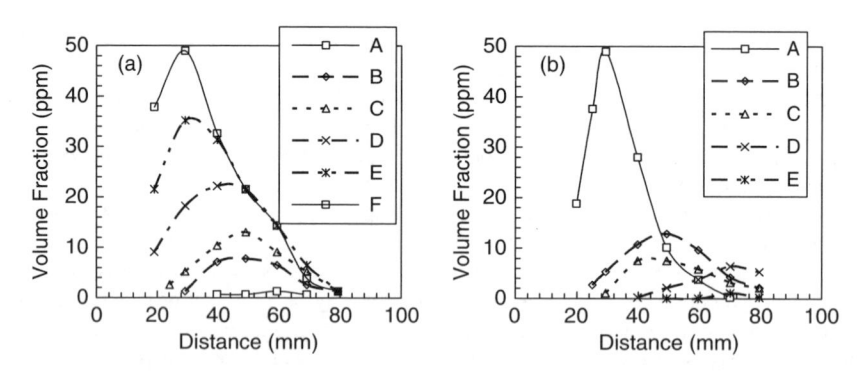

Figure 7.12 Average soot volume fraction across the combusting plume as a function of distance from the injection nozzle, as measured by light extinction in a constant-volume combustion chamber. Fuel injection, common rail; fuel, no. 2 diesel (aromatics 33.8 %, cetane no. 46). (a) Effect of charge temperature (K), lift-off length (mm) and estimated lambda at premixed flame, as follows: A (900, 29.9, 0.50); B (950, 22.4, 0.37); C (1000, 18.3, 0.30); D (1100, 12.5, 0.19); E (1200, 9.3, 0.14); F (1300, 7.2, 0.10). Charge density, 14.8 kg/m^3; injection pressure, 138 MPa. (b) Effect of charge density (kg/m^3), charge temperature (K), lift-off length (mm) and estimated lambda at premixed flame, as follows: A (30.0, 950, 11.8, 0.32); B (14.8, 1000, 18.3, 0.29); C (14.8, 950, 22.4, 0.37); D (7.3, 1050, 26.9, 0.25); E (7.3, 1000, 35.5, 0.34). Injection pressure, 138 MPa (Pickett and Siebers, 2004).

utilisation begins to outweigh the greater presence of oxygen. Again, this is where laser imaging of the combusting plume has helped enormously: some data are reproduced in Figure 7.12(b) (Pickett and Siebers, 2004). A higher charge density yielded a shorter lift-off length, and a richer (admittedly still hypothetical) premixed flame; the in-plume soot rose to a higher peak, and the peak shifted further upstream. But soot survival beyond the end of the plume, which is what really counts, was not straightforwardly related to charge density.

7.6 Lubrication

Engine lubrication is of twofold relevance to particulate emissions. Firstly, impish escape of lubricant into the exhaust constitutes a particulate emission in its own right. Secondly, particles generated in the combustion – principally soot – escape into the lubricant, and thereby degrade the performance of the lubricating system. There is thus a dual carriageway, so to speak, but one on which both sides are subtly interrelated: for example, soot, when present in the lubricant, accelerates wear in the piston rings; this wear then allows more lubricant to pass into the exhaust. There is also an underappreciated *indirect* effect on emissions, via the fuel. For example, adulteration of the oil and the consequent degradation in fuel economy require, to maintain the same operating point, the consumption of more fuel. If one considers, for the sake of argument, that emissions scale with the quantity of fuel consumed by the engine, then there is some justification for claiming that the *fuel*-derived particulate emissions are a function of *oil* condition (Manni *et al.*, 1997). Clearly there is great potential here for a vicious cycle, or even several closely interwoven vicious cycles.

7.6.1 Oil in Particulate

The significance of engine lubrication for the particulate emission is easily underestimated. In the field, exceptional circumstances are seen in which gross emissions of lubricant occur through malmaintenance or vehicle neglect. But, even in unexceptional circumstances, the possibility that lubricant emissions *alone* might account for the entire statutory limit could not be overlooked at one time (Munro, 1990), and, in the intervening years, these limits have been tightened considerably. There is little point, then, in restricting fuel-derived particulate emissions while ignoring that fraction which is oil-derived. Of course, solely from the emissions standpoint, much rests on what percentage of this escaping oil is successfully burned, as emergence from the engine as 'particulate' is not a foregone conclusion.

Fugitive Escape

This fugitive emission is fundamentally controlled by engine architecture, and by the design of four salient escape routes: the piston rings (Jakobs and Westbrooke, 1990), the turbocharger seals (Ariga *et al.*, 1992), the valve stem seals (Mori *et al.*, 1990), and the positive crankcase ventilation (PCV) system (Koch *et al.*, 2002). Some idea of the importance of one of these routes in oil control can be gauged from the wide range of research, encompassing piston-ring design (Signer and Steinke, 1987; Jakobs and Westbrooke, 1990; Munro, 1990; Essig *et al.*, 1990; Mihara and Inoue, 1995); piston design (Munro, 1990); honing of the cylinder liner (Essig *et al.*, 1990); inter-ring gas pressure and piston-ring kinematics (Richardson, 1996); piston-ring oil transport and film thickness (Wong V.W. and Hoult, 1991); soot deposition into piston-ring grooves (Kim J.-S. *et al.*, 1998) and the implications of such for piston-ring lubrication (Urabe *et al.*, 1998) and bore polishing (Guertler, 1986); and bore surface and cylindricity (Klein *et al.*, 1990). Finally, it is worth recalling that a throttled gasoline engine experiences a cylinder depression (below crankcase pressure), likely to draw oil in (Baumbach *et al.*,

1995), whereas a diesel engine is not customarily throttled. These aspects of engine design are not reviewed here, although one pre-eminent point is that reductions in oil consumption cannot be pursued indiscriminately: for example, this would introduce risks of scuffing at the piston ring pack.

The development of apparatus capable of *real-time* oil consumption measurement (Froelund *et al.*, 2001b) has confirmed long-standing suspicions as to the importance of transient engine operation: significant short-lived increases are seen during accelerations (Yilmaz *et al.*, 2001); these are tentatively adduced to temporary reversal of the gas flow at the top-ring groove; ring motion and blow-by are other facets (Bailey B.K. and Ariga, 1992). But one must also remember that emissions of particulate during acceleration tend to be dominated by the fuel-derived carbonaceous fraction (Section 7.5); so that improvements in oil control, on a *transient* basis, might not, therefore, be significant in the total picture. The converse transient, deceleration, these days, is met by 'fuel cut'; hence, emissions during these moments – except for confounding by outgassing deposits – *must* be oil-derived, and these particles are unlikely to be soot (Andersson *et al.*, 2001). This mode of emission becomes significant in duty cycles containing substantial numbers of engine-braking manoeuvres (Cartellieri and Herzog, 1988). Naturally, the escape of lubricant from a motored engine is higher when the engine is hot (Andrews *et al.*, 1993).

A common metric is oil consumption referenced to fuel consumption: at one time, this may well have exceeded 1 %; today, a preferable maximum is 0.1 % (Andrews *et al.*, 1998a), i.e. a reduction of more than an order of magnitude. But this escaping oil is not necessarily discharged by the engine as particulate: *emission* is the net result of *consumption* and *combustion*. Thus, oil consumption, on its own, is unrevealing as to the true state of affairs. In fact, consumption and combustion are curious bedfellows, exhibiting, across the engine operating map, trends directionally consistent in one region, and diametrically opposed in another (Essig *et al.*, 1990), with different maxima and different minima that may or may not coincide (Shore, 1988). Some data along these lines are given in Figure 7.13 (Ariga *et al.*, 1992). This wanton or capricious behaviour, with few unambiguous or explicable trends, is unsurprising in view of the various escape routes available to lubricant (Inoue T. *et al.*, 1997),

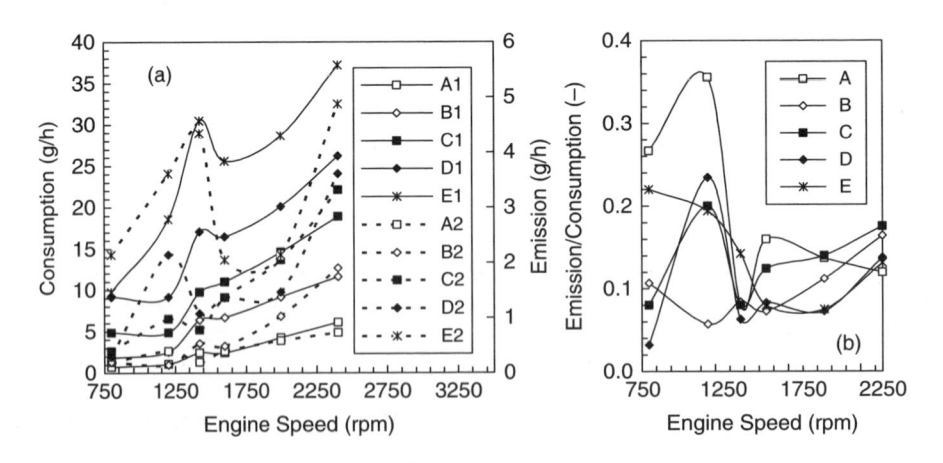

Figure 7.13 As measured by the SO_2 tracer technique: (a) oil consumption (1, solid line), and oil consumed but emitted unburned, i.e. captured as particulate (2, dotted line); (b) ratio of oil emission to oil consumption. Engine load (per cent of maximum): A, 0; B, 25; C, 50; D, 75; E, 100. Engine: 7.6-litre, six-cylinder heavy-duty truck engine, turbocharged and aftercooled, rated power 188 kW, 2400 rpm. Oil: sulphur, 1.07 %wt. Fuel: sulphur, 1 ppm; aromatics, 28.3 % (Ariga *et al.*, 1992).

the relative importance of which must vary with engine operation, and the disparate ways in which consumption and combustion are affected.

Obviously, certain trends might be supposed: for example, a dependency on engine speed, tempered by the combustion duration; and a dependency on engine load, tempered by the combustion temperature (Kawatani *et al.*, 1993). But, less conjecturally, data such as those in Figure 7.13 explain why placing greater restrictions, especially blanket restrictions, on oil consumption does not always pay dividends in restricting oil's contribution to particulate. Indeed, oil's greatest contribution to the particulate emission is sometimes attained when oil consumption is actually at its lowest ebb (Shore, 1988); and, conversely, an engine of high oil consumption may actually be rather clean in terms of oil emission (Andrews *et al.*, 1998a). The same situation is encountered with gasoline engines (Andrews and Ahamed, 1999). Occasionally, the promotion of oxidation seems a better strategy; this is where oxidation catalysts are great levellers, ensuring all engines reach the same touchline, whatever their starting point.[3]

One perhaps not immediately obvious consequence of the escape *route* arises from the temperature to which the escaping oil is exposed, and also, perhaps, from the divided state of the oil, as decided by the mechanical agitation provided by the engine: this is the mode or *phase* of escape, whether as a *vapour* or as a *liquid*: for example, the oil exists within the lubricating system as fluid films and as aerosol mists. From this observation it seems reasonable to suppose that, among the various oil *components*, liquid-phase escape is *unbiased*, whereas the same suggestion seems implausible for vapour-phase escape, as some components are more volatile than others (Takeuchi *et al.*, 2003). This understanding explains, for example, why certain additives or metals become, over time, more concentrated in the remaining oil. The mode of escape, then, has a direct bearing on which oil components are lost.

The escape route is equally critical in deciding the extent of oxidation: for example, a small leak through the exhaust valve generates more particulate than a far larger leak through the inlet valve, simply because, in the former case, oil is oxidised less effectively (Mori *et al.*, 1990; Inoue T. *et al.*, 1997). Within the cylinder itself, movement of the burning gases, and the proximity of the flame to the cylinder walls, is relevant for similar reasons (Cartellieri and Herzog, 1988). Lubricant particularly evaporates unscathed from this region during the expansion stroke, when the likelihood of oxidation is rapidly declining. Local wall temperatures evaporate the lightest lubricant components (Yilmaz *et al.*, 2002), and vary with engine design and operating point (Taylor G.W.R., 2001). Spatially, the hottest zone is towards the flamedeck, where, although oil evaporation is swifter, oxidation is also more thorough; conversely, the coldest zone is towards the crankcase, where the opposite situation applies; the region in between contributes disproportionately to the oil-derived particulate.

The consequences of this escape for the combustion are not necessarily insignificant: if the air – fuel charge is adulterated (say) to a thousand ppm (Dowling, 1992)[4], and, if this lubricant burns fully, then the effects on heat release or ignition delay might well be nonnegligible. The *fate* of lubricant, however, within the combustion chamber, is oddly under-reported and somewhat open to conjecture. Lubricant compounds are *not* subjected to the same conditions as fuel compounds: they are, after all, at least initially, *outside* the combusting plume, in the surrounding air, from which one might suppose, for example, less likelihood of assimilation by carbonaceous spherules. A couple of observations here are intriguing. First, more than two-thirds of piston-crownland deposits were, in fact, organic compounds derived not from fuel but from lubricant; and similar compounds were also found residing on injector tips (Andrews *et al.*, 1992). Lubricant-derived compounds are unsurprising in the ring-pack region; why they should appear on injector tips is unknown, as no obvious transport route exists. Second, if a

---

[3] Oil consumption is still important for *other* reasons affecting the customer's perception of the vehicle: for example, engine performance ('top-ups') and catalyst aging (ash deposition) (Section 7.8.1).

[4] To place this figure in perspective, treat rates for certain fuel additives are often, in fact, considerably lower (tens of ppm).

lubricant – fuel *mixture* is burned, the resulting ash differs in its morphology from that observed when lubricant is instead entrained into the air intake as a *mist* (Sutton *et al.*, 2004). Thus, investigations in which oil consumption is increased deliberately, that is, by artificial means, might be relying on precarious assumptions as to the combustion mode of this oil. These remarks highlight the need to learn much more about the combustion of escaping lubricant.

Considerable emissions abatement is attainable via oil *reformulation*: the escape (evaporation) and the burn-up (oxidation) arguably correspond, respectively, to the physical properties and the chemical composition of the lubricant. Of the two salient *property* characteristics, the reason for decreasing the volatility (after the above discussion) will be self-evident (Dowling, 1992); the reason why increasing the viscosity is also beneficial (Arai, 1992) remains unclear, but relates possibly to oil film thickness (Dowling, 1992). Synthetic oils, of higher viscosity and lower volatility than mineral oils, reportedly contribute less to the particulate emission (Froelund *et al.*, 2001a). Chiefly through volatility and viscosity, particulate emissions were cut by a third during motoring (Cartellieri and Herzog, 1988), and, over complete test cycles, results compared favourably with the alternative strategy of *fuel* reformulation (Manni *et al.*, 1997). The improvements possible through oil reformulation should not, then, be dismissed too lightly; although, of course, numerous *other* criteria, not germane to the present section, are also important in oil formulation, so that the room for manoeuvre is, actually, smaller than might appear from this discussion vis-à-vis particulate.

Before passing to the implications for individual particulate fractions, a further, less delineated aspect of oil formulation is worthy of note, namely *fuel – oil interactions* (Andrews *et al.*, 1993). For example, evidence exists that differences in emissions exhibited by various oil formulations are not, in fact, retained across different fuel formulations (Manni *et al.*, 1997). Moreover, oil consumption is often disguised in some measure by oil dilution, wherein fuel enters the lubricating system; and this adulteration, in turn, modifies the lubricating properties of the oil, with suspected effects on the oil emission. The incestuousness of this relationship cannot be unravelled here in any detail; but it arises, presumably, from aspects such as the diffusion or solubility of oil in fuel (or *vice versa*).

Particulate Characterisation

With regard to particulate *composition*, lubricant is predominantly emitted in the *organic* fraction – if one takes the vast literature devoted to this subject as unwitting testimony; in fact, the contribution is *so* biased that lubricant is often perceived as *only* being emitted in this fraction: an erroneous perception, as we shall see shortly. More to the point, however, is that efforts to restrict the organic fraction via the fuel-derived compounds are obviously misplaced if oil-derived compounds dominate; hence, the two contributors should be resolved according to the procedures outlined in Section 5.2.6. This also explains why – unlike for soot – empirical relationships between the emissions of organic particulate and combustion parameters (such as adiabatic flame temperature) invariably fail (Ahmad and Plee, 1983): simply, oil-derived organic compounds represent a significant and independently varying emission with the ability to confound such empiricisms. Moreover, the two sources compete on an unequal footing: for an equivalent rate of consumption, oil supplies one or two orders of magnitude more organic particulate than does fuel (Mayer W.J. *et al.*, 1980). This disparity is what makes oil significant as an emission.

The lubricant contribution to the carbonaceous fraction is potentially confused by limitations in the organic extraction (Sections 5.2.5 and 6.3.1), as some compounds may remain bound to the soot. The contribution nevertheless is *probably* small, although this area is woefully unexplored. Experiments with radioactive tracers suggest that very few of the carbon atoms originally in oil compounds are actually emitted as soot – at most, a few per cent (Hilden and Mayer, 1984; Dowling, 1992; Buchholz *et al.*, 2003) – and full oxidation to CO_2 is negligible. On the basis that susceptibility to hydrocarbon pyrolysis increases with molecular weight, and the molecular weight of oil (~ 400) is more than twice that of fuel (~ 170), a marked contribution to the carbonaceous fraction might be expected (Mayer W.J. et al., 1980). This discrepancy points to an almost complete escape of organic compounds in *uncombusted* form. Perhaps this supposition is a little simplistic: oil might escape predominantly under conditions where its pyrolysis is less likely. On the other hand, partial combustion of oil is known to create hard, carbonaceous deposits within the engine, such as at the topland (Guertler, 1986): these deposits may eventually fracture and release atypical coarse-mode particles.

The contribution to the ash fraction, although minuscule in homologation terms, is, nonetheless, a continuing embarrassment, insofar as it shortens the life of aftertreatment devices (Section 7.8). A mixed bag of oil-derived ashing elements appears in the exhaust (Lim M.C.H. et al., 2007). Other than sulphur, which is discussed below, the four most prominent elements – present intentionally as additives – are phosphorus, zinc, calcium and magnesium (Vaaraslahti et al., 2005), all of them only too familiar from catalyst deactivation studies (Section 7.8.1) and ash build-up in the DPF (Sections 6.3.2 and 7.8.2); there is thus pressure to reduce these concentrations, moreover, without compromising lubricant performance (Arrowsmith, 2003). The phosphorus and zinc are present at a few tens to a hundred ppm (Andersson et al., 2000), and calcium at a couple of thousand ppm (Vouitsis et al., 2007). Less cited elements are sodium and chlorine, present at, say, a few ppm and one ppm, respectively (Mohr et al., 2003b). Chlorine has no function, and is simply a relic of the manufacturing process (Stunnenberg et al., 2001), but, because of its potential to form dioxins, there are moves to restrict its presence in lubricants; recent work, however, was unable to establish a clear link (Dyke et al., 2007).

Ashing elements, if characteristic of and unique to the oil, serve as useful tracers with which to estimate, by extrapolation, oil consumption. The result, however, of this extrapolation should be factored carefully, according to the remaining fraction in the oil, since the consumption rates of inorganic compounds and of organic compounds differ discernibly (Cooke, 1990) – the reason for which was already mentioned. That any specific additive, element or compound is uniformly lost with respect to temperature, throughout the distillation range of the oil, is an inconvenient assumption and, as we have seen, vapour losses are not the same as liquid losses. Thus, oil consumption estimates, based, say, on measurements of calcium emission rates are underestimated (Nemoto et al., 2004) without appropriate adjustment. These caveats notwithstanding, *calcium* does enjoy widespread use as a tracer (Andrews et al., 1998a). Phosphorus and zinc are held to be somewhat less useful, owing to their formation, *inter alia*, of zinc pyrophosphate ($Zn_2P_2O_7$), which, in the exhaust, evades easy quantification by partitioning between the particulate phase and gas phase (Elamir et al., 1991).

Tracer studies, although aimed initially at estimating the *consumption* of oil, have proven quite revealing also about the *mechanisms* by which these ashing elements are lost. Figure 7.14 shows correlations between emission rates of inorganic elements measured *directly* in the exhaust gas, and emission rates estimated *indirectly* from oil consumption and oil composition (Givens et al., 2003). That direct measurements are consistently lower than indirect measurements is easily ascribable to losses experienced *en route*, between oil source and filter collection; this is likely to involve storage somewhere within the exhaust system. But less obvious, and certainly intriguing, are variations in loss rates *among* these elements.

At a first level, these variations are explicable in terms of volatility, with zinc-containing compounds perhaps evaporating more easily than others. But volatility is not all, because these elements fulfil different *roles* in the oil, resulting from their different chemical forms: the original additive compounds break down or decompose, after which the products reside, as will be explained in the next section, (a) on soot surfaces, (b) on metal surfaces and (c) in the bulk oil, from which it follows that emission rates must be in some way affected. Mo, S, P and B, for instance, are sorbed onto/into the protective anti-wear layers on metal surfaces, whereas Zn, Mg and Ca stay in the bulk; there are thus, in loss behaviour, two elemental groups (Givens et al., 2003). But, taken in aggregate, such investigations

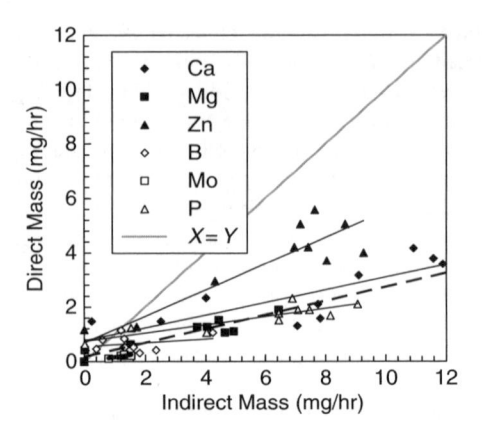

Figure 7.14 Direct measurements of ashing elements in exhaust compared with indirect measurements, based on oil concentrations and oil consumption. Duty cycle: various steady states, 25 h duration. Eight test oils of various inorganic loadings; see original paper for full speciations. Engine: 2.15-litre, four-cylinder, direct-injection, common-rail, turbocharged, model year 2001. Aftertreatment: oxidation catalyst and NO_x storage catalyst. Fuel: sulphur, 3 ppm. Elements analysed by mass spectrometry; oil consumption measured by tritium technique (Givens et al., 2003).

highlight the dangers of estimating the total emission of oil from the emission of certain inorganic elements carried by that oil.

As highlighted in Section 3.5.1, the elemental composition of the ash fraction is not the only story, as chemical reactions occur between these elements (and with oxygen). The chemical compounds, e.g. $CaSO_4$ and $Zn(PO_3)_2$, that appear in the exhaust reflect the relative abundance of these elements, i.e. Ca, Zn, P and S, in the oil, or rather in the escaped products (Nemoto et al., 2004). Of course, these compositional trends among oil-derived compounds are somewhat disguised by the confounding action of fuel sulphur, through which metals are also sulphated.

The contribution oil makes to the sulphate fraction is traditionally small, but becoming ever more prominent with the demise of *fuel* sulphur in conventional diesel, and the rise of sulphur-free oxygenated fuels such as biodiesel (Durán et al., 2006). The fuel sulphur content is correlated to the sulphate particulate; yet if fuel sulphur is exceptionally low, this correlation is confounded by oil sulphur (Engler et al., 1993). Indeed, where fuel sulphur levels are exceptionally low and oxidation catalysts are exceptionally active, this secondary source of sulphur cannot nowadays be ignored. Sulphur is present in oil at several thousand parts per million (Mohr et al., 2003b): for example, it is estimated that, if the concentration were one per cent, and oil consumption were 0.2 % of the fuel consumption, this would add an additional 20 ppm of sulphur to the exhaust (Khalek et al., 2000) – an addition which is certainly significant in the light of sulphuric acid's role in instigating nucleation (Section 3.7.2 and 7.2.1). To complain vociferously, therefore, about *fuel* sulphur, at today's levels of a few tens of parts per million, seems somewhat hypocritical.

But an important distinction should be made here, in the propensity of either source of sulphur to form sulphates (Stunnenberg et al., 2001): oil is less likely to be exposed to such high combustion temperatures as fuel, and hence sulphur-bearing compounds in oil, upon evaporation, are better placed, at least partially, to escape from the engine unscathed in their *native* forms – better placed, that is, than sulphur-bearing compounds in fuel, which experience the full brunt of the flame, leading to their oxidation, and emission as SO_2, SO_3 and SO_4^{2-}. These ideas are bolstered by efforts to use oil-derived sulphur in the exhaust as a tracer for oil consumption, in a like manner to metals, as discussed above: this is possible if fuel sulphur is reduced to negligible levels. The sulphur-containing compounds, or organosulphur compounds, and sundry other organic compounds in the oil are not necessarily lost at equivalent rates, *inter alia*, because of differences in their susceptibility to evaporation or decomposition (Bailey B.K. and Ariga, 1990; Givens *et al.*, 2003).

The identities of these emitted sulphur-bearing oil compounds do not seem to have been comprehensively reported. But an extremely interesting observation nevertheless emerges: sulphur-bearing oil compounds are suspected to pose fewer dangers for catalysts than sulphur-bearing fuel compounds (Asanuma *et al.*, 2003). In fact, sulphur-bearing oil compounds in the *base* components were more dangerous to catalysts than sulphur-bearing oil compounds supplied as *additives*, and for just the same reason. It seems that the volatility of these compounds, or their disinclination to adsorb onto catalysts, allows their safe passage to the tailpipe, where SO_2 and SO_3 would otherwise be captured or interact. This suggests that catalysts might be protected by making these sulphur-bearing oil compounds more stable, rather than limiting bulk oil loss *per se*.

Lastly, some mention should be made of oil's implications for particle demography: a topic which has been poorly addressed. The prominence of nucleation modes during deceleration (fuel cut) speaks for itself (Andersson *et al.*, 2004a), but the trends, vis-à-vis one oil versus another, do not always stand out (Andersson *et al.*, 2004c). So, to obtain unambiguous results, fuel was doped to the tune of 2% (Kytö *et al.*, 2002): for reasons already cited, some artificiality in the combustion of this oil cannot be ruled out, and the artificial treat rate was also higher than the likely reality. Nevertheless, lubricants producing the highest particle numbers produced also the lowest particulate mass. Others have performed the same experiment and observed an order-of-magnitude increase in particle numbers, mainly in the nucleation mode (Jung H. *et al.*, 2003). The difference could only be explained in *part* by ash content (Kytö *et al.*, 2002); a possibility is that high-molecular-weight organic compounds from the oil contribute preferentially to the nucleation mode (Inoue M. *et al.*, 2006). On the other hand, in Sections 3.7.2 and 7.2.1 a wealth of evidence was presented indicting sulphuric acid as a nucleating agent, raising the suspicion that oil sulphur, rather than fuel sulphur, is, on occasion, the source of this nucleation mode (Hall and Dickens, 2003). Should this be the case, the headlong pursuit of low-sulphur fuel is unlikely to pay demographic dividends, if oil is really the source of the nucleating sulphates.

7.6.2 Particulate in Oil

The lubricating system is, obviously, compromised and perhaps endangered when successfully penetrated by foreign material from without – always a risk for engines operating in excessively dusty environments, or where failure of the air filter has passed unobserved. These are cases where foreign material is brought in via the air. But in cases more germane to the current topic, where this foreign material arises in the combustion chamber, special instances exist for which penetration by the ash fraction may become a nuisance: fuel-borne organometallic additives (Vincent and Richards, 2000). The implications here for the lubricating system, being so unreported, are perhaps small, and the treat rates in the fuel are, nowadays, only a few parts per million. Penetration by the sulphate fraction is not an issue *as such*; but this fraction's gaseous precursors (SO_2 and SO_3) are a nuisance insofar as the ensuing sulphuric acid requires neutralising compounds in the oil. Penetration by the organic fraction, or rather *fuel* hydrocarbons in the upper distillation range (Andrews *et al.*, 1993), is not, again, an issue *as such*, but these compounds are undesirable insofar as they reduce the viscosity of the lubricant; and ongoing in-oil reactions might also convert them into particulate of a secondary nature.

With diesel engines – possibly with direct-injection gasoline engines not too far behind (Bardasz *et al.*, 1999) – these issues are considerably overshadowed by the litany of engine lubrication problems created by *soot*. This field is too multitudinously reported to review in detail, and the present section can only take a brisk canter through the tribology and rheology of soot-in-oil. (It should, however, be noted that, in the open literature, the same issues have repeatedly resurfaced for several decades.) These particles of soot and the lubricant which carries them constitute a two-phase system, the properties of which are actually quite interesting from a scientific standpoint. Moreover, aerosol science parallels colloid science: particles collide and agglomerate in lubricant, just as they collide and agglomerate in exhaust gas, although the particle – liquid boundary in lubricant is not, perhaps, as equivocal as the particle – vapour boundary in exhaust gas.

Two aspects of soot-in-oil are worth mentioning, for which space does not allow a detailed review. The first is quantifying the amount of soot carried by the oil. The obvious method, of course, is centrifuging; others are thermogravimetric analysis (Andrews *et al.*, 1993; 2000a), infrared spectrometry (McGeehan and Fontana, 1980; Ryason *et al.*, 1994) and quasi-elastic light scattering (Bardasz *et al.*, 1996). A pressing task, though, is the realisation of an on-line, real-time instrument suitable for production vehicles, as part of engine management: this would obviate the need for various software fudges and unnecessary oil changes. Some possibilities for sensing are the effect of soot on the thermal (Lockwood F.E. *et al.*, 2001), acoustic (Agoston *et al.*, 2005) and electrochemical (Goodlive *et al.*, 2004) properties of the oil. The second issue is on-vehicle maintenance of oil quality (Frehland *et al.*, 1999; Andrews *et al.*, 1999a,b; Andrews *et al.*, 2001c,e,f). It is increasingly recognised that 'full-flow' filtration is insufficient, as to maintain an acceptable pressure drop the filter must be large, making it expensive and difficult to package. These systems are therefore augmented nowadays by 'by-pass' filtration, where a small amount of the total oil flow is processed and then returned to the sump: this can also use either barrier filters or centrifuges (Cox and Samways, 1999). Widespread collateral benefits are expected from improvements in oil quality: for example, reductions in oil consumption and oil emission (Andrews *et al.*, 2001f).

The *rate* of soot infiltration is expressible as a certain fraction, say, a few per cent, of the engine-out soot, although, as a cautionary remark, the two parameters show some independence, and in fact may even be directionally opposed; moreover, the retention is not permanent and some soot can, in fact, escape into the exhaust (Andrews *et al.*, 1999a,b). Contemporary oils indisputably experience greater levels of soot than in former times, mass concentrations exceeding 5 % being no longer considered exceptional (Lockwood F.E. *et al.*, 2001). This adulteration is the simple corollary of a surprisingly large number of factors (McGeehan *et al.*, 1999). First, EGR ensures a sootier charge and a sootier combustion. Second, timing retardation[5] exposes a larger area of naked cylinder wall to soot-laden combustion gases. Third, high piston rings scrape more soot-laden oil from the top of the liner, towards the sump. Fourth, high cylinder pressures promote the blow-by of soot-laden gas. Fifth – actually quite ironically – engine manufacturers face considerable market pressures to extend oil-change intervals, and the laudably vigorous pursuit of low emissions, is, in this instance, counterproductive: as erstwhile engines consumed more oil, they benefited, correspondingly, from more frequent fresh additions of the same, by which the soot was *diluted*. Reductions in oil consumption and restrictions on soot entry should, therefore, go hand-in-hand.

Although the *consequences* of soot entry into oil are vastly reported, the *transport* of soot *into* oil is not. This situation is a little strange insofar as measures necessary to handle the soot would be less exigent if the lubricating system was better sealed. As described extensively in Section 3.4, good grounds have been established for the belief that thermophoresis is the dominant particle transport

---

[5] Like EGR, adopted as an NO_x countermeasure.

mechanism *to* the cylinder walls. But the question, really, is what happens then: whether these particles are carried past the piston by the blow-by gases, or by the oil film. According to a lonely experimental and modelling study (Tokura *et al.*, 1982), only a small fraction of the soot, less than 3 %, is actually carried into the oil by the blow-by. The oil *itself* accounted for the majority of the transport, through the scraping action of the piston rings. On the other hand, particle transport via blow-by might become prominent as engines age and piston rings gradually wear (Lockwood F.E. *et al.*, 2001) – incidentally, as exacerbated by soot!

Ideally, these soot particles would not interfere with any of the desired properties of a lubricant, of which there are many, e.g. viscosity; wear, rust and corrosion control; acid neutralisation; minimal deposit formation; engine cooling; the maintenance of combustion chamber pressure; low foaming proclivity; and oxidation resistance. Two lesser-known implications of soot entry are modifications to an oil's thermal (Lockwood F.E. *et al.*, 2001) and electrical (Kornbrekke *et al.*, 1998) conductivities. But by far and away the two largest and most frequently cited implications of soot entry are the fillip given to *wear* (discussed below) and the uplift given to *viscosity*.[6]

Viscosity

In a colloid, the viscosity displays an ever-increasing sensitivity to the rising fraction of space occupied by the particles, leading ultimately to a gelled state (Gopalakrishnan and Zukoski, 2006). Thus, the impact of soot on viscosity can be profound: mass concentrations of $\sim 5\%$ involve factors of 10^4 (cited in McGeehan *et al.*, 1991), or certainly several hundred per cent. Of course, this statement is slightly disingenuous, since the worst effects occur when soot particles have opportunity to *agglomerate*, something which oils are expressly formulated to impede: in the converse sense, this property is described as 'dispersancy'. If the particles remain well dispersed, then the increase in viscosity need only be a few per cent.

The rheology of soot-laden oils is immensely complex and can only be cursorily described here. Behaviour is 'thixotropic', meaning the assumption of a gel-like state when left standing, and immediate liquefaction on agitation; and it is also non-Newtonian, i.e. shear stress is not proportional to shear rate. Given this awkwardness, the inadequacy of conventional kinematic viscometric apparatus – employing constant, low shear rates – for characterising in-service rheology will be obvious (George *et al.*, 1997); other techniques are under discussion (l'Hermine *et al.*, 2000). The kinematic and rotational viscosities diverge to an extent dependent on the extremity of non-Newtonian behaviour (Parry *et al.*, 2001) Several such characterisations are given in Figure 7.15 (Selby, 1998).

Satisfactory analytical expressions for viscosity exist for colloids containing independent, well-dispersed, noninteracting spheres. The understanding of soot-laden oils, which is more qualitative and empirical (Covitch *et al.*, 1985), is that agglomerates – or perhaps more accurately, 'flocculates' – occupy large effective volumes (Ryason and Hansen, 1991), and that the volume of oil entrapped within the pores and interstices or immobilised on the surfaces of these agglomerates is, in these circumstances, more rheologically relevant than simply the volume or mass fraction of soot – which, in the extreme case of a gelled oil, might account for just two per cent of the whole structure (McGeehan *et al.*, 1984). (Electroviscous effects may have some adjunctive role (Ryason and Hansen, 1991).) Thus, agglomerate morphology, shape, size and size distribution are all factors. These networks of interconnected flocculates are not, though, stable structures as such, but loose and fragile curlicues, the sizes of which vary continually during fluid motion, according to the shifting balance, or rather

---

[6] Soot is not the only reason why viscosity can increase: oil oxidation is another. Actually, soot is suspected of fortuitously adsorbing the polar species that initiate oil oxidation (McGeehan and Fontana, 1980).

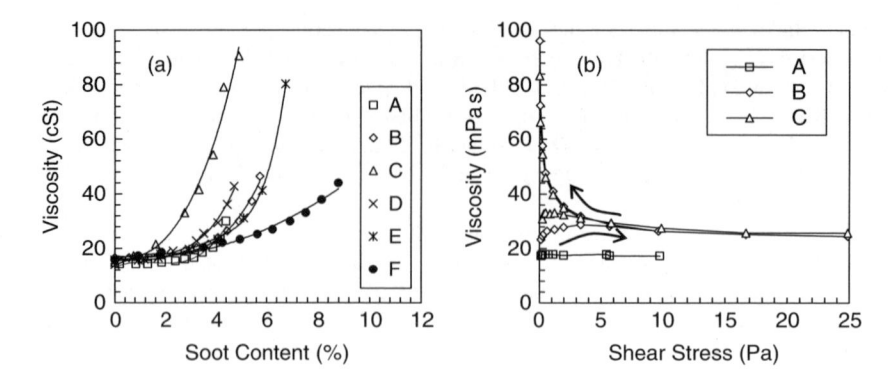

Figure 7.15 Effect of soot on viscosity of lubricant (100 °C): (a) *kinematic*, as measured by conventional means (narrow capillary; reverse-flow viscometer; low, constant shear stress; soot-laden oils of various formulations, A – F); (b) *rotational*, as measured by a newly proposed means (concentric-cylinder rheometer; each point held for two minutes; forward and reverse sweeps as indicated by arrows; soot concentration A, 2.8 %; B, 4.9 %; C, 4.9 % following one hour of thermal equilibration; oil B of (a) was used) (Selby, 1998).

competition, between *agglomeration* – for example, as determined by temperature (Bardasz *et al.*, 1996) – and *fragmentation* – for example, as determined by shearing in the base oil (McGeehan *et al.*, 1984). Faced with these shifting and transitory states, analytical expressions of practical utility seem unlikely to emerge.

Colloidal agglomeration in the oil parallels aerosol agglomeration in the combustion chamber (Section 3.2.4), and the agglomerative mechanisms in both situations are studied using identical fractal concepts (Bezot *et al.*, 1997; Wedlock *et al.*, 1999). Still, a distinction arises between 'primary' agglomeration, or what takes place in the combustion chamber, and 'secondary' agglomeration, or what takes place in the oil. Primary agglomeration, if early enough, is partially coalescent, affecting spherules in their pre-solidification stages; and inter-spherular bonding is stronger if the spherules have fused – stronger, one might suppose, than the shearing forces likely to be experienced in the oil – but whether these bonds withstand mechanical crushing remains to be ascertained. It is the secondary bonds that are repeatedly broken and remade.

Soot particles agglomerate by virtue of van der Waals forces, while adjunctive ionic or electrostatic effects, and interactions with aromatic compounds in the oil, are also suspected (Kornbrekke *et al.*, 1998). But whatever the nature of these attractive forces, agglomeration is obviously highly undesirable, and, to impart soot tolerance, i.e. to impede agglomeration, oil additives known as 'dispersants' are introduced, of which succinimides are an example; these have affinities for particle surfaces and, when adsorbed, generate repulsive forces of a steric, osmotic or electrostatic character. The polarity of the additive molecule (Jonkman *et al.*, 1987), and its size in relation to the spherule (Ruot *et al.*, 2000) are critical, while detergents can also act as dispersants (Covitch *et al.*, 1985). It should, however, be noted that deliberate dispersal, in order to limit the impact on viscosity, runs counter to the need to filter particles out of the oil (Miyahara *et al.*, 1991).

Wear

Purely on a *phenomenological* basis, the wear-promoting effects of soot in oil are clear enough: for example, engines experience greater rates of soot entry into oil and greater rates of wear at metal

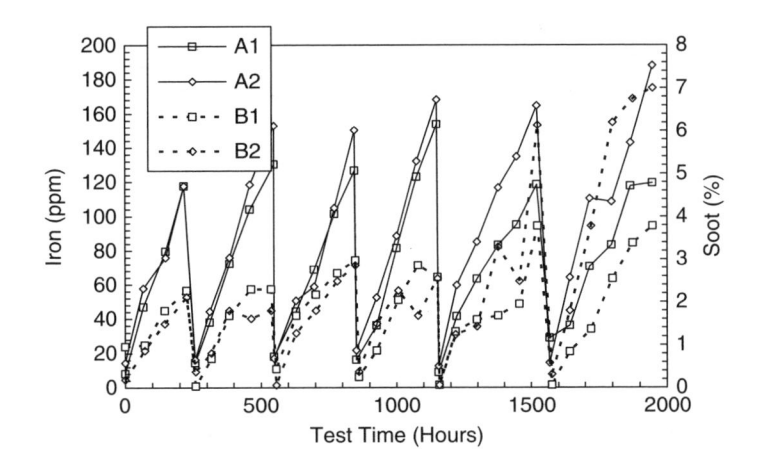

Figure 7.16 Effect of EGR on engine wear rate and soot accumulation in lubricant. Engine: 4-litre, four-cylinder, direct-injection. Duty cycle: 1600 rpm, 70 % maximum load. Abrupt decreases are caused by oil changes. 1, without EGR; 2, with EGR at 28 %. A, iron in oil; B, soot in oil (Dennis *et al.*, 1999).

surfaces upon the implementation of EGR.[7] As shown in Figure 7.16 (Dennis *et al.*, 1999), this wear is, ferruginously speaking, eminently measurable in the accumulation of *iron* particles. This situation creates a vicious cycle, in which the released metal particles are themselves wear-promoting, and lead, in turn, to more metal particles: for example, with EGR, the rate at which iron particles accumulated in the oil *doubled* (Ryan *et al.*, 1999). Other wear metals found to increase with EGR are copper, chromium, aluminium and manganese (Singh *et al.*, 2006).

Of course, this indictment might conceivably fail under artful cross-examination, on the grounds that the observed association between soot and wear is, in fact, circumstantial rather than causative. To gainsay this counter-intuitive augment, plausible wear mechanisms must be posited – of which there are *several*, any one of which might predominate at one location or another, or in one operating mode or another (van Dam *et al.*, 1999), with strong mediation by the nature of the soot particle surface, as no single explanation appears universally applicable (Gautam *et al.*, 1998). Four posited mechanisms are as follows.

A. *Blockage and oil starvation.* A less abstruse mechanism is the accumulation of soot particles at critical points in the lubricating system, such as within narrow channels, leading to a damming or blocking effect, by which the flow of oil to critical components is impeded. The bulk pumpability of oil may also be affected if flocculates accumulate as a sludge (McGeehan *et al.*, 1999). The oil filter experiences an increase in pressure differential, ultimately leading to plugging. And, following a cold start, the length of time needed for oil to reach outlying parts of the engine is protracted (Stehouwer *et al.*, 2002). The oil, though, must first circulate, if flocculates are to be sheared down (McGeehan *et al.*, 1984).

---

[7] Strictly speaking, some 'wear' could arise from corrosion by sulphuric acid, carried either by the oil or by the recirculated exhaust gas (Takakura *et al.*, 2005), although these possibilities are admittedly adjunctive in the data of Figure 7.16. The wear is a function of the increased soot in the combustion chamber just as much as of increased soot in the oil.

B. *Direct abrasion of metal surfaces.* Particles trapped between two surfaces in relative motion engender so-called 'three-body wear'. This may be an issue when contact zones lack sufficient lubricant, such as at low speed or high load, or during a cold start; while at TDC and BDC, the relative motion between liner and piston slows and briefly stops, during which the boundary lubrication layer narrows to less than 50 nm. Wear scars generated in a three-body-wear machine (Sato *et al.*, 1999) and on crossheads and rocker arms (Kuo *et al.*, 1998) are comparable with spherule sizes, and distinguishable from the results of other wear mechanisms (McGeehan at al, 1999). How exactly these scar dimensions impute spherules, rather than agglomerates, is uncertain: fragmentation might arguably occur within actual contacts. Spherules might also become harder through microstructural modifications induced by repeated entrapment between two surfaces at high load. Piston – liner friction increases when rings slide on soot captured by the oil film (Urabe *et al.*, 1998); and there are grounds for believing that the abrasive properties of the metal surface itself are promoted when lubricant and soot combine to act as a cutting fluid (Ishiki *et al.*, 2000). Hardness, size and angularity, as factors in wear, evaluated in the three-body-wear machine, were apparently exonerated (Rounds, 1981); but particles of soot and of alumina generated scars of similar dimensions (Gautam *et al.*, 1999).

C. *Interference with anti-wear compounds on metal surfaces.* The decomposition products of anti-wear compounds such as zinc dithiophosphate (ZDP or ZDTP) and zinc dialkyldithiophosphate (ZDDP), in adsorbing on metal surfaces, act as wear-protecting layers, through preventing metal-to-metal contact. An equilibrium, or balance, exists between the removal of these protective coatings by relative motion between surfaces, and re-formation by adsorption (van Dam *et al.*, 1999). With time, a sudden increase occurs in cam wear rate, suggestive of a threshold; the wear rate decreases again with the incorporation of fresh additive (Gauthier and Delvigne, 2000). The soot particles may (a) preferentially adsorb on metal surfaces, blocking formation of anti-wear films; (b) restrict access of oxygen to metal surfaces, inducing a transition in the oxide film from anti-wear Fe_3O_4 to pro-wear FeO (Corso and Adamo, 1984); (c) engender carbon diffusion into metal surfaces, causing hardening and embrittlement (Corso and Adamo, 1985); (d) participate in tribochemical reactions with anti-wear additives and metal surfaces; and (e) remove or abrade anti-wear films on metal surfaces (Nagai *et al.*, 1983). The liner surface, with its greater affinity for ZDDP products, is potentially more affected than the ring surface (Truhan *et al.*, 2005).

D. *Interference with anti-wear compounds in the lubricant.* ZDP decomposition products might not have opportunity to form the intended anti-wear films, if soot adsorbs ZDP; adsorbs ZDP decomposition products; modifies the rates or products of ZDP decomposition reactions; or modifies chemical interactions amongst anti-wear compounds, detergents and dispersants (Corso and Adamo, 1984). Studies in a three-body-wear machine point to the preferential adsorption onto soot of ZDP decomposition products; adsorption of ZDP and interference with decomposition reactions were ruled out (Rounds, 1977). Yet the elements characteristic of ZDP decomposition (P and Zn) are not always found on soot surfaces (Gautam *et al.*, 1999). There were indications in a three-body-wear machine that highly polar materials are the first to adsorb on virgin soot surfaces, thereby blocking adsorption of ZDP decomposition products; this 'polar reservoir' must be depleted before soot surfaces begin purloining anti-wear compounds (Rounds, 1981).

The soot surface is paramount: it becomes enriched, relative to the bulk oil, with oil additives, the decomposition products of these additives, and the products of oil oxidation. Some compositional analyses are shown in Table 7.1 (Rounds, 1981): actual compositions range markedly, but, relative to the surrounding oil, soot particles are enriched in O, N, S, P, Zn and Ca by factors of ~ 20–90, suggestive of preferential adsorption. Some of these adsorbed compounds are water-soluble. Suggested formulae are $CaSO_4$, $CaCO_3$, $ZnSO_4$, $Zn_3(PO4)_2$ and ZnS. The C/H ratio of oil-recovered, organic-extracted soot is one-third that of exhaust soot, suggesting tenaciously bound detergents or dispersants

Table 7.1 Elemental compositions of soot taken from a selection of 21 engines and 51 samples, and compared with graphite and carbon black. Extraction was by centrifuging and three washes in isooctane. Samples were about 15 mg, with up to a gallon of used oil. (Only about half the soot was removed from the oil.) Analysis: C, H, N, S by combustion; O by pyrolysis; P by photometry; Zn and Ca by atomic absorption (n.d. = not determined). (Test results were similar to those obtained with thermogravimetry.) (Rounds, 1981.)

Element (Mass %)	Graphite	Carbon black	Oil soot Typical	Oil soot Range	Exhaust soot Range
C	99.93	97.7	86.6	38.0–93.7	48.6–87.1
H	0.01	0.19	2.4	1.33–5.23	1.47–10.82
O	0.06	0.9	5.4	3.40–15.8	4.0–22.2
N	n.d.	0.04	0.30	0.02–1.1	0.1–1.49
S	n.d.	0.83	1.30	0.82–4.26	0.70–7.0
P	n.d.	0.01	0.44	0.11–2.29	0.021–0.95
Zn	n.d.	0.004	0.63	0.15–3.99	0.061–0.69
Ca	n.d.	n.d.	0.78	0.36–4.86	0.066–1.43

(Nagai et al., 1983). The wear-inducing nature of this surface is captured in a key metric, the 'ZDP requirement', meaning the mass of phosphorus needed per unit mass of soot to generate similarly sized wear scars for different soot loadings. There is little systematic understanding of why this requirement varies (by factors of ten to thirty) between soots generated over *different* duty cycles. Apparently, soot generated at high loads is more wear-inducing than soot generated at low loads. The differently structured surfaces could not alone, it seems, account comprehensively for variation amongst soots in their ZDP requirements; but there *were* indications that higher oxygen presence on the surface rendered the soot less wear-promoting (Rounds, 1981).

7.7 Alternative Combustion Systems

The lion's share of the text is given over – for obvious reasons, shall we say – to the *diesel* engine, but other forms of internal combustion engine *also* emit particles. Of these other forms, there is, of course, the time-honoured gasoline or spark-ignition engine, to which Chapter 8 is exclusively devoted. The current section focuses, instead, on various modifications, or refinements, to the conventional heterogeneous-charge diesel engine, and also on considerably more drastic measures, wherein the combustion departs markedly from conventionality. These alternative combustion systems have been awarded various appellations and acronyms, the commonest of which, although not necessarily the most accurate from a technical standpoint, is homogeneous charge compression ignition (HCCI); others are premixed compression ignition (PCI), partially premixed compression ignition (PPCI), and low-temperature combustion (LTC). This list is not complete. The aim of these alternative combustion systems is not always the improvement of engine-out emissions: regeneration of aftertreatment devices furnishes another (Hountalas et al., 2007). Demarcations amongst these numerous alternative combustion systems are often uncertain; indeed, such distinctions are, in some instances, more semantical than technical. Fortunately, as combustion is not our subject, but rather particulate emissions, the present discussion is confinable to a far narrower compass than would otherwise be necessary.

As we have learned, soot is a product of rich combustion or, more correctly, regions of the charge that are *locally* rich, even though, overall, or globally, air–fuel ratios are in fact lean. However, it appears possible, within certain constraints, to burn fuel at rich air–fuel ratios *without* soot, the formation of which, according to considerations of a fundamental nature, assumes a bell-shaped curve, peaking, say, somewhere between 2000 K and 2400 K (Uyehara, 1980), depending *inter alia* on flame structure, pressure and fuel formulation (Frenklach *et al.*, 1983). This thermal window for soot formation is adduced on the low-temperature side to effete pyrolysis *kinetics*, and on the high-temperature side to *thermodynamic* instability in soot precursors (Kitamura *et al.*, 2001). The concept is best illustrated by a graph on which air–fuel ratio is plotted against temperature, whereon the soot-prone region appears as a kind of peninsula (Akihama *et al.*, 2001). The task is to burn the fuel in a manner that avoids this peninsula: that is, by careful management of air–fuel ratio, temperature and pressure (and residence time at each combination of the three); a suitable course is charted from start to end of combustion. By a similar ploy it is possible, in theory at any rate, to negotiate a route that avoids the NO_x-forming regions.

The practical realisation of this concept, inside internal combustion engines, is not, of course, trivial, because of the many trade-offs, entailing unavoidable compromises which cannot be explored here; although it should be said that operating envelopes are smaller than for conventional diesel engines, maximum load being a significant constraint, and fuel economy also suffers. Passing over these difficulties, to avoid soot formation, flame temperature might be *increased*, say to >2300 K, for example by reducing heat rejection or raising the compression ratio; but if the pressure shifts, the soot formation window also shifts, and, in any case, the NO_x formation window must be avoided (Kamimoto and Bae, 1988). Alternatively, flame temperature might be *decreased*, say to <1600 K, by high rates of EGR (Akihama *et al.*, 2001) – a practice which overturns convention, as contemporary understanding is that this always exacerbates emissions of soot (Section 7.9). But if EGR is increased to rates far larger than those conventionally realised, the soot emission in fact declines. This effect is not held to be one of improved fuel–air mixing: rich-burning pockets of fuel still arise; the reduction in flame temperature is what counts. It was once suggested, for instance, that below 1600 K, the retardation in carbonisation kinetics would aid the oxidation of soot-*precursor* PAH, instead of, as now, permitting the same precursors to form less readily oxidised soot (Dobbins *et al.*, 1996). Many alternative combustion systems use a start of injection that is either very retarded or very advanced, so that the burn is mostly premixed, rather than mixing-controlled, indeed, the ignition delay might actually exceed the fuel-injection period (Klingbeil *et al.*, 2003). Engine manufacturers have devised various proprietary schemes for burning lean, well-mixed charges (Besson *et al.*, 2003); 'modulated kinetics', for example, uses a low-temperature premixed (but not necessarily homogeneous) charge, high rates of EGR, retarded timing and a lower compression ratio (Kawamoto *et al.*, 2004).

The HCCI engine (in theory) completely eschews mixing-controlled combustion, for example by deliberately prolonging the ignition delay and by foreshortening the injection duration (Lejeune *et al.*, 2004). Soot, if it emerges, appears to relate to instances of incomplete mixing, such as slowly evaporating fuel (Kim D.S. and Lee, 2005); this is betrayed in the combustion chamber by incandescence (Wagner U. *et al.*, 2003), and remedied by delaying the combustion, for example via the charge temperature (Lee K. *et al.*, 2005). In a hybrid configuration, the HCCI is *partial*: some fuel is injected into the inlet manifold (premixed burn), and the remainder injected subsequently into the cylinder (mixing-controlled burn). Counter-intuitively, a greater premixed fraction can lead to more, rather than less soot: perhaps the balance between formation and oxidation is adversely affected at the start of the mixing-controlled burn (Kim D.S. and Lee, 2006).

Some overmixed charge may not burn fully, if beyond the lean flammability limit; and some hydrocarbons may shelter in the thermal boundary layer at the combustion chamber wall, or within crevices (Gray and Ryan, 1997), in a manner akin to what occurs in the gasoline engine. Emissions of HC are inconveniently high, and many of these compounds are ultimately captured as organic particulate. Moreover, lubricant is still discharged (Kawano *et al.*, 2004a). Particulate deposits collected on filters are grey, rather than black, evidencing some compositional difference (Wagner R.M. *et al.*, 2003): logically, a greater organic and a lesser carbonaceous presence than for conventional combustion.

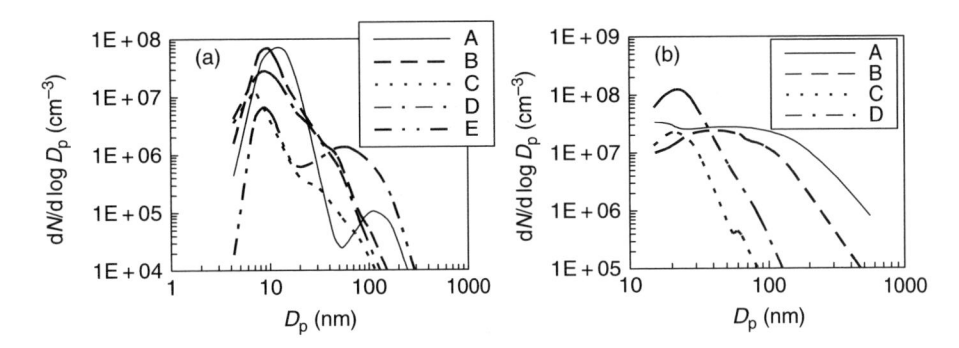

Figure 7.17 Number distributions. (a) As determined by various combustion systems: A, fumigation; B, port fuelling with hot inlet air; C, early homogenisation in cylinder; D, late homogenisation in cylinder; E, conventional diesel. Engine: 1.8-litre, single-cylinder. Operating point: 1200 rpm, 50 kPa BMEP (40 kPa for B). Dilution: two-stage ejector, with ratio of 10 in each stage. Instrumentation: SMPS, 4–150 nm (Raatz and Mueller, 2001). (b) As determined by air – fuel ratio: A, 50; B, 70; C, 100; D, 200. Engine: 0.67-litre, single-cylinder, direct-injection. End of injection, 320°BTDC; start of combustion, ~ TDC. Operating point: 1500 rpm (Kaiser E.W. *et al.*, 2005).

But that is not all: in low-temperature, rich combustion, soot precursors are still formed. Thus, the organic fraction, according to the ideas related in Section 3.6.1, consists not just of fuel hydrocarbons, but also of partial-reaction products (Sluder *et al.*, 2004), making emissions of PAH a potential issue, inasmuch as these novel combustion modes prematurely terminate soot-forming reactions (Merritt *et al.*, 2006).

Combustion schemes along the lines described often emit so little soot that mass measurements are not always reported, while smoke is unmeasurable. This does not, however, necessarily render particulate emissions negligible, in mass terms (Kawano *et al.*, 2004b) or in number terms (Price *et al.*, 2007). Greater organic presence on the one hand and loss of solid carbonaceous surface on the other are circumstances likely to foment nucleation: nanoparticles are indeed prominent with HCCI – plausibly, these are nucleated sulphates or organics derived from the fuel and lubricant (Kawano *et al.*, 2004b). Some demographic characterisations are provided in Figure 7.17(a) (Raatz and Mueller, 2001): these show fewer accumulation-mode particles compared with conventional diesel combustion, as one might expect from the reduction in soot, but more abundant nanoparticles, presumably through higher saturation, i.e. as a necessary concomitant. Intriguingly, however, these nanoparticles seemed relatively *nonvolatile*, being uncharacteristically resistant to evaporation; thus, a possibility is that the absence of soot, in the *combustion chamber*, promoted an earlier nucleation of ash, according to ideas sketched in Section 3.5. In another investigation, shown in Figure 7.17(b), at an air–fuel ratio of ~ 70 the number distribution changed sharply, from domination by accumulation-mode soot to domination by nucleation-mode volatiles (Kaiser E.W. *et al.*, 2005). Obviously, such a shift betrays some corresponding (but unknown) transition in the combustion process; but, whatever the cause, even very lean and ostensibly homogeneous charges are capable of generating soot in some circumstances.

7.8 Aftertreatment

The technology of 'aftertreatment' – an exhaust gas clean-up between the engine and the tailpipe – is well established, and, famously, concerns the catalytic converter, which today is found, in one form or another, on almost all motor vehicles. The primary aim of the catalytic converter is to address the emissions of HC, CO and NO_x; but there are certain benefits of an indirect nature and occasional drawbacks for emissions of particulate, as reviewed here. The device that, by contrast, addresses particulate *directly* is the DPF, which, at the time of writing, is undergoing, in preparation for imminent emissions control legislation, extremely rapid development; indeed, forthcoming statutory limits are unlikely to be met by any other means.

7.8.1 Catalytic Converters

The original reason for installing catalytic converters on diesel engines was to control the emissions of HC and CO: hardly a misapplication, since, unlike on gasoline engines, exhaust gas is always rich in oxygen, so the task is, straightforwardly, one of accelerating the oxidation reactions, the corollary of which – destruction of the organic fraction – came as an obvious bonus. However, the precise implications for particulate emissions, as a collective group, very much depend on the fraction of particulate, all four of which interact in *some* way with catalysts, and not necessarily in a manner conducive to abatement. These undesirable interactions are twofold: obfuscating *storage effects* and unpropitious *side reactions*.

The destruction of organic compounds, probably as gas-phase precursors, takes place in a manner akin to that of HC. Yet, as with HC, the conversion efficiencies *within* the group vary widely: speciative measurements upstream and downstream of the catalytic converter are, unfortunately, infrequent, but, from the work conducted, oil organics oxidise more readily than fuel organics (Voss *et al.*, 1997); alkanes and aromatics oxidise more readily than partially oxidised transitionals, oxygenates and acidics (Hunter *et al.*, 1981); native fuel compounds oxidise more readily than pyrosynthesised polar compounds (Narusawa *et al.*, 1995); heavier hydrocarbons oxidise more readily than lighter hydrocarbons (Kawatani *et al.*, 1993); and some PAH (*BbF*, *BkF* and *BaP*) oxidise slightly more readily than other PAH (*Flu*, *Pyr*, *BaA* and *Chr*) (Pataky *et al.*, 1994). This picture is tempered by the suspected *formation* within the converter of *secondary* organic compounds, such as transitionals and oxygenates, perhaps through partial oxidation of primary compounds, such as alkanes and aromatics (Hunter *et al.*, 1981). The suspicion also lingers that large PAH molecules are catalytically formed from small PAH molecules (Janakiraman *et al.*, 2002). The conversion efficiency is, then, not only a function of what is oxidised, but also a function of what is synthesised; although it should be said that, *overall*, a net reduction takes place in the organic fraction.

Catalysts also store organic compounds: this seems inevitable when temperatures are low, that is, *below* light-off: HC 'conversion efficiencies' of 70% (Andrews *et al.*, 2000d) are, in this regime, obviously factitious. Naturally, the period of this storage is dependent, also, on the rate of throughput, but is not, apparently, exhausted even after several tens of minutes of idling (Andrews *et al.*, 2000d). This engenders some difficulty in distinguishing between genuine oxidation and storage. The mode of storage is, perhaps, one of adsorption or condensation of gas-phase precursors onto the cold catalyst surface; although, in exceptionally cold conditions, since some gas-to-particle conversion takes place within the exhaust system, in extreme instances with the formation of white smoke (Section 3.6.5), fuel droplets may simply collide with the surface. The fates of these stored compounds are undefined: some are oxidised at a later stage, when high temperatures return; others, depending on the vagaries of desorption, evaporation, oxidation and rate of temperature rise, escape unscathed to form a fugitive emission. The conditions of release do not necessarily coincide with a resumption of catalytic activity. It should be noted that fuel, if excessively stored within catalysts and ignited *in situ*, is thermally destructive to the monolith.

The relevance of oxidation catalysts to sulphate particulate is in the precursive step (Section 3.7.1), conversion of SO_2 to SO_3 – an inevitable consequence, it seems, of the *oxidation* function. This side reaction is highly undesirable, as the formation of H_2SO_4 is, thereafter, virtually assured. The diesel oxidation catalyst is, in this respect, identical to the gasoline oxidation catalyst (Section 8.2.2). The conversion, being strongly driven by temperature, worsens markedly for heavily loaded duty cycles (Bielaczyc et al., 2002); although, at truly high temperatures (>700 °C), the equilibrium reverses, so to speak, and favours instead SO_2 over SO_3. These temperatures are, however, less likely on diesel than on gasoline engines.

Catalysts avidly store sulphur. The precious metals appear less important here, although some storage probably takes place, for example as $PdSO_4$ (Neyestanaki et al., 2004). Otherwise, the storage takes two forms. At higher temperatures (>400 °C), sulphuric acid reacts with the washcoat, with storage as a sulphate of the surface or perhaps of the bulk: aluminium oxide, a common washcoat constituent, plays a large role here. At lower temperatures (<350 °C), sulphuric acid is merely stored in condensed form (Horiuchi et al., 1991). At some stage, the catalyst stops storing, not necessarily because its capacity is satiated; the stored sulphur compounds and the sulphur compounds passing through might simply fall into equilibrium. Post-mortem examinations, which probably focus on acid–washcoat compounds rather than the condensed acid, show a relatively even distribution of this element lengthwise (Bardasz et al., 2003), and also with depth into the washcoat (Lampert et al., 1996). This pattern of storage distinguishes sulphur from phosphorus, calcium and zinc (see below), and accords, plausibly, with its arrival as a gas (SO_2) rather than as a particle, and also with its site-specific, rather than indiscriminate, adsorption. According to the duty cycle, this stored sulphur is ejected in sudden bursts – usually measured as sulphate particulate. The condensed form is given up more easily, by evaporation, whereas thermodynamics predict the decomposition of aluminium sulphate at >650 °C (Neyestanaki et al., 2004).

From the foregoing, a key operating characteristic of the oxidation catalyst will be obvious: the *trade-off* between, on the one hand, the organic fraction, the oxidation of which is the intended function, and the sulphate fraction, the formation of which is an unwanted side effect. A common experience is one where the mass of sulphate particulate formed erodes, and occasionally surpasses, the mass of organic particulate oxidised, so that the conversion efficiency of the catalyst for particulate, taken as a whole, becomes, in fact, *negative*. In fact, because of 'sulphate make', the mass emission across the catalyst can increase infelicitously by severalfold (Ketcher and Horrocks, 1990).

It need hardly be stated that Herculean efforts have been devoted to ameliorating this trade-off. In the catalyst formulation, there are two approaches. The first is through *selectivity*, i.e. oxidising HC, CO and organic particle precursors *without* oxidising SO_2, for which there are plausible grounds in catalytic chemistry, as catalysts do not promote all reactions equally. For example, selectivity is imparted by special ageing processes and washcoat additives (Wyatt et al., 1993). The other approach is to seek formulations, particularly washcoat formulations, with poorer affinities for sulphur storage, silica being much reported at one time. Even so, neither of these two methods has been wholly successful; and a major enabler for oxidation catalysts came from a wholly different quarter – the introduction of low-sulphur diesel.

Catalysts are unable, as a rule, to oxidise the carbonaceous fraction or soot; formulations *with* this ability have long been the philosopher's stone in aftertreatment (e.g. Watabe et al., 1983), and the frenetic pace of research in this area shows no sign of abating (e.g. Liu S. et al., 2001; An et al., 2004; Biamino et al., 2005; Cauda et al., 2005; López-Fonseca et al., 2005; Liu J. et al., 2005; Uner et al., 2005; An and McGinn, 2006; Villani et al., 2006).[8] There are numerous contenders (Neeft et al., 1996a), but, as yet, no one has succeeded in this quest – commercially, at any rate. The failure of these formulations is not, actually, purely a chemical aspect of catalytic oxidation as such: the fact that the reactant (soot) is not gaseous, like the other emissions, but in the *solid phase* is the principal problem:

---

[8] An obvious parallel arises here with the need to regenerate a DPF (Section 7.8.2).

on the catalyst surface, the burn, as it were, on atomic scales, once localised reactions have completed, 'goes out', where self-propagating reactions are really required. And even isolated spherules are still too large to enter the pores of conventional catalysts: surface atoms on nanoparticle catalysts are, however, more labile, helping to maintaining contact (Liu J. *et al.*, 2007). This 'contact' problem has been circumvented by formulating catalysts such that they 'fluidise' at typical exhaust temperatures, thus rendering them mobile, or labile, with the ability to maintain contact by diffusing into agglomerate microstructures (Genc *et al.*, 2005). Amongst these formulations are certain chlorides. But since such catalysts when hot readily vaporise or sublime, they are easily lost; and this is the principal stumbling block with fluidising formulations.

Soot is, however, pertinent to oxidation catalysts for an altogether different reason: it is stored, to some extent, the deposition mechanisms for which were described thoroughly in Section 4.2.2. There is also a possibly that stored hydrocarbons may be coked *in situ* (Nakane *et al.*, 2005). This storage impedes the intended function by masking catalytic sites, in a manner akin to the ash particulate (see below). Stored soot *is* oxidisable – as might happen, say, intermittently and noncatalytically, in fortuitous, high-temperature duty cycles. But since the oxidation of soot is highly exothermic, thermal damage to the catalyst might well result. The periodic release rather than oxidation of soot is occasionally advanced to explain otherwise anomalous test results (Arai, 1991), but no systematic study of this has been published. Intriguingly, catalysts are also suspected of *forming* soot: this phenomenon is not widely reported either, but there is, apparently, some theoretical justification for suspecting the action of benzene radicals and hydrocarbon dehydrogenation. If soot formation really does occur within catalysts, this formation is a bit-player in the overall production, while the feedstock hydrocarbons are probably heavy, and not, therefore, easily distinguishable from soot itself (Johnson J.H. *et al.*, 1981).

The propensity of catalysts to store ash is regrettable. With the possible exception of zinc compounds, which are, apparently, sufficiently volatile to escape to *some* extent at high temperatures (Andersson *et al.*, 2004c), and other captured ash which might escape by abrasive mechanisms (Section 9.6), this material cannot be evicted, at least in service, although partial recovery is possible via acid treatments (Christou *et al.*, 2006). The accumulation of this ash within catalysts is, therefore, as inexorable as the degradation this accumulation brings to conversion efficiencies (Neyestanaki *et al.*, 2004). Ash, in smothering the pores in the washcoat, forms a 'diffusion barrier', the action of which restricts the access of gases to catalytic sites; this interference is physical. There are also chemical aspects of this interference, involving electronic interactions and the formation of alloys. The predominant element in this diffusion barrier is normally phosphorus, but zinc, calcium and magnesium make regular appearances; silicon is a minor contaminant in normal circumstances (Galisteo *et al.*, 2005). This suggests that the escape of oil into the exhaust is the main culprit in catalyst deactivation; contamination by phosphorus is, logically enough, linearly dependent on mileage and oil consumption (Ball and Kirby, 1997). Dissimilar patterns of deposition between gasoline and diesel catalysts perhaps reflect dissimilar washcoats or operating conditions (Engler *et al.*, 1993). For example, in gasoline catalysts, the higher temperatures form glassy pyrophosphates, $Zn_2P_2O_7$, and cerium in the washcoat forms $CePO_4$, (López Granados *et al.*, 2005). It should be said that virtually nothing has been published about the chemical forms or particle sizes in which these ashing elements *arrive* in the catalyst.

Deposits of calcium, phosphorus and zinc decline with increasing distance along the catalyst, being thickest at the inlet (Lampert *et al.*, 1996; Bardasz *et al.*, 2003). Thus, this distribution, which is uneven, although displaying a trend, differs markedly from what is observed with sulphur, as aforesaid. The likely explanation is that whereas sulphur is choosy as to deposition site, the adsorption being essentially chemical in nature, the other ashing elements are not choosy, being prey to transport mechanisms of a physical or mechanical nature, outside their control. It is tempting, on this basis, to argue that these elements, arriving perhaps as particles, experience the strongest depositional transport mechanisms when in the *inlet*; but no specific study has been published along these lines. Incidentally, this pattern of deposition, biased towards the inlet of the catalytic converter, is really quite intriguing insofar as incombustible ash, trapped within a DPF, tends to accumulate in the *rearmost* section. This discrepancy does not appear to have been addressed in the literature.

These factors taken together, the conversion efficiency for particulate taken as a whole varies considerably over the engine operating map, not only because of catalytic activity, but also because of particulate *composition*. At high load, engine-out particulate emissions are dry and predominantly carbonaceous, so that although catalysts are active towards organic compounds, only a small amount of such compounds is available to oxidise. A high production of sulphate particulate is, however, not unexpected. At low load, plenty of organic compounds are emitted; but, if temperatures are *too* low, the catalyst is inactive, and stores rather than oxidises. This explains instances where high 'conversions' are seen in the organic fraction even below light-off temperature (Unsworth *et al.*, 1996). Moreover, in-engine emission control strategies which make the particulate emissions more organic and less carbonaceous – even when not actually changing the mass emission – are still worthwhile insofar as they aid the performance of the catalyst. For this reason, one must distinguish carefully between engine-out and tailpipe locations.

Even though organic compounds are undoubtedly oxidised, quite often effectively, the effects of catalysts on mutagenicity are less easily established, for this scales unstraightforwardly with the mass emission. What is intriguing (and critical) is the difference between the *total* mutagenicity of the organic fraction, expressed simply as revertants, and the *specific* mutagenicity, expressed as revertants per unit mass of organic particulate. Even if the mass emission of organic particulate declines significantly, the compounds within this emission, per unit mass, are sometimes disobligingly *more* mutagenic, such as by an order of magnitude (Johnson J.H. *et al.*, 1981). An example of this phenomenon is shown in Figure 7.18 (Hunter *et al.*, 1981). Others have measured a decline of severalfold (Vanrullen *et al.*, 2000); and the effects do seem to vary, one way or the another, depending, *inter alia*, on engine operating point. A definitive explanation for these mutagenic effects goes unreported; but they arise, obviously, from subtle speciative changes induced in the organic fraction, as already mentioned. The phenomenon is, however, difficult to extricate from measurement artefacts: nitration, by which NPAH form from PAH, as detailed in Section 6.4, is clearly implicated (Pataky *et al.*, 1994); particularly so,

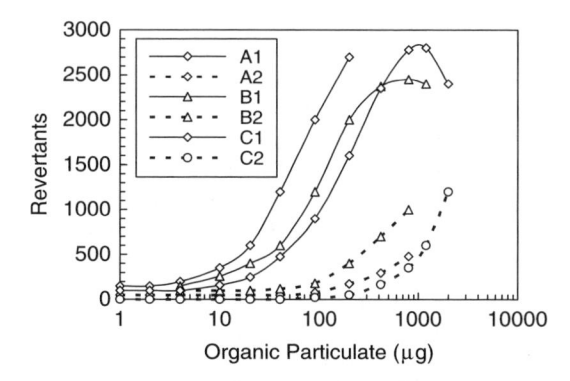

Figure 7.18 Mutagenicity as determined in the Ames test. The following legend includes the mass concentration of organic particulate in parentheses: A1, No. 2 diesel, with catalyst (6.0 mg/m^3); A2, No. 2 diesel, without catalyst (26 mg/m^3); B1, shale oil with catalyst (7.3 mg/m^3); B2, shale oil without catalyst (49.7 mg/m^3); C1, No. 1 diesel, with catalyst (7.3 mg/m^3); C2, No. 1 diesel, without catalyst (25.8 mg/m^3). Catalyst: platinum, <3500 g/m^3, cordierite monolith. Engine: 1.4-litre V8 diesel, direct-injection, naturally aspirated; rated power 157 kW at 2800 rpm. Fuel (aromatics %, sulphur ppm): No. 1, 16.7, 400; No. 2, 24.3, 2600; shale oil, 21.0, 5 (Hunter *et al.*, 1981).

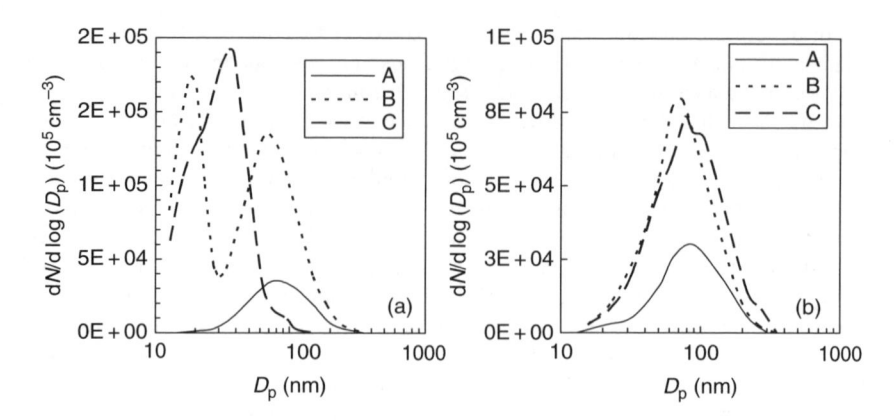

Figure 7.19 Number distribution, as affected by an oxidation catalyst. Tests conducted in wind tunnel; the wind speed was the vehicle speed. (a) Active catalyst, 2.5 mg/cm³ platinum with alumina – zeolitic washcoat, relative humidity in ambient 45 % (NB: curve C is shown reduced by a factor of ten). (b) Blank monolith, relative humidity in ambient 14 %. (Note different scales on y-axes.) Vehicle speeds (km/h): A, 40; B, 70; C, 70 + 3 % grade. Vehicle: 2.5-litre diesel, direct-injection, turbocharged, intercooled, light-duty truck, model year 1997, Stage II. Fuel: aromatics, 30 %; cetane number, 46, sulphur, 350 ppm. Dilution: as realised in exhaust plume; the sampling point was 5.5 m behind the tailpipe, 0.8 m above ground level. Ambient temperature 20 °C (Maricq *et al.*, 2002b).

as oxidation catalysts are active in the conversion of NO to NO_2. But the moment of nitration – in the exhaust gas, in the catalyst, in the dilution tunnel or on the filter – is unclear; and this is the perennial problem with artefactual transformations, as already discussed (Sections 3.6.2 and 4.6).

The nucleation mode is very different: sulphuric acid, through mechanisms described at length in Section 3.7.2, is a notable nucleating agent (Schneider *et al.*, 2005), and this acid is avidly formed by oxidation catalysts. Should the surface area in the accumulation mode prove inadequate to the task of suppressing saturation by adsorbing this acid, then nanoparticle formation seems inevitable – as strongly suggested by Figure 7.19, in which an active catalyst is compared with a blank monolith (Maricq *et al.*, 2002b). This issue might be exacerbated were an oxidation catalyst to release, suddenly, stored sulphuric acid or even stored organic compounds, the latter of which, as growth species in the nucleation mode, were also described in Section 3.7.2. Hence, plausible reasons exist for supposing that oxidation catalysts considerably worsen the emission of *volatile* nucleation-mode particles under certain conditions. *Nonvolatile* nucleation-mode particles are different still: if present already, in the feed gas, as might be expected of ash, these will, according to the conventional understanding of particle transport, be captured relatively easily by the catalyst monolith.

DeNO$_x$ Catalysts

Some mention of so-called 'deNO$_x$ catalysts' is apposite in this section since, although the laudable intention here is to address the emissions of NO$_x$, certain side effects are nonetheless seen on the emissions of particulate, albeit through a different set of circumstances than for oxidation catalysts. In 'active deNO$_x$', hydrocarbons are injected intentionally into the exhaust stream, to act as chemical reductants for NO$_x$. However, the use that deNO$_x$ catalysts make of these hydrocarbons is poor, and many of them escape, to be measured, if still in the gaseous state, as HC, or, following gas-to-particle conversion, as organic particulate. And, in both major branches of deNO$_x$ catalysis, 'active' and 'passive', stored hydrocarbons are released from time to time just as from oxidation catalysts, and measured as bursts of organic particulate (Kawanami et al., 1996).

Hydrocarbons are not the only possible reductant: ammonia and urea have been investigated extensively for the same purpose (Hug et al., 1993), in which case the reducing chemistry, if improperly contrived, *also* leads to particle emissions – albeit of an entirely different chemical composition from those discussed throughout this book. The ammonia reacts with SO_3 to form particles of ammonium *sulphate*, according to (Lüders et al., 1995)

$$SO_3 + 2NH_3 + H_2O \longrightarrow (NH_4)_2 SO_4(s) \qquad (7.1)$$

and

$$SO_3 + NH_3 + H_2O \longrightarrow NH_4HSO_4(s). \qquad (7.2)$$

These particulate-phase compounds are mostly studied for the way they impede proper operation of the catalyst by depositing on, and thereby masking, the reactive sites – this is particularly acute whenever SO_3, rather than SO_2, predominates in the feed gas, such as when there is a preceding oxidation catalyst. But otherwise, very little is known about these particles compared with others more characteristic of exhaust gas.

This particle formation chemistry highlights a frustrating, but nonetheless seemingly inevitable conundrum. If an oxidation catalyst is *not* added downstream of the deNO$_x$ catalyst, then ammonia, should its dosage exceed the requirements of the deNO$_x$ chemistry, escapes into the environment – this is seen from time to time, especially during transient engine operation, because the metering is still performed open-loop – and ongoing atmospheric reactions eventually form *secondary* particles of ammonium sulphate. On the other hand, if an oxidation catalyst *is* added downstream, then, rather than being eradicated, this ammonia reacts with SO_3 – which is, after all, being formed from SO_2 in the same location – and exhaust-gas reactions form *primary* particles of ammonium sulphate (Havenith and Verbeek, 1997). And, in both cases – in the exhaust and in the atmosphere – particles of ammonium *nitrate* are alternative possibilities (Saito S. et al., 2003). Nature, it seems, is utterly determined, one way or another, to turn ammonia into particles: sulphate or nitrate, primary or secondary (see the final points made in Sections 2.3.3 and 2.3.4)

Demographic measurements of particles emitted in deNO$_x$ catalysis are insufficiently reported to allow any general picture to be delineated. In one study of urea as a reductant, particles of size >100 nm were not, apparently, affected, whereas those of size <30 nm were fewer in number, by an order of magnitude or more, than without the aftertreatment; and urea, or its breakdown products, did not, evidently, instigate formation of a nucleation mode (Gekas et al., 2002). In another study, though, an order-of-magnitude increase was observed in the nucleation mode; and, significantly, there was *no* corresponding decrease in the accumulation mode, thus suggesting surface area not to be an aspect (Kati et al., 2004b). The nucleation mode was volatile in that it disappeared at 250 °C. When NH_4HSO_4 and $(NH_4)_2SO_4$ are emitted in extreme quantities, they become manifested as smoke (Morimune et al., 1998). With deNO$_x$ catalysts – in *contradistinction* to oxidation catalysts – the preparatory conversion of SO_2 to SO_3, by which sulphates form, might be helpfully put under some restraint, if the exhaust chemistry is now less oxidising, and this restraint might oppose, to some extent, the tendency of ammonia to form sulphate particles; this idea is conjectural. Finally, mutagenicity – also poorly reported – is not consistently affected one way or the other (Hori M. and Oguchi, 2004); whether such trends relate solely to the organic component or to some additional aspect introduced by the ammonia chemistry is unknown.

7.8.2 Particulate Filters

The diesel particulate filter is *not* a new idea: relevant publications in the open literature first emerged as far ago as the late 1970s, and have been appearing with great regularity ever since. Throughout this period, whenever any new (and tighter) statutory emissions regulations for particulate were introduced, various pundits, more learned than the present author, declared the 'absolute necessity' of the DPF 'in a few years' – these regulations not being thought attainable by any other means. But this did not happen: engineers are, after all, a resourceful community, and were spurred to look elsewhere by the formidable complexity and expense of the DPF. Hence, this technology never really made it to mass production.[9] Meanwhile, a quarter of a century or more elapsed, and only now is the DPF finally stepping into the limelight. There are, of course, important enabling technologies in *other* fields, without which the DPF would not be able to assume centre stage: for example, certain Olympic-class gymnastics are required in fuel-injection schedules for which the necessary technology was simply unavailable prior to the 1990s.

If one takes a broader view, capturing the particles in some form of exhaust-mounted device is an approach with an even lengthier history: this technology was avidly researched for *gasoline* engines in the early 1970s, with a view to controlling particle emissions of *lead* (Section 8.2.1). Of course, in great distinction to lead, soot is eminently *combustible*, by which process the filter is periodically cleaned or purged, *on vehicle*. The operating principle of the DPF, therefore, unlike the catalytic converter, is *discontinuous*: periods of particle capture or filtering (several hundred miles), punctuated by periods of particle burning or incineration (a few minutes).[10] In the *capture* phase, the already solid particulate fractions (carbonaceous and ash) are detained to high efficiencies; the volatile particulate fractions (sulphates and organics) are also detained to some extent; and that which is still in the gas phase can pass through. The *incineration* phase is commonly perceived as essentially *oxidative* in nature, but this can only apply in the purest sense to the carbonaceous fraction. Ideally, the organic fraction is *also* oxidised, but some species revaporise, and, rather than await destruction, make good their escape. The sulphate fraction also vaporises, but with consequent escape, partially as sulphuric acid. The ash fraction, being essentially nonvolatile and incombustible (although it may partially break down), remains behind as ash.

Filtering

There are many, many, many types of DPF, but grouped according to filtering *principle*, just two variations are made on one theme. Deep-bed filters, such as wool and foam, collect particles throughout the whole of their structures; surface filters collect particles on their porous walls. Two operational differences separate the filter types: deep-bed filters, upon the sudden application of a high flow rate, are more likely to eject their contents as a 'blow-off', whereas surface filters, if not cleared with sufficient regularity, are more easily clogged.

Capture mechanisms are diverse, but what can be safely dispensed with is the prevalent misconception of filter operation as a kind of 'sieve': exhaust gas streams through, whereas particles are physically blocked. This is correct only for the very largest particles (several microns at least). Filters, if designed purely on this principle, would be impractical because the pressures required to force through the exhaust gas would be untenable to the engine. In reality, the voids or pores in the filter are one or even

---

[9] There were some small-scale field applications, but the results were not encouraging.
[10] There are, moreover, acoustic benefits, as the noise-attenuating effect of the DPF resembles that of the conventional muffler (Katari *et al.*, 2004).

two orders of magnitude *larger* than the particles, so that capture is *not* purely an operation of sieving (Mayer A. *et al.*, 2002).

Filtering, as described in Section 2.2.3, operates through a variety of *other* particle transport mechanisms, namely diffusion, inertia, interception and thermophoresis; and, actually, an instructive question arises here that does not seem to have been pursued in the literature. The need to filter particles from gases is common enough (e.g. in air conditioning), and, in this technology, shortfalls in trapping efficiency at a few hundred nanometres or so are generic (Konstandopoulos *et al.*, 2000): this is ascribed to the absence of significant particle transport mechanisms between the inertial and diffusional regimes (Section 2.2.3). But, in filtering particles from *exhaust gas*, these shortfalls are not always seen: a result, *perhaps*, of thermal gradients, i.e. thermophoresis, or simply because the combination of hydrodynamics and filter properties (tortuousity, porosity and permeability) shifts this gap outside the size range of the measurement (Wirojsakunchai *et al.*, 2007). Still, whatever the reason, the augmentation of naturally arising particle transport mechanisms in this 'blind spot', such as by electrostatic attraction (Xiaoguang *et al.*, 2001), has not yet been necessary, as capture efficiencies are already high enough to meet current statutory requirements.

A further and interesting twist is the diminishing importance of underlying filter structures once particles have begun to accumulate in any great quantity (Tessier *et al.*, 1980). In surface filters, particles accumulate as a contiguous but *porous* layer, or 'cake', through which exhaust gas continually streams, and the filtering efficiency of this layer often exceeds that of the underlying filter. In this situation the filter acts predominantly, and secondarily, as a porous support for the overlying deposit, to which the filtering action has now passed. But in deep-bed filters also, captured particles resting, say, on strands of wire also assist the filtering action through acting as excrescences, or 'collectors', to which yet more particles attach themselves gregariously. A fouled filter is therefore more effective in particle capture than a recently regenerated one (Montajir *et al.*, 2007).

Regeneration

In the incinerating, burn-off or purging procedure, known in contemporary parlance as 'regeneration', an illuminating question is whether soot can be eradicated while in transit through the exhaust system, *without* a DPF. As it turns out, the trade-off between exhaust temperature and residence time is brutally unforgiving: several workers have shown the need for a temperature in excess of 1000°C, *sustained* for the whole of the journey, from engine to tailpipe (Fujiwara and Fukazawa, 1980; Otto *et al.*, 1980; Murphy *et al.*, 1981). Obviously, this thermal extremity is impractical, if only because of the energy demand: there is the need to heat all of the exhaust gas, rather than just the particles, and also the need to arrest or reverse the rapidly plunging temperature. The solution, then, is not so much to increase the temperature as to increase the residence time, by *capturing* the particles. Moreover, the oxidation of carbon is comfortably exothermic, and, if the particles are laid down in a contiguous layer, their combustion becomes self-sustaining, thus assuring an energy input that is even more practical. This is the prime purpose of the DPF: to hold the particles securely in place for long enough to ensure incineration.

A precise qualification of ignition temperature is impossible, as the physicochemical properties of the particulate deposit are so diverse and ill-defined. The following aspects are thought relevant: dendritic or fluffy deposits, the interstices of which entrap oxygen (Gabathuler *et al.*, 1991); the organic fraction, which acts as a kind of 'lighter-fluid' (Pattas *et al.*, 1998); the ash fraction, the metals of which act as catalysts (Bardasz *et al.*, 2005); spherule microstructure, which controls the density of carbon atoms exposed to oxidants (Boehman *et al.*, 2005); graphitisation (Fang H.L. and Lance, 2004; Vander Wal *et al.*, 2007); and carbonisation (Yamanaka *et al.*, 2005). It should, moreover, be noted that such physicochemical properties are *not* stable: ongoing carbonisation during residence, for example, renders the deposit progressively more difficult to burn. For all these reasons, analytically predicting the conditions required for ignition are nontrivial (Zheng and Keith, 2004); but, experimentally, temperatures of >550 °C must be exceeded.

Unfortunately, diesel engines – according to typical customer duty cycles, that is – usually discharge exhaust gas cooler than this figure; indeed, conditions are considerably cooler in the *underbody* location, where, owing to packaging constraints, the DPF is often located, so that some method of *forcing* regeneration is absolutely necessary. Many, many, many such methods have been devised; we shall not review these methods in detail, but they are essentially of two types: *thermal*, i.e. raising the temperature of the exhaust, such as by an electrical heater, auxiliary burner or throttling (Persiko-Karakash and Sher, 2006); and *catalytic*, i.e. lowering the ignition temperature of the particulate, such as by a catalytic coating on the DPF, or a catalytic fuel additive. In this last-mentioned method, some recapitulation of Section 3.5.2 is worthwhile: close contact or integration between metal and spherule is essential, and if treat rates are increased beyond this point, ash forms distinct nucleation modes independently of accumulation-mode soot, of little benefit to regeneration.

Once ignited, the problems are only just beginning: the burn-off is exceedingly exothermic, and this is why, to avoid thermal damage to the DPF, the mass of the particulate deposit must never be allowed to exceed a certain well-defined limit. Moreover, some method of restraint is advisable: a particular embarrassment is raised by 'return-to-idle', where the DPF is suddenly flooded with oxygen, yet is also deprived of the convective cooling imparted by a high exhaust flow rate. During these moments, the exhaust oxygen and flow rate are manipulated by, *inter alia*, throttling the engine's air intake (Flörchinger *et al.*, 2004). The difficulty, though, of starving the regeneration of oxygen is the conversion of carbon to CO, rather than to CO_2 (Biancotto *et al.*, 2004). It should be also mentioned that regeneration involves running the wrong way up an escalator, so to speak: the engine, after all, is still running, so that burn-off rates must exceed, by a considerable margin, the rate at which soot arrives (Bruetsch and Bloom, 1991), otherwise insufficient headway is made.

The Four Fractions

As aforesaid, the pre-eminent advantage of the DPF lies in its remarkably efficient capture of 'already solid' particles, meaning, chiefly, the carbonaceous fraction or soot, present as accumulation-mode agglomerates: thus, the mass emission of particulate declines from several tens of milligrams, or a few hundreds of milligrams per kilometre, to just a few milligrams.

Capture is not, however, restricted to solid particles: organic compounds including PAH are also retained. Now, the intuitive understanding is that these compounds, being in the gas phase, are able to pass straight through a DPF, and thus evade capture. This is true to some extent, but as, on occasion, capture efficiencies are several tens of per cent, including for PAH (Howitt *et al.*, 1983), the action of a DPF towards organic compounds should not be dismissed too lightly. Some small amount of gas-to-particle conversion probably takes place in the exhaust system, upstream of the filter: the significance of this capture mechanism depends on the exhaust temperature, as determined by duty cycle, and the location of the DPF with respect to the engine. But the capture of organic compounds is not, by any means, dependent on gas-to-particle conversion upstream: this relates not to the DPF *per se*, but to the particulate deposit within it; if the retained mass is (say) 50 g, then the total surface area may be (using information in Section 6.2.4) 15 000 m^2. The adsorbing affinity of carbon for organic compounds is well known – indeed, this is the principle of many scrubbers. So, although organic compounds may well *enter* the DPF in the gas phase, their successful penetration is by no means assured.

Left to itself, a *simple* DPF would interact only weakly with sulphur, which arrives, more probably, as gas-phase particle precursors (SO_2 and SO_3) rather than as sulphuric acid. The interactions between these gas-phase precursors and the particulate deposit or the underlying filter structure remain wholly unreported. But the source of the problem lies elsewhere: to force temperatures up during regeneration, some oxidation function is *added*:[11] either as a separate, upstream catalyst (Salvat *et al.*, 2000), or as a catalytic coating within the DPF itself (Warner *et al.*, 2003); in either case, acid formation proceeds whenever exhaust temperatures enter the catalytic window (Sections 3.7 and 7.8.1). Hence, although operation in the regeneration mode is not an essential precondition, the hotter conditions encountered during this time tend to exacerbate the sulphate emission (Bickel and Majewski, 1993). Once this acid has formed, it finds a high surface area within the DPF on which to condense or adsorb, from which it is intermittently evicted.

And finally to the ash fraction, which, as said, remains behind, as it is incombustible. The complexity of ash retention mechanisms is often glossed over: for example, why zinc should deposit evenly, yet strontium preferentially at the rear (Obiols *et al.*, 2005), is not known. Ash retention has various, but always deleterious consequences for the working life of the DPF. Fuel economy suffers, because the (baseline) exhaust back-pressure creeps inexorably upwards; the precise impact depends on the composition, morphology and distribution (Gaiser and Mucha, 2004) of the ash. The storage capacity for what is combustible declines correspondingly; and, ultimately, a point is reached where regeneration is being demanded too frequently.[12] Structurally weakening solid-state reactions occur between ash and the filter structure (Child and Cioffi, 1994). And a lesser-known aspect is ash-promoted formation of sulphuric acid (Czerwinski *et al.*, 2000), as the metals are catalytic. For these and various other reasons, the working life of the DPF is traded very carefully against the ash content of the fuel, particularly when a fuel additive is used, and also the ash content of the *oil* (McGeehan, 2004). Some vehicle manufacturers choose to replace the DPF at specified vehicle service intervals; the ash is removed with purpose-built equipment, and the filters returned to service. The only method to prevent, or at least to delay, this ash build-up *on-vehicle* is to flush it out, with compressed air; this is an alternative method of regeneration (Khalil and Levendis, 1992).

Demography

The demographic impact of the DPF is profound yet complex, and not always beneficial. In the first instance, one should distinguish carefully between normal running and periods of regeneration. During regeneration, some soot may not actually, as intended, be incinerated or gasified to CO_2 and H_2O, but instead may be released (Laymac *et al.*, 1991): plausibly, the particulate deposit burning in the DPF disintegrates, whereupon some debris is released which successfully negotiates a safe passage through the filter structure (Cauda *et al.*, 2006; Montajir *et al.*, 2007), briefly raising, also, the emission of accumulation-mode particles (Baumgard and Kittelson, 1985). The evaporation and escape of particle precursors during this time adds to the mass emission, and also creates nucleation-mode particles (Bikas and Zervas, 2007). These momentary emissions are styled 'release events'. All is not lost; this worsening of the particle emission during regeneration, and *only* during regeneration, is reasonable enough insofar as it occurs only every few hundred miles, in which case the DPF is still very beneficial in overall terms. It pays, however, not to become complacent when the mass emission during regeneration is one hundredfold that between regenerations (Kono *et al.*, 2005).

Contrastingly, the ramifications for *volatile* material are highly equivocal. True, some of the organic fraction is, like soot, oxidised to CO_2 and H_2O during regeneration; escape happens (Kantola *et al.*,

---

[11] With which to burn hydrocarbons, the concentrations of which are, during this time, intentionally raised artificially.
[12] Regeneration strategies, virtually without exception, degrade the fuel economy. And for strategies in which late post-injections are used, there is a raised risk of oil adulteration by fuel. So, the preference is to regenerate as infrequently as possible.

1992) when temperatures are high enough to evaporate, yet insufficient to oxidise. It remains unknown whether sulphuric acid decomposes into its gas-phase precursors, but its straightforward release seems to be the more common experience (Sappok and Wong, 2007). Yet *this* form of release event, i.e. of volatiles, is less easily dismissed than the preceding fragmentation phenomenon, *not* being restricted to regeneration (Mohr *et al.*, 2003a), but likely whenever lengthy, low-temperature storage periods give way to brief high-temperature excursions (Liu Z.G. *et al.*, 2002), or where cold starts are followed by high-speed running (Richards P. and Rogers, 2002). The DPF, then, acts as a kind of reservoir for volatile material (Lepperhoff and Kroon, 1985) which, when emptying, erodes, or even reverses, the intended effect. During these moments, mass emissions of particulate increase and, because volatile hydrocarbons are also discharged, momentary increases are measurable in HC (Hardenberg *et al.*, 1987b), sometimes with the manifestation of white smoke (Ullman *et al.*, 1984; Kodama *et al.*, 2005). Moreover, these released volatile compounds act as *particle precursors*, as evidenced by the more pronounced nucleation-mode particles not infrequently seen when vehicles are fitted with a DPF (Campbell B. *et al.*, 2006). This explains the otherwise unphysical result of greater particle numbers downstream than upstream. These particles then continue growing: the pronounced accumulation-mode particles during release events are probably not solid or carbonaceous (see Figure 6.13).

Particle number is perhaps the greatest bone of contention with the DPF, but it should not be understood that this bone relates solely to release events (Kati *et al.*, 2004b): the whole operating envelope is potentially affected. This is because gas-phase particle precursors which pass through a DPF experience a very different situation from that pertaining when *no* DPF is fitted: little solid surface area now exists on which gas-to-particle conversion might take place, the carbonaceous agglomeration mode having been preferentially captured; hence, the conditions for nucleation have ripened correspondingly; this was lucidly explained many years ago (MacDonald J.S., 1983). These nucleation-mode particles are readily demonstrated as volatile: they disappear if the exhaust gas is reheated (Warner *et al.*, 2003). It should not, therefore, be too surprising if particle numbers are boosted by an order of magnitude (Baumgard and Kittelson, 1985), or, at 10 nm, by two orders of magnitude (Suresh and Johnson, 2001). Nor are these releases necessarily short-lived: they can last for half an hour (Hall and Dickens, 2003). Thus, although the DPF may well impart significant reductions in particulate mass – in the carbonaceous accumulation mode – there may occur, conversely, significant increases in particle number – in the volatile nucleation mode. Of course, it seems unfair from one standpoint to castigate the DPF for failing to filter material from the exhaust stream which was, in fact, in the *gas* phase when it passed through, and only *subsequently* formed particles; but the real point is whether these volatile compounds are now emitted (and inhaled) in the nucleation mode, where previously, that is *without* the DPF, they would have been emitted in the (less respirable) accumulation mode.

There is also some discussion as to whether ash is *also* emitted as a nucleation mode. This question needs answering, since fuel additives used to regenerate the DPF may engender precisely such a scenario, for which some data are reproduced in Figure 7.20 (Mayer A. *et al.*, 1998). The figure shows the presence of an ash nucleation mode in the exhaust with no DPF, and also, but less strongly, *with* a DPF. This is quite interesting, since the general understanding of the ash fraction is that gas-to-particle conversion is swiftly completed in the exhaust system, and the filtering efficiency of the DPF in the nanorange is very high. So, Figure 7.20 suggests either that the DPF was overwhelmed by huge particle numbers, or that nucleation was in operation downstream. Some heart, however, can be taken from the fact that although pronounced ash nucleation modes do arise with such fuel additives, they only do so upon *over*dosing (Cook and Richards, 2002). Moreover, any escape of these additive-derived products into the atmosphere should be balanced against the emission of other ash, and much else besides, that takes place *without* a DPF. But, before any additive is used in combination with a DPF, demographic characterisations such as that in Figure 7.20 are advisable.

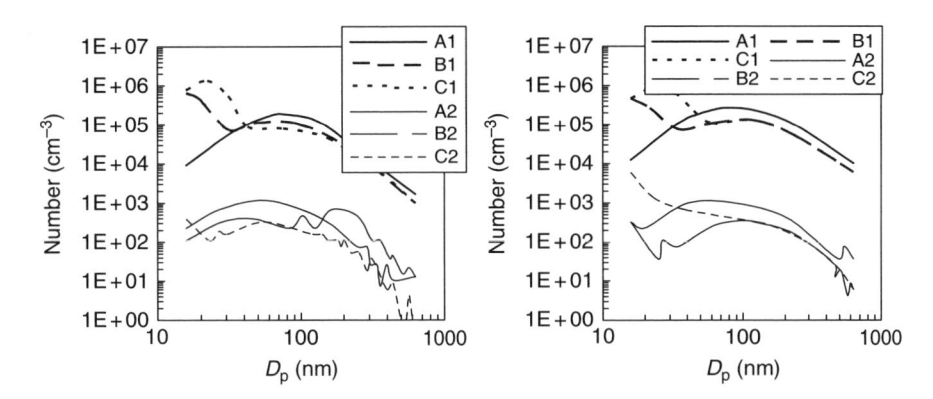

Figure 7.20 Number emissions, as affected by fuel-borne catalyst and DPF. (a) Iron additive: A, none; B, 18 ppm; C, 36 ppm; 1, without DPF; 2, with DPF. Duty cycle: 1400 rpm, 50 % maximum load. (b) Cerium additive: A, none; B, 50 ppm; C, 100 ppm; 1, without DPF; 2, with DPF. Duty cycle: 1400 rpm, full load. (NB: ppm levels are for the additive and not the element.) (Mayer A. *et al.*, 1998.)

Toxicology

As in the case of demography, toxicological characterisations should also take due cognisance of the two operating modes. The prognosis *during* regeneration is presumably favourable, inasmuch as the far hotter conditions are conducive to effective incineration of mutagens that otherwise might have been emitted (Laymac *et al.*, 1991). *Between* regenerations, however – the considerably longer operating mode – the toxicological implications should be examined extremely carefully, and on two grounds.

Firstly, a DPF interferes so extensively with gas–particle partitioning, including the partitioning of PAH and NPAH (Gratz *et al.*, 1991), that some parallel phase-related interference in mutagenicity is only to be expected. Logically, preferential removal of solid carbonaceous particles must shift some organic compounds accordingly, from the particulate phase into the gas phase (Kantola *et al.*, 1992). Now, it could be argued that greater dispersal into the gas phase engenders a reduced risk, inasmuch as toxicology is driven by particle surfaces on which mutagens are locally concentrated. On the other hand, since nucleation comes into operation, as already described, these mutagens might instead shift from the accumulation mode into the nucleation mode.

The second ground for concern is that emissions of certain species, otherwise *foreign* to exhaust gas, might be created within the DPF (Mayer A. *et al.*, 2003), which is, after all, a chemical reactor, in which thousands of compounds are exposed to hot exhaust gas for many hours. This situation – which resembles the 'shake and bake' methods of the early chemists, or even their alchemical forebears – provides ample opportunity even for slowly proceeding chemical reactions to take effect. The toxicological implications then depend on whether these compounds are effectively incinerated during regeneration, or allowed to evaporate and escape. Very similar concerns with the dilution tunnel filter have already been mentioned (Section 3.6.2); and there is no less reason for supposing that, *mutatis mutandis*, clandestine transformations of this nature occur also inside the DPF. Indeed, transformations in the DPF may be more extreme than on the dilution tunnel filter: the particulate deposit is aged for far longer periods, the total surface area is much higher, the temperature is much hotter, and the exhaust gas has not been diluted with air (Howitt *et al.*, 1983).

In the parallel case of the dilution tunnel filter, the principal transformation indicted in enhancing mutagenicity, namely nitration of PAH into NPAH, is designated 'artefactual', in that it takes place *on* the dilution tunnel filter, i.e. the mutagens were not previously present – although it is difficult to be conclusive on this point. The same argument of artefactuality can hardly be applied to the DPF: if the compounds are there, then they are there, the aim being not measurement, but public protection. More-volatile compounds have the potential to escape (Mayer A. *et al.*, 2003); for example, during regeneration, the emission of *1-NP* at ten times the rates seen without a DPF (Draper *et al.*, 1987) argues against successful incineration. Generally speaking, mutagenicities with a DPF are higher than without, although factors of two (Draper *et al.*, 1987) are not easily reconciled with orders of magnitude (Bagley *et al.*, 1991), and decreases are also known, depending on operating point (McMahon *et al.*, 1984). This, not implausibly, reflects daunting complexity, inasmuch as each compound has its own history of feed gas precursors, synthesis, evaporation, oxidation and gas-to-particle conversion. An important distinction should be made, however: very little mass is actually emitted. This is why an increase in mutagenicity per unit mass of several orders of magnitude becomes a decrease when expressed in the alternative metric, per unit volume of exhaust gas (Bagley *et al.*, 1991).

We should point out that researchers have not taken full cognisance of two shortfalls in conventional protocols: samples of exhaust gas (say, taken from downstream of a DPF) are passed through a dilution tunnel filter, and the particulate deposit thereon is analysed for mutagenicity. Yet the dilution tunnel filter is *itself* likely to instigate the very transformations under investigation! To accurately gauge the mutagenicity, therefore, analyses of particulate deposits extracted from within a DPF are surely a more rigorous approach. A similar question is posed in relation to gas-to-particle conversion: in the dilution tunnel, organic compounds are partitioned between the particulate phase (filter) and vapour phase (adsorbent). This protocol is not a true representation of partitioning in the ambient atmosphere, nor does it capture the aforementioned nucleation of mutagens, since there is no size fractionation.

Returning, however, to DPF-mediated fermentation, this extends to two highly undesirable chlorine-containing compound groups, namely dioxins and furanes, emissions of which, upon the installation of a DPF, have been seen to increase by three orders of magnitude (Mayer A. *et al.*, 1999). The fuel (7 ppm) and the oil (300 ppm) were both sources of chlorine. But, even though a DPF inopportunely provides a convenient vessel for these reactions, and chlorine is regrettably available, the formation of dioxins and furanes does not follow inevitably: this appears to require also certain metals, such as copper, to act as catalysts. The formation of dioxins by heterogeneous catalytic reactions at comparatively low temperatures (300–400 °C), well known in other fields, is thought to occur on soot particle surfaces (Huang H. and Buekens, 1995). But, whatever the mechanisms, some fuel additives cannot, because of these unwanted side reactions, be used in combination with a DPF – even though such additives might well aid, quite admirably, the process of regeneration.

7.9 Closure

In this chapter, several long-standing strategies for abating the emission of particulate were examined. Most of them, in one way or another, directly address the combustion: what is burned, and how it is burned. Central is the rate of soot formation versus the rate of soot oxidation, each being affected differently. Another section addressed the escape of engine oil into the exhaust, which is a kind of fugitive particulate emission, and the entry of soot into the lubricant. A final section concerned the post-combustion clean-up of the exhaust gas, known as 'aftertreatment', which involved two devices, the catalytic converter and the diesel particulate filter (DPF).

7.9.1 Fuel Formulation

Reformulation encompasses both physical and chemical aspects of the fuel, which, generally speaking, influence the process of air–fuel mixing, and the pyrolytic soot-forming reaction paths, respectively. The approach engenders three distinct possibilities: an immediate emissions benefit when current-technology vehicles are refuelled at the filling station; an easing of the NO_x–soot trade-off, allowing, for example, the implementation of more EGR; and a shifting of particulate composition towards organic and away from carbonaceous, assisting operation of the oxidation catalyst. Against this there are the need for changes, perhaps quite drastic ones, in fuel-supply infrastructure; the modest benefits in comparison with new vehicle technology; the interdependencies among numerous fuel parameters; the difficulty of understanding the direct effects on combustion; and the hidden or partly hidden interactions with engine calibration.

Sulphur

Sulphur, the salient aspect in fuel reformulation, contributes directly to the particulate emission via the sulphate fraction. However, fuel desulphurisation is subject to practical limits, and the sulphur compounds confer on the fuel important self-lubricating properties. Through some covert cross-influence, the organic fraction is also reduced: this might be a side effect of the desulphurisation, for example through modifications to the hydrocarbon speciation in the fuel, or an aspect of the filter, for example if sulphuric acid acts as a scrubber. Higher levels of some PAH and higher mutagenicities have been observed with low-sulphur fuels, the reasons for which are unreported; but altered gas–particle partitioning is implicated. More clearly demonstrated is that low-sulphur fuels engender fewer nanoparticles, sulphuric acid being imputed in nucleation. The accumulation mode is mostly unaffected.

Hydrocarbons

With hydrocarbon speciation, a well-recognised need is to restrict aromaticity, these compounds being strongly associated with sooting combustion. But some in-engine studies are ambiguous; the net benefit is easily disguised, so it seems, by engine model, engine calibration and duty cycle. Better established is the importance of multiple-ringed rather than single-ringed aromatics in soot production. No clear effect on the organic fraction is seen. That emitted PAH decline with fuel PAH is difficult to demonstrate conclusively, because of pyrosynthesis, and the release of stored PAH by lubricant and wall deposits. But reductions in mutagenicity are nonetheless observed. One study, into aliphatics, suggests that double-bonded molecular structures are undesirable.

Oxygenates

It is possible to supply oxygen to the combustion via the fuel molecule, in the form of 'oxygenates'. Dozens of such compounds have been tested; examples are esters, alcohols and ethers. The main target of oxygenates is soot; indeed, the particulate emitted is usually reported as 'wetter'. Soot declines in proportion to the oxygen content in the fuel, in many studies disappearing at 30%. This 'internal' oxygen discourages soot formation by leaning-out the combustion zone, but this is not, by any means, the whole explanation; indeed, given the enormous diversity of oxygenates, the existence of one common, unifying mechanism seems rather unlikely. Some studies point to molecular structure and the carbon–oxygen bond; others to fuel evaporation, air entrainment and the structure of the combusting plume. According to the 'sequestration' theory, when an oxygen atom is bound to a carbon atom, the latter cannot participate in soot-forming reactions. Demographically, the higher organic–carbonaceous ratio leads to greater nanoparticle abundance.

Additives

Soot-suppressing organometallic fuel additives are well known: some promote soot oxidation and others inhibit soot formation. Barium is the most commonly reported additive, appearing in the literature as far ago as the 1970s. The soot emission is reduced, and, since the metal compounds interact comparatively weakly with light, the smoke emission is also reduced. But the particulate mass is not much reduced, and in fact can increase; in which case the improvement is largely cosmetic. Besides, the emission of these metals is unlikely to prove acceptable today. Nowadays, the strategy is to use organometallic fuel additives to assist in regenerating the DPF; hence the metals are not emitted, but retained as an incombustible ash. Treat rates are also much lower than in the original soot-suppressing application (by one or two orders of magnitude). Unfortunately, *nonmetallic* fuel additives of comparable performance are rarely reported.

Volatility, Cetane Number and Density

The effect of cetane number on soot is commonly ascribed to the ignition delay, and the duration of the soot-forming mixing-controlled burn. But longer ignition delay, since it allows greater opportunity for mixing and wall impingement, can favour the emission of organic particulate. (In the extreme case, low cetane number causes white smoke.) One should take into account the engine load, as this also decides the proportion of premixed and mixing-controlled burn. Demographics are seldom reported, but the higher organic–carbonaceous ratio exhibited with fuels of low cetane number is suspected to encourage nucleation. Fuel volatility is generally associated with soot because it influences mixing: detrimentally, if the charge is overmixed or undermixed. Correlations exist between soot and T90, or 'back-end' volatility. A well-mixed charge avoids rich, soot-forming pockets. The implications for organic particulate are less demonstrated, but overmixed charges tend to result in the emission of unburned fuel compounds. Investigations into fuel density must take into account interactions with the fuel-metering strategy used by the engine management system, which is volumetrically based. When this is done, the decline in particulate emissions brought about by fuels of lesser density is largely apparent.

7.9.2 Fuel Injection

The Injector Nozzle

A significant factor in the emission of organic particulate used to be fuel entrapment in the nozzle sac at the end of injection: this fuel survived combustion by evaporating or dribbling into the combustion chamber, in an uncontrolled manner. At low load this was a particular problem, as the volume of the sac represented a significant percentage of the fuel quantity. The valve-covered orifice (VCO) eliminated this problem. Orifice diameters have been gradually reduced, with a view to more effective atomisation and mixing, which is of benefit for emissions of soot. However, the effects of small orifices are not always beneficial: this is ascribed to the lower momentum of smaller droplets, and consequently poorer penetration of the fuel spray into the surrounding air.

Injection Pressure

The low-sooting operation of today's diesel engines, in large measure, is due to inexorably rising injection pressures. At a first level, atomisation and mixing are improved, with benefit for emissions of soot, but this explanation is known to gloss over many complexities of the combusting plume. High injection pressure can be detrimental if the fuel spray penetrates too far and strikes the wall of the combustion chamber, or if the fuel droplets are too small and do not penetrate far into the air charge. There is ongoing discussion about whether soot oxidation or soot formation is the more greatly affected. Overmixing also causes emissions of organic particulate. Although the mass of accumulation-mode soot is regularly reported as being reduced by high injection pressure, a concomitant generation of nanoparticles has been observed.

Injection Scheduling

Great gains in emissions reduction have been made since diesel engines were operated using just one injection event, the timing of which was the only calibratable. Modern fuel injection equipment (FIE) provides enormous choice in injection schedules, i.e. pilot, main and post-injections. Such schedules help ensure that fuel no longer has to be injected into a hot, rich, soot-producing mixture. But the timing and the brief interruption between injections must be carefully optimised, as the subsequent injection can be detrimental, for example if, in imparting a cooling effect, it hinders soot oxidation. A post-injection, if timed correctly, might encourage oxidation of the soot cloud left by the main injection, if the combustion temperature is raised or additional air is entrained. Effects on particle demographics are mixed.

7.9.3 Exhaust Gas Recirculation (EGR)

EGR requires careful scheduling, according to engine operating point, as emissions of NO_x and soot must be traded against one another. The soot-producing effect of EGR restricts its use to regions of comfortably lean air–fuel ratios, i.e. low speed/low load. Opinions still vary about how EGR affects combustion, but broadly there are three aspects: dilution (replacing oxygen), thermal (lower heat capacity) and chemical (supply of radicals). Experiments with diluents, designed to isolate the three effects, show that the dilution effect predominantly controls soot. There is still discussion, though, about whether soot oxidation or soot formation is predominantly affected. In stark contrast to the mass emission, little literature exists about how EGR affects particle demographics. Some workers have measured fewer nanoparticles, perhaps because of higher exhaust temperatures and the consequent improved oxidation of organic particle precursors; improved oxidation of HC is indeed sometimes reported with EGR, but the effect on organic particulate is often equivocal.

7.9.4 Induction

The so-called 'turbo-lag', and the brief dip in air–fuel ratio during acceleration, is responsible for emissions spikes, or puffs of soot. These spikes are difficult to control, because a clear conflict of interest exists between driveability and emissions: to avoid soot spikes, engine management systems must restrict the rate at which fuelling is increased, and this compromises the acceleration of the vehicle. Today's fuel-metering software is open-loop, and calibrated for steady-state operation: it thus approximates transient manoeuvres by a locus of steady-state points, something that is expedient rather than accurate. A related aspect is that for steady-state engine operation at low speed/high load, the turbocharger is unable to supply sufficient air – again, soot is emitted. Boost pressure is not always beneficial in controlling the soot emission, as the density of the charge is an aspect of fuel-spray penetration. Rarefied air, at altitude, needs to be compensated in the fuel metering, otherwise the charge is enriched, leading to more soot. One effect of charge temperature is to modify the ignition delay. Swirl is imparted to inducted air by the inlet ducts; this prevents localised regions becoming too rich during injection. But overswirl and underswirl both produce more soot, so that careful optimisation is required at each operating point.

7.9.5 Lubrication

Oil in Particulate

Lubricant adds to the particulate emission by escaping through the valve stem and turbocharger seals, through the piston ring-pack, from the cylinder walls, and around the positive crankcase ventilation system. Local temperature and lubricant volatility drive this escape, as does wear. Escape as a vapour is biased towards the more volatile components; escape as a liquid is not. Transient operation of the engine is an important factor. The lubricant contribution to particulate should not be confused with lubricant consumption, as significant oxidation, prior to emission, can occur; the two parameters are mutually independent, and restricting consumption might only pay dividends at certain operating points. There is, potentially at any rate, a contribution to all four of the fractions, but the largest is generally held to be to the organic fraction, in which, on an equal-mass basis, lubricant seems to contribute more readily than fuel. The contribution to the sulphate fraction, with the decline in *fuel* sulphur levels, is becoming more noticeable; but if they are unburned, sulphur-containing oil compounds are not necessarily quantified in the final analysis as 'sulphates'. The contribution to the ash fraction is small, but nonetheless a nuisance for aftertreatment devices, the durability of which is compromised. The contribution to soot is not well investigated, but thought relatively small.

Particulate in Oil

The reason for soot entry into the oil is not well elucidated, but the oil film residing on the cylinder wall, rather than the blow-by gas, is thought to act as a carrier. Various properties of the oil are affected, or rather degraded, although chiefly this concerns the viscosity and anti-wear performance. Today's engines are more vulnerable to soot, and for a gamut of reasons, e.g. greater use of EGR, timing retardation for NO_x control, fewer 'top-ups', and ever-widening oil-change intervals. Oil consumption should not be restricted without also limiting soot entry into the oil. If soot particles succeed in agglomerating, then the impact on viscosity is enormous. Dispersants are added to the oil to restrict this agglomeration; these adsorb onto particle surfaces, and oppose the attractive van der Waals forces by conferring repulsive steric, osmotic or electrostatic forces. Evaluation of soot-contaminated oils is complicated owing to thixotropicity and non-Newtonian rheology, and alternative characterisation methods are used. The particle morphology resembles that in the exhaust; the primary agglomerates, with partially fused spherules, do not appear to be fragmented by the forces of fluid shear. The similarity to exhaust soot does not, though, extend to the chemistry of particle surfaces: for this reason, one should be chary of using carbon black as a surrogate soot. The literature is extensive about the wear-promoting effects of soot: it seems probable that several mechanisms operate, e.g. direct abrasion, filching of anti-wear compounds by soot, and local blockage and oil starvation.

7.9.6 Alternative Combustion Systems

In principle, combustion can be carried out at temperatures either too low or too high to form soot, even though the air–fuel ratios are rich. In the first case, soot-forming pyrolytic reactions are thought to be rate-controlling; in the second, soot precursors are thought to be unstable. Both of these concepts have been tested in engines. One difficulty is recovering the lost fuel economy. A completely different alternative uses a homogeneous charge and lean air–fuel ratios. Soot emissions are negligible from such engines in mass terms, but nanoparticles are reported. Particulate deposits on filters are grey rather than black, reflecting a higher organic component. As soot-forming reactions are interrupted, this allows the release of partial-reaction products.

7.9.7 Aftertreatment

Catalytic Converters

A catalyst oxidises organic compounds, probably as gas-phase particle precursors. Ash is stored irreversibly and deactivates the device through the formation of a 'diffusion barrier', preventing the ingress of reactants. Soot is not oxidised, at least to any useful degree. Sulphates are formed, because SO_2 is oxidised as a side effect. The conversion efficiency is poor for 'dry' particulate. Some soot is stored, but this is reversible through its subsequent oxidation. Sulphate and organic particulate are subject to long-term storage–release effects. Storage of organic particulate takes place when the catalyst is too cold for oxidation, but this can be beneficial, provided these compounds are not released too early before higher temperatures return. The great restriction with oxidation catalysts for many years was that the formation of sulphate particulate easily outweighed the oxidation of organic particulate. The mass emission of PAH is reduced, but some studies show an increase in mutagenicity: this could be because of nitration of PAH to NPAH, but no conclusive proof has been reported. The formation of sulphates engenders a more prominent nucleation mode. The deNO$_x$ catalyst presents an entirely new set of problems in its generation of ammonium sulphate particles, as either a primary or a secondary pollutant.

Particulate Filters

The DPF is chiefly designed to address the carbonaceous particulate, but as it traps whatever is in the solid phase, the ash is also captured. Some ability to capture the organic and sulphate fractions is generally assumed, but the quantity cannot be specified *a priori*, as it depends on temperature and the adsorbing properties of the particulate deposit already captured. The mechanical process of sieving is not responsible for this capture: many particles are considerably smaller than the voids in the filter; other implicated mechanisms are inertia, diffusion and thermophoresis. The DPF must be periodically (i.e. every few hundred miles) 'regenerated' by incinerating the particulate deposit. The ignition temperature is poorly defined: this depends on the variable physical and chemical nature of the particulate deposit, but temperatures in excess of 550 °C are generally cited. Unfortunately, this temperature is not realised in many duty cycles, and so some means of intervention is required. In thermal regeneration, burners or electrical heaters are used to raise the temperature of the DPF. In catalytic regeneration, the ignition temperature is lowered by fuel additives or a catalytic coating. During regeneration, the soot is oxidised; the organic fraction might be oxidised, but some species escape as a fugitive emission. The sulphate fraction evaporates and might decompose to SO_2 or SO_3. The ash fraction remains behind; this reduces the working life of the DPF. A DPF can considerably boost the nucleation-mode particles, because the volatile species that pass through, and especially those that volatilise after a brief period of storage, according to the duty cycle, are deprived of solid particle surface area; these species may thus gain in toxicological significance. Some studies show the mutagenicity to increase, a finding that is ascribed to the DPF being, in effect, a chemical reactor, in which the particulate deposit is held for protracted periods in a hot exhaust stream, thus allowing otherwise unfeasible reactions to take place. Any rise in mutagenicity is probably offset by the considerably smaller quantity of organic compounds emitted.

8
Gasoline Engines

8.1 Introduction

The long exemption enjoyed by gasoline engines from statutory regulations is a natural consequence of their low particulate emissions when compared with diesel engines: *gravimetrically*, this difference was, originally, as much as two orders of magnitude.[1] Understandably, therefore, the eyes of legislators did not come to rest on gasoline engines. The particulate mass in diesel engine exhaust has, however, over the last three decades, because of increasing stringency in statutory regulations, declined so enormously that this difference has now dwindled to a few tens of per cent (Andrews and Ahamed, 1999). Indeed, widespread introduction of the diesel particulate filter (DPF), as seems only too likely at the time of writing, could well turn the tables completely (Cadle *et al.*, 1997). Consequently, in a few years most countries will probably move to regulate the particulate mass in gasoline engine exhaust.

Research into gasoline engine particulate has been lent further urgency by a burgeoning literature in the biomedical field, the findings of which point to particle *size* as a more relevant toxicological metric than the total mass of particulate. (These studies will be reviewed in Chapter 10.) Such a reorientation would, very probably, change the gasoline – diesel picture quite considerably. Particles emitted by gasoline engines are undeniably just as respirable as, if not more than, those emitted by diesel engines, while their *numbers* might not compare so favourably (Mohr *et al.*, 2006): and even if vehicles are lower emitters on an *individual* basis, this might be negated, in the overall pollution inventory, by the all-pervading presence of gasoline vehicles.[2] Should gasoline vehicles turn out to be more responsible for nanoparticles than are diesel engines, they may be more rather than less culpable in public health terms – notwithstanding their lower mass emissions.

We thus arrive at an unfortunate twist in the debate about gasoline engines and the particles they emit. Currently, *direct injection* is widely touted as providing a superior power plant to port

---

[1] Obviously, this remark excludes the (now dead) issue of lead particles resulting from leaded gasoline.
[2] This highlights a significant difference between the USA, where diesel passenger cars are still unusual, and Europe, where they now constitute >40 % of the population.

Particulate Emissions from Vehicles   P. Eastwood
© 2008 John Wiley & Sons, Ltd

injection. This alternative combustion system, which uses a heterogeneous or stratified charge, rather than the conventional homogeneous charge, potentially provides, depending on the duty cycle, an improvement in fuel economy of 5–10 %. Yet the heterogeneous nature of the charge inevitably fathers certain diesel-like features, and thus the particulate emissions from direct-injection gasoline engines are inconveniently higher than from conventional port-injection gasoline engines.

The literature on gasoline engine particulate falls naturally into two periods. The earliest investigations were conducted in the USA in the 1970s, partly because of sulphuric acid aerosols, partly because of leaded gasoline, and partly as comparisons were needed in expectation of a diesel passenger car market. Many years of indifference followed this initial surge of interest. In the second period, i.e. the mid 1990s onwards, interest grew again, partly because of the narrowing gap between gasoline engines and diesel engines, and partly because of the aforementioned biomedical research.

Many aspects of gasoline engine particulate are common with those of diesel engine particulate, and, as such, they have already been discussed. This is because the diesel engine was used only as a *template*, so to speak, for particulate formation (Chapters 3 and 4), particulate characterisation (Chapter 6) and particulate abatement (Chapter 7). For example, with gasoline engines there seems no reason to adjust or reinterpret the classical understanding of soot formation related at length in Section 3.2: this delineation was generalised. Obviously, it would be clumsy and pedantic to re-present all this earlier material, from the groundwork upwards, merely for the satisfaction of isolating gasoline engines within a self-contained chapter. The reader is therefore left to interpret the earlier discussions accordingly; confident or precise reorientation to gasoline engines is not, however, possible in many areas, as the necessary research has not been done. The present chapter focuses only on whatever distinguishes gasoline engines from diesel engines.

8.2 A Historical Perspective

Gasoline engines emit particulate in all four fractions, carbonaceous, organic, ash and sulphate, the relative proportions of which have changed markedly over the years. Two major areas of technological innovation are responsible for these shifts. One is *fuel metering*, for which a three-level transition took place, from the carburettor, to single-point injection, and then to multipoint injection. Various reasons for this transition are evident but, arguably, the most important was the quest for ever-greater fuel-metering precision – as demanded, particularly, by the second major area of innovation, *aftertreatment*, for which a three-level transition also took place, from no aftertreatment, to oxidation catalysts (mid 1970s in the USA), and then to three-way catalysts (late 1980s in the USA). (The three levels of technology in the two fields were not contemporaneous.)

Of a more immediate and clear-cut bearing on particulate composition is the effect of aftertreatment. The original, and, at times, heatedly controversial issue concerned particles of lead discharged by vehicles running on leaded gasoline; these engines had no aftertreatment. The switch to unleaded fuel, and the introduction of oxidation catalysts thereby facilitated, saw a decline in organic particulate and, of course, a virtual disappearance of lead particulate, but a significant worsening of sulphate particulate. These are the two historical issues discussed here. The *present-day* configuration of the powertrain, incorporating the three-way catalyst, is addressed in Section 8.3.

8.2.1 Organometallic Fuel Additives and Ash

A generic drawback of the conventional gasoline engine, or perhaps of gasoline, is premature ignition in the end gas, ahead of the propagating flame front: this is the phenomenon of 'knock', in which reverberating shockwaves inflict various sorts of engine damage. The problem is that compression ratio, and hence efficiency, are very much restricted, as knock, being so destructive to the engine, must be absolutely avoided. So, tackling the problem from the other side, lead alkyls (tetraethyl lead (TEL, $(C_2H_5)_4Pb$) and tetramethyl lead (TML, $(CH_3)_4Pb$)), i.e. organometallic compounds, were, for several decades, habitually added to gasoline as anti-knock agents, or *octane enhancers*, thus allowing access to greater efficiency. (There were also beneficial side effects in arresting valve-seat wear.) The designation *leaded fuel* was, however, somewhat loosely applied, and the benefits were seen across a wide range of treat rates (0.1–1.2 g Pb/litre as TEL; see Heywood, 1988, p. 476). Commercial levels were about 0.6 g Pb/litre, although maxima of ∼2.5 g Pb/litre were also known (Lynam *et al.*, 1994).

Unfortunately, leaded gasoline ultimately proved untenable in two critical respects: the health risks airborne lead particles posed to the general public (Smith R.G., 1971; Russell Jones, 1987), and the potent poison lead represented to catalytic converters (Gandhi, 1987), both highly negative corollaries which forced, eventually, an outright ban; since which, correspondingly, the prevalence of airborne lead has measurably declined (Graedel, 1988; Watson and Lu, 1993), and, of course, catalytic converters have seen immeasurably widespread application throughout the world. The issue of leaded gasoline has, therefore, been consigned, so to speak, to the dustbin of history. Still, a brief discussion is apposite notwithstanding, as lead is a useful exemplar with which to elucidate gas-to-particle conversion in the ash fraction. It should also be noted that lead is still present even in 'unleaded' gasoline, albeit at trace levels (a few milligrams Pb/litre), and so, in the air pollution inventory, motor vehicles may not be wholly negligible as a source of lead (Pacyna, 1998), even though the anti-knock properties are now nonexistent.

Lead is a useful exemplar because, if admixed on its own, or rather as an organometallic, the resulting vapour of lead, or more probably of lead oxide (PbO),[3] following decomposition of the additive, undergoes gas-to-particle conversion *prematurely*, i.e. within the engine; and this is inconvenient, as the ensuing deposition, *inter alia*, fouls the spark-plug electrodes. The necessary countermeasure will be obvious from Section 3.5: gas-to-particle conversion is an aspect not only of the metal *per se*, but also of metal *compounds*, and this principle can be used to delay or to arrest this transition. Hence, in addition to lead alkyls, gasoline also contained, as 'scavenging agents', ethylene dibromide and dichloride, the chlorine and bromine of which reacted avidly with lead oxide, forming compounds of greater volatility, allowing, in turn, lead to persist in the vapour state, thus keeping the engine free of deposits. The lead compounds did, however, undergo gas-to-particle conversion in the exhaust system, and so showed strong proclivities to deposit onto exhaust-system walls. This should be expected, if only because lead would not be such an effective catalyst poison if it simply passed through to the tailpipe.

As also described in Section 3.5, falling temperatures during the emission process force compositional shifts, but, the cooling being so swift, thermodynamic states are probably precluded, allowing lead compounds to persist in nonequilibrated states: indeed, if these compounds are collected and reheated, there takes place a certain amount of microstructural reconstruction, segregation and crystallisation (Steiner *et al.*, 1992). Gas-to-particle conversion seemed to take place at around 300 °C (Treuhaft and Wisnewski, 1977). The lead is found in a wide variety of chemical compounds, of which the three most characteristic are lead bromide ($PbBr_2$), lead chloride ($PbCl_2$) and lead bromochloride (PbBrCl).[4] Studies of atomic ratios suggest predominant emission as bromochloride (Braddock and Perry, 1986), corresponding roughly to fuel treat rates (Pierson and Brachaczek, 1976). Other forms encountered are PbO, Pb(OH)Cl and Pb(OH)Br (van Malderen *et al.*, 1996b). Further compositional

---

[3] Lead alkyls are generally supposed first to decompose, with the anti-knock effect arising indirectly, via the formation of lead oxide, but this oxide may be present in the combustion chamber as a vapour or as particles (Heywood, 1988, p. 475).

[4] The chlorine and bromine are also emitted as gas-phase haloalkanes (Atkinson R., 1988).

modifications, courtesy of ongoing atmospheric reactions, generate complex, hybridised particles, such as $PbSO_4(NH_4)_2SO_4$ (Claes *et al.*, 1998), with the halogens being preferentially lost during ageing.

Mass emission rates[5] for lead should not be taken wholly out of context, as, at one time, gasoline vehicles, whether running on leaded or unleaded gasoline, emitted much else besides. For example, for a selection of automobiles (model years 1970–81) running on the FTP-75, total particulate emission rates were ∼100 mg/mile, burning *leaded* gasoline at 1.4 g/gallon (US), and ∼20 mg/mile, burning *unleaded* gasoline at 1 mg/gallon (US) (Lang *et al.*, 1981). Similarly, for an automobile from the pre-emission control era (model year 1968), running on the FTP-75, total particulate emission rates were 140 mg/mile, burning *leaded* gasoline at 0.8 g/litre, and 40 mg/mile, burning *unleaded* gasoline (Storey *et al.*, 2000). When the vehicle was running on leaded gasoline, the lead *alone* accounted for 40 mg/mile: higher, by four orders of magnitude, than when running on unleaded gasoline. Thus, it seems, a lead emission rate on the order of several tens of milligrams per mile was typical with leaded gasoline.

Particle size distributions in the accumulation mode do not necessarily differ appreciably between leaded and unleaded gasoline (Ristovski *et al.*, 1998). But although particles of size 250 nm or so were commonly found (Cantwell *et al.*, 1972), a remarkable tendency was exhibited for growth, either through agglomeration or, more probably, deposition onto and re-entrainment from exhaust system walls: more than half of the lead consumed by the engine failed to appear at the tailpipe (Beltzer, 1976). In accordance, presumably, with the vagaries and randomicity of storage and release, clumps, exceeding even 10μm, were emitted intermittently (Treuhaft and Wisnewski, 1977). This strong proclivity for storage and release precluded easy statements as to how much of this lead was respirable, how much remained airborne and how much quickly deposited onto neighbouring roads, pavements, buildings and vegetation. In one estimate, a third of the mass was submicron and one-fifth smaller than 200 nm, while one half of the total emission was arguably 'airborne' (Cantwell *et al.*, 1972).

The physical form in which the lead appears is quite interesting insofar as autonomous particles are not, generally, the rule – plausibly an aspect, in part, of the vehicle technology of the time (e.g. carburettored fuel metering) – conditions which were more favourable than today for the co-generation of soot. The lead was embedded within spherules as nanodomains or islands of size ∼10 nm, identifiable by virtue of their greater densities compared with the surrounding carbon under the electron microscope; aside from lead, these islands were rich also in bromine and chlorine. (This segregation provides a clue as to the formation mechanisms, as discussed in Sections 3.5 and 6.3.2.) Such sequestrations were absent in particles discharged by vehicles burning unleaded gasoline; the carbonaceous agglomerates otherwise appeared very similar (Storey *et al.*, 2000).

Except in a few specialist fields, and in developing countries that are now, in any case, rapidly following suit, leaded gasoline has now been universally banned. However, research was, at one time, vigorously conducted into the *capture* of lead particles within the exhaust system – a strategy currently being deployed on diesel engines (Section 7.8.2). Unlike soot, as already said, lead was not necessarily in the particulate phase within the exhaust system, at least not fully. But the distinguishing disadvantage of lead, in great contradistinction to soot, is that gasification by some periodic burn-off procedure is utterly impossible; and this difficulty greatly intensified the need to avoid compromising vehicle service intervals with such a device. This was done by designing three stages into the aftertreatment (Treuhaft and Wisnewski, 1977). (a) Particles were first caught on a loosely packed bed of glass or alumina beads, whereon they agglomerated, and, when large enough, and vulnerable to fluid shear, they became re-entrained, quite naturally, into the flow: by this process, growth was deliberately fomented. (b) The re-entrained particles then entered a second section, where an intense swirl threw them onto

---

[5] One should distinguish carefully between the mass of particulate, as a total; the mass of lead compounds, in the particulate (including Cl, Br etc.); and the mass of lead, as an element.

the surrounding walls, the point being that, with their original sizes, the inertial forces would have been too weak for this deposition principle to work. (c) In a third and final section, pleated glass fibres caught the smallest particles surviving transit through the first two stages. Various other designs were researched (Cantwell *et al.*, 1972).

Methylcyclopentadienyl manganese tricarbonyl (MMT, $C_9H_7MnO_3$) has also been used as an antiknock agent, particularly, with the advent of unleaded gasoline, as an alternative to lead alkyls. In the USA, though, this additive has had something of a chequered history, with discussions pivoting on whether emission control systems for regulated emissions (HC, CO and NO_x) are compromised. The technical community was, however, divided as to the precise implications, with various posited effects in the engine, as well as in the catalytic converter – some of them possibly beneficial, rather than detrimental. The manganese particles discharged by the engine are, of course, a form of pollution in their own right, for which the implications for public health require proper assessment.

At treat rates of 30 mg Mn/gallon (US), vehicles emitted manganese at a few hundreds of micrograms per mile; to place this in perspective, vehicles fuelled with MMT-free gasoline *still* emitted manganese, but only at a few micrograms per mile (Hammerle *et al.*, 1992). Three-quarters (Lynam *et al.*, 1994) or 99.5 % (Lenane, 1990) of the manganese may well lodge within the engine or be retained by exhaust systems. Some of this retention undoubtedly takes place in catalysts, mostly as Mn_3O_4; deposits were rust-coloured, fluffy and porous in appearance (Hurley *et al.*, 1989); other compounds to have been detected were amorphous (Zayed *et al.*, 1999). $MnSO_4.H_2O$ and $Mn_5(PO_4)[PO_3(OH)]_2.4H_2O$ (Ressler *et al.*, 2000) reveal reactions with sulphur (from fuel, and possibly from oil) and phosphorus (from oil). Phosphates form first, in the exhaust system, followed by sulphates in the dilution tunnel; once these elements are exhausted, oxides are formed (Colmenares *et al.*, 1999).

A third anti-knock agent is ferrocene (iron dicyclopentadiene, $(C_5H_5)_2Fe$), the behaviour of which in diesel engines and hydrocarbon flames was used in Section 3.5 as an exemplar for ash formation. The treat rate in gasoline is 20–40 mg Fe/litre (Kameoka and Tsuchiya, 2006). Following combustion and decomposition of the additive, gas-to-particle conversion occurs within the engine, as evidenced by iron oxide deposits on the spark plugs, which compromise ignition; the oxide also deposits within the exhaust system, where it forms a reddish-brown powder. The composition of the oxide is, however, variable, according to temperature, being predominantly Fe_3O_4 at 300 °C and Fe_2O_3 at 800 °C; oxygen availability is also a factor.

8.2.2 Oxidation Catalysts and Sulphates

Although sulphates are nowadays strongly, indeed virtually exclusively, associated with diesel engines, they did, in fact, first rise to prominence with *gasoline* engines (Herling *et al.*, 1977). But knowledge of just how plague-ridden gasoline engines once were by this emission is probably restricted to the more mature members of the technical community. Unlike the situation for diesel engines, the law laid down no statutory requirements for the mass of emitted particulate. The sulphates, or rather sulphuric acid, were more an issue of customer perception, and indeed of public nuisance. The acid, moreover, attacked exhaust system components, leading to rust, and, following exfoliation of this rust, the emission of coarse-mode metal-containing particles and 'acid smut' (Section 9.5).

At this time sulphuric acid accounted for half, and sometimes virtually the entire particulate emission. The undoubted trigger for this acid was the introduction of *oxidation catalysts*. The evidence for this indictment is incontrovertible: the sulphuric acid, virtually overnight, had become strongly dependent on fuel sulphur content, whereas previously it was largely independent; and on vehicles without catalysts, less than two per cent of the fuel sulphur was being converted to sulphuric acid, whereas on vehicles with catalysts, the figure was more like 30–40 % (Beltzer *et al.*, 1974).

The exhaust gas chemistry behind the formation of sulphates has already been described in Section 3.7, and, in this respect, the commonality between diesel engines and gasoline engines is straightforward. The sulphur is emitted from the engine predominantly as SO_2. In the exhaust, SO_2 converts to SO_3; the SO_3 swiftly converts to H_2SO_4; and the H_2SO_4, depending on the acid dew point, readily undergoes gas-to-particle conversion. The acid droplets were virtually all emitted at submicron sizes (Beltzer, 1976). The acid attaches to surfaces, including particle surfaces, but when present in high concentrations, forms an aerosol of its own. The bottleneck, or rate-limiting step, in this reaction chain is oxidation of SO_2 to SO_3, which, kinetically, is negligible without a catalyst. Contrarily, high conversion is predicted above catalyst light-off temperature, and in the vicinity of the thermodynamic limit (Griffing et al., 1975).

Unlike in diesel engines, the air–fuel ratios of which are always lean, gasoline engines operate on rich, stoichiometric and lean mixtures. The oxidising chemistry therefore required by the oxidation catalyst is not inherently available with the gasoline engine; *rich* air–fuel ratios, for example, are used to meet high power demand. The exhaust gas is, in this condition, *reducing*, and to maintain oxidising conditions in the oxidation catalyst, secondary air was pumped into the exhaust. For proper operation of the catalyst, this secondary air, metered to the exhaust, and the fuel, metered to the engine, required appropriate matching. The sulphate problem may have been inadvertently exacerbated, for example, by fuel-metering strategies adopted to prevent the stalling or hesitation experienced by some customers driving vehicles fitted with early catalysts (Gibbs et al., 1977).

Gasoline oxidation catalysts, like diesel oxidation catalysts, exhibit remarkable propensities for sulphur storage, and this storage considerably worsened the problem, as the acid could be discharged in sudden bursts, during which transient emissions exceeded vastly the steady-state emissions. Driving mode and catalyst formulation were equally significant factors in deciding the pattern of storage and release. Storage periods were extremely protracted: following changes in fuel sulphur content or engine operating point, a whole hour of steady-state running was required to stabilise the acid emission (Beltzer et al., 1975).

Sulphuric acid emissions from gasoline vehicles fitted with oxidation catalysts were never actually solved as such; rather, they disappeared fortuitously with the advent of three-way catalysts. Air–fuel ratios in the automotive industry, from this moment, changed to stoichiometric or, more correctly, were controlled to tight fluctuations about stoichiometry. The exhaust gas was no longer oxidising, or not predominantly so; the chemistry, therefore, became unfavourable for acid formation.

8.3 Port-injection Engines

In the *conventional* realisation of the gasoline engine, which incidentally still accounts for the majority of the market for gasoline vehicles, fuel and air are mixed in the intake system, and inducted together into the cylinder: historically, this charge preparation was the task of the carburettor, in which fuel became entrained, via a venturi, into the airstream. This older form of technology is not explicitly addressed in the current section, our aim being, rather, to discuss the implications for particulate emissions of *current* technology. Charge preparation is nowadays carried out with a fuel injector mounted in the inlet port; indeed, contemporary emissions legislation could not be met with the carburettor, which lacks sufficient fuel-metering precision; this was the reason for its desuetude. Both types of charge preparation are, though, assignable to the same category of engine if one adopts the nature of the charge, rather than the method of charge preparation, as the criterion. This is because, immediately prior to initiation of the spark, the charge is *premixed* and, in ideal circumstances, if not actually in practice – an issue to which we shall turn shortly – *homogeneous*. Thus, 'premixed-charge engine' and 'premixed homogeneous-charge engine' serve as alternative appellations for conventional gasoline engine technology. This distinguishes it from direct injection, to be discussed in Section 8.4.

8.3.1 Formation

The conventional gasoline engine is not particularly renowned for the emission of soot. That partial-combustion products of a carbonaceous nature are formed is evident in other quarters, namely, spark-plug fouling (Quader and Dasch, 1992) and lubricant adulteration. In 'normal' circumstances, when engine operation is properly optimised and maintenance is satisfactorily attended to, visible emissions of smoke (Quader, 1989) are unlikely. Nevertheless, *some* soot is still emitted. This soot forms predominantly because of practical limitations in fuel management, and, although not immediately introduced for this reason, the fuel injector is to be particularly welcomed, as it offers far greater fuel-metering precision than could ever have been realised with the carburettor. But this is not the place to review fuel-metering hardware.

Emissions of HC, NO_x and CO are the prime concern with the conventional gasoline engine. The *minutiae* of soot formation have mostly been neglected: it is unknown whether the generalised mechanisms (Section 3.2) differ in their essentials, and a conceptual model derived from laser imaging, akin to that devised for the diesel engine (Section 3.3), has not appeared. But soot formation is easy enough to explain in terms of a wider vista. Following the electrical discharge, a turbulent flame forms at the spark plug, the outward propagation of which consumes fresh (unburned) charge in front, and leaves spent (burned) charge behind. Insofar as rich air–fuel ratios are a necessary prerequisite for soot formation, and this premixed, homogenous charge is eminently amenable to *lean* air–fuel ratios, the conventional gasoline engine is not, unlike its diesel counterpart, inherently soot-prone. This 'lean-burn' mode of combustion is not, though, customarily adopted with conventional port injection, but, rather, more associated with direct injection (Section 8.4), although the reasons for this preference would take us too far from the current topic. What concerns us here is *why* air–fuel ratios become *rich*.

For the present, it is sufficient to observe that the *three-way catalyst*, with which all port-injection engines are now virtually without exception fitted, requires, for its proper operation, precise control of exhaust gas chemistry, which requires, in turn, precise control of air–fuel ratio: that is, never deviating too far from *stoichiometry*. In actuality, the air–fuel ratio, under steady-state engine operation, oscillates around stoichiometry at a few hertz: this is known as 'limit cycling', which is carried out by the engine management system, using an oxygen sensor in the exhaust as feedback. Whether or not the brief excursions to the rich side of stoichiometry during *mild* steady-state limit cycling generate soot is unreported; but this mechanism, if it exists, is certainly overshadowed by rich excursions in air–fuel ratio resulting from imprecisions and compromises in fuel metering. These rich air–fuel ratios may be *global*, relating to the entire charge as a whole, or *local*, relating to pockets or regions of charge. It should be recognised that locally rich pockets of charge can still exist even when global air–fuel ratios are, in fact, comfortably lean, and therefore outwardly plentiful in oxygen. In this respect, even ostensibly lean air–fuel ratios can cause the emission of soot. And with single-point injection, there are practical limitations in subdividing the fuel, so that poor distribution, or 'maldistribution', allows one cylinder, say, to run rich when others in fact run lean.

For reasons of performance, driveability and customer expectations, globally rich air–fuel ratios are used in some operating conditions. During cold starting and powertrain warm-up, rich air–fuel ratios ensure the realisation of an ignitable mixture, evaporation being essentially incomplete and unreliable – engine running would otherwise be rough and subject to stalling. Unfortunately, during these moments, films of liquid fuel accumulate on combustion chamber walls, ignite capriciously, and burn protractedly long into the expansion stroke, as sooty pool fires (Witze and Green, 1997), and in a fashion controlled by oxygen availability (Kayes and Hochgreb, 1999a). A second circumstance of globally rich air–fuel ratios is the need for acceptable vehicle acceleration, as the powertrain response would otherwise be too sluggish. A third circumstance is the need to restrain exhaust temperatures at high load in order not to overheat the catalyst. In the second two cases the fuel may well be fully vaporised, but it burns rich nonetheless, and this creates soot.

Some fuel, rather than entering the cylinder as intended, accumulates as reservoirs or fuel films on the walls of inlet ports and on the rears of the inlet valves (Skippon et al., 1996). This 'hang-up', between what the fuel system *supplies* and what the combustion process really *needs*, is not necessarily a problem during steady-state operation, as the fuel-metering system will compensate via the oxygen sensor. Significant metering problems, though, arise during acceleration, or transient operation, as these reservoirs supply additional fuel or filch required fuel, so that one cycle runs excessively lean, say, and the next runs excessively rich. Upon shifts in engine operation, the temperature of the inlet valve – since this controls the accumulation and evaporation of fuel reservoirs – resembles the transient response in the soot emission (Kayes et al., 1999). Nor is the fuel always successfully vaporised prior to combustion (Maly, 1994): liquid fuel ligaments or fuel droplets, on the order of 100–200 μm in size (Gold M.R. et al., 2000), are able to enter the cylinder, and, if these evaporate too slowly, they burn with sooty flames. This seems to be the predominating factor (Kayes and Hochgreb, 1999c). The paramountcy of fuel evaporation and charge inhomogeneities (i.e. droplets) in the formation and emission of soot will be apparent (Kayes and Hochgreb, 1999c). This remark holds across a wide range of parameters: injection timing (open/closed valves), spark timing, EGR, coolant temperature, etc. (Kayes and Hochgreb, 1999a).

Consequently, global air–fuel ratios, although conveniently computed, lay greater emphasis on soot as a global phenomenon than is perhaps justified. This problem notwithstanding, the strength of the relationship between global air–fuel ratio and soot is really quite remarkable, involving high orders, or strong exponents. A 2.3-litre, four-cylinder engine, running at a stoichiometric air–fuel ratio ($\lambda = 1$), was estimated to emit particulate mass at $30\,\mu g/m^3$ – this figure is *lower*, in fact, than the ambient background of PM10 in many regions, and indeed *was* lower than in the air intake of the engine (Abdul-Khalek and Kittelson, 1995). Purely on this basis, one would perceive the engine as a vacuum cleaner! However, a similar estimate for a rich air–fuel ratio ($\lambda = 0.4$) arrived at a considerably larger emission, $50\,mg/m^3$ – *higher* than for many diesel engines. This was admittedly a long extrapolation, based upon the measured *eighth-order* dependence; but it strongly suggests the likely significance of rich air–fuel ratios. The same workers made similar computations for particle numbers: the lowest, for lean air–fuel ratios, were well below concentrations in the ambient air; the highest, for rich air–fuel ratios, approached those of diesel engines. The particle number $N\,(s^{-1})$ has been related to λ according to an exponential relationship (Maricq et al., 1999c):

$$N = 8.1 \times 10^8 \exp\left[(160-\lambda)^2\right]. \tag{8.1}$$

Of course, purely in *field* terms what really counts is the emission rate, as controlled by driving behaviour, expressed, say, as a function of vehicle speed or perhaps of acceleration; and these patterns are revealing. Initially, the particulate emission increases fairly moderately, as a function of vehicle speed, but then gathers pace, and begins climbing ever more vertiginously. This effect, being so strong, cannot be attributed simply to increased exhaust flow rates; the mechanisms of particle production must also become more fecund (Maricq et al., 1999a). Although other factors, such as spark timing, presumably have some effect, the trend depicted in Figure 8.1 (Maricq et al., 1999c) strongly imputes air–fuel ratio, customary practice being to enrich the charge as driver demand increases, for reasons already described. Consequently, the carbonaceous fraction, for lightly-loaded duty cycles, may account for one-third of the particulate mass (Bosteels et al., 2006), but dominate overwhelmingly at WOT (Andrews and Ahamed, 1999). Given the strength and nature of this dependency, the gentleness of certain homologation duty cycles, in comparison with typical customer duty cycles likely to be experienced in the field (i.e. harsh accelerations), should be noted.

We have dwelled at some length on soot, but the organic fraction is nonnegligible. There is a parallel here with gas-phase hydrocarbons, commonly designated (and regulated) as 'HC' – although the

Gasoline Engines

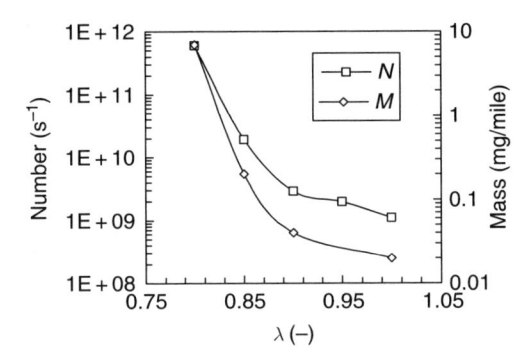

Figure 8.1 Number and mass emissions plotted against air–fuel ratio (normalised here as λ). Vehicle: six-cylinder light-duty truck, model year 1995; odometer 2600 miles, automatic transmission. Duty cycle: 50 mph cruise. Fuel: 'EPA certified'. Instrumentation: particle number by ELPI; particulate mass estimated thereby (Maricq *et al.*, 1999c).

provisos relating to such a comparison were mentioned as long ago as Section 3.6.4. Identification of *unburned fuel* with the organic fraction, and not just HC, seems less appropriate with gasoline engines than with diesel engines, as gasoline fuel is more volatile than diesel fuel and thus, presumably, less susceptible to gas-to-particle conversion. But some unburned fuel does indeed appear in the organic fraction, especially during cold starting and, less expectedly, fuel enrichment (Andrews and Ahamed, 1999).

So, it is worth restating briefly the already well-understood reasons for emissions of HC (Heywood, 1988, pp. 601–619). (a) Fuel hydrocarbons are sheltered from the flame, such as by the threads of the spark plug, the gap at the piston top-land and the micropores in combustion chamber deposits, and through absorption into lubricant (Andrews *et al.*, 2000b); during the exhaust stroke, the hydrocarbons simply emerge from these refuges. (b) There are inhomogeneities and partial burning, resulting from imperfections in air–fuel mixing or improper control of EGR and spark timing during transient operation: some regions of the charge are too lean to support the flame or, rather, the flame propagates through these regions too slowly, in extreme instances causing misfires; and some regions are too rich for all hydrocarbons to be fully oxidised, in a manner akin to soot. The escape of unburned hydrocarbons with global charge enleanment may explain why particulate emissions are at a minimum near stoichiometry (Kayes and Hochgreb, 1999b). For all these reasons, hydrocarbons emerge from the engine unburned during the exhaust stroke, although some late-phase oxidation admittedly takes place if temperatures remain sufficiently high.

The carbonaceous and organic fractions, with respect to air–fuel ratio, appear to vary relatively independently, being merely superimposed on one another, suggesting different causal mechanisms (Shin and Cheng, 1997). Hence, whereas the carbonaceous fraction, as already described, rises precipitously with charge enrichment, the organic fraction remains comparatively indifferent. This picture accords with the findings of other investigators (Abdul-Khalek and Kittelson, 1995). The characteristics depicted in Figure 8.1, for example, are plausibly adducible to the carbonaceous rather than the organic fraction.

Fuel is unlikely to be the only source of the organic fraction (Kayes and Hochgreb, 1999b), but the emission of lubricant remains largely equivocal, because, until fairly recently, little incentive has existed for such research – unlike with the diesel engine, for which the corresponding literature is intimidatingly voluminous (Section 7.6.1). Logically, this is matter of burned-gas temperature in the vicinity of the cylinder walls, insofar as this drives evaporation, as observed for the diesel engine. Subsequent burn-up of this vaporised lubricant may be poor (Puffel et al., 1998). Compared with the diesel engine, though, there are two potentially significant differences: the oxidation of this oil is limited during periods of enrichment, as oxygen is unavailable (Andrews and Ahamed, 1999); and the load delivered by the conventional gasoline engine is regulated via air path throttling, by which the escape of lubricant is, intuitively at any rate, *assisted*, insofar as the ensuing depression encourages losses from the lubricating system. Under *normal* operation, unburned fuel could be the minor partner in the organic fraction (Shin and Cheng, 1997).

Gasoline engines, notwithstanding their stoichiometric fuelling schedules, also emit sulphates. From this, one would surmise that, during lean excursions, the three-way catalyst is furnished with sufficient opportunity to convert SO_2 into SO_3 via exhaust gas chemistry similar to that of diesel engines. But the chemistry differs from that of diesel exhaust in one particular: rich excursions generate ammonia, and sufficient quantities of this compound may exist for full neutralisation of the acid to ammonium sulphate (Cadle et al., 1999a). On a different tack, the reason why sulphates become more prominent during cold starts (Andrews and Ahamed, 1999) is unclear: perhaps, at low temperatures, dissolution of SO_2 is followed by aqueous-phase conversion to SO_3. It will be apparent that little research has been reported in this area.

8.3.2 Characterisation

Accumulation-mode particles discharged by gasoline engines are polydisperse agglomerates of (relatively) monodisperse spherules, differing little morphologically from what is observed with diesel engines. The interiors of the spherules are poorly probed, but, if one speculates that carbonaceous, in-engine deposits represent what is emitted, then the microstructures are, similarly, quasi-graphitic: partly amorphous and partly crystalline (Zerda et al., 1999). The colour range exhibited by deposits of particulate as collected on filters parallels also what is observed for diesel engines, insofar as this reflects the notional compositional trend, driven principally by air–fuel ratio: for charges richer than $\lambda = 0.7$, the deposits are black, reflecting a predominating carbonaceous fraction, and for charges leaner than $\lambda = 0.8$, the deposits are yellow – white, reflecting a predominating organic fraction (Shin and Cheng, 1997). Such a trend is fully expected from the discussion in Section 8.3.1. The carbonaceous agglomerate is the particle *archetype*; other material, of atypical morphology and presumably also of different chemical composition, is certainly emitted, and seemingly more noticeable than with diesel engines, because archetypal carbonaceous particles are not so overwhelmingly abundant. Indistinct features, possibly of a fluffy, dendritic nature, are attached to the agglomerates (Shin and Cheng, 1997).

Ash elements, not discernibly different from those emitted by diesel engines, are more prominent, making up several per cent (Cadle et al., 1999a) or as much as several tens of per cent (Andrews and Ahamed, 1999) of the whole mass: this reflects the low emissions in other fractions, especially on recent-model vehicles (Cadle et al., 1999a). In the organic fraction, PAH are well represented, although the absence of sufficient material for analysis tends to force pooling of gas-phase and particulate-phase compounds; the first group probably preponderate, and significant compounds were *Ant*, *Flu* and *Pyr* (Cadle et al., 1999a). All PAH were, however, dominated by a little-monitored compound, 2-methylnaphthalene. A speciative breakdown for the 16 so-called 'priority' compounds did not seem to show any systematic differences between a gasoline vehicle and a diesel vehicle; *Phe* and *Pyr* were more prominent on the latter, and *Ace* and *Nap* were more prominent on the former (Collier et al., 1998). Most PAH lighter than *Phe* are predominantly in the gas phase; those heavier are predominantly in the particulate phase; as much as one-fifth of those in the particulate phase are, in fact, oxygenated; and six NPAH are measurable, one-third of which are in the particulate phase (Knapp et al., 2003).

Mutagenic characterisations are rare and, in view of the small quantities of organic material available for collection and insufficient knowledge of standard deviations, it is unsafe to make definitive pronouncements. The organic fraction is, however, seemingly mutagenic. What stands out is the considerable abatement brought about by catalysts, in comparison with the pre-catalyst era (Kokko et al., 2000). Whether or not the mutagenicity is comparable (Shore et al., 1987) or greater by tenfold (Lang et al., 1981) compared with diesel particulate, this is compensated by the considerably lower rate at which mutagens are emitted.

Plenty of demographic characterisations[6] are now found in the open literature, where, a decade ago, very little information was available. Emissions, as already detailed in the previous section, while comparable to the particle background for steady-state, low-load operation, are rather pronounced during high-speed driving and acceleration. But although evaluations have been conducted on statutory test cycles, it should be remembered that, as gasoline vehicles were not formerly subject to statutory constraints for particulate emissions, they were not, correspondingly, designed and calibrated by the manufacturer with these cycles in mind.

The statutory test cycle evaluations of particulate mass to have been reported, with model year or homologation level specified in parentheses, run as follows. (**A**) NEDC: (a) five vehicles (model year 1998) and two fuels, 0.5–5.8 mg/km (Kokko et al., 2000); (b) two vehicles (model years 1995 and 1997), 0.9–1.1 mg/km (Mohr et al., 2000); (c) two vehicles (Stage II and III) and three fuels, <1 mg/km (Andersson et al., 2001); (d) one vehicle (model year 2001), ~1 mg/km (Aakko and Nylund, 2003); (e) one vehicle, 0.4 mg/km (Bosteels et al., 2006); (f) four vehicles (Stage III) and three fuels, 0.1–5 mg/km (Ntziachristos et al., 2004b). (**B**) F T P-75: (a) fifteen vehicles (1990s models), 0.8–8.5 mg/km (Maricq et al., 1999a).

From these studies, mass emissions for gentle or lightly loaded driving schedules do not usually exceed a few milligrams per kilometre. As said, the load expected of the engine makes a difference, but even on a 'motorway' schedule, mass emissions were still restricted to 1–2 mg/km (Ntziachristos et al., 2004b). It should be noted that such mass emissions are low enough to pose measurement difficulties in current test protocols, and the standard deviations are inconveniently large. Moreover, with such limited amounts of material available for size fractionation, modes do not stand out unambiguously. One such attempt for the NEDC is given in Figure 8.2(a) (Andersson et al., 2001). This figure seems to contain no systematic trend according to vehicle or fuel formulation. Whatever the size classification, a mass of, say, 0.25–1.0 mg/km is the probable result. Supermicron particles are evidently well represented.

The statutory test cycle evaluations of particle numbers to have been reported, with model year or homologation level specified in parentheses, run as follows. (**A**) NEDC: (a) one vehicle, $\sim 5 \times 10^{12}$/km (Aakko and Nylund, 2003); (b) two vehicles (Stage II and III) and three fuels, $2 \times 10^{11} - 2 \times 10^{12}$/km (Andersson et al., 2001); four vehicles (Stage III) and three fuels, 7×10^{11}/km $- 4 \times 10^{12}$/km (Ntziachristos et al., 2004b); (c) one vehicle, 4×10^{11}/km (Bosteels et al., 2006). (**B**) FTP-75: fifteen vehicles (1990s models), $5.1 \times 10^{12} - 6.5 \times 10^{13}$/km (Maricq et al., 1999a). From these studies, number emissions would appear to fall somewhere around 10^{12}/km, but there is wide variation, even for vehicles of ostensibly the same certification level: for example, on a 'motorway' schedule, four vehicles (Stage III), tested on three fuels, emitted at 1×10^{11}/km $- 2 \times 10^{14}$/km (Ntziachristos et al., 2004b).

Distributions of particle number for two vehicles, operated at three steady-state conditions, are given in Figure 8.2(b) (Andersson et al., 2001). An extremely large range in concentration was traversed, from 10^5 cm^{-3} at idle, to 10^{12} cm^{-3} at 120 kph. Faster-moving vehicles generated greater concentrations of particles in an unbiased fashion, until the highest speeds were attained, when nanoparticles came to dominate. It should be admitted that Figure 8.2(b) shows little evidence of fuel enrichment according to the causal mechanisms discussed in the previous section, insofar as the distributions are not

---

[6] See also the provisos listed in Section 6.5.

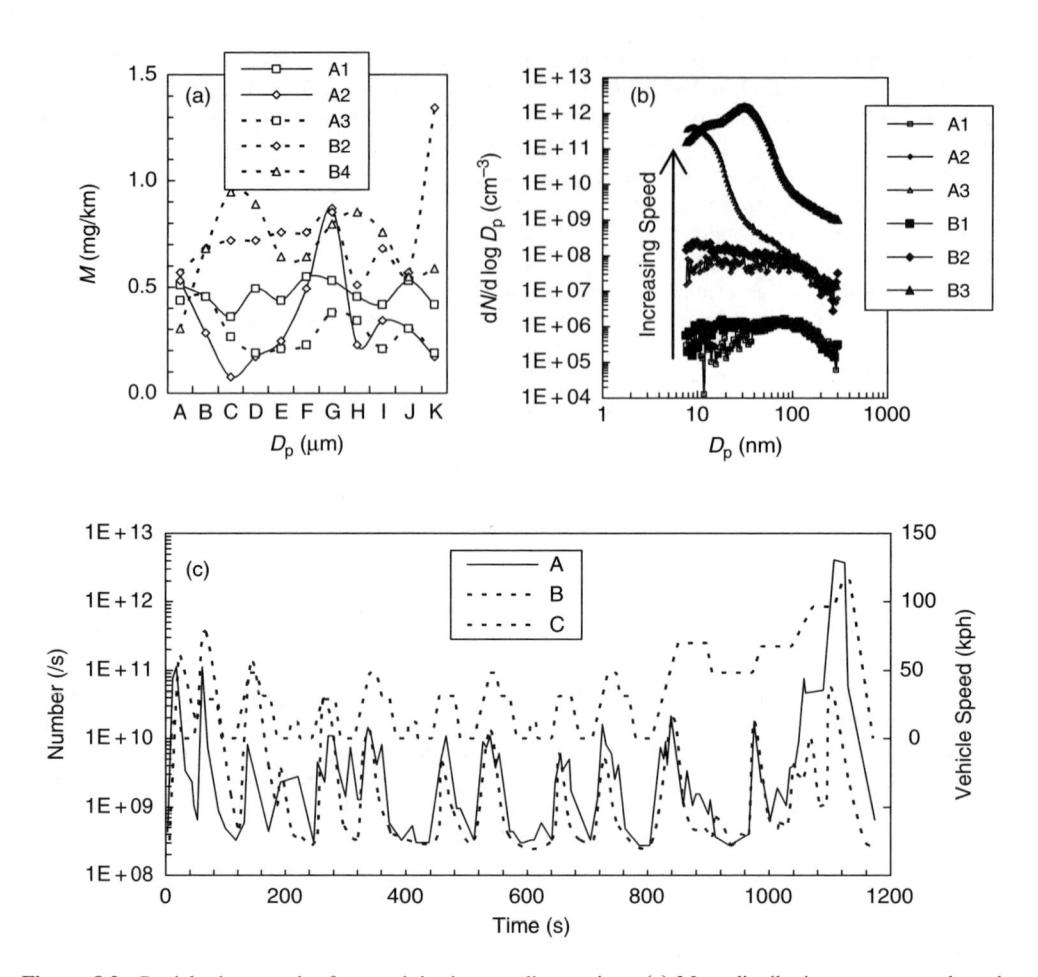

Figure 8.2 Particle demography for port-injection gasoline engines. (a) Mass distribution, as measured on the NEDC, cold start. Vehicles: A, 1.8-litre, three-way catalyst, Stage II; B, 1.8-litre, 'bifuel', three-way catalyst, Stage II. Fuels (aromatics, %, and sulphur, ppm): 1, 39.7 and 480; 2, 36.1 and 130; 3, 28.8 and 22 ppm; 4, liquefied petroleum gas (for full fuel specifications, see original reference). Instrument: MOUDI. Size classifications: A, <0.056 µm; B, 0.056–0.1 µm; C, 0.1–0.18 µm; D, 0.18–0.32 µm; E, 0.32–0.56 µm, F, 0.56–1.0 µm; G, 1.0–1.8 µm; H, 1.8–3.2 µm; I, 3.2–5.6 µm; J, 5.6–10 µm, K, >10 µm (Andersson *et al.*, 2001). (b) Number distribution, as measured for steady states: 1, idle (particles per second); 2, 50 kph (particles per kilometre); 3, 120 kph (particles per kilometre). Vehicles: A, B as for (a). Fuel: aromatics, 36.1 %; sulphur, 130 ppm. Instrument: SMPS, 7–320 nm (Andersson *et al.*, 2001; data courtesy of J. Andersson, Ricardo Consulting Engineers). (c) Particle numbers during the NEDC. Vehicles: A, 1.4-litre, four-cylinder, single-point injection, three-way catalyst, model year 1997; B, 1.8-litre, four-cylinder, multipoint injection, three-way catalyst, model year 1995. Vehicle speed: C. Dilution ratio: primary (constant-volume sampler), >9; secondary (ejector), 14. Fuel: RON 95. Instrument: CPC, >7 nm (Mohr *et al.*, 2000).

overwhelmingly dominated by carbonaceous accumulation modes; but this is a matter also of volatile material in relation to soot, and the fuel-metering strategy adopted by the manufacturer.

Figure 8.2(c) illustrates instantaneous particle numbers for two vehicles, as recorded throughout the NEDC (Mohr *et al.*, 2000). It should be noted that all particles larger than 7 nm were included (summed). The importance of accelerations in particle generation is clear: close examination of the trace suggests that emission rates decline once the vehicle has reached a steady cruise, i.e. the brief period of fuel enrichment has ended and stoichiometric fuelling has been re-established (Aakko and Nylund, 2003). On one of the vehicles, the highest emission rates were experienced during the high-speed cruise, suggesting sustained fuel enrichment. Less expected is the generation, during cruises, of particle numbers at similar rates to those at seen during idling (see $t = 920-960$ s).

8.3.3 Abatement

Emission control problems for port-injection gasoline engines, stoichiometrically fuelled and aftertreated with three-way catalysts, do not generally concern particulate, at least from the present-day type-approval standpoint and current statutory test cycles. The focus is, instead, on HC, CO and NO_x; and, consequently, no emission control technologies are directed *specifically* at particulate. However, certain *indirect* benefits accrue, as technology successful in controlling emissions of HC, CO and NO_x, is, typically, also advantageous in controlling emissions of particulate. As said, this is the present-day situation; in forthcoming years, legislation may be expanded to include operating conditions currently 'off-cycle', for example harsher accelerations, for which special measures might be required to address particulate emissions.

The control of particulate via fuel reformulation is obvious enough, and a noteworthy trend in recent times has been the relentless reduction in fuel sulphur content (FSC) – although here, again, the underpinning motivation lies elsewhere. Sulphur, for a variety of reasons, impedes the action of the catalyst towards HC, CO and NO_x; the benefits of low FSC for particulate abatement are indirect, first in reducing the feedstock for sulphates, especially insofar as this fraction is exacerbated by oxidation catalysis, and second in improving the oxidation of organic particulate in a manner akin to HC.

Some effort has been directed at the hydrocarbon speciation of the fuel: for example, a reduction in heavy alkenes ($>C_6$) was only marginally beneficial in terms of particulate-phase PAH and mutagenicity (Pentikäinen *et al.*, 2004); and gasoline – oxygenate blends (e.g. with ethanol) have been particularly aimed at easing the enrichment used during cold starts (Mulawa *et al.*, 1997). More drastically, it is proposed to move away from gasoline altogether. Particle numbers with liquefied petroleum gas (LPG) were higher (Ristovski *et al.*, 1998), and differed little in their demography from the case with gasoline (Li L., *et al.*, 2004). Compressed natural gas (CNG), on the other hand, generated exhaust gas relatively devoid of soot (Okamoto *et al.*, 2006), including during more susceptible operating conditions, namely cold starts and low ambient temperatures (Aakko and Nylund, 2003); this is a feature of improved mixing and greater fuel-metering precision than is possible with liquid fuels, which must first be vaporised. The particles were, however, still mutagenic (Kado *et al.*, 2005), while soot and heavy organic compounds, including aromatics, were still generated if air–fuel ratios were too rich (Sidhu *et al.*, 2001). Last but not least, oxygenates, the principle motivation for which is to improve octane number, have some effect: for example, ethanol significantly reduced particulate mass (Aakko and Nylund, 2003), and marginally reduced particulate PAH and mutagenicity (Pentikäinen *et al.*, 2004).

Turning to combustion, the facetious advice, in view of what was related in Section 8.3.1, is to avoid rich air–fuel ratios; but, as equally said, this is easier said than done. Not unexpectedly, in view of the causal mechanisms, any technology able to improve fuel vaporisation and the destruction of fuel droplets or ligaments of liquid fuel concomitantly obviates rich-burning pockets, and thereby also combats emissions of soot. Hotter intake valves are beneficial for precisely this reason: this is why exhaust gas recirculation (EGR) can actually decrease engine-out soot (Kayes *et al.*, 2000), where increases are seen on diesel engines (Section 7.4). On the other hand, EGR may affect the particulate emission indirectly, since, to compensate for the power loss, the throttle is opened, giving, for example, higher in-cylinder pressure (less lubricant drawn into the combustion chamber) and higher exhaust temperature (more sulphate formation in the catalyst).

One extremely long-standing issue is the need to pass swiftly through the cold-start period, as such operation is characterised, inherently, by high emissions of CO and HC; and, as hinted, what is detrimental in these terms is detrimental also for emissions of particulate, both in the carbonaceous fraction (badly burning fuel) and in the organic fraction (unburned fuel). In the FTP-75, mass emissions at a cold start at $-15\,°C$ exceed those at $22\,°C$ by an order of magnitude (Ahlvik *et al.*, 1997); similar differences are seen in the NEDC (Kokko *et al.*, 2000), and in particle numbers between *Phase 1* (the cold start) and *Phase 2* (the warm start) (Maricq *et al.*, 1999a). There are essentially two aspects to this cold-start problem: the engine requires considerable enrichment, during which time the fuel burns poorly, and the catalyst is inactive when below its light-off temperature. A review of cold-start emission control technology would take us too far off course; but one approach, among many, is to select the appropriate injection timing, i.e. in order to vaporise the liquid fuel fully, prior to initiation of the spark (Kayes *et al.*, 2000).

The conversion efficiency of the three-way catalyst for HC, CO and NO_x is sharply peaked within a narrow range of air–fuel ratio centred on stoichiometry. Thus, an enormous incentive already exists to avoid embarrassingly lean or embarrassingly rich excursions; and in either case, there are positive corollaries for particulate. On the lean side, favourable conditions in the *exhaust* for the catalytic production of sulphate particulate are avoided; and on the rich side, favourable conditions in the *engine* for the pyrolytic production of soot are avoided. And throughout this operation, if the HC and CO can be oxidised, then, logically, so can organic compounds which act as particle precursors. This, it seems, is a no-lose situation, although two caveats are evident. First, the storage of sulphur by three-way catalysts is invariably followed by a release of some sort, although, fortunately (in terms of *particulate* control, that is), this release takes place as SO_2 or H_2S, rather than as SO_3 or a sulphated nucleation mode (Maricq *et al.*, 2002a). Second, a little-investigated area is the generation by three-way catalysts of ammonia; this gas, a well-known neutralising agent in the atmosphere, leads, via ongoing atmospheric reactions, to secondary sulphate particles.

But although one may theorise significant benefits from the three-way catalyst, the demographic implications of these are poorly delineated; some available data are presented in Figure 8.3(Graskow *et al.*, 1999a): particle numbers at low engine load decreased by almost 80 %, but at intermediate load by only 10 %, the smallest (<70 nm), and the largest (>200 nm) sizes were most affected. Interestingly, although particulate emissions were markedly lower following a warm start than following a cold start, it was the warm engine, rather than the warm catalyst, which seemed predominantly responsible (Maricq at al, 2002a). Some soot is also stored in catalysts (Andrews and Ahamed, 1999). Obviously, there is much that cannot be determined from such a regrettably slender database: whether particles are captured, created or destroyed, is a matter for careful compositional investigations. But the question that certainly begs itself is what happens within the catalyst during brief periods of enrichment, when engine-out emissions of soot are at their peak, for which one might speculate efficient deposition, mediated by strong thermophoretic forces (Section 4.2.2).

8.4 Direct-injection Engines

The injection of gasoline *directly* into the cylinder, rather than into the inlet manifold, is not a recent idea (Harrington and Yetter, 1981); what *is* recent is the existence of engine management systems capable of supporting the greater fuel-metering complexity. This was, principally, why research into direct-injection gasoline engines suddenly mushroomed in the mid 1990s and continues to thrive up to the present day. Actually, there is an instructive distinction to be made here between the *combustion* system and the *injection* system. At *low* load, fuel is injected during the compression stroke, the charge

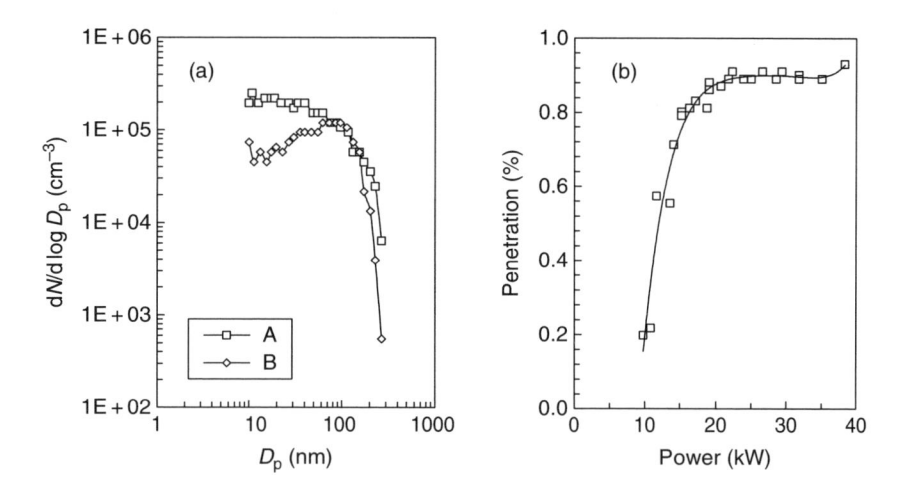

Figure 8.3 Particle removal by three-way catalytic converter. (a) Number distribution. The operating point was 2500 rpm, 60 kPa MAP; A, upstream; B, downstream. (b) Penetration (by number). The duty cycle was 24 different (steady-state) operating points. Engine: 2.3-litre, four-cylinder, port fuel-injected, model year 1993. Fuel: aromatics, 40.5 %; sulphur, 317 ppm, RON, 92.2. Instrumentation: CNC and SMPS, 7 to 420 nm. Dilution: single-stage (15:1) ejector diluter, transit time <50 ms (Graskow *et al.*, 1999a).

is stratified, inhomogeneous or heterogeneous (take your choice), and the air–fuel ratio is ∼30–40. At *medium* load, fuel is injected during the intake stroke, the charge is homogeneous, and the air–fuel ratio is ∼20–25. At *high* load, injection takes place during the intake stroke, the charge is homogeneous, and the air–fuel ratio is under closed-loop stoichiometric control (Graskow *et al.*, 1999b). Thus, even the direct-injection engine adopts the homogeneous combustion mode used in the port-injection engine, where appropriate. From this description it becomes obvious that 'stratified-charge engine' is really a common-or-garden term for a combustion mode restricted to a certain range of engine operation, whereas the *principle* of injection is a generic feature of the engine. Given this distinction, *direct-injection engine* seems an apter term than 'stratified-charge engine'. But this is not just pedantry: the mode of combustion profoundly influences the particulate emission, as we shall see.

8.4.1 Formation

The explanation for emissions of soot, at least at a first level, is hardly arcane: stratification of the charge inevitably confers certain diesel-like characteristics: a rich mixture is used in the vicinity of the spark plug, even if the remainder of the charge is comfortably lean, and rich-burning pockets are obviously conducive to the formation of soot (Yang J. and Kenney, 2002). There is also less time available for fuel evaporation; and surviving fuel droplets produce, when engulfed by the flame, luminous streaks characteristic of incandescent soot.

The in-cylinder soot concentration, as a function of crank angle, increases and then decreases – behaviour logically attributed to shifting contention between *formation* and *oxidation*. Some results along these lines are given in Figure 8.4 (Kubach *et al.*, 2001). As in diesel engines, the amount of soot discharged depends not just on the maximum in-cylinder concentration, but also on the extent of burn-up in the later stages of the combustion. But although this pattern of rise and fall is similar to that

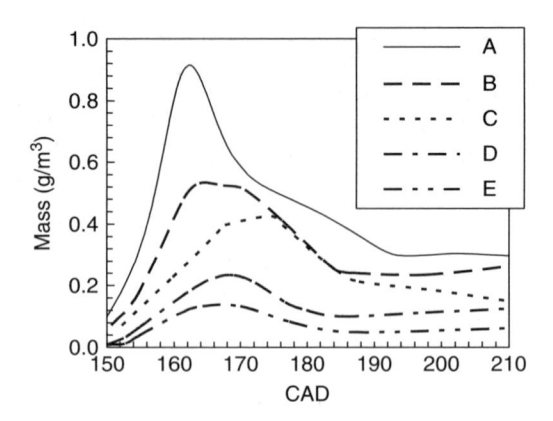

Figure 8.4 In-cylinder soot concentration as measured by two-colour pyrometry. Injection pressure (bar): A, 60; B, 85; C, 100; D, 120 (with swirl); E: 120 (without swirl). Operating point: 1000 rpm, 2 bar. Injection timing: 52°BTDC. Spark timing: 42°BTDC. Engine: 0.5-litre, single-cylinder, common-rail direct-injection (Kubach *et al.*, 2001).

in diesel engines, the combustion itself is obviously different. Here it is impossible to accommodate all variants of stratified combustion; but in one such engine, combustion proceeds in *two* stages: first, a soot-forming flame front propagates through the charge, and second, a soot-oxidising flame front propagates through the spent gases left by the first flame front (Kuwahara *et al.*, 1998). The first flame is detectable in the ultraviolet through chemiluminescence (OH, CH and CO–O), and the second is detectable in the infrared through soot incandescence.

On a second level of understanding, soot arises in direct-injection gasoline engines through a decidedly undiesel-like mechanism, for when fuel sprays strike the piston, droplets accumulate on the piston surface to form liquid pools, films or puddles – fractionally, this piston wetting may represent a few per cent of the total injected fuel quantity (Karlsson and Heywood, 2001). Then, following initiation of the spark, the flame spreads through the charge and, in catching, so to speak, the descending piston, ignites these puddles, which now burn belatedly via turbulent diffusion flames. These 'pool fires' are highly sooting (Mehta D. *et al.*, 2001), and readily identified as such by their bright yellow flames. Pool-fire combustion is self-sustaining insofar as the heat transfer from the burning gasoline vapour above to the fire below drives the evaporation of more fuel. These fires are long-lived: they survive well into the expansion stroke (Karlsson and Heywood, 2001), and are still smouldering even in the exhaust stroke (Stevens and Steeper, 2001). The importance of pool-fire combustion in the emission of soot is ardently debated. But pool fires do seem to be a primary controlling factor (Drake *et al.*, 2003), inasmuch as engine-out soot becomes undetectable in their absence (Stevens and Steeper, 2001): the effects are switched on and off by the deliberate injection of liquid fuels onto the piston of an engine running on gaseous fuel (Warey *et al.*, 2002).

These pools do not necessarily turn into fully fledged pool fires. An *early* injection (−300CAD) produced no pool fires, even though fuel pools were present at initiation of the spark; perhaps the pools had mostly evaporated prior to ignition, or perhaps they had lost their lighter components, leaving heavier, unignitable ones (Stevens and Steeper, 2001). A *late* injection (−150CAD), which allowed substantial pools to persist until initiation of the spark, *did* give rise to pool fires. Of course, injection timing not only determines the time available for pool evaporation; it also determines the distance traversed by the spray, and hence the amount of fuel that impinges on the piston in the first place.

Pool-fire combustion is actually quite an interesting phenomenon, since piston surfaces are, superficially at any rate, too hot to harbour liquid pools of unburned fuel. This apparently paradoxical situation is adduced to the Leidenfrost effect (Huang Y. et al., 2001b): should the temperature of a surface significantly exceed the boiling point of a liquid resting on that surface, then an intervening layer of vapour is generated. This, for example, is the situation pertaining when water is poured onto a hot metal surface, and droplets hover on cushions of their own vapour.[7] This affects splash dynamics (Han Z. et al., 2000): in one circumstance the fuel wets the surface, to accumulate as a pool; in another, nonwetting droplets, upon impaction, fragment and bounce off. The intervening cushion of vapour, moreover, has an *insulating* effect, such that heat transfer to the liquid from the surface is inhibited; hence there is prolongation of the evaporation. Perversely, therefore, fuels of greater volatility persist as liquid pools, where those of *lesser* volatility experience *faster* rates of evaporation: 'wrens make prey where eagles dare not perch'. The strength of this effect should not be underestimated: the cushion of vapour slows evaporation by as much as two orders of magnitude.

Some researchers argue against the Leidenfrost effect in direct-injection engines (Han Z. et al., 2000). This is where out-of-engine experiments are intriguing, as they point to adjunctive factors (Stanglmaier et al., 2002). The cushion of insulating vapour underneath the fuel is compressed at high cylinder pressure; this increases the heat transfer, and thus diminishes the Leidenfrost effect. Leidenfrost-inhibited evaporation seems restricted to cases where the in-cylinder pressure is low, such as during the intake stroke or early in the compression stroke. According to this understanding, the rate of evaporation accelerates as compression proceeds, not because of increasing temperature but because of increasing pressure. One should also note that, in an engine, the cylinder pressure changes much more rapidly than does the surface temperature of the piston.

In direct-injection engines the Leidenfrost effect is, with considerable justification, blamed not just for soot, but also for a large proportion of HC (Huang Y. et al., 2001a). The problem is that hydrocarbons escape from these pools in a manner wholly unsuited to their oxidation, before the fires have ignited, or after they have extinguished. Indeed, it is not unknown for the fires to become extinguished before consuming all of their fuel, so that pools survive into the *next* stroke. In this case the charge motion during engine breathing might assist in sweeping fuel back into the bulk gas. The underpinning physics is obfuscated by multicomponent fuels with no single defined boiling point, and differential evaporation undoubtedly occurs. Experiments in which fuel droplets are allowed to impact on hot surfaces show that evaporation rates of multicomponent fuels cannot be easily predicted from the behaviour of individual fuel components in isolation (Stanglmaier et al., 2002). But there is an expectation that high-volatility fuels may actually generate *more* HC, since if evaporation from the surface is *slowed*, the hydrocarbons will be released too late for their effective oxidation.[8] The final twist is modelling studies which indicate the rate-controlling factor to be not evaporation *per se*, but diffusion of this fuel vapour away from the surface into the bulk gas (Stanglmaier et al., 2002). This might explain instances where the accumulated mass of fuel, rather than the time available for evaporation or the piston surface temperature, was the important factor.

Lest the impression be given that the particulate discharged by direct-injection engines is simply soot, a nonnegligible organic fraction is expected, first, because emissions of HC are higher when running in stratified mode (Castagné et al., 2000), and second, *some* hydrocarbons inevitably undergo gas-to-particle conversion, notwithstanding the greater volatility of gasoline (compared with diesel fuel,

---

[7] The same effect is thought to explain how people can walk barefoot over hot coals without, apparently, burning their soles: either they wet their feet prior to the stunt, or they become so nervous that the perspiration on their feet is sufficient. (The reader is not encouraged to test this theory.)

[8] Of course, droplets of a high-volatility fuel might not actually reach the surface in the first place. The physics of evaporation for surface-residing droplets is completely different from that of droplets suspended in the bulk gas.

that is). Three explanations advanced for the greater survival of fuel hydrocarbons are undermixing (too rich to burn), overmixing (too lean to burn) and poor late oxidation (too cool to burn). These three categories have been advanced, principally, with the bulk charge in mind, but there seems good reason to suppose that pools whose fires are extinguished or which fail to ignite also contribute to HC emissions in these ways.

Poor burn-up is a reminder that dilution of the combustion products with unused air in a lean-burn engine inevitably lowers late-cycle temperatures – this aspect has been used to explain why the HC speciation in stratified mode resembles the fuel speciation more closely than in homogenous mode: hydrocarbons are less likely to undergo late-cycle partial oxidation (Castagné *et al.*, 2000). This is a timely reminder that the speciation of HC is not a straightforward reproduction of the fuel speciation. (Here it may be useful to recall Figure 3.15 and the definition of 'direct and indirect' organic compounds.) Partial combustion may also explain why the HC speciation differs from that of port-injection engines (Villinger *et al.*, 2002a). Presumably, these remarks apply also, in some measure, to the organic fraction, which contains partially oxidised and pyrolysed hydrocarbons (Owrang *et al.*, 2004). One need only consider the action of pool fires, and the dark tide marks they leave behind, to suggest how a welter of new organic compounds (alien to the fuel) may be formed.

There are, then, a host of factors driving particulate formation, but the central feature of the emission – one that stands out beyond anything else – is the switch between a low-load stratified charge and a high-load homogeneous charge. The general pattern, which, it should be noted, is completely different from that for port-injection engines, is easily stated: with increasing load, the mass of emitted particulate rises gradually and then, at a certain point, falls back suddenly (Andersson *et al.*, 2001). This happens because of the greater sooting propensity of stratified operation compared with homogeneous operation. And since this change occurs in the soot emission, a corresponding change occurs also in the accumulation mode, this being where most of the soot resides. Of course, one should carefully qualify such statements by size range, as a sudden loss of the accumulation-mode soot is logically expected to increase the saturation of volatiles, and thus to promote nucleation. For this reason, particle numbers may become *higher* when in homogeneous mode, even though the particulate mass is, in fact, lower. How exactly this characteristic is connected to the duty cycle would require some tangential explanation of vehicle calibration, but it seems that vehicle acceleration, rather than speed, has the greater influence on emission rates (Smallwood *et al.*, 2001).

8.4.2 Characterisation

A reasonable body of demographic data[9] exists for the particles emitted by direct-injection gasoline engines, although the range of engines, or rather the range of vehicles, is understandably narrow, as, at present, the power plant is but little commercialised in comparison with traditional port-injection engines. The data are nevertheless sufficient to highlight informative similarities to (and differences from) diesel engines and port-injection engines, and also to show that particulate emissions are an issue. As explained in the previous section, the chief demographic feature of particulate emissions from direct-injection engines lies in the switch from one operating mode to the other (Hall and Dickens, 1999). The greater responsibility for the emission (notably soot) is safely apportionable to the stratified charge, although it should be said that 'whole-cycle' summations necessarily conflate the contributions made by the two operating modes.

The statutory test cycle evaluations of particulate mass to have been reported, with model year or homologation level specified in parentheses, run as follows. (**A**) NEDC: (a) one vehicle (model year

---

[9] See also the provisos listed in Section 6.5.

1997), 9 mg/km (Mohr *et al.*, 2000); (b) one vehicle (Stage III) and two fuels, 6.8 g/km and 14.5 g/km (Andersson *et al.*, 2001); (c) one vehicle (model year 1998) and two fuels, 15 mg/km and 30 mg/km (Kokko *et al.*, 2001); (d) two vehicles, 3.5 mg/km and 4 mg/km (Mohr *et al.*, 2003a); (e) one vehicle (model year 2002), 10 mg/km (Aakko and Nylund, 2003); (f) three vehicles (Stage III) and three fuels, 2–15 mg/km (Ntziachristos *et al.*, 2004b). (**B**) FTP-75: (a) one vehicle (model year 1997) and two fuels, 6 mg/mile and 22 mg/mile (Cole *et al.*, 1999); (b) one vehicle (model year *c.*2001) and two tests, 3.5 mg/mile and 14.5 mg/mile (Smallwood *et al.*, 2000); (c) one vehicle (model year *c.*2000), ~10 mg/mile (Kwon *et al.*, 1999a).

From these studies, mass emissions of 15 mg/km or so on statutory test cycles would seem to be typical, although instances are known of lower figures. More highly loaded duty cycles with direct-injection engines produce, in a manner converse to other powertrains, not more but *less* particulate mass: on a 'motorway' schedule, emissions for three vehicles (Stage III) and three fuels were 1.5–1.8 mg/mile, compared with the NEDC result of 2–15 mg/km cited above (Ntziachristos *et al.*, 2004b). This trend is the natural consequence of switching from stratified charge to homogeneous charge. As the conditions for switching are a central feature of the vehicle calibration adopted by the manufacturer, it is unreasonable to ascribe low statutory emissions to successful stratification without first verifying whether or not the stratification envelope is smaller, i.e. whether a homogeneous charge is being used over a wider range of the cycle (Aakko and Nylund, 2003).

The distribution of this mass has received less attention. Figure 8.5(a) shows two such distributions, as cumulatively measured over the NEDC (Andersson *et al.*, 2001); these are broadly corroborated by other researchers (Aakko and Nylund, 2003). Generally speaking, the accumulation mode is well represented at a few hundred nanometres; yet the contribution of massive particles is nonnegligible, with some suggestion of a coarse mode at several microns.

The statutory test cycle evaluations of particle numbers to have been reported, with model year or homologation level specified in parentheses, run as follows. These were all on the NEDC: (a) one vehicle (Stage III) and two fuels, $2 \times 10^{13} - 9 \times 10^{13}$/km (Andersson *et al.*, 2001); one vehicle (model year 2002), $\sim 1 \times 10^{13}$/km (Aakko and Nylund, 2003); (b) two vehicles, 2×10^{12}/km and 1×10^{13}/km (Mohr *et al.*, 2003a); (c) three vehicles (Stage III) and three fuels, $2 \times 10^{12} - 4 \times 10^{13}$/km (Ntziachristos *et al.*, 2004b).

From these studies, it seems reasonable to conclude that particle numbers are on the order of 10^{13}/km, although there is a wide range, with some results higher and some lower by an order of magnitude or so. This time, the switch from stratified to homogeneous charge is disadvantageous: on a 'motorway' schedule, number emissions for three vehicles (Stage III) and three fuels were $7 \times 10^{12} - 4 \times 10^{14}$/km, as compared with the NEDC result mentioned immediately above, of $2 \times 10^{12} - 4 \times 10^{13}$/km (Ntziachristos *et al.*, 2004b). The appearance of more particles on more highly loaded duty cycles is wholly consistent with the lower mass emission: conversion from stratified to homogeneous charge has encouraged nucleation (Mohr *et al.*, 2003a) by depriving volatile particle precursors of soot surfaces.

As shown in Figure 8.5(b) (Andersson *et al.*, 2001), the switch is equally apparent in the number distribution. At low to medium load, the accumulation mode is well represented, and even dominates when an engine is idling, while at 50 kph, particle concentrations reach $10^{10} - 10^{11}$ cm^{-3}, with the higher end of this range still representing soot. At still greater speed, however, the engine now has a homogeneous charge, whereupon the accumulation-mode soot is significantly attenuated, even to the level seen at idle, just as the nucleation mode is correspondingly amplified, exceeding $> 10^{12}$ cm^{-3}. This pattern, though, is very much a matter of the ratio of gas-phase particle precursors to solid particles: cases where nucleation-mode particles are at first prominent at light load, and then decrease with rising load, yet operation is nevertheless still stratified (Graskow *et al.*, 1999b), are also explainable: more soot surface is becoming available for the accretion of volatiles, by which saturation is depressed and nucleation discouraged.

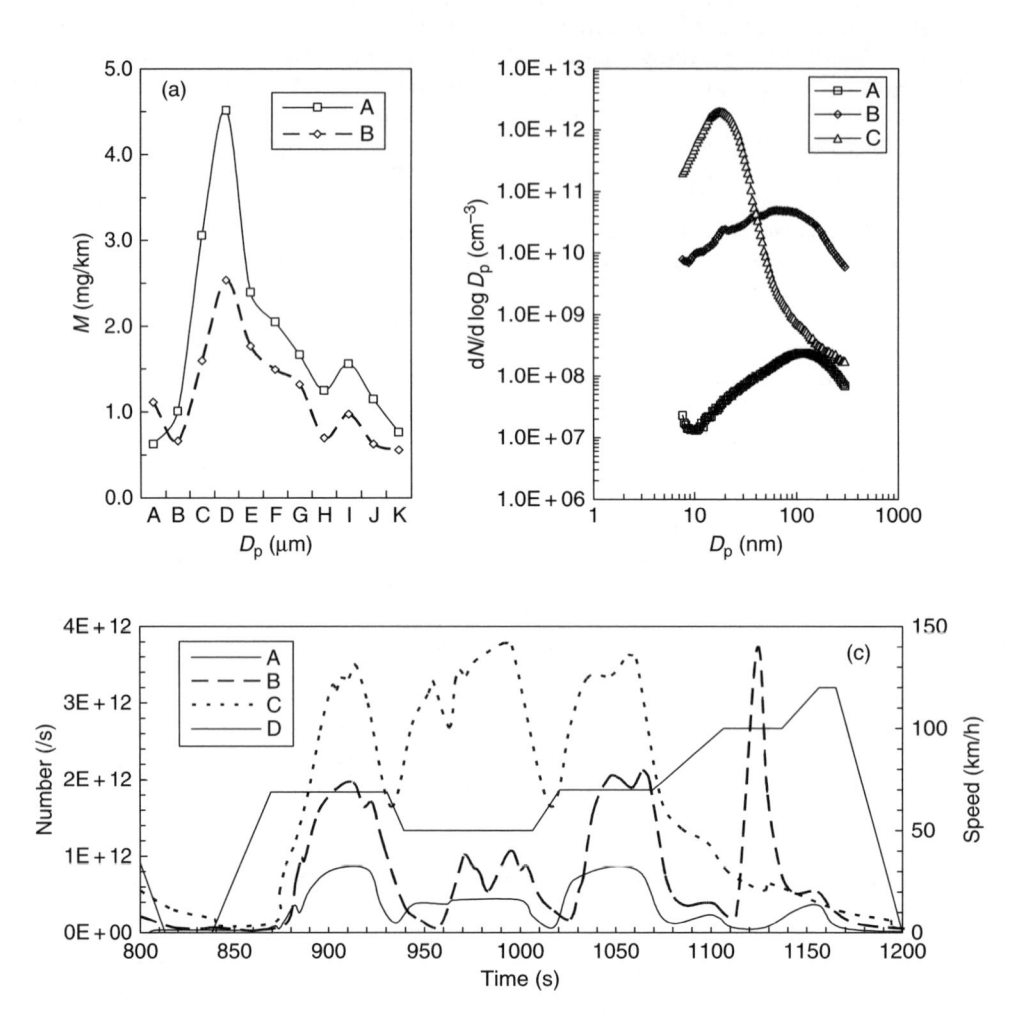

Figure 8.5 Particle demography for direct-injection gasoline engines. (a) Mass distribution, as measured on the NEDC. Vehicle: 1.8-litre, two oxidation catalysts, lean NO$_x$ trap, Stage III. Fuels: A, aromatics 39.7 %, sulphur 480 ppm; B, aromatics 36.1 %, sulphur 130 ppm. Instrument: MOUDI. Size classifications: A, <0.056 µm; B, 0.056–0.1 µm; C, 0.1–0.18 µm; D, 0.18–0.32 µm; E, 0.32–0.56 µm; F, 0.56–1.0 µm; G, 1.0–1.8 µm; H, 1.8–3.2 µm; I, 3.2–5.6 µm; J, 5.6–10 µm; K, >10 µm (Andersson *et al.*, 2001). (b) Number distribution, as measured for steady-state operation: A, idle (particles per second); B, 50 kph (particles per kilometre); C, 120 kph (particles per kilometre). Vehicle: as for (a). Fuel: sulphur 22 ppm, aromatics 28.8 %. Instrument: SMPS, 7–320 nm (Andersson *et al.*, 2001; data courtesy of J. Andersson, Ricardo Consulting Engineers). (c) Particle numbers during final, high-speed section of NEDC (EUDC). Vehicle: 1.8-litre, lean NO$_x$ catalyst. Particle sizes: A, 25 nm; B, 50 nm; C, 100 nm. Vehicle speed: D. Fuel: aromatics, 44 %; RON, 98; sulphur, 148 ppm. Instrument: SMPS, as locked on the stated sizes for each test run (Hall and Dickens, 1999).

Finally, as revealed by Figure 8.5(c) (Hall and Dickens, 1999), which portrays the final section of the NEDC, the switch in operating mode is discernible as well in transient duty cycles. Particle numbers at 100 nm during the initial cruises are on the order of $3-4 \times 10^{12}$/s. However, immediately upon accelerating to the highest speed, at around 1075 s, the emission of 100 nm particles declines swiftly, signifying conversion from stratified mode to homogeneous mode – and corresponding to a sudden but momentary surge in 50 nm particles. These sub-100 nm particles may well be volatiles released in such large quantities that they have agglomerated significantly, thus passing the 25 nm size, which shows no such spike. But it is not possible to be conclusive on this point, since chemical compositions were not directly verified.

8.4.3 Abatement

Emissions abatement via fuel formulation is little reported with direct-injection engines, but, as one might surmise, the necessary strategies parallel those used for diesel engines (Kwon et al., 1999b). Reductions of a third or more in the mass emission, brought about by oxygenates, are reported (Kokko et al., 2000), although this was not validated as being in the carbonaceous fraction; PAH were also reduced. Comparable benefits are obtained with reformulated gasolines, i.e. those of lower aromaticity and higher volatility (Hashimoto et al., 2000). Two *nonmetallic* additives exerted no consistently strong demographic influence (Graskow et al., 2000): a polyether amine reduced particle numbers throughout, whereas a polyolefin amine reduced only the accumulation-mode particles. Another additive, this time proprietary, reduced particle numbers by a third. The results suggested that additives could themselves produce particles at some operating points.

The central abatement effort with direct-injection engines is directed at *soot*, a reduction in which might be accomplished within the engine, either through impeding the formation or by encouraging the oxidation; both processes are usually modified concurrently, but *not* equally, and this is the basis of the control method. The practical realisation of these measures will, in most cases, be suggested by analogy to the diesel engine (Chapter 7): for example, the spray geometry needed to avoid impingement on the piston (Yang J. and Kenney, 2002), as determined by the injector nozzle (Nogi et al., 1998); and the flow field (tumble and swirl), as determined by the induction process (Mehta D. et al., 2001). Other measures are unique to the direct-injection engine: in *two-stage* combustion, the soot formed initially by the *first* flame is subsequently consumed by a *second* flame (Kuwahara et al., 1998). Actually, from the opposite standpoint, the soot particles within the burned zone of the first flame are put to good use, insofar as they act as *ignition sites*, facilitating burn-up in the remaining mixture, which is lean.

Other possibilities for emission abatement lie in engine calibratables (Price et al., 2006). Higher injection pressures are beneficial in controlling the mass emission of soot (Kubach et al., 2001), probably through greater atomisation of the fuel. Both particle numbers and particulate mass increased by a factor of 20 as the injection timing was retarded from the homogeneous mode (>200°BTDC) into the stratified mode (<150°BTDC) (Maricq et al., 1999d): gradually intensifying charge stratification (steepening gradients in local air–fuel ratio) is the probable cause. A somewhat lesser sensitivity was exhibited to spark timing, with the stratified mode more affected than the homogeneous mode. As shown in Figure 8.6, this time for an engine with an air-forced fuel-injection system, upon retardation in spark timing, a nucleation mode progressively develops, while, correspondingly, the accumulation mode declines (Maricq et al., 2000a). Again, this suggests lessening charge inhomogeneity. Nucleated volatiles are the likely consequence of less carbonaceous surface area.

Aftertreatment for direct-injection engines requires two sets of hardware, one for each of the two operating modes. The stratified mode, with a lean air–fuel ratio, uses an oxidation catalyst (for HC and CO) and a deNO$_x$ catalyst (for NO$_x$); the homogeneous mode, with an air–fuel ratio cycled about stoichiometry between slightly lean and slightly rich, uses a three-way catalyst (for HC, CO and NO$_x$),

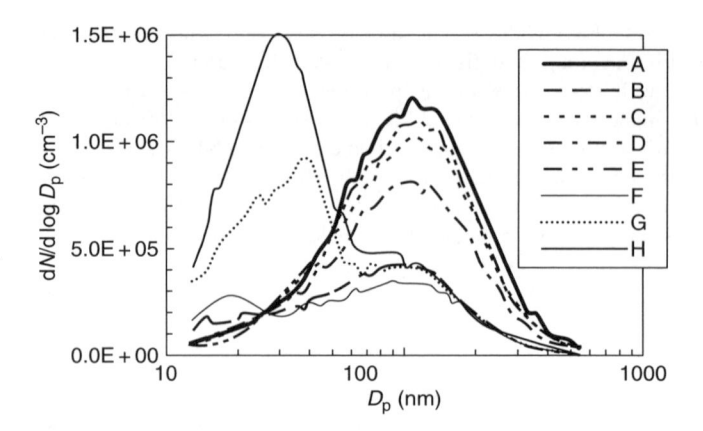

Figure 8.6 Number distribution, as affected in stratified operation by spark timing (degrees BTDC). A – H: 32, 28, 26, 24, 22, 20, 18, 16. Engine: 0.3-litre, single-cylinder, with air-forced fuel injection. Operating point: 1500 rpm/3.2 bar BMEP, air–fuel ratio 31, no EGR, start of injection 80° BTDC. Fuel: alkanes 71 %, aromatics 28 %, alkenes 1 %; 6000 ppm polyether amine as injector cleaner. Dilution: ejector pump diluter, dilution ratio (two stages) approximately 64. Instrument: SMPS (Maricq *et al.*, 2000a).

the operation of which follows precepts already discussed with respect to the port-injection engine (Section 8.3.3). In the former case, low-sulphur fuel is advisable (Cole *et al.*, 1999), as the oxidising exhaust chemistry reinstates, potentially, the sulphate-producing problems once experienced with port-injection engines (Section 8.2.2). However, the reduction these catalysts impart to organic particulate, in either operating mode, is welcome.

Whether or not the soot emitted by direct-injection engines will ultimately be conquered via methods internal to the engine is a moot point. Should combustion engineers be unsuccessful in this quest, then some form of particulate *aftertreatment* will be necessary: but with gasoline engines, there is barely any experience, at the time of writing, in this technology. As with diesel engines, there are two sequential (and usually alternating) steps: particle capture and particle incineration. Thermophoretic capture using exhaust gas cooling was ineffective compared with electrostatic capture via a centrally mounted electrode operating at up to 20 kV (Crane *et al.*, 2002). There seems little reason, though, to pursue such methods, because so much experience already exists with exhaust-mounted filters for *diesel* engines (Section 7.8.2). The principal *operational* difference from diesel engines, and the greater technical stumbling block, concerns regeneration: at low load and a lean air–fuel ratio, the temperatures are insufficient; at high load and a stoichiometric air–fuel ratio, the exhaust gas oxygen is insufficient. Thus, novel catalysts, with the ability to exploit H_2O and CO_2 rather than O_2 as oxidants, are required (Crane *et al.*, 2002). Catalyst instability and poor contact with the soot are other difficulties (Section 7.8.1).

8.5 Two-stroke Engines

The profuse discharge by two-stroke engines of greyish-white or bluish clouds of smoke is a matter of common observation, and indeed of notoriety. These smoke clouds, along with equally profuse emissions of HC and CO, are the bane of developing countries, where mopeds, scooters and motorbikes,[10] which represent the predominant affordable mode of motorised personal transport, reach extraordinarily high densities. Since these dense clouds of liquid-phase droplets are hardly stable entities as such, the implications for passers-by are difficult to gauge: in close proximity, compounds are inhaled as an aerosol; further away, the droplets may well have evaporated, rather than simply dispersed, so that inhalation takes place as a gas instead. This volatility also creates problems in measurement akin to those encountered with white smoke discharged by diesel engines (Section 3.6.5).

Related concerns are evident in the developed world, where the pollution from sundry utility engines is now coming under regulation. These utility engines are stupefyingly diverse, being used domestically (e.g. lawnmowers), municipally (e.g. chainsaws) and recreationally (e.g. snowmobiles). The last-cited example serves as a reminder of the need to protect not only the air of urban and suburban areas, but also the soil and water of pristine ecosystems. Indeed, outboard motors engender less-appreciated concerns, where the exhaust gas, as an anti-noise measure, is routed underwater, so that rivers and lakes inopportunely act as scrubbers (Wasil and Montgomery, 2003); and when cooling water is injected directly into the exhaust system so that it intermingles with exhaust gas, the pollutants emerge as an exhaust – water aerosol (Carroll *et al.*, 2004).

It should be stressed, however, that two-stroke engines are currently undergoing, so to speak, something of a renaissance, as significant emissions benefits accrue by replacing the somewhat limited carburettor with modern direct-injection systems, and catalytic converters are now appearing on many applications. It remains to be seen just how much mileage (literally as well as figuratively) there is in these innovations. For example, the impingement of liquid fuel on the piston is one aspect of particulate emissions with direct-injection two-strokes that is in common with direct-injection four-strokes (Cromas and Ghandhi, 2005). But the exact implications for particulate emissions, in each and every twist and turn of this bewilderingly diverse technology, cannot be discussed in this short review.

Incisively, though, the particulate emission is governed by one dominant aspect: two-stroke engines burn, designedly, a mixture of fuel and lubricant; this lubricant is either premixed with the gasoline or injected into the airstream. However, as much as one-third of the charge escapes combustion, or 'short-circuits', during scavenging (the period of valve overlap), to be emitted, unburned – a problem often exacerbated by unstable combustion at idle and low speed. Simply, the escape of unburned charge is the reason for high emissions of HC: but since many of these hydrocarbons – especially those from the oil – have large molecular masses, they readily undergo gas-to-particle conversion to form aerosols of liquid-phase droplets: this is the smoke. The smoke emission is therefore, arguably, to *some* extent controlled indirectly via legislation for HC (Rijkboer *et al.*, 2005), a category which covers the more volatile components, although the two emissions do not necessarily display similar trends with respect to engine operation. Not should the oil components be overstressed: the particulate emission also contains a nonnegligible fuel component, possibly a third or more (Carroll *et al.*, 2000).

Not unexpectedly, then, oil–fuel ratio (OFR), rather than air–fuel ratio (AFR), provides the principal index: this is *inversely* (i.e. nonlinearly) related to the density of the smoke (Yashiro, 1987). An OFR (by mass) of 50–100 is typical:[11] emissions of smoke (and particulate) tend to have levelled out by this stage, although the overarching significance of oil is demonstrated in the observation that trends in *combustion-derived* particulate are masked until an OFR as weak as 450 is reached (Cromas and Ghandhi, 2004). To restrict these oil emissions, the obvious strategy is to devise engines requiring less oil. This does not simply concern the thermal constraints and mechanical clearances of the engine: the manner of oil metering is also a factor, perhaps erring on the excess side if safeguarding engine operation is allowed to take precedence over emissions. If the pump is able to meter oil more precisely,

---

[10] Four-stroke motorbikes are covered implicitly in Section 8.3. See also Andersson *et al.* (2003).
[11] 1:400 is typical of four-stroke engines.

it is possible to meet more appropriately the engine's varying requirements, say, as a function of load and speed, hence realising ratios as weak as 200 (Laimböck *et al.*, 1999).

Oil reformulation is an alternative anti-smoke measure. The unaesthetic aspects, of course, should not be confused with the implications for public health: for example, oil properties (e.g. viscosity) appear to have some effect on droplet size (Sugiura and Kagaya, 1977), but reducing the offending visibility by such means may bring purely cosmetic improvements. Reformulation for greater susceptibility to oxidation is surely the better method. Actually, two-stroke oils are *already* formulated very much with their combustibility in mind, but from a different perspective: should they burn improperly, deposits will form in combustion chambers and exhaust ports, compromising engine performance. The oils are often ashless, consisting of additives, solvents, a base oil and a highly lubricious component known as 'brightstock', the latter of which not only consists of the heaviest aromatics, but also represents the poorest-burning component, thus contributing disproportionately to the particulate emission (Wasil *et al.*, 2004). However, as always, additives can compensate to some extent for lost hydrocarbons (Lai *et al.*, 1991). Synthetic molecules, such as high-molecular-weight esters and polyisobutylenes, have been developed to replace the base stock (Broun *et al.*, 1989).

Exhaust systems exhibit a remarkable predilection for storing the hydrocarbons emitted by two-stroke engines: a reflection, perhaps, of the large molecular masses involved. Particulate deposits extracted from a muffler, unsurprisingly, contain oil-derived hydrocarbons of the highest boiling points (Yashiro, 1987); the retention mechanisms for these are unreported, but, should oil droplets form *within* the exhaust system, these all too probably are captured by inertial impaction. The hydrocarbons from which visible smoke forms possess *intermediate* boiling points, logically, because they remain in the gas phase, until entering the exhaust plume; the lightest hydrocarbons resolutely remain in the gas phase, to be measured as HC. The temperature of the exhaust gas, as decided by the distance from the engine and the duty cycle, controls this threefold partitioning between gas phase, smoke plume and wall deposit.

Short-lived emission events, then, are not so much an aspect of the engine: hydrocarbon storage by the exhaust system (especially the muffler) makes smoke a particular characteristic of, and indeed nuisance during, acceleration, wherein rapid increases in temperature and in flow rate promote abrupt evaporation and release (Yashiro and Takahashi, 1991), and this causes, briefly, a massive upsurge in saturation, with nucleation following unavoidably. The extremity of a smoke burst depends, also, on the duration of the preceding storage. To ensure test consistency, purging or preconditioning procedures are essential, otherwise the exhaust system enters the test in an undefined state. Acceleration-related smoke bursts are ameliorated through discouraging this deposition, either by curbing the high-boiling-point hydrocarbons in the oil or by raising the temperature in the muffler (Yashiro and Takahashi, 1991). The impact of catalytic converters is variable: there may be little demographic effect (Czerwinski *et al.*, 2002), although others report a halving of particulate mass (Palke and Tyo, 1999).

In fractionation terms, the fact that organic particulate predominates will be obvious: it is readily apparent in the *colour* of the filter deposit (not to be confused with the colour of the airborne smoke), which is quite different from that found with diesel engines, being light-coloured, and tan or brown (White *et al.*, 1991), or grey–brown and yellow (Kojima *et al.*, 2002), rather than black. This light-coloured smoke aerosol, which adsorbs on filter fibres as oleaginous liquid films rather than as discrete particles, is >95 % (Sakai *et al.*, 1999) or 98–99 % (Kojima *et al.*, 2002) organic-extractable. That said, some operating conditions – such as following a cold start, when the choke forces considerable enrichment of the air–fuel ratio – may still generate black smoke, dominated by carbonaceous particulate (Palke and Tyo, 1999).

All-encompassing statements about particulate mass are impossible, since two-stroke engines enjoy such wide-ranging applications: metrics such as 'grams per km' are hardly relevant to chainsaws, and the duty cycle of a lawnmower is not that of a leaf-blower (Volckens *et al.*, 2007). This means that comparisons are not easily drawn: it just depends; but the problem faced by regulators is another topic. To speak, therefore, with sweeping generality, two-strokes tend to discharge particulate mass at rates comparable to diesel engines in the pre-emission-control era (Carroll *et al.*, 2000). That said, the aforementioned technological innovations have evidently been effective: when evaluated in the WMTC, a carburettored, two-stroke motorcycle emitted at 80 mg/km; another, with electronic fuel injection, emitted at 10 mg/km (Etheridge *et al.*, 2003). In-use 'three-wheelers', of the type commonplace in the developing world, not infrequently emit at several hundred milligrams per kilometre (Kojima *et al.*, 2002).

As one would expect, nucleation-mode particles are well represented – sometimes, even, below 10 nm (Cromas and Ghandhi, 2004). But intense, and sometimes more dominant, accumulation-mode particles are also seen. As the particulate emission contains little material that is solid or carbonaceous, the logical inference is that, following nucleation, vigorous agglomeration creates liquid-phase accumulation-mode particles. This understanding seems borne out in Figure 8.7(a): as OFR increases, the particle emission shifts from the nucleation mode into an accumulation mode (Czerwinski *et al.*, 2005). Figure 8.7(b) compares the number distributions for two scooters, one two-stroke, the other four-stroke (Czerwinski *et al.*, 2002): the size-range envelope displayed is similar for the two engines, and both distributions are monomodal; this finding is not immediately obvious, since solid particles were more preponderant with the four-stroke scooter. Moreover, particle numbers were orders-of-magnitude greater for the *two-stroke* scooter. The total counts are comparable to the present-day diesel engine, although possibly not one fitted with a DPF. For example, a carburettored motorcycle with catalyst emitted, on the WMTC, at $\sim 3 \times 10^{13}$/km; and a fuel-injected motorcycle without catalyst emitted, on the WMTC, at $\sim 10^{13}$/km, and, on the NEDC, $\sim 2 \times 10^{14}$/km (Andersson *et al.*, 2003).

Organic-fraction bioassays give positive results (Carroll *et al.*, 2000) in which mutagenicity appears to be controlled by PAH (Laanti *et al.*, 2001) – perversely, perhaps, when these compounds are predominantly *gasoline*-derived: this suggests inactivity in the considerably greater mass of lubricant-derived organic compounds, but the adjunctive importance, if any, of pyrosynthesised PAH goes unreported. Catalytic converters and synthetic oils are effective countermeasures to this mutagenicity

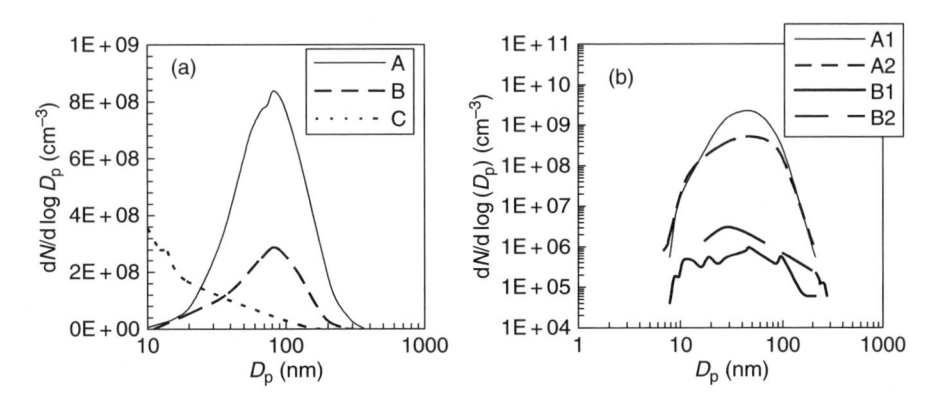

Figure 8.7 Demographic characterisations for two-stroke engines. (a) Effect of oil – fuel ratio (OFR): A, twice nominal; B, nominal; C, one-third nominal. Vehicle: scooter, 50 cm³, direct-injection, oxidation catalyst, model year 2002. Oil: 6250 ppm sulphur. Duty cycle: 30 kph; hot dilution (150 °C) (Czerwinski *et al.*, 2005). (b) Number distributions for A, two-stroke scooter (50 cm³, maximum power 2.5 kW at 6800 rpm, model year 2000), and B, four-stroke scooter (124 cm³, maximum power 8.5 kW at 9000 rpm, model year 1999). Aftertreatment: 1, with catalyst; 2, without catalyst. Vehicle speed: 30 kph. Instrument: SPMS (Czerwinski *et al.*, 2002).

(Sakai *et al.*, 1999). Intriguingly, the mutagenicity with two-stroke engines is reportedly lower than with diesel engines, and, on the basis that less NO_x is also emitted, one might presuppose some restriction on the conversion of PAH to NPAH (Section 3.6.2) (Sakai *et al.*, 1999). The implications of direct injection and of a stratified charge for mutagenicity are also unclear: the late-phase burn-up of hydrocarbons might well, on occasion, be less effective, but, from a converse standpoint, greater precision is available in oil metering (Wasil *et al.*, 2004). Unfortunately, insufficient investigations have been mounted to reach any definitive conclusions about any these widely ramifying talking points.

8.6 Closure

This chapter surveyed the particulate emissions of three types of gasoline engine: port-injection, direct-injection and two-stroke. The port-injection engine is found in almost all of today's gasoline automobiles. The direct-injection engine, while not yet widely commercialised, is currently being intensively researched with this expectation. The two-stroke engine is omnipresent for utility purposes, and also in mopeds and motorbikes.

8.6.1 Port-injection Engines

Formation

Port-injection engines are low emitters of particulate, but only during steady-state driving, cruises or light loads; they become prolific emitters during brief periods of driving requiring fuel enrichment, such as cold starts, hard acceleration or full load. Even when they are operating on globally stoichiometric or lean air–fuel ratios, inadequate fuel vaporisation allows droplets to enter the cylinder, and these droplets furnish the locally rich conditions required for soot production. Fuel pools within the inlet ports are a generic limitation; the accumulation and depletion of these pools, according to engine operation, interferes with proper fuel management. Rich air–fuel ratios are sometimes required to ensure acceptable driveability; the soot is strongly dependent on the degree of enrichment, with exponents as large as *eight* reported. The greater volatility of gasoline, relative to diesel fuel, implies a lesser contribution to the organic fraction. Instead, this fraction is suspected to consist predominantly of unburned lubricant. This emission, which changes little with air–fuel ratio, is thought driven by the burned-gas temperature, and the evaporation of oil from the cylinder wall.

Characterisation

Definitive statements about physical and chemical characteristics are not possible, since gasoline-engine particulate emissions are not nearly so well delineated in the literature as diesel-engine particulate. Accumulation-mode particles are agglomerates of relatively monodisperse spherules. The colour of the particulate deposit on the dilution tunnel filter is black (carbonaceous) at rich air–fuel ratios and yellow–white at lean air–fuel ratios (organic). The organic fraction is mutagenic, albeit emitted at low rates compared with diesel engines. On statutory test cycles, vehicles emit particulate mass at one or two milligrams per kilometre, but, for more aggressive duty cycles with higher loads and harsher accelerations, figures reach several milligrams per kilometre. This mass is distributed fairly evenly across the size spectrum. On statutory test cycles, particle numbers are on the order of $10^{12}-10^{13}$/km; there is wide variation. Instantaneous measurements show particle numbers to increase during acceleration, and then to fall back once the cruise condition has been reached.

Abatement

Emission control has not traditionally been directed at particulate emissions with this form of engine, but certain benefits are available though efforts to control HC, CO and NO_x. Reductions in fuel sulphur help to control sulphate particulate. Of greater impact is conversion to natural gas, as the improved fuel – air mixing virtually eliminates soot. With gasoline, the strategy is to improve the homogeneity of the charge (avoiding the entry of liquid fuel drops into the combustion chamber). The cold-start period must be passed through as swiftly as possible to limit emissions of organic and carbonaceous particulate. To obtain the best conversion efficiencies in the three-way catalyst, air–fuel ratios are being controlled with greater precision, closer to stoichiometry, with benefit for emissions of soot (in rich excursions).

8.6.2 Direct-injection Engines

Formation

The stratified or inhomogeneous charge of the direct-injection gasoline engine provides diesel-like conditions favouring the generation of soot. In fact, the emissions fall between those of diesel engines and port-injection gasoline engines. Switching between homogeneous and stratified charge at mid-load engenders a step change in the nature of the particulate emission. Some fuel strikes the piston, and accumulates as liquid films or pools; these pools ignite and burn with sooty flames. Some of this soot burns in a secondary combustion and so is not emitted; the flow field can promote this burn-up. The reason why pools survive on the piston is thought to be because of the Leidenfrost effect, i.e. the formation of an insulating layer of fuel vapour between the liquid and the piston surface that impedes heat transfer. The poor combustion in pool fires is also responsible for organic particulate, derived either directly from the fuel or from its pyrolysis. The problems more generally are overmixing, undermixing and insufficient temperature.

Characterisation

Particle demography is governed by the switch in operating mode between homogeneous charge and stratified charge. On statutory test cycles, vehicles emit 1–15 mg/km; when low figures are seen, these may relate to instances where the stratified-charge operation envelope is smaller. Mass emissions decline at higher engine load, commensurate with homogeneous operation. The mass is present as an accumulation mode, although the coarse mode is nonnegligible. On statutory test cycles, particle numbers are $\sim 10^{13}$/km; more highly loaded cycles generate greater numbers and a more prominent nucleation mode, logically, because there is less carbonaceous surface area available when in homogeneous mode.

Abatement

The central abatement effort is directed at soot, for example through spray geometry and flow field. The soot formed initially may aid a later combustion or burn-up phase. Experiments with spark timing and fuel-injection timing show that accumulation-mode soot and nucleation-mode volatiles are affected in diametrically opposed ways. The oxidising exhaust in stratified mode favours the formation of sulphates. The conditions for regeneration of a particulate filter are unfavourable: in stratified mode, there is oxygen but temperatures are too low; the converse situation occurs in homogeneous mode.

8.6.3 Two-stroke Engines

Two-stroke engines are notorious for the emission of white or grey smoke, colours which signify a strongly organic, rather than carbonaceous nature. These engines are designed to burn a mixture of fuel and oil, and much of this mixture can escape uncombusted. The heavy hydrocarbons from the oil then condense into a grey aerosol. The oil – fuel ratio is an important factor, but any reduction in oil consumption must be accompanied by design changes to reduce the lubrication requirement. Alternatively, the oil formulation can be adjusted so that the hydrocarbons are more susceptible to burn-up. Long-term storage takes place in the exhaust system, with subsequent release of these deposits in sudden bursts; this considerably exacerbates the smoke. Preconditioning procedures are therefore essential in evaluations of the particle emission. There are substantial amounts of nucleation-mode particles; these seem to agglomerate into liquid-phase accumulation-mode particles. The organic fraction is mutagenic.

9
Disintegration

9.1 Introduction

Material disintegration happens through the combined action of thermal and mechanical stresses, corrosion, abrasion and erosion: taken together, these processes are described colloquially as 'wear and tear'. From the standpoint of composition or of origin, there seems little reason to suspect these particles of any lesser diversity than others generated by the combustion process. The information assembled under this 'disintegrative' banner, therefore, is disparate, and arguably represents a cat's cradle of loose ends; indeed, as far as the present author is aware, a coherent review of particles generated by material disintegration for which motor vehicles can be safely castigated has not hitherto appeared.

The sources, then, are disparate, but several distinct categories are evident. The boldest evidence that material disintegration has taken place resides simply in surface recession. Saliently, there is the intriguing question of what happens to so much tyre material: where does it all go? Recession in the tyre surface is obvious to any vehicle user, if only because tyres must be periodically replaced. Brake linings also recede, as do the surfaces of roads on which vehicles travel. But not all disintegrative processes are associated with the action of one surface on another: particles of disintegration are also released by exhaust systems and, more especially, by catalysts; these are somewhat less studied, but this tide is currently reversing. All of these particles, and many others besides (unconnected with motor vehicles), reside in the detritus found on every road surface.

What chiefly distinguishes the disintegrative process from the combustion process is not so much particle composition as particle *size*: the broken-off debris resides primarily in the *coarse mode*. Because of their large sizes, these particles are more readily rejected by the respiratory tract; they are, moreover, considerably more prone to gravitational settling, hence their comparatively limited itinerancy, from which one might suspect fewer implications for public health. However, unlike other coarse-mode particles in the atmospheric aerosol, which are disturbed only by the wind, road dust is repeatedly raised through continual agitation by passing vehicles. It should, moreover, be noted that this raised road dust contains particulate matter such as soot and organic compounds discharged originally from tailpipes.

Once particles have had opportunity to mix freely with the wider atmospheric aerosol, apportionment of responsibility is much more difficult. One way around this problem is to compare the environment within road tunnels with that outside, in the open air. Some of these studies are quite long-standing, and there seems no reason for revising the belief that, by mass, most particles associated with motor vehicles are heavy and settle rapidly, outweighing by, say, two orders of magnitude those light enough to remain airborne (Pierson and Brachaczek, 1976). Of course, as we have learned, the mass of a collection of particles is hardly the same index as the number of particles. But, importantly, this rapidly settled mass is not emitted by tailpipes. This leads us to the notion that particulate generated by wear and particulate generated by combustion are quite distinct entities.

9.2 Roads

Constant pounding of road surfaces by motor vehicles generates a never-ending supply of road-wear particles, the dangers of which to the engine, if inducted with the air, were recognised in the earliest days (Summers, 1925). Aside from debris derived specifically from roads, this dust carries a variety of other, nonroad material, i.e. tyre-wear particles; spent combustion-generated particles such as soot, PAH (Johnsen *et al.*, 2006) and lead (Chow *et al.*, 2003); road salt (Owega *et al.*, 2004); and last but not least, vegetative remains, especially during leaf fall in autumn.

The flotsam and jetsam are continually entrained and pulverised between road surfaces and the wheels of passing vehicles. Much of this dust settles readily; but since passing vehicles repeatedly raise it, gradual accumulation occurs at points where it is less agitated, such as on road shoulders, alongside kerbs and on central reservations. A large part of the coarse mode in the urban environment is probably composed of repeatedly resuspended particles, particularly through the action of heavy-duty traffic, since rates of road-surface abrasion do not seem to account for what is airborne (Charron and Harrison, 2005). For example, construction work in the vicinity, unrelated to motor traffic, may actually generate the particles that vehicles repeatedly disturb.

Wear patterns are various yet well understood: fundamentally, the degradation is cyclic, as particles act as grinding material that promotes wear, leading to yet more particles and more grinding. In winter months, this wear is significantly exacerbated by studded tyres and traction sand (Kupiainen *et al.*, 2005); in damp conditions, dust is sequestered until released by the return of dry weather; other wear-controlling factors are the frequency of street cleaning undertaken by local authorities, the presence of nearby construction sites or farmland, and the course taken by rainwater run-off.

The wear resistance of road surfaces, especially in terms of mineralogy, is quantified under simulated conditions in the laboratory, with a view to improving resistance to fragmentation (Kupiainen *et al.*, 2005). Airborne wear particles emitted from various anti-skid and asphalt aggregates were typically a few microns in size (Räisänen *et al.*, 2003). Traction sand, via the 'sandpaper' effect, greatly increased, and studded tyres doubled, the particle emission; but the first seemed relatively uninfluential when combined with the second (Kupiainen *et al.*, 2003).

A paramount distinction exists between *unpaved* and *paved* roads (Chow *et al.*, 2003): in the case of the former, the prolific production of dust plumes, predominantly of mineral particles, is notorious: this concerns not just public roads, but any activities in which overlying vegetation has been stripped, such as opencast mines and construction sites, where there are obvious implications for occupational health and machinery maintenance alike. But even in the case of paved roads, the public health implications of dust raised in heavily trafficked areas cannot be ignored; hence, the settled dust, especially in gutters, is periodically collected by municipal authorities with purpose-built trucks. Damp conditions temporarily restrict this resuspension (Lough *et al.*, 2005); the dust gradually worsens with time from the previous rainfall, and unpaved roads at least have the advantage of drying out more slowly (Kuhns *et al.*, 2003).

Given so many controlling factors, closely defined figures for the quantity of dust raised by passing vehicles cannot be universally applicable; and it is unclear, moreover, how 'mass per unit distance' translates into 'mass per unit volume', the usual air quality metric (although many particles are larger than the commonly used PM10 designation). From kerbside sampling on public roads, encompassing high- and low-speed driving, PM10 and PM2.5 from passenger cars were respectively 140–780 mg/km and 2–25 mg/km; and from heavy-goods vehicles, they were respectively 230–7800 mg/km and 15–30 mg/km (Abu-Allaban et al., 2003). From a laboratory test facility simulating low-speed driving, the contribution to PM10 without traction sand was 10–20 mg/km, and with traction sand (900 g/m^2), 30–50 mg/kg (Kupiainen et al., 2003).

Direct measurements of road dust are made, traditionally, through collection: by vacuuming and sweeping the road surface. But these methods are labour-intensive, and do not quantify that fraction which is readily raised. An alternative method is to mount an optical probe immediately behind the tyre of a moving vehicle (Etyemezian et al., 2003b), thus measuring, in *real time*, by light scattering, the particles raised from the road surface during actual driving. The mass of raised particles, as a function of vehicle speed v, follows a polynomial expression

$$M = av^b \tag{9.1}$$

where a and b are determined by regression (Etyemezian et al., 2003a), and M is actually a difference, since the signal from a similar device mounted in front of the vehicle, measuring the particle background, must be subtracted. Some results are depicted in Figure 9.1 (Etyemezian et al., 2003a). In this analysis, the exponent b was a function of particle size, being ∼3 for the less-settled particles (<4 μm, PM2.5, PM10), and declining to 2 or even 1 for much larger, easily settled particles.

The contributions made by high-speed and low-speed roads to PM10 seemed of the same order, once total traffic flow was taken into account. This apparatus, moreover, led to an unexpected conclusion (Etyemezian et al., 2003b). For a distance of one kilometre on a 3 m-wide lane, i.e. an area of 3000 m^2, carrying 1000 vehicles per day, the particle turnover was 0.3–3 g/m^2/day. Assuming a dust loading of <3 mg/m^2, for a particle size of <75 μm, and assuming also that 10 % of this dust is PM10, this leads to residence times of just a few hours. Thus it seems that settled PM10 is completely replaced

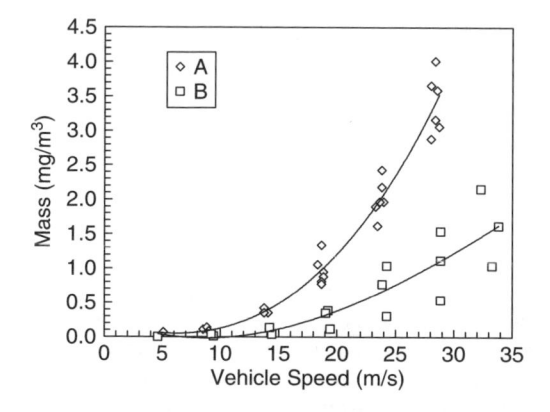

Figure 9.1 The dust (PM10) raised from paved roads by a passing vehicle. A: Treasure Valley, ID, USA. B: Fort Bliss, TX, USA. Instrument: light scattering, calibrated with Arizona Road Dust, range 0.001–150 mg/m^3; frequency 1 Hz (Etyemezian et al., 2003a).

several times in one day. Should this rapid turnover be the case, a regular and proximate source of replenishment must exist: perhaps dust is disturbed by vehicles travelling slightly outside their lanes, or by the turbulent wakes from trucks. This suggests that lane cleaning would have only short-term benefits, and that a wiser approach might be to eliminate particle reservoirs in kerbs, gutters and shoulders.

Mere quantification, though, along these lines, although indicating the environmental burden, is nevertheless insufficient for source apportionment, which really requires measurements of chemical composition: these reveal what fraction of dust comes directly from the road, what fraction from motor vehicles, and what fraction from elsewhere. The following sections discuss the second category; the third is not the immediate concern of this text. Most (95 %) of the asphalt is mineral (stone); the rest is bitumen (Lindgren, 1996). Bitumen and diesel-combustion products were equally represented in suspended particles collected at a busy road (Moriyoshi *et al.*, 2002); road tar ($>C_{50}$) is chromatographically distinguishable from lubricating oil (Neukom *et al.*, 2002). The bitumen is composed of four generic groups, saturates, aromatics, resins and asphaltenes, the latter of which are solid black particles, surrounded by the resins and dispersed into an oily matrix (García-Morales *et al.*, 2006). The compounds are mainly high-molecular-weight aromatic hydrocarbons (in the region of 1000 to 100 000 amu), for which the carbon–hydrogen atomic ratio is ~0.8–0.9; there are also, at concentrations of 1–4 wt %, sulphur, nitrogen and oxygen, and mineral grains containing sodium, potassium, calcium (limestone fragments), aluminium and silicon (Nord A.G., 2000). The ratio of calcium to aluminium in street dusts derived from asphalt pavements is several times that in average soil (Kadowaki, 2000).

This discussion of road dust provides a suitable moment to return to the ash fraction of the particulate emission, from the tailpipe, that is; as whatever enters the engine along with the inducted air presumably stands some chance of being re-emitted. This is evident if only because an engine whose intake air is deliberately seeded with particles of aluminium oxide discharges these same particles in its exhaust stream (Stevenson, 1984). So, the ratios of silicon, iron and calcium discharged at the tailpipe are consistent with the ingestion of road dust by the engine (Cadle *et al.* 1999a); and, as sodium is characteristic of sea-salt ingestion by vehicles driven in coastal areas (Pattas *et al.*, 1992), the same ingestion presumably happens in winter months because of road de-icing salt. *Historically*, the focus on such ingested debris has been engine wear (Truhan *et al.*, 1995), as this debris (silica, quartz and iron) possesses a high degree of hardness (Poon *et al.*, 1997). As a particulate *emission*, ingested dust is, even today, a small player – unless, that is, the air filter has failed structurally, or the operating environment of the vehicle happens to be exceptionally dirty. The introduction of exhaust-mounted particulate filters (Section 7.8.2) represents a newer vulnerability, not only because the dust adds to the incombustible ash, but also because some elements (sodium especially) undergo embrittling reactions with the substrate.

9.3 Brakes

A braking system consists of a pad and a metallic disc, the former being a soft polymer matrix impregnated with small, hard particles or fibres. During braking, the hard particles embedded in the pad wear the disc, generating additional particles which play adjunctive roles in friction and wear (Ostermeyer, 2003), and which, because they are loosely held, escape into the environment.

The controlling factors in the sizes and morphologies of these particles are, first, wear *modes* (i.e. adhesion, fatigue, delamination and thermal degradation): for example, sheet-like particles imply delamination and fatigue; small aspect ratios imply abrasion (Mosleh *et al.*, 2004).[1] Second, the braking

---

[1] Note: this paper was reviewed 'in press' in 2003.

style (i.e. the pressure and duration of the contact) is a factor (Talib *et al.*, 2003). Agglomeration and fragmentation occur concurrently, and the particle composition, by virtue of frictional heating and consequent tribochemical reactions, can differ from that of the progenitor material.

Brake-wear particles inherently reflect, compositionally, the evolution of brake materials. The longstanding historical concern was asbestos (Cha *et al.*, 1983), present in brake linings, prior to the 1980s, to about 10–70 % (Spurny, 2000d). The chrysotile form was a major component, but heat energy generated by friction during braking converted this into olivine or forsterite (Jacko *et al.*, 1973). Asbestos has now been replaced by organic, semimetallic and metallic materials. Three characteristic metals are Cu, Ba and Sb, of which the last, as Sb(III), is suspected of carcinogenicity, and the second, at concentrations of <40 %, is commonly used in fillers. Other elements found in brake-wear dust are Si, S, Al, Fe, Mo and C (Mosleh *et al.*, 2004). Of course, none of these elements is unique to brakes; and chemical compounds, while probably unique, and so exploitable as markers to some extent, are poorly reported, pointing, probably, to the proprietary nature of braking materials. Found in road tunnels, and ascribed to brake-wear debris, are Sb_2S_3 and $BaSO_4$ (Sternbeck *et al.*, 2002); other metal sulphides such as Cu_2S and PbS might also be present (Gudmand-Høyer *et al.*, 1999). The elements Fe, Cu, Si, Ba, K and Ti are seemingly unsegregated with respect to particle size, and reflect those in the parent material (Sanders *et al.*, 2003).

The particles, although able to escape from the immediate vicinity, are not necessarily airborne: deposition occurs on internal surfaces, such as on wheel rims, and also on road surfaces; dampness (relative humidity) is thought a factor in retention, but particle size and local air currents are plausible adjuncts. Because of these other particle destinations, the total mass as determined from measurements of the recessed surface does not reveal the mass escaping as airborne particles; comprehensive inventories require purpose-built laboratory test rigs (Jacko *et al.*, 1973). The airflow on an engine dynamometer, though, is unrepresentative, and so unlikely to reveal the real suspension rate; a wind tunnel provides a controlled environment, and mounting probes under the wheel arches of moving vehicles is an alternative (Sanders *et al.*, 2003).

The total loss from vehicle brakes in a large city may exceed 100 tonnes annually (Nord A.G., 2000); and in the USA, the annual loss of asbestos from brakes and clutches was once estimated at ~33 000 tonnes – but most of this settled quickly on roads or deposited within braking systems: that fraction able to remain airborne was small, 3.7 % (Jacko *et al.*, 1973). A recent estimate obtained in a wind tunnel put the airborne fraction at one-half (Sanders *et al.*, 2003). Estimates obtained via kerbside sampling on public roads of the brake-wear contribution to PM10 and PM2.5, per light-duty vehicle, were <80 mg/km and <5 mg/km, respectively, and, per heavy-duty truck, <610 mg/km and <5 mg/km, respectively (Abu-Allaban *et al.*, 2003).

It seems logical to expect some correlation between brake-wear particles and braking *frequency*; for example, brake dust is strongly represented at freeway exits where vehicles decelerate (Abu-Allaban *et al.*, 2003). Direct observations of brake-light illumination or less direct observations of traffic density provide similar quantitative information. In road tunnels, a correlation was found between airborne Cu and traffic density (vehicles per second) (Sternbeck *et al.*, 2002); two outliers related to damp days, when particles are thought to adhere more avidly to wheel rims. Cu and Ba correlated, but not Cu and Sb; perhaps the latter elements are not necessarily found together in brake linings (Sternbeck *et al.*, 2002). A correlation was found between Ba and the preponderance of heavy-duty traffic (no Ba smoke-suppressing fuel additive was expected).

Several demographic characterisations of brake-wear particles, given in Figure 9.2, show accumulation modes (350 nm) and coarse modes (2–15 μm) of comparable prominence (Mosleh *et al.*, 2004). Two aspects of these data are intriguing. First, coarse-mode particle sizes are strong functions of braking style; indeed, continuous braking and high contact pressures seemed to facilitate growth or, rather, to favour agglomeration over fragmentation. The accumulation modes, by contrast,

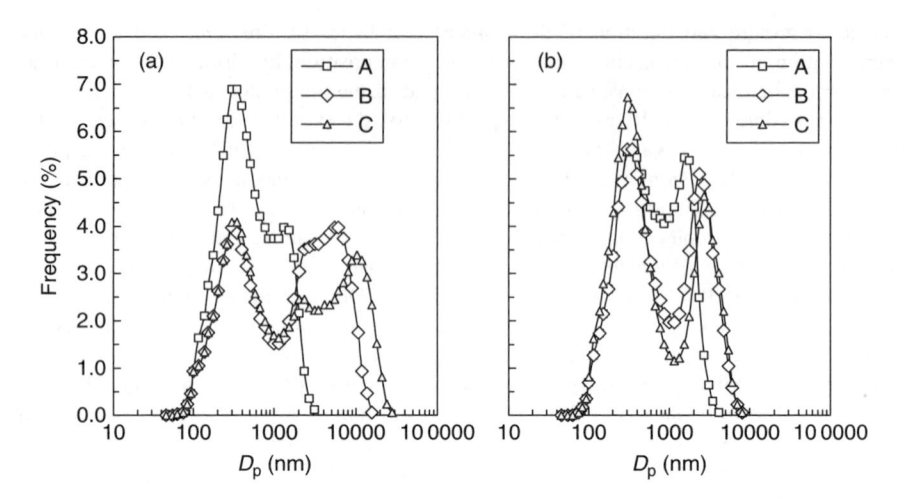

Figure 9.2 Wear particles generated during braking: (a) low-speed, continuous (sliding speed 0.275 m/s); (b) low-speed, discontinuous (sliding speed 0.275 m/s). Contact pressures (MPa): A, 0.125; B, 0.375; C, 0.625. Commercial truck brake pad, pressed against cast iron. Relative humidity, 70–77 %; air temperature, 20–24°C. Instrument: laser scattering, 40 nm–262 μm (Mosleh *et al.*, 2004).

were relatively indifferent to braking style. Secondly, accumulation modes are unexpected, insofar as these do not, like coarse modes – according to classical notions of aerosols, that is – stem from material disintegration. Conjecturally, frictional heating brought about decomposition, and the resulting emission of some unknown vapour, the nucleation of which formed nanoparticles, and the agglomeration of which formed the accumulation modes. Measurements of particle composition pointed to the brake pad as the parent material for these accumulation modes. Others find increased numbers of submicron particles when brakes exceed 500°C, suggesting accclerating material breakdown (Sanders *et al.*, 2003); and significant amounts of particulate mass at sizes below 100 nm are found, the emission of which continues between braking events if temperatures remain hot enough (Garg *et al.*, 2000). But, as always, the implications here for public exposure need further study, since nucleation events are so much dependent on the interplay of dilution and saturation.

9.4 Tyres

In fractional terms, ~10 % of the tyre and ~30 % of the tyre tread are reportedly lost in the form of wear particles, corresponding to ~800 g for automobile tyres and ~8000 g for truck tyres (Camatini *et al.*, 2001). The *rate* of wear, though, is not constant, but rather is a strong function of driving behaviour: logically, it is highest for the greatest tread forces, i.e. during acceleration, deceleration and manoeuvring. Wear rates compiled from the literature show wide variations, ranging from 10 μg/km for gentle driving, through 70 μg/km for aggressive driving, to 490 μg/km for rapid cornering (various references cited by Councell *et al.*, 2004). Of course, these are instantaneous rates, and what counts is the driving pattern taken as a whole, in which case 25–50 μg/km seems typical. This lost mass, estimated dimensionally using tread recession, agrees quite satisfactorily with direct gravimetric measurements (Councell *et al.*, 2004).

Tyre-wear particles are partly organic and partly inorganic. The greater proportion of the composition is taken up by carbon black, embedded in an organic, elastomeric polymer consisting of styrene–butadiene rubber (SBR), polybutadiene and natural rubber (Camatini *et al.*, 2001). SBR, being a compound characteristic of tyres, is used as a tracer in source apportionment studies: airborne particles within a road tunnel, and deposits scraped from various surfaces within this tunnel contained <2.5 % SBR (Pierson and Brachaczek, 1976). Organic compounds from tyre wear are chromatographically distinguishable: molecular weights above 600 are less represented than in diesel particulate or road bitumen (Moriyoshi *et al.*, 2002). Particles also carry a range of inorganic elements, of which only a salient few will be mentioned. Less usually, tungsten is generally indicative of studded tyres (Sternbeck *et al.*, 2002); and more characteristically, vulcanised rubber contains sulphur, a cross-linking agent; zinc, an activator for the sulphur (Heideman, 2004); and silicon, as a reinforcing filler. Zinc and sulphur, in combination, are characteristic of tyres, and can be used for source apportionment studies (Councell *et al.*, 2004).

Studies of tyre-wear particles in road simulators show that interactions between studded or nonstudded tyres and quartzite or granite pavements are significant sources of *submicron* particles, including nanoparticles, and, on the basis that demographics are determined by the type of tyre rather than the type of pavement, the former is adduced as the source (Dahl *et al.*, 2006).[2] Some number distributions are given in Figure 9.3(a). With *studded* tyres, the overall shape of the distribution was unchanged by vehicle speed, but particle numbers increased, and there was little evidence of a nucleation mode. With *nonstudded* tyres, a nucleation mode appeared, the magnitude of which was a strong function of vehicle speed. When the particles generated by the studded tyres were subjected

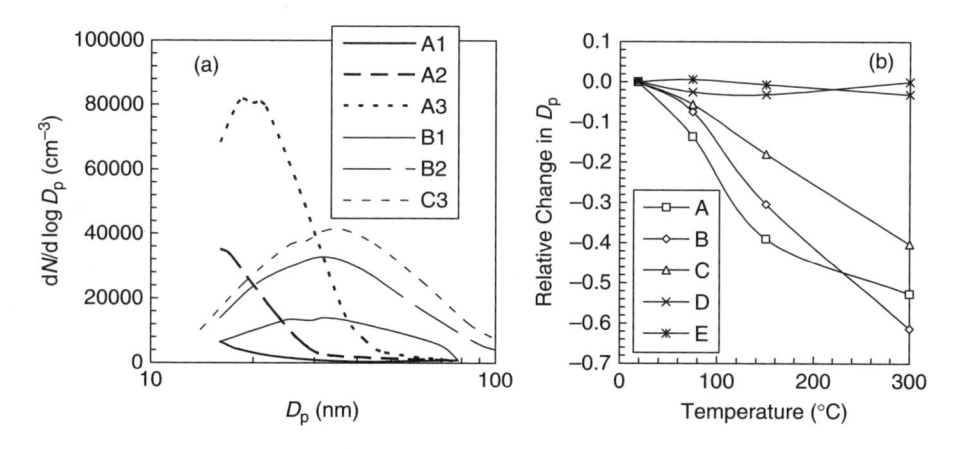

Figure 9.3 Particles generated at a pavement – tyre interface in a road simulator, four wheels negotiating a circular track. (a) Number distribution. Vehicle speed (kph): 1, 30; 2, 50; 3, 70. Pavements and tyres: A, nonstudded and quartzite; B, studded and quartzite: C, studded and granite. The original data points have been omitted. (b) Relative change in particle size, following passage through a thermodesorber. Size (nm): A, 40; B, 80; C, 160; D, 320; E, 640. Instrument: SMPS, 15–700 nm (Dahl *et al.*, 2006).

---

[2] Note: this paper was reviewed 'in press' in 2005, i.e. prior to acceptance by the journal. A speedy re-perusal of the finalised version, on the eve of publishing this book, shows the basic tenets of the paper to be the same as originally stated.

to heat treatments, the smallest, according to Figure 9.3(b), were found to be volatile, and the largest nonvolatile.

The appearance of accumulation-mode particles and, especially, of nanoparticles is surprising in view of the expected emission of coarse-mode particles through what is known about wear and attrition of surfaces. These smaller particles have, perhaps, been overlooked because of historical obsessions with particulate mass. It does seem plausible, though, that when tyres are hot, certain material, perhaps organic in nature, vaporises; and equally plausible that this outgassed material subsequently nucleates. Vehicle speed would then be a critical factor, inasmuch as this determines the temperature in the tyre, through the rate of work done in elastic compression and deformation. If this is the case, tyres release not only coarse-mode particles through mechanical stress, but also nucleation-mode particles through thermal stress, with subsequent agglomeration into the accumulation mode. Emission rates estimated for particles of size 15–700 nm are 3.7×10^{11}/km at 50 kph and 3.2×10^{12}/km at 70 kph (Dahl *et al.*, 2006). These figures are comparable with particle emission rates at the tailpipe for conventional diesel vehicles, and, for the cleanest vehicles, such as those fitted with particle traps, the road – tyre interface may become a significant contributor to the total particle burden in heavily trafficked areas. But the general significance of these nanoparticles is unknown, since nucleation is very much driven by local saturation.

Tyre-wear particles, in the coarse mode, are readily distinguishable under the electron microscope. Topologically, they resemble the parent material: the surfaces are warped, and pock-marked with numerous pores or scales (Camatini *et al.*, 2001). Morphologically, they are monolithic – rather than consisting of agglomerations of smaller particles, that is – and elongated, with aspect ratios ranging from 2:1 to 5:1, and they are massive, with sizes of 20–40 μm (cited in Weingartner *et al.*, 1997) or 5–200 μm (Rauterberg-Wulff *et al.*, 1995). The newly discovered accumulation- and nucleation-mode particles display several morphologies, differing distinctly from the coarse mode, and also from one another: isolated, near-spherical particles; agglomerates of near-spherical particles; particles apparently containing occlusions; and particles seemingly in the liquid phase at the time of collection (Dahl *et al.*, 2006). Plausibly, the provenances differ in each case, but the ideas here are still conjectural: for example, the solidified, erstwhile liquid-phase particles may be fillers or mineral oils that have outgassed and nucleated, and the occlusions may be ZnO or ZnS.

The broader environmental concern with tyre-wear particles is not so much the rubber matrix as zinc, present in tyres to 0.5–1.5 % (Councell *et al.*, 2004), and concentrations of which in lake sediments are increasing, where, for example, the concentrations of lead (from leaded gasoline) are now measurably declining. The zinc thereby released to the environment is thought comparable to that from waste incineration, and to be surpassed only by zinc production; a correlation has been observed between zinc concentrations in watersheds and the intensity of neighbouring traffic (Councell *et al.*, 2004). For the USA in 1999, the amount of zinc annually released from tyres was estimated at a staggering 10 000 tonnes; such estimates are rising year on year. Considerable evidence exists that much of this zinc is not stable within host particles, but rather is *leachable*, by which it then becomes *bioavailable*: the element is taken up, for instance, by plants. The cytotoxicity of tyre-wear particles in bioassays is related not just to the organic component but also, in some way, to zinc (Gualtieri *et al.*, 2005); in fact, the toxicity of this element or its oxide, in particle form, to the lung is more generally noted (Amdur, 1996).

9.5 Exhausts

That exhaust systems release particles of rust and scale is common currency. Particle sizes are thoroughly in accordance with processes of disintegration, i.e. comfortably in the coarse mode; and since internal surfaces inevitably degrade with age, the mass of material larger than PM2.5, or even

PM10, may increase with mileage (Cadle et al., 1999a). Particle colours are also in accordance with what is expected of iron oxide (haematite, Fe_2O_3): dark red or brown (Hardenberg et al., 1987a), as commonly found on corroded iron or steel structures. Formation of this iron oxide probably induces microstresses, aiding destruction of the surface. The regular appearance of iron particles of several microns or more in size in urban dust (such as that deposited onto tree leaves; Davila et al., 2006) is plausibly enough ascribed to motor vehicles, but difficult to ascribe conclusively, because *direct* tailpipe measurements in the supermicron range are so seldom undertaken.

The causes of this emission are quite revealing. An association apparently exists between emissions of iron and transient driving modes, thus engendering an obvious supposition that surface encrustations are weakened by the stresses and strains imparted by rapid thermal cycling, leading to exfoliation (Beltzer, 1976). This release is particularly fostered, or mediated, by the co-presence of sulphuric acid, a notion consistent with the observation that emissions of iron suddenly became more prominent when oxidation catalysts were fitted to gasoline vehicles. The chemistry here was explained in Section 3.7: if the acid dew point is crossed within the exhaust system, the acid condenses onto exhaust system walls, or perhaps nucleates into droplets which impact on these walls. The metal walls, moreover, may themselves have catalytic effects promoting the conversion of SO_2 and SO_3 into H_2SO_4. The various metals in these surfaces – chiefly iron – are then attacked, creating a friable surface that eventually exfoliates, exposing virgin surface on which the process begins anew.

Emissions of iron and emissions of acid show interdependencies or divergent trends that are not immediately obvious (Beltzer, 1976). Higher emissions of iron in the presence of oxidation catalysts – which, after all, manufacture the acid – are intuitively correct; yet, puzzlingly, the fuel sulphur content was uninfluential. The acid, moreover, was predominantly emitted during cruises, or steady-state driving, whereas the iron was predominantly emitted during transient driving. This observation implies that the *availability* of iron at the wall of the exhaust system is rate-controlling, rather than the amount of sulphuric acid *per se,* or that what counts might be the availability of virgin surface; and this, in turn, depends on the rate at which corroded, friable material exfoliates.

Material losses from exhaust systems are not restricted to iron; a less usual emission, for example, is hafnium, an element attributed to specialised steels (McDonald J.D. et al., 2004). But little work is reported in this area other than for zinc, thought to be from the muffler, a source difficult to distinguish from fugitive lubricant, which contains the same element (Beltzer, 1976) – see Section 7.7.1. Interestingly, the two sources of zinc seemed to be comparable only in the *absence* of an oxidation catalyst; in the *presence* of such, the total emission of zinc was higher, by one order of magnitude, than what was conceivable from the escape of lubricant. Faced with this observation, it seems only reasonable to impeach sulphuric acid in the loss of zinc from the muffler, in a manner akin to iron. Supporting evidence for this conjecture is that calcium, another common oil constituent, was emitted at similar rates for catalyst and noncatalyst vehicles alike. As with iron, the zinc emission failed to correlate with fuel sulphur content, leading to similar suppositions about the availability of freshly exposed surface.

9.6 Catalysts

In-service attrition in all catalytic converters engenders a quiescent particle emission. Some components vaporise at the highest temperatures, and perhaps subsequently nucleate, but fragmentation, or abrasion, seems the more dominant mechanism of attrition. Rapid thermal cycling and severe vibration foster ideal conditions for microcracks, and these microcracks compromise the structural integrity of the substrate, the washcoat, the precious metals, and the union between all three, leading, eventually, to particle release. This is where (present-day) monolithic catalysts are thought superior to (first-generation) pelleted catalysts (Beltzer et al., 1974). But examinations of aged monoliths still show evidence of abrasion under the microscope (Zotin et al., 2005). Purely in catalyst ageing terms, material attrition is probably small in relation to other mechanisms such as sintering and the deposition of poisons (Section 7.8.1). It is not, then, an aspect of catalyst performance, but it is undesirable for its wider environmental repercussions: platinum group metals, especially soluble platinum salts, cause allergic reactions, dermally, as well as inhalationally, i.e. asthma, rhinorrea and the condition known as 'platinosis' (Brubaker et al., 1975). The detection and quantification of platinum group metals at trace levels in the environment poses particular problems to the analytical chemist (Bosch Ojeda and Sánchez Rojas, 2005).

Microcracking is not perhaps the sole mechanism: abrasion or erosion might also occur when catalysts are struck by particles carried by the feed gas – potentially important for locations close to the engine, where particle velocities, and hence kinetic energies, are higher. Surface erosion by particle impaction is well known in wear science generally (Clark H.M., 1995); corresponding experiments on cordierite monoliths, using bench-top apparatus and artificial particles (of aluminium oxide and silicon carbide) at two sizes (200 μm and 500 μm), have been reported (Day, 2001). Wear rates based on monolith weight loss using the larger particles were around twice those for the smaller; yet, in either case, *low* gas flows could produce more damage than high gas flows. The silicon carbide particles, of angular morphology, were more damaging than the aluminium oxide particles, which were rounded (Hampton et al., 2003). The cell walls at the entrance to the substrate were chipped, with breakage occurring, presumably, along grain boundaries. Fundamentally, the relative positioning of the substrate geometry with respect to the particle trajectory is what counts. But the relevance of these bench-top experiments to in-service erosion remains uncertain. The test particles are untypical, being far larger and harder than those likely to predominate in exhaust gas. The experiments were conducted at room temperature, whereas at typical exhaust temperatures, cordierite monoliths become stronger and more rigid, with likely modifications to surface–particle interactions.

Material lost from catalytic converters has been studied since the devices were first introduced: an early investigation reported that aluminium emissions increased sixfold on the introduction of catalysts; the element was presumably derived from the washcoat (Beltzer, 1976). In addition to precious metals (see below), washcoats and substrates, material is released from intumescent mats (Pattas et al., 1992), of an organic and inorganic nature, which relates to the curing process in new catalytic converters (Bagley et al., 1991). All in all, recently quoted rates for emissions of Pt and Pd are 10–15 ng/km from gasoline catalysts and 75–200 ng/km from diesel catalysts (Rauch et al., 2002); the reason for the discrepancy is not reported. Interestingly, an order-of-magnitude increase in the emission rate was observed in a road tunnel following passage of 1970s automobiles, demonstrating the importance of catalyst mileage or the earlier level of technology (Lough et al., 2005).

The chemical forms in which the precious metals are released are critical to bioavailability, and hence also to environmental impact. The fraction of water-soluble Pt is small (Shafer et al., 2006). Pt is primarily metallic, but attached to alumina in the washcoat (Tuit et al., 2000), perhaps in the form of nanocrystals (Barefoot, 1997); its association with ceria, a common washcoat component characteristic of gasoline catalysts, provides a useful tracer or fingerprint (Silva and Prather, 1997). The particles are, not unexpectedly, in the coarse mode, i.e. 1–5 μm; Pd is reportedly resident in slightly smaller particles than Pt. The precious metal group includes Ir, Os and Ru: these are also present in particles, but as impurities; the first two have been discovered in road dust recovered from tunnels (Rauch et al., 2004).

A growing body of evidence suggests that particles lost from catalytic converters are gradually accumulating in the environment; these sequestrations are observable in air, soil, road dust, the sedimentary record in rivers, and the bodies of aquatic invertebrates. Typical concentrations in urban air are Pt 5–25 pg/m^3, Pd 2–15 pg/m^3 and Rh 0.5–4 pg/m^3 (Rauch et al., 2005b): that is, about an order of magnitude higher than in the pre-catalyst era. Concentrations in air (Zereini et al., 2004) and soil (Cicchella *et al.*, 2003) correlate with traffic density, but a comparison of weekends with weekdays suggests meteorological factors also (Rauch *et al.*, 2005b). In soils, a background of 3 ng/g suggests no pollution by Pt prior to 1992, whereas 11 ng/g, by 2001, is considered noticeable (Cinti *et al.*, 2002). Soil samples collected from flower beds in an urban area contain concentrations of Pt and Pd at <2−52 ng/g and <10−110 ng/g, respectively (Cicchella *et al.*, 2003). (By comparison, lead was present in soils to the tune of 0.1−1 mg/g; Cinti *et al.*, 2002.) Other accumulations are road dust, 200 ng/g, and river sediments, 50 ng/g. At a typical urban site, the daily deposition rate of Pt in airborne dust was 23 ng/m^2 (Schäfer *et al.*, 1999); this is somewhat higher than would be expected, based on measurements conducted at vehicle tailpipes, and perhaps reflects other sources. A large fraction (>95%) seems, in fact, to be transported regionally or even globally, rather than being deposited in the immediate vicinity (Rauch *et al.*, 2005a).

Catalyst formulations are not disclosed for proprietary reasons; they are, however, always changing, in order to optimise cost and performance, and they vary also across vehicle models; this seems to be subtly reflected in particle compositions. A gradual shift from Pt to Pd throughout the 1990s corresponds nicely to changes in catalyst formulations undertaken during the same period; and the Pt/Pd/Rh ratio reflects that in catalysts (Rauch *et al.*, 2002). As shown in Figure 9.4 (Cicchella *et al.*, 2003), the ratios of Pt to Pd correspond quite closely to what one would expect if these metals had been discharged by catalytic converters – a conclusion reached by others (Rauch *et al.*, 2005b). This is also the case for Pt, Pd and Rh in lake sediments, for which concentrations appear now to be levelling off, in a manner tracking the penetration into the market of catalyst-equipped vehicles (Rauch *et al.*, 2004). It does not follow, of course, that a particle's composition represents the catalyst formulation *as a whole*: the precious metals are nonuniformly distributed; for example, where one metal lies predominantly at the surface of the washcoat, another might lie deep inside. This inhomogeneity leads to a 'nugget' effect in subsequent chemical analysis, and confounds inventories.

An emerging issue is the release of osmium, an element with no intended function in catalytic converters, but present, as said, as an impurity along with other precious metals. There are reasonable grounds for supposing the release of osmium as OsO_4, which, being more volatile at typical operating temperatures than other precious metals and compounds thereof, is preferentially lost (Poirer and

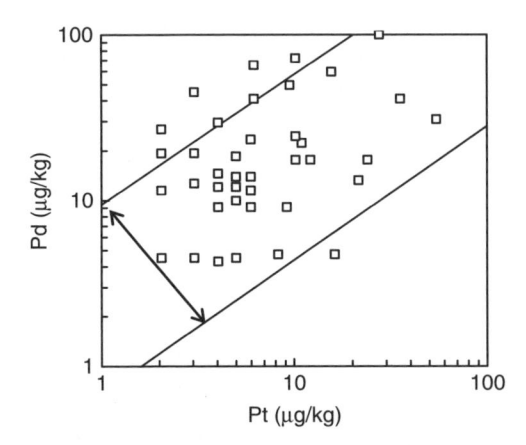

Figure 9.4 Correlation between palladium and platinum, as found in soils taken from nonfertilised flower beds established in 1992 in a nonresidential urban area (Naples, Italy). The area enveloped by the two lines signifies the ranges expected from typical catalyst formulations (Cicchella *et al.*, 2003).

Gariépy, 2005). The airborne oxide is probably converted subsequently to the metal or the dioxide. That Os fails to correlate with Pt, Pd or Rh implies differing emission mechanisms or environmental behaviour (Rauch *et al.*, 2005b). On a large scale this release does not seem too significant, but locally, in densely trafficked areas, catalytic converters probably represent an important anthropogenic source of osmium.

9.7 Closure

The present chapter examined a range of particles for which motor vehicles are held responsible: these are *not* combustion-derived, but generated instead by various processes of attrition or material disintegration. Nature, as it were, favours the break-up of bulk matter into particulate matter. That such particles have gone substantially less reported is notable. This is perhaps because of the relative superabundance and omnipresence of combustion-derived particles, but also because, being in the coarse mode, particles from material disintegration inevitably travel shorter distances from the road, and are also less respirable. It should be noted that, in great contrast to particles emitted by tailpipes, very little information exists as to the toxicological implications of these wear particles.

9.7.1 Roads

All road surfaces are covered with dust of a diverse nature. Some particles are from the road itself; some are from vehicles (tyres, brakes, exhausts and catalysts); and some are exogenous (e.g. vegetative). Dust is far worse on unpaved roads. A grinding action between wheels and roads causes cyclic degradation of the surface. Repeated disturbance by passing vehicles is responsible for dust accumulation on kerbs, etc. Other controlling factors are studded tyres, street cleaning and rainwater run-off. Estimates of g/km are highly context-dependent. Road erosion is simulated in the laboratory in order to identify mineralogies less susceptible to dust formation. The quantity of dust has traditionally been determined by vacuuming or sweeping. Real-time measurements, using light scattering, show that the mass of dust raised from the road surface is a straightforward polynomial function of vehicle speed. Recent evaluations imply that the residence time of road dust is only a few hours; the turnover mechanism is probably agitation by passing vehicles. Chromatography can distinguish between road-surface bitumen and diesel particulate. Dust entrained into vehicle intakes can be emitted from the tailpipe as inorganic particulate; this particularly concerns silica.

9.7.2 Brakes

Particles are generated by wear between a brake pad and a brake disc; their nature is dependent on the mode of wear (e.g. fatigue or delamination) and the pattern of braking (e.g. the contact pressure and stop – start nature). The particles may be retained in the housing of the braking system, settle onto the road or become airborne. Local air currents, particle size and dampness decide this fate. Metals are well represented in brake-wear particles, especially Sb, Cu and Ba. Correlations have been demonstrated between brake-wear particles and braking frequency. Measurements of size distributions show equally dominant accumulation and coarse modes. The size of the coarse-mode particles is a function of contact pressure and braking pattern (continuous/discontinuous): this could reflect the relative rates of agglomeration and fragmentation at the rubbing surfaces.

9.7.3 Tyres

The average tyre on a light-duty vehicle loses about 800 g over its usable life, or about 10 % of the tyre weight. Accelerating, decelerating and manoeuvring promote tyre wear. Tyre-wear particles are both inorganic and organic. They have warped surfaces with numerous pores. The organic component contains styrene – butadiene rubber (SBR), natural rubber and polybutadiene. Characteristic inorganic elements are Zn, Si and S. Studded tyres release W. Chromatography can easily distinguish between tyre-wear particles and combustion particles. Recent research strongly suggests outgassing of volatile material from the tyre, with the consequent formation of nucleation- and accumulation-mode particles; these particles show various morphologies and may, on the latest vehicles, be generated at rates comparable to emissions from the tailpipe.

9.7.4 Exhausts

Rust and scale form on all internal surfaces of exhaust systems; these encrustations are then released owing to rapid thermal cycling and vibration. The particles are usually dark red or brown. The degradation is exacerbated by sulphuric acid, which may condense onto metal walls. Experimental investigations imply that the availability of iron, rather than the amount of sulphuric acid, is what counts; encrustations must first exfoliate to expose virgin surface. The muffler is suspected to release zinc; this, again, is probably provoked by sulphuric acid.

9.7.5 Catalysts

Material lost from catalytic converters forms a slowly growing field of research; this concern arises not so much from performance degradation as from the wider environmental implications. The interest lies more in precious metals (Pt, Pd and Rh) than in the accompanying washcoat compounds such as alumina and ceria, the cordierite substrate, or the intumescent mat. Rapid thermal cycling and vibration weaken the unity between support, washcoat and precious metals. Some volatilisation at high temperature, followed by subsequent nucleation, is not ruled out. It is suspected that impaction by particles arriving in the feed gas causes additional attrition; bench experiments have shown that the leading edge of the monolith becomes chipped, but the particles used were unrepresentative of those in the exhaust. Direct measurements of precious metal releases from catalysts show emission rates of nanograms per kilometre. Accumulation of these metals is undoubtedly taking place in the environment (in soils and river sediments). Current concentrations in the urban atmosphere are a few pg/m^3; this is an order of magnitude higher than in the pre-catalyst era. The concentrations and relative ratios of precious metals in the environment follow the number of catalyst-equipped vehicles and the traffic density, respectively. But a catalyst does not necessarily lose particles in a manner that reflects its formulation in its entirety.

10
Toxicology

10.1 Introduction

Strictly within the framework of today's statutory regulations, particulate mass, and *only* particulate mass, as captured by the dilution tunnel filter, is what counts. But since this quantity is so equivocally and so ambivalently related to particle *sizes*, then the protection afforded to members of the public by such legislation is open to debate. Occupational health in this respect differs noticeably from public health: for example, concerning the notoriety of asbestos, Spurny (2000d) writes:

> Pollution measurements made in terms of the mass of asbestos [per unit volume] of air are useless, as they include thick, non-respirable and minute non-carcinogenic fibres.

To illustrate further, consider a bowl of fruit in which there rest nine grapes and one apple (Hinds, 1982, p. 78). This bowl, *by number*, is 90 % grapes, and *by mass*, 90 % apples: thus, the fructuosity of the impression is quite different depending on whether the orchard or the vineyard is adopted as the metric. Similarly, with aerosols and the mass metric, a few large particles (the apples) easily dominate many small particles (the grapes) – an understanding clearly evident in Figure 2.17. Hence, the scarcity or abundance of nanoparticles emitted by internal combustion engines goes mostly unrecorded in current statutory practice (Mayer A. et al., 1998).

From this it is but a short step to supposing that the fraction of particles able to reach the deep lung, expressed in number terms, is far higher than when expressed in mass terms: large particles are less respirable than small particles, and the deftness with which nanoparticles reach the pulmonary region is well known (Figure 2.12). It is thus hardly a counter-intuitive proposition that the number of particles rather than the mass of particles, per unit volume of air, should correlate with the prevalence and extent of respiratory disease.

Following deposition in the respiratory tract, a welter of diseases may arise, amongst which, in the biomedical field, concerns have shifted somewhat over the years. Traditionally, the disquiet over motor vehicle particulate or, more specifically, diesel engine particulate concerned cancer; and, not unexpectedly, the pulmonary system received the greatest attention, but other organs were also implicated, such the bladder. More recently, i.e. into the 1990s, a sea change took place in which cardiovascular and cardiopulmonary complaints rose to prominence. And, in contemporary society, at least in the developed world, a virtual epidemic has arisen in asthma – especially childhood asthma. Of course, particles, if inhaled, do not in all cases bear sole responsibility: they might simply exacerbate diseases with unrelated aetiologies, or allow opportunistic infections such as pneumonia to take hold (Dockery and Pope, 1996).

Epidemiological studies of the general population repeatedly find positive and statistically significant associations between day-to-day fluctuations in ambient particles and various health statistics, e.g. hospitalisations for respiratory or cardiovascular conditions; these associations remain, at the time of writing, unexplained, and some scientists are still suspicious, for example, of the statistical methods employed; nor is an association between two parameters conclusive proof of causality, i.e. a firm scientific corollary linking cause to effect. Batteries of tests, both *in vivo* and *in vitro*, show a range of effects on biological systems. The mutagenicity and cytotoxicity towards bacteria, although well demonstrated, are of uncertain relevance to humans; and other bioassays, more representative of mammalian biology, are under development. Laboratory animals invariably exhibit adverse health effects; but here, the issue is whether the high exposures[1] – much higher than the human experience, even in the most polluted environments – generate unrepresentative health outcomes.

Inhaled particles and human health are, in totality, a vast subject; more than five hundred scientific studies were published between 1996 and 2000 (AML, 2000). The present review will, however, be more modest in scope, and encompass material relating more specifically to motor vehicle particulate. This is a compromise because, as already detailed extensively in Chapter 2, the atmospheric aerosol is a cocktail, although in urban environments, motor vehicles are undoubtedly a large ingredient – probably the largest. For example, in expectation of a substantial diesel passenger car market, a comprehensive review of the biomedical literature was once published in the USA (Griffin *et al.*, 1981). It concluded that much was not known; and although much has been elucidated since, definitive answers are unlikely to emerge for some time yet.

10.2 Public Exposure

Legislators and policymakers seek to safeguard public health by establishing certain expectations, or 'standards', for ambient air quality – framed in terms of maximum recommended exposures. Such standards are expressed in two ways, depending on the timescale of exposure, a strategy which aims to capture the dual impact on public health of short-lived and long-lived pollution, i.e. acute and chronic effects. By 2010 in the EU, annual levels of PM10 should not exceed $20\,\mu g/m^3$, and daily levels should not exceed, on more than seven days of the year, $50\,\mu g/m^3$ (Maynard and Cameron, 2001). Table 10.1 quotes the long-term and short-term US National Ambient Air Quality Standard (NAAQS) for PM2.5 and PM10. (The figures in this table are currently under review.) In view of what is to be related, it is perhaps worth observing that, at one time, concentrations of $75\,\mu g/m^3$ (Barth, 1971) or even $250\,\mu g/m^3$ (Maynard, 1999) were once deemed to provide comfortable margins of safety.

---

[1] Strictly speaking, a distinction exists between 'exposure', relating to the concentration of a particular pollutant in the immediate environment, and 'dose', meaning the concentration received at any point in the body, such as in the respiratory tract or wherever the damage occurs (Bates and Watson, 1988). For example, a person exercising vigorously, and so breathing rapidly and deeply, will likely receive in their lung tissue a higher dose than will a sedentary person residing in the same environment, but both individuals have the same exposure (Sexton and Ryan, 1988). The two terms are, however, used more loosely in the present text.

Table 10.1 National Ambient Air Quality Standards (NAAQS) adopted in the USA (EPA, 2004).

Size class	Concentration (24 hour)	Concentration (annual)
PM2.5	<65 µg/m^3	<15 µg/m^3
PM10	<150 µg/m^3	<50 µg/m^3

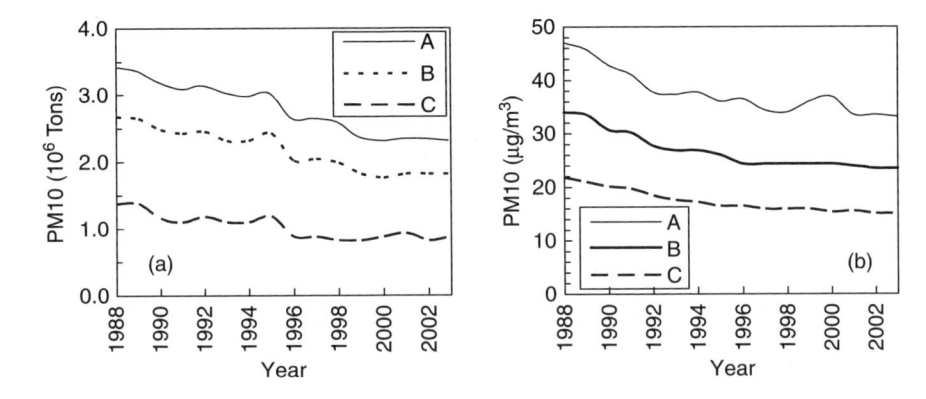

Figure 10.1 USA, national trends in PM10 emissions. (a) As estimated from *direct* emissions: A, transport; B, industrial; C, fossil fuel combustion. (NB: the EPA refined its estimation methods in 1996). (b) Measurements of air quality taken from 434 monitoring sites: A, 90 % of sites below this line; B, average; C, 90 % of sites above this line (EPA, 2004).

Regulatory authorities routinely monitor particles in the ambient atmosphere: from the US record, up to 2003, there is good news, and there is not-so-good news. The good news is that, *nationally*, measurements of PM2.5 were at their lowest since monitoring began in 1999, and measurements of PM10 were at their second lowest since monitoring began in 1988. Figure 10.1 shows yearly estimates of the PM10 inventory, corroborated by actual measurements: evidently, the downward trend is no mere artefact of accounting methods. These improvements have been brought about, not just by more effective emission control in motor vehicles: other measures include fuel reformulation in all types of fossil fuel combustion, wiser practices in agriculture, paving of unpaved roads, and treatment of flue gases discharged by coal-fired power stations.

This, however, as said, is the national picture; the not-so-good news is that, *regionally*, PM10 exceedances in the USA continue to affect 21 million people, and PM2.5 exceedances continue to affect 53 million people (EPA, 2004), so there is no room for complacency. Presumably, motor vehicles are to blame in part for these regional exceedances, although the pollution inventory is not reviewed here in detail. Of course, locally high particle burdens are experienced in industrial and manufacturing centres (Morawska *et al.*, 2006), and even in bucolic or rustic settings; what are germane are *urban* areas, where motor vehicles are always found at high densities; and for these areas, the need to integrate air-quality management into municipal transport policy is now increasingly recognised (Parkhurst, 2004). On still smaller scales, i.e. 'microenvironments', the action of street canyons in containing and channelling vehicle pollution (Jiménez-Horrnero *et al.*, 2007) and thereby exacerbating public exposure (Samson, 1988) is well known.

Regional variations in public exposure follow regional variations in the vehicle fleet: for example, in Europe, sales of light-duty diesels have grown inexorably in the last decade, to the point where, in some countries, they now threaten to oust gasoline engines as the preferred *automobile* power plant – unlike in the US market, where they still have only a toehold. In the USA, then, an interesting question is whether the lower particulate emissions per gasoline engine are, in fact, negated by the considerably greater presence of gasoline automobiles. The majority of motor vehicles and vehicle-miles travelled are light-duty, and some research indicates that gasoline engines might actually be the majority contributor to ambient PM10 (Gertler, 2005). This possibility is difficult to assess because diesels have always received the greater attention, although, admittedly, this situation is now changing. In the US, the annual emission of particulate attributable to diesel vehicles is allegedly 10^8 kg (cited by Sidhu *et al.*, 2001); and, although heavy-duty diesels are, on the roads, proportionally minuscule (2 %), they nevertheless reportedly dominate the inventory (65 %) (Jacobs *et al.*, 1998), annually emitting, in the state of California, some 27×10^6 kg (cited in Walsh, 2001). However, regions still exist where gasoline engines do seem the larger contributor, especially at weekends (Kim E. *et al.*, 2005b), owing, not implausibly, to partial cessation of goods transport by heavy-duty trucks, during which time particles become slightly less carbonaceous and slightly more organic (Harley *et al.*, 2005).

Some heavy-duty diesels are more equal than others: buses, through their close proximity to people, their preponderance in certain congested urban areas and their start–stop duty cycles, represent the greater part of *public* exposure (Walsh, 1985), whereas in goods transport, the all-pervasive presence of heavy-duty diesel trucks seems unassailable, and, while switching (back!) this transport from road to rail will no doubt transfer the particulate emissions from diesel trucks to diesel locomotives (Jørgensen and Sorenson, 1998), public exposures in urban areas seem unlikely to benefit if such transitions are restricted to inter-city haulage. Heavy-duty diesels are, equally obviously, responsible for elevated *occupational* exposures: in wide-ranging off-road applications particularly, where maintenance is carried out with widely varying degrees of assiduity, these engines have attained great notoriety as emitters of soot (Samaras and Zierock, 1995). Indeed, diesel engines of all shapes and sizes are found in numerous stationary, semi-stationary or off-road applications, including the ubiquitous generator set, which generates particles as well as electricity (Liu Z. *et al.*, 2005). Regrettably, workplace exposures also include habitual engine operation in *enclosed* environments, where these particles are dispersed less readily.

Reference to enclosed environments is a timely reminder of the central distinction, in air quality, between exposure *indoors* and exposure *outdoors* (Harrison P.T.C., 1998). Less obviously, individuals stir up 'personal clouds', inside of which particle concentrations are increased, i.e. static measurements in unoccupied environments fail to capture the continual disturbance humans impart to settled particles (Yakovleva *et al.*, 1999). *Personal* exposure, therefore – which should not be confused with indoor or outdoor exposure (Monn, 2001) – is only measurable by equipping subjects, throughout the course of their day, with portable air samplers (Sexton and Ryan, 1988; Mark, 1998). Such a strategy also captures the effects of lotions, perfumes, etc., the application of which to the face or hair creates further sources of particles in the near-head region (Corsi *et al.*, 2007).

Particles are adept at penetrating building envelopes:[2] PM10 and PM2.5 found inside buildings in fact originate, to several tens of per cent, from outdoor sources (Özkaynak and Spengler, 1996), while *indoor* soot and PAH reflect, quite closely, traffic densities in the immediate neighbourhood (Fischer *et al.*, 2000; Funasaka *et al.*, 2000), so that 'indoor air is outdoor air' (Ormstad *et al.*, 1997). Coarse-mode particles are less able to penetrate building envelopes than accumulation-mode particles (Özkaynak and Spengler, 1996). Logically, this just reflects how weakly the latter interact with bulk matter: a curious irony, as staying indoors and closing the windows – as often enough advised during

---

[2] Not forgetting that persons themselves (or rather their clothes) act as vehicles for particle transport into buildings.

pollution episodes – can be a more effective countermeasure against gases than against particles, if the building fabric efficiently adsorbs these gases upon passage of the air through niches in the building envelope. Because most infiltration studies have used mass and not number, penetration by nucleation-mode particles is less clear: these do, however, seem more readily filtered than the accumulation mode (Zhu Y. et al., 2005), and this accords with the received understanding of diffusional deposition. Probably for these reasons, a considerable fraction of indoor particles are, in fact, *soot*: fluctuations in light absorption, inside and outside buildings, correlate closely with fluctuations outdoors, but with a lower amplitude, and a lag of an hour or so, while the advisability or inadvisability of window ventilation depends on the time of day (Horvath H. et al., 1997). Of course, in some buildings, particle ingress may simply be aided by air-conditioning systems lacking appropriate filtration systems (Ekberg, 1995).

Occasions when indoor PM is poorly correlated to outdoor PM (Yakovleva et al., 1999) are ascribable to major sources of indoor particles, three obvious activities being vacuuming, cooking (especially frying) and cigarette smoking – the first-mentioned is, logically, responsible for resuspension; the ability of the other two to generate prodigious numbers of nanoparticles should be noted. During cooking, PM2.5 reaches a couple of hundred micrograms per cubic meter, and a thousand or so if food is burned (Olson and Burke, 2006); during frying, particle numbers of size <100 nm reach 10^5 cm^{-3} (Wallace et al., 2004). Less obvious are secondary aerosols formed by reactions between ozone and chemicals released from building materials (Aoki and Tanabe, 2007) or household cleaning fluids (Sarwar and Corsi, 2007). Indoor exposures must be carefully assessed since, if they are significant in overall health terms and predominantly responsible for episodic peaks (Jones N.C. et al., 2000), then little will be accomplished by controlling outdoor exposures. A converse argument is that epidemiological studies have highlighted sulphates (Section 10.5), a particulate fraction which largely originates outdoors.

Corresponding, if rather less appreciated, health risks are carried by air entering the cabins or compartments of motor vehicles. Aside from obvious occupational exposures for truck drivers and so forth, lengthy commutes are nowadays a common experience for many persons. CARB has estimated that although an average person spends 6% of his time inside vehicles, this period accounts for around one-third of all particle exposure (cited by Edgar et al., 2003). Vehicles, when driven on busy roads, in effect pass through 'pollution tunnels', or, perhaps more accurately, pollution columns, inside of which the occupants inhale particles from preceding vehicles – thus bringing a new meaning to the phrase 'keeping a safe distance'. Moreover, in the phenomenon of 'self-pollution', occupants inhale particles from their own vehicles, depending on the external flow field (and whether the exhaust system has leaks). Thus, it may be immaterial whether the vehicle windows are open or closed (Marr et al., 2004). In some markets, the absence of closed venting systems for crankcase emissions may actually be the more significant source of occupant exposure (Clark N.N. et al., 2007). Air filters installed in cabin air intakes have been traditionally designed with a view to retaining coarse-mode particles (e.g. pollen and road dust): filters with greater submicron efficiencies (e.g. using electrostatic charge) are under development (Frédéric et al., 2007), and the degree of protection against vehicle-associated nanoparticles (see below) should be particularly ascertained. Taking typical lifestyles, a regular commute of an hour or so on a busy freeway may account for 10–50% of daily exposure to traffic-related nanoparticles (Zhu Y. et al., 2007).

Studies of airborne particles in school buses[3] have proven somewhat controversial, and are still disputed (Marshall and Behrentz, 2006; Valberg and Long, 2006), but the effects of cabin ventilation are not always intuitive (Kuritz et al., 2004): the highest particle concentrations were observed with the windows closed; fully open and half-open windows were equivalent; no significant interactions were found with vehicle speed; and, finally, particle concentrations were lower at idle with the

---

[3] Of the design prevalent in the USA.

windows closed, but also lower when driving with the windows open. School buses, it is argued, are perhaps more effective than other vehicles at delivering their own emissions to the occupants, for whom particle inhalation during short bus rides might actually exceed the daily inhalation pertaining elsewhere (Marshall and Behrentz, 2005). This underscores the difference between (and different cost-effectiveness of) two strategies: avoiding particle inhalation and restricting particle emission.

A relatively unaddressed aspect of public exposure is particle *ageing*, i.e. during suspension in the atmosphere (Section 4.3). Particles act as carriers of other pollution by accreting gas-phase material; undergo mutagenicity-modifying photochemical reactions in their organic coatings; and gradually increase in hygroscopicity, with implications for particle deposition in the atmosphere and also in the moisture-laden respiratory tract. Capillary action and surface tension in accreted liquid films force carbonaceous agglomerates into tighter, more compacted structures (Kütz and Schmidt-Ott, 1992); on this basis, freshly emitted particles, by virtue of their significantly more curlicued or convoluted structures, are thought to be more effective carriers of volatile compounds and more deeply penetrating into the lung than their aged counterparts (Dye *et al.*, 1998).

The particle concentrations people are likely to inhale courtesy of internal combustion engines is not a straightforward question to answer, since ambient particles are so diversely sourced; 'source apportionment' techniques, by which the contribution from motor vehicles is extricated from the pollutant mix, and diesel vehicles are divorced from gasoline vehicles, cannot be discussed here. In heavily trafficked urban areas, a significant fraction of the air quality standard for PM10 is consumed by diesels, say $5-10\,\mu g/m^3$ (according to various citations by Cook and Richards, 2002). Source apportionment is less of an issue in occupational exposure, as there are fewer particle sources: for example, inside a bus station containing 50 buses during the morning warm-up period, PM10 was $\sim 200\,\mu g/m^3$ (Briedé *et al.*, 2005), and inside a locomotive repair shop, $150-200\,\mu g/m^3$ (Huang W. *et al.*, 2007). Highway construction personnel working in a tunnel experienced elemental carbon concentrations of $4-178\,\mu g/m^3$ (Blute *et al.*, 1999). Miners, and other workers in underground construction sites, appear to experience the highest exposures to diesel particulate: $\sim 1-2\,mg/m^3$ (Cantrell and Watts, 1997; Haney *et al.*, 1997; Pratt *et al.*, 1997), to which cutting operations add further (admittedly coarse-mode) particles (Dahmann and Bauer, 1997). Peak concentrations of diesel particulate reached $1\,mg/m^3$ for some crew members of ships and ferries (Ulfvarson *et al.*, 1987). In traffic, passing diesel vehicles can generate emission spikes of several seconds' duration, during which concentrations reach a few hundred micrograms per cubic metre (Walsh *et al.*, 2005).

10.2.1 Nanoparticles

Studies of public exposure expressed in terms of particle *numbers* – especially in the nanorange – are far fewer than for mass, but at the time of writing are multiplying with great rapidity (Pitz *et al.*, 2003). Examples, with particle sizes stated in parentheses, are as follows. (**A**) An urban 'background', $2-3 \times 10^4\,cm^{-3}$ ($\gtrsim 7\,nm$) (Harrison R.M. and Jones, 2005). (**B**) At the kerbside of a street canyon, carrying 80 000 vehicles per day, $1.2 \times 10^5\,cm^{-3}$ ($\gtrsim 7\,nm$) (Harrison R.M. and Jones, 2005). (**C**) In the wintertime, 30 m from a highway, carrying 200–270 vehicles per minute, $4 \times 10^4\,cm^{-3}$ (6–12 nm) and $9 \times 10^2\,cm^{-3}$ (100–200 nm) (Zhu Y. *et al.*, 2004). (**D**) At the kerbside of a street canyon: morning, carrying 1300 vehicles per hour, $2 \times 10^4\,cm^{-3}$; afternoon, carrying 1400 vehicles per hour, $1.2 \times 10^4\,cm^{-3}$; nighttime, carrying 50 vehicles per hour, $4 \times 10^3\,cm^{-3}$ (18–50 nm) (Imhof *et al.*, 2005). From these and various other studies, the particle concentrations per cubic centimetre experienced by many city dwellers would, on a regular basis, appear to fall somewhere between the fourth and fifth orders of magnitude – and predominantly for sizes below 100 nm.

Detailed measurements have been made of diurnal variations in particle numbers, segregated according to size: one such recording, made in an urban setting, is depicted in Figure 10.2(a) (Jeong

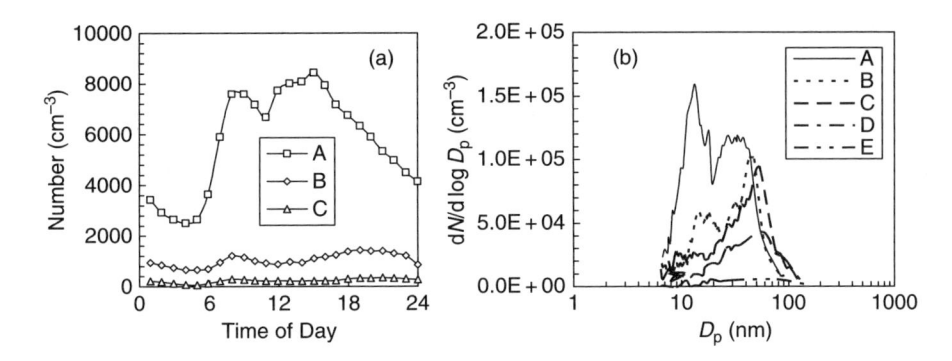

Figure 10.2 Particles in the ambient. (a) Diurnal variations during 2003 at Rochester, NY, USA, expressed as number, and in three size ranges: A, 11–50 nm; B, 50–100 nm; C, 100–470 nm. The location was approximately 50 m from major roads, atop a 10 m-high building. December 2001 to December 2002. Standard deviations in the original have been omitted (Jeong *et al.*, 2004). (b) Number distributions in proximity to a freeway passing through Los Angeles, CA, at various distances: A, 30 m; B, 60 m; C, 90 m; D, 150 m; E, upwind. Summertime; 200–270 vehicles per minute; 5 % heavy-duty trucks (Zhu Y. *et al.*, 2004).

et al., 2004). Whereas the largest measured size range (100–470 nm) was comparatively staid throughout the course of the whole day, the smallest measured size range (11–50 nm) varied severalfold, peaking somewhere in the morning rush hour, 8.00am–9.00am, and then again in the afternoon, 1.00pm–5.00pm, before declining to its lowest ebb in the earliest hours. The morning peak seemed to relate directly to the number of motor vehicles on the roads. The afternoon peak was of less certain provenance, i.e. less easily ascribed to the population of motor vehicles. This difference betrays the action of so-called nucleation 'events', during which atmospheric processes manufacture nanoparticles from their gas-phase particle precursors (Zhang Q. *et al.*, 2004). Similar measurements also showed how number distributions shift towards larger particles with distance from a road, until, at 300 m, particle numbers approached those of the background; see Figure 10.2(b) (Zhu Y. *et al.*, 2004). The nanoparticles were particularly prominent within a few metres of the road.

As one might expect from studies already cited (Section 4.3), the population of nanoparticles in the atmosphere very much depends on the prevailing meteorology and climatology, i.e. temperature and relative humidity (saturation and gas-to-particle conversion), solar intensity (photochemical reactions and the creation of new particle precursors), and wind currents (mixing and pollutant dispersal) (Kerminen *et al.*, 2007; Olivares *et al.*, 2007). Inversion layers, which occur more frequently at night and in the wintertime, are well known to restrict pollutant dispersal: these can raise the saturation of volatile particle precursors. Clearly, many extraneous factors have a hand in determining the appearance of nanoparticles (Li X.L. *et al.*, 2007), so that abundance does not automatically scale in any obvious way with traffic density (Zhu Y. *et al.*, 2006). But, generally speaking, the emerging understanding is that although motor vehicles are strong sources of nanoparticles, an impish duality is displayed, that is, some nanoparticles are primary and others secondary. Indeed, the boundary between nanoparticles 'as-emitted' by tailpipes or formed immediately thereafter within exhaust plumes, and nanoparticles formed much later, through ongoing atmospheric reactions or fluctuations in temperature and humidity, is hugely problematical.

Clearly, persons who drive regularly on heavily trafficked roads in vehicles with inadequate cabin ventilation systems are exposed to such nanoparticles. But one cannot, it seems, argue, on the grounds that lifetimes in this mode are so short – agglomeration being so rapid, and evaporation on dispersal being a distinct possibility – that zones of exposure are restricted to the immediate vicinity of busy roads. Nanoparticles are formed wherever natural atmospheric processes have access to the relevant gas-phase particle precursors. A reservoir of undetected particle kernels (<3 nm) probably exists at all times and in many areas; observations are linked to the growth of such kernels into fully fledged particles, whereafter detection is possible (Holmes, 2007). Primary particles may, it is true, be restricted to the vicinity of busy roads; but the implications for public health of secondary particles are less clear, as these may form elsewhere, whenever natural processes are infelicitously furnished with the prerequisite conditions.

The chemical composition of these nanoparticles is an equally critical question for public health. According to studies already cited (Section 3.7.2), it appears plausible that nanoparticles, at least those for which motor vehicles are *immediately* responsible, consist of sulphate kernels, around which organic compounds have gathered. Sulphates, for example, reside disproportionately in the nucleation mode (Gertler et al., 2002, p. 29). Thus, freshly emitted nanoparticles may well be acidic: yet subsequent neutralisation by ammonia seems to be rapid, requiring a few tens of seconds (Grose et al., 2006). The composition of *later-forming* nanoparticles is less clear, as various possible nucleating agents are found in the atmosphere – not necessarily, it should be said, those discharged by motor vehicles. For example, nucleation events may well be strongly associated with the prevalence of SO_2; but this gas can travel, also, from coal-fired power stations (Jeong et al., 2004). Other possible participants in nucleation events are nitrates, ammonia and secondary organics (Zhang Q. et al., 2004). Nanoparticles collected from the ambient atmosphere consisted mainly of organic compounds, but there were some sulphates, and iron was the most prominent transition metal (Hughes L.S. et al., 1998).

Pivotally, these nanoparticles are the reason why correlations between number concentration and mass concentration (as PM2.5 or PM10) are poor or nonexistent (Monn, 2001; Jeong et al., 2004; Harrison R.M. and Jones, 2005). This observation will not be a startling revelation for those already familiar with the current text. The explanation is obvious enough from Figure 2.17: simply, such small particles, although present in large numbers, possess, on aggregate, little mass. For example, ambient particles in the range 17–100 nm were present at number concentrations of 10^4 cm^{-3}, yet their corresponding mass concentrations were only ~1μg/m^3 (Hughes L.S. et al., 1998) – comfortably or, rather, uncomfortably below current air quality standards for PM2.5 and PM10. From this, it is difficult to avoid concluding that large fluctuations in the abundance of ambient nanoparticles pass largely unobserved by current mass-based assessments of air quality.

10.3 Public Health

The health risk *traditionally* associated with diesel engines is cancer, for which the long-suspected carcinogenic agents, or 'carcinogens', are the particulate-phase organic compounds. As generally defined, a carcinogen is what increases the incidence of malignant tumours, but, unfortunately, this black-and-white distinction is inadequate to the vagaries of cancer. Hence, four classifications for an agent, or suspected agent, have been defined (Rushton, 2001), paraphrased thus: (i) it is carcinogenic to humans; (ii) the degree of evidence for carcinogenicity in humans is almost sufficient; (iii) the carcinogenic risk to humans is not classifiable; and (iv) it is probably not carcinogenic to humans.

The International Agency for Research on Cancer (IARC) concluded in 1988 that there is sufficient evidence, in experimental animals, for the carcinogenicity of whole diesel exhaust; inadequate evidence, in experimental animals, for the carcinogenicity of filtered diesel exhaust; sufficient evidence, in experimental animals, for the carcinogenicity of extracts of diesel particles; and limited evidence, in humans, for the carcinogenicity of diesel exhaust (cited in Walsh and Bradow, 1991). The particulate emissions were assigned to group (ii), and designated 'probably carcinogenic'. Various national and international agencies have followed suit: these include the state of California, the US National

Institute for Occupational Health (NIOSH), the US National Institute for Environmental Health Sciences (NIEHS), the World Health Organization (WHO), the California Air Resources Board (CARB) and the US Environmental Protection Agency (EPA) (Walsh, 2001). Hence, diesel particulate or diesel exhaust is regularly branded a 'potential', 'possible', 'probable' or 'likely' human carcinogen: these qualifiers tend to suggest some equivocation, but they are not synonyms (although the semantics of legalese need not detain us). Such pronouncements are based on a mixture of toxicological and epidemiological data, especially that which relates specifically to the lung; the *quantification* of this risk is, however, still uncertain and controversial (Rogers A. and Davies, 2005).

For reasons to be examined presently, the focus in recent times has shifted from cancer to various respiratory and cardiovascular illnesses. The posited health risks do not concern so much healthy adults as 'at risk' groups, i.e. infants, pre-adolescent children (<13 years), the elderly (>65 years) and the infirm, i.e. those with pre-existing respiratory or cardiovascular diseases.[4] Thus, the difference between occupational health and public health concerns rather more than what is inhaled: the 'general public' includes a greater fraction of susceptible or vulnerable individuals. It could be, for example, that for those persons whose cardiovascular systems are already seriously compromised or impaired, particle inhalation is just sufficient to push the body to its tipping point. This advancement of death for those already about to die, albeit through other causes, is in biomedical circles inelegantly known as 'harvesting'. Thus, when concentrations of ambient particles briefly rise, life expectancies are correspondingly foreshortened. But if harvesting is truly in operation, an immediately following dip in death rates is to be expected; the fact that these dips are not always observed remains a puzzle (Wilson R., 1996). Still, one might argue that a shortening of lifespan by one week is less serious than by three months, which, in turn, is less serious than by several years; but one cannot put a figure on this advancement (Wilson R. and Spengler, 1996); and how to balance harvesting and premature deaths against the costs of pollution abatement is an ethical question outside the scope of this book.

Sundry other effects on the respiratory system are linked to motor vehicle pollution more generally, as distinct from particulate emissions specifically. This distinction is important, since ozone (O_3), nitrogen dioxide (NO_2) and sulphur dioxide (SO_2) are always part of the pollutant mix, but none of these species are exactly benign. These health effects encompass a temporary decrease in lung function (functional vital capacity (FVC) and forced expiratory volume (FEV)), eye irritation, laboured breathing, coughing, chest tightness, nasal discharge, prevalence of colds, and wheezing. Implicated diseases here are bronchitis and two prominent allergic reactions, rhinitis and asthma.

Interestingly, asthma in the developed world has in the last generation become a virtual plague, having increased in childhood by a factor of 50 % (BMA, 1997), whereas, in the undeveloped world, this illness continues to be virtually unknown. This trend cannot be ascribed entirely to shifts in reporting or diagnostic criteria; some shift in lifestyle or behaviour is strongly suspected. Motor vehicle pollution has been widely mooted as a possible explanation, although this is not by any means the only hyopothesis.[5] Investigations into childhood asthma sometimes do and sometimes do not point

---

[4] The point is that the epithelium of the immature lung is more permeable to pollutants, and children consume 50 % more air per unit of body weight than do adults (Schwartz, 2004). Lung function (FEV and FVC) attains a maximum in the early twenties, and thereafter declines, *inter alia* because of progressive inelasticity in lung tissue, becoming clinically apparent at 40 years; the rate is swifter for persons exposed to air pollution, and much swifter for smokers. This 'available reserve' serves as a reasonable predictor for mortality; indeed, a threshold is eventually crossed in old age, whereafter lung function becomes inadequate (Evans and Wolff, 1996).

[5] The alternative 'dirt theory' holds that modern-day high standards of domestic hygiene, which reduce infantile infections, also deprive the developing immune system of critical stimuli, resulting in sensitisation to allergens (Karol, 2002). In fact, humans have evolved with germs, and it could be that such infections are a necessary precondition for the immune system to develop properly.

to traffic pollution (Elliott and Briggs, 1999). There are logical grounds for supposing that pollutants sensitise the airways of asthmatics and rhinitics to inhaled allergens (of which domestic dust mites are perhaps the best-known example), and thus exacerbate the symptoms of the disease (Devalia *et al.*, 1996); and the ability of allergens to become airborne is paramount, for which suspended particles provide ideal vehicles. But it is unclear whether pollution could initiate asthma in healthy individuals or just aggravate the symptoms in persons already asthmatic.

A wide range of airborne allergens, or 'aeroallergens', relating to animals, insects and fungi, is found in the domestic sphere (Jones A.P., 1999). Such aeroallergens, asthma and particles are probably connected (Ormstad, 2000). Canine, feline and birch allergens are almost exclusively attached to one type of particle, namely *soot*, or the accumulation-mode carbonaceous agglomerates. These allergen-carrying particles are even found, albeit to lesser extents, in households with *no* canine occupant, thus demonstrating their peripatetic nature. But mite allergen was *not* found; this suggests a necessary precursive collision of allergen with airborne particles, possibly through the ejection of saliva droplets from the mouth of the pet. Road dust from sundry sources, e.g. soil, exhausts, tyres, brake linings and plant fragments, also contains pollen, pollen fragments, moulds and animal dander, and when entrained into the air by passing vehicles, becomes a virtually unavoidable source of allergen exposure (Miguel *et al.*, 1999).

10.4 Pathogenesis

In the first instance, particle-induced diseases are governed by the point of deposition in the respiratory tract, and secondarily by the body's ability to expel the offending particles from this tract. Particles that successfully evade these clearance mechanisms then cause harmful chemical and biochemical reactions, from which the various diseases arise.

10.4.1 Particle Deposition and Clearance

The necessary precursive step in deposition is particle *transport*; the fluid mechanics and hydrodynamics of this transport are governed, in turn, by many well-chronicled anatomical and physiological aspects of the respiratory tract (Smith J.R., 1971). This tract consists, sequentially (according to the passage of air), of three distinct regions: the *oral–nasal*, *tracheobronchial* and *pulmonary* (alveolar) regions. From the point of entry, air passes through the nose or mouth, turns immediately downward into the trachea, divides repeatedly, according to successive bifurcations in the bronchial tree, enters the bronchioles (with diameters of $\sim$500 nm), and finally arrives in the alveoli (with diameters of $\sim$200 nm). Gas exchange (CO_2 and O_2) is restricted to the pulmonary region, the finely divided structure of which, consisting of several hundred million alveoli, secures a remarkable surface area ($\sim$50 m^2). Some selected statistics appertaining to lung geometry and airflow are presented in Table 10.2 (Hinds, 1999, p. 236).

Since the complexity of the flow field in the human lung is not satisfactorily resolved (Kennedy, 2007), neither are the finer details of particle deposition. The exact site a particle will deposit depends on its size, the airflow rate and the airway geometry. The flow, according to simple computations of airway diameters and flow rates, should be laminar, but bends, bifurcations, surface roughness and convolutions engender localised turbulence (Schlesinger, 1988), especially at peak inspiration and expiration, and the flow is undeveloped. The deposition mechanisms are interception, impaction, sedimentation and diffusion (Seinfeld, 1986, pp. 59–67); the predominant one, owing to continually shifting hydrodynamics, changes with depth of penetration. This remark can best be illustrated with a few particulars.

Table 10.2 Characteristics of selected regions of the respiratory tract, abstracted from Hinds (1999, p. 236).

Airway	Diameter (mm)	Length (mm)	Total cross section (cm^2)	Residence Time[a] (ms)
Trachea	18	120	2.5	30
Terminal bronchus	0.60	1.6	180	31
Alveolar sac	0.41	0.5	72 000	550

[a] Assuming a flow rate of one litre per second.

The diameter of individual airways decreases with depth, but this decrease is outweighed by the significantly increasing total cross-section, so that air velocities drop rapidly. If particles are hygroscopic, they will grow by water take-up, since the airways are always moisture-laden. Local air velocities are governed by rate and depth of breathing, particularly in the upper reaches of the respiratory tract. A swifter rate of inspiration leads to more inertial impaction in the upper respiratory tract. Transition from expiration to inspiration, and *vice versa*, causes a momentary cessation and then reversal of the flow, twice in each cycle. Inertial deposition is more likely in the nasal than the oral path, as air velocities are higher, the nose is lined with hairs, and the passageways are narrower and more tortuous. Exercise marks a transition from nasal to oral breathing, so that the lower reaches of the tract are then less protected. Deposition in the head is almost total for particles of more than a few microns. At the opposite end of the respiratory tract, no tidal flow survives by the time air reaches the alveoli; hence mass transport in this region is entirely diffusional.

Deposition efficiency as a function of particle size is well characterised (Seinfeld, 1986, pp. 59–); a generic example was presented in Figure 2.12(b). As with filters, there is a mid-range size window at 0.1–1.0 μm, in which particles are too large to experience diffusional deposition, yet too small to experience inertial deposition. The accumulation-mode particles, therefore, stand the greatest chance of escaping deposition, i.e. of being exhaled. The deposition efficiency is approximately 90 % for 10 μm particles, 50 % at 2.0 μm, and 20–30 % at a few hundred nanometres. But, owing to the increasing importance of diffusion, this efficiency rises again for the smallest particles, passing 60 % at 200 nm; this characteristic is highly pertinent for nanoparticles, the deposition of which throughout, including in the tracheobronchial region, is predominantly diffusional (Smith S. *et al.*, 2001). Yet it is not quite correct that the smaller the particle, the deeper the penetration: alveolar deposition peaks at ∼20 nm; below this, particles diffuse so rapidly that they now deposit in the upper airways (i.e. the nose) (Maynard, 2001).

The detailed characteristics of this curve vary: between men and women, between children and adults, and between the healthy and the sick. For example, certain respiratory ailments, if they cause slower and shallower breathing, lead to longer residence times, and more effective diffusional deposition. The pattern of deposition also changes in illnesses where fluid build-up reduces the cross-section or where airway geometry is distorted.

Retained particles are an unfortunate corollary of the need to inhale fresh air; these particles are not wanted, and nature has, therefore, bestowed on the respiratory tract a formidable range of defence mechanisms. In the nasal passage, particles are evicted by sneezing or by the discharge of mucus. In the tracheobronchial region, particles lodge on a layer of mucus which, through the coordinated, wave-like action of cilia, is continually moved upward, towards the throat: this is the 'mucociliary escalator'. Residence time is a function of ciliary activity and the rheological properties of the mucus, but particles are generally removed in minutes or hours; deeper particles take longer to evict, owing to

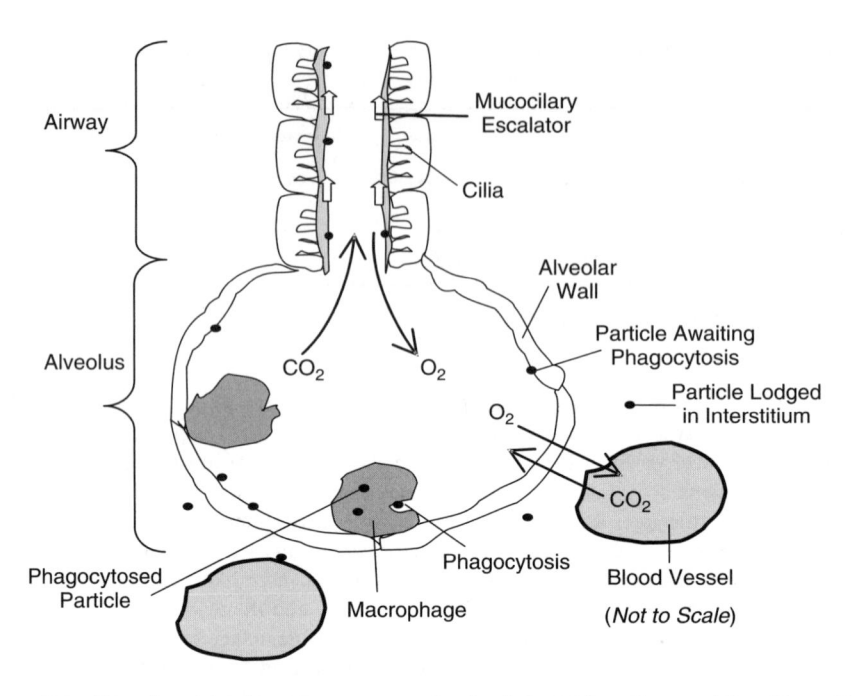

Figure 10.3 Fate of particles that gain entry to an alveolus (adapted from Stone and Donaldson, 1998).

the greater distance, and the slower rate of mucus transport with depth. Some inhaled pollutants slow down the escalator. Upon reaching the top of the escalator, particles are swallowed along with the mucus, and hence cleared to the digestive tract, usually subconsciously; or, social niceties permitting, they are ejected from the mouth by expectoration. The mucociliary escalator is assisted by the cough reflex.

Clearance is far slower and less straightforward in the *deep* lung, where alveoli lack the protection of cilia, and particles are able to encamp with impunity for weeks or months – or even as long as couple of years. Various possible fates, illustrated schematically in Figure 10.3 (adapted from Stone and Donaldson, 1998), await particles in an alveolus. Clearance is mediated by white blood cells called 'macrophages', which, being mobile, approach, either by chance or through chemotaxis, the particles, and engulf them in a process known as 'phagocytosis'. One macrophage may engulf many particles. Macrophages themselves are cleared by many routes (Lehnert, 1992), but often when they enter the lower terminus of the mucociliary escalator. They can also pass through the pulmonary epithelium, to enter the bloodstream or the lymphatic system. Of course, some particles are cleared without intervention by macrophages, because they are soluble in lung fluid.

The relative kinetics of *deposition* and *clearance* are absolutely critical in pathogenesis, as they determine particle residence time, that is, *retention* (Gradon et al., 1996): simply, the longer particles are permitted to reside, the greater the likelihood of harmful side effects, which we shall come to shortly. It is possible to estimate, via some simple assumptions, the rate at which an alveolus might receive particles (cited in Utell and Samet, 1996):[6] for example, a particle concentration of 10^5cm^{-3},

---

[6] Original reference Oberdörster (1995).

a particle size of 20 nm, an alveolar population per lung of 3×10^8 and an 'at-rest' ventilation rate of 8 litres/min suggest that 80 particles are deposited per alveolus per hour. A similar figure was reached by others (Seaton *et al.*, 1995; McAughey, 1999). With 70–80 macrophages per alveolus, such deposition rates would seem to be tenable, based solely on a consideration of what particle volume can be phagocytosed (McAughey, 1999). This deposition will not, however, be even, with some alveoli receiving more particles than others, especially in individuals where pre-existing lung diseases have altered local lung ventilation.

Such a calculation gives an idea of the macrophage-mediated clearance rate required in order to avoid particle accumulation, this rate being obviously finite, and, if it is insufficient to the task, lung *overload* can occur. It is, though, difficult to avoid some disingenuousness here, since nucleation-mode particles are the obvious focus of the current remarks, whereas most work in this area relates to accumulation-mode particles, for which deposition rates are unlikely to exceed clearance rates for the general population on a routine basis[7] – although it is noteworthy that particle exposures used in animal studies considerably exceed what humans experience under normal circumstances, leading, perhaps, to unrepresentative aetiologies (Stöber *et al.*, 1998). Of course, if particles cannot be phagocytosed, or if particles are toxic to macrophages, then they cannot be evicted. There is good evidence that macrophages are less able to remove nanoparticles and, indeed, are more likely to be harmed by them.

The phenomenon of *overload* is not simply one of an overwhelming particle burden, but seems to involve a slowing down in macrophage-mediated clearance (Donaldson and MacNee, 1998). The original hypothesis was that macrophages gradually lose their mobility with continuing phagocytosis, expressed as a certain volume of particles. Recently, however, overload has been related also to the surface area of the phagocytosed particles (Donaldson *et al.*, 1998). This seems a likely hypothesis for toxic particles, because interaction with cells occurs via the surface, and leaching of soluble components is presumably a function of surface area; but it is unclear why nontoxic particles mediate their effects similarly. Macrophage dysfunction is, moreover, still seen with nanoparticles, when the phagocytosed particle volume is small in relation to the cell.[8] And during overload, macrophages might themselves release species such as oxygen radicals that injure the lung. This is because one unfortunate particularity of the lung's defence systems is that chemical substances released to kill invading microorganisms are potentially toxic to the cells that are being protected (Seaton, 1999), and this is known to be an aspect of particle ingress.

10.4.2 The Chemistry and Biochemistry of Particle-induced Reactions

The next step is what happens to particles if they are not properly cleared. Pulmonary epithelium is not impenetrable to particles, especially nanoparticles, some of which, for instance, catalyse cell-damaging reactions leading to exposure of the delicate, underlying tissues. Consequently, particles are able to infiltrate the 'interstitium', that is, the delicate connective lung tissue permeated by blood and lymphatic vessels (Donaldson and MacNee, 1998), the thickness of which is only one or two microns. This penetration, or 'interstitialisation', allows particles *to escape normal lung clearance mechanisms*.

The general response of lung tissue is *inflammation*, a phenomenon engendering an unfortunate circularity, inasmuch as the permeability of the pulmonary epithelium increases still further, thus promoting more interstitialisation. Inflammation of the interstitium is thought more harmful than inflammation of the epithelium. Clearance of particles, once interstitialisation has taken place,

---

[7] Coal miners in the past may have experienced overload.
[8] It should be mentioned that, within cell tissue like anywhere else, nanoparticles will rapidly agglomerate, i.e. their effects are not necessarily expressed individually (Limbach *et al.*, 2005).

is complicated, but, essentially, particles can travel, or 'translocate', bringing *extrapulmonary* toxicological risks, i.e. risks to remote regions of the body, including other organs; particle accumulation, both inside and outside macrophages, in the lymph nodes of laboratory animals is often noted. Accumulation happens where clearance is less efficient, and where ongoing transport is inadequate.

Nanoparticles are not restricted to obvious conduits or extracellular pathways: they possess the ability to move around the body virtually with impunity;[9] their membrane-crossing mechanism is as yet unknown, but the nature of the particle surface is certainly a key factor. Nanoparticles penetrate red blood cells (Rothen-Rutishauser *et al.*, 2006) and hamster ovary cells (Suzuki H. *et al.*, 2007), and partition into subcellular compartments, including mitochondria, where they cause structural damage (Li N. *et al.*, 2003). In rats, nanoparticles that deposit in the nasopharyngeal region appear able to translocate along axons of the olfactory nerve into the central nervous system, engendering the potential to cross the blood – brain barrier (Oberdörster *et al.*, 2004). This nonlung clearance route may have been overlooked because of obsessional concerns with alveoli.

Translocation is not, however, restricted to whole particles, and much rests on 'bioavailability', i.e. those components that are soluble in lung fluid. In these instances, whatever resides at particle *surfaces* might be more toxicologically relevant than the total or average particle composition. Particles of silica and asbestos tend to persist, and hence to retain their identities in the lung for lengthy periods; the fate of particles discharged by internal combustion engines is still being established, but some components undoubtedly translocate through dissolution; and, indeed, organic compounds are obvious candidates, as we shall see shortly. Metals also dissolve to some extent (Schlesinger, 1988). It has been suggested, but not, apparently, widely digested, that low-sulphur fuels might, perversely, cause metals to be emitted not as sulphates but in more soluble, and thus more bioavailable, chemical forms (Weaver *et al.*, 1986). Nevertheless, the epithelial damage incurred by exposure to sulphuric acid aerosols, *especially* when this acid is sorbed onto solid particles (Amdur, 1996), remains to be elucidated, insofar as translocation cannot be the only factor.

Particles, or their dissolved components, instigate chemical reactions that drive inflammation: free radicals and 'reactive oxygen species' (ROS) are generated, leading to oxidant – antioxidant imbalance, or oxidative stress (Donaldson *et al.*, 2003) – a condition thought to underlie many diseases.[10] Indeed, some scientists believe that the ability of the body to handle this stress is a central aspect of why we all age: it should be said that such species are also generated naturally, as part of the body's normal metabolism. Free radicals and ROS damage cells, e.g. by oxidising lipids and proteins, by breaking strands of DNA, by generating further deleterious compounds and generally by consuming the very antioxidants that form part of the body's natural defences: for example, antioxidant enzymes catalyse opposing reactions that convert free radicals to harmless or less toxic compounds; cellular damage occurs once these antioxidant defences are overwhelmed. This is the basis for the notion that *dietary* antioxidants bolster the body's defences against pollutant oxidants, as is apparently the case with vitamins A, C and E, which reputedly confer protection against O_3 and NO_2 (Holgate, 1999).

Two chemical mechanisms are commonly invoked (Briedé *et al.*, 2005): redox cycling of organic, quinone-like molecules such as PAH, and the catalytic action of metals, notably iron, in Fenton-type (and other) reactions (Imlay and Linn, 1988). (The two mechanisms are not necessarily independent.) Oxidant generation leading to the damage of bacterial DNA depends on the co-presence of iron, which mediates a Fenton reaction (Imlay *et al.*, 1988). Iron is particularly interesting inasmuch as it purposefully fulfils special biological roles in the body, including, *inter alia*, haemoglobin; hence its handling as an inhaled pollutant may also differ from that of other metals (Wilson M.R. *et al.*,

---

[9] They may even be able to penetrate dermally from skin care products (Reijnders, 2006).
[10] Exposure to tobacco smoke is thought to cause similar oxidative stress (Maskos *et al.*, 2005).

2002). Fortunately, indices do exist for ROS-generating ability (per unit mass of substance): although uncorrelated with PM10, such an index was significantly higher for the particles immediately discharged by gasoline and diesel engines than for ambient particles generally present in the urban environment (Briedé *et al.*, 2005). Even so, PAH, total metal content and total transition metal content were *not* implicated, and the generation of ROS seemed to be the work of copper and nickel, rather than of iron. Particles – especially nanoparticles, the unusual catalytic properties of which are well known – are thought particularly effective in the generation of ROS (Donaldson *et al.*, 1996).

10.4.3 Particle-induced Diseases

With actual diseases to have been associated with motor vehicle particulate, the largest question mark, historically speaking at any rate, concerns *cancer*, especially of the lung, for those persons occupationally exposed to diesel exhaust (HEI, 2002). Of course, the dangers of *all* soots and tars – and cigarette smoke, for that matter – are well known in this respect; here, PAH are widely held to be the key cancer-causing agents. In the case of pitch, this was demonstrated as long ago as the 1930s (Hecht, 1988). The impeachment of NPAH on the same charge is more recent.

Now, the view generally entertained in relation to carcinogenesis is that information encoded within DNA is somehow corrupted, very likely by chemical reactions with exogenous compounds, and the offspring of these reactions – mutated cells – then proliferate out of control (Kaufman, 1988). But for this to happen, the cancer-causing agents, and PAH are no exception, must first gain access to body tissue. And it turns out that carcinogens are often more strongly carcinogenic when inhaled as sorbates on particle surfaces than when inhaled as vapours. This may logically be ascribed to the localisation of the dose, and the *rate of uptake* into body tissue, which, from a particle surface, is slower, yet more prolonged (Sun J.D. *et al.*, 1988); indeed, this prolongation might even assume more significance than the actual quantity *per se*. Conversely, however, the organic compounds must at least be bioavailable; for if molecules are bound stubbornly to particle surfaces, they may well remain essentially nonreactive and benign. Particles well-covered with PAH and other organic compounds exhibit gradually declining release rates, with the freer compounds being denuded first.

The presence of an organic compound in cell tissue does not necessarily constitute a problem. What counts in mutagenesis and carcinogenesis is the subsequent fate, or *metabolism*, of the organic compounds in (or by) various organs (i.e. the liver, lung, skin, intestine or kidney), and interactions with certain other compounds able to act as 'promoters', 'co-carcinogens' and 'inhibitors' (Kaufman, 1988; Hecht, 1988). PAH – or their metabolised products, the so-called toxic metabolites, which are chemically reactive intermediates – are able to bind covalently to macromolecules such as proteins, RNA and DNA (Sun J.D. *et al.*, 1998), leading *inter alia* to 'PAH – DNA adducts'. Several studies demonstrate significantly higher levels of these adducts in persons occupationally exposed to diesel exhaust: mechanics, traffic police, taxi drivers, road workers, bus depot workers (Okona-Mensah *et al.*, 2005) and miners (Qu *et al.*, 1997); such adducts correlate with cell mutations (Lambert B. *et al.*, 1995). The effects are difficult to unravel because metabolic interactions engender synergistic or antagonistic effects. This should be borne in mind whenever *BaP* is studied in isolation to suggest the total risk arising from PAH, as is often done for expediency.

From what has already been related, and indeed commonsensically, adverse effects on the lungs are only to be expected; less readily explained are adverse effects on the cardiovascular system (*heart attacks* and *strokes*). There is more than one hypothesis (Seaton, 1999). A well-respected suspicion is that particles promote haemostasis and thrombosis, or blood clotting, which in the coronary arteries causes heart attacks, and in the cerebral microcirculation causes strokes. Two posited mechanisms are sketched in Figure 10.4 (Donaldson and MacNee, 1998). The first imputes polymorphonuclear neutrophil leukocytes (PMN), white blood cells which, because they are slightly larger (7μm) than

```
                    ┌─────────────────┐
                    │ Particle deposition │
                    └─────────┬───────┘
                              ▼
    ┌──────────────┐    ┌──────────────┐
    │  Release of  │◄───│  Pulmonary   │
    │  cytokines   │    │ inflammation │
    └──────┬───────┘    └──────┬───────┘
           ▼                    │
    ┌──────────────┐            │
    │    Liver     │            │
    └──────┬───────┘            │
           ▼                    ▼
    ┌──────────────┐    ┌──────────────┐
    │  Release of  │    │ Decreased PMN│
    │pro-coagulant │    │ deformability│
    │factors, e.g. │    │in the pulmonary│
    │Factor VII,   │    │microcirculation│
    │ fibrinogen   │    │              │
    └──────┬───────┘    └──────┬───────┘
           ▼                    ▼
    ┌────────────────────────────────┐
    │Coronary and cerebral microcirculation│
    └────────────────┬───────────────┘
                     ▼
    ┌────────────────────────────────┐
    │  Haemostasis and clot formation │
    └────────────────┬───────────────┘
                     ▼                      │
            ╱─────────────╲          ╱─────────────╲
           │ Heart attack  │         │Exacerbation of│
           │  and stroke   │         │airways disease│
            ╲─────────────╱          ╲─────────────╱
```

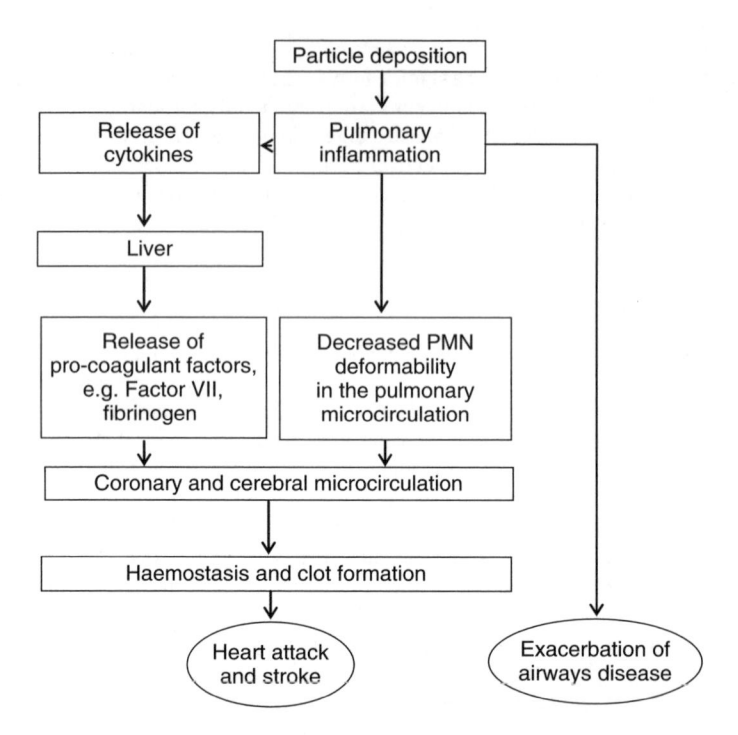

Figure 10.4 Hypothetical scheme for the events leading from particle deposition in the lung to health outcomes (morbidity and mortality) (Donaldson and MacNee, 1998).

the alveolar capillaries (5μm), must deform in order to negotiate the microcirculation (O'Byrne and Postma, 1999). This delays their transit even in normal circumstances, so that they tend to accumulate, or *sequester*, at various points – a condition that might be worsened during periods of inflammation and oxidative stress. The second mechanism is enhanced production of fibrinogen (a precursor to fibrin) – a protein of major importance in controlling blood viscosity and coagulability.

Three diseases to be mentioned concern the upper airways. The first, *asthma*, an inflammatory disease characterised by wheezing, panting and shortness of breath, is an allergic reaction in which the lining of the bronchial tree is inflamed; the release of histamine causes mucus hypersecretion, and contraction of the airway-encircling musculature, so that airway resistance is increased. Particles increase levels of IgE, a class of antibody involved in allergic reactions (Salvi and Holgate, 1999), and environmental allergens penetrate because the integrity of the epithelium is compromised (Ormstad *et al.*, 1998). The immune response is, though, reversible. The pathogenesis of asthma in relation to diesel exhaust is given in detail elsewhere (Pandya *et al.*, 2002). The second disease, *bronchitis*, is particularly characterised by persistent coughing, sputum production and expectoration; it can be acute, such as through the action of a chemical stimulus, or chronic, involving reduced or absent ciliary action. ROS are implicated in both asthma and bronchitis (Sagai *et al.*, 1996). A third frequently mentioned illness, but one more typically associated with cigarette smoking, is *chronic obstructive pulmonary disease* (COPD), which in fact is not one distinct illness, but an umbrella term encompassing chronic (obstructive) bronchitis, emphysema (destructive enlargement of the alveoli) and small airways disease, in which lung function is permanently compromised owing to a fixed airway obstruction (O'Byrne and Postma, 1999). It should be noted, however, that clear lines of demarcation do not always exist: diagnostic practices vary from one country to another, and have changed over the years; physicians are not always in agreement; and diagnoses can be anatomically as well as symptomatically based.

10.5 Epidemiology

Observations of the association between exposure and illness considerably pre-date modern epidemiology: the inimical nature of combustion's unwanted by-products – sooty and tarry deposits resulting from incomplete burning – was famously recognised as long ago as the Enlightenment, when, in 1775, the physician Percival Pott, arguably a proto-epidemiologist, famously pointed to an unusually high incidence of scrotal cancer in chimney sweeps (Hecht, 1988). And although chimneys are no longer an issue as such, some present-day workers are still exposed to fumes associated with similar substances: coke, tar, bitumen, asphalt, pitch, etc.

Equally demonstrated in modern times is the influence of locale: among the general public, incidences of lung cancer in metropolitan districts, small towns and rural communities all lie along a decreasing scale of frequency (Rushton, 2001); and, in fact, data even suggest an antagonism, insofar as cigarette smoking is riskier for city dwellers (Coffin, 1971). Such findings cannot be labelled as *solely* a consequence of *occupational* exposure to diesel engines – some unknown aspect of population density, conceivably relating to ambient air pollution, is implicated (Russell Jones, 1987). Of course, this topic broadens out into lung cancer or cancers generally, and airborne particles or air pollution generally, and it is not the purpose of the present section to review this enormous topic: the reader is referred elsewhere (Utell and Samet, 1996; Lippmann, 1998; Ayres, 1998).

Epidemiological studies show associations between levels of suspended particles in the ambient air, and various health statistics. But what, exactly, do individuals inhale? Some pollutants have strong spatial gradients (Monn *et al.*, 1997): poorly correlated variations in exposure from one location to another in the same airshed potentially distort epidemiological findings when averages are used from several monitors, or data from a single, central site within a community (Kim E. *et al.*, 2005a); and the strength of an association could be underestimated without consideration of spatial variations in large cities (Chen L. *et al.*, 2007). And, from Section 10.2, there is every reason to suppose that exposures to nucleation-mode particles are strongly localised (Monn, 2001), whereas exposures to longer-lived accumulation-mode particles are more uniform. Moreover, building interiors are where most people, at least in the developed world, predominantly reside – this is paradoxical, insofar as epidemiological studies of public health rely habitually on air-quality-monitoring networks situated outdoors. As we also learned in Section 10.2, personal or indoor exposures are not unambiguous functions of outdoor exposures, and in epidemiological studies, these relationships are potential sources of bias (Meng *et al.*, 2005) or noise (Özkaynak and Spengler, 1996).

With other pollutants such as CO, for example, it is possible to state an exposure threshold, below which no discernible health effects are seen. For carcinogens this is not possible, as, theoretically at any rate, no identifiable threshold exists;[11] and this is also the understanding with diesel particulate (Walsh, 1995). Thresholds observed in animal studies may instead be artefacts of the high doses used (Klingenberg and Winneke, 1990); we shall return to this topic in Section 10.7. Some researchers believe that just one molecule of an offending substance is capable of causing cancer, although this has

---

[11] This is for genotoxic carcinogens – it is believed that thresholds can be established for carcinogens acting via nongenotoxic mechanisms (Larsen and Larsen, 1998).

not been proven. So, instead, data are expressed in terms of *risk*.[12] Of course, this approach appears unreassuring to members of the public, encouraged by the gutter press to believe in 'zero risk' – an obviously unreal concept. A *lifetime* risk of one in 10^6, corresponding to an *annual* risk of one in 10^8 (assuming for convenience a lifespan of 100 years), is used in setting air quality standards. Placing these figures in perspective (Harrison R.M., 1998), five annual risks of death are as follows: (a) through smoking ten cigarettes per day, one in 200; (b) through contracting influenza, one in 5000; (c) through playing soccer, one in 25 000; (d) through homicide, one in 10^5 (UK); (e) through being struck by lightning, one in 10^7.

Risk, it is supposed, follows a linear dose–response relationship supportive of extrapolation (Larsen and Larsen, 1998), normally being represented by measurements of particulate mass (PM10 or PM2.5, or sometimes compositions or fractions derived therefrom). With carcinogens, the risk of cancer does rise with dose, but, at low doses, the relationship is often unclear (Kaufman, 1988). The assumption of linearity and the presence (or absence) of a threshold both remain under discussion (Wilson R., 1996). It has not, even for cigarette smoking, been possible to prove the existence of a threshold, even though far more deaths are involved than conceivably could be the case with air pollution. Within a large city, variations in susceptibility or exposure from one individual to another, may mask any threshold, if it exists (Harrison R.M., 1998). A decidedly important clue, however, is surely the temporally close nature of the association between particle concentrations and mortality – this closeness throws a spanner in the works, so to speak, as the latency period for cancer is years or even decades, rather than a few days – hence the posited cardiovascular effects, which are more immediately felt, and which we shall come to shortly.

It should not go unmentioned that epidemiological findings pointing to causality are not infrequently disputed: some scientists are still suspicious, for example, of the statistical analyses; others cite flawed methodologies or point to misinterpretation (Maynard, 2001; Phalen, 2002), with insufficient weight being accorded to what is *statistically significant* (Stöber, 1987; Morgan *et al.*, 1997). Saliently, epidemiology must account for significant confounding by cigarette smoking;[13] other difficulties are obtaining knowledge of (or accounting for) true exposure, occupation, income, locale, ethnicity, diet, alcohol consumption, body mass index, cause of death and the weather (extreme heat and cold). The historical record is something of a moving target: a decades-long latency must be set against the fact that, during the same period, the technology of the diesel engine has undoubtedly changed out of all recognition; while emissions data prior to 1970, i.e. before the emission control era, are scant (Fritz *et al.*, 2001). Estimating year-by-year worker exposure from, say, 1950 to the present requires painstaking considerations of nondiesel particles, vehicle-miles travelled, real-world emission rates, the turnover rate in vehicle technology, etc. (Bailey C.R. *et al.*, 2003). Whereas the epidemiological cases against cigarette smoking and asbestos, for example, are indisputable, the lung cancer risks posed by occupational exposure to diesel particulate do not universally, consistently and unambiguously stand out above the noise. This, for some, is sufficient grounds for stating that such health risks, if they exist at all, are minuscule, and comparable, for example, to death by lightning strike.

Two types of group form the subject of epidemiological investigation: workers and the public. Risk assessment is less problematical in occupational health, for here, the substance to which people are exposed bears clearer definition. Some risks posed by airborne dust in the workplace are well known, e.g. pneumoconiosis, silicosis and asbestosis. (The practice of naming the disease after the causative agent testifies to the surety with which that agent has been identified.) Similarly, therefore, as diesel

---

[12] The *relative* risk, namely the proportional change associated with an increment of pollution, should not be confused with the *attributable* risk, namely the absolute increase associated with an increment – in turn, a function of the baseline incidence and the relative risk (Anderson *et al.*, 2003).

[13] Not forgetting passive smoking (Jones A.P., 1999), risk factors for which continue to be controversial.

engines are characteristic of certain occupations, the workers one might select for epidemiological studies are obvious, e.g. garage workers, railway workers, dock workers, etc. Of course, exhaust gas is hardly a closely defined substance like asbestos, although occupational exposures are often somewhat higher than for the general population, so that the health effects may be more pronounced. Extrapolation to the general public is uncertain; but the protection of persons in the workplace is surely sufficient justification in itself. A lung cancer risk was found in the US trucking industry (Steenland *et al.*, 1998), and in fact approximately thirty epidemiological studies have shown, for various occupations, a consistent association between lung cancer and diesel exhaust, the risks ranging from 20 to 89 % (Walsh, 2001). Other cancers to have been associated with exposures to motor vehicle exhaust (in various occupations) are of the bladder, colon, kidney and gastrointestinal tract (Rushton, 2001).

Because of the great diversity of airborne pollution – not forgetting noxious gases – *public health* studies focus, understandably, on the total quantity or concentration of particles. This is a problem, as pollutants are unlikely to act on the body as separate entities, and will more probably *interact*, yet the synergisms, if any, are poorly understood, and unhelpful co-variation occurs in the environment (although secondary pollutants such as ozone may differ). There are therefore problems in untangling the various effects: it is unclear from public health studies how much blame can be apportioned to diesel engines specifically or, for that matter, internal combustion engines generally, as suspended particles are of such diversity. All one can say is that, in urban areas, motor vehicles contribute significantly to the entire particle burden. Nevertheless, a consultation of death certificates shows that high-pollution days ($141\mu g/m^3$) are riskier than low-pollution days ($47\mu g/m^3$) (Schwartz, 1994).

Four major US epidemiological studies devoted to public health were published in the 1990s. (**A**) In the Harvard Six Cities Study, 8000 people were monitored for a period of 14–16 years. As depicted in Figure 10.5 (Pope A. and Dockery, 1996), mortality risk was directly proportional to ambient particle concentration, and residence in the most polluted city was riskier, by 26 %, as compared with the least polluted city – approximating to a shortened life expectancy of one to two years. As shown, the correlation was closer for fine particles and sulphates than for coarse particles. (**B**) In a study conducted by the American Cancer Society (ACS), statistics were compiled for 500 000 adults living in 151 cities over the period 1982–89. Again, the levels of fine particles experienced in the most polluted city were riskier, by 15–17 % as compared with the least polluted city; and again, fine particles and sulphates were implicated (Pope A. and Dockery, 1996). (**C**) In the National Morbidity, Mortality and Air Pollution Study (NMMAPS), which concerned 90 of the largest US cities, a particle-associated increase was found in relative mortality (0.51 % per $10\mu g/m^3$ PM10) (cited by Walsh, 2001).[14] (**D**) In the Multiple Air Toxics Exposure Study (MATES-II), 30 pollutants were tracked in the South Coast Air Basin (California): about 70 % of carcinogenic risk was attributable to diesel particulate (cited by Walsh, 2001). Sixteen other studies show similar statistically significant associations (a range of 0.5–1.6 % per $10\mu g/m^3$ PM10, with a mean of 0.8 % per $10\mu g/m^3$ PM10) (Dockery and Pope, 1996).

With cause of death, there are less strongly displayed or even insignificant associations between ambient particles and cancer mortality; yet repeatedly implicated are cardiovascular mortality (1.4 % per $10\mu g/m^3$ PM10), and, stronger still, respiratory mortality (3.4 % per $10\mu g/m^3$ PM10) (Pope A. and Dockery, 1996). NMMAPS, for example, found 'consistent evidence that the level of PM10 is associated with the rate of death from cardiovascular and respiratory illnesses.' This association does not solely concern the elderly: it extends to infant mortality (Kaiser R. *et al.*, 2004), especially respiratory-related mortality (Ha *et al.*, 2003). And an analysis of 190 000 stroke deaths between 1990 and 1992, among persons over 45 years of age, found a significant (but small) association with distance of residence from main roads (Maheswaran and Elliott, 2003), although this does not, of course, implicate particles as such.

---

[14] Original reference Samet *et al.* (2000).

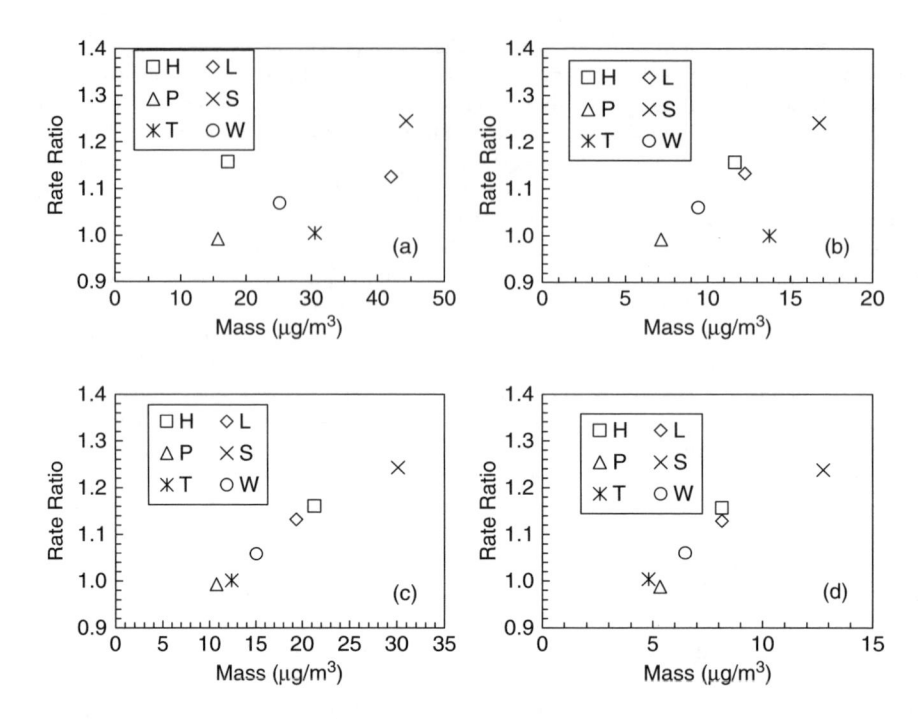

Figure 10.5 Estimated mortality rates (expressed as ratios) obtained from the Harvard Six Cities Study: (a) noninhalable fraction, i.e. inhalable particles subtracted from total suspended particles; (b) coarse fraction, i.e. fine particles subtracted from inhalable particles; (c) fine particles (PM2.5); (d) sulphate particles. Cities: H, Harriman, TN; L, St Louis, MO; C, Portage, WI; S, Steubenville, OH; T, Topeka, KS; W, Watertown, MA (Pope A. and Dockery, 1996).

Whereas death is an objective and indisputable health end point, diagnoses for disease states are more complex and can be subjective, leaving room for bias and misclassification. Still, various measures of morbidity also show plausible associations with ambient particles. These include modifications to blood rheology, including raised plasma viscosity (Peters et al., 1997); restricted heart-rate variability (a predictor for cardiovascular conditions) (Gold D.R. et al., 2000); myocardial infarctions (Peters et al., 2001); atherosclerosis (Künzli et al., 2005); hospitalisations for respiratory conditions in the elderly (Anderson et al., 2003); greater occurrences of COPD (Ye et al., 1999); and hospitalisations for asthma (Devalia et al., 1996; Claiborn et al., 2002). In Austria, France and Switzerland, motor vehicles were implicated, annually, in 20 000 deaths, 25 000 new cases of chronic bronchitis in adults, 290 000 episodes of bronchitis in children, more than half a million asthma attacks and more than 16 million man-days of restricted activity (Künzli et al., 2000).

Something undeniable is the great dependency of epidemiology on lengthy records of PM10, and less lengthy records of PM2.5, rather than other particle measurements; and all the above studies have been based on these metrics. Yet it appears that health statistics such as mortality and morbidity correlate somewhat more strongly with PM2.5 than they do with PM10. This is certainly not discordant with the understanding that particle size is paramount. PM10 can be, and indeed often is, dominated by natural, coarse-mode particles of a benign, nonrespirable nature, whereas PM2.5 awards greater emphasis to smaller, combustion-derived particles, and, *ipso facto,* focuses more intently on motor vehicles

(Mark, 1998). Nor is it surprising when, in tailpipe measurements, PM10 and PM2.5 correspond so closely: little material is actually emitted between 2.5 μm and 10 μm (Cadle *et al.*, 1999a; Knapp *et al.*, 2003; Lev-On and Zielinska, 2004). Even so, 2.5 μm is still very large compared with what is commonly emitted by internal combustion engines – both gasoline and diesel. It thus seems probable that epidemiological studies impugning particle size will continue to accumulate.

Concluding, even small shifts in levels of suspended particles ($\pm 5\mu g/m^3$ or so) would seem to have discernible effects on public health (Medina *et al.*, 2004); and variations of several tens of $\mu g/m^3$ are not unknown. No obvious or convenient threshold has been demonstrated which might serve as an air quality standard below which, as in the case of CO, for example, adverse heath effects disappear. Without a threshold, commonly agreed standards are not so easily set (Wilson R. and Spengler, 1996). The above associations are still discernible even when exposures are actually at and below *current* air quality standards. This suggests that either these adverse health effects arise at exposures lower than those at which they were previously thought possible or, alternatively, that PM10 and PM2.5 act as weak surrogates for some other aspect of ambient air pollution inadequately captured in current standards. Intriguingly, even though the USA largely lacks the light-duty diesel passenger car now commonplace in Europe, the epidemiological associations between suspended particles and mortality are still seen – and these are reportedly *stronger* than in Europe, rather than weaker (Maynard, 2000). This last observation provokingly suggests either that the heavy-duty diesel engine is a dominant factor or that the gasoline engine requires more assessment.

10.6 *In Vitro*

As already related (Section 5.2.7), the *Ames* test has gradually established itself as a standard bioassay in toxicological evaluations of motor vehicle particulate – and of many chemicals besides, including pharmaceuticals. In this test, organic extracts of particulate, virtually without exception, cause mutations in the *Salmonella* bacterium. There are, however, other aspects of particle–cell interactions that demand consideration. The Ames test uses a bacterial rather than a mammalian system. The nitroreductase enzyme may give greater and undue prominence to NPAH (Griffin *et al.*, 1981). The quantity and nature, and hence mutagenicity, of the extracted organics are solvent-dependent; and in the lung, the release of organic compounds from carbonaceous agglomerates is mediated not by solvents such as DCM or cyclohexane, but rather by body fluids of considerably greater complexity. Lastly, the formation of mutant cells is just one aspect of pathogenesis.

To meet these needs, alternative bioassays using various cell cultures and tissue samples have been devised; these may one day replace, or at least routinely complement, the Ames test. This is not the place to review bioassays, but some examples, taken from a multitude, are mutations in yeast cells (Griffin *et al.*, 1981), single-strand breaks in the DNA of human liver cells (Hasspieler *et al.*, 1995), enzyme activity in the human nasal epithelium (Schmidt *et al.*, 1996), antioxidant depletion in the epithelial fluid of the respiratory tract (Greenwell *et al.*, 2002), the formation of PAH–DNA adducts in calf thymus (Kuljukka *et al.*, 1998), mutations in human lymphoblasts (Pedersen *et al.*, 1999), chromosomal damage in rat liver cells (Shore *et al.*, 1987), mutations in the bacterium *Escherichia coli* (Shore *et al.*, 1987), the proliferation of white blood cells (Shore *et al.*, 1987), cytokine production in rat lung tissue (le Prieur *et al.*, 2000), intercellular communication in the liver and lung cells of rats (Vanrullen *et al.*, 2000), and the inhibition of enzymes involved in oxidative stress (Hatzis *et al.*, 2006).

Cell viability and function are affected by particulate in many ways. Enzyme activity in human nasal epithelial cells is modified, but why this modification is more pronounced for unextracted particles than organic-extracted ones remains to be ascertained (Schmidt *et al.*, 1996). There are posited effects on the immune system, as the proliferation of white blood cells is arrested, although this appears to be through some nonspecific cytotoxicity, rather than a true immunosuppressive effect (Shore et al., 1987). Adenosine triphosphate (ATP), a general health indicator of cells, is released (Nikula et al., 1999), as are cytokines – indicative of an inflammatory response. For example, tumour necrosis factor (TNF) is found in rat lung tissue exposed to whole diesel exhaust, but not when the exhaust is filtered, prior to exposure – pointing to the importance of particles in generating the inflammatory response. In cytokine expression, the inflammatory response was similar for particles of sizes 54 nm and 130 nm, but the allergy response was stronger for the smaller size (Papaioannou et al., 2006). Glutathione (GSH), an antioxidant which serves as a measure of oxidative stress, is depleted in rat lung tissue (Morin et al., 2002); the effects are attenuated significantly when exhaust gas is filtered (Morin et al., 1999). The deleterious effects extend to a tissue's ability to self-correct or repair alterations induced by chemical agents, and to suppress the growth of mutated cells; this is a key aspect of tumour growth, and is expressed as gap junction intercellular communication (GJIC), which seems compromised by organic extracts or inhibited in a dose-dependent manner, in the liver and lung cells of rats (Vanrullen et al., 2000). Repair mechanisms are instigated in human bronchial epithelial cells by both organic and aqueous extracts of particulate collected in the vicinity of a busy road (Rumelhard et al., 2007). Last but not least, diesel particles not only are cytotoxic towards macrophages, but also cause, through unknown mechanisms, the production of NO, opening up even wider implications for respiratory diseases (Hälinen et al., 1999).

There are, then, a range of adverse effects over and above that which receives the greatest attention of all, namely corruption of cellular DNA. Diesel exhaust, as a whole, alters the DNA in rat lung tissue; these alterations are less pronounced when exhaust gas is diluted or when particles are filtered out prior to exposure (le Prieur et al., 2000). The chromosomes in rat liver cells are damaged by organic extracts of particulate, and in a dose-dependent manner (Shore et al., 1987), while bacterial DNA is altered by water-extracted components, with a linear dose response (Greenwell et al., 2002). Diesel particulate emissions are mutagenic towards mammalian cells in a dose-dependent manner that begins with phagocytosis, whereafter the agglomerates lodge within vacuoles (Bao et al., 2007).

The sixty-four thousand dollar question is how DNA is damaged; and there are three salient comments to be made here. First, there are reactions between PAH and DNA, as demonstrated by incubating organic extracts of diesel particulate with calf thymus (Kuljukka et al., 1998): the frequency with which DNA–PAH adducts appeared corroborated the results for Ames mutagenicity. NPAH formed more adducts than PAH, and a reformulated fuel was able to reduce adduct formation by 80%. Second, there is oxidative stress. Diesel particulate catalyses the formation of a strong but unidentified oxidant, the potency of which rivals that of the OH radical (Vogl and Elstner, 1989); and ROS, provocatively generated by particulate in biological systems, are themselves implicated in DNA damage. Nanoparticles of carbon black, for instance, generate significantly more ROS than do fine particles (Wilson M.R. et al., 2002). The generation of ROS points to the catalytic action of metals. Third, PAH and ROS are suspected to interact in their corrupting effects on DNA, perhaps when the first, through redox cycling, generates the second (Kumagai et al., 1997). Organic compounds such as NPAH might well promote the formation of ROS which subsequently react with DNA (Hasspieler et al., 1995). Apparently, O_2^- can arise from metabolised *BaP* and *1-NP, without* any biochemical activating system, perhaps through the action of quinone-like compounds (Sagai et al., 1993). The finding that PAH, rather than just metals, are responsible for ROS is convenient, since the ROS-forming substance, whatever its nature, is lost by organic extraction of the particulate (Sagai et al., 1993).

Cells are naturally equipped with antioxidants that confer some measure of protection against oxidants, and these defences are depleted by exposure to particulate. An assay able to measure the antioxidant defence remaining in plasma showed depletion in persons suffering from asthma and COPD, and significant depletion during actual attacks (Donaldson and MacNee, 1998). The notion that

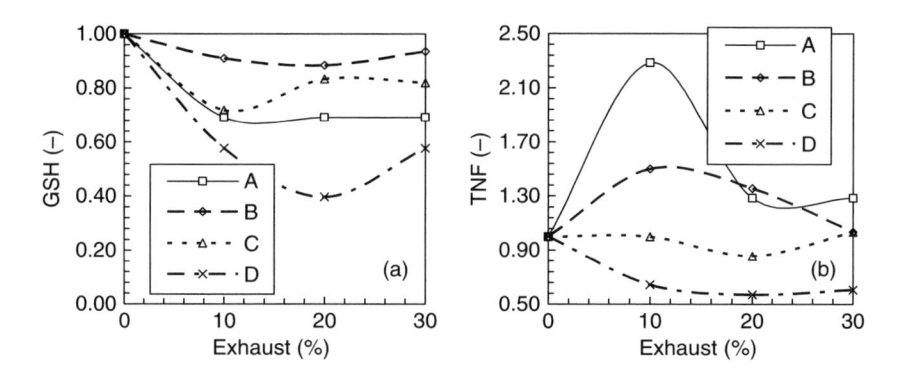

Figure 10.6 Fractional change in (a) glutathione (GSH) and (b) tumour necrosis factor (TNF) after exposure of rat lung slices to diluted engine exhaust, following incubation for two hours with isoflavones (genistein and daidzen), at the following concentrations (μmol/litre): A, control (zero); B, 0.3; C, 1.0; D: 3.0. Engine: 1.9-litre diesel, direct-injection, turbocharged, intercooled. Operating point: 2500 rpm, 10 bar BMEP (Morin *et al.*, 2002).

certain foodstuffs confer protection against the multifarious effects of oxidants by supplying additional antioxidants has widespread support. For example, soy-derived isoflavones protected rat lung tissue against the depredations of diesel exhaust, since they restrained oxidative stress, the inflammatory response, secretion of TNF and GSH, and DNA damage; the benefit was dependent on dose. Selected results are presented in Figure 10.6 (Morin *et al.*, 2002).

10.7 *In Vivo*

Laboratory animals used in toxicological investigations are invariably mammals, rats being the most favoured species, but mice, rabbits, guinea pigs and hamsters are also known. This is the first area of compromise: for example, lung retention of and tissue reactions to diesel particulate in rats differ from those in primates (monkeys), with a greater proportion interstitialised in the former, yet greater inflammatory responses in the latter (Nikula *et al.*, 1997). Such tests can be long-term (chronic) or short-term (acute). Animals either die during the period of exposure or are killed at predetermined points, and, following dissection, their lungs are examined for histological changes indicative of tumours, particle sequestrations, and various aspects of tissue damage and repair (Richards R.J. *et al.*, 1999).

There are two methods of particle administration: inhalation as an aerosol, and *intracheal instillation* as a colloid (such as particle suspensions in saline). This is not the place to review the relative merits of each method, but neither is wholly immune to artefacts. Intracheal instillation is nonphysiological, i.e. bypasses the filtering action of the upper airways; a major difference between humans and other mammals lies in the pattern of bronchial airway branching (Schlesinger, 1988). Yet it has been argued that the doses are not that out of keeping with what animals might actually inhale in polluted environments (Nemmar *et al.*, 2003b). Inhalation, on the other hand, raises three easily overlooked stumbling blocks: there may be reaction of animal-derived ammonia with exhaust components, to form secondary sulphate and nitrate particles; there may be day-to-day variations in engine emissions, especially in particle-bound organic compounds (McDonald J.D. *et al.*, 2004); and animals, when resting, can place their noses in fur, the hairs of which impart an additional filtering action.

These practicalities notwithstanding, the tallest question mark in animal inhalation studies undoubtedly hangs over the unrepresentative doses – one or even two orders of magnitude higher than realised even in the most heavily polluted human environments: for example, *a few milligrams per cubic meter* rather than *a few tens of micrograms* per cubic metre. This is a natural corollary of the statistical nature of disease, particularly of cancer. For example, if the lifetime risk of exposure to particles in ambient air is 10^{-5} and the test group contains 100 animals, then the laboratory exposures must exceed those pertaining in the real environment by a factor of 10^3 – otherwise the disease may fail to manifest itself (Stöber, 1987). Animals, moreover, exhibit a statistical 'background' of spontaneous (naturally occurring) tumours, by which the effects of the test substance may be disguised. And some compensation is required for the natural lifespans of these animals (two or three years), which are short compared with those of humans (Kaufman, 1988). Since these constraints immediately suggest exposure to undiluted rather than diluted exhaust, the compromises will be apparent.

Health effects manifested at high doses might be unrepresentative, arising, for example, through acute toxicity, excessive inflammation, or the crossing of some threshold in the aforementioned phenomenon of 'overload'. Tumour induction might then simply be 'epigenetic' rather than genetic, i.e. arising not through the metabolism of organic compounds associated with the particulate, but merely from an unrealistically high lung burden, for which the aetiology is not (or less) substance-specific. Overload appears to be a particular characteristic of the rat, as high particle doses of TiO_2, carbon black, coal dust and diesel soot all tend to generate similar, and possibly epigenetic, lung tumours. This renders extrapolation to low doses fatuous, and has no immediate bearing on the human situation in the real world. On this basis, it is not inappropriate to scrutinise the relevance of *in vivo* tests to public health (Klingenberg and Winneke, 1990).

With these caveats in mind, the animal studies, grouped according to type of exposure, are as follows.

A. *Exposure to diesel exhaust*. During long-term exposure, the frequency of lung tumours in rats is dose-dependent; the tumours are not induced when particles are filtered out prior to exposure. Mice and hamsters are noticeably less susceptible to these tumours (Mauderly *et al.*, 1996; cited by Baker, 1998). But many other deleterious effects are seen. Rats were exposed to diluted diesel exhaust for 16 hours per day, six days per week: hypersecretion of airway mucus (the basic pathology of chronic bronchitis) developed by six months at $\sim 2,900\,\mu g/m^3$, and by 18 months at $1100\,\mu g/m^3$; both doses provided indications of incipient emphysema; no untoward effects were, however, observed at $210\,\mu g/m^3$ (cited by Kagawa, 2002). Rats were exposed to diesel exhaust for 12 hours per day, seven days per week for four weeks: marked histological changes, including lesions, appeared in bronchial epidermal cells at $3000\,\mu g/m^3$, but not at $300\,\mu g/m^3$ or in controls exposed to filtered exhaust; the changes were adduced to oxidative stress (Sato *et al.*, 2001). Rats were reared for 60 weeks beside a high-traffic trunk road, for which exposure was $63\,\mu g/m^3$: by 48 weeks, swelling was observed in the alveolar epithelium, thickening in the interstitium, cellular modifications in the bronchiolar region, and carbon translocation to the lymph nodes; the prominence of these effects depended on the duration of the exposure (cited by Kagawa, 2002).

B. *Exposure to particulate*. Mice were instilled intracheally with 0.9 mg of diesel particulate. All died within 24 hours; the cause of death seemed to be pulmonary oedema, as mediated by epithelial cell damage (Sagai *et al.*, 1993). Death rates were reduced markedly by prior injection of chemicals expected to promote or to replace the action of natural antioxidant defences. Rats were instilled intracheally with 1 mg or 5 mg of diesel particulate; at 24 hours there was evidence of pulmonary oedema and haemorrhaging (Hälinen *et al.*, 1999). Hamsters intracheally instilled with $50\,\mu g$ of diesel particulate experienced considerably increased venous thrombosis in an hour, extending for

24 hours (Nemmar *et al.*, 2003a); 60 nm polystyrene particles administered intravenously had similar effects (Nemmar *et al.*, 2002a).

C. *Exposure to extracted organics.* The PAH, being infamous as carcinogens and suspected carcinogens, and omnipresent wherever organic compounds are burned, have motivated considerable research, far beyond the bounds of motor vehicle particulate, and this extensive body of research cannot be reviewed here. Administered orally, PAH produce tumours of the forestomach, liver, lungs, mammary glands, pituitary glands and colon (Larsen and Larsen, 1998). Administered dermally, organic extracts of diesel particulate produce skin cancers in mice (Griffin *et al.*, 1981). Administered intracheally, *BaP* produces lung tumours in 90 % of rats at 30 mg, and in 25 % at 15 mg (dose subdivided into 16 weekly applications) (Dasenbrock *et al.*, 1996). Rat embryos at ten days, exposed to organic extracts for 48 hours, show abnormalities in growth and development (Shore *et al.*, 1987); reductions in rump and head length were observed for gasoline engine particulate, but the results were not significant for diesel engine particulate.

D. *Exposure to organic-extracted particulate.* Exposure to particulate denuded of its sorbed organic compounds assists in partitioning the relative roles of each. Mice instilled intracheally with 1.0 mg methanol-extracted diesel particulate survived those receiving unextracted particulate (Sagai *et al.*, 1993). Rats were intracheally instilled with 15 mg of particulate, subdivided into 16 weekly applications; one group received unextracted particulate, another received toluene-extracted particulate (Dasenbrock *et al.*, 1996). The tumour rate was higher in the former (17 %) than in the latter (4 %). The carbonaceous core (agglomerate) and the adsorbed organics were suggested to have a *combined* tumour-inducing effect.

E. *Exposure to organically recoated particulate.* Extracted particulate, *recoated* with BaP at 11 µg/mg, caused, following intracheal instillation into rats, a lower tumour rate than did 'as-emitted' particulate, which contained only 0.9 ng/mg *BaP* but a variety of other PAH and NPAH, thus demonstrating the importance of sundry other organic compounds in tumour induction (Dasenbrock *et al.*, 1996). Recoating experiments have also been used to measure the release rate of *BaP*, following deposition in the lung of the dog: starting from 25 % of a monolayer, some was desorbed from the particulate and absorbed into the circulating blood in just a few minutes. After this point, absorption rates declined drastically; 16 % of a monolayer remained even after five and a half months – this fraction, presumably, not being bioavailable (Gerde *et al.*, 2001).

F. *Exposure to carbon black.* Particles of carbon black, which are, relative to diesel particulate, deficient in organic compounds, still induce tumours at similar rates in rats, when intracheally instilled at 15 mg, subdivided into 16 weekly applications (Dasenbrock *et al.*, 1996). This is the case even for toluene-extracted carbon black, which can be as tumour-inducing as unextracted diesel particulate. On this basis, it was thought that adsorbed organic compounds make a comparatively minor contribution, and that the carbon core, via size and surface area, is the major causative factor in the tumours.

G. *Exposure to sulphuric acid.* This extensive corpus of research, which has been reviewed (Amdur, 1996), strongly suggests sulphuric acid to have been overlooked. Two metrics are airway resistance and mucociliary clearance; data for the latter are shown in Figure 10.7. Guinea pigs (rats are insensitive to sulphuric acid) experience a discernible increase in airway resistance. There are, moreover, three intriguing findings. First, animals whose airway resistance is already in the upper range are more sensitive. Second, this sensitivity relates *not* to the acid administered as gas or vapour, but to the acid as an *aerosol*. Third, greater sensitivity to the acid is less marked at high doses – arguing against the practice of extrapolation to low doses. Donkeys exposed for one hour per day, five days per week, for several months, to repeated doses of $100 \mu g/m^3$ displayed profound changes in mucociliary clearance rates, persisting for three months after cessation. Rabbits experienced accelerated clearance at $200-300 \mu g/m^3$, but retarded clearance at $1000-2000 \mu g/m^3$.

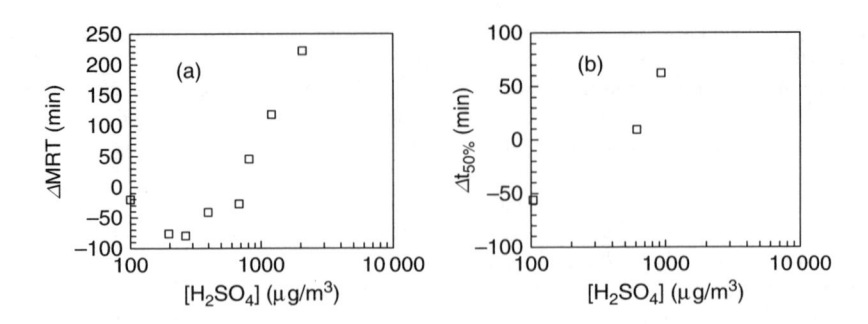

Figure 10.7 Exposure-dependent changes in bronchial mucociliary clearance: (a) rabbits, expressed as mean residence time (MRT); (b) humans, expressed as halftime. Exposure was one hour, to submicrometer sulphuric acid aerosols. (Cited by Amdur, 1996; original reference Schlesinger *et al.*, 1984.)

H. *Exposure to metals.* The deleterious effects of metals, through the generation of ROS and consequent oxidative stress, are difficult to resolve in motor vehicle particulate, and it is not inappropriate to mention research on residual oil fly ash (ROFA), where metals predominate; in this case the biochemistry, following lung deposition, is complex (Donaldson and MacNee, 1998). Fly ash particles are of especial concern because toxic metals (Li, Na, S, K, V, Cr, Mn, Fe, Tl and Pb) are present on their surfaces, which, often soluble, are bioavailable, with enhanced concentrations at the point of contact (Injuk *et al.*, 1998). Experiments on rats have linked these metals to hypertension and pulmonary injury (Kodavanti *et al.*, 2000). Inhalation by rats of an admittedly high ROFA dose, 15 mg/m^3, for six hours per day over three days resulted in irregular heartbeats (arrhythmias). Other observations are depletion of antioxidants (Jiang *et al.*, 2000) and sensitisation to the house dust mite (Lambert A.L. *et al.*, 2000). The effects are linked in some way to sulphuric acid (Gavett *et al.*, 1997).

I. *Exposure to sulphuric acid sorbed on metal oxides.* Extensive research, which has been reviewed (Amdur, 1996), highlights the considerably worsened effects when metals and sulphuric acid act in concert. In guinea pigs, impaired oxygenation of the blood across the alveolar – capillary membrane and damage to airway epithelial cells correlated with the amount of sulphuric acid sorbed on the particles. There was evidence of pulmonary oedema and distension of the alveolar interstitium. These effects were clear at exposures of 60 µg/m^3, following a single three-hour exposure, but slight at 30 µg/m^3. Lungs exhibited some recovery, but there was greater sensitivity to re-exposure. Expressed in terms of impaired oxygenation of the blood, sulphuric acid adsorbed onto zinc oxide particles was a respiratory irritant ten times more potent than the acid aerosol acting alone. Research into particles discharged in combustion suggests the effects arise not from the sulphate *per se* but from acidity, on the grounds that neutralisation of this acid brings about some amelioration.

J. *Exposure to nanoparticles.* Materials hitherto found to be inert or relatively benign when administered as *fine* particles (hundreds of nanometres) prove significantly more toxic as *ultrafine* particles (<100 nm); the impact, not unexpectedly, is in the lower reaches of the respiratory tract, especially in the alveolar region. Rats were intracheally instilled with TiO$_2$ and carbon black, both as fine and as ultrafine particles, and at equal mass (Renwick *et al.*, 2004). The ultrafine particles were markedly more harmful in terms of inflammation, damage to the epithelium, cytotoxicity, and inhibition of phagocytosis; and there were indications that macrophages would more likely be retained in the lung, rather than migrate along clearance routes. The inflammatory response induced by intracheal installation of ultrafine and fine particles of TiO$_2$ correlates well with surface area, but poorly with mass (Oberdorster et al., 1992). Following intracheal instillation, the greatest incidence of lung tumours in rats was observed for a carbon black of the highest surface area (270 m^2/g) and the smallest spherules (14 nm) (Dasenbrock et al., 1996). Similar results are found with inhalation experiments. Rats inhaling ultrafine (20 nm) and fine (250 nm) TiO$_2$ particles, at a mass concentration of 23 μg/m^3, experience significantly retarded rates of macrophage-mediated clearance for the ultrafine particles, but not for the fine particles (Oberdorster et al., 1992). Teflon, if inhaled for a *few tens of minutes* by rats as 10 nm particles at concentrations of 9 μg/m^3 and 10^5 cm^{-3}, causes extensive pulmonary inflammation, and even haemorrhaging (cited by Utell and Samet, 1996, and Amdur, 1996). The point is that Teflon is considered as essentially an inert material. These and other findings suggest the primacy of particle size in toxicity, perhaps involving surface area rather than some chemical aspect of the material. It should be noted that nanoparticle-induced effects are seen at mass concentrations of *micro*grams per cubic metre; and so, unlike many of the aforementioned studies, which used *milli*grams per cubic metre, these effects cannot be so easily dismissed as mere artefacts of unrepresentative doses.

10.8 Humans

In so-called 'human challenge' or 'clinical' studies, healthy volunteers or persons with pre-existing but mild diseases are exposed, usually in a special chamber, to test aerosols for short periods (minutes). Subjects are always adults, ethical considerations generally ruling out the use of children, and they also tend to be from younger age groups, rather than the elderly; hence, those persons most at risk in the human population at large are not evaluated. The study may include the effects of respiration rate, determined, for example, by exercising on a treadmill. Outcomes are only mild and transient, typically defined in terms of 'lung function', for example FEV or FVC; bronchoscopy and bronchiolar lavage are also used.

The largest relevant body of work, relevant, that is, to motor vehicles, some of which is quite long-standing (Higgins I.T.T., 1971), relates to aerosols of sulphuric acid (Utell and Samet, 1996). Asthmatics, especially adolescent asthmatics, are more sensitive to these aerosols than healthy persons, experiencing measurable increases in airway resistance even for tens of minutes at <100 μg/m^3 (cited by Amdur, 1996). Mucocilary clearance rates accelerate at 100 μg/m^3, but decelerate at 1000 μg/m^3 (see Figure 10.7); and the longer the exposure, the more protracted, following removal of the subject, is the period of slowed clearance. Upon re-exposure, individuals show an increased sensitivity to these aerosols. Few effects are observed at the alveolar level, and individuals with COPD are less affected than asthmatics. Neutralisation of the acid with ammonia mitigates the effects; yet it remains unestablished whether the important factor is the acidity (H$^+$) or the sulphate ion (SO$_4^{2-}$).

Radioactive labelling has been used to study particle translocation within the body, or clearance kinetics, rather than toxicity (Spurny, 2000b). Radioactively labelled carbon particles of size 5–10 nm were inhaled by human subjects, and the radioactivity monitored in the body. This provided plausible evidence that nanoparticles are able to pass rapidly from the lungs into the systemic circulation – having penetrated the gas – blood barrier and translocated to the liver and other regions of the body, in just *a few minutes* (Nemmar et al., 2002b). The radioactivity seemed to relate to particles that had penetrated intact, rather than through dissolution, and the rapidity of the translocation tends to rule out the work of macrophages.

Clinical studies using *real* motor vehicle particulate are rarer, and, obviously, focus only on the less inimical aspects of particle exposure, meaning acute rather than chronic health effects, and the lowest exposures. After one hour's exposure to diesel particulate, a proliferation is seen in the lung of macrophages and other white blood cells associated with increased production of proinflammatory cytokines (Holgate, 1999). Such changes are not, though, apparent in standard cardiovascular parameters (Nightingale *et al.*, 2000) or measurements of lung function (Salvi *et al.*, 1999); nor are they eliminated by applying a particulate filter to an idling engine (Rudell *et al.*, 1999). In healthy, nonsmoking persons exposed for an hour to diesel particulate at $200\,\mu g/m^3$, increases in nasal-cavity ascorbic acid appear sufficient to prevent further oxidant stress in the respiratory tract (Blomberg *et al.*, 1998). Changes in the production of IgE upon exposure to a combination of ragweed allergen and diesel particulate suggest synergy to be a key feature in the expression of respiratory allergy (Diaz-Sanchez *et al.*, 1997).

10.9 Closure

The mechanisms leading to particle-induced diseases are obviously extremely complicated, and the present chapter has only been able to sketch a cursory outline of the salient ideas. We began with some disparate thoughts about public health issues, including how persons are exposed to motor vehicle particulate in various environments and microenvironments. Then various aetiologies were traced, starting from the moment of inhalation, and proceeding to the manifestation of actual diseases. In the subsequent section, the study of disease statistics, or epidemiology, highlighted some concerns raised by practitioners in occupational and public health. Finally, the findings of scientists working with tissue samples and cell cultures (*in vitro*), laboratory animals (*in vivo*) and human subjects were discussed.

10.9.1 Public Exposure

Typical urban environments carry motor vehicle particulate at mass concentrations (PM10) of $\sim 10-20\,\mu g/m^3$, with diesel engines making a significant contribution. Conventional understanding holds that rapid transfer takes place from the nucleation mode into the accumulation mode; thus the dangers posed by nanoparticles might well diminish rapidly with distance from roadways. Motor vehicles are responsible for primary nanoparticles, and indirectly responsible for secondary nanoparticles formed subsequently by atmospheric processes. Accumulation-mode particles travel the largest distances, yet these deposit less efficiently in the lung. Particle hygroscopicity is enhanced by long-term ageing in the atmosphere. Penetration of building envelopes by the accumulation mode is efficient; hence exposure might not be greatly influenced by the proportion of time spent indoors. The air quality within vehicle compartments is often overlooked, but some people spend large portions of their lives in such microenvironments.

10.9.2 Public Health

Several national and international organisations have branded diesel particulate a 'probable' or 'potential' human carcinogen. It has been noticed that particles act as carriers for allergens. This is intriguing, as asthma, especially childhood asthma, is now extensive in the developed world, whereas in the Third World, this disease is still relatively unknown.

10.9.3 Pathogenesis

The first step in pathogenesis is particle deposition in the lung: this follows well-established fluid-mechanical laws. Coarse-mode particles deposit in the oral and nasal cavities; accumulation-mode particles stand the best chance of avoiding deposition completely, and hence of being exhaled; and nucleation-mode particles are adept at reaching the alveoli. Sneezing ejects particles from the nasal passages; others deposited in the trachea and bronchi are removed in a few hours by the mucociliary escalator, after which they are swallowed or expectorated. The alveoli are unciliated, and particle clearance, undertaken by white blood cells called macrophages, is slow, taking weeks or even months. The macrophages, which engulf, or phagocytose, the particles, make their way to the escalator. Clearance is hindered if particles are cytotoxic to macrophages. Particles, if they pass through the pulmonary epithelium, can become lodged in the interstitium, and thereby escape the normal lung clearance mechanisms. Particles can travel, or translocate, to other sites in the body, such as the liver and lymph nodes. Lung fluid dissolves material from particle surfaces, thus aiding assimilation. The generation of reactive oxygen species (ROS), i.e. O_2^- and OH, and consequent oxidative stress, is suspected to underpin more than one disease. ROS could arise through the catalytic action of finely divided metals. The general response to particle residence in the lung is inflammation. Asthma is an inflammatory disease of the upper airways caused by the production of histamine, in which there occurs excessive mucus secretion and contraction of the airways. Bronchitis is another inflammation of the upper airways, characterised by coughing and expectoration. It is less easy to explain the effect of particles on the cardiovascular system, i.e. heart attacks and strokes, but there is a suspicion of enhanced blood clotting, as mediated by release of cytokines (chemical messengers), or loss of deformability in leukocytes (white blood cells) as they negotiate the microcirculation. The view generally entertained of cancer is that PAH, or rather their metabolised products, react with DNA to form 'adducts'. However, the first step is the release of PAH from particle surfaces. PAH interact metabolically, and engender synergisms that are poorly understood.

10.9.4 Epidemiology

The oldest fears about motor vehicle particulate, and indeed combustion-derived particulate generally (e.g. chimney soot), concern cancer. This must be gauged as a 'risk', as no identifiable threshold exists for carcinogens. The dose – response relationship is assumed to be linear, and the low-level risk simply computed by extrapolation. Many epidemiological studies are still disputed, not just by vested interests, but also by fellow epidemiologists. It is difficult to account for confounders such as cigarette smoking, diet and ethnicity, and knowledge of real exposures is insufficient. Since cause and effect are often separated by many years, or even decades, this adds an additional layer of complexity, as risks have probably changed during the selected periods of analysis, in accordance with the many technological improvements brought to diesel engines in the same period. Two groups of people are studied in epidemiology: workers and the public. Exposure is more tightly circumscribed in the former, i.e. those who work in close proximity to diesel engines, such as on the railways. Many epidemiological studies have indeed shown associations between occupational exposure to diesel engines and lung cancer. But more recently, concerns have shifted to cardiovascular effects. Parallel studies of the general public have persistently shown an association between the mass of suspended particulate and statistics for mortality and morbidity – not for cancer, but rather for cardiovascular and respiratory ailments such as heart attacks, strokes, asthma and bronchitis. Associations with PM2.5 are stronger than with PM10; this tends to imply the greater significance of anthropogenic, combustion-derived particles, and the lesser relevance of coarse, naturally arising particles. From the compositional standpoint, sulphates are particularly implicated. However, the risks, if any, are small compared with those associated with cigarette smoking.

10.9.5 In Vitro

The *in vitro* touchstone for motor vehicle particulate, for many years, has been the Ames bioassay for mutagenicity. Alternative but less used bioassays have been developed: these include other microbes such as yeast, the detection of strand breaks in DNA, formation of DNA–PAH adducts in calf thymus, and the depletion of antioxidants in plasma. *In vitro* investigations reveal corruption of DNA in mammalian cells; alterations in rat lung tissue are fewer when particles are filtered from the exhaust prior to exposure. The ability of tissue to suppress the growth of mutated cells is impeded. ROS, not just PAH, are implicated in this damage to DNA; the two are in some way connected: for example, the metabolism of PAH may generate ROS. Bioassays have shown that persons suffering from asthma and COPD have depleted levels of antioxidants. The administration of antioxidants restricts oxidative stress, inflammatory responses and ROS-induced damage to DNA.

10.9.6 In Vivo

Rats and mice and, to a lesser extent, hamsters, guinea pigs, rabbits, monkeys and dogs have been used to probe the health effects of particulate. Exposure is via polluted air or intracheal instillation. Inhalation exposures, expressed as milligrams per cubic metre of air, are one or sometimes two orders of magnitude higher than those humans are likely to experience, even in heavily polluted environments. This is because of the limited lifespan of laboratory animals, and because the statistical nature of a disease may afford no opportunity for its occurrence in a small group. Unfortunately, high exposures might lead to unrepresentative health outcomes, for example a breakdown in alveolar clearance. With these provisos, motor vehicle particulate does induce cancer in laboratory animals; there are, moreover, various other effects: pulmonary oedema, mucus hypersecretion, carbon accumulation in lymph nodes and a reduction in the population of macrophages. The depletion of reductants in the lung fluid of rats, and tissue damage, i.e. lesions of the epithelium, suggest the action of ROS. Sulphuric acid may have been improperly downplayed: this is associated with increased airway resistance and slowed mucociliary clearance. The acid seems particularly potent when administered as an adsorbate on other particles. PAH have adverse effects on the reproductive performance of mice and rats, and cause skin cancer in mice. Tumours are observed in several organs. The rate of release is a factor in carcinogenesis; it is swiftest in the minutes following deposition, but then slows down, and some PAH do not appear particularly bioavailable. The carcinogenicity of PAH is in some way enhanced through attachment to particles, perhaps because of the inflammatory effects. Because particles relatively free of organics (carbon black) and inorganic particles (TiO_2) also induce cancer, it is suggested that PAH have a lesser role, and that the underlying carbonaceous agglomerates are really responsible. Particles in the nanorange generate considerably greater inflammatory effects than the same material in the micron range, implicating particle size rather than composition. The ability of macrophages to clear the deep lung is related to particle surface area and volume; the cells become dysfunctional for reasons that are only just being elucidated.

10.9.7 Humans

Human challenge studies show that asthmatics are more sensitive to sulphuric acid aerosols. There are measurable increases in airway resistance, mucociliary clearance rates are altered, and individuals become sensitised on re-exposure. Radioactively labelled particles easily translocate from the lung to distant regions in the body. Exposure to diesel engine particulate, as an aerosol, causes a proliferation of macrophages.

10.9.8 Which Particulate Fraction?

A generation ago it seemed indisputable that the health risks stemmed from organic compounds, since many of the PAH residing in, or rather on, the particles were known carcinogens. However, extensive in vitro and in vivo research has been unable to exculpate the carbonaceous, sulphate and ash fractions; and epidemiology now suggests cardiac and pulmonary diseases to be the risks. In the face of so many biochemical interactions and the great chemical complexity of particulate, it seems unlikely that any Grand Unified Theory for these particle-induced health effects will emerge. This understanding contrasts, for example, with the once-held notion that sulphuric acid is a pollutant of the past, that the battle with this pollutant has been fought and won, and that the health risks to which the public are exposed have now disappeared (Higgins I.T.T., 1971; Wilson R., 1996); indeed, the concern is now with invisible rather than visible particles.

10.9.9 Healthy Antioxidant Diet

The notion that antioxidant-rich diets confer, in some measure, protection against cancer, and indeed many other diseases, has gained a considerable following, far outside the current topic of discussion, and indeed into the public arena; for example, we are constantly urged to consume more fresh fruit and vegetables. Hence there is strong *scientific* evidence, rather than pseudo-scientific mumbo-jumbo, that those individuals who consume oranges, spinach, etc. protect themselves against various diseases contracted from or accentuated by inhaled particles.

10.10 Glossary of Biomedical Terms

Acute. Relating to the short term.
Adenosine triphosphate. A compound essential in energy transfer in the cell; part of the metabolism.
Aetiology. The cause of a disease.
Atherosclerosis. Hardening and furring of the arteries by deposits (which can later be released, creating a thrombus).
Axon. The nerve fibre along which electrical impulses are transmitted.
Chronic. Relating to the long term.
Cytokine. Signalling molecules that affect interactions between cells in the immune response.
Cytotoxic. Toxic to cells.
Emphysema. Enlargement of the alveoli, which causes breathlessness.
Fibrin. The fibrous, insoluble protein that forms the structure of a blood clot.
Fibrinogen. A plasma protein synthesised by the liver.
Forced Expiratory Volume (FEV). The amount of air expelled from the lungs in a certain time (usually one second, FEV_1).
Functional vital capacity (FVC). The maximum amount of air expelled from the lungs, following full inspiration.
Glutathione (GSH). A naturally occurring antioxidant.
Haemostasis. The stopping or slowing of blood (or of the circulation).
Hyperplasia. An abnormal increase in the number of normal cells in normal arrangements within a tissue.
IgE. Immunoglobulin E, a class of antibody involved in allergic reactions; associated with hypersensitivity.
Inflammation (inflammatory response). A nonspecific defensive reaction of the body to invasion by a foreign substance.
Interstitium. Small region of connective tissue (in an organ).
Instillation (intracheal). Administration of a test substance by bypassing the upper respiratory tract.
In vitro. In an artificial container.

In vivo. In a living organism.
Ischaemic. Localised tissue anaemia due to obstruction of the blood supply.
Isoflavones. A type of plant oestrogen found chiefly in soybeans.
Lesion. An adverse change in the texture or functioning of an organ.
Lipid. Organic compounds insoluble in water but soluble in organic solvents (fatty acids, oils, waxes and steroids). Includes cell membranes.
Lymphoblast. A dividing cell that gives rise to a mature lymphocyte (white blood cell).
Macrophage. A mobile scavenging cell representing part of the cell-mediated immune system.
Oedema. The accumulation of abnormal amounts of fluid in the tissues and lungs.
Oxidative stress. See under 'Reactive oxygen species'.
Pathogenesis. The study of the origin, cause and development of disease.
Phagocytosis. The engulfing of foreign particles by certain cells, including macrophages.
Proteases. Enzymes that digest proteins.
Reactive oxygen species (ROS). These cause damaging biochemical reactions leading to oxidative stress, thought to underlie many diseases.
Thrombosis. Coagulation or clotting of the blood (in a vessel). Formation of a thrombus.
Tumour necrosis factor (TNF). A type of cytokine.

11

Closure

11.1 Recommendations for Research

The number of papers consulted in preparing this monograph comfortably exceeds three thousand, and, within this considerable corpus of knowledge, certain concerns are repeatedly raised, while the reporting of certain other areas is manifestly poor. The salient issue, perhaps, is whether a headlong rush towards lower particulate mass has resulted in larger numbers of smaller particles; this question will be considered separately in Section 11.2. Various other issues are enumerated in this section, as follows.

11.1.1 Signal-to-noise Ratios

Conventional particulate measurement protocols, as used in homologation or type approval, use gravimetric assessments of dilution tunnel filters. These protocols are reasonably reproducible and well established, and comfortably underwritten by many years of hard-won experience. However, they were originally designed for vehicles with far higher emissions than those of today. Particulate emission rates have declined to such an extent that signal-to-noise ratios are becoming, frankly, an embarrassment; errors introduced by extraneous factors represent ever-greater proportions of the result; and sensitivity towards confounders has swelled correspondingly. The technical hurdles to be overcome in particulate measurement have therefore multiplied; and merely continuing to weigh dilution tunnel filters to ever-greater precision, on its own, will be insufficient, while subtleties in measurement protocols and nuances in instrument design must be subjected to the minutest scrutiny. Indeed, one could argue that progress in particulate abatement is hampered for want of appropriate instruments, just as much as through limitations in the control methods. As Vuk *et al.* (1976) remarked, '... meaningful control standards are related to the extent of development of measurement techniques. Measurement aids understanding of behaviour in the environment and particle formation processes'. This statement, admittedly made quite some years ago, is not for that reason any less relevant today. Statutory tailpipe limits for particulate, or for that matter any pollutant, cannot be reduced until measurements become unreliable, otherwise these legal restrictions are utterly meaningless. Statutory limits and the technology of measurement must proceed hand in hand.

Particulate Emissions from Vehicles P. Eastwood
© 2008 John Wiley & Sons, Ltd

11.1.2 Statutory Test Cycles and Real Emission Rates

A long-standing question, not solely with particulate but with most pollutants is whether statutory test cycles used in type approval really reflect true emission rates in the field. These are the so-called 'off-cycle' emissions. The accelerations used on the NEDC, for example, are notable for their gentleness, being, to all intents and purposes, quasi-steady states when compared with customer 'foot-to-floor' driving. While this discrepancy does not hold automatically for all pollutants, one particularity of soot is its great sensitivity to transient manoeuvres, during which momentary enrichment of the air – fuel ratio is a strong causal factor. There is, therefore, some justification for supposing that field emissions are higher than supposed from statutory test cycles To improve the database of 'real-world' particulate emissions, miniaturised *on-vehicle* instrumentation is desperately required (Booker *et al.*, 2007).

11.1.3 Inspection and Maintenance

Individual vehicles that pollute disproportionately, i.e. 'gross polluters', vitiate substantially the gains made on aggregate with new emission control technologies (Weaver *et al.*, 2000); a smoking vehicle discharges the same mass of particulate as one hundred nonsmoking ones (Cadle *et al.*, 1999a), and a small proportion (5 %) of passenger cars accounts for a large proportion (43 %) of the soot (Kurniawan and Schmidt-Ott, 2006). Relevant here are component wear, customer neglect, inexpert maintenance and, on occasion, downright tampering (Jacobs *et al.*, 1998). This issue is known as 'in-use compliance'. Catching out-of-compliance vehicles is arguably an easier task with publicly owned fleets than those in private ownership (Reul-Chen *et al.*, 2004). The gross polluter is detectable in statutory (e.g. annual) inspections, and also by pulling vehicles over for roadside checks. But far greater numbers of vehicles can be speedily surveyed by 'remote sensing' (Li W. *et al.*, 2007), i.e. roadside units that monitor tailpipe emissions from vehicles as they pass (Moosmüller *et al.*, 2003); and civic duty, where wanting, can be aided by the imposition of fines, *pour encourager les autres*. But should particle numbers also become an aspect, then in-use compliance is orders-of-magnitude more difficult than in the old days of regulation by smoke (Klausmeier *et al.*, 1985). Particle-sizing instruments are delicate and costly units unsuited to garage environments, and they require levels of scientific skill and understanding not expected of the average garage mechanic. At least, this want of ruggedness is the general expectation: low-cost, compact nanoparticle units are under development.

11.1.4 The Soot Sensor in Engine and Aftertreatment Management

The introduction of the DPF means that the smoke emissions historically used to detect problems with diesel engines will no longer be available, except in instances were this unit has already failed. One great difficulty with this technology is that engine management systems lack direct knowledge of the engine-out soot emission: whether or not regeneration should take place is a decision based on indirect or inferred knowledge about the state of the DPF. An on-board exhaust sensor, capable of reporting the soot emission rate to the engine management system and thereby furnishing feedback, would be absolutely invaluable. Of course, there are many *other* ways in which an engine-out soot sensor could be used (air path management, etc.).

11.1.5 Surface Area Distribution

In the last ten years, two particle-sizing instruments, the ELPI and the SMPS, have seen widespread use in the automotive industry, and an extensive database of demographic measurements has been built, where previously very little was known. Not unnaturally, in view of the operating principles of these instruments, eyes have focused, virtually to the exclusion of all other demographic aspects, on number distribution and mass distribution. Yet *surface area* is central to toxicology – it governs interactions with cells, and the ability of particles to carry volatile material, for example – and hence it seems appropriate to measure the surface area distribution as well. This parameter has, however, been mostly ignored by researchers in the automotive industry, even though suitable instruments are certainly available (Kasper *et al.*, 2001).

11.1.6 Instrumentation for Number-based Legislation

In a few years, an additional layer of legislation will likely be enacted, proscribing particles by *number*. Any such shift will increase measurement complexity; the deficiencies of conventional protocols, involving dilution tunnels, become especially manifest in particle enumeration. The problem is irreproducibility in the nucleation mode: nucleation among volatile particle precursors is highly nonlinear, while the take-up and release of volatile material from or to sample-line walls, etc. has promotional or suppressing effects, the control of which is notoriously difficult. Yet, if nucleation cannot be controlled appropriately, the results reported by particle-counting instruments will be unpredictable, meaningless, misleading or downright bogus. This calls for substantial precautionary safeguards in measurement protocols (Holmén and Qu, 2004). The developing policy is to legislate for particle number according to what is solid or *nonvolatile*, i.e. following desorption and evaporation of what is volatile (Montajir *et al.*, 2007) – as realised by carefully selecting the conditions of dilution (Kasper, 2004) and by thermally conditioning the diluted exhaust to, say, 300 °C (Herner *et al.*, 2007). Even so, avoiding volatile nucleation modes under all conditions is not easy; the distinction between what is solid, volatile and semivolatile is not straightforward, being strongly dependent on the chosen protocol (Ntziachristos *et al.*, 2005), and consistent measurement in the face of 'renucleation' or variations in vehicle operation (e.g. lubricant consumption and storage – release effects in aftertreatment) is nontrivial (Bernemyr and Ångström, 2007).

11.1.7 Nanoparticles in Real Exhaust Plumes (and the Ambient)

The database devoted to nanoparticles in the ambient atmosphere is in its infancy, and must be considerably expanded. Epidemiology is still based almost exclusively on PM10, and to a lesser extent on PM2.5 – two groupings which are very large compared with the sizes at which most particles discharged by motor vehicles are manifested, and which may simply be acting as surrogates for number, or other, stronger and more meaningful toxicological metrics. Some have argued that nanoparticles formed in motor vehicle exhaust are simply artefacts of measurement protocols, as used in the laboratory – unlikely to form in real exhaust plumes, as discharged into the ambient atmosphere. Yet measurements in real, on-road exhaust plumes have belied this claim. The database of 'chase' experiments must now be expanded, so that the true public health risks can be fully established. This will facilitate the development of laboratory instruments with the ability to simulate the real exhaust process more faithfully. Noticeable also is how far theories of nucleation lag behind experimental investigations. This is not the case for airliner exhaust, where, owing to concerns over climate change, theoreticians have been especially active, and some of their ideas might be advantageously redeployed to motor vehicles.

11.1.8 How Should Primary and Secondary Particles Be Demarcated?

Particle formation within exhaust plumes, whether we like it or not, in effect forces a *modus vivendi*: where is the line of demarcation to be drawn between a primary particle and a secondary particle? The pollution is emitted from the tailpipe, yet only at some later point is it actually inhaled. The automotive engineer has little control over the mixing an exhaust plume undergoes in the ambient, and no control whatsoever once wider dispersal, mingling with other pollutants and subsequent atmospheric processes (which may worsen the environmental impact) have taken place (Qian *et al.*, 2007). There seems no hard and fast rule for dividing primary nanoparticles from secondary nanoparticles. Of course, from the particle standpoint, we have actually been here before: photochemical smog consists of secondary particles, yet these are legislated and controlled via their gas-phase progenitors, NO_x and HC. Perhaps nanoparticles could simply be legislated prior to dilution, via their precursors?

11.1.9 Will Gas–particle Partitioning in the Wider Environment Be Affected?

Fears have been voiced that low ambient particulate levels, as vigorously pursued, might inadvertently shift certain volatile pollutants into the *gas* phase, for which uptake by plants, for example, is actually more efficient than for the particulate phase. These pollutants then become concentrated in the bodies of animals that consume the plants; and humans eat the animals. For example, the route from grass to cattle to humans already accounts for the majority of people's exposure to PCDDs, PCDFs and PCBs (Harrad, 1998). Direct inhalation here is *not* the issue; it is indirect ingestion of pollutants, through their entry into the food chain.

11.1.10 The Chemical Compositions of Individual Particles

'Particulate', in the statutory sense, simply refers to whatever a filter collects. Compositional investigations are then conducted on this entire collection of motley material, taken as a whole. These are, in effect, *summed* or *aggregate* evaluations. A similar comment could be made of investigations into ambient air quality, for which aggregate evaluations are often reported for PM10 or PM2.5 (Cecinato *et al.*, 1999). This practice of collectively grouping the particulate loses sight of *individual particles* as they might deposit, for example, in the lung. Studies into the chemical composition of particles *as a function of size* are uncommon, and the compositional database for nanoparticles is still in its infancy. Of course, the practical difficulties such studies pose to analytical chemists are hardly trivial. The recognition that particles of ostensibly the same 70 nm size class can be of two distinct types, 'less volatile' and 'more volatile', is a step in the right direction (Sakurai *et al.*, 2003b).

11.1.11 Toxicity as a Function of Particle Size

As in the case of the preceding comment, toxicological evaluations are commonly conducted on aggregate samples of particulate – there is little systematic knowledge of toxicity as a function of particle size. Studies of ultrafine particles have used unrepresentative materials, such as TiO_2, or those hopefully representative, such as carbon black. Sorely needed are toxicological studies of *size-segregated* samples of real particulate emissions. The same case can be made in the field of air quality monitoring, where PM10 is dominated by relatively benign particles generated from material disintegration, whereas the submicron particles are those generated by combustion, and are thought to be the more inimical ones. Toxicology studies of volatile or soluble nanoparticles, for example, are in their infancy (Fujitani *et al.*, 2006). This observation leads us conveniently to the next section.

11.2 Smaller Particles in Larger Numbers; or Larger Particles in Smaller Numbers

A vital question repeatedly raised in this book is whether current emission control strategies aimed at reducing mass emissions are also, inadvertently, increasing number emissions, particularly in the nanorange; in this section, a few additional points are made along these lines. Partly, this question is answerable from trends in ambient air quality: measurements conducted inside a road tunnel, where motor vehicles are relatively isolated from other sources of pollution, show, for 2004, half as much particulate mass (PM2.5) yet severalfold more particle numbers (7–270 nm) than for seven years previously (Geller M.D. *et al.*, 2005). This finding is certainly consistent with the notion that, whenever solid-phase particle surfaces are less available, the nucleation of gas-phase particle precursors is rendered more likely. Such a conclusion should be carefully tempered, however, with the proviso that saturations are very likely to be elevated artificially by the restricted dispersal available to pollutants in an enclosed tunnel environment. The absence of sunlight, moreover, will very likely impose some additional artificiality on the formation of secondary particles. Thus, the broad applicability of this work, i.e. to urban environments, remains uncertain.

Direct measurements of exhaust gas are the other side of the coin: in a widely cited study, two engines of the same model but different model years (1988 and 1991) were compared (Baumgard and Johnson, 1996). The 1991 engine emitted less than half the particulate mass of the 1988 engine, in accordance with the greater stringency of the mass-based legislation; yet the exhaust gas contained more particles per unit volume – by two orders of magnitude. Others have observed factors of severalfold (Mayer A. *et al.*, 1998). The notion that particle numbers in the nucleation mode increase, *irrespective of whatever happens to particulate mass as a whole*, seems borne out in the data of Figure 11.1 (Wedekind *et al.*, 2000). Two heavy-duty diesels (certified to Stage I and Stage II) were tested using three different fuels. The regulated mass emission did indeed fall from Stage I to Stage II, as suggested

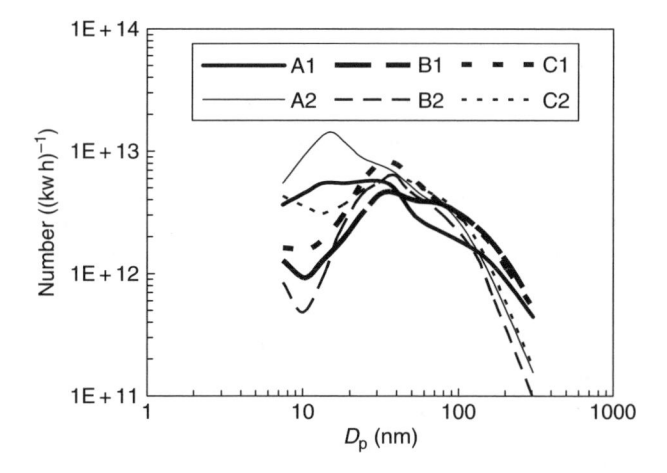

Figure 11.1 Cumulative number distribution over test cycle for two heavy-duty diesels and three fuels. Fuels (cetane no., T95 (°C), sulphur (ppm), aromatics (%)): A, 52, 340, 300, 21.2; B, 52, 340, <10, 4.4; C, 54, 338, 53, 16.4. Test cycle and engine: 1, 2 litres per cylinder, direct-injection, turbocharged, no EGR, Stage I; 2, 1 litre per cylinder, direct-injection, no EGR, Stage II. Instrumentation: SMPS, 7–320 nm. Regulated particulate emissions (conventional dilution tunnel, g/kW h): A1, 0.215; B1, 0.219; C1, 0.187; A2, 0.124; B2, 0.155; C2, 0.149 (Wedekind *et al.*, 2000).

also in the number distribution (see the consistency in the trend for particles of >100 nm). But the nucleation modes varied widely, over more than an order of magnitude; and, *below* 30 nm, the Stage II engine actually emitted particles in greater abundance. Two engines (certified to Stage II and Stage III), similarly differentiated in the accumulation mode according to mass, were approximately on a par in the nucleation mode.

Unfortunately, measures which reduce the accumulation mode tend to boost the nucleation mode: this was shown in the discussions of fuel additives (Section 7.2.4), oxygenates (Section 7.2.3), injection pressure (Section 7.3.2), particulate filters (Section 7.8.2) and alternative combustion systems (Section 7.7). Logically, there is just one underpinning cause: this behaviour is wholly commensurate with a loss of carbonaceous surface area, and correspondingly more vigorous nucleation. If engineers have been more successful in reducing carbonaceous particulate than in reducing organic or sulphate particle precursors, then this situation will, plausibly, increase the likelihood of nucleation. This may have happened: in one study, mass emissions of carbonaceous particulate, from 1988 to 1995, decreased by a factor of five, whereas mass emissions of organic particulate decreased at best by a factor of two (Kreso *et al.*, 1998). It should also be noted that the masses of emitted organics and sulphates, even if extremely low, will nonetheless influence the population of nanoparticles quite profoundly – this is a consequence of the finely divided state, and the highly nonlinear nature of nucleation. But again, the implications these trends hold for public health are difficult to gauge: nucleation exhibited in laboratory tests inevitably reflects the vagaries of exhaust gas dilution and the specific sampling conditions. On this basis – it being utterly impracticable to cover all possible permutations that may instigate nucleation amongst volatiles – the current inclination is to proscribe only that component which is *nonvolatile*, i.e. still exists in the particulate phase after, say, heating the diluted exhaust to 300 °C. Of course, such a distinction may not be pertinent to particle toxicology in the lung.

The proviso that laboratory dilution might provoke artificial nucleation – of a type unmanifested in the real-world emission process – is surely inapplicable to the ash fraction, which transfers into the particulate phase long before the exhaust gas is diluted – probably, even within the engine. But similar concerns may still apply: when less soot is available in the combustion chamber to take up metal vapours, and thereby to depress the saturation of metals, ash nucleation could become more likely: in the converse case, when organometallic fuel additives are administered at high treat rates, this phenomenon does indeed occur (Burtscher and Matter, 2000). Supersaturation of sulphates and organics occurs in the exhaust, so why not supersaturation of metal vapours during the combustion process? Engine-out concentrations of soot from diesel engines have declined enormously over the past generation: have metal concentrations in fuel remained the same? But whatever the nanoparticle composition – ash, organic or sulphate – the inadvisability of preferential reductions in accumulation-mode soot is clear.

Such terminological distinctions may in any case be irrelevant: a core or kernel of refractory material (Abdul-Khalek *et al.*, 1998) – soot, metal or low-volatility organic (Sakurai *et al.*, 2003a) – could be surrounded by volatile coatings of organics and sulphates. Such hybridised nanoparticles, if they exist, differ markedly from the historical archetype, namely the carbonaceous agglomerate – and presumably pose, also, quite different toxicological risks. Intriguingly, a 'thermodesorber' operating at 250 °C reduced particle numbers, for *gasoline* vehicles, by only a third or a half (Li W. *et al.*, 2006). Unfortunately, the size range was wide (7 nm–3 μm, although 'most' particles were smaller than 150 nm), but this result does tend to suggest the presence of refractory kernels. This is a deeply ramifying irony, since the DPF, at the time of writing, is seeing widespread introduction on diesel engines, and ash kernels will be captured efficiently therein, whereas those formed in gasoline engines will be allowed to enter the atmosphere with impunity!

Because the DPF preferentially traps accumulation-mode soot, it will change rather markedly the particulate emission from diesel engines, as traditionally experienced. One should point out, again, that nucleation-mode particles which form *upstream* of the DPF (say, ash) are seldom an issue, according to received wisdom, since capture efficiencies in the nanorange are high, although it should be noted that diesel engines equipped in such a way can still emit substantial numbers of sub-20 nm ostensibly 'solid' particles (several tens of per cent of the total count) (Herner *et al.*, 2007). There is also sufficient evidence to conclude that gas-phase particle precursors which successfully penetrate the DPF will subsequently nucleate, where previously they would have sorbed onto soot particles. The implications here for public health are proximate and regional. In the immediate vicinity of the tailpipe and the exhaust plume, some persons are likely to be exposed to nanoparticles in greater abundance, although the locations of greatest exposure are difficult to define, these being dependent on vehicle driving history and duty cycle, as these determine, in turn, the pattern of storage in and release from the DPF. However, it is conceivable that volatile compounds will suddenly be released at certain characteristic moments (say, when accelerating to high speed after a lengthy period of idling). On larger, regional scales, diesel engines are largely to blame for much of the soot found in urban settings; the widespread disappearance of this soot will, perhaps, bring about more regular nucleation events, wherein unfavourable meteorology and ongoing atmospheric reactions create *secondary* nanoparticles – as already implied by irradiation experiments in smog chambers (Lee S.-B. *et al.*, 2007). This underscores the need for a two-pronged pollution abatement strategy: one focusing on emissions of primary particles *as well as* particle precursors (Kleeman and Cass, 1999).

It should be emphasised that such nanoparticles are not those customarily investigated in nanotoxicology: distinctions between solid and liquid are, perhaps, misleading at these minute scales, as we shall see in the next section, and it may be more helpful, i.e. pragmatic, to talk, again, of what is 'volatile' and 'nonvolatile', or even of 'soluble' and 'insoluble'. Nanotoxicology has undoubtedly focused on nonvolatile or insoluble substances, such as particles of carbon (Hurt *et al.*, 2006) and TiO_2 (Long *et al.*, 2006). The cytotoxicity of nanoparticles, though, has two strands: physical, or irritations imparted by intruding particles; and chemical, or the release of particle components to the body (Brunner T.J. *et al.*, 2006).

Nucleated volatiles will very probably behave quite differently from insoluble nanoparticles once they deposit in the respiratory tract, perhaps dissipating speedily into lung surfactant (Kasper, 2004) and so resembling the distributed take-up of a gas, rather than surviving as coherent particles awaiting macrophage clearance or interstitialisation. But their components are still released to the body: this question does not seem to have been addressed as vigorously as it might. The intermolecular or interatomic bonding within the nanoparticle in relation to the dissolution properties of the surfactant is, naturally, pivotal. But, on the basis that what survives protractedly in the lung as a coherent particle is more likely to be inimical, it is argued that statutory regulations need only be framed in terms of what is nonvolatile, and *will* survive. The counter-argument is that such volatiles will now enjoy, as nanoparticles, high deposition efficiencies in the deep lung, where previously they would have sorbed onto less penetrating accumulation-mode soot, and stood a far greater chance of being exhaled.

The twist is that, since relative humidity is an adjunctive factor in the emergence and growth of these volatile nucleation modes, such nanoparticles are probably hygroscopic (Grose *et al.*, 2006), in which case they may grow in passing down the moisture-laden respiratory tract, thus losing their predisposition for deep penetration: the point of deposition cannot, however, be generalised, because alveolar deposition peaks at $\sim$10–20 nm, and the sizes may be either side of this peak. Moreover, dissolution in the mucous membranes is quite different from in the deep lung. The alveoli are comparatively dry, and the behaviour of water-soluble material therein is poorly understood, but it seems unlikely that nonsolid substances, as *particles*, would be able to pass through cellular membranes into the cytoplasm: they would have to dissolve first (Maynard, 2001). This is quite an interesting line of thought: nanoparticles, according to recent research, are believed to consist of sulphuric acid kernels, around which organic compounds have gathered. Now, in the London smog of 1952, the current hypothesis is that acidity was responsible for the adverse health effects. And, in contemporary epidemiology, associations between cardiorespiratory ailments and sulphates have been found. Might sulphates simply be a measurement of acidity (H^+)? The acidity in the atmospheric aerosol does seem to be biased towards the submicron particles (Yao *et al.*, 2007). Of course, the amount of sulphuric acid deposited in the lung today bears no relation to the 1952 incident, and the buffering capacity of the lung lining fluid may well confer sufficient protection in today's environments.

Much of this is unproven and conjectural, but two key points emerge nonetheless. First, there is indeed some justification for suggesting, tentatively, that current emission control strategies have demographic effects that might prove prejudicial to public health, i.e. smaller particles are created in larger numbers. (Certainly, low-sulphur fuels are a strongly advisable countermeasure.) Secondly, since the accumulation mode contains most of the particulate mass, and since current regulations are defined in terms of mass, then it follows, almost inevitably, that fluctuations in nanoparticles remain largely unquantified in current type-approval procedures, i.e. smaller particles in larger numbers will pass unobserved. The efficacy of current type approval, in number terms, rests entirely on mass acting as a surrogate for number. In certain circumstances there is a correlation between number and mass (Rickeard *et al.*, 1996; Lehmann *et al.*, 2003), say, for the same model of engine on the same test cycle, in the same why that, for an ensemble of identical spheres, the number of individuals must correspond to the total combined mass. Ultimately, this surely relies on one or other of the two modes consistently dominating the emission, as such correlations are invalidated when these modes shift in relative proportion to one another – as is often the case, because of the interaction between solid surface area (accumulation mode) and saturation (nucleation mode).

But instead of making particles ever smaller, why not make them *larger*? Ironically, engineers have already investigated such technologies and abandoned them: growing particles seems an unnecessary complication. If particles attach to surfaces and are allowed to accumulate as deposits, these deposits can be subsequently encouraged to exfoliate, to re-enter the exhaust stream as *supermicron* particles: this, after all, is how coarse modes are created. Such particles, if emitted, fall to earth more rapidly, and are less, rather than more, respirable; they will also furnish surfaces on which volatile particle precursors will attach, thus depressing saturation and discouraging nucleation. Moreover, it would now be feasible to capture them in an aftertreatment device using inertial impaction (Springer and Stahman, 1977) – cyclones being, unless the pressure is reduced considerably (Hsu *et al.*, 2005), ineffective for *as-emitted* particles, yet strong for supermicron particles (Arcoumanis *et al.*, 1994). Such methods were proposed for lead particles generated in the combustion of leaded gasoline (Section 8.2.1). Updating the same strategy, it might be possible to capture nanoparticles this way – provided, of course, they have had the opportunity to form (Wright *et al.*, 2003).

There are several ways of growing particles. (**A**) Within deep-bed filters (Fang C.P. and Kittelson, 1984), dendrite-like agglomerates grow which, having reached a certain size, become unstable, owing to the drag imparted by the flow, and so escape into the exhaust stream (Mayer A. and Buck, 1992). Deep-bed filters have, however, lost out to wall filters, in which supermicron particles are not released; and this technology is being pursued, instead, as simply a means of capturing the accumulation-mode soot prior to incineration (Section 7.8.2). (**B**) Electrostatic attraction, as already practised in powder technology (Riehle and Wadenpohl, 1996), and investigated for motor vehicles (Thimsen *et al.*, 1990), causes similar dendritic growth in agglomerates, the chains of which orientate themselves along the field lines; soot particles have thereby been successfully grown from a few hundred nanometres to several microns (Kittelson *et al.*, 1992). (**C**) High-intensity sound waves, at sonic and ultrasonic wavelengths, although wholly unexplored in the automotive industry, have been used to force particle agglomeration in the flues of coal-fired combustion plants – with no need for interactions with a surface (Gallego-Juárez *et al.*, 1999). (**D**) Thermophoretic forces in the exhaust system become exceptionally strong when a vehicle is accelerated (during which time internal combustion engines are often at their most soot-producing); this might be used to promote thermophoretic deposition in 'flow-through' aftertreatment (Messerer *et al.*, 2006a).

11.3 Smaller and Smaller and Smaller

How small can particles be? Preining (1995, 1998) coined the term 'very, very small particle' to denote, particularly, sizes *smaller than 5 nm*. This is a world still relatively unexplored in motor vehicle emissions, and still outside the range of many emissions laboratories. A schematic depiction of a very, very small particle appears in Figure 11.2 (Preining, 1998). The particle is enveloped by a molecular cloud, the members of which, on scales of nanoseconds, are continually interchanging with the surrounding gas; some molecules are permanently adsorbed; and the particle kernel possesses an outer electronic structure.

The chief peculiarity of nanoparticles lies in their high proportion of *surface* atoms. At one nanometre, a particle contains perhaps only ten atoms, practically all of which reside at the surface; at 5 nm, there may now be a few hundred molecules; yet, even at 10 nm, as much as a quarter or a fifth of the molecules, or about 8000, are still found at the surface. (Obviously, such estimates do not encompass large organic molecules, but only more mundane ones of size ~0.1−0.5nm.) At such small scales, traditional concepts of 'surface' break down, as no such clearly demarcated topology is determinable; and exact values for 'diameter' or 'size' applicable under all conditions are not assignable, as these properties, in the limit, depend on *electronic states*, i.e. the distribution of charge in the electron cloud, and this is a matter of *probability*.

That macro notions of 'surface' break down at minute dimensions is an aspect also of molecules and of atoms; but good grounds exist for regarding very, very small particles as quite different. The mutual interactions between the electronic structures of individual molecules, and the relevant energy states (electronic, vibrational and rotational), are not those of the bulk (solid or liquid), but nor can they be those of isolated molecules (gas); and, since the chemical properties are the natural corollary of these energy states, then very, very small particles inevitably differ in these respects from individual atoms or molecules and from bulk matter, and also from particles as conventionally conceived. Among other things, there must be a transition somewhere from the *elastic* collisions of molecules to the *inelastic* collisions of particles (Tammet, 1995). On the one hand, the terms 'liquid' and 'solid' − and, indeed, 'particle', in the conventional sense − cannot be applied; yet on the other hand, 'atom' and 'molecule' are also improper terms. Very, very small particles are neither fish nor fowl.

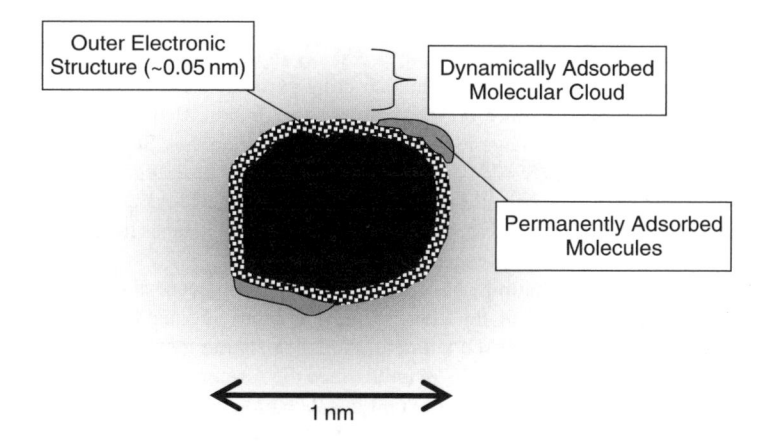

Figure 11.2 Schematic depiction of a very, very small particle (Preining, 1998).

Until fairly recently, it was believed that nanoparticle structures were those of the equivalent bulk material, the surfaces of which terminated similarly (Jefferson and Tilley, 1999). If this were the case, properties dependent on surface area – such as catalysis – would scale inversely with diameter, whereas this is true only up to a certain point, with stronger or weaker dependences being seen for the smallest particles. In nanoparticles, some structures are modified versions of the bulk; others have no bulk counterpart. Moreover, the structure is not endlessly 'periodic' like in the bulk, but varies from surface to centre; and particle shapes can be distorted by adsorbents. The density of reactive sites is greater, because the lattice structure must incorporate more edges and corners, while strange alterations are seen in the energy distribution between surface and interior as particles attempt to maintain their shape and stability (Al-Abadleh and Grassian, 2003). Hence, there are excellent theoretical grounds for regarding very, very small particles as an entirely different form of matter *sui generis*.

Because of these altered energy states, nanoparticles possess unusual electronic, optical, electrical, chemical, magnetic and mechanical properties: hence their suitability for novel applications (Kruis *et al.*, 1998), and hence also the avalanche of interest in nanotechnology. For example, novel adsorptive properties make nanoparticles profoundly effective as catalysts (Sun Y. *et al.*, 2006); these properties differ appreciably from those of micron-sized particles. And it is precisely such unusual properties that are opening up the field of nanomedicine (Emerich and Thanos, 2006). So, from the vantage point of nanochemistry, it would indeed be surprising if the tiniest particles discharged by motor vehicles did *not* possess unusual properties.

Moreover, electronic states are paramount in determining interactions with other systems, *including biological ones* (i.e. *cells*), thus engendering different toxicological risks – not just for so-called 'incidental' nanoparticles, as emitted by tailpipes, but also intentionally engineered nanoparticles in the area of nanotechnology, where workers, the public (Reijnders, 2006; Soto *et al.*, 2006) and the wider environment (Guzmán *et al.*, 2006) require adequate protection. Consequently there have been calls for a moratorium on research in some areas of nanomaterials (Brumfiel, 2003). The role of the media in promoting public fears of 'grey goo' and encouraging obscurantism has been unfortunate. But humans have been exposed for rather longer than commonly supposed: glazes in medieval and Renaissance pottery, after all, derive their attractive lustre from nanoparticles of silver and copper (Padovani *et al.*, 2003).

11.4 Broader Questions of Policy

Since the beginning of the modern emission control era, say in the early 1970s or perhaps late 1960s, engineers have undoubtedly made huge strides in restricting particulate emissions from motor vehicles. In the 1970s, emissions from light-duty diesel vehicles were hundreds of milligrams per kilometre (French and Pike, 1979). In Europe, Stage I (1992) specified 140 g/km, and Stage IV (2005), 25 mg/km, while at the time of writing, Stage V (2010), at 5 mg/km, has been promulgated. This two-orders-of-magnitude reduction has been matched in the heavy-duty sector (McGeehan, 2004). Such progress is really quite remarkable, when one considers that, at each successive stage, many pundits viewed the *next* stage as impracticable and virtually inaccessible.

But since what emerges from the tailpipe has been the principal subject of this text, it is worth pausing to reflect, briefly, on the other side of the coin, namely ambient air quality. *Air quality standards*, which, incidentally, should not be confused with emissions legislation (Maynard and Cameron, 2001), specify particulate mass also; the difference is in units, since the former prescribes mass per unit volume of air ($\mu g/m^3$), and the latter prescribes mass per unit distance travelled by the vehicle (mg/km). Motor vehicles, of course, are just one culprit in particulate pollution, albeit a prominent one; and a detailed consideration of the particulate inventory in the wider environment, the contribution motor vehicles make to this inventory, and the broader implications for air quality could justifiably have formed a separate chapter in this book, as sufficient material certainly exists. (As it was, this text did not proceed far beyond the tailpipe.) The relationship between tailpipe emissions and ambient air quality is a large subject in its own right.

One hundred years have witnessed a gradual but arguably ironic shift in the way societies or governments regulate polluting particles. In former times only smoke was regulated – certainly because visibility furnished an obvious and readily tangible method of measurement, but also because the offensive choking properties and generally unaesthetic nature were unacceptable. These asphyxiating and unaesthetic aspects have improved markedly, almost disappearing from view, in the developed world; yet, in tandem with this improvement, concerns have shifted to what is *not* visible. So, while coarse particles have undoubtedly decreased over the decades, suspicions are emerging that exposures to fine or ultrafine particles have increased throughout the same period (Spurny, 2000a).

Air quality standards say little about the *composition* of these particles. Currently, the only element singled out is *lead*; this proscription dates from the days of leaded gasoline, and so is now of questionable relevance. Even so, it seems inevitable that legislators and policymakers will have to look more closely at *what* is present in suspended particles, rather than just the total mass. Failure to do so could be a problem, if only because of the ingress of natural aerosols. In southern Europe, for example, elevated levels of ambient particles arise episodically, courtesy of African dust storms. Stationary high-pressure regions exacerbate this ingress. Contrariwise, westerly winds and eastward-moving depressions bring rain and cold fronts favouring pollutant dispersal in northern Europe. Thus, without taking into account particle composition, it may prove difficult to meet forthcoming standards for PM10 even at rural sites (Rodriguez *et al.*, 2003). Moreover, recent investigations in metropolitan London indicate that international abatement of particle precursors (SO_x, NO_x and organic compounds) is more logical than controlling primary particles from tailpipes: this is because exceedances of the PM10 standard are often caused by secondary aerosols advected in large air masses from the continent (Charron *et al.*, 2007).

Ambient air quality standards, like emission control laws, as yet make no reference to the number of particles or, for that matter, to the number distribution. Concerns, nevertheless, about particle size led eventually to the introduction of a PM2.5 air quality standard, as an adjunct to the PM10 standard. (Compositionally, this measure also helped to quantify naturally arising coarse-mode dust.) A large database of air quality measurements already exists for PM10, and to a lesser extent for PM2.5; data on particle numbers, especially of nanoparticle abundance, in the environment are scarcer and more controversial. We have learned, on various occasions in this text, that particle number is a considerably more slippery metric than mass; and the nano-database in air quality monitoring is still in its infancy. But, as anyone working in the emissions field knows, the dimension of 2.5 μm is still enormous – by two orders of magnitude – when compared with the sizes of typical particles emitted by motor vehicles, and also other, secondary particles which form in the atmosphere from primary pollutants.

In the USA, the implementation of the PM2.5 standard, and the ensuing costs to meet it, provoked lively and spirited debate between industry and the legislature, and the EPA eventually acted primarily owing to a lawsuit against them issued by the American Lung Association (BMA, 1997). The costs accrued in meeting this standard, which have presumably been passed on to the customer, should be wisely offset against certain credits accrued through abating the deleterious effects, particularly on public health, i.e. health care costs. This is where the balance sheet becomes formidably contentious. Measures to control emissions of particulate in the USA were once estimated as having brought a net financial *benefit*, running into billions of dollars (Walsh, 1983). And the health costs incurred by traffic pollution in Austria, France and Switzerland are put at 1.7 % of the GDP (Künzli *et al.*, 2000).

Members of the public might not wait for engineers to develop new emission control technologies, or, for that matter, for scientists to explain the aetiologies of particle-induced diseases. The litigious nature of contemporary society is no secret, and the public are now so much more conscious of health risks. This is parched brushwood, for which the compensation culture and the pecuniary designs of certain lawyers might act as powerful incendiaries. Unlike the situation for other particle emitters such as fossil-fuel power stations, for example, large numbers of people live and work in close proximity to large numbers of motor vehicles negotiating heavily trafficked arterial routes. In Japan, exceedance of prescribed limits for diesel particulate in the vicinity of a busy road resulted in damages being awarded to asthma sufferers (Walsh, 2001). In 2000, the Kobe district court fined the Japanese government and the Hanshin Expressway Public Corporation 200 million yen, the onus being on these two bodies to keep down levels of airborne particles. The court acknowledged the existence of a relationship between the asthma experienced by the plaintiffs and the suspended particles, and made particular mention of diesel engines, ruling that concentrations within a 50 m distance of the road ought not to exceed $150\,\mu g/m^3$.

Reference to the compensation culture highlights an illuminating (albeit partial) parallel: much ink has been spilt over the health risks of passive smoking. Is it acceptable to pursue one's own interests by degrading the amenity for others? Employers, spurred by the fear of class-action lawsuits, have pushed their smoking employees into 'smoking rooms'; and a growing conundrum, alike, faces restaurateurs, publicans and many owners or managers of places of entertainment. Evidently, a complete ban is required in order to head off the danger that *passive* smokers – customers and employees – will subsequently sue the owners of these establishments for illnesses contracted therein. With cigarette smokers, and the money paid out by the tobacco industry in compensation for illnesses that were self-evident, much of the legal case rests on the narcotic properties of the smoke. Of course, it is impossible to argue that motor vehicle exhaust is narcotic... but neither is this the point with *passive* smoking. The nonnarcotic argument, then, is unlikely to head off legal disputes concerning passive inhalation of motor vehicle particulate.

Further Reading

Anon. (1977) *Fine Particulate Pollution*. United Nations Economic Commission for Europe, Pergamon Press.
Anon. (1997) *Road Transport and Health*. British Medical Association.
Baron P.A., Willeke K. (editors) (2001) *Aerosol Measurement – Principles, Techniques and Applications*. John Wiley & Sons, Inc.
Bohren C.F., Huffman D.R. (1998) *Adsorption and Scattering of Light by Small Particles*. John Wiley & Sons, Inc.
Borman G.J., Ragland K.W. (1998) *Combustion Engineering*. McGraw-Hill.
Brimblecombe P., Maynard R.L. (editors) (2001) *The Urban Atmosphere and Its Effects*, Air Pollution Reviews, Vol. 1. Imperial College Press.
Caines A.J., Haycock R.F., Hillier J.E. (2004) *Automotive Lubricants Reference Book*. Society of Automotive Engineers.
Davies C.N. (editor) (1966) *Aerosol Science*. Academic Press.
Degobert P. (1995) *Automobiles and Pollution*. SAE and Editions Technio.
Edwards J.B. (1977) *Combustion – The Formation and Emission of Trace Species*. Ann Arbor Science Publishers.
Flagan R.C., Seinfeld J.H. (1988) *Fundamentals of Air Pollution Engineering*. Prentice Hall.
Frederick E.W. (1980) *The Technical Basis for a Size Specific Particulate Standard Parts 1 and 2*. Air Pollution Control Association.
Friedlander S.K. (1977) *Smoke, Dust and Haze – Fundamentals of Aerosol Behavior*. John Wiley & Sons, Inc.
Fuchs N.A. (1959) *Evaporation and Droplet Growth in Porous Media*. Pergamon Press.
Fuchs N.A. (1964) *The Mechanics of Aerosols*. Pergamon Press.
Glassman I. (1996) *Combustion*. Academic Press.
Griffin H.E., Brusick D.J., Castagnoli N., Crump K.S., Goldschmidt B.M., Higgins I.T., Hoffmann D., Horvath S.M., Nettesheim P., Stuart B.O., Witschi H. (1981) *Health Effects of Exposure to Diesel Exhaust*, Report of the Health Effects Panel of the Diesel Impacts Study Committee, National Research Council. National Academy Press, USA.
Griffiths J.F., Barnard J.A. (1995) *Flame and Combustion*. Blackie.
Harrison R.M., van Grieken R.E. (editors) (1998) *Atmospheric Particles*. IUPAC Series on Analytical and Physical Chemistry of Environmental Systems, Vol. 5. John Wiley & Sons, Inc.
Harrison R.M., Hester R.E. (editors) (1998) *Air Pollution and Health*, Issues in Environmental Science and Technology, Vol. 10. Royal Society of Chemistry.
Hayward D.O., Trapnell B.M. (1964) *Chemisorption*. Butterworths, London.

Heywood J.B. (1988) *Internal Combustion Engine Fundamentals*. McGraw-Hill.
Hidy G.M. (editor) (1972) *Aerosols and Atmospheric Chemistry*. Academic Press.
Hinds W.C. (1982) *Aerosol Technology – Properties, Behavior, and Measurement of Airborne Particles*. John Wiley & Sons, Inc.
Hinds W.C. (1999) *Aerosol Technology – Properties, Behavior, and Measurement of Airborne Particles*. John Wiley & Sons, Inc.
Kerker M. (1969) *The Scattering of Light and Other Electromagnetic Radiation*. Academic Press.
Kohse-Höinghaus K., Jeffries J.B. (editors) (2002) *Applied Combustion Diagnostics*. Taylor and Francis.
Kouimtzis T., Samara C. (editors) (1995) *Airborne Particulate Matter*. Springer.
Lahaye J., Prado G. (editors) (1983) *Soot in Combustion Systems and Its Toxic Properties*. Plenum Press.
Lawton J., Weinberg F.J. (1969) *Electrical Aspects of Combustion*. Clarendon Press.
Marsh H., Heintz E.A., Rodríguez-Reinoso F. (editors) (1997) *Introduction to Carbon Technologies*. University of Alicante.
Maynard R.L., Howard C.V. (editors) (1999) *Particulate Matter: Properties and Effects upon Health*. Bios Scientific Publishers.
Owen K., Coley T. (1995) *Automotive Fuels Reference Book*. Society of Automotive Engineers.
Phalen R.F. (2002) *The Particulate Air Pollution Controversy*. Kluwer Academic.
Revoir W.H., Bien C.-T. (1997) *Respiratory Protection Handbook*. Lewis.
Schuetzler D. (editor) (1979) *Monitoring Toxic Substances*. ACS Symposium Series 94. American Chemistry Society.
Seinfeld J.H. (1986) *Atmospheric Chemistry and Physics of Air Pollution*. John Wiley & Sons, Inc.
Song C., Hsu C.S., Mochida I. (editors) (2000) *Chemistry of Diesel Fuels*. Taylor and Francis.
Spurny K.R. (editor) (1986) *Physical and Chemical Characterization of Individual Airborne Particles*. Ellis Horwood.
Spurny K.R. (editor) (1999) *Analytical Chemistry of Aerosols*. CRC Press LLC.
Spurny K.R., Hochrainer D. (editors) (2000) *Aerosol Chemical Processes in the Environment*. CRC Press LLC.
Strauss W., Mainwaring S.J. (1984) *Air Pollution*. Edward Arnold.
Torvela H. (1994) *Measurement of Atmospheric Emissions*. Springer-Verlag.
Twomey S. (1977) *Atmospheric Aerosols*. Developments in Atmospheric Science, Vol. 7. Elsevier.
Watson A.Y., Bates R.R., Kennedy D. (editors) (1988) *Air Pollution, the Automobile, and Public Health*. National Academy Press.
Wellburn A. (1994) *Air Pollution and Climate Change – the Biological Impact*. Longman.
Wilson R., Spengler J. (editors) (1996) *Particles in Our Air – Concentrations and Health Effects*. Harvard University Press.
Zhao F., Harrington D.L., Lai M.C. (2002) *Automotive Gasoline Direct-Injection Engines*. Society of Automotive Engineers.
Zhao H., Ladommatos N. (2001) *Engine Combustion Instrumentation and Diagnostics*. Society of Automotive Engineers.

Literature Cited (Cross-referenced Against the Text)

(*Note*: surnames carrying detached prefixes such as 'de' and 'van' are listed under the main root).

A

Aakko P., Nylund N.-O. (2003) 'Particle emissions at moderate and cold temperatures using different fuels'. Society of Automotive Engineers 2003-01-3285. [Cited on pp. 120, 166, 242, 243, 244, 323, 325, 331]

Abbass M.K., Williams P.T., Andrews G.E., Bartle K.D. (1987) 'The aging of lubricating oil, the influence of unburnt fuel and particulate SOF contamination'. Society of Automotive Engineers 872085. [Cited on p. 108]

Abbass M.K., Andrews G.E., Williams P.T., Bartle K.D., Davies I.L., Tanui L.K. (1988) 'Diesel particulate emissions: Pyrosynthesis of PAH from hexadecane'. Society of Automotive Engineers 880345. [Cited on pp. 108, 139]

Abbass M.K., Andrews G.E., Williams P.T., Bartle K.D. (1989a) 'Diesel particulate composition changes along an air cooled exhaust pipe and dilution tunnel'. Society of Automotive Engineers 890789. [Cited on pp. 117, 118, 119, 139]

Abbass M.K., Andrews G.E., Asadi-Aghdam H.R., Lalah J.O., Williams P.T., Bartle K.D., Davies I.L., Tanui L.K. (1989b) 'Pyrosynthesis of PAH in a diesel engine operated on kerosene'. Society of Automotive Engineers 890827. [Cited on p. 108]

Abbass M.K., Andrews G.E., Williams P.T., Bartle K.D. (1989c) 'The influence of diesel fuel composition on particulate PAH emissions'. Society of Automotive Engineers 892079. [Cited on pp. 108, 117]

Abbass M.K., Andrews G.E., Kennion S.J., Williams P.T., Bartle K.D. (1991a) 'The survivability of diesel fuel components in the organic fraction of particulate emissions from an IDI diesel'. Society of Automotive Engineers 910487. [Cited on p. 108]

Abbass M.K., Andrews G.E., Ishaq R.B., Williams P.T., Bartle K.D. (1991b) 'A comparison of the particulate composition between turbocharged and naturally aspirated DI diesel engines'. Society of Automotive Engineers 910733. [Cited on p. 108]

Abbass M.K., Shen Y., Abdelhaleem S.M., Andrews G.E., Williams P.T. (1991c) 'Comparison of methods for the determination of the SOF of diesel particulates and development of a rapid TGA method for the estimation of the unburnt fuel and lubricating oil fractions'. In *Experimental Methods in Engine Research and Development*, Institution of Mechanical Engineers, December 11–12, pp. 109–114. [Cited on pp. 141, 169]

Abdul-Khalek I.S., Kittelson D.B. (1995) 'Real time measurement of volatile and solid exhaust particles using a catalytic stripper'. Society of Automotive Engineers 950236. [Cited on pp. 119, 140, 219, 320, 321]

Abdul-Khalek I.S., Kittelson D.B., Graskow B.R., Wei Q., Brear F. (1998) 'Diesel exhaust particle size: Measurement issues and trends'. Society of Automotive Engineers 980525. [Cited on pp. 96, 97, 392]

Abel E. (1946) 'The vapor phase above the system sulfuric acid – water'. Journal of Physical Chemistry 50:260–283. [Cited on p. 174]

Abraham J. (1996) 'Thermophoretic effects on soot distribution in a direct-injection diesel engine'. Society of Automotive Engineers 960320. [Cited on p. 93]

Abu-Allaban M., Gillies J.A., Gertler A.W., Clayton R., Proffitt D. (2003) 'Tailpipe, resuspended road dust, and brake-wear emission factors from on-road vehicles'. Atmospheric Environment 37:5283–5293. [Cited on pp. 343, 345]

Abu-Qudais M., Kittelson D.B. (1995) 'Experimental and theoretical study of particulate re-entrainment from the combustion chamber walls of a diesel engine'. Insitution of Mechanical Engineers: Proceedings, Part D 211:49–57. [Cited on pp. 93, 136]

Adams K.M., Davis L.I., Japar S.M., Pierson W.R. (1989) 'Real-time, in situ measurements of atmosphere optical absorption in the visible via photoacoustic spectroscopy – II. Validation for atmospheric elemental carbon aerosol'. Atmospheric Environment 23:693–700. [Cited on p. 193]

Adams K.M., Davis L.I., Japar S.M., Finley D.R., Cary R.A. (1990a) 'Measurement of atmospheric elemental carbon: Real-time data for Los Angeles during summer 1987'. Atmospheric Environment 24A:597–604. [Cited on p. 193]

Adams K.M., Davis L.I., Japar S.M. (1990b) 'Real-time, in situ measurements of atmospheric optical absorption in the visible via photoacoustic spectroscopy – IV. Visibility degradation and aerosol optical properties in Los Angeles'. Atmospheric Environment 24A:605–610. [Cited on p. 193]

Agoston A., Ötsch C., Jakoby B. (2005) 'Viscosity sensors for engine oil condition monitoring – application and interpretation of results'. Sensors and Actuators A 121:327–332. [Cited on p. 286]

Ahlvik P., Erlandsson L., Laveskog A. (1997) 'The influence of block heaters on the emissions from gasoline fueled cars with varying emission control technology at low ambient temperatures'. Society of Automotive Engineers 970747. [Cited on p. 326]

Ahlvik P., Ntziachristos L., Keskinen J., Virtanen A. (1998) 'Real time measurements of diesel particulate size distribution with an electrical low pressure impactor'. Society of Automotive Engineers 980410. [Cited on pp. 148, 221]

Ahmad T. (1980) 'Effect of particulates on the measurement of oxides of nitrogen in diesel exhaust'. Society of Automotive Engineers 800189. [Cited on p. 46]

Ahmad T., Plee S.L. (1983) 'Application of flame temperature correlations to emissions from a direct-injection diesel engine'. Society of Automotive Engineers 831734. [Cited on pp. 66, 282]

Ahmed S.H., Rikeit H.E., Hayat S.O. (2000) 'Effect of a non-metallic combustion enhancer diesel additive on mass and number particulate emissions from light duty vehicles and heavy duty engines'. Society of Automotive Engineers 2000-01-1910. [Cited on p. 262]

Akard M., Oestergaard K., Chase R.E., Richert J.F.O., Fukushima H., Adachi M. (2004) 'Comparison of an alternative particulate mass measurement with advanced microbalance analysis'. Society of Automotive Engineers 2004-01-0589. [Cited on pp. 165, 167, 170]

Akihama K., Takatori Y., Inagaki S., Dean A.M. (2001) 'Mechanism of the smokeless rich diesel combustion by reducing temperature'. Society of Automotive Engineers 2001-01-0655. [Cited on pp. 68, 72, 105, 292]

Akinlua A., Torto N., Ajayi T.R., Oyekunle J.A.O. (2007) 'Trace metals characterisation of Niger delta kerogens'. Fuel 86:1358–1364. [Cited on p. 50]

Al-Abadleh H.A., Grassian V.H. (2003) 'Oxide surfaces as environmental interfaces'. Surface Science Reports 52:63–161. [Cited on p. 396]

Ålander T.J.A., Leskinen A.P., Raunemaa T.M., Rantanen L. (2004) 'Characterization of diesel particles: Effect of fuel reformulation, exhaust aftertreatment, and engine operation on particle carbon composition and volatility'. Environmental Science and Technology 38:2707–2714. [Cited on p. 146]

Alatas B., Pinson J.A., Litzinger T.A., Santavicca D.A. (1993) 'A study of NO and soot evolution in a DI diesel engine via planer imaging'. Society of Automotive Engineers 93973. [Cited on p. 106]

Albers P.W., Klein H., Lox E.S., Seibold K., Prescher G., Parker S.F. (2002) 'INS-, SIMS- and XPS-investigations of diesel engine exhaust particles'. Physical Chemistry Chemical Physics 2:1051–1058. [Cited on pp. 229, 230]

d'Alessio A., Gambi G., Minutolo P., Russo S., D'Anna A. (1994) 'Optical characterization of rich premixed CH_4/O_2 flames across the soot formation threshold'. Twenty-Fifth Symposium (International) on Combustion, The Combustion Institute, pp. 645–651. [Cited on p. 69] d'Alessio A., D'Anna A., Gambi G., Minutolo P. (1998) 'The spectroscopic characterisation of UV absorbing nanoparticles in fuel rich soot forming flames'. Journal of Aerosol Science 29:397–409. [Cited on p. 106]

Alfè M., Apicella B., Barbella R., Tregrossi A., Ciajolo A. (2007) 'Distribution of soot molecular weight/size along premixed flames as inferred by size exclusion chromatography'. Energy and Fuels 21:136–140. [Cited on p. 72]

Alkidas A.C. (1984) 'Relationships between smoke measurements and particulate measurements'. Society of Automotive Engineers 840412. [Cited on pp. 203, 204]

Allen J.O., Fergenson D.P., Gard E.E., Hughes L.S., Morrical B.D., Kleeman M.J., Gross D.S., Gälli M.E., Prather K.A., Cass G.R. (2000) 'Particle detection efficiencies of aerosol time of flight mass spectrometers under ambient sampling conditions'. Environmental Science and Technology 34:211–217. [Cited on p. 209]

Amann C.A., Stivender D.L., Plee S.L., MacDonald J.S. (1980) 'Some rudiments of diesel particulate emissions'. Society of Automotive Engineers 800251. [Cited on pp. 63, 81, 83, 114, 115, 161]

Amdur M. (1996) 'Animal toxicology'. In *Particles in Our Air – Concentrations and Health Effects*, Wilson R., Spengler J. (editors), Harvard University Press, pp. 85–121. [Cited on pp. 348, 368, 379, 380, 381]

Ames B.N. (1979) 'The detection and hazards of environmental carcinogens/mutagens'. In *Monitoring of Toxic Substances*, Schuetzle D. (editor), ACS Symposium Series 94, American Chemical Society, pp. 1–11. [Cited on p. 180]

Ammann M., Burtscher H., Siegmann H.C. (1990) 'Monitoring volcanic activity by characterization of ultrafine aerosol emissions'. Journal of Aerosol Science 21:S275–S278. [Cited on p. 30]

Ammann M., Kalberer M., Jost D.T., Tobler L., Rössler E., Piguet D., Gäggeler H.W., Baltensberger U. (1998) 'Heterogeneous production of nitrous acid on soot in polluted air masses'. Nature 395:157–160. [Cited on p. 46]

AML (2000) *Selected Key Studies on Particulate Matter and Health: 1997–2000*. American Lung Association. [Cited on p. 356]

An H., McGinn P.J. (2006) 'Catalytic behavior of pottasium containing compounds for diesel soot combustion'. Applied Catalysis B: Environmental 62:46–56. [Cited on p. 295]

An H., Kilroy C., McGinn P.J. (2004) 'Combinatorial synthesis and characterization of alkali metal doped oxides for diesel soot combustion'. Catalysis Today 98:423–429. [Cited on p. 295]

Anastopoulos G., Lois E., Karonis D., Kalligeros S., Zannikos F. (2005) 'Impact of oxygen and nitrogen compounds on the lubrication properties of low sulfur diesel fuels'. Society of Automotive Engineers 30:415–426. [Cited on p. 252]

Anderson H.R., Atkinson R.W., Bremner S.A., Marston L. (2003) 'Particulate air pollution and hospital admissions for cardiorespiratory diseases: Are the elderly at greater risk'. European Respiratory Journal 21:39S–46S (Supplement). [Cited on pp. 372, 374]

Andersson J.D., Wedekind B.G.A., Hall D., Stradling R., Barnes C., Wilson G. (2000) 'DETR/SMMT/CONCAWE particulate research program: Sampling and measurement experiences'. Society of Automotive Engineers 2000-01-2850. [Cited on p. 283]

Andersson J.D., Wedekind B.G.A., Hall D., Stradling R., Wilson G. (2001) 'DETR/SMMT/CONCAWE particulate research program: Light duty results'. Society of Automotive Engineers 2001-01-3577. [Cited on pp. 186, 241, 242, 243, 244, 280, 323, 324, 330, 331, 332]

Andersson J.D., Lance D.L., Jemma C.A. (2003) 'DfT motorcycle emissions measurement programmes: Unregulated emissions results'. Society of Automotive Engineers 2003-01-1898. [Cited on p. 337]

Andersson J., Preston H., Warrens C., Brett P. (2004a) 'Fuel and lubricant effects on nucleation mode particle emissions from a Euro III light duty diesel vehicle'. Society of Automotive Engineers 2004-01-1989. [Cited on pp. 160, 285]

Andersson J.D., Clarke D.P., Watson J.A. (2004b) 'UK Particulate Measurement Program (PMP): A near US 2007 approach to heavy duty diesel particulate measurements – comparison with the standard European method'. Society of Automotive Engineers 2004-01-1990. [Cited on p. 150]

Andersson J.D., Preston H., Warrens C., Brett P., Payne M. (2004c) 'Lubricant composition impact on the emissions from a heavy duty diesel engine equipped with a diesel particulate filter'. Society of Automotive Engineers 2004-01-3012. [Cited on pp. 150, 285, 296]

Andreae M.O. (1995) 'Climate effects of changing aerosol levels'. In *Future Climates of the World: A Modelling Perspective*, Henderson-Sellers A. (editor), Elsevier. [Cited on p. 24]

Andrews G.E., Ahamed F.M. (1999) 'The composition of spark ignition engine steady state particulate emissions'. Society of Automotive Engineers 1999-01-1143. [Cited on pp. 178, 281, 313, 320, 321, 322, 326]

Andrews G.E., Charalambous L.A. (1991) 'An organic diesel fuel additive for the reduction of particulate emissions'. Society of Automotive Engineers 912334. [Cited on p. 262]

Andrews G.E., Abdelhalim S., Abbass M.K., Asadi-Aghdam H.R., Williams P.T., Bartle K.D. (1992) 'The role of exhaust pipe and incylinder deposits on diesel particulate composition'. Society of Automotive Engineers 921648. [Cited on pp. 109, 230, 281]

Andrews G.E., Abdelhalim S., Williams P.T. (1993) 'The influence of lubricating oil age on emissions from an IDI diesel'. Society of Automotive Engineers 931003. [Cited on pp. 179, 280, 282, 285, 286]

Andrews G.E., Elamir I.E., Abdelhalim S., Ahmed F.M., Shen Y. (1998a) 'The measurement of lubricationg oil combustion efficiency using diesel particulate analysis'. Society of Automotive Engineers 980523. [Cited on pp. 178, 280, 281, 283]

Andrews G.E., Ishaq R.B., Farrar-Khan J.R., Shen Y., Williams P.T. (1998b) 'The influence of speciated diesel fuel composition on speciated particulate SOF emissions'. Society of Automotive Engineers 980527. [Cited on pp. 108, 178, 255, 266]

Andrews G.E., Abdelhalim S.M., Li H. (1999a) 'The influence of lubricating oil age on oil quality and emissions from IDI passenger car diesels'. Society of Automotive Engineers 1999-01-1135. [Cited on p. 286]

Andrews G.E., Li H., Xu J., Jones M.H., Hall J., Rahman A., Saykali S. (1999b) 'Oil quality in diesel engines with on line oil cleaning using a heated lubricating oil recycler'. Society of Automotive Engineers 1999-01-1139. [Cited on p. 286]

Andrews G.E., Li H., Jones M.H., Hall J., Rahman A.A., Saykali S. (2000a) 'The influence of an oil recycler on lubricating oil quality with oil age for a bus using in service testing'. Society of Automotive Engineers 2000-01-0234. [Cited on pp. 178, 197, 286]

Andrews G.E., Abbass M.K., Abdelhalim S., Farrar-Khan J., Ghazikhani M., Ounzain A., Salih F.M., Shen Y. (2000b) 'Unburned liquid hydrocarbons using differential temperature hydrocarbon analysers'. Society of Automotive Engineers 2000-01-0506. [Cited on pp. 48, 105, 118, 197, 266, 267, 321]

Andrews G.E., Clarke A.J., Rojas N.Y., Gragory D., Sale T. (2000c) 'Diesel particle size distribution changes along a practical exhaust system during cold start in a passenger car IDI diesel'. Society of Automotive Engineers 2000-01-0514. [Cited on p. 138]

Andrews G.E., Clarke A.J., Rojas N.Y., Gregory D., Sale T. (2000d) 'Particulate mass accumulation and release in practical diesel engine exhaust systems under cold start conditions'. Society of Automotive Engineers 2000-01-2983. [Cited on p. 294]

Andrews G.E., Clarke A.J., Rojas N.Y., Gregory D., Sale T. (2001a) 'The transient storage and blow-out of diesel particulate in practical exhaust systems'. Society of Automotive Engineers 2001-01-0204. [Cited on p. 137]

Andrews G.E., Clarke A.J., Rojas N.Y., Gregory D., Sale T. (2001b) 'The transient deposition and particle changes across a combined oxidation and storage catalyst under diesel cold start conditions'. Society of Automotive Engineers 2001-01-1951. [Cited on p. 138]

Andrews G.E., Xu J., Jones M.H., Hall J., Rahman A.A., Mawson P. (2001c) 'Oil quality with oil age in an IDI diesel passenger car using an on line lubricating oil recycler under real world driving'. Society of Automotive Engineers 2001-01-1898. [Cited on p. 286]

Andrews G.E., Clarke A.G., Rojas N.Y., Sale T., Gregory D. (2001d) 'Diesel particle size distribution: The conversion of particle number size distribution to mass distribution'. Society of Automotive Engineers 2001-01-1946. [Cited on pp. 220, 221, 222]

Andrews G.E., Xu J., Jones M.H., Hall J., Rahman A.A., Mawson P. (2001e) 'The influence of an on line oil recycler on oil quality from a bus in service using synthetic oil'. Society of Automotive Engineers 2001-01-1969. [Cited on pp. 179, 286]

Andrews G.E., Xu J., Hall J., Rahman A.A., Mawson P. (2001f) 'The influence of an on line oil recycler on emissions from a low emission DI diesel engine as a function of oil age'. Society of Automotive Engineers 2001-01-3617. [Cited on p. 286]

Andrews G.E., Xu J., Sale T. (2002) 'Influence of catalyst and exhaust system on particulate deposition and release from an IDI diesel passenger car under real world driving'. Society of Automotive Engineers 2002-01-1006. [Cited on p. 138]

d'Anna A., Kent J.H. (2006) 'Modeling particulate carbon and species formation in coflowing diffusion flames of ethylene'. Combustion and Flame 144:249–260. [Cited on p. 107]

d'Anna A., Violi A., D'Alessio A. (2000) 'Modeling the rich combustion of aliphatic hydrocarbons'. Combustion and Flame 121:418–429. [Cited on p. 68]

Antoni C., Peters N. (1997) 'Cycle resolved emission spectroscopy for IC engines'. Society of Automotive Engineers 972917. [Cited on p. 91]

Aoki T., Tanabe S.-i. (2007) 'Generation of sub-micron particles and secondary pollutants from building materials by ozone reaction'. Atmospheric Environment 41:3139–3150. [Cited on p. 359]

Aoyagi Y., Kamimoto T., Matsui Y., Matsuoka S. (1980) 'A gas sampling study on the formation processes of soot and NO in a DI diesel engine'. Society of Automotive Engineers 800254. [Cited on p. 84]

Appel J., Bockhorn H., Frenklach M. (2000) 'Kinetic modelling of soot formation with detailed chemistry and physics: Laminar premixed flames of C_2 hydrocarbons'. Combustion and Flame 121:122–136. [Cited on pp. 74, 75]

Arai M. (1991) 'SOF reduction and sulfate formation characteristics by diesel catalyst'. Society of Automotive Engineers 910328. [Cited on p. 296]

Arai M. (1992) 'Impact of changes in fuel properties and lubrication oil on particulate emissions and SOF'. Society of Automotive Engineers 920556. [Cited on pp. 263, 264, 282]

Arana C.P., Pontoni M., Sen S., Puri I.K. (2004) 'Field measurements of soot volume fractions in laminar partially premixed coflow ethylene/air flames'. Combustion and Flame 138:362–372. [Cited on p. 68]

Arcoumanis C., Barbaris L.N., Crane R.I., Wisby P. (1994) 'Evaluation of a cyclone-based particulate filtration system for high-speed diesel engines'. Institution of Mechanical Engineers: Part D: Journal of Automobile Engineering 203:269–279. [Cited on p. 394]

Arcoumanis C., Bae C., Nagwaney A., Whitelaw J.H. (1995) 'Effect of EGR on combustion development in a 1.9L DI diesel optical engine'. Society of Automotive Engineers 950850. [Cited on p. 272]

Arhami M., Kuhn T., Fine P.M., Defino R.J., Sioutas C. (2006) 'Effects of sampling artifacts and operating parameters on the performance of a semicontinuous particulate elemental carbon/organic carbon monitor'. Environmental Science and Technology 40:945–954. [Cited on p. 168]

Ariga S., Sui P.C., Shahed S.M. (1992) 'Instantaneous unburned oil consumption measurement in a diesel engine using SO_2 tracer technique'. Society of Automotive Engineers 922196. [Cited on pp. 252, 279, 280, 285]

Armas O., Ballesteros R., Gómez A. (2001) 'Morphological analysis of particulate matter emitted by a diesel engine using digital image analysis algorithms and scanning mobility particle sizer'. Society of Automotive Engineers 2001-01-3618. [Cited on p. 215]

Arnold F., Stilp T., Busen R., Schumann U. (1998) 'Jet engine exhaust chemiion measurements: Implications for gaseous SO_3 and H_2SO_4'. Atmospheric Environment 32:3073–3077. [Cited on p. 128]

Arnott W.P., Moosmüller H., Rogers C.F., Jin T., Bruch R. (1999) 'Photoacoustic spectrometer for measuring light absorption by aerosol: Instrument description'. Atmospheric Environment 33:2845–2852. [Cited on p. 193]

Arnott W.P., Zielinska B., Rogers C.F., Sagebiel J., Park K., Chow J., Moosmüller H., Watson J.G., Kelly K., Wagner D., Sarofim A., Lighty J., Palmer G. (2005) 'Evaluation of 1047-nm photoacoustic instruments and photoacoustic aerosol sensors in source-sampling of black carbon aerosol and particle-bound PAHs from gasoline and diesel powered vehicles'. Environmental Science and Technology 39:5398–5406. [Cited on pp. 192, 193, 195]

Arrowsmith S. (2003) 'Challenges for future heavy duty diesel lubricant development: PC-10/DX-2/Euro IV'. Society of Automotive Engineers 2003-01-1964. [Cited on p. 283]

Arsie I., di Genova F., Pianese C., Sorrentino M., Rizzo G., Caraceni A., Cioffi P., Flauti G. (2004) 'Development and identification of phenomenological models for combustion and emissions of common-rail multi-jet diesel engines'. Society of Automotive Engineers 2004-01-1877. [Cited on p. 67]

Asanuma T., Hirota S., Yanaka M., Tsukasaki Y., Tanaka T. (2003) 'Effect of sulfur-free and aromatics-free diesel fuel on vehicle exhaust emissions using simultaneous PM and NOx reduction system' Society of Automotive Engineers 2003-01-1865. [Cited on p. 285]

Atkinson C.M., Thompson G.J., Traver M.L., Clark N.N. (1999) 'In-cylinder combustion pressure characteristics of Fischer – Tropsch and conventional diesel fuels in a heavy duty CI engine'. Society of Automotive Engineers 1999-01-1472. [Cited on p. 251]

Atkinson R. (1988) 'Atmospheric transformations of automotive emissions'. In *Air Pollution, the Automobile, and Public Health*, Watson A.Y., Bates R.R., Kennedy D. (editors), National Academy Press, pp. 99–132. [Cited on p. 315]

Aubin D.G., Abbatt J.P. (2006) 'Laboratory measurements of thermodynamics of adsorption of small aromatic gases to *n*-hexane soot surfaces'. Environmental Science and Technology 40:179–187. [Cited on p. 114]

Ausset P., Crovisier J.L., de Monte M., Furlan V., Girardet F., Hammecker C., Jeanette D., Lefevre R.A. (1996) 'Experimental study of limestone and sandstone sulphation in polluted realistic conditions: The Lausanne Atmospheric Simulation Chamber (LASC)'. Atmospheric Environment 30:197–3207. [Cited on p. 38]

Ayala A., Olson B., Cantrell B., Drayton M., Barsic N. (2003) 'Estimation of diffusion losses when sampling diesel aerosol: A quality assurance measure'. Society of Automotive Engineers 2003-01-1896. [Cited on p. 146]

Ayers J.P., Keywood M.D., Gras J.L. (1999) 'TEOM vs. manual gravimetric methods for determination of PM2.5 aerosol mass concentrations'. Atmospheric Environment 33:3717–3721. [Cited on p. 198]

Ayres J.G. (1998) 'Health effects of gaseous air pollutants'. In *Air Pollution and Health*, Issues in Environmental Science and Technology, Volume 10, Hester R.E., Harrison R.M. (editors), Royal Society of Chemistry, pp. 1–20. [Cited on p. 371]

Axelsson B., Witze P.O. (2001) 'Qualitative laser-induced incandescence measurements of particulate emissions during transient operation of a TDI diesel engine'. Society of Automotive Engineers 2001-01-3574. [Cited on p. 188]

Axenfeld F., Munch J., Pacyna J.M., Duiser J.A., Veldt C. (1992) 'Test-Emissionsdatenbasis der Spurenelements As, Cd, Hg, Pb, Zn und der speziellen Organischen Verbindungen y-HCH (Lindan), HCB, PCB and PAK für Modellrechnungen in Europa', Umweltforschungsplan des Bundesministers für Umwelt Naturschutz und Reaktorsicherheit, Forschungsbericht 104 02 0588, Dornier GmbH, Friedrichshafen. [Cited on p. 32]

Axetell K. (1980) 'Assessment of nontraditional source impacts in the western United States'. In *The Technical Basis for a Size Specific Particulate Standard Parts I & II*, E.R. Frederick (editor), Speciality Conference, Air Pollution Control Association, pp. 223–229. [Cited on p. 23]

B

Badami M., Mallamo F., Millo F., Rossi E.E. (2002) 'Influence of multiple injection strategies on emissions, combustion noise and BSFC of a common rail diesel engine'. Society of Automotive Engineers 2002-01-0503. [Cited on pp. 269, 270]

Badr T., Harion J.-L. (2007) 'Effect of aggregate storage piles configuration on dust emissions'. Atmospheric Environment 41:360–368. [Cited on p. 31]

Bae C., Yu J., Kang J., Kong J., Lee K.O. (2002) 'Effect of nozzle geometry on the common-rail diesel spray'. Society of Automotive Engineers 2002-01-1625. [Cited on p. 266]

Bae M.-S., Hong C.-S., Kim Y.J., Han J.-S., Moon K.-J., Kondo Y., Komazaki Y., Miyazaki Y. (2007) 'Intercomparison of two different thermal-optical elemental carbons and optical black carbon during ABC-EAREX2005'. Atmospheric Environment 41:2791–2803. [Cited on p. 169]

Bagley S.T., Baumgard K.J., Gratz L.D., Bickel K.L., Watts W.F. (1991) 'Effects of a catalyzed diesel particulate filter on the chemical and biological'. Society of Automotive Engineers 911840. [Cited on pp. 173, 238, 306, 350]

Bailey B.K., Ariga S. (1990) 'On-line diesel engine oil consumption measurement'. Society of Automotive Engineers 902113. [Cited on pp. 280, 285]

Bailey C.R., Somers J.H., Steenland K. (2003) 'Exposures to diesel exhaust in the International Brotherhood of Teamsters, 1950–1990'. American Industrial Hygiene Association Journal 64:472–479. [Cited on p. 372]

Baker J.A. (1998) 'Particulate matter regulation and implications for the diesel engine'. Society of Automotive Engineers 981174. [Cited on p. 378]

Bakrania S.D., Miller T.A., Perez C., Wooldridge M.S. (2007) 'Combustion of multiphase reactants for the synthesis of nanocomposite materials'. Combustion and Flame 148:76–77. [Cited on p. 98]

Ball D.J., Kirby C.W. (1997) 'A survey of automotive catalyst technologies using rapid aging test schedules which incorporate engine oil derived poisons'. Society of Automotive Engineers 973050. [Cited on p. 296]

Balthasar M., Frenklach M. (2005) 'Detailed kinetic modeling of soot aggregate formation in laminar premixed flames'. Combustion and Flame 149:130–145. [Cited on p. 77]

Balthasar M., Mauss F., Knobel A., Kraft M. (2002a) 'Detailed modeling of soot formation in a partially stirred plug flow reactor'. Combustion and Flame 128:395–409. [Cited on pp. 62, 78]

Balthasar M., Mauss F., Wang H. (2002b) 'A computational study of the thermal ionization of soot particles and its effect on their growth in laminar premized flames'. Combustion and Flame 129:204–216. [Cited on pp. 74, 78, 225]

Bamford H.A., Bezabeh D.Z., Schantz M.M., Wise S.A., Baker J.E. (2003) 'Determination and comparison of nitrated-polycyclic aromatic hydrocarbons measured in air and diesel particulate reference materials'. Chemosphere 50:575–587. [Cited on pp. 111, 177, 235]

Bao L., Chen S., Wu L., Hei T.K., Wu Y., Yu Z., Xu A. (2007) 'Mutagenicity of diesel exhaust particles mediated by cell – particle interaction in mammalian cells'. Toxicology 229:91–100. [Cited on p. 376]

Barale R., Bulleri M., Cornetti G., Loprieno N., Wachter W.F. (1992) 'Preliminary investigation on genotoxic potential of diesel exhaust'. Society of Automotive Engineers 920397. [Cited on pp. 110, 238]

Bardasz E.W., Carrick V.A., Ebeling V.L., George H.F., Graf M.M., Kornbrekke R.E., Pocinki S.B. (1996) 'Understanding soot mediated oil thickening through designed experimentation – Part 2: GM 6.5L'. Society of Automotive Engineers 961915. [Cited on pp. 286, 288]

Bardasz E.W., Arters D.C., Schiferl E.A., Righi D.W. (1999) 'A comparison of gasoline direct injection and port fuel injection vehicles: Part II – Lubricant oil performance and engine wear'. Society of Automotive Engineers 1999-01-1499. [Cited on p. 286]

Bardasz E., Mackney D., Britton N., Leinschek G., Oloffson K., Murray I., Walker A.P. (2003) 'Investigations of the interactions between lubricant-derived species and aftertreatment systems on a state-of-the-art heavy duty diesel engine'. Society of Automotive Engineers 2003-01-1963. [Cited on pp. 295, 296]

Bardasz E.A., Cowling S., Panesar A., Durham J., Tadrous T.N. (2005) 'Effects of lubricant derived chemistries on performance of the catalyzed diesel particulate filters'. Society of Automotive Engineers 2005-01-2168. [Cited on p. 301]

Barefoot R.R. (1997) 'Determination of platinum at trace levels in environmental and biological materials'. Environmental Science and Technology 31:309–314. [Cited on p. 350]

Barth D.S. (1971) 'Effects of air pollution on man and his environment'. Society of Automotive Engineers 710326. [Cited on p. 356]

Bascom R.C., Chiu W.S., Padd R.J. (1973) 'Measurement and evaluation of diesel smoke'. Society of Automotive Engineers 730212. [Cited on p. 202]

Bassoli C., Conetti G.M., Biaggini G., di Lorenzo A. (1979) 'Exhaust emissions from a light duty turbocharged diesel'. Society of Automotive Engineers 790316. [Cited on pp. 168, 172]

Bates R.R., Watson A.Y. (1988) 'Motor vehicle emissions: A strategy for quantifying risk'. In *Air Pollution, the Automobile, and Public Health*, Watson A.Y., Bates R.R., Kennedy D. (editors), National Academy Press, pp. 17–36. [Cited on p. 356]

Baum M.M., Kiyomiya E.S., Kumar S., Lappa A.M., Lord H.C. (2000) 'Multi-component remote sensing of vehicle exhaust by dispersive absorption spectroscopy'. Society of Automotive Engineers 2000-01-3103. [Cited on p. 37]

Baumbach G., Vogt U., Hein K.R.G., Oluwole A.F., Ogunsola O.J., Olaniyi H.B., Akeredolu F.A. (1995) 'Air pollution in a large tropical city with a high traffic density – results of measurements in Lagos, Nigeria'. Science of the Total Environment 169:25–31. [Cited on pp. 119, 279]

Baumgard K.J., Johnson J.H. (1996) 'The effect of fuel and engine design on diesel exhaust particle size distributions'. Society of Automotive Engineers 960131. [Cited on pp. 126, 127, 220, 391]

Baumgard K.J., Kittelson D.B. (1985) 'The influence of a ceramic particle trap on the size distribution of diesel particles'. Society of Automotive Engineers 850009. [Cited on pp. 158, 303, 304]

Baxter L.L., Mitchell R.E., Fletcher T.H. (1997) 'Release of inorganic material during coal devolatization'. Combustion and Flame 108:494–502. [Cited on p. 98]

Baxter L.L., Miles T.R., Jenkins B.M., Milne T., Dayton D., Bryers R.W., Oden L.L. (1998) 'The behavior of inorganic material in biomass-fired power boilers: Field and laboratory experiences'. Fuel Processing Technology 54:47–78. [Cited on p. 97]

Bazari Z. (1990) 'The transient performance analysis of a turbocharged vehicle diesel engine with electronic fuelling control'. Society of Automotive Engineers 900236. [Cited on p. 276]

Beaver H. (1953) *Interim Report of the Committee on Air Pollution* (on London air pollution incident), Cmd 9011. Her Majesty's Stationery Office, London. [Cited on p. 3]

Bechtold R.L., Timbario T.J., Miller M.T., Urban C. (1991) 'Performance and emissions of a DDC 8V-71 transit bus engine using ignition-improved methanol and ethanol'. Society of Automotive Engineers 912356. [Cited on p. 257]

Becker S., Halsall C.J., Tych W., Hung H., Attewell S., Blanchard P., Li H., Fellin P., Stern G., Billeck B., Friesen S. (2006) 'Resolving the long-term trends of polycyclic aromatic hydrocarbons in the Canadian Arctic atmosphere'. Environmental Science and Technology 40:3217–3222. [Cited on p. 35]

van Beckhoven L.C. (1991) 'Effects of fuel properties on diesel engine emissions – a review of information available to the EEC-MVEG Group'. Society of Automotive Engineers 910608. [Cited on p. 255]

Belardini P., Bertoli C., del Giacomo N., Iori B. (1993) 'Soot formation and oxidation in a DI diesel engine: A comparison between measurements and three dimensional computations'. Society of Automotive Engineers 932658. [Cited on p. 66]

Beltzer M., Campion R.J., Petersen W.L. (1974) 'Measurement of vehicle particulate emissions'. Society of Automotive Engineers 740286. [Cited on pp. 317, 349]

Beltzer M. (1976) 'Non-sulfate particulate emissions from cars'. Society of Automotive Engineers 760038. [Cited on pp. 316, 318, 349, 350]

Beltzer M., Campion R.J., Harlan J., Hochhauser A.M. (1975) 'The conversion of SO_2 over automotive oxidation catalysts'. Society of Automotive Engineers 750095. [Cited on p. 318]

Berger C., Horvath H., Schindler W. (1998) 'The deposition of soot particles from hot gas streams through pipes'. Journal of Aerosol Science 26:211–217. [Cited on p. 146]

Bergstrand P. (2004) 'The effects of orifice shape on diesel combustion'. Society of Automotive Engineers 2004-01-2920. [Cited on p. 265]

Bergstrand P., Denbratt I. (2001) 'Diesel combustion with reduced nozzle orifice diameter'. Society of Automotive Engineers 2001-01-2010. [Cited on p. 266]

Bergstrand P., Denbratt I. (2003) 'The effects of multirow nozzles on diesel combustion'. Society of Automotive Engineers 2003-01-0701. [Cited on pp. 265, 266]

Bernemyr H., Ångström H.-E. (2007) 'Number measurements of diesel exhaust particles – influence of dilution and fuel sulphur content'. Society of Automotive Engineers 2007-01-0064. [Cited on p. 389]

Bertola A., Schubiger R., Kasper A., Matter U., Forss A.M., Mohr M., Boulouchos K., Lutz T. (2001) 'Characterization of diesel particulate emissions in heavy-duty DI-diesel engines with common rail fuel injection influence of injection parameters with fuel composition'. Society of Automotive Engineers 2001-01-3573. [Cited on pp. 260, 268, 270]

Bertola A., Li R., Boulouchos K. (2003) 'Influence of water – diesel fuel emulsions and EGR on combustion and exhaust emissions of heavy duty DI-diesel engines equipped with common-rail injection system'. Society of Automotive Engineers 2003-01-3146. [Cited on p. 251]

Bertoli C., del Giacomo N., Caprotti R., Smith A.K. (1991) 'The influence of automotive diesel back-end volatility and new fuel additive technology on regulated emissions'. Institution of Mechanical Engineers C427/34/218. [Cited on pp. 105, 263]

Bertoli C., del Giacomo N., Beatrice C., na Migliaccio M. (1998) 'Evaluation of combustion behavior and pollutants emission of advanced fuel formulations by single cylinder engine experiments'. Society of Automotive Engineers 982492. [Cited on p. 272]

BéruBé K.A., Jones T.P., Williamson B.J., Richards R.J. (1999) 'The physicochemical characterization of urban airborne particulate matter'. In *Particulate Matter: Properties and Effects upon Health*, Maynard R.L., Howard C.V. (editors), Bios Scientific Publishers, pp. 39–62. [Cited on p. 215]

Bessagnet B., Rosset R. (2001) 'Fractal modelling of carbonaceous aerosols – application to car exhaust plumes'. Atmospheric Environment 35:4751–4762. [Cited on p. 143]

Besson M., Hilaire N., Lahjaily H., Gastaldi P. (2003) 'Diesel combustion study at full load using CFD and design of experiments'. Society of Automotive Engineers 2003-01-1858. [Cited on p. 292]

Bester K, Hühnerfuss H. (2000) 'Transport and chemistry of pesticides in the atmosphere'. In *Aerosol Chemical Processes in the Environment*, Spurny K.R., Hochrainer D. (editors), CRC Press, pp. 577–600. [Cited on p. 30]

Bezot P., Hesse-Bezot C., Diraison C. (1997). 'Aggregation kinetics of colloidal suspensions of engine soots: Influence of polymeric lubricant additives'. Carbon 35:53–60. [Cited on p. 288]

Bi X., Sheng G., an Peng P., Zhang Z., Fu J. (2002) 'Extractable organic matter in PM10 from LiWan district of Guangzhou City, PR China'. Science of the Total Environment 300:213–228. [Cited on p. 29]

Biamino S., Fino P., Fino D., Russo N., Badini C. (2005) 'Catalyzed traps for diesel soot abatement: In situ processing and deposition of perovskite catalyst'. Applied Catalysis B: Environmental 61:297–305. [Cited on p. 295]

Bianchi D., Jean E., Ristori A., Vonarb R. (2005) 'Catalytic oxidation of a diesel soot formed in the presence of a cerium additive. III. Microkinetic-assisted method for the improvement of the ignition temperature'. Energy and Fuels 19:1453–1461. [Cited on p. 99]

Bianchi G.M., Pelloni P., Corcione F.E., Luppino F. (2001) 'Numerical analysis of passenger car HSDI diesel engines with the 2nd generation of common rail injection systems: The effect of multiple injections on emissions'. Society of Automotive Engineers 2001-01-1068. [Cited on p. 269]

Biancotto D., Erol T., Georges H., Lavy J., Martin B., Blanchard G., Macaudiere P., Grellier J.-M., Seguelong T. (2004) 'Retrofit program of a Euro 1 and Euro 2 urban bus fleet in La Rochelle using the ceria-based fuel-borne catalyst for diesel particulate filter regeneration (phase #1)'. Society of Automotive Engineers 2004-01-0821. [Cited on p. 302]

Bickel K., Majewski W.A. (1993) 'Evaluation of a catalyzed ceramic diesel particulate filter and catalytic converter on an underground mine vehicle'. Society of Automotive Engineers 932493. [Cited on p. 303]

Bielaczyc P., Merkisz J., Kozak M. (2002) 'Analysis of the influence of fuel sulphur content on diesel engine particulate emissions'. Society of Automotive Engineers 2002-01-2219. [Cited on pp. 252, 295]

Bielaczyc P., Kozak M., Merkisz J. (2003) 'Effects of fuel properties on exhaust emissions from the latest light-duty DI diesel engine'. Society of Automotive Engineers 2003-01-1882. [Cited on pp. 250, 263]

Bikas G., Zervas E. (2007) 'Regulated and non-regulated pollutants emitted during the regeneration of a diesel particulate filter'. Energy and Fuels 21:1543–1547. [Cited on p. 303]

Binder K., Hilburger W. (1981) 'Influence of the relative motions of air and fuel vapor on the mixture formation processes of the direct injection diesel engine'. Society of Automotive Engineers 810831. [Cited on pp. 267, 277]

Black F., High L. (1979) 'Methodology for determining particulate and gaseous diesel hydrocarbon emissions'. Society of Automotive Engineers 790422. [Cited on pp. 235, 236]

Blanchard G., Seguelong T., Michelin J., Schuerholz S., Terres F. (2003) 'Ceria-based fuel-borne catalysts for series diesel particulate filter regeneration'. Society of Automotive Engineers 2003-01-0378. [Cited on p. 222]

Blomberg A., Sainsbury C., Rudell B., Frew A.J., Holgate S.T., Sandstrom T., Kelly F.J. (1998) 'Nasal cavity lining fluid asorbic acid concentration increases in healthy human volunteers following short term exposure to diesel exhaust'. Free Radical Research 29:59–67. [Cited on p. 382]

Blute N.A., Woskie S.R., Greenspan C.A. (1999) 'Exposure characterisation for highway construction. Part I. Cut and cover and tunnel finish stages'. Applied Occupational and Environmental Hygiene 14:632–641. [Cited on p. 360]

BMA (1997) *Road Transport and Health*. British Medical Association. [Cited on pp. 363, 397]

Boehman A.L., Song J., Alam M. (2005) 'Impact of biodiesel blending on diesel soot and the regeneration of particulate filters'. Energy and Fuels 19:1857–1864. [Cited on pp. 81, 215, 259, 301]

Bohren C.F., Huffman D.R. (1998) *Adsorption and Scattering of Light by Small Particles*. John Wiley & Sons, Inc. [Cited on p. 189]

Boman C., Nordin A., Boström D., Öhman M. (2004) 'Characterization of inorganic particulate matter from residential combustion of pellitized biomass fuels'. Energy and Fuels 18:338–348. [Cited on p. 97]

Bond T.C., Sun H. (2005) 'Can reducing black carbon emissions counteract global warming?'. Environmental Science and Technology 39:5921–5925. [Cited on p. 42]

Bond T.C., Streets D.G., Yarber K.F., Nelson S.M., Woo J.-H., Klimont Z. (2004) 'A technology-based global inventory of black and organic carbon emissions from combustion'. Journal of Geophysical Research 109:1–43. [Cited on p. 32]

Booker D.R., Giannelli R.A., Hu J. (2007) 'Road test of an on-board particulate mass measurement system'. Society of Automotive Engineers 2007-01-1116. [Cited on pp. 199, 388]

Borman G.J., Ragland K.W. (1998) *Combustion Engineering*. McGraw-Hill. [Cited on pp. 98, 255]

Bosch Ojeda C., Sánchez Rojas F. (2005) 'Determination of rhodium: Since the origins until today spectrophotometric methods'. Talanta 67:1–19. [Cited on p. 350]

Bosteels D., May J., Karlsson M., de Serves C. (2006) ' "Regulated" and "unregulated" emissions from modern European passenger cars'. Society of Automotive Engineers 2006-01-1516. [Cited on pp. 243, 320, 323]

Bouffard R.A., Beltzer M. (1981) 'Light duty diesel particulate emissions – fuel and vehicle effects' Society of Automotive Engineers 811191. [Cited on p. 263]

Bouris D., Crane R.I., Evans J.M., Tippayawong N. (2000) 'An approach to characterization and after-treatment of particulate emissions from gasoline engines'. International Journal of Engine Research 1:291–300. [Cited on p. 138]

Brace C.J., Cox A., Hawley J.G., Vaughan N.D., Wallace F.W., Horrocks R.W., Bird G.L. (1999) 'Transient investigation of two variable geometry turbocharge for passenger vehicle diesel engines'. Society of Automotive Engineers 1999-01-1241. [Cited on p. 277]

Braddock J.N., Gabele P.A. (1977) 'Emission patterns of diesel-powered passenger cars – Part II'. Society of Automotive Engineers 770168. [Cited on p. 212]

Braddock J.N., Perry N.K. (1986) 'Gaseous and particulate emissions from gasoline- and diesel-powered heavy-duty trucks'. Society of Automotive Engineers 860617. [Cited on p. 315]

Bradow R.L., Zweidinger R.B., Black F.M., Dietzmann H.M. (1982) 'Sampling diesel engine particle and artifacts from nitrogen oxide interactions'. Society of Automotive Engineers 820182. [Cited on pp. 110, 111]

Brahma I., Rutland C.J., Foster D.E., He Y. (2005) 'A new approach to system level soot modeling'. Society of Automotive Engineers 2005-01-1122. [Cited on p. 89]

Brandenberger S., Nohr M., Grob K., Neukom H.P. (2005) 'Contribution of unburned lubricating oil and diesel fuel to particulate emission from passenger cars'. Atmospheric Environment 39:6985–6994. [Cited on p. 178]

Brandes J.G. (1970) 'Diesel fuel specification and smoke suppressant additive evaluations'. Society of Automotive Engineers 700522. [Cited on p. 261]

Brandl F., Reverencic I., Cartellieri W., Dent J.C. (1979) 'Turbulent air flow in the combustion bowl of a D.I. diesel engine and its effect on engine performance'. Society of Automotive Engineers 790040. [Cited on p. 277]

Braun A., Huggins F.E., Seifert S., Ilavsky J., Shah N., Kelly K.E., Sarofim A., Huffman G.P. (2004) 'Size-range analysis of diesel soot with ultra-small angle X-ray scattering'. Combustion and Flame 137:63–72. [Cited on pp. 214, 223]

Braun A., Shah N., Huggins F.E., Kelly K.E., Sarofim A., Jacobsen C., Wirick S., Francis H., Ilavsky J., Thomas G.E., Huffman G.P. (2005a) 'X-ray scattering and spectrocsopy studies on diesel soot from oxygenated fuel under various engine load conditions'. Carbon 43:2588–2599. [Cited on pp. 215, 216]

Braun A., Huggins F.E., Shah N., Chen Y., Wirick S., Mun S.B., Jacobsen C., Huffman G.P. (2005b) 'Advantages of soft X-ray absorption over TEM-EELS for solid carbon particles – a comparative study on diesel soot with EELS and NEXAFS'. Carbon 43:117–124. [Cited on p. 259]

Braun A., Huggins F.E., Kelly K.E., Mun B.S., Ehrlich S.N., Huffman G.P. (2006) 'Impact of ferrocene on the structure of diesel exhaust soot as probed with wide-angle X-ray scattering and C(s) NEXAFS spectroscopy'. Carbon 44:2904–2911. [Cited on p. 231]

Brezinsky K. (2002) 'Opportunities for diagnostics in the combustion synthesis of materials'. In *Applied Combustion Diagnostics*, Kohse-Höinghaus K., Jeffries J.B. (editors) Taylor and Francis, pp. 587–605. [Cited on p. 98]

Brichard J., Cabane M., Madelaine G., Vigla D. (1972) 'Formation and properties of neutral ultrafine particles and small ions conditioned by gaseous impurities in the air'. In *Aerosols and Atmospheric Chemistry*, Hidy G.M. (editor), Academic Press, pp. 27–43. [Cited on p. 125]

Bricklemyer B.A., Spindt R.S. (1978) 'Measurement of polynuclear aromatic hydrocarbons in diesel exhaust gases'. Society of Automotive Engineers 780115. [Cited on p. 178]

Briedé J.J., de Kok T.M.C.M., Hogervorst J.G.F., Moonen E.J.C., Op den Campl C.L.P., Kleinjans J.C.S. (2005) 'Development and application of an electron spin resonance spectrometry method for the determination of oxygen free radical formation by particulate matter'. Environmental Science and Technology 39:8420–8426. [Cited on pp. 360, 368, 369]

Brimblecombe P. (2001) 'Urban air pollution'. In *The Urban Atmosphere and Its Effects*, Brimblecome P., Maynard R.L. (editors), Air Pollution Reviews, Vol. 1, Imperial College Press, pp. 1–20. [Cited on p. 1]

Bristol C.W. (1971) 'Gas turbine engine emission characteristics and future outlook'. Society of Automotive Engineers 710319. [Cited on p. 4]

Broun R.M., Fog D.A., Gorland D.H. (1989) 'Smoke reduction in two-stroke gasoline engine'. Society of Automotive Engineers 891338. [Cited on p. 336]

Brubaker P.E., Moran J.P., Bridbord K., Hueter F.G. (1975) 'Noble metals: A toxicological appraisal of potential new environmental contaminants'. Environmental Health Perspectives 10:39–56. [Cited on p. 350]

Bruetsch R.I., Bloom R. (1991) 'Low mileage evaluation of an electrically regenerable diesel particulate trap'. Society of Automotive Engineers 912669. [Cited on p. 302]

Brumfiel G. (2003) 'A little knowledge . . . '. Nature 424:246–428. [Cited on p. 396]

Brunauer S., Emmett P.H., Teller E. (1938) 'Adsorption of gases in multimolecular layers'. Journal of the American Chemistry Society 60:309–319. [Cited on p. 222]

Bruneaux G., Verhoeven D., Baritaud T. (1999) 'High pressure spray and combustion visualisation in a transparent model diesel engine'. Society of Automotive Engineers 1999-01-3648. [Cited on p. 272]

Brunner N.R. (1995) 'Test equipment and methods at the 3M Company diesel filter products laboratory'. Society of Automotive Engineers 950516. [Cited on p. 151]

Brunner T.J., Wick P., Manser P., Spohn P., Grass R.N., Limbach L.K., Bruinink A., Stark W.J. (2006) 'In vitro cytotoxicity of oxide nanoparticles: Comparison to asbestos, silica, and the effect of particle solubility'. Environmental Science and Technology 40:4374–4381. [Cited on p. 393]

Bryzik W., Smith C.O. (1977) 'Relationships between exhaust smoke emissions and operating variables in diesel engines'. Society of Automotive Engineers 770718. [Cited on p. 65]

Buchholz B.A., Cheng A.S., Dibble R.W., Mueller C.J., Martin G.C. (2002) 'Isotopic tracing of fuel component carbon in the emissions from diesel engines'. Society of Automotive Engineers 2002-01-1942. [Cited on p. 259]

Buchholz B.A., Dibble R.W., Rich D., Cheng A.S. (2003) 'Quantifying the contribution of lubrication oil carbon to particulate emissions from a diesel engine'. Society of Automotive Engineers 2003-01-1987. [Cited on p. 282]

Buchholz B.A., Mueller C.J., Upatnieks A., Martin G.C., Pitz W.J., Westbrook C.K. (2004) 'Using carbon-14 isotope tracing to investigate molecular structure effects of the oxygenate dibutyl maleate on soot emissions from a DI diesel engine'. Society of Automotive Engineers 2004-01-1849. [Cited on p. 259]

Buckley S.G., Sawyer R.F., Koshland C.P., Lucas D. (2002) 'Measurements of lead vapor and particulate in flames and post-flame gases'. Combustion and Flame 128:435–446. [Cited on p. 98]

Bukowiecki N., Kittelson D.B., Watts W.F., Burtscher H., Weingartner E., Baltensperger U. (2002) 'Real-time characterization of ultrafine and accumulation mode particles in ambient combustion aerosols'. Journal of Aerosol Science 33:1139–1154. [Cited on pp. 143, 195]

Bukowiecki N., Gehrig R., Hill M., Lienemann P., Zwicky C.N., Buchmann B., Weingartner E., Baltensperger U. (2007) 'Iron, manganese and copper emitted by cargo and passenger trains in Zürich (Switzerland): Size-segregated mass concentrations in ambient air'. Atmospheric Environment 41:878–889. [Cited on p. 31]

Bünger J., Krahl J., Frank H.-U., Munack A., Hallier E. (1998) 'Mutagenic and cytotoxic effects of exhaust particulate matter of biodiesel compared to fossil diesel fuel'. Mutation Research 415:13–23. [Cited on p. 260]

Burley H.A., Rosebrook T.L. (1979) 'Automotive diesel engines – fuel composition vs particulates'. Society of Automotive Engineers 790923. [Cited on pp. 171, 254, 263, 264]

Burtscher H. (1992) 'Measurement and characteristics of combustion aerosols with special consideration of photoelectric charging and charging by flame ions'. Journal of Aerosol Science 23:549–595. [Cited on pp. 192, 193, 195, 202, 218, 227]

Burtscher H. (2000) 'Characterization of ultrafine particle emissions from combustion systems'. Society of Automotive Engineers 2000-01-1997. [Cited on p. 194]

Burtscher H. (2001) 'Literature study on tailpipe particulate measurement for diesel engines'. Fachhochschule Aargau, University of Applied Science, Windisch, Switzerland (http://www.akpf.org/pub/burtscher_bericht.pdf). [Cited on pp. 198, 199]

Burtscher H., Matter U. (2000) 'Particle formation due to fuel additives'. Society of Automotive Engineers 2000-01-1883. [Cited on pp. 103, 231, 292]

Burtscher H., Matter U., Skillas G. (1999) 'The effect of fuel additives on diesel engine particulate emissions'. Journal of Aerosol Science 30:S851–S852. [Cited on pp. 104, 262]

C

Cachier H., Liousse C., Pertuisot M.H., Guadichet A., Echlara F., Lacaux J.P. (1996) In *Biomass Burning and Global Change*, Levine J.S. (editor), MIT Press, Cambridge, MA, pp. 428–440. [Cited on p. 33]

Cachier H. (1998) 'Carbonaceous combustion aerosols'. In *Atmospheric Particles*, Harrison R.M., van Grieken R.E. (editors), IUPAC Series on Analytical and Physical Chemistry of Environmental Systems, John Wiley & Sons Ltd, pp. 295–348. [Cited on pp. 33, 41, 145, 168]

Cadle S.H., Mulawa P.A., Ball J., Donase C., Weibel A., Sagebiel J.C., Knapp K.T., Snow R. (1997) 'Particulate emission rates from in-use high-emitting vehicles recruited in Orange County, California'. Environmental Science and Technology 31:3405–3413. [Cited on p. 313]

Cadle S.H., Mulawa P., Ragazzi R.A., Knapp K.T., Norbeck J.M., Durbin T.D., Truex T.J., Whitney K.A. (1999a) 'Exhaust particulate matter emissions from in-use passenger vehicles in three locations: CRC Project E-24'. Society of Automotive Engineers 1999-01-1545. [Cited on pp. 169, 173, 235, 322, 344, 345, 375, 388]

Cadle S.H., Mulawa P.A., Hunsanger E.C., Nelson K., Ragazzi R.A., Barrett R., Gallagher G.L., Lawson D.R., Knapp K.T., Snow R. (1999b) 'Composition of light-duty motor vehicle exhaust particulate matter in the Denver, Colorado Area'. Environmental Science and Technology 33:2328–2339. [Cited on p. 235]

Caines A.J., Haycock R.F., Hillier J.E. (2004) *Automotive Lubricants Reference Book*. Society of Automotive Engineers [Cited on p. 95]

Calcote H.F. (1957) 'Mechanisms for the formation of ions in flames'. Combustion and Flame 1:385–403. [Cited on p. 225]

Calcote H.F. (1983) 'Ionic mechanisms of soot formation'. In *Soot in Combustion Systems and its Toxic Properties*, Lahaye J., Prado G. (editors), Plenum Press, pp. 197–215. [Cited on pp. 70, 225]

Calcote H.F. (2001) 'Comments on "The origin of soot in flames: Is the nucleus an ion?". by V.J. Hall-Roberts, A.N., Hayhurst, D.E. Knight, and S.G. Taylor'. Combustion and Flame 126:1607–1610. [Cited on p. 71]

Camatini M., Crosta G.F., Dolukhanyan T., Sung C., Giuliani G.P., Corbetta G.M., Cencetti S., Regazzoni C. (2001) 'Microcharacterization and identification of tire debris in heterogeneous laboratory and environmental specimens'. Materials Characterization. 46:271–283. [Cited on pp. 346, 347, 348]

Campbell B., Peckham M., Symonds J., Parkinson J., Finch A. (2006) 'Transient gaseous and particulate emissions measurements on a diesel passenger car including a DPF regeneration event'. Society of Automotive Engineers. 2006-01-1079. [Cited on p. 304]

Campbell J., Scholl J., Hibbler F., Bagley S., Leddy D., Abata D., Johnson J. (1981) 'The effect of fuel injection rate and timing on the physical, chemical and biological character of particulate emissions from a direct injection diesel'. Society of Automotive Engineers 810996. [Cited on pp. 180, 239]

Campbell P.H., Sinko K.M., Chehroudi B. (1995) 'Liquid and vapor phase distributions in a piloted diesel fuel spray'. Society of Automotive Engineers 950445. [Cited on p. 270]

Campenon T., Wouters P., Blanchard G., Macaudiere P., Seguelong T. (2004) 'Improvement and simplification of DPF system using a ceria-based fuel-borne catalyst for diesel particulate filter regeneration in serial applications'. Society of Automotive Engineers 2004-01-0071. [Cited on p. 231]

Camuffo D., Bernardi A. (1996) 'Deposition of urban pollution on the Ara Pacis, Rome'. Science of the Total Environment 189/190:235–245. [Cited on p. 39]

Cantrell B.K., Watts W.F. (1997) 'Diesel exhaust aerosol: Review of occupational exposure'. Applied Occupational and Environmental Hygiene 12:1019–1027. [Cited on p. 360]

Cantwell E.N., Jacobs E.S., Kunz W.G., Liberi V.E. (1972) 'Control of particulate lead emissions from automobiles'. Society of Automotive Engineers 720672. [Cited on pp. 316, 317]

Caprotti R., Fowler W.J., Lepperhoff G., Houben M. (1993) 'Diesel additive technology effects on injector hole erosion/corrosion, injector fouling and particulate traps'. Society of Automotive Engineers 932739. [Cited on p. 261]

Caprotti R., Field I., Michelin J., Schuerholz S., Terres F. (2003) 'Development of a novel DPF additive'. Society of Automotive Engineers 2003-01-3165. [Cited on p. 222]

Carpenter K., Johnson J.H. (1979) 'Analysis of the physical characteristics of diesel particulate matter using transmission electron microscope techniques'. Society of Automotive Engineers 790815. [Cited on pp. 215, 217, 219]

Carroll J.N., White J.J., Khalek I.A., Kado N.Y. (2000) 'Characterization of snowmobile particulate emissions'. Society of Automotive Engineers 2000-01-2003. [Cited on pp. 335, 337]

Carroll J.N., White J.J., Khalek I.A., de Vita J., Bekken M., Lenieux S. (2004) 'Marine outboard and personal watercraft engine gaseous emissions, and particulate emission test procedure development'. Society of Automotive Engineers 2004-32-0093. [Cited on p. 335]

Cartellieri W.P., Herzog P.L. (1988) 'Swirl supported or quiescent combustion for 1990's heavy-duty DI diesel engines – an analysis'. Society of Automotive Engineers 880342. [Cited on pp. 277, 280, 281, 282]

Cartellieri W.P., Wachter W.F. (1987) 'Status report on a preliminary survey of strategies to meet US-1991 HD diesel emission standards without exhaust gas aftertreatment'. Society of Automotive Engineers 870342. [Cited on p. 266]

Castagné M., Chevé E., Dumas J.P., Henriot S. (2000) 'Advanced tools for analysis of gasoline injection engines'. Society of Automotive Engineers 2000-01-1903. [Cited on pp. 329, 330]

Cauda E., Mescia D., Fino D., Saracco G., Specchia V. (2005) 'Diesel particulate filtration and combustion in a wall-flow trap hosting a $LiCrO_2$ catalyst'. Industrial and Engineering Chemistry 44:9549–9555. [Cited on p. 295]

Cauda E., Hernandez S., Fino D., Saracco G., Specchia V. (2006) 'PM0.1 emissions during diesel trap regeneration'. Environmental Science and Technology 40:5532–5537. [Cited on p. 303]

Cavaliere A., Barbella R., Ciajolo A., D-Anna A., Ragucci R. (1994) 'Fuel and soot oxidation in diesel-like conditions'. Twenty-Fifth Symposium (International) on Combustion, The Combustion Institute, pp. 167–174. [Cited on pp. 79, 80]

Cecinato A., Marino F., di Filippo P., Lepore L., Possanzini M. (1999) 'Distribution of n-alkanes, polynuclear aromatic hydrocarbons and nitrated polynuclear aromatic hydrocarbons between the fine and coarse fractions of inhalable atmospheric particles'. Journal of Chromatography A 846:255–264. [Cited on p. 390]

Cerru F.G., Kronenburg A., Lindstedt R.P. (2006) 'Systematically reduced chemical mechanisms for sulfur oxidation and pyrolysis'. Combustion and Flame 146:437–455. [Cited on p. 123]

Cha S., Carter P., Bradow R.L. (1983) 'Simulation of automobile brake wear dynamics and estimation of emissions'. Society of Automotive Engineers 831036. [Cited on p. 345]

Chaffin C.A., Ullman T.L. (1994) 'Effects of increased altitude on heavy-duty diesel engine emissions'. Society of Automotive Engineers 940669. [Cited on p. 276]

Chan Y., He Y., Ge Y. (1999) 'An innovative carbon-atom-balance-based method for diesel particulate measurement'. Measurement Science and Technology 10:N93–N100. [Cited on p. 168]

Chandler M.F., Teng Y., Koylu U.O. (2007) 'Diesel engine particulate emissons: A comparison of mobility and microscopy size measurements'. Proceedings of the Combustion Institute 11:2971–2979. [Cited on p. 218]

Charalampopoulos T.T., Hahn D.W., Chang H. (1992) 'Role of metal additives in light scattering from flame particulates'. Applied Optics 31:6519–6528. [Cited on p. 189]

Charmley W.J. (2004) 'The Federal Government's role in reducing heavy duty diesel emissions'. Society of Automotive Engineers 2004-01-2708. [Cited on p. 6]

Charron A., Harrison R.M. (2005) 'Fine ($PM_{2.5}$) and coarse ($PM_{2.5-10}$) particulate matter on a heavily trafficked London highway: Sources and processes'. Environmental Science and Technology: 39:7768–7776. [Cited on p. 342]

Charron A., Harrison R.M., Quincey P. (2007) 'What are the sources and conditions responsible for exceedences of the 24h PM_{10} limit value ($50\,\mu gm^{-3}$) at a heavily trafficked London site?'. Atmospheric Environment 41:1960–1975. [Cited on p. 397]

Chase R.E., Schamp D.H. (2007) 'Measuring the electrostatic charge on filter'. Society of Automotive Engineers 2007-01-0323. [Cited on p. 166]

Chase C.L., Peterson C.L., Lowe G.A., Mann P., Smith J.A., Kado N.Y. (2000) 'A 322,000 kilometer (200,000 mile) over the road test with HySEE biodiesel in a heavy duty truck'. Society of Automotive Engineers 2000-01-2647. [Cited on p. 260]

Chase R.E., Duszkiewicz G.J., Richert J.F.O., Lewis D., Maricq M.M., Xu N. (2004) 'PM measurement artifact: Organic vapor deposition on different filter media'. Society of Automotive Engineers 2004-01-0967. [Cited on pp. 151, 161, 165]

Chase R.E., Duszkiewicz G.J., Lewis D., Podsiadlik D.H. (2005) 'Reducing PM measurment variability by controlling static charge'. Society of Automotive Engineers 2005-01-0193. [Cited on p. 167]

Chatterjee S., McDonald C., Conway R., Windawi H., Vertin K., le Tavec C.A., Clark N.N., Gautam M. (2001) 'Emission reductions and operational fleet experiences with heavy duty diesel fleet vehicles retrofitted with continuously regenerated diesel particulate filters in southern California'. Society of Automotive Engineers 2001-01-0512. [Cited on p. 253]

Chatterjee S., Conway R., Lanni T., Frank B., Tang S., Rosenblatt D., Bush C., Lowell D., Evans J., McLean R., Levy S.J. (2002) 'Performance and durability evaluation of continuously regenerating particulate filters on diesel powered urban buses at NY City Transit – Part II'. Society of Automotive Engineers 2002-01-0430. [Cited on p. 238]

Chen D.-R., Pui D.Y.H., Hummes D., Fissan H., Quant F.R., Sem G.J. (1998) 'Design and evaluation of a nanometer aerosol differential mobility analyzer (Nano-DMA)'. Journal of Aerosol Science 29:497–509. [Cited on p. 186]

Chen H.X., Dobbins R.A. (2000) 'Crystallogenesis of particles formed in hydrocarbon combustion'. Combustion Science and Technology 159:109–128. [Cited on pp. 71, 212]

Chen L., Mengersen K., Tong S. (2007) 'Spatiotemporal relationship between particle air pollution and respiratory emergency hospital admissions in Brisbane, Australia'. Science of the Total Environment 373:57–67. [Cited on p. 371]

Chen S.K. (2000) 'Simultaneous reduction of NOx and particulate emissions by using multiple injections in a small diesel engine'. Society of Automotive Engineers 2000-01-3084. [Cited on p. 269]

Chen S.K., Yanakiev O. (2005) 'Transient NOx emission reduction using exhaust oxygen concentration based control for a diesel engine'. Society of Automotive Engineers 2005-01-0372. [Cited on pp. 272, 277]

Chernich D.J., Jacobs P.E., Kowalski J.D. (1991) 'A comparison of heavy-duty diesel truck engine smoke opacities at high altitude and at sea level'. Society of Automotive Engineers 911671. [Cited on p. 276]

Cheung S.K., Elder S.T., Raine R.R. (2000) 'Diesel particulate measurements with a light scattering photometer'. Society of Automotive Engineers 2000-01-1136. [Cited on p. 190]

Chi Y., Cheong J., Kim C., Choi K. (2002) 'Effects of VGT and injection parameters on performance of HSDI diesel engine with common rail FIE system'. Society of Automotive Engineers 2002-01-0504. [Cited on p. 270]

Chikahisa T., Araki T. (1996) 'In-cylinder control of smoke and NOx by high turbulent two-stage combustion in diesel engines'. Society of Automotive Engineers 962113. [Cited on p. 91]

Child D., Cioffi J. (1994) 'An electronically controlled diesel particulate filter system employing electrically regenerated ceramic fiber wound cartridges'. Society of Automotive Engineers 942265. [Cited on p. 303]

Chow J.C., Watson J.G., Ashbaugh L.L., Magliano K.L. (2003) 'Similarities and differences in PM_{10} chemical source profiles for geological dust from the San Joaquin Valley, California'. Atmospheric Environment 37:1317–1340. [Cited on p. 342]

Christoforou C.S., Salmon L.G., Cass G.R. (1996) 'Fate of atmospheric particles within the Buddhist cave temples at Yuang, China'. Environmental Science and Technology 30:3425–3434. [Cited on p. 38]

Christou S.Y., Birgersson H., Fierro J.L.G., Efstathiou A.M. (2006) 'Reactivation of an aged commecial three-way catalyst by oxalic and citric acid washing'. Environmental Science and Technology 40:2030–2036. [Cited on p. 296]

Chu R.R. (2004) 'Issues and behaviors of airborne particulate matters under microgravity environment'. Society of Automotive Engineers 2004-01-2328. [Cited on p. 17]

Chughtai A.R., Atteya M.O., Kim J., Konowalchuk B.K., Smith D.M. (1998) 'Adsorption and adsorbate interaction at soot particle surfaces'. Carbon 36:1573–1589. [Cited on pp. 124, 125]

Ciajolo A., D'Anna A., Barbbella R., Tregrossi A. (1994) 'The formation of aromatic carbon in sooting ethylene flames'. Twenty-Fifth Symposium (International) on Combustion, The Combustion Institute, pp. 679–685. [Cited on p. 106]

Cicchella D., de Vivo B., Lima A. (2003) 'Palladium and platinum concentrations in soils from the Napoli metropolitan area, Italy: Possible effects of catalytic exhausts'. Science of the Total Environment 308:121–131. [Cited on p. 351]

Cinti D., Angelone M., Masi U., Cremisini C. (2002) 'Platinum levels in natural and urban soils from Rome and Latium (Italy): Significance for pollution by automobile catalytic converter'. Science of the Total Environment 293:47–57. [Cited on p. 351]

Claes M., Gysels K., van Grieken R., Harrison R.M. (1998) 'Inorganic composition of atmospheric particles'. In *Atmospheric Particles*, Harrison R.M., van Grieken R.E. (editors), IUPAC Series on Analytical and Physical Chemistry of Environmental Systems, John Wiley & Sons Ltd, pp. 95–146. [Cited on pp. 30, 31, 36, 37, 46, 124, 165, 316]

Clague A.D.H., Donnet J.B., Wang T.K., Peng J.C.M. (1999) 'A comparison of diesel engine soot with carbon black'. Carbon 37:1553–1565. [Cited on p. 214]

Claiborn C.S., Larson T., Sheppard L. (2002) 'Testing the metals hypothesis in Spokane, Washington'. Environmental Health Perspectives 110:547–552. [Cited on p. 374]

Clark H.M. (1995) 'A comparison of particle impact in gas–solid and liquid–solid erosion'. Wear 186–187:465–472. [Cited on p. 350]

Clark N.N., Atkinson C.M., McKain D.L., Nine R.D., El-Gazzar L. (1996) 'Speciation of hydrocarbon emissions from a medium duty diesel engine'. Society of Automotive Engineers 960322. [Cited on p. 175]

Clark N.N., Gautam M., Boyce J., Xie W., Mehta S., Jarett R., Rapp B. (2003) 'Heavy duty vehicle exhaust plume study in the NASA/Langley wind tunnel'. Society of Automotive Engineers 2003-01-1895. [Cited on pp. 142, 143]

Clark N.N., Tatli E., Barnett R., Wayne W.S., McKain D.L. (2007) 'Characterization and abatement of diesel crankase emissions'. Society of Automotive Engineers 2007-01-3372. [Cited on p. 359]

Clarke A.G., Chen J.-M., Pipitsangchand S., Azadi-Bougar G.A. (1996) 'Vehicular particulate emissions and roadside air pollution'. Science of the Total Environment 189/190:417–422. [Cited on p. 95]

Clerc J.C., Johnson J.H. (1982) 'A computer heat transfer and hydrocarbon adsorption model for predicting diesel particulate emissions in dilution tunnels'. Society of Automotive Engineers 821218. [Cited on pp. 114, 116, 162]

Coffin D.L. (1971) 'Health aspects of airborne polycyclic hydrocarbons'. Society of Automotive Engineers 710301. [Cited on p. 371]

Colbeck I. (1995) 'Particle emission from outdoor and indoor sources'. In *Airborne Particulate Matter*, Volume 4, Part D, Kouimtzis T., Samara C. (editors) Springer, pp. 1–34. [Cited on p. 24]

Cole R.L., Poola R.B., Sekar R.R. (1999) 'Gaseous and particulate emissions from a vehicle with a spark-ignition direct-injection engine'. Society of Automotive Engineers 1999-01-1282. [Cited on pp. 331, 334]

Cole R.L., Poola R.B., Sekar R., Schaus J.E., McPartlin P. (2001) 'Effects of ethanol additives on diesel particulate and NO_X emissions'. Society of Automotive Engineers 2001-01-1937. [Cited on p. 257]

Collier A.R., Wedekind B. (1997) 'The effect of hydrocarbon composition on lean NOx catalysis'. Society of Automotive Engineers 973000. [Cited on p. 110]

Collier A.R., Jemma C.A., Wedekind B., Hall D.E., Heinze P. (1998) 'Sampling and analysis of vapour-phase and particulate-bound PAH from vehicle exhaust'. Society of Automotive Engineers 982727. [Cited on p. 322]

Collin F., Gommord M.F., Momique J.C., Monier R., Walter C. (2001) 'Particulate matter size distribution and associated polycyclic aromatic hydrocarbon content from indirect and direct injection diesel engines'. International Journal of Engine Research 2:23–31. [Cited on p. 237]

Collings N., Baker N., Wolber W.G. (1986) 'Real-time smoke sensor for diesel engines'. Society of Automotive Engineers 860157. [Cited on pp. 196, 225]

Collings N., Reavell K., Hands T., Tate J. (2003) '194 roadside aerosol measurements with a fast particulate spectrometer'. Japanese Society of Automotive Engineers: Annual Congress Proceedings, No. 41–03. [Cited on p. 186]

Collura S., Chaoui N., Azambre B., Finqueneisel G., Heintz O., Krzton A., Koch A., Weber J.V. (2005) 'Influence of the soluble organic fraction on the thermal behaviour, texture and surface chemistry of diesel exhaust soot'. Carbon 43:605–613. [Cited on pp. 168, 169, 229, 233]

Colmenares C., Deutsch S., Evans C., Nelson A.J., Terminello L.J., Reynolds J.G., Roos J.W., Smith I.L. (1999) 'Analysis of manganese particulates from automotive decomposition of methylcyclopentadienyl manganese tricarbonyl'. Applied Surface Science 151:189–202. [Cited on p. 317]

Cook S.L., Richards P.J. (2002) 'An approach towards risk assessment for the use of a synergistic metallic diesel particulate filter (DPF) regeneration additive'. Atmospheric Environment 36:2955–2954. [Cited on p. 304]

Cooke V.B. (1990) 'Lubrication of low emissions diesel engines Part 1 and Part 2'. Society of Automotive Engineers 900814. [Cited on p. 283]

Cooper B.H., Donnis B.B.L. (1996) 'Aromatic saturation of distillates: An overview'. Applied Catalysis A: General 137:203–223. [Cited on p. 254]

Cooper M.G. (1986) 'Air treatment design for clean rooms'. Institution of Mechanical Engineers MEP-254. [Cited on p. 209]

Corcione F.E., Vaglieco B.M. (1994) 'Cycle resolved measurements of diesel particulate by optical techniques'. Society of Automotive Engineers 941948. [Cited on p. 136]

Corcione F.E., Merola S.S., Vaglieco B.M. (2001) 'Nanometric particle formation in optically accessible diesel engine'. Society of Automotive Engineers 2001-01-1258. [Cited on p. 86]

Corcione F.E., Merola S.S., Vaglieco B.M. (2002) 'Evaluation of temporal and spatial distribution of nanometric particles in a diesel engine by broadband optical techniques'. International Journal of Engine Research 3:93–101. [Cited on p. 86]

Corn M. (1966) 'Adhesion of particles'. In *Aerosol Science*, Davies C.N. (editor), Academic Press, pp. 359–392. [Cited on p. 18]

Corsi R.L., Siegel J., Karamalegos A., Simon H., Morrison G.C. (2007) 'Personal reactive clouds: Introducing the concept of near-head chemistry'. Atmospheric Environment 41:3161–3165. [Cited on p. 358]

Corso S., Adamo R. (1984) 'The effect of diesel soot on reactivity of oil additives and valve train materials'. Society of Automotive Engineers 841369. [Cited on p. 290]

Corso S., Adamo R. (1985) 'Incipient scuffing detection by ferrography in a diesel valve train system'. Society of Automotive Engineers 852124. [Cited on p. 290]

Councell T.B., Duckenfield K.A., Landa E.R., Callender E. (2004) 'Tire-wear particles as a source of zinc to the environment'. Environmental Science and Technology 38:4206–4214. [Cited on pp. 346, 347, 348]

Covitch M.J., Humphrey B.K., Ripple D.E. (1985) 'Oil thickening in the mack T-7 engine test – fuel effects and the influence of lubricant additives on soot aggregation'. Society of Automotive Engineers 852126. [Cited on pp. 287, 288]

Cox I.M., Samways A.L. (1999) 'Diesel lube oil conditioning – the systems approach'. Society of Automotive Engineers 1999-01-1218. [Cited on p. 286]

Crane R.I., Rubino L., Arcoumanis C., Golunski S.E., McNamara J.M., Poulston S., Rajaram R.R. (2002) 'Strategies for gasoline particulate emission control – a "foresight vehicle" project'. Society of Automotive Engineers 2002-01-1894. [Cited on pp. 138, 334]

Cromas J., Ghandhi J.B. (2004) 'Lubricating oil contribution to direct-injection two stroke engine particulate emissions'. Society of Automotive Engineers 2004-32-0012. [Cited on p. 337]

Cromas J., Ghandhi J.B. (2005) 'Particulate emissions from a direct-injection spark-ignition engine'. Society of Automotive Engineers 2005-01-0103. [Cited on p. 335]

Crookes R.J., Sivalingham G., Nazha M.A.A., Rajakaruna H. (1999) 'Analysis of soot particulate formation in a high-pressure confined spray-flame'. Society of Automotive Engineers 1999-01-3488. [Cited on pp. 86, 90]

Cucchi C., Hublin M. (1989) 'Evolution of emissions legislation in Europe and impact on technology'. Society of Automotive Engineers 890487. [Cited on p. 6]

Cunningham L.J., Henly T.J., Kulinowski A.M. (1990) 'The effects of diesel ignition improvers in low-sulfur fuels on heavy-duty diesel emissions'. Society of Automotive Engineers 902173. [Cited on p. 264]

Cunningham P.J., Meckl P.H. (2006) 'Measuring particulate load in a diesel particulate filter'. Society of Automotive Engineers 2006-01-0686. [Cited on p. 159]

Curran H.J., Fisher E.M., Glaude P.-A., Marinov N.M., Pitz W.J., Westbrook C.K., Layton D.W., Flynn P.F., Durrett R.P., zur Loye A.O., Akinyemi O.C., Dryer F.L. (2001) 'Detailed chemical kinetic modeling of diesel combustion with oxygenated fuels'. Society of Automotive Engineers 2001-01-0653. [Cited on pp. 68, 258, 259]

Currie L.A., Eglinton T.I., Benner B.A., Pearson A. (1997) 'Radiocarbon "dating" of individual chemical compounds in atmospheric aerosol: First results comparing isotopic and multivariate statistical apportionment of specific polycyclic aromatic hydrocarbons'. Nuclear Instruments and Methods in Physics Research B 123:475–486. [Cited on p. 35]

Cuthbertson R.D., Stinton H.C., Wheeler R.W. (1979) 'The use of a thermogravimetric analyser for the investigation of particulates and hydrocarbons in diesel engine exhaust'. Society of Automotive Engineers 790814. [Cited on pp. 168, 169, 170]

Czerwinski J., Jaussi F., Wyser M., Mayer A. (2000) 'Particulate traps for construction machines properties and field experience'. Society of Automotive Engineers 2000-01-1923. [Cited on p. 303]

Czerwinski J., Comte P., Napoli S., Wili P. (2002) 'Summer cold start and nanoparticles of small scooters'. Society of Automotive Engineers 2002-01-1096. [Cited on pp. 336, 337]

Czerwinski J., Comte P., Reutimann F. (2005) 'Nanoparticle emissions of a DI 2-stroke scooter with varying oil- & fuel quality'. Society of Automotive Engineers 2005-01-1101. [Cited on p. 337]

Czerwinski J., Comte P., Astorga C., Rey M., Mayer A., Reutimann F. (2007) '(Nano) particles from 2-S scooters: SOF/INSOF; improvements of aftertreatment; toxicity'. Society of Automotive Engineers 2007-01-1089. [Cited on p. 169]

D

Dahl A., Gharibi A., Swietlicki E., Gudmundsson A., Bohgard M., Ljungman A., Blomqvist G., Gustafsson M. (2006) 'Traffic-generated emissions of ultrafine particles from pavement–tire interface'. Atmospheric Environment 40:1314–1323. [Cited on pp. 347, 348]

Dahmann D., Bauer H.-D. (1997) 'Diesel particulate matter (DPM) in workplaces in Germany'. Applied Occupational and Environmental Hygiene 12:1028–1031. [Cited on p. 360]

van Dam W., Willis W.W., Cooper M.W. (1999) 'The impact of additive chemistry and lubricant rheology on wear in heavy duty diesel engines'. Society of Automotive Engineers 1999-01-3575. [Cited on pp. 289, 290]

Dani A.D., Nagpurkar U.P., Lakshminarayanan (1990) 'Universal mixing correlations for the performance and emission of open chamber diesel combustion supported by air swirl'. Society of Automotive Engineers 900446. [Cited on p. 277]

Dasenbrock C., Peters L., Creutzenberg O., Heinrich U. (1996) 'The carcinogenic potency of carbon particles with and without PAH after repeated intratracheal administration in the rat'. Toxicology Letters 88:15–21. [Cited on pp. 223, 379, 381]

Davila A.F., Rey D., Mohamed K., Rubio B., Guerra A.P. (2006) 'Mapping the sources of urban dust in a coastal environment by measuring magnetic parameters of *Platanus hispanica* leaves'. Environmental Science and Technology 40:3922–3928. [Cited on p. 349]

Day J.P. (2001) 'Particulate erosion of automotive catalyst supports'. Society of Automotive Engineers 2001-01-1995. [Cited on p. 350]

Dec J.E. (1992) 'Soot distribution in a D.I. diesel engine using 2-D imaging of laser-induced incandescence, elastic scattering, and flame luminosity'. Society of Automotive Engineers 920115. [Cited on pp. 84, 89]

Dec J.E. (1997) 'A conceptual model of DI diesel combustion based on laser-sheet imaging'. Society of Automotive Engineers 970873. [Cited on pp. 84, 86, 87, 88]

Dec J.E., Coy E.B. (1996) 'OH radical imaging in a DI diesel engine and the structure of the early diffusion flame'. Society of Automotive Engineers 960831. [Cited on pp. 80, 84]

Dec J.E., Espey C. (1995) 'Ignition and early soot formation in a DI diesel engine using multiple 2-D imaging diagnostics'. Society of Automotive Engineers 950456. [Cited on pp. 80, 88]

Dec J.E., Kelly-Zion P.L. (2000) 'The effects of injection timing and diluent addition on late combustion soot burn-out in a DI diesel engine based on simultaneous 2-D imaging of OH and soot'. Society of Automotive Engineers 2000-01-0238. [Cited on pp. 84, 90, 91]

Dec J.E., Tree D.R. (2001) 'Diffusion-flame/wall interactions in a heavy-duty DI diesel engine'. Society of Automotive Engineers 2001-01-1295. [Cited on p. 84]

Dec J.E., zur Loye A.O., Siebers D.L. (1991) 'Soot distribution in a D.I. diesel engine using 2-D laser-induced incandescence imaging'. Society of Automotive Engineers 910224. [Cited on p. 84]

DeCarlo P.F., Slowik J.G., Worsnop D.R., Davidovits P., Jimenez J.L. (2004) 'Particle morphology and density characterisation by combined mobility and aerodynamic diameter measurements. Part 1: Theory'. Aerosol Science and Technology 38:1185–1205. [Cited on p. 219]

Degobert P. (1995) *Automobiles and Pollution*. SAE and Editions Technio. [Cited on pp. 197, 239]

Demirbas A. (2005) 'Biodiesel production from vegetable oils via catalytic and non-catalytic supercritical methanol transesterification methods'. Progress in Energy and Combustion Science 31:466–487. [Cited on p. 257]

Dennis A.J., Garner C.P., Taylor D.H.C. (1999) 'The effect of EGR on diesel engine wear'. Society of Automotive Engineers 1999-01-0839. [Cited on p. 289]

Dent J.C. (1980) 'Turbulent mixing rate – and its effect on smoke and hydrocarbon emissions from diesel engines'. Society of Automotive Engineers 800092. [Cited on p. 91]

Desai R.R., Watson H.C. (1997) 'Effects of fuel type and engine operation on diesel particulates and NO_x emissions'. Institution of Mechanical Engineers S496/006/97. [Cited on pp. 263, 265]

Desantes J.M., Arrégle J., Molina S., Lejeune M. (2000) 'Influence of the EGR rate, oxygen concentration and equivalent fuel/air ratio on the combustion behaviour and pollutant emissions of a heavy-duty diesel engine'. Society of Automotive Engineers 2000-01-1813. [Cited on pp. 66, 272]

Desantes J.M., Benajes J., Molina S., González C.A. (2004) 'The modification of the fuel injection rate in heavy-duty diesel engines. Part 1: Effects on engine performance and emissions'. Applied Thermal Engineering 24:2701–2714. [Cited on p. 269]

Desantes J.M., Payri R., Salvador F.J., Gil A. (2006) 'Development and validation of a theoretical model for diesel spray penetration'. Fuel 85:910–917. [Cited on p. 265]

Devalia J.L., Rusznak C., Eang J., Khair O.A., Abdelaziz M., Calderon M.A., Davies R.J. (1996) 'Air pollutants and respiratory hypersensitivity'. Toxicology Letters 86:169–176. [Cited on pp. 364, 374]

Diaz-Sanchez D., Tsien A., Fleming J. (1997) 'Combined diesel exhaust particulate and ragween allergen challenge markedly enhances human in vivo nasal ragweed-specific IgE and skew cytokine production to a T helper cell 2-type pattern'. Journal of Immunology 158:2406–2413. [Cited on p. 382]

Dietzmann H.E. (1984) 'Emissions characterization of diesel forklift trucks'. Society of Automotive Engineers 841396. [Cited on p. 173]

Dietzmann H.E., Parness M.A., Bradow R.L. (1980) 'Emissions from trucks by chassis version of 1983 transient procedure'. Society of Automotive Engineers 801371. [Cited on p. 172]

Dillner A.M., Schauer J.J., Christensen W.F., Cass G.R. (2005) 'A qualitative method for clustering size distributions of elements'. Atmospheric Environment 39:1525–1537. [Cited on p. 35]

Dobbins R.A. (2002) 'Soot inception temperature and the carbonization rate of precursor particles'. Combustion and Flame 130:204–214. [Cited on p. 70]

Dobbins R.A. (2007) 'Hydrocarbon nanoparticles formed in flames and diesel engines'. Aerosol Science and Technology 41:485–496. [Cited on p. 107]

Dobbins R.A., Megaridis C.M. (1987) 'Morphology of flame-generated soot as determined by thermophoretic sampling'. Langmuir 3:254–259. [Cited on p. 215]

Dobbins R.A., Megaridis C.M. (1991) 'Absorption and scattering of light by polydisperse aggregates'. Applied Optics 30:4747–4754. [Cited on p. 189]

Dobbins R.A., Fletcher R.A., Lu W. (1995) 'Laser microprobe analysis of soot precursor particles and carbonaceous soot'. Combustion and Flame 100:301–309. [Cited on pp. 70, 71]

Dobbins R.A., Govatzidakis G.J., Lu W., Schwartzman A.F., Fletcher R.A. (1996) 'Carbonization rate of soot precursor particles'. Combustion and Flame 121:103–121. [Cited on pp. 70, 106, 292]

Dobbins R.A., Fletcher R.A., Chang H.-C. (1998) 'The evolution of soot precursor particles in a diffusion flame'. Combustion and Flame 115:285–298. [Cited on pp. 62, 70, 107]

Dobbins R.A., Fletcher R.A., Benner B.A., Hoeft S. (2006) 'Polycyclic aromatic hydrocarbons in flames, in diesel fuels, and in diesel emissions'. Combustion and Flame 144:773–781. [Cited on pp. 70, 105]

Dockery D., Pope A. (1996) 'Epidemiology of acute health effects: Summary of time-series studies'. In *Particles in Our Air – Concentrations and Health Effects*, Wilson R., Spengler J. (editors), Harvard University Press, pp. 123–147. [Cited on pp. 356, 373]

Dodge L.G., Simescu S., Neeley G.D., Maymar M.J., Dickey D.W., Savonen C.L. (2002) 'Effect of small holes and high injection pressures on diesel engine combustion'. Society of Automotive Engineers 2002-01-0494. [Cited on pp. 66, 266]

Dolan D.F., Kittelson D.B. (1978) 'Diesel exhaust aerosol particle size distributions – comparison of theory and experiment'. Society of Automotive Engineers 780110. [Cited on p. 76]

Dolan D.F., Kittelson D.B. (1979) 'Roadway measurements of diesel exhaust aerosols'. Society of Automotive Engineers 790492. [Cited on p. 142]

Dolan D.F., Kittelson D.B., Pui D.Y.H. (1980) 'Diesel exhaust particle size distribution measurement techniques'. Society of Automotive Engineers 800187. [Cited on pp. 158, 215]

Donahue N.M., Robinson A.L., Stanier C.O., Pandis S.N. (2006) 'Coupled partitioning, dilution, and chemical aging of semivolatile organics'. Environmental Science and Technology 40:2635–2643. [Cited on p. 145]

Donaldson K., MacNee W. (1998) 'The mechanism of lung injury caused by PM_{10}'. In *Air Pollution and Health*, Hester R.E., Harrison R.M. (editors), Issues in Environmental Science and Technology, Volume 10, Royal Society of Chemistry, pp. 21–32. [Cited on pp. 367, 369, 370, 376, 380]

Donaldson K., Beswick P.H., Gilmour P.S. (1996) 'Free radical activity associated with the surface of particles: A unifying factor in determining biological activity?'. Toxicology Letters 88:293–298. [Cited on p. 369]

Donaldson K., Li X.Y., MacNee W. (1998) 'Ultrafine (nanometre) particle mediated lung injury'. Journal of Aerosol Science 29:553–560. [Cited on p. 367]

Donaldson K., Stone V., Borm P.J.A., Jimenez L.A., Gilmour P.S., Schins R.P.F., Knappen A.M., Rahman I., Faux S.P., Brown D.M., MacNee W. (2003) 'Oxidative stress and calcium signaling in the adverse effects of environmental particles (PM_{10})'. Free Radical Biology and Medicine 34:1369–1382. [Cited on p. 368]

Dorie L.D., Bagley S.T., Woon P.V., Leddy D.G., Johnson J.H., Wiczynski P.D., Lu J. (1987) 'Collection and characterization of particulate and gaseous-phase hydrocarbons in diesel exhaust modified by ceramic particulate traps'. Society of Automotive Engineers 870254. [Cited on pp. 165, 180, 240]

Dowling M. (1992) 'The impact of oil formulation on emissions from diesel engines'. Society of Automotive Engineers 922198. [Cited on pp. 281, 282]

Drake M.C., Fansler T.D., Solomon A.S., Szekely G.A. (2003) 'Piston fuel films as a source of smoke and hydrocarbon emissions from a wall-controlled spark-ignited direct-injection engine'. Society of Automotive Engineers 2003-01-0547. [Cited on p. 328]

Draper W.M., Hartmann H., Kittelson D.B., Watts W.F., Baumgard K.J. (1987) 'Impact of a ceramic trap and manganese fuel additive on the biological activity and chemical composition of exhaust particles from diesel engines used in underground mines'. Society of Automotive Engineers 871621. [Cited on pp. 112, 168, 261, 306]

Draper W.M., Phillips J., Zeller H.W. (1988) 'Impact of a barium additive on the mutagenicity and polycyclic aromatic hydrocarbon content of diesel exhaust particulate emissions'. Society of Automotive Engineers 881651. [Cited on pp. 112, 181, 238]

Du C.-J., Kittelson D.B., Zweidinger R.B. (1984) 'Measurements of polycyclic aromatic compounds in the cylinder of an operating engine'. Society of Automotive Engineers 840364. [Cited on pp. 93, 105, 109, 110, 116]

Du C.-J., Kracklauer J., Kittelson D.B. (1998) 'Influence of an iron fuel additive on diesel combustion'. Society of Automotive Engineers 980536. [Cited on pp. 96, 99, 103]

Dunnu G., Hilber T., Schnell U. (2006) 'Advanced size measurements and aerodynamic classification of solid recovered fuel particles'. Energy and Fuels 29:1685–1690. [Cited on p. 217]

Durán A., de Lucas A., Carmona M., Ramos M.J., Armas O. (2002) 'Accuracy of the European standard method to measure the amount of DPM emitted into the atmosphere'. Fuel 81:2053–2060. [Cited on pp. 48, 115, 223]

Durán A., Monteagudo J.M., Armas O., Hernández J.J. (2006) 'Scrubbing effect on diesel particulate matter from transesterified waste oils blends'. Fuel 85:923–928. [Cited on p. 284]

Dusek U., Reischl G.P., Hitzenberger R. (2006) 'CCN activation of pure and coated carbon black particles'. Environmental Science and Technology 40:1223–1230. [Cited on p. 145]

Dye A., Rhead M., Trier C. (1998) 'The physical nature of airborne particulate matter (PM10) in Plymouth, UK'. International Journal of Vehicle Design 20:3–9. [Cited on p. 360]

Dyke P.H., Sutton M., Wood D., Marshall J. (2007) 'Investigations on the effect of chlorine in lubricating oil and the presence of a diesel oxidation catalyst on PCDD/F releases from internal combustion engines'. Chemosphere 67:1275–1286. [Cited on pp. 231, 283]

E

Easley W.L., Mellor A.M. (2001) 'Flame temperature correlation of emissions from diesels operated on alternative fuels'. Society of Automotive Engineers 2001-01-2014. [Cited on p. 66]

Edgar B., Rumminger M., Streichsbier M. (2003) 'A framework for evaluating aftertreatment PM control strategies'. Society of Automotive Engineers 2003-01-2306. [Cited on p. 359]

Edwards J.B. (1977) *Combustion – The Formation and Emission of Trace Species*. Ann Arbor Science Publishers, Inc. [Cited on pp. 52, 72, 124, 125]

Egsgaard H. (1996) 'Investigation of the initial reactions of the Calcote mechanism for soot formation'. Journal of the American Society for Mass Spectrometry 7:559–564. [Cited on p. 77]

Ehrman S.H., Friedlander S.K., Zachariah M.R. (1998) 'Characteristics of SiO_2/TiO_2 nanocomposite particles formed in a premixed flat flame'. Journal of Aerosol Science. 29:687–706. [Cited on p. 98]

Eiglmeier C., Lettmann H., Stiesch G., Merker G.P. (2001) 'A detailed phenomenological model for wall heat transfer prediction in diesel engines'. Society of Automotive Engineers 2001-01-3265. [Cited on p. 92]

Eisenberg W.C., Schuetzle D., Williams R.L. (1984) 'Cooperative evaluation of methods for the analysis of PAH in extracts from diesel particulate emissions'. Society of Automotive Engineers 840414. [Cited on p. 178]

Ekberg L.E. (1995) 'Concentrations of NO_2 and other traffic related contaminants in office buildings located in urban environments'. Building and Environment 30:293–298. [Cited on p. 359]

Elamir I.E., Andrews G.E., Williams P.T. (1991) 'Determination of diesel engine lubricating oil consumption through analysis of the calcium in diesel particulates'. In *Experimental Methods in Engine Research and Development*, Institution of Mechanical Engineers, December 11–12, pp. 171–180. [Cited on pp. 165, 283]

Eldabbagh F., Ramesh A., Hawari J., Hutney W., Kozinski J.A. (2005) 'Particle–metal interactions during combustion of pulp and paper biomass in a fluidized bed combustor'. Combustion and Flame 142:249–257. [Cited on p. 97]

El-Hannouny E.M., Lee T.W., Farrell P.V., Reitz R.D. (2003) 'An experimental and numerical study of injector behavior for HSDI diesel engines'. Society of Automotive Engineers 2003-01-0705. [Cited on pp. 266, 267]

Elliott P., Briggs D. (1999) 'Traffic related pollution and chronic health effects'. In *Health Effects of Vehicle Emissions*, Feb. 16–17, Royal Society of Medicine. [Cited on p. 364]

Emerich D.F., Thanos C.G. (2006) 'The pinpoint promise of nanoparticle-based drug delivery and molecular diagnosis'. Biomolecular Engineering 23:171–184. [Cited on p. 396]

Engler B.H., Lox E.S., Ostgathe K., Cartellieri W., Zelenka P. (1993) 'Diesel oxidation catalysts with low sulfate formation for HD-diesel engine application'. Society of Automotive Engineers 932499. [Cited on pp. 152, 284, 296]

Enya T., Suzuki H., Watanabe T., Hirayama T., Hisamatsu Y. (1997) '3-nitrobenzanthrone, a powerful bacterial mutagen and suspected human carcinogen found in diesel exhaust airborne particulates'. Environmental Science and Technology 31:2772–2776. [Cited on pp. 112, 240]

EPA (2004) 'The particle pollution report – current understanding of air quality and emissions through 2003'. Contract No. 68-D-02-065, Work Assignment No. 2-01, U.S.Environmental Protection Agency, December 2004. [Cited on p. 357]

Erdemir A., Ozturk O., Alzoubi M., Woodford J., Ajayi L., Fenske G. (2000) 'Near-frictionless carbon coatings for use in fuel injectors and pump systems operating with low-sulfur diesel fuels'. Society of Automotive Engineers 2000-01-0518. [Cited on p. 252]

Espey C., Dec J.E. (1993) 'Diesel engine combustion studies in a newly designed optical-access engine using high-speed visualization and 2-D laser imaging'. Society of Automotive Engineers 930971. [Cited on pp. 84, 91]

Espey C., Dec J.E., Litzinger T.A., Santavicca D.A. (1994) 'Quantitative 2-D fuel vapor concentration imaging in a firing D.I. diesel engine using planar laser-induced rayleigh scattering'. Society of Automotive Engineers 940682. [Cited on p. 84]

Essig G., Kamp H., Wacker E. (1990) 'Diesel engine emissions reduction – the benefits of low oil consumption design'. Society of Automotive Engineers 900591. [Cited on pp. 279, 280]

Etheridge P., Hews I.M., Wearn S.J., Andersson J.D. (2003) 'DfT motorcycle emissions measurement program: Regulated emissions results'. Society of Automotive Engineers 2003-01-1897. [Cited on p. 337]

Etyemezian V., Kuhns H., Gillies J., Green M., Pitchford M., Watson J. (2003a) 'Vehicle-based road dust emission measurement: I – methods and calibration'. Atmospheric Environment 37:4559–4571. [Cited on p. 343]

Etyemezian V., Kuhns H., Gillies J., Chow J., Hendrickson K., McGown M., Pitchford M. (2003b) 'Vehicle-based road dust emission measurement (III): Effect of speed, traffic volume, location, and season on PM_{10} road dust emissions in the Treasure Valley, ID'. Atmospheric Environment 37:4583–4593. [Cited on p. 343]

Evans J., Wolff S. (1996) 'Modeling of air pollution impacts: One possible explanation of the observed chronic mortality'. In *Particles in Our Air – Concentrations and Health Effects*, Wilson R., Spengler J. (editors), Harvard University Press, pp. 189–204. [Cited on p. 363]

Ezekoye O.A., Zhang Z. (1997) 'Soot oxidation and agglomeration modeling in a microgravity diffusion flame'. Combustion and Flame 110:127–139. [Cited on pp. 79, 81]

F

Fan Z., Chen D., Birla P., Kamens R.M. (1995) 'Modeling of nitro-polycyclic aromatic hydrocarbon formation and decay in the atmosphere'. Atmospheric Environment 29:1171–1181. [Cited on p. 112]

Fang C.P., Kittelson D.B. (1984) 'The influence of a fibrous diesel particulate trap on the size distribution of emitted particles'. Society of Automotive Engineers 840362. [Cited on p. 394]

Fang H.L., Lance M.J. (2004) 'Influence of soot surface changes on DPF regeneration'. Society of Automotive Engineers 2004-01-3043. [Cited on p. 301]

Fanick E.R., Schubert P.F., Russel B.J., Freerks R.L. (2001) 'Comparison of emission characteristics of conventional, hydrotreated, and Fischer–Tropsch diesel fuels in a heavy-duty diesel engine'. Society of Automotive Engineers 2001-01-3519. [Cited on p. 242]

Farfaletti A., Astorga C., Martini G., Manfredi U., Mueller A., Rey M., de Santi G., Krasenbrink A., Larsen B.O. (2005) 'Effect of water/fuel emulsions and a cerium-based combustion improver additive on HD and LD diesel exhaust emissions'. Enviromental Science and Technology. 39:6792–6799. [Cited on p. 261]

Färnlund J., Holman C., Kågeson P. (2001) 'Emissions of ultrafine particles from different types of light duty vehicles', Swedish National Road Administration, January, 2001:10. [Cited on p. 240]

Farrar-Khan J.R., Andrews G.E., Williams P.T., Bartle K.D. (1992) 'The influence of nozzle sac volume on the composition of diesel particulate fuel derived SOF'. Society of Automotive Engineers 921649. [Cited on pp. 266, 267]

Farrell A., Keating T.J. (2000) 'The globalization of smoke: co-evolution in science and governance of a commons problem'. Eighth Biennial Conference, International Association for the Study of Common Property, June, Bloomington, IN, USA. [Cited on pp. 2, 3]

Feilberg A., Kamens R.M., Strommen M.R., Nielsen T. (1999) 'Modeling the formation, decay, and partitioning of semivolatile nitro-polycyclic aromatic hydrocarbons (nitronaphthalenes) in the atmosphere'. Atmospheric Environment 33:1231–1243. [Cited on pp. 110, 112]

Ferge T., Karg E., Schröppel A., Coffee K.R., Tobias H.J., Frank M., Gard E.E., Zimmermann R. (2006) 'Fast determination of the relative elemental and organic carbon content of aerosol samples by on-line single-particle aerosol time-of-flight mass spectrometry'. Environmental Science and Technology 40:3327–3335. [Cited on p. 209]

Fialkov A.B. (1997) 'Investigations of ions in flames'. Progress in Energy and Combustion Science 23:399–528. [Cited on pp. 71, 72, 98, 197]

Figler B., Sahle W., Krantz S., Ulfvarson U. (1996) 'Diesel exhaust quantification by scanning electron microscope with special emphasis on particulate size distribution'. Science of the Total Environment 193:77–83. [Cited on p. 233]

Fily M., Bourdelles B., Dedieu J.P., Sergent C. (1997) 'Comparison of *in situ* and Landsat thematic mapper derived snow grain characteristics in the Alps'. Remote Sensing of the Environment 59:452–460. [Cited on p. 42]

Fine P.M., Cass G.R., Simoneit B.R.T. (1999) 'Characterization of fine particle emissions from burning church candles'. Environmental Science and Technology 33:2352–2362. [Cited on p. 38]

Fine P.M., Chakrabarti B., Krudysz M., Schauer J.J., Sioutas C. (2004) 'Diurnal variations of individual organic compound consituents of ultrafine and accumulation mode particulate matter in the los angeles basin'. Environmental Science and Technology 38:1296–1304. [Cited on p. 176]

Fiorello S.C. (1968) 'The Navy's smoke abatement program'. Society of Automotive Engineers 680345. [Cited on p. 4]

Fischer P.H., Hoek G., van Reeuwijk H., Briggs D.J., Lebret E., van Wijnen J.H., Kingham S., Elliott P.E. (2000) 'Traffic-related differences in outdoor and indoor concentrations of particles and volatile organic compounds in Amsterdam'. Atmospheric Environment 34:3713–3722. [Cited on p. 358]

Fishenden M.W. (1964) 'Smoke and smoke prevention'. Encyclopaedia Britannica 20:840–842. [Cited on p. 2]

Flagan R.C., Seinfeld J.H. (1988) *Fundamentals of Air Pollution Engineering*. Prentice Hall. [Cited on pp. 15, 16, 43, 98]

Fletcher T.H., Ma J., Rigby J.R., Brown A.L., Webb B.W. (1997) 'Soot in coal combustion systems'. Progress in Energy and Combustion Science 23:283–301. [Cited on p. 62]

Flörchinger P., Zink U., Cutler W., Tomazic D. (2004) 'DPF regeneration-concept to avoid uncontrolled regeneration during idle'. Society of Automotive Engineers 2004-01-2657. [Cited on p. 302]

Florig H.K., Sun G., Song G. (2002) 'Evolution of particulate regulation in china – prospects and challenges of exposure-based control'. Chemosphere 49:1163–1174. [Cited on p. 33]

Fløysand S.Å., Kvinge F., Betts W.E. (1993) 'The influence of diesel fuel properties on particulate emissions in a catalyst equipped european car'. Society of Automotive Engineers 932683. [Cited on pp. 264, 265]

Fraenkle G., Hardenberg H.O. (1975) 'New methods for reducing visible emissions of diesel engines'. Society of Automotive Engineers 750772. [Cited on p. 120]

Fraser M.P., Cass G.R. (1998) 'Detection of excess ammonia emissions from in-use vehicles and the implications for fine particle control'. Environmental Science and Technology 32:1053–1057. [Cited on p. 37]

Fraser M.P., Cass G.R., Simoneit B.R.T. (1998) 'Gas-phase and particulate-phase organic compounds emitted from motor vehicle traffic in a Los Angeles roadway tunnel'. Environmental Science and Technology 32:2051–2060. [Cited on p. 29]

Frédéric L., Shinichi H., Jean-Pascal S., Laurent T. (2007) 'In-car air quality – a global approach to enhance comfort'. Society of Automotive Engineers 2007-01-1185. [Cited on p. 359]

Frehland P., Fischer H., Paffrath H. (1999) 'The free jet driven centrifugal cleaner – a present and future concept for oil filtration'. Society of Automotive Engineers 1999-01-0824. [Cited on p. 286]

French C.C.J., Pike D.A. (1979) 'Diesel engined, light duty vehicles for and emission controlled environment'. Society of Automotive Engineers 790761. [Cited on p. 396]

Frenklach M., Taki S., Matula R.A. (1983) 'A conceptual model for soot formation in pyrolysis of aromatic hydrocarbons'. Combustion and Flame 49:275–282. [Cited on p. 292]

Friedlander S.K. (1977) *Smoke, Dust and Haze – Fundamentals of Aerosol Behavior*. John Wiley & Sons, Inc. [Cited on p. 37]

Friedlander S.K., Ogawa K., Ullman M. (2001) 'Elasticity of nanoparticle chain aggregates: implications for polymer fillers and surface coatings'. Powder Technology 118:90–96. [Cited on p. 219]

Frisch L.E., Johnson J.H., Leddy D.G. (1979) 'Effect of fuels and dilution ratio on diesel particulate emissions'. Society of Automotive Engineers 790417. [Cited on pp. 51, 52, 125, 142, 165]

Fritz S.G., Bailey C.R., Scarbro C.A., Somers J.H. (2001) 'Heavy-duty diesel truck in-use emission test program for model years 1950–1975'. Society of Automotive Engineers 2001-01-1327. [Cited on p. 372]

Froelund K., Owens E.C., Frame E., Buckingham J.P., Garbak J., Tseregounis S., Jackson A. (2001a) 'Impact of lubricant oil on regulated emissions of a light-duty Mercedes-Benz OM611 CIDI-engine'. Society of Automotive Engineers 2001-01-1901. [Cited on p. 282]

Froelund K., Menezes L., Johnson H.R., Rein O. (2001b) 'Real-time transient and steady-state measurement of oil consumption for several production SI-engines'. Society of Automotive Engineers 2001-01-1902. [Cited on p. 280]

Frølund K., Schramm J. (1997) 'PAH-transport in diesel engines'. Society of Automotive Engineers 972960. [Cited on p. 108]

Fuchs N.A. (1964) *The Mechanics of Aerosols*. Pergamon Press. [Cited on pp. 18, 21, 22, 143]

Fujii T., Ikezawa H., Kotani Y. (2002) 'A study of the analysis of PM components with CR-DPF'. Society of Automotive Engineers 2002-01-1686. [Cited on pp. 174, 175]

Fujimoto H., Kurata K., Asai G., Senda J. (1998) 'OH radical generation and soot formation/oxidation in a DI diesel engine'. Society of Automotive Engineers 982630. [Cited on p. 91]

Fujimoto H., Senda J., Kawano D., Wada Y. (2004) 'Exhaust emission through diesel combustion of mixed fuel oil compose of fuel with high volatility and that with low volatility'. Society of Automotive Engineers 2004-01-1845. [Cited on p. 263]

Fujitani Y., Ideno Y., Fushimi A., Tanabe K., Obayashi S., Kobayashi T. (2006) 'Generation of nanoparticles of lubricating motor oil for inhalation studies'. Aerosol Science and Technology 41:14–23. [Cited on p. 390]

Fujiwara Y., Fukazawa S. (1980) 'Growth and combustion of soot particulates in the exhaust of diesel engines'. Society of Automotive Engineers 800984. [Cited on pp. 216, 301]

Fujiwara Y., Fukazawa S., Tosaka S., Muryama T. (1984) 'Formation of soot particulates in the combustion chamber of a precombustion chamber type diesel engine'. Society of Automotive Engineers 840417. [Cited on p. 86]

Fujiwara Y., Tosaka S., Murayama T. (1990) 'The microcrystal structure of soot particulates in the combustion chamber of prechamber type diesel engines'. Society of Automotive Engineers 901579. [Cited on p. 214]

Fujiwara Y., Tosaka S., Murayama T. (1993) 'Formation process of SOF in the combustion chamber of IDI diesel engines'. Society of Automotive Engineers 932799. [Cited on p. 109]

Fukushima H., Asano I., Nakamura S., Ishida K., Gregory D. (2000) 'Signal processing and practical performance of a real-time PM analyzer using fast FIDs'. Society of Automotive Engineers 2000-01-1135. [Cited on p. 197]

Fukushima H., Uchihara H., Asano I., Adachi M., Nakamura S., Ikeda M., Ishida K. (2001) 'An alternative technique for low particulate measurement'. Society of Automotive Engineers 2001-01-0218. [Cited on pp. 169, 170]

Fukushima H., Asano I., Hill L., Adachi M., Ishida K., Onishi T. (2003a) 'Measurement of nitrogen compounds in diesel particulate matter'. Society of Automotive Engineers 2003-01-2019. [Cited on p. 171]

Fukushima H., Asano I., Nakamura S., Adachi M (2003b) 'New techniques for measurement of particulate emission from advanced vehicles'. Society of Automotive Engineers 2003-26-0005. [Cited on p. 170]

Funasaka K., Miyazaki T., Tsuruho K., Tamura K., Mizuno T., Kuroda K. (2000) 'Relationship between indoor and outdoor carbonaceous particulates in roadside households'. Environmental Pollution 110:127–134. [Cited on p. 358]

Funkenbusch E.F., Leddy D.G., Johnson J.H. (1979) 'The characterization of the soluble organic fraction of diesel particulate matter'. Society of Automotive Engineers 790418. [Cited on pp. 111, 171, 175, 176, 234, 239]

G

Gabathuler J.P., Mizrah T., Fischer A., Käser P., Maurer A. (1991) 'New developments of ceramic foam as a diesel particulate filter'. Society of Automotive Engineers 910325. [Cited on p. 301]

Gabele P.A., Black F.M., King F.G., Zweidinger R.B., Brittain R.A. (1981) 'Exhaust emission patterns from two light-duty diesel automobiles'. Society of Automotive Engineers 810081. [Cited on p. 180]

Gaiser G., Mucha P. (2004) 'Prediction of pressure drop in diesel particulate filters considering ash deposit and partial regenerations'. Society of Automotive Engineers 2004-01-0158. [Cited on p. 303]

Galisteo F.C., Mariscal R., Granados M.L., Fierro J.L.G., Brettes P., Salas O. (2005) 'Reactivation of a commercial diesel oxidation catalyst by acid washing'. Environmental Science and Technology 39:3844–3848. [Cited on p. 296]

Gallego-Juárez J.A., Riera-Franco de Sarabia E., Rodríguez-Corral G., Hoffmann T.L., Gálvez-Moraleda J.C., Rodríguez-Maroto J.J., Gómez-Moreno F.J., Bahillo-Ruiz A., Martín-Espigares M., Acha M. (1999) 'Application of acoustic agglomeration to reduce fine particle emissions from coal combustion plants'. Environmental Science and Technology 33:3482–3489. [Cited on p. 394]

Gallorini M. (2000) 'Trace elements in atmospheric pollution processes: The contribution of neutron activation analysis'. In *Aerosol Chemical Processes in the Environment*, Spurny K.R., Hochrainer D. (editors), CRC Press LLC, pp. 431–455. [Cited on pp. 35, 36]

Gandhi H.S. (1987) 'Technical considerations for catalyst for European market'. Institution of Mechanical Engineers C344/87. [Cited on p. 315]

Garciá-Morales M., Partal P., Navarro F.J., Gallegos C. (2006) 'Effect of waste polymer addition on the rheology of modified bitumen'. Fuel 85:936–943. [Cited on p. 344]

Garg B.D., Cadle S.H., Mulawa P.A., Groblicki P.J., Laroo C., Parr G.A. (2000) 'Brake wear particulate matter emissions'. Environmental Science and Technology 34:4463–4469. [Cited on p. 346]

Gautam M., Kelly B., Gupta D., Clark N., Atkinson R., El-Gazzar L., Lyons D.W. (1994) 'Sampling strategies for characterisation of reactive components of heavy duty diesel exhaust emissions'. Society of Automotive Engineers 942263. [Cited on pp. 165, 172]

Gautam M., Durbha M., Chitoor K., Jaraiedi M., Mariwalla N., Ripple D. (1998) 'Contribution of soot contaminated oils to wear'. Society of Automotive Engineers 981406. [Cited on p. 289]

Gautam M., Chitoor K., Durbha M., Summers J.C. (1999) 'Effect of diesel soot contaminated oil on engine wear – investigation of novel oil formulations'. Tribology International 32:687–699. [Cited on p. 290]

Gautam M., Clark N.N., Mehta S., Boyce J.A., Rogers F., Gertler A. (2003) 'Concentrations and size distributions of particulate matter emissions from a Class-8 heavy-duty diesel truck tested in a wind tunnel'. Society of Automotive Engineers 2003-01-1894. [Cited on pp. 142, 143, 144, 183]

Gauthier A., Delvigne T. (2000) 'Soot induced cam wear in diesel engines: An investigation using thin layer activation'. Society of Automotive Engineers 2000-01-1990. [Cited on p. 290]

Gavett S.H., Madison S.L., Dreher K.L., Winsett D.W., McGee J.K., Costa D.L. (1997) 'Metal and sulfate composition of residual oil fly ash determines airway hyperreactivity and lung injury in rats'. Environmental Research 72:162–172. [Cited on p. 380]

Gekas I., Gabrielsson P., Johansen K., Bjørn I., Kjaer J.H., Reczek W., Cartellieri W. (2002) 'Performance of a urea SCR system combined with a PM and fuel optimized heavy-duty diesel engine able to achieve the Euro V emission limits'. Society of Automotive Engineers 2002-01-2885. [Cited on p. 299]

Gelfand B.E. (1996) 'Droplet breakup phenomena in flow with velocity lag'. Progress in Energy and Combustion Science 22:201–265. [Cited on p. 267]

Geller D.P., Goodrum J.W. (2004) 'Effects of specific fatty acid methyl esters on diesel fuel lubricity'. Fuel 83:2351–2356. [Cited on p. 257]

Geller M.D., Sardar S.B., Phuleria H., Fine P.M., Sioutas C. (2005) 'Measurements of particle number and mass concentrations and size distributions in a tunnel environment'. Environmental Science and Technology 39:8653–8663. [Cited on p. 391]

Genc V.E., Altay F.E., Uner D. (2005) 'Testing molten metal oxide catalysts over structured ceramic substrates for diesel soot oxidation'. Catalysis Today 105:537–543. [Cited on p. 296]

George H.F., Bardasz E.A., Soukup B.L. (1997) 'Understanding soot mediated oil thickening through designed experimentation part 3: An improved approach to drain oil viscosity measurements – rotational rheology'. Society of Automotive Engineers 971692. [Cited on p. 287]

Gerde P., Muggenburg B.A., Lundborg M., Tesfaigzi Y., Dahl A.R. (2001) 'Respiratory epithelial penetration and clearance of particle-borne benzo[a]pyrene'. Report No. 101, Health Effects Institute. [Cited on pp. 181, 379]

Gertler A.W. (2005) 'Diesel vs. gasoline emissions: Does PM from diesel or gasoline vehicles dominate in the US'. Atmospheric Environment 39:2349–2355. [Cited on p. 358]

Gertler A.W., Gillies J.A., Pierson W.R., Rogers C.F., Sagebiel J.C., Abu-Allaban M., Coulombe W., Tarnay L., Cahill T.A. (2002) 'Emissions from diesel and gasoline engines measured in highway tunnels'. Report No. 107, Health Effects Institute. [Cited on pp. 169, 233, 238, 362]

Geyer S.M., Lestz S.S., Litzinger T.A. (1987) 'Chemical and biological character of particulate matter for a variety of oxidants in a constant-volume combustion bomb'. Society of Automotive Engineers 872135. [Cited on p. 110]

Ghan S.J., Guzman G., Abdul-Razzak H.A. (1998) 'Competition between sea salt and sulfate particles as cloud condensation nuclei'. Journal of the Atmospheric Sciences 55:3340–3347. [Cited on pp. 25, 27]

Gibbs R., Wotzak G., Byer S., Johnson R., Hill B., Werner P. (1977) 'Emissions from in-use catalyst vehicles'. Society of Automotive Engineers 770064. [Cited on p. 318]

Giechaskiel B., Ntziachristos L., Samaras Z., Scheer V., Casati R., Vogt R. (2005) 'Formation potential of vehicle exhaust nucleation mode particles on-road and in the laboratory'. Atmospheric Environment 39:3191–3198. [Cited on p. 144]

Gierens K. (2007) 'Are fuel additives a viable contrail mitigation option?'. Atmospheric Environment 41:4548–4552. [Cited on p. 41]

Givens W.A., Buck W.H., Jackson A., Kaldor A., Hertzberg A., Moehrmann W., Mueller-Lunz S., Pelz N., Wenninger G. (2003) 'Lube formulation effects on transfer of elements to exhaust after-treatment system components'. Society of Automotive Engineers 2003-01-3109. [Cited on pp. 97, 233, 283, 284, 285]

Glassman I. (1996) *Combustion*. Academic Press. [Cited on p. 255]

Gleitsmann G., Zellner R. (1998) 'The effects of ambient temperature and relative humidity on particle formation in the jet regime of commercial aircrafts: A modelling study'. Atmospheric Environment 32:3079–3087. [Cited on pp. 126, 128]

Gogou A., Stephanou E.G. (2000) 'Urban and rural organic fine aerosols: Components source reconciliation using and organic geochemical approach'. In *Aerosol Chemical Processes in the Environment*, Spurny K.R., Hochrainer D. (editors). CRC Press LLC, pp. 457–486. [Cited on pp. 29, 34]

Gold D.R., Litonjua A., Schwartz J., Lovett E., Larson A., Nearing B., Allen G., Verrier M., Cherry R., Verrier R. (2000) 'Ambient pollution and heart rate variability'. Circulation 101:1267–1273. [Cited on p. 374]

Gold M.R., Arcoumanis C., Whitelaw J.H., Gaade J., Wallace S. (2000) 'Mixture preparation strategies in an optical four-valve port-injected gasoline engine'. International Journal of Engine Research 1:41–56. [Cited on p. 320]

Gomes P.C.F., Yates D.A. (1992) 'The influence of some engine operating parameters on particulate emissions'. Society of Automotive Engineers 922222. [Cited on p. 125]

Gómez-Moreno F.J., Núñez L., Plaza J., Alonso D., Pujadas M., Artíñano B. (2007) 'Annual evolution and generation mechanisms of particulate nitrate in Madrid'. Atmospheric Environment 41:394–406. [Cited on p. 46]

González D.M.A., Piel W., Asmus T., Clark W., Garbak J., Liney E., Natarajan M., Naegeli D.W., Yost D., Frame E.A., Wallace J.P. (2001) 'Oxygenates screening for advanced petroleum-based diesel fuels: Part 2. The effect of oxygenate blending compounds on exhaust emissions'. Society of Automotive Engineers 2001-01-3632. [Cited on pp. 258, 259]

Goodlive S.A., Lvovich V.F., Humphrey B.K., Boyle F.P. (2004) 'On-board sensor systems to diagnose condition of diesel engine lubricants – focus on soot'. Society of Automotive Engineers 2004-01-3010. [Cited on p. 286]

Gopalakrishnan V., Zukoski C.F. (2006) 'Viscosity of hard-sphere suspensions: Can we go lower?'. Industrial and Engineering Chemistry Research 45:6906–6914. [Cited on p. 287]

Gosse K., Paranthoën P., Patte-Rouland B., Gonzalez M. (2006) 'Dispersion in the near wake of idealized car model'. International Journal of Heat and Mass Transfer 49:1747–1752. [Cited on p. 142]

Goto Y., Kawai T. (2004) 'Real-time measurement of particle size distribution from diesel engines equipped with continuous regenerative DPF under a transient driving condition'. Society of Automotive Engineers 2004-01-1984. [Cited on p. 186]

Gradon L., Pratsinis S.E., Podgorski A., Scott S.J., Panda S. (1996) 'Modeling retention of inhaled particles in rat lungs including toxic and overloading effects'. Journal of Aerosol Science 27:487–503. [Cited on p. 366]

Graedel T.E. (1988) 'Ambient levels of anthropogenic emissions and their atmospheric transformation products'. In *Air Pollution, the Automobile, and Public Health*, Watson A.Y., Bates R.R., Kennedy D. (editors) National Academy Press, pp. 133–160. [Cited on p. 315]

Graskow B.R., Kittelson D.B., Ahmadi M.R., Morris J.E. (1999a) 'Exhaust particulate emissions from two port fuel injected spark ignition engines'. Society of Automotive Engineers 1999-01-1144. [Cited on pp. 326, 327]

Graskow B.R., Kittelson D.B., Ahmadi M.R., Morris J.E. (1999b) 'Exhaust particulate emissions from a direct injection spark ignition engine'. Society of Automotive Engineers 1999-01-1145. [Cited on pp. 327, 331]

Graskow B.R., Ahmadi M.R., Morris J.E., Kittelson D.B. (2000) 'Influence of fuel additives and dilution conditions on the formation and emission of exhaust particulate matter from a direct injection spark ignition engine'. Society of Automotive Engineers 2000-01-2018. [Cited on p. 333]

Gratz L.D., Bagley S.T., King K.S., Baumgard K.J., Leddy D.G., Johnson J.H. (1991) 'The effect of a ceramic particulate trap on the particulate and vapor phase emissions of a heavy-duty diesel engine'. Society of Automotive Engineers 910609. [Cited on pp. 173, 305]

Gray A.W., Ryan T.W. (1997) 'Homogeneous charge compression ignition (HCCI) of diesel fuel'. Society of Automotive Engineers 971676. [Cited on p. 292]

Green G.L., Wallace D. (1980) 'Correlation studies of an in-line, full-flow opacimeter'. Society of Automotive Engineers 801373. [Cited on p. 292]

Green R.M. (2000) 'Measuring the cylinder-to-cylinder EGR distribution in the intake of a diesel engine during transient operation'. Society of Automotive Engineers 2000-01-2866. [Cited on p. 202]

Greenfield C., Quarini G. (1998) 'A Lagrangian simulation of particle deposition in a turbulent boundary layer in the presence of thermophoresis'. Applied Mathematical Modelling 22:759–771. [Cited on p. 272]

Greenfield S. (1957) 'Rain scavenging of radioactive particulate material from the atmosphere'. Journal of Meteorology 14:115–125. [Cited on p. 92]

Greenwell L.L., Moreno T., Jones T.P., Richards R.J. (2002) 'Particle-induced oxidative damage is ameliorated by pulmonary oxidants'. Free Radical Biology & Medicine 32:898–905. [Cited on pp. 375, 376]

Greeves G., Tullis S., Barker B. (2003) 'Advanced two-actuator EUI and emission reduction for heavy-duty diesel engines'. Society of Automotive Engineers 2003-01-0698. [Cited on p. 269]

Griffin H.E., Brusick D.J., Castagnoli N., Crump K.S., Goldschmidt B.M., Higgins I.T., Hoffmann D., Horvath S.M., Nettesheim P., Stuart B.O., Witschi H. (1981) *Health Effects of Exposure to Diesel Exhaust*, Report of the Health Effects Panel of the Diesel Impacts Study Committee, National Research Council. National Academy Press, USA. [Cited on pp. 238, 356, 375, 379]

Griffing M.E., Gilbert L.F., Haar G.T., Immethun P.A., Zutaut D.W. (1975) 'Exhaust sulfur oxide measurement using air dilution'. Society of Automotive Engineers 750697. [Cited on pp. 173, 318]

Griffiths J.F., Barnard J.A. (1995) *Flame and Combustion*. Blackie. [Cited on pp. 49, 71, 72, 99, 225, 254]

Grigg H.C. (1976) 'The role of fuel injection equipment in reducing 4-stroke diesel engine emissions'. Society of Automotive Engineers 760126. [Cited on p. 65]

Grillot J.M., Icart G. (1997) 'Fouling of a cylindrical probe and a finned tube bundle in a diesel exhaust environment'. Experimental Thermal and Fluid Science 14:442–454. [Cited on p. 275]

Grimaldi C.N., Postrioti L., Battistoni M., Millo F. (2002) 'Common rail HSDI diesel engine combustion and emissions with fossil/bio-derived fuel blends'. Society of Automotive Engineers 2002-01-0865. [Cited on p. 258]

Grose M., Sakurai H., Savstrom J., Stolzenburg M.R., Watts W.F., Morgan C.G., Murray I.P., Twigg M.V., Kittelson D.B., McMurry P.H. (2006) 'Chemical and physical properties of ultrafine diesel exhaust particles sampled downstream of a catalytic trap'. Environmental Science and Technology 40:5502–5507. [Cited on pp. 362, 393]

Gualtieri M., Andrioletti M., Mantecca P., Vismara C., Camatini M. (2005) 'Impact of tire debris on *in vitro* and *in vivo* systems'. Particle and Fibre Toxicology 2005, 2:1 (*Creative Commons*). [Cited on p. 348]

Gudmand-Høyer L., Bach A., Nielsen G.T., Morgen P. (1999) 'Tribological properties of automotive disc brakes with solid lubricants'. Wear 232:168–175. [Cited on p. 345]

Guerrassi N., Dupraz P. (1998) 'A common rail injection system for high speed direct injection diesel engines'. Society of Automotive Engineers 980803. [Cited on p. 265]

Guerrieri D.A., Rao V., Caffrey P.J. (1996) 'An investigation of the effect of differing filter face velocities on particulate mass weight from heavy-duty diesel engines'. Society of Automotive Engineers 960253. [Cited on p. 151]

Guertler R.W. (1986) 'Excessive cylinder wear and bore polishing in heavy duty diesel engines: Causes and proposed remedies'. Society of Automotive Engineers 860165. [Cited on pp. 279, 283]

Gui B., Chan T.L., Leung C.W., Xiao J., Wang H., Zhao L. (2004) 'Modeling study on the combustion and emissions characteristics of a light-duty DI diesel engine fueled with dimethyl ether (DME) using a detailed chemical kinetics mechanism'. Society of Automotive Engineers 2004-01-1839. [Cited on p. 259]

Guo H., Liu F., Smallwood G.J., Gülder Ö. (2006) 'Numerical study on the influence of hydrogen addition on soot formation in a laminar ethylene – air diffusion flame'. Combustion and Flame 145:324–338. [Cited on p. 74]

Gupta A.K. (1997) 'Gas turbine combustion: Prospects and challenges'. Energy and Conversion Management 38:1311–1318. [Cited on pp. 4, 62]

Gupta S., Poola R., Sekar R. (2001) 'Issues for measuring diesel exhaust particulates using laser induced incandescence'. Society of Automotive Engineers 2001-01-0217. [Cited on p. 188]

Gupta S., Hillman G., El-Hannouny E., Sekar R. (2003) 'Transient particulate emission measurements in diesel engine exhausts'. Society of Automotive Engineers 2003-01-3155. [Cited on p. 188]

Gustafson K.E., Dickhut R.M. (1997) 'Particle/gas concentrations and distributions of PAHs in the atmosphere of southern Chesapeake Bay'. Environmental Science and Technology 31:140–147. [Cited on p. 117]

Gustafsson Ö., Gschwend P.M., Buesseler K.O. (1997) 'Using the ^{234}Th disequilibria to estimate the vertical removal rates of polycyclic aromatic hydrocarbons from the surface area'. Marine Chemistry 57:11–23. [Cited on p. 35]

Guzmán K.A.D., Taylor M.R., Banfield J.F. (2006) 'Environmental risks of nanotechnology: National Nanotechnology Initiative funding, 2000–2004'. Environmental Science and Technology 40:1401–1407. [Cited on p. 396]

H

Ha E.-E., Lee J.-T., Kim H., Hong Y.-C., Lee B.-E., Park H.-S., Christiani D.C. (2003) 'Infant susceptibility of mortality to air pollution in Seoul, South Korea'. Pediatrics 111:284–290. [Cited on p. 373]

Haagen-Smit A.G. (1970) 'A lesson from the smog capital of the world'. Proceedings of the National Academy of Sciences 67:887–897. [Cited on p. 36]

Hackenberg S., Ranalli M. (2007) 'Ammonia on a LNT: Avoid the formation or take advantage of it'. Society of Automotive Engineers 2007-01-1239. [Cited on p. 37]

Hagena J.R., Filipi Z.S., Assanis D.D.N. (2006) 'Transient diesel emissions: Analysis of engine operation during a tip-in'. Society of Automotive Engineers 2006-01-1151. [Cited on p. 276]

Hälinen A.J., Komulainen H., Salonen R.O., Ruotsalainen M., Hirvonen M.-R. (1999) 'Diesel particles induce nitric oxide production in murine alveolar macrophages and rat airways'. Environmental Toxicology and Pharmacology 7:11–18. [Cited on pp. 376, 378]

Hall D.E., Dickens C.J. (1999) 'Measurement of the number and size distribution of particles emitted from a gasoline direct injection vehicle'. Society of Automotive Engineers 1999-01-3530. [Cited on pp. 330, 332, 333]

Hall D.E., Dickens C.J. (2000) 'Measurement of the numbers of emitted gasoline particles: Genuine or artefact?'. Society of Automotive Engineers 2000-01-2957. [Cited on p. 139]

Hall D., Dickens C. (2003) 'The effect of sulphur-free diesel fuel on the measurement of the number and size distribution of particles emitted from a heavy-duty diesel engine equipped with a catalyzed particulate filter'. Society of Automotive Engineers 2003-01-3167. [Cited on pp. 125, 139, 253, 285, 304]

Hall D.E., King D.J., Morgan T.D.B., Baverstock S.J., Heinze P., Simpson B.J. (1998) 'A review of recent literature investigating the measurement of automotive particulate; the relationship with environmental aerosol, air quality and health effects'. Society of Automotive Engineers 982602. [Cited on pp. 4, 6]

Hall D.E., Stradling R.J., Zemroch P.J., Rickeard D.J., Mann N., Heinze P., Martini G., Hagemann R., Rantanen L., Szendefi J. (2000) 'Measurement of the number and size distribution of particle emissions from heavy duty engines'. Society of Automotive Engineers 2000-01-2000. [Cited on p. 242]

Hall-Roberts V.J., Hayhurst A.N., Knight D.E., Taylor S.G. (2000) 'The origin of soot in flames: Is the nucleus an ion?'. Combustion and Flame 120:578–584. [Cited on pp. 71, 98]

Halsall R., McMillan M.L., Schwartz B.J. (1987) 'An improved method for determining the hydrocarbon fraction of diesel particulates by vacuum oven sublimation'. Society of Automotive Engineers 872136. [Cited on pp. 168, 171, 172]

Hammerle R.H., Korniski T.J., Weir J.E., Chladek E., Gierczak C.A., Chase R.E., Hurley R.G. (1992) 'Effect of mileage accumulation on particulate emissions from vehicles using gasoline with methylcyclopentadienyl manganese tricarbonyl'. Society of Automotive Engineers 920731. [Cited on p. 317]

Hampson G.J., Reitz R.D. (1998) 'Two-color imaging of in-cylinder soot concentration and temperature in a heavy-duty DI diesel engine with comparison to multidimensional modeling for single and split injections'. Society of Automotive Engineers 980524. [Cited on pp. 62, 83, 270]

Hampton L.E., Shanley D.L., Miao W., Tilgner I.-C., Lê Bich-van, Hodjati S. (2003) 'Erosion mechanisms and performance of cellular ceramic substrates'. Society of Automotive Engineers 2003-01-3071. [Cited on p. 350]

Han H.-S., Chen D.-R., Pui D.Y.I., Anderson B.E. (2000) 'A nanometer aerosol size analyzer (nASA) for rapid measurement of high-concentration size distributions'. Journal of Nanoparticle Research 2:43–52. [Cited on p. 186]

Han S.B., Hinze P.C., Kwon Y.J., Mun S.S. (1997) 'Experimental investigation of smoke emission dependent upon engine operating conditions'. Society of Automotive Engineers 971658. [Cited on p. 65]

Han Z., Uludogan A., Hampson G.J., Reitz R.D. (1996) 'Mechanism of soot and NOx emission reduction using multiple-injection in a diesel engine'. Society of Automotive Engineers 960633. [Cited on p. 270]

Han Z., Xu Z., Trigui N. (2000) 'Spray/wall interaction models for multidimensionalo engine simulation'. International Journal of Engine Research 1:127–146. [Cited on p. 329]

Haney R.A., Saseen G.P., Waytulonis R.W. (1997) 'An overview of diesel particulate exposures and control technology in the U.S. mining industry'. Applied Occupational and Environmental Hygiene 12:1013–1018. [Cited on p. 360]

Hanisch S., Jander H., Pape T., Wagner H.G. (1994) 'Soot mass growth and coagulation of soot particles in C_2H_4/air-flames at 15 bar'. Twenty-Fifth Symposium (International) on Combustion, The Combustion Institute. [Cited on pp. 74, 76]

Hansen K.F., Jensen M.G. (1997) 'Chemical and biological characteristics of exhaust emissions from a DI diesel engine fuelled with rapeseed oil methyl ester (RME)'. Society of Automotive Engineers 971689. [Cited on p. 260]

Hara H., Itoh Y., Henein N.A., Bryzik W. (1999) 'Effect of cetane number with and without additive on cold startability and white smoke emissions in a diesel engine'. Society of Automotive Engineers 1999-01-1476. [Cited on pp. 120, 121]

Haralampous O.A., Koltsakis G.C., Samaras Z.C., Vogt C.-D., Ohara E., Watanabe Y., Mizutano T. (2004) 'Modeling and experimental study of uncontrolled regenerations in SiC filters with fuel borne catalysts'. Society of Automotive Engineers 2004-01-0697. [Cited on p. 104]

Hardenberg H.O. 'Testing the smoke formation properties of compression ignition engine fuels in premixed open flames – a promising approach'. Society of Automotive Engineers 811194. [Cited on p. 65]

Hardenberg H.O., Daudel H.L., Erdmannsdörfer H.J. (1987a) 'Experiences in the development of ceramic fiber coil particulate traps'. Society of Automotive Engineers 870015. [Cited on pp. 233, 349]

Hardenberg H.O., Daudel H.L., Erdmannsdörfer H.J. (1987b) 'Particulate trap regeneration induced by means of oxidizing agents injected into the exhaust gas'. Society of Automotive Engineers 870016. [Cited on p. 304]

Hare C.T., Springer K.J., Bradow R.L. (1976) 'Fuel and additive effects on diesel particulate – development and demonstration of methodology'. Society of Automotive Engineers 760130. [Cited on pp. 233, 261]

Harley R.A., Marr L.C., Lehner J.K., Giddings S.N. (2005) 'Changes in motor vehicle emissions on diurnal to decadal time scales and effects on atmospheric composition'. Environmental Science and Technology 39: 5356–5362. [Cited on p. 358]

Harner T., Bidleman T.F. (1998) 'Octanol–air partition coefficient for describing particle/gas partitioning of aromatic compounds in urban air'. Environmental Science and Technology 32:1494–1502. [Cited on p. 117]

Harrad S.J. (1998) 'Dioxins, dibenzofurans and PCBs in atmospheric aerosols'. In *Atmospheric Particles*, Harrison R.M., van Grieken R.E. (editors), IUPAC Series on Analytical and Physical Chemistry of Environmental Systems. John Wiley & Sons, Ltd, pp. 233–251. [Cited on p. 390]

Harrington J.A., Yetter R.A. (1981) 'Application of a mini-dilution tube in the study of fuel effects on stratified charge engine emissions'. Society of Automotive Engineers 811198. [Cited on p. 326]

Harris S.J., Maricq M.M. (2001) 'Signature size distributions for diesel and gasoline engine exhaust particulate matter'. Journal of Aerosol Science 32:749–764. [Cited on pp. 76, 77, 78]

Harrison P.T.C. (1998) 'Health effects of indoor air pollutants'. In *Air Pollution and Health*, Hester R.E., Harrison R.M. (editors), Issues in Environmental Science and Technology, Volume 10. Royal Society of Chemistry, pp. 101–125. [Cited on pp. 29, 358]

Harrison R.M. (1998) 'Setting health-based air quality standards'. In *Air Pollution and Health*, Hester R.E., Harrison R.M. (editors), Issues in Environmental Science and Technology, Volume 10. Royal Society of Chemistry, pp. 57–73. [Cited on p. 372]

Harrison R.M., Jones A.M. (2005) 'Multisite study of particle number concentrations in urban air'. Environmental Science and Technology 39:6063–6070. [Cited on pp. 360, 362]

Hashimoto K., Inaba O., Akasaka Y. (2000) 'Effects of fuel properties on the combustion and emission of direct-injection gasoline engine'. Society of Automotive Engineers 2000-01-0253. [Cited on p. 333]

Hasse C., Bikas G., Peters N. (2000) 'Modeling DI-diesel combustion using the eulerian particle flamelet model (EPFM)'. Society of Automotive Engineers 2000-01-2934. [Cited on p. 79]

Hasspieler B.M., Ali F.N., Alipour M., Haffner G.D., Adeli K. (1995) 'Human bioassays to assess environmental genotoxicity: Development of a DNA break bioassay in HepG2 cells'. Clinical Biochemistry 28:113–116. [Cited on pp. 375, 376]

Hatzis C., Godleski J.J., González-Flecha B., Wolfson J.M., Koutrakis P. (2006) 'Ambient particulate matter exhibits direct inhibitory effects on oxidative stress enzymes'. Environmental Science and Technology 40:2805–2811. [Cited on p. 375]

Haudiquert M., Cessou A., Stepowski D., Coppalle A. (1997) 'OH and soot concentration measurements in a high-temperature laminar diffusion flame'. Combustion and Flame 111:338–349. [Cited on p. 79]

Haupt D., Nord K., Tingvall B., Ahlvik P., Egebäck K.-E., Andersson S., Blomquist M. (2004) 'Investigating the potential to obtain low emissions from a diesel engine running on ethanol and equipped with EGR, catalyst and DPF'. Society of Automotive Engineers 2004-01-1884. [Cited on p. 260]

Hauser G. (2006) 'Smoke particulate sensors for OBD and high precision measuring'. Society of Automotive Engineers 2006-01-3549. [Cited on pp. 195, 196]

Havenith C., Verbeek R.P. (1997) 'Transient performance of a urea deNOx catalyst for low emissions heavy-duty diesel engines'. Society of Automotive Engineers 970185. [Cited on p. 299]

Hawthorne S.B., Grabanski C.B., Martin E., Miller D.J. (2000) 'Comparisons of Soxhlet extraction, pressurized liquid extraction, supercritical fluid extraction and subcritical water extraction for environmental solids: Recovery, selectivity and effects on sample matrix'. Journal of Chromatography A. 892:421–433. [Cited on p. 172]

Haynes B.S., Wagner H.G. (1981) 'Soot formation'. Progress in Energy and Combustion Science 7:229–273. [Cited on pp. 65, 68, 72, 75, 123]

Hays M.D., Vander Wal R.L. (2007) 'Heterogeneous soot nanostructure in atmospheric and combustion source aerosols'. Energy and Fuels 21:801–811. [Cited on p. 212]

Hayward D.O., Trapnell B.M. (1964) *Chemisorption*. Butterworths, London. [Cited on p. 114]

Hecht S.S. (1988) 'Potential carcinogenic effects of polynuclear aromatic hydrocarbons and nitroaromatics in mobile source emissions'. In *Air Pollution, the Automobile, and Public Health*, Watson A.Y., Bates R.R., Kennedy D. (editors). National Academy Press, pp. 555–578. [Cited on pp. 369, 371]

Heddling G.H., Kittelson D.B., Scherrer H., Liu X., Dolan D.F. (1981) 'Total cylinder sampling from a diesel engine'. Society of Automotive Engineers 810257. [Cited on p. 84]

HEI (2002) 'Research directions to improve estimates of human exposure and risk from diesel exhaust'. Health Effects Insitute. [Cited on p. 369]

Heideman G. (2004) 'Reduced zinc oxide levels in sulphur vulcanisation of rubber compounds'. PhD Thesis, University of Twent, Enschede, Netherlands. [Cited on p. 347]

Heintzenberg J. (1998) 'Condensed water aerosols'. In *Atmospheric Particles*, Harrison R.M., van Grieken R.E. (editors), IUPAC Series on Analytical and Physical Chemistry of Environmental Systems. John Wiley & Sons, Ltd, pp. 509–542. [Cited on pp. 25, 38]

Heinze P., Hutcheson R., Kapus P., Cartellieri W. (1996) 'The interaction between diesel fuel density and electronic engine management systems'. Society of Automotive Engineers 961975. [Cited on p. 265]

Heisey J.B., Lestz S.S. (1981) 'Aqueous alcohol fumigation of a single-cylinder DI diesel engine'. Society of Automotive Engineers 811208. [Cited on p. 260]

Henderson R.D. (1970) 'Air pollution and construction equipment'. Society of Automotive Engineers 700551. [Cited on p. 201]

Hentschel W., Richter J.-U. (1995) 'Time-resolved analysis of soot formation and oxidation in a direct-injection diesel engine for different EGR-rates by an extinction method'. Society of Automotive Engineers 952517. [Cited on p. 272]

Hepp H., Siegmann K. (1998) 'Mapping of soot particles in a weakly sooting diffusion flame by aerosol techniques'. Combustion and Flame 115:275–283. [Cited on pp. 71, 106, 230]

Herling R.J., Gafford R.D., Carlson R.R., Lyles A., Bradow R.L. (1977) 'Characterization of sulfate and gaseous emissions from California consumer-owned, catalyst-equipped vehicles'. Society of Automotive Engineers 770062. [Cited on p. 317]

l'Hermine E., Hipeaux J.C., Faure D., Pariset T., Chambard L., Tiquet L. (2000) 'Rheology of used diesel lubricants when contaminated with soot – the brookfield measurement opportunity'. Society of Automotive Engineers 2000-01-1808. [Cited on p. 287]

Herner J.D., Green P.G., Kleeman M.J. (2006) 'Measuring the trace elemental composition of size-resolved airborne particles'. Environmental Science and Technology 40:1925–1933. [Cited on p. 23]

Herner J.D., Robertson W.H., Ayala A. (2007) 'Investigation of ultrafine particle number measurements from a clean diesel truck using the European PMP protocol'. Society of Automotive Engineers 2007-01-0114. [Cited on pp. 253, 389, 393]

Herr J.D., Dukovich M., Lestz S.S., Yergey J.A., Risby T.H., Tejada S.B. (1982) 'The role of nitrogen in the observed direct microbial mutagenic activity for diesel engine combustion in a single-cylinder DI engine'. Society of Automotive Engineers 820467. [Cited on pp. 110, 111]

Hessami M.-A., Child C. (2004) 'Statistical analysis of diesel vehicle exhaust emission: Development of empirical equations to calculate particulate emission'. Society of Automotive Engineers 2004-01-3066. [Cited on p. 203]

Hewitt C.N., Davison B.M. (1998) 'Formation of aerosol particles from biogenic precursors'. In *Atmospheric Particles*, Harrison R.M., van Grieken R.E. (editors), IUPAC Series on Analytical and Physical Chemistry of Environmental Systems. John Wiley & Sons, Ltd, pp. 369–383. [Cited on pp. 27, 29]

Heyder J., Gebhart J. (1986) 'Light scattering by single particles'. In *Physical and Chemical Characterization of Individual Airborne Particles*, Spurny K.R. (editor). Ellis Horwood, pp. 86–100. [Cited on p. 182]

Heymann D., Yancey T.E., Wolbach W.S., Thiemens M.H., Johnson E.A., Roach D., Moecker S. (1998) 'Geochemical markers of the Cretaceous–Tertiary boundary event at Brazos River, Texas, USA'. Geochimica et Cosmochimica Acta 62:173–181. [Cited on p. 41]

Heywood J.B. (1988) *Internal Combustion Engine Fundamentals*. McGraw-Hill. [Cited on pp. 5, 65, 80, 85, 136, 176, 315, 321]

Higgins B., Siebers D., Aradi A. (2000) 'Diesel-spray ignition and premixed-burn behavior'. Society of Automotive Engineers 2000-01-0940. [Cited on pp. 84, 86]

Higgins I.T.T. (1971) 'Sulfur oxides and particulates'. Society of Automotive Engineers 710299. [Cited on pp. 2, 3, 381]

Hilden D.L., Mayer W.J. (1984) 'The contribution of engine oil to particulate exhaust emissions from light-duty, diesel-powered vehicles'. Society of Automotive Engineers 841395. [Cited on p. 282]

Hills F.J., Wagner T.O., Lawrence D.K. (1969) 'CRC correlation of diesel smokemeter measurements'. Society of Automotive Engineers 690493. [Cited on p. 201]

Hinds W.C. (1982) *Aerosol Technology – Properties, Behavior, and Measurement of Airborne Particles*. John Wiley & Sons, Inc. [Cited on pp. 17, 18, 20, 22, 23, 26, 76, 143, 355]

Hinds W.C. (1999) *Aerosol Technology – Properties, Behavior, and Measurement of Airborne Particles*. John Wiley & Sons, Inc. [Cited on pp. 11, 13, 15, 16, 19, 20, 21, 23, 24, 37, 183, 184, 189, 364, 365]

Hirasawa T., Sung C.-J., Yang Z., Joshi A., Wang H. (2004) 'Effect of ferrocene addition on sooting limits in laminar premixed ethylene – oxygen – argon flames'. Combustion and Flame 139:288–299. [Cited on pp. 99, 100]

Hiroyasu H., Kadota T. (1976) 'Models for combustion and formation of nitric oxide and soot in direct injection diesel engines'. Society of Automotive Engineers 760129. [Cited on pp. 67, 81, 85]

Hiroyasu H., Arai M., Nakanishi K. (1980) 'Soot formation and oxidation in diesel engines'. Society of Automotive Engineers 800252. [Cited on pp. 86, 91]

Hiroyasu H., Kadota T., Arai M. (1983) 'Development and use of a spray combustion modeling to predict diesel engine efficiency and pollutant emissions. Part 1: Combustion modeling'. Bulletin of the JSME 26(214):569–575. [Cited on p. 81]

Ho T.C. (2004) 'Deep HDS of diesel fuel: Chemistry and catalysis'. Catalysis Today 98:3–18. [Cited on p. 252]

Hoard J., Bretz R.L., Ehara Y. (2003) 'Diesel exhaust simulator: Design and application to plasma discharge testing'. Society of Automotive Engineers 2003-01-1184. [Cited on p. 211]

Hofeldt D.L., Chen G. (1996) 'Transient particulate emissions from diesel buses during the central business district cycle'. Society of Automotive Engineers 960251. [Cited on pp. 138, 275, 276]

Holgate S. (1999) 'Mechanisms of air pollutants and the health impacts of ground level ozone'. Health Effects of Vehicle Emissions, Feb. 16–17, Royal Society of Medicine. [Cited on pp. 368, 382]

Holländer W. (1995) 'Particle counting and particle size analysis'. In *Airborne Particulate Matter*, Volume 4, Part D, Kouimtzis T., Samara C. (editors). Springer, pp. 253–278. [Cited on p. 38]

Holmén B.A., Qu Y. (2004) 'Uncertainty in particle number modal analysis during transient operation of compressed natural gas, diesel, and trap-equipped diesel transit buses'. Environmental Science and Technology 38:2413–2423. [Cited on pp. 240, 389]

Holmes N.S. (2007) 'A Review of particle formation events and growth in the atmosphere in the various environments and discussion of mechanistic implications'. Atmospheric Environment 41:2183–2201. [Cited on p. 362]

Homan H.S. (1985) 'Conversion factors among smoke measurements'. Society of Automotive Engineers 850267. [Cited on pp. 201, 203]

Hong G., Collings N. (1989) 'Design of diesel smoke feedback control using a combination of PI control algorithm and performance optimisation'. Society of Automotive Engineers 890387. [Cited on p. 197]

Hong G., Collings N., Baker N.J. (1987) 'Diesel smoke transient control using a real-time smoke sensor'. Society of Automotive Engineers 871629. [Cited on p. 196]

Hong S., Wooldridge M.S., Im H.G., Assanis D.N., Pitsch H. (2005) 'Development and application of a comprehensive soot model for 3D CFD reacting flow studies in a diesel engine'. Combustion and Flame 143:11–26. [Cited on pp. 72, 74]

Hoornaert S., van Malderen H., van Grieken R. (1996) 'Gypsum and other calcium-rich aerosol particles above the North Sea'. Environmental Science and Technology 30:1515–1520. [Cited on p. 27]

Horakouchi N., Fukano I., Shoji T. (1989) 'Measurement of diesel exhaust emissions with mini-dilution tunnel'. Society of Automotive Engineers 890181. [Cited on p. 160]

Hori M., Oguchi M. (2004) 'Feasibility study of urea SCR systems on heavy duty commercial vehicles'. Society of Automotive Engineers 2004-01-1944. [Cited on p. 299]

Hori S., Narusawa K. (1998) 'Fuel composition effects on SOF and PAH exhaust emissions from DI diesel engines'. Society of Automotive Engineers 980507. [Cited on p. 152]

Hori S., Narusawa K. (2001) 'The influence of fuel components on PM and PAH exhaust emissions from a DI diesel engine – effects of pyrene and sulfur contents'. Society of Automotive Engineers 2001-01-3693. [Cited on pp. 108, 151]

Horiuchi M., Saito K., Ichihara S. (1991) 'Sulfur storage and discharge behavior on flow-through type oxidation catalysts'. Society of Automotive Engineers 910605. [Cited on p. 295]

Horvath J. (1998) 'Influence of atmospheric aerosols upon the global radiation balance'. In *Atmospheric Particles*, Harrison R.M., van Grieken R.E. (editors), IUPAC Series on Analytical and Physical Chemistry of Environmental Systems. John Wiley & Sons, Ltd, pp. 543–596. [Cited on pp. 23, 41]

Horvath H., Catalan L., Trier A. (1997) 'A study of aerosol of Santiago de Chile II: Light absorption measurements'. Atmospheric Environment 31:3737–3744. [Cited on p. 359]

Hou Z.-X., Abraham J. (1995) 'Three-dimensional modeling of soot and NO in a direct-injection diesel engine'. Society of Automotive Engineers 950608. [Cited on p. 84]

Hountalas D.T., Papagiannakis R.G. (2002) 'Theoretical and experimental investigation of a direct injection dual fuel diesel–natural gas engine'. Society of Automotive Engineers 2002-01-0868. [Cited on p. 251]

Hountalas D.T., Maragiannis G.T., Arapatsakos C. (2007) 'Evaluation of various rich combustion techniques for diesel engines using modeling'. Society of Automotive Engineers 2007-01-0671. [Cited on p. 291]

Howard J.B., Bittner J.D. (1983) 'Structure of sooting flames'. In *Soot in Combustion Systems and its Toxic Properties*, Lahaye J., Prado G. (editors). Plenum Press, pp. 57–93. [Cited on p. 106]

Howard J.B., Kausch W.J. (1980) 'Soot control by fuel additives'. Progress in Energy and Combustion Science 6:263–276. [Cited on pp. 78, 98]

Howe F.D. (1910) 'Smoky and ill-smelling exhausts'. Society of Automotive Engineers 100015. [Cited on p. 4]

Howitt J.S., Montierth M.R. (1981) 'Cellular ceramic diesel filter'. Society of Automotive Engineers 810114. [Cited on p. 220]

Howitt J.S., Elliott W.T., Mogan J.P., Dainty E.D. (1983) 'Application of a ceramic wall flow filter to underground diesel emissions reduction'. Society of Automotive Engineers 830181. [Cited on pp. 302, 305]

Hsu Y.-D., Chein H.M., Chen T.M., Tai C.-I. (2005) 'Axial flow cyclone for segregation and collection of ultrafine particles: Theoretical and experimental study'. Environmental Science and Technology 39:1299–1306. [Cited on p. 394]

Huai T., Durbin T.D., Younglove T., Scora G., Barth M., Norbeck J.M. (2005) 'Vehicle specific power approach to estimating on-road NH_3 emissions from light-duty vehicles'. Environmental Science and Technology 39:9595–9600. [Cited on p. 37]

Huang H., Buekens A. (1995) 'On the mechanisms of dioxin formation in combustion processes'. Chemosphere 31:4099–4117. [Cited on p. 306]

Huang W., Smith T.J., Ngo L., Wang T., Chen H., Wu F., Herrick R.F., Christiani D.C., Ding H. (2007) 'Characterizing and biological monitoring of polycyclic aromatic hydrocarbons in exposures to diesel exhaust'. Environmental Science and Technology 41:2711–2716. [Cited on p. 360]

Huang X.-F., Yu J.Z., He L.-Y., Hu M. (2006) 'Size distribution characteristics of elemental carbon emitted from Chinese vehicles: Results of a tunnel study and atmospheric implications'. Environmental Science and Technology 40:5355–5360. [Cited on p. 254]

Huang Y., Alger T., Matthews R., Ellzey J. (2001a) 'The effects of fuel volatility and structure on HC emissions from piston wetting in DISI engines'. Society of Automotive Engineers 2001-01-1205. [Cited on p. 329]

Huang Y., Matthews R., Ellzey J., Dai W. (2001b) 'Effects of fuel volatility, load, and speed on HC emissions due to piston wetting'. Society of Automotive Engineers 2001-01-2024. [Cited on p. 329]

Huang Z., Lu H., Jiang D., Zeng K., Liu B., Zhang J., Wang X. (2005) 'Performance and emissions of a compression ignition engine fueled with diesel/oxygenate blends for various fuel delivery advance angles'. Energy and Fuels 29:403–410. [Cited on p. 259]

Hueglin C., Scherrer L., Burtscher H. (1996) 'Description and application of a dilution method for the characterization of particles from emission sources'. Journal of Aerosol Science 27: S311–S312. [Cited on p. 165]

Huffman H.D. (1996) 'The reconstruction of aerosol light absorption by particle measurements at remote sites: An independent analysis of data from the Improve network – II'. Atmospheric Environment 30:85–99. [Cited on p. 169]

Hug H.T., Mayer A., Hartenstein A. (1993) 'Off-highway exhaust gas after-treatment: Combining urea-SCR, oxidation catalysis and traps'. Society of Automotive Engineers 930363. [Cited on p. 299]

Huggins F.E., Huffman G.P., Robertson J.D. (2000) 'Speciation of elements in NIST particulate matter SRMs 1648 and 1650'. Journal of Hazardous Materials 74:1–23. [Cited on pp. 231, 232]

Hughes L.S., Cass G.R., Gone J., Ames M., Olmez I. (1998) 'Physical and chemical characterization of atmospheric ultrafine particles in the Los Angeles area'. Environmental Science and Technology 32:1153–1161. [Cited on p. 362]

Hughes T.J., Lewtas J., Claxton L.D. (1997) 'Development of a standard reference material for diesel mutagenicity in the salmonella plate incorporation assay'. Mutation Research 391:243–258. [Cited on p. 180]
Human D.M., Ullman T.L., Baines T.M. (1990) 'Simulation of high altitude effects on heavy-duty diesel emissions'. Society of Automotive Engineers 900883. [Cited on p. 276]
Hunter G., Scholl J., Hibbler F., Bagley S., Leddy D., Abata D., Johnson J. (1981) 'The effects of fuels on diesel oxidation catalyst performance and the physical, chemical, and biological character of diesel particulate emissions'. Society of Automotive Engineers 811192. [Cited on pp. 294, 297]
Hurley R.G., Watkins W.L.H., Griffis R.C. (1989) 'Characterization of automotive catalysts exposed to the fuel additive MMT'. Society of Automotive Engineers 890582. [Cited on p. 317]
Hurt R.H., Monthioux M., Kane A. (2006) 'Toxicology of carbon nanomaterials: Status, trends, and perspectives on the special issue'. Carbon 44:1028–1033. [Cited on p. 393]
Hwang J.Y., Lee W., Kang H.G., Chung S.H. (1998) 'Synergistic effect of ethylene–propane mixture on soot formation in laminar diffusion flames'. Combustion and Flame 114:370–380. [Cited on p. 69]
Hyde J.D., Gibbs R.E., Whitby R.A., Byer S.M., Hill B.J., Hoffman T.E., Johnson R.E., Werner P.L. (1982) 'Analysis of particulate and gaseous emissions data from in-use diesel passenger cars'. Society of Automotive Engineers 820772. [Cited on pp. 180, 181, 238, 239]

I

Ikegami M., Li X.-h., Nakayama Y., Miwa K. (1983) 'Trend and origins of particulate and hydrocarbon emission from a direct-injection diesel engine'. Society of Automotive Engineers 831290. [Cited on pp. 109, 277]
Ikegami M., Shioji M., Kimoto T. (1988) 'Diesel combustion and the pollutant formation as viewed from turbulent mixing concept'. Society of Automotive Engineers 880425. [Cited on pp. 91, 267, 277]
Imhof D., Weingartner E., Ordóñez C., Gehrig R., Hill M., Buchmann B., Baltensperger U. (2005) 'Real-world emission factors of fine and ultrafine aerosol particles for different traffic situations in Switzerland'. Environmental Science and Technology 39:8341–8350. [Cited on p. 360]
Imlay J.A., Linn S. (1988) 'DNA damage and oxygen radical toxicity'. Science 240:1302–1309. [Cited on p. 368]
Imlay J.A., Chin S.M., Linn S. (1988) 'Toxic DNA damage by hydrogen peroxide through the Fenton reaction in vivo and in vitro'. Science 240:640–642. [Cited on p. 368]
Ingham M.C., Warden R.B. (1987) 'Cost-effectiveness of diesel fuel modifications for particulate control'. Society of Automotive Engineers 870556. [Cited on p. 173]
Injuk J., de Bock L., van Grieken R. (1998) 'Structural heterogeneity within airborne particles'. In *Atmospheric Particles*, Harrison R.M., van Grieken R.E. (editors), IUPAC Series on Analytical and Physical Chemistry of Environmental Systems. John Wiley & Sons, Inc., pp. 173–202. [Cited on pp. 145, 380]
Inoue M., Murase A., Yamamoto M., Kubo S. (2006) 'Analysis of volatile nanoparticles emitted from diesel engine using TOF-SIMS and metal-assisted SIMS (MetA-SIMS)'. Applied Surface Science 252:7014–7017. [Cited on p. 285]
Inoue T., Masuda Y.-h., Yamamoto M. (1997) 'Reduction of diesel particulate matter by oil consumption improvement utilizing radioisotope tracer techniques'. Society of Automotive Engineers 971630. [Cited on pp. 280, 281]
International Commission on Radiological Protection (1966) 'Task Force on Lung Dynamics'. Health Physics 12:173–207. [Cited on p. 40]
Ishida M., Ueki H., Yoshimura Y., Matsumura N. (1990) 'Studies on combustion and exhaust emissions in a high speed DI diesel engine'. Society of Automotive Engineers 901614. [Cited on p. 66]
Ishiguro T., Suzuki N., Fujitani Y., Morimoto H. (1991) 'Microstructural changes of diesel soot during oxidation'. Combustion and Flame 85:1–6. [Cited on pp. 81, 224]
Ishiki K., Oshida S., Takiguchi M., Urabe M. (2000) 'A study of abnormal wear in power cylinder of diesel engine with EGR – wear mechanism of soot contaminated in lubricating oil'. Society of Automotive Engineers 2000-01-0925. [Cited on p. 290]
de Iuliis S., Barbini M., Benecchi S., Cignoli F., Zizak G. (1998) 'Determination of the soot volume fraction in an ethylene diffusion flame by multiwavelength analysis of soot radiation'. Combustion and Flame 115:253–261. [Cited on p. 182]

Iyer S.S., Litzinger T.A., Lee S.-Y., Santoro R.J. (2007) 'Determination of soot scattering coefficient from extinction and three-angle scattering in a laminar diffusion flame'. Combustion anf Flame 149:206–216. [Cited on p. 190]

Iyer V., Abraham J. (1998) 'The computed structure of a combusting transient jet under diesel conditions'. Society of Automotive Engineers 981071. [Cited on p. 83]

J

Jacko M.G., DuCharme R.T., Somers J.H. (1973) 'Brake and clutch emissions generated during vehicle operation'. Society of Automotive Engineers 730548. [Cited on p. 345]

Jacobs P.E., Chernich D.J., Miller E.F., Cabrera R.P., Baker M., Gaslan D.P., Duleep K.G., Meszler D. (1998) 'California's revised heavy-duty vehicle smoke and tampering inspection program'. Society of Automotive Engineers 981951. [Cited on pp. 358, 388]

Jacobson M.Z. (2000) 'Reversible chemical reactions in aerosols'. In *Aerosol Chemical Processes in the Environment*, Spurny K.R., Hochrainer D. (editors). CRC Press LLC, pp. 135–156. [Cited on p. 19]

Jacobson M.Z., Seinfeld J.H. (2004) 'Evolution of nanoparticle size and mixing state near the point of emission'. Atmospheric Environment 38:1839–1850. [Cited on p. 144]

Jacobson M.Z., Kittelson D.B., Watts W.F. (2005) 'Enhanced coagulation due to evaporation and its effect on nanoparticle evolution'. Environmental Science and Technology 39:9486–9492. [Cited on p. 144]

Jaeger L.W., Boulouchos K., Mohr M. (1999) 'Analysis of factors influencing particulate matter emissions of a compression ignition direct injection diesel engine'. Society of Automotive Engineers 1999-01-3492. [Cited on pp. 270, 271, 274]

Jaenicke R. (1998) 'Atmospheric aerosol size distribution'. In *Atmospheric Particles*, Harrison R.M., van Grieken R.E. (editors), IUPAC Series on Analytical and Physical Chemistry of Environmental Systems. John Wiley & Sons, Inc., pp. 1–28. [Cited on p. 25]

Jakobs R.J., Westbrooke K. (1990) 'Aspects of influencing oil consumption in diesel engines for low emissions'. Society of Automotive Engineers 900587. [Cited on p. 279]

Janakiraman G., Johnson J.H., Bagley S.T., Warner J.R., Huynh C.T., Khan A., Leddy D.G. (2002) 'Oxidation catalytic converter and emulsified fuel effects on neavy-duty diesel engine emission'. Society of Automotive Engineers 2002-01-1277. [Cited on pp. 124, 230, 294]

Jang H.-N., Seo Y.-C., Lee J.-H., Hwang K.-W., Yoo J.-I., Sok C.-H., Kim S.-H. (2007) 'Formation of fine particles enriched by V and Ni from heavy oil combustion: Anthropogenic sources and drop-tube furnace experiments'. Atmospheric Environment 41:1053–1063. [Cited on p. 98]

Jang M., Kamens R.M., Leach K.B., Strommen M.R. (1997) 'A thermodynamic approach using group contribution methods to model the partitioning of semivolatile organic compounds on atmospheric particulate matter'. Environmental Science and Technology 31:2805–2811. [Cited on p. 117]

Jang M., Czoschke N.M., Northcross A.L. (2005) 'Semiempirical model for organic aerosol growth by acid-catalyzed heterogeneous reactions of organic carbonyls'. Environmental Science and Technology 39:164–174. [Cited on p. 129]

Jang M.J., Kamens R.M. (1998) 'A thermodynamic approach for modeling partitioning of semivolatile organic compounds on atmospheric particulate matter: Humidity effects'. Environmental Science and Technology 32:1237–1243. [Cited on p. 117]

Jang M.J., Kamens R.M. (1999) 'A predictive model for adsorptive gas partitioning of SOCs on fine atmospheric inorganic dust particles'. Environmental Science and Technology 33:1825–1831. [Cited on p. 117]

Jaoui M., Corse E., Kleindienst T.E., Offenberg J.H., Lewandowski M., Edney E.O. (2006) 'Analysis of secondary organic aerosol compounds from the photooxidation of *d*-limonene in the presence of NO$_X$ and their detection in ambient PM$_{2.5}$'. Environmental Science and Technology 40:3819–3828. [Cited on p. 29]

Japar S.M., Szkarlat A.C. (1981) 'Real-time measurements of diesel vehicle exhaust particulate using photoacoustic spectroscopy and total light extinction'. Society of Automotive Engineers 811184. [Cited on p. 192]

Jarrett R.P., Clark N.N. (2001) 'Evaluation methods for determining continuous particulate matter from transient testing of heavy-duty diesel engines'. Society of Automotive Engineers 2001-01-3575. [Cited on p. 200]

Jefferson D.A., Tilley E.E.M. (1999) 'The structural and physical chemistry of nanoparticles'. In *Particulate Matter: Properties and Effects upon Health*, Maynard R.L., Howard C.V. (editors). Bios Scientific Publishers, pp. 63–84. [Cited on p. 396]

Jeguirim M., Tschamber V., Brilhac J.F., Ehrburger P. (2005) 'Oxidation mechanism of carbon black by NO_2: Effect of water vapour'. Fuel 84:1949–1956. [Cited on p. 46]

Jenkin M.E., Clemitshaw K.C. (2000) 'Ozone and other secondary photochemical pollutants: Chemical processes governing their formation in the planetary boundary layer'. Atmospheric Environment 34:2499–2527. [Cited on p. 173]

Jennings S.G. (1998) 'Wet processes affecting atmospheric aerosols'. In *Atmospheric Particles*, IUPAC Series on Analytical and Physical Chemistry of Environmental Systems, Harrison R.M., van Grieken R.E. (editors). John Wiley & Sons, Inc., pp. 475–508. [Cited on p. 25]

Jensen D. (1971) 'From air quality criteria to control regulations'. Society of Automotive Engineers 710303. [Cited on p. 145]

Jeong C.-H., Hopke P.K., Chalupa D., Utell M. (2004) 'Characteristics of nucleation and growth events of ultrafine particles measured in Rochester, NY'. Environmental Science and Technology 38:1933–1940. [Cited on pp. 360, 361, 362]

Jeuland N., Dementhon J.B., Gagnepain L., Plassat G., Coroller P., Momique J.C., Belot G., Dalili D. (2004) 'Performance and durability of DPF (diesel particulate filter) tested on a fleet of Peugeot 607 taxis: Final results'. Society of Automotive Engineers 2004-01-0073. [Cited on pp. 242, 243]

Jiang N., Dreher K.L., Dye J.A., Li Y., Richards J.H., Martin L.D., Adler K.B. (2000) 'Residual oil fly ash induces cytotoxicity and mucin secretion by guinea pig tracheal epithelial cells via an oxidant-mediated mechanism'. Toxicology and Applied Pharmacology 163:221–230. [Cited on p. 380]

Jiao K., Lafleur A.L. (1997) 'Improved detection of polycyclic aromatic compounds in complex mixtures by liquid chromatographic fractionation on poly(divinylbenzene) prior to gas chromatography – mass spectrometry. Application to the analysis of diesel particulates'. Journal of Chromatography A 791:203–211. [Cited on p. 175]

Jiménez S., Ballester J. (2005) 'Influence of operating conditions and the role of sulfur in the formation of aerosols from biomass combustion'. Combustion and Flame 140:346–358. [Cited on p. 97]

Jiménez-Horrnero F.J., Giráldez J.V., Gutiérrez de Rave E., Moral F.J. (2007) 'Description of pollutant dispersion in an urban street canyon using a two-dimensional lattice model'. Atmospheric Environment 41:221–226. [Cited on p. 357]

Johansen K., Gabrielsson P., Stavnsbjerg P., Bak F., Andersen E., Autrup H. (1997) 'Effect of upgraded diesel fuels and oxidation catalysts on emission properties, especially PAH and genotoxicity'. Society of Automotive Engineers 973001. [Cited on p. 256]

Johnsen A.R., de Lipthay J.R., Reichenberg F., Sørensen S.J., Andersen O., Christensen P., Binderup M.-L., Jacobsen C.S. (2006) 'Biodegradation, bioaccessibility, and genotoxicity to diffuse polycyclic aromatic hydrocarbon (PAH) pollution at a motorway site'. Environmental Science and Technology 40:3293–3298. [Cited on pp. 40, 342]

Johnson J.E., Kittelson D.B. (1994) 'Physical factors affecting hydrocarbon oxidation in a diesel oxidation catalyst'. Society of Automotive Engineers 941771. [Cited on pp. 119, 139, 140, 141, 236, 237]

Johnson J.H. (1988) 'Automotive emissions'. In *Air Pollution, the Automobile, and Public Health*, Watson A.Y., Bates R.R., Kennedy D. (editors). National Academy Press, pp. 39–75. [Cited on pp. 176, 234, 239]

Johnson J.H., Reinbold E.O., Carlson D.H. (1981) 'The engineering control of diesel pollutants in underground mining'. Society of Automotive Engineers 810684. [Cited on pp. 296, 297]

Johnson J.H., Bagley S.T., Gratz L.D., Leddy D.G. (1994) 'A review of diesel particulate control technology and emissions effects'. Society of Automotive Engineers 940233. [Cited on pp. 178, 179]

Johnson T., Caldow R., Pöcher A., Mirme A., Kittelson D. (2004) 'A new electrical mobility particle sizer spectrometer for engine exhaust particle measurements'. Society of Automotive Engineers 2004-01-1341. [Cited on p. 186]

Jones A.P. (1999) 'Indoor air quality and health'. Atmospheric Environment 33:4535–4564. [Cited on pp. 29, 364, 372]

Jones B.L., Stollery D.J., Clifton J.M., Wylie T.F. (1997) 'In-service smoke and particulate measurements'. Society of Automotive Engineers 970748. [Cited on p. 202]

Jones C.C., Chugtai A.R., Murugaverl B., Smith D.M. (2004) 'Effects of air/fuel ratio combustion ratio on the polycyclic aromatic hydrocarbon content of carbonaceous soots from selected fuels'. Carbon 42:2471–2484. [Cited on p. 172]

Jones N.C., Thornton C.A., Mark D. (2000) 'Indoor/outdoor relationships of particulate matter in domestic homes with roadside, urban and rural locations'. Atmospheric Environment 34:2603–2612. [Cited on p. 359]

Jones T.D. (1995) 'Use of bioassays in assessing health hazards from complex mixtures: A rash analysis'. Chemosphere 31:2475–2484. [Cited on p. 182]

Jonker M.T.O., Hawthorne S.B., Koelmans A.A. (2005) 'Extremely slowly desorbing polycyclic aromatic hydrocarbons from soot and soot-like materials: Evidence by supercritical fluid extraction'. Environmental Science and Technology 39:7889–7895. [Cited on pp. 172, 181]

Jonkman L., de Jong F., Sant P., Barnes J.R. (1987) 'Relation between molecular structure of succinimide dispersants and dispersancy of soot in a carbon black dispersancy test'. Society of Automotive Engineers 871274. [Cited on p. 288]

Jørgensen M.W., Sorenson S.C. (1997) 'A 2-dimensional simulation model for a diesel particulate filter'. Society of Automotive Engineers 970471. [Cited on p. 80]

Jørgensen M.W., Sorenson S.C. (1998) 'Estimating emissions from railway traffic'. International Journal of Vehicle Design 20:210–218. [Cited on p. 358]

de Juan J., de la Mora J.F. (1998) 'High resolution size analysis of nanoparticles and ions: Running a Vienna DMA of near optimal length at Reynolds numbers up to 5000'. Journal of Aerosol Science 29:617–626. [Cited on p. 186]

Jung D., Assanis D.N. (2001) 'Multi-zone DI diesel spray combustion model for cycle simulation studies of engine performance and emissions'. Society of Automotive Engineers 2001-01-1246. [Cited on pp. 67, 81, 82]

Jung H., Kittelson D.B. (2005) 'Measurement of electrical charge on diesel particles'. Aerosol Science and Technology 39:1129–1135. [Cited on pp. 128, 228]

Jung H., Kittelson D.B., Zachariah M.R. (2003) 'The influence of engine lubricating oil in diesel nanoparticle emissions and kinetics of oxidation'. Society of Automotive Engineers 2003-01-3179. [Cited on p. 285]

Jung H., Kittelson D.B., Zachariah M.R. (2004) 'Kinetics and visualisation of soot oxidation using transmission electron microscopy'. Combustion and Flame 136:445–456. [Cited on pp. 80, 81]

Jung H., Kittelson D.B., Zachariah M.R. (2005) 'The influence of a cerium additive on ultrafine diesel particle emissions and kinetics of oxidation'. Combustion and Flame 142:276–288. [Cited on pp. 99, 101]

Junge C.G. (1952) Arch. Met. Geophys. Biokdim. 5:44. [Cited on p. 24]

K

Kaario O., Larmi M., Tanner F. (2002) 'Relating integral length scale to turbulent time scale and comparing $k\text{-}\varepsilon$ and RNG $k\text{-}\varepsilon$ turbulence models in diesel combustion simulation'. Society of Automotive Engineers 2002-01-1117. [Cited on pp. 67, 83]

Kado N.Y., Okamoto R.A., Kuzmicky P.A., Kobayashi R., Ayala A., Gebel M.E., Riger P.L., Maddox C., Zafonte L. (2005) 'Emissions of toxic pollutants from compressed natural gas and low sulfur diesel-fueled heavy-duty transit buses tested over multiple driving cycles'. Environmental Science and Technology 39:7638–7649. [Cited on pp. 22, 235, 325]

Kadowaki S. (2000) 'Characterization of urban aerosols in the Nagoya area'. In *Aerosol Chemical Processes in the Environment*, Spurny K.R., Hochrainer D. (editors). CRC Press LLC, pp. 379–404. [Cited on pp. 27, 30, 33, 36, 344]

Kagawa J. (2002) 'Health effects of diesel exhaust emissions – a mixture of air pollutants of worldwide concern'. Toxicology 181–182:349–353. [Cited on p. 378]

Kageyama K., Kinehara N. (1982) 'Characterization of particulate emission from swirl chamber type light-duty diesel engine as a function of engine parameters'. Society of Automotive Engineers 820181. [Cited on p. 178]

Kaiser E.W., Maricq M.M., Xu N., Yang J. (2005) 'Detailed hydrocarbon species and particulate emissions from a HCCI engine as a function of air–fuel ratio'. Society of Automotive Engineers 2005-01-3749. [Cited on p. 293]

Kaiser R., Romieu I., Medina S., Schwartz J., Krzyzanowski M., Künzli N. (2004) 'Air pollution attributable postneonatal infant mortality in U.S. metropolitan areas: A risk assessment study'. Environmental Health: A Global Access Science Source 3:4. [Cited on p. 373]

Kajitani S., Usisaki H., Clasen E., Campbell S., Rhee K.T. (1994) 'MTBE for improved diesel combustion and emissions?'. Society of Automotive Engineers 941688. [Cited on p. 259]

Kameoka A., Tsuchiya K. (2006) 'Influence of ferrocene on engine and vehicle performance'. Society of Automotive Engineers 2006-01-3448. [Cited on p. 317]

Kamimoto T., Bae M.-h. (1988) 'High combustion temperature for the reduction of particulate in diesel engines'. Society of Automotive Engineers 880423. [Cited on p. 292]

Kamimoto T., Yokota H., Kobayashi H. (1987) 'Effect of high pressure injection on soot formation processes in a rapid compression machine to simulate diesel flames'. Society of Automotive Engineers 871610. [Cited on pp. 267, 268]

Kamm S., Möhler O., Naumann K.-H., Saathoff H., Schurath U. (1999) 'The heterogeneous reaction of ozone with soot aerosol'. Atmospheric Environment 33:4651–4661. [Cited on p. 42]

Kammler H.K., Pratsinis S.E., Morrison P.W., Hemmerling B. (2002) 'Flame temperature measurements during electrically assisted synthesis of nanoparticles'. Combustion and Flame 128:369–381. [Cited on p. 98]

Kantola T.C., Bagley S.T., Gratz L.D., Leddy D.G., Johnson J.H. (1992) 'The influence of a low sulfur fuel and a ceramic particle trap on the physical, chemical, and biological character of heavy-duty diesel emissions'. Society of Automotive Engineers 920565. [Cited on pp. 303, 305]

Kärcher B. (1998) 'On the potential importance of sulfur-induced activation of soot particles in nascent jet aircraft exhaust plumes'. Atmospheric Research 46:293–305. [Cited on p. 128]

Karlsson R.B., Heywood J.B. (2001) 'Piston fuel film observations in an optical access GDI engine'. Society of Automotive Engineers 2001-01-2022. [Cited on p. 328]

Karol M.H. (2002) 'Respiratory allergy: What are the uncertainties'. Toxicology 181–182:305–310. [Cited on p. 363]

Kasper M. (2003) 'Nanoparticle exhaust gas measurement: On-line response, high sensitivity, low cost'. Society of Automotive Engineers 2003-01-0286. [Cited on pp. 125, 161, 209]

Kasper M. (2004) 'The number concentration of non-volatile particles – design study for an instrument according to the PMP recommendations'. Society of Automotive Engineers 2004-01-0960. [Cited on pp. 389, 393]

Kasper M., Siegmann K. (1998) 'The influence of ferrocene on PAH synthesis in acetylene and methane diffusion in flames'. Combustion Science and Technology 140:333–350. [Cited on p. 99]

Kasper M., Sattler K., Siegmann K., Matter U., Siegmann H.C. (1999) 'The influence of fuel additives on the formation of carbon during combustion'. Journal of Aerosol Science 30:217–225. [Cited on pp. 100, 101, 102]

Kasper M., Matter U., Burtscher H. (2000) 'NanoMet: On-line characterization of nanoparticle size and composition'. Society of Automotive Engineers 2000-01-1998. [Cited on pp. 164, 195]

Kasper M., Matter U., Burtscher H., Bukowiecki N., Mayer A. (2001) 'NanoMet, a new instrument for on-line size- and substance-specific particle emission analysis'. Society of Automotive Engineers 2001-01-0216. [Cited on p. 389]

Katari A., Syed M., Sickels M., Wahl T., Rajadurai S. (2004) 'Effect of aspect ratio on pressure drop and acoustics in diesel particulate filters'. Society of Automotive Engineers 2004-01-0695. [Cited on p. 300]

Kathuria V. (2004) 'Impact of CNG on vehicular pollution in Delhi: A note'. Transportation Research Part D 9:409–417. [Cited on p. 252]

Kati V., Virtanen A., Ristimäki J., Keskinen J. (2004a) 'Nucleation mode formation in heavy-duty diesel exhaust with and without a particulate filter'. Environmental Science and Technology 38:4884–4890. [Cited on p. 117]

Kati V., Annele V., Jyrki R., Jorma K. (2004b) 'Effect of after-treatment systems on size distribution of heavy duty diesel exhaust aerosol'. Society of Automotive Engineers 2004-01-1980. [Cited on pp. 253, 299, 304]

Kato M., Masunaga K., Hoshi H. (1987) 'The influence of fuel properties on white smoke emissions from light-duty diesel engine'. Society of Automotive Engineers 870341. [Cited on pp. 121, 122]

Kato S., Takayama Y., Sato G.T., Tanabe H., Sato T.G. (1997) 'Investigation of particulate formation of DI diesel engine with direct sampling from combustion chamber'. Society of Automotive Engineers 972969. [Cited on p. 109]

Kaufman D.G. (1988) 'Assessment of carcinogenicity: Generic issues and their application to diesel exhaust'. In *Air Pollution, the Automobile, and Public Health*, Watson A.Y., Bates R.R., Kennedy D. (editors). National Academy Press, pp. 519–553. [Cited on pp. 369, 372, 378]

Kaufmann R.L. (1986) 'Laser-microprobe mass spectrometry (LAMMA) of particulate matter'. In *Physical and Chemical Characterization of Individual Airborne Particles*, Spurny K.R. (editor). Ellis Horwood, pp. 226–250. [Cited on p. 209]

Kavouras I.G., Mihalopoulos N., Stephanou E.G. (1999) 'Secondary organic aerosol formation vs primary organic aerosol emission: In situ evidence for the chemical coupling between monoterpene acidic photooxidation products and new particle formation over forests'. Environmental Science and Technology 33:1028–1037. [Cited on p. 29]

Kawaguchi A., Sato Y., Yanagihara H.l, Ishiguro T. (1992) 'A study of soot formation processes in dual fueled compression ignition engine'. Society of Automotive Engineers 922304. [Cited on pp. 213, 214]

Kawai T., Iuchi Y., Nakamura S., Ishida K. (1998) 'Real time analysis of particulate by flame ionisation detection'. Society of Automotive Engineers 980048. [Cited on pp. 197, 198]

Kawamoto K., Araki T., Shinzawa M., Kimura S., Koide S., Shibuya M. (2004) 'Combination of combustion concept and fuel property for ultra-clean DI diesel'. Society of Automotive Engineers 2004-01-1868. [Cited on p. 292]

Kawanami M., Okumura A., Horiuchi M., Schäfer-Sindlinger A., Zerafa K. (1996) 'Advanced catalyst studies of diesel NOx reduction for heavy-duty diesel trucks'. Society of Automotive Engineers 961129. [Cited on p. 299]

Kawano D., Naito H., Suzuki H., Ishii H., Hori S., Goto Y., Odaka M. (2004a) 'Effects of fuel properties on combustion and exhaust emissions of homogeneous charge compression ignition (HCCI) engine'. Society of Automotive Engineers 2004-01-1966. [Cited on p. 292]

Kawano D., Kawai T., Naito H., Goto Y., Odaka M., Bachalo W.D. (2004b) 'Comparative measurement of nano-particulates in diesel engine exhaust gas by laser-induced incandescence (LII) and scanning mobility particle sizer'. Society of Automotive Engineers 2004-01-1982. [Cited on pp. 271, 274, 293]

Kawatani T., Mori K., Fukano I., Sugawara K., Koyama T. (1993) 'Technology for meeting the 1994 USA exhaust emission regulations on a heavy-duty diesel engine'. Society of Automotive Engineers 932654. [Cited on pp. 281, 294]

Kayes D., Hochgreb S. (1998) 'Investigation of the dilution process for measurement of particulate matter from spark-ignition engines'. Society of Automotive Engineers 982601. [Cited on pp. 160, 162]

Kayes D., Hochgreb S. (1999a) 'Mechanisms of particulate matter formation in spark-ignition engines. 1. Effect of engine operating conditions'. Environmental Science and Technology 33:3957–3967. [Cited on pp. 319, 320]

Kayes D., Hochgreb S. (1999b) 'Mechanisms of particulate matter formation in spark-ignition engines. 2. Effect of fuel, oil, and catalyst parameters'. Environmental Science and Technology 33:3968–3977. [Cited on p. 321]

Kayes D., Hochgreb S. (1999c) 'Mechanisms of particulate matter formation in spark-ignition engines. 3. Model of PM formation'. Environmental Science and Technology 33:3978–3992. [Cited on p. 320]

Kayes D., Liu H., Hochgreb S. (1999) 'Particulate matter emission during start-up and transient operation of a spark-ignition engine'. Society of Automotive Engineers 1999-01-3529. [Cited on pp. 161, 320]

Kayes D., Hochgreb S., Maricq M.M., Podsiadlik D.H., Chase R.E. (2000) 'Particulate matter emission during start-up and transient operation of a spark-ignition engine (2): Effect of speed, load, and real-world driving cycles'. Society of Automotive Engineers 2000-01-1083. [Cited on pp. 325, 326]

Kazakov A., Foster D.E. (1998) 'Modelling of soot formation during DI diesel combustion using a multistep phenomenological model'. Society of Automotive Engineers 982463. [Cited on pp. 76, 78]

Kennedy I.M. (1997) 'Models of soot formation and oxidation'. Progress in Energy and Combustion Science 23:95–132. [Cited on pp. 63, 65, 72, 73]

Kennedy I.M. (2007) 'The health effects of combustion-generated aerosols'. Proceeding of the Combustion Institute 31:2757–2770. [Cited on p. 364]

Kennedy I.M., Yam C., Rapp D.C., Santoro R.J. (1996) 'Modeling and measurements of soot and species in a laminar diffusion flame'. Combustion and Flame 107:368–382. [Cited on p. 79]

Kenney T.E., Gardner T.P., Low S.S., Eckstrom J.C., Wolf L.R., Korn S.J., Szymkowicz P.G. (2001) 'Overall results: Phase I ad hoc diesel fuel test program'. Society of Automotive Engineers 2001-01-0151. [Cited on p. 250]

Kerker M. (1969) *The Scattering of Light and Other Electromagnetic Radiation*. Academic Press. [Cited on pp. 37, 189]

Kerminen V.-M., Mäkelä T.W., Ojanen C.H., Hillamo R.E., Vilhunen J.K., Rantanen L., Havers N., von Bohlen A., Klockow D. (1997) 'Characterization of the particulate phase in the exhaust from a diesel car'. Environmental Science and Technology 31:1883–1889. [Cited on p. 237]

Kerminen V.-M., Pakkanen T., Mäkelä T.W., Hillamo R.E., Sillanpää T., Rönkkö T., Virtanen A., Keskinen J., Pirjola L., Hussein T., Hämeri K. (2007) 'Development of particle number size distribution near a major road in Helsinki during an episodic inversion situation'. Atmospheric Environment 41:1759–1767. [Cited on p. 361]

Keshtkar H., Ashbaugh L.L. (2007) 'Size distribution of polycyclic aromatic hydrocarbn particulate emission factors from agricultural burning'. Atmospheric Environment 41:2729–2739. [Cited on p. 116]

Keskinen J., Marjamäki M., Virtanen A., Mäkelä T., Hillamo R. (1999) 'Electrical calibration method for cascade impactors'. Journal of Aerosol Science 30:111–116. [Cited on p. 184]

Ketcher D.A., Horrocks R.W. (1990) 'The effect of fuel sulphur on diesel particulate emissions when using oxidation catalysts'. In *Fuels for Automotive and Diesel Engines*, Nov. 19–20. Institution of Mechanical Engineers. [Cited on pp. 174, 295]

Keyser P.I., Howard G.W. (1991) 'The application of sterile filtration technology in the environmental control and life support systems of Space Station Freedom'. Society of Automotive Engineers 911518. [Cited on p. 209]

Khair M.K., Bykowski B.B. (1992) 'Design and development of catalytic converters for diesels'. Society of Automotive Engineers 921677. [Cited on p. 253]

Khalek I.A. (2000) 'Characterization of particle size distribution of a heavy-duty diesel engine during FTP transient cycle using ELPI'. Society of Automotive Engineers 2000-01-2001. [Cited on p. 284]

Khalek I.A., Kittelson D.B., Brear F. (2000) 'Nanoparticle growth during dilution and cooling of diesel exhaust: Experimental investigation and theoretical assessment'. Society of Automotive Engineers 2000-01-0515. [Cited on pp. 127, 128]

Khalek I.A., Ullman T.L., Shimpi S.A., Jackson C.C., Dharmawardhana B., Silvis W.M., Kreft N., Harvey R.N., Munday D., Yamagishi Y., Graze R., Smitherman J., Adkins J. (2002) 'Performance of partial flow sampling systems relative to full flow CVS for determination of particulate emissions under steady-state and transient diesel engine operation'. Society of Automotive Engineers 2002-01-1718. [Cited on pp. 159, 160]

Khalek I.A., Fritz S.G., Paas N. (2003) 'Particle size distribution and mass emissions from a mining diesel engine equipped with a dry system technologies emission control system'. Society of Automotive Engineers 2003-01-1893. [Cited on p. 151]

Khalil N., Levendis Y.A. (1992) 'Development of a new diesel particulate control system with wall-flow filters and reverse cleaning regeneration'. Society of Automotive Engineers 920567. [Cited on p. 303]

Khalili N., Scheff P.A., Holsen T.M. (1995) 'PAH source fingerprints for coke ovens, diesel and gasoline engines, highway tunnels, and wood combustion emissions'. Atmospheric Environment 29:33–542. [Cited on p. 35]

Khatri N.J., Johnson J.H., Leddy D.G. (1978) 'The Characterization of the hydrocarbon and sulfate fractions of diesel particulate matter'. Society of Automotive Engineers 780111. [Cited on pp. 112, 123, 124, 125, 173, 221]

Kim C.H., El-Leathy A.M., Xu F., Faeth G.M. (2004) 'Soot surface growth and oxidation in laminar diffusion flames at pressures of 0.1–1.0atm'. Combustion and Flame 136:191–207. [Cited on pp. 74, 75, 79, 81]

Kim D.-H., Gautam M., Gera D. (2002a) 'Effect of ambient dilution on coagulation of particulate matter in a turbulent dispersing plume'. Society of Automotive Engineers 2002-01-0652. [Cited on pp. 143, 144]

Kim D.-H., Gautam M., Gera D. (2002b) 'Prediction of pollutant concentration variation inside a turbulent dispersing plume using PDF and Gaussian models'. Society of Automotive Engineers 2002-01-0654. [Cited on p. 143]

Kim D.S., Lee C.S. (2005) 'Effect of n-heptane premixing on combustion characteristics of diesel engine'. Energy and Fuels 19:2240–2246. [Cited on p. 292]

Kim D.S., Lee C.S. (2006) 'Improved emission characteristics of HCCI engine by various premixed fuels and cooled EGR'. Fuel 85:695–704. [Cited on p. 292]

Kim E., Hopke P.K., Pinto J.P., Wilson W.E. (2005a) 'Spatial variability of fine particle mass, components, and source contributions during the regional air pollution study in St. Louis'. Environmental Science and Technology 39:4172–4179. [Cited on p. 371]

Kim E., Hopke P.K., Kenski D.M., Koerber M. (2005b) 'Sources of fine particles in a rural Midwestern U.S. area'. Environmental Science and Technology 39:4953–4960. [Cited on p. 358]

Kim J.H., Ma X., Zhou A., Song C. (2006) 'Ultra-deep desulfurization and denitrogenation of diesel fuel by selective adsorption over three different adsorbents: A study on adsorptive selectivity and mechanism'. Catalysis Today 111:74–83. [Cited on p. 252]

Kim J.-S., Min B.-S., Lee D.-S., Oh D.Y., Choi J.-K. (1998) 'The characteristic of carbon deposit formation in piston top ring groove of gasoline and diesel engine'. Society of Automotive Engineers 980526. [Cited on p. 279]

Kim S.H., Fletcher R.A., Zachariah M.R. (2005) 'Understanding the difference in oxidative properties between flame and diesel soot nanoparticles: The role of metals'. Environmental Science and Technology 39:4021–4026. [Cited on pp. 99, 101, 220]

Kimura K., Lynskey M., Corrigan E.R., Hickman D.L., Wang J., Fang H.L., Chatterjee S. (2006) 'Real world study of diesel particulate filter ash accumulation in heavy-duty diesel trucks'. Society of Automotive Engineers 2006-01-3257. [Cited on p. 95]

Kirchstetter T.W., Novakov T. (2007) 'Controlled generation of black carbon particles from a diffusion flame and applications in evaluating black carbon measurement methods'. Atmospheric Environment 41:1874–1888. [Cited on p. 202]

Kitagawa H., Murayama T., Tosaka S., Fujiwara Y. (2001) 'The effect of oxygenated fuel additive on the reduction of diesel exhaust particulates'. Society of Automotive Engineers 2001-01-2020. [Cited on pp. 258, 259]

Kitamura T., Ito T., Senda J., Fujimoto H. (2001) 'Detailed chemical kinetic modeling of diesel spray combustion with oxygenated fuels'. Society of Automotive Engineers 2001-01-1262. [Cited on pp. 259, 292]

Kitamura T., Ito T., Kitamura Y., Ueda M., Senda J., Fujimoto H. (2003) 'Soot kinetic modeling and empirical validation on smokeless diesel combustion with oxygenated fuels'. Society of Automotive Engineers 2003-01-1789. [Cited on p. 258]

Kittelson D.B., Collings N. (1987) 'Origin of the response of electrostatic particle probes'. Society of Automotive Engineers 870476. [Cited on pp. 225, 226, 227, 228]

Kittelson D.B., Dolan D.F. (1978) 'Dynamics of sampling and measurement of diesel engine exhaust aerosols'. Presented at the Conference on Carbonaceous Particles in the Atmosphere, Berkeley, CA, 20–22 March 1978. [Cited on p. 113]

Kittelson D.B. (1998) 'Engines and nanoparticles: A review'. Journal of Aerosol Science 29:575–588. [Cited on pp. 53, 54]

Kittelson D.B., Johnson J.H. (1991) 'Variability in particulate emission measurements in the heavy duty transient test'. Society of Automotive Engineers 910738. [Cited on pp. 148, 160, 162]

Kittelson D.B., Dolan D.F., Verrant J.A. (1978a) 'Investigation of a diesel exhaust aerosol'. Society of Automotive Engineers 780109. [Cited on p. 183]

Kittelson D.B., Dolan D.F., Diver R.B., Aufderheide E. (1978b) 'Diesel exhaust particle size distribution – fuel and additive effects'. Society of Automotive Engineers 780787. [Cited on pp. 98, 103, 104, 220, 221, 231, 233, 261, 262, 264]

Kittelson D.B., Pui D.Y.H., Moon K.C. (1986a) 'Electrostatic collection of diesel particles'. Society of Automotive Engineers 860009. [Cited on pp. 224, 225, 226, 227, 228]

Kittelson D.B., Pipho M.J., Ambs J.L., Siegla D.C. (1986b) 'Particle concentrations in a diesel cylinder: Comparison of theory and experiment'. Society of Automotive Engineers 861569. [Cited on pp. 81, 82]

Kittelson D.B., Pipho M.J., Ambs J.L., Luo L. (1988) 'In-cylinder measurements of soot production in a direct-injection diesel engine'. Society of Automotive Engineers 880344. [Cited on p. 85]

Kittelson D.B., Ambs J.L., Hadjkacem H. (1990) 'Particulate emissions from diesel engines: Influence of in-cylinder surface'. Society of Automotive Engineers 900645. [Cited on pp. 91, 93, 94, 95, 136]

Kittelson D.B., Reinertsen J., Michalski J. (1991) 'Further studies of electrostatic collection and agglomeration of diesel particles'. Society of Automotive Engineers 910329. [Cited on pp. 225, 228]

Kittelson D.B., Sun R., Blackshear P.L., Brehob D.D. (1992) 'Oxidation of soot agglomerates in a direct injection diesel engine'. Society of Automotive Engineers 920111. [Cited on pp. 79, 80, 274, 394]

Kittelson D., Johnson J., Watts W., Wei Q., Drayton M., Paulsen D., Bukowiecki N. (2000) 'Diesel aerosol sampling in the atmosphere'. Society of Automotive Engineers 2000-01-2212. [Cited on pp. 142, 144]

Kitto A.-M.N., Colbeck I. (1999) 'Filtration and denuder sampling techniques'. In *Analytical Chemistry of Aerosols*, Spurny K.R. (editor). CRC Press LLC, pp. 103–132. [Cited on p. 146]

Kittredge G.D., McNutt B.D. (1971) 'Role of NAPCA in controlling aircraft pollutant emissions'. Society of Automotive Engineers 710338. [Cited on p. 4]

Klausmeier R., Gallagher J., Barrett R., Holman T. (1985) 'The Colorado diesel emissions control program – Stage I'. Society of Automotive Engineers 850149. [Cited on p. 388]

Kleeman M.J., Cass G.R. (1999) 'Effect of emissions control strategies on the size and composition distribution of urban particulate air pollution'. Environmental Science and Technology 33:177–189. [Cited on p. 393]

Klein R.F., Brunner N.R., Hughes J.C., Pahl B.A., Turpin M.E. (1990) 'The Cummins A3.4–125: A charge cooled IDI turbo diesel for the 1991 US light–heavy duty market'. Society of Automotive Engineers 901570. [Cited on p. 279]

Kleindienst T.E., Edney E.O., Lewandowski M., Offenberg J.H., Jaoui M. (2006) 'Secondary organic carbon and aerosol yields from the irradiations of isoprene and α-pinene in the presence of NO_X and SO_2'. Environmental Science and Technology 40:3807–3812. [Cited on p. 29]

Klingbeil A.E., Juneja H., Ra Y., Reitz R.D. (2003) 'Premixed diesel combustion analysis in a heavy-duty diesel engine'. Society of Automotive Engineers 2003-01-0341. [Cited on p. 292]

Klingen H.-J., Roth P. (1991) 'Real-time measurements of soot particles at the exhaust valve of a diesel engine'. Society of Automotive Engineers 912667. [Cited on pp. 136, 137, 190, 218]

Klingenberg H., Winneke H. (1990) 'Studies on health effects of automotive exhaust emissions – how dangerous are diesel emissions'. Science of the Total Environment 93:95–105. [Cited on pp. 371, 378]

Knapp K.T., Tejada S.B., Cadle S.H., Lawson D.R., Snow R., Zielinska B., Sagebiel J.C., McDonald J.D. (2003) 'Central carolina vehicle particulate emissions study'. Society of Automotive Engineers 2003-01-0299. [Cited on pp. 6, 322, 375]

Knutson E.O. (1999) 'History of diffusion batteries in aerosol measurements'. Aerosol Science and Technology 31:83–128. [Cited on p. 158]

Kobayashi S., Sakai T., Nakahira T., Komori M., Tsujimura K. (1992) 'Measurement of flame temperature distribution in D.I. engine with high pressure fuel injection'. Society of Automotive Engineers 920692. [Cited on p. 267]

Koch F., Haubner F.G., Orlowsky K. (2002) 'Lubrication and ventilation system of modern engines – measurements, calculations and analysis'. Society of Automotive Engineers 2002-01-1315. [Cited on p. 279]

Kocis D., Song K., Lee H., Litzinger T. (2000) 'Effects of dimethoxymethane and dimethylcarbonate on soot production in an optically-accessible DI diesel engine'. Society of Automotive Engineers 2000-01-2795. [Cited on p. 259]

Kodama K., Hiranuma S., Doumeki R., Takeda Y., Ikeda T. (2005) 'Development of DPF system for commercial vehicles (second report) – active regenerating function in various driving condition'. Society of Automotive Engineers 2005-01-3694. [Cited on p. 304]

Kodavanti U.P., Schladweiler M.C., Ledbetter A.D., Watkinson W.P., Campen M.J., Winsett D.W., Richards J.R., Crissman K.M., Hatch G.E., Costa D.L. (2000) 'The spontaneously hypertensive rat as a model of human cardiovascular disease: Evidence of exacerbated cardiopulmonary injury and oxidative stress from inhaled emission particulate matter'. Toxicology and Applied Pharmacology 164:250–263. [Cited on p. 380]

Kojima M., Bacon R.W., Shah J., Mainkar M.S., Chaudhari M.K., Bhanot B., Iyer N.V., Smith A., Atkinson W.D. (2002) 'Measurement of mass emissions from in-use two-stroke engine three-wheelers in south Asia'. Society of Automotive Engineers 2002-01-1681. [Cited on pp. 336, 337]

Kokko J., Rantanen L., Pentikäinen J., Honkanen T., Aakko P., Lappi M. (2000) 'Reduced particulate emissions with reformulated gasoline'. Society of Automotive Engineers 2000-01-2017. [Cited on pp. 323, 326, 331, 333]

Kolb C.E., Herndon S.C., McManus J.B., Shorter J.H., Zahniser M.S., Nelson D.D., Jayne J.T., Canagaratna M.R., Worsnop D.R. (2004) 'Mobile laboratory with rapid response instruments for real-time measurements of urban and regional trace gas and particulate distributions and emission source characteristics'. Environmental Science and Technology 38:5694–5703. [Cited on p. 142]

Kolesnikov E.M., Boettger T., Kolesnikova N.V. (1999) 'Finding of probable Tunguska cosmic body material: Isotopic anomalies of carbon and hydrogen in peat'. Planetary and Space Science 47:905–914. [Cited on p. 41]

Koltsakis G.C., Stamatelos A.M. (1997) 'Catalytic automotive exhaust aftertreatment'. Progress in Energy and Combustion Science 23:1–39. [Cited on p. 123]

Koltsakis G.C., Konstantinou A., Haralampous O.A., Samaras Z.C. (2006) 'Measurement and intra-layer modeling of soot density and permeability in wall-flow filters'. Society of Automotive Engineers 2006-01-0261. [Cited on p. 212]

Kong S.-C., Ricart L.M., Reitz R.D. (1995) 'In-cylinder diesel flame imaging compared with numerical computations'. Society of Automotive Engineers 950455. [Cited on pp. 269, 270]

Kono N., Kobayashi Y., Takeda H. (2005) 'Fuel effects on emissions from diesel vehicles equipped with aftertreatment devices'. Society of Automotive Engineers 2005-01-3700. [Cited on p. 303]

Konstandopoulos A.G., Kostoglou M., Skaperda E., Papaioannou E., Zarvalis D., Kladopoulou E. (2000) 'Fundamental studies of diesel particulate filters: Transient loading, regeneration and aging'. Society of Automotive Engineers 2000-01-1016. [Cited on pp. 222, 301]

Korn S.J. (2001) 'An advanced diesel fuels test program'. Society of Automotive Engineers 2001-01-0150. [Cited on p. 263]

Kornbrekke R.E., Patrzyk-Semanik P., Kirchner-Jean T., Raguz M., Bardasz E.A. (1998) 'Understanding soot mediated oil thickening Part 6: Base oil effects'. Society of Automotive Engineers 982665. [Cited on pp. 287, 288]

Koshland C.P., Fischer S.L. (2002) 'Diagnostic requirements for toxic emission control'. In *Applied Combustion Diagnostics*, Kohse-Höinghaus K., Jeffries J.B. (editors) Taylor and Francis, pp. 606–626. [Cited on p. 1]

Koutrakis P., Sioutas C. (1996) 'Physico-chemical properties and measurements of ambient particles'. In *Particles in Our Air – Concentrations and Health Effects*, Wilson R., Spengler J. (editors). Harvard University Press, pp. 15–39. [Cited on pp. 150, 165]

Koyanagi K., Öing H., Renner G., Maly R. (1999) 'Optimizing common rail-injection by optical diagnostics in a transparent production type diesel engine'. Society of Automotive Engineers 1999-01-3646. [Cited on pp. 266, 277]

Köylü Ü.Ö., Faeth G.M. (1994) 'Optical properties of soot in buoyant laminar diffusion flames'. Journal of Heat Transfer 116:971–979. [Cited on p. 189]

Kraft J., Lies K.-H. (1981) 'Polycyclic aromatic hydrocarbons in the exhaust of gasoline and diesel vehicles'. Society of Automotive Engineers 810082. [Cited on p. 177]

Kraft J., Hartung A., Sculze J., Lies K.-H. (1982) 'Determination of polycyclic aromatic hydrocarbons in diluted and undiluted exhaust gas of diesel engines'. Society of Automotive Engineers 821219. [Cited on pp. 114, 177]

Krahl J., Munack A., Schröder O., Stein H., Bünger J. (2003) 'Influence of biodiesel and different designed diesel fuels on the exhaust gas emissions and health effects'. Society of Automotive Engineers 2003-01-3199. [Cited on pp. 259, 260]

Krämer L., Bozoki Z., Niessner R. (2000) 'Setup, calibration and characterization of a mobile photoacoustic soot sensor'. Journal of Aerosol Science 31:S72–S73. [Cited on p. 192]

Krenn M., Kampelmühler F., Weidinger C., Mariani G., Masera F. (2000) 'Evaluation of a new diesign for CVS-systems meeting the requirements of S-ULEV and Euro IV'. Society of Automotive Engineers 2000-01-0800. [Cited on p. 160]

Kreso A.M., Johnson J.H., Gratz L.D., Bagley S.T., Leddy D.G. (1998) 'A study of the effects of exhaust gas recirculation on heavy-duty diesel engine emissions'. Society of Automotive Engineers 981422. [Cited on pp. 271, 274, 392]

Krestinin A.V. (2000) 'Detailed modeling of soot formation in hydrocarbon pyrolysis'. Combustion and Flame 121:513–524. [Cited on pp. 68, 72]

Kreyling W.G., Tuch T., Peters A., Pitz M., Heinrich J., Stötzel M., Cyrys J., Heyder J., Wichmann H.E. (2003) 'Diverging long-term trends in ambient urban particle mass and number concentrations associated with emission changes caused by the German unification'. Atmospheric Environment 37:3841–3848. [Cited on p. 40]

Kroll J., Ng N.L., Murphy S.M., Flagan R.C., Seinfeld J.H. (2006) 'Secondary organic aerosol formation from isoprene photooxidation'. Environmental Science and Technology 40:1869–1877. [Cited on p. 29]

Kruis F.E., Fissan H., Peled A. (1998) 'Synthesis of nanoparticles in the gas phase for electronic, optical and magnetic applications – a review'. Journal of Aerosol Science 29:511–535. [Cited on p. 396]

Krutzsch B., Wenninger G. (1992) 'Effect of sodium- and lithium-based fuel additives on the regeneration efficiency of diesel particulate filters'. Society of Automotive Engineers 922188. [Cited on p. 233]

Kubach H., Gindele J., Spicher U. (2001) 'Investigations of mixture formation and combustion in gasoline direct injection engines'. Society of Automotive Engineers 2001-01-3647. [Cited on pp. 327, 328, 333]

Kubach H., Velji A., Spicher U., Fischer W. (2004) 'Ion current measurement in diesel engines'. Society of Automotive Engineers 2004-01-2922. [Cited on p. 225]

Kuhns H., Etyemezian V., Green M., Hendrickson K., McGown M., Barton K., Pitchford M. (2003) 'Vehicle-based road dust emission measurement – Part II: Effect of precipitation, wintertime road sanding, and street sweepers on inferred PM10 emission potentials from paved and unpaved roads'. Atmospheric Environment 37:4573–4582. [Cited on p. 342]

Kuljukka T., Savela K., Peltonen K., Mikkonen S., Rantanen L. (1998) 'Effect of fuel reformulation in diesel particulate emissions – application of DNA adduct test'. Society of Automotive Engineers 982650. [Cited on pp. 375, 376]

Kulmala M., Vesala T., Laaksonen A. (2000) 'Physical chemistry of aerosol formation'. In *Aerosol Chemical Processes in the Environment*, Spurny K.R., Hochrainer D. (editors). CRC Press LLC, pp. 23–46. [Cited on pp. 19, 41]

Kumagai Y., Arimoto T., Shinyashiki M., Shimojo N., Nakai Y., Yoshikawa T., Sagai M. (1997) 'Generation of reactive oxygen species during interaction of diesel exhaust particle components with Nadph-cytochrome P450 reductase and involvement of the bioactivation in the DNA damage'. Free Radical Biology and Medicine 22:479–487. [Cited on p. 376]

Kumata H., Uchida M., Sakum E., Uchida T., Fujiwara K., Tsuzuki M., Yonedea M., Shibata Y. (2006) 'Compound class specific ^{14}C analysis of polycyclic aromatic hydrocarbons associated with PM_{10} and $PM_{1.1}$ aerosols from residential areas of suburban Tokyo'. Environmental Science and Technology 40:3474–3480. [Cited on p. 35]

Künzli N., Kaiser R., Medina S., Studnicka M., Chanel O., Filliger P., Herry M., Horak F., Puybonnieux-Texier V., Quénel P., Schneider J., Seethaler R., Vergnaud J.-C., Sommer H. (2000) 'Public-health impact of outdoor and traffic-related air pollution: A European assessment'. The Lancet 356:795–801. [Cited on pp. 374, 397]

Künzli N., Jerrett M., Mack W.J., Beckermann B., LaBree L., Gilliland F., Thomas D., Peters J., Hodis H.N. (2005) 'Ambient air pollution and atherosclerosis in Los Angeles'. Environmental Health Perspectives 113:201–206. [Cited on p. 374]

Kuo C.C., Passut C.A., Jao T.-C., Csontos A.A., Howe J.M. (1998) 'Wear mechanism in Cummins M-11 high soot diesel test engines'. Society of Automotive Engineers 981372. [Cited on p. 290]

Kupiainen K., Tervahattu H., Räisänen M. (2003) 'Experimental studies about the impact of traction sand on urban road dust composition'. Science of the Total Environment 308:175–184. [Cited on pp. 342, 343]

Kupiainen K.J., Tervahattu H., Räisänen M., Mälelä T., Aurela M., Hillamo R. (2005) 'Size and composition of airborne particles from pavement wear, tires, and traction sanding'. Environmental Science and Technology 39:699–706. [Cited on p. 342]

Kuritz B., Hearne J., Toback A., Hesketh R.P., Marchese A.J., Gephardt Z.O. (2004) 'Application of experimental design in the steady state particulate exposure levels in a 1992 international school bus'. Society of Automotive Engineers 2004-01-1088. [Cited on p. 359]

Kurniawan A., Schmidt-Ott A. (2006) 'Monitoring the soot emissions of passenger cars'. Environmental Science and Technology 40:1911–1915. [Cited on pp. 195, 388]

Kurtz E.M., Mather D.K., Foster D.E. (2000) 'Parameters that affect the impact of auxiliary gas injection in a DI diesel engine'. Society of Automotive Engineers 2000-01-0233. [Cited on p. 91]

Kütz S., Schmidt-Ott A. (1992) 'Characterization of agglomerates by condensation-induced restructuring'. Journal of Aerosol Science 23:S357–S360. [Cited on pp. 20, 219, 360]

Kuwahara K., Ueda K., Ando H. (1998) 'Mixing control strategy for engine performance improvement in a gasoline direct injection engine'. Society of Automotive Engineers 980158. [Cited on pp. 328, 333]

Kweon C.-B., Foster D.E., Schauer J.J., Okada S. (2002) 'Detailed chemical composition and particle size assessment of diesel engine exhaust'. Society of Automotive Engineers 2002-01-2670. [Cited on p. 109]

Kweon C.-B., Okada S., Foster D.E., Bae M.-S., Schauer J.J. (2003a) 'Effect of engine operating conditions on particle-phase organic compounds in engine exhaust of a heavy-duty direct-injection (D.I.) diesel engine'. Society of Automotive Engineers 2003-01-0342. [Cited on p. 109]

Kweon C.-B., Okada S., Stetter J.C., Christenson C.G., Shafer M.M., Schauer J.J., Foster D.E. (2003b) 'Effect of injection timing on detailed chemical composition and particulate size distributions of diesel exhaust'. Society of Automotive Engineers 2003-01-1794. [Cited on pp. 176, 269, 271]

Kweon C.-B., Okada S., Stetter J.C., Christenson C.G., Shafer M.M., Schauer J.J., Foster D.E. (2003c) 'Effect of fuel composition on combustion and detailed chemical/physical characteristics of diesel exhaust'. Society of Automotive Engineers 2003-01-1899. [Cited on pp. 109, 256]

Kwon Y., Stradling R., Heinze P., Broekx W., Esmilaire O., Martini G., Bennett P.J., Rogerson J., Kvinge F., Lien M. (1999a) 'The effect of fuel sulphur content on the exhaust emissions from a lean burn gasoline direct injection vehicle marketed in Europe'. Society of Automotive Engineers 1999-01-3585. [Cited on p. 331]

Kwon Y.K., Bazzani R., Bennett P.J., Esmilaire O., Scorletti P., Morgan T.D.B., Goodfellow C.L., Lien M., Broeckx W., Liiva P. (1999b) 'Emissions response of a European specification direct-injection gasoline vehicle to a fuels matrix incorporating independent variations in both compositional and distillation parameters'. Society of Automotive Engineers 1999-01-3663. [Cited on p. 333]

Kwon Y., Mann N., Rickeard D.J., Haugland R., Ulvund K.A., Kvinge F., Wilson G. (2001) 'Fuel effects on diesel emissions – A new understanding'. Society of Automotive Engineers 2001-01-3522. [Cited on pp. 250, 251, 252, 256, 264, 265]

Kyriakides S.C., Dent J.C., Mehta P.S. (1986) 'Phenomenological diesel combustion model including smoke and NO emission'. Society of Automotive Engineers 860330. [Cited on pp. 67, 82, 272]

Kytö M., Aakko P., Nylund N.-O., Niemi A. (2002) 'Effect of lubricant on particulate emissions of heavy duty diesel engines'. Society of Automotive Engineers 2002-01-2770. [Cited on p. 285]

L

Laanti S., Sorvari J., Elonen E., Pitkänen M. (2001) 'Mutagenicity and PAH:S of particulate emissions of two-stroke chainsaw engines'. Society of Automotive Engineers 2001-01-1825. [Cited on p. 337]

Labeckas G., Slavinskas S. (2006) 'Performance of direct-injection off-road diesel engine on rapeseed oil'. Renewable Energy 31:849–863. [Cited on p. 257]

Ladommatos N., Balian R., Horrocks R., Cooper L. (1996a) 'The effect of exhaust gas recirculation on combustion and NOx emissions in a high-speed direct-injection diesel engine'. Society of Automotive Engineers 960840. [Cited on p. 273]

Ladommatos N., Balian R., Horrocks R., Cooper L. (1996b) 'The effect of exhaust gas recirculation on soot formation in a high-speed direct-injection diesel engine'. Society of Automotive Engineers 960841. [Cited on pp. 273, 278]

Ladommatos N., Abdelhalim S.M., Zhao H., Hu Z. (1996c) 'The dilution, chemical, and thermal effects of exhaust gas recirculation on diesel engine emissions – Part 1: Effect of reducing inlet charge oxygen'. Society of Automotive Engineers 961165. [Cited on p. 273]

Ladommatos N., Abdelhalim S.M., Zhao H., Hu Z. (1996d) 'The dilution, chemical, and thermal effects of exhaust gas recirculation on diesel engine emission – Part 2: Effects of carbon dioxide'. Society of Automotive Engineers 961167. [Cited on p. 273]

Ladommatos N., Rubenstein P., Bennett P. (1996e) 'Some effects of molecular structure of single hydrocarbons on sooting tendency'. Fuel 75:114–124. [Cited on pp. 65, 254]

Ladommatos N., Abdelhalim S.M., Zhao H., Hu Z. (1997a) 'The dilution, chemical, and thermal effects of exhaust gas recirculation on diesel engine emissions – Part 3: Effects of water vapour'. Society of Automotive Engineers 971659. [Cited on p. 273]

Ladommatos N., Abdelhalim S.M., Zhao H., Hu Z. (1997b) 'The dilution, chemical, and thermal effects of exhaust gas recirculation on diesel engine emissions – Part 4: Effects of carbon dioxide and water vapour'. Society of Automotive Engineers 971659. [Cited on p. 273]

Ladommatos N., Rubenstein P., Harrison K., Xiao Z., Zhao H. (1997c) 'The effect of aromatic hydrocarbons on soot formation in laminar diffusion flames in a diesel engine'. Journal of the Institute of Energy 70:84–94. [Cited on p. 255]

Ladommatos N., Abdelhalim S., Zhao H. (2000) 'The effects of exhaust gas recirculation on diesel combustion and emissions'. International Journal of Engine Research 1:107–126. [Cited on p. 273]

Lafleur A.L., Taghizadeh K., Howard J.B., Anacleto J.F., Quilliam M.A. (1996) 'Characterization of flame-generated C10 to C160 polycyclic aromatic hydrocarbons by atmospheric-pressure chemical ionization mass spectrometry with liquid introduction via heated nebulizer interface'. Journal of the American Society for Mass Spectrometry 7:276–286. [Cited on p. 70]

Lahaye J. (1992) 'Particulate carbon from the gas phase'. Carbon 3:309–314. [Cited on p. 81]

Lahaye J., Ehrburger-Dolle F. (1994) 'Mechanisms of carbon black formation. Correlation with the morphology of aggregates'. Carbon 32:1319–1324. [Cited on pp. 68, 71, 73, 77]

Lahaye J., Boehm S., Chambrion P., Ehrburger P. (1996) 'Influence of cerium oxide on the formation and oxidation of soot'. Combustion and Flame 104:1996–2007. [Cited on pp. 216, 231]

Lai S.-T., Sheng J.P.M., Lin C.-S., Lin R.S., Lin C.-Y., Shiau Y.-F., Huang G.G.S. (1991) 'Study of exhaust emission reduction and lubricity of two-stroke engine'. Society of Automotive Engineers 911276. [Cited on p. 336]

Laimböck F.J., Trigg R.V., Kirchberger R.S., Meister G.F., Dorfstätter M.J., Brasseur G. (1999) 'HYC – a hybrid concept with small lean burn engine, electrically heated catalyst and asynchronous motor for enhanced performance and ULEV emissions'. Society of Automotive Engineers 1999-01-1330. [Cited on p. 336]

Lambert A.L., Dong W., Selgrade M.J.K., Gilmour M.I. (2000) 'Enhanced allergic sensitization by residual oil fly ash particles is mediated by soluble metal constituents'. Toxicology and Applied Pharmacology 165:84–93. [Cited on p. 380]

Lambert B., Bastlova T., Österholm A.-M., Hou S.-M. (1995) 'Analysis of mutation at the *hprt* locus in human T lymphocytes'. Toxicology Letters 82/83:323–333. [Cited on p. 369]

Lampert J.K., Kazi M.S., Farrauto R.J. (1996) 'Methane emissions abatement from lean burn natural gas vehicle exhaust: Sulfur's impact on catalyst performance'. Society of Automotive Engineers 961971. [Cited on pp. 295, 296]

Lamprecht A., Eimer W., Kohse-Höinghaus K. (1999) 'Dynamic light scattering in sooting premixed atmospheric-pressure methane –, propane –, ethene –, and propene – Oxygen flames'. Combustion and Flame 118:140–150. [Cited on p. 182]

Lang J.M., Snow L., Carlson R., Black F., Zweidinger R., Tejada S. (1981) 'Characterization of particulate emissions from in-use gasoline-fueled motor vehicles'. Society of Automotive Engineers 811186. [Cited on pp. 316, 323]

Lanni T., Chatterjee S., Conway R., Windawi H., Rosenblatt D., Bush C., Lowell D., Evans J., McLean R. (2001) 'Performance and durability evaluation of continuously regenerating particulate filters on diesel powered urban buses at NY City Transit'. Society of Automotive Engineers 2001-01-0511. [Cited on p. 219]

Lanning L.A., Smith K.W., Tennant C.J. (2001) 'A new method for diesel HC collection and speciation'. Society of Automotive Engineers 2000-01-2951. [Cited on p. 175]

Lapuerta M., Armas O., Ballesteros R., Carmona M. (2000) 'Fuel formulation effects on passenger car diesel engine particulate emissions and composition'. Society of Automotive Engineers 2000-01-1850. [Cited on pp. 46, 240]

Lapuerta M., Armas O., Ballesteros R., Fernández J. (2005) 'Diesel emissions from biofuels derived from Spanish potential vegetable oils'. Fuel 84:773–780. [Cited on p. 257]

Lapuerta M., Ballesteros R., Martos F.J. (2006) 'A method to determine the fractal dimension of diesel soot agglomerates'. Journal of Colloid and Interface Science 303:149–158. [Cited on p. 218]

Larsen J.C., Larsen P.B. (1998) 'Chemical carcinogens'. In *Air Pollution and Health*, Hester R.E., Harrison R.M. (editors), Issues in Environmental Science and Technology, Volume 10. Royal Society of Chemistry, pp. 33–56. [Cited on pp. 35, 371, 372, 379]

Larsson A.-C., Einvall J., Sanati M. (2007) 'Deactivation of SCR catalysts by exposure to aerosol particles of potassium and zinc salts'. Aerosol Science and Technology 41:369–379. [Cited on p. 97]

Last R.J., Krüger M., Dürnholz M. (1995) 'Emissions and performance characteristics of a 4-stroke, direct injected diesel engine fueled with blends of biodiesel and low sulfur diesel fuel'. Society of Automotive Engineers 950054. [Cited on p. 250]

Lawton J., Weinberg F.J. (1969) *Electrical Aspects of Combustion*. Clarendon Press. [Cited on p. 225]

Laymac T.D., Johnson J.H., Bagley S.T. (1991) 'The measurement and sampling of controlled regeneration emissions from a diesel wall-flow particulate trap'. Society of Automotive Engineers 910606. [Cited on pp. 221, 237, 303, 305]

Lazar A.C., Reilly P.T.A., Whitten W.B., Ramsey J.M. (1999) 'Real-time surface analysis of individual airborne environmental particles'. Environmental Science and Technology 33:3993–4001. [Cited on p. 209]

Lee K.B., Thring M.W., Beer J.M. (1962) 'On the rate of combustion of soot in a laminar soot flame'. Combustion and Flame 6:137–145. [Cited on p. 82]

Lee K., Lee C., Ryu J., Kim H. (2005) 'An experimental study on the two-stage combustion characteristics of a direct-injection-type HCCI engine'. Energy and Fuels 19:393–402. [Cited on p. 292]

Lee K.-O., Zhu J. (2005) 'Effects of exhaust system components on particulate morphology in a light-duty diesel engine'. Society of Automotive Engineers 2005-01-0184. [Cited on pp. 137, 215, 217]

Lee K.-O., Cole R., Sekar R., Choi M.Y., Zhu J., Kang J., Bae C. (2001) 'Detailed characterization of morphology and dimensions of diesel particulates via thermophoretic sampling'. Society of Automotive Engineers 2001-01-3572. [Cited on pp. 77, 137, 214, 215, 218]

Lee K.O., Zhu J., Ciatti S., Yozgatligil A., Choi M.Y. (2003) 'Sizes, graphitic structures and fractal geometry of light-duty diesel engine particulates'. Society of Automotive Engineers 2003-01-3169. [Cited on pp. 216, 218]

Lee R., Pedley J., Hobbs C. (1998) 'Fuel quality impact on heavy duty diesel emission: – A literature review'. Society of Automotive Engineers 982649. [Cited on p. 256]

Lee S., Baumann K., Schauer J.J., Sheesley R.J., Naeher L.P., Meinardi S., Blake D.R., Edgerton E.S., Russell A.G., Clements M. (2005) 'Gaseous and particulate emissions from prescribed burning in georgia'. Environmental Science and Technology 39:9049–9056. [Cited on pp. 33, 35]

Lee S.-B., Bae G.-N., Moon K.-C., Choi M. (2007) 'Effect of diesel particles on the photooxidation of a diluted diesel exhaust – toluene mixture'. Society of Automotive Engineers 2007-01-0315. [Cited on p. 393]

Lee T., Reitz R.D. (2003) 'The effects of split injection and swirl on a HSDI diesel engine equipped with a common rail injection system'. Society of Automotive Engineers 2003-01-0349. [Cited on p. 269]

Lehmann U., Mohr M., Schweizer T., Rütter J. (2003) 'Number size distribution of particulate emissions of heavy-duty engines in real world test cycles'. Atmospheric Environment 37:5247–5259. [Cited on p. 394]

Lehmann U., Niemelä V., Mohr M. (2004) 'New method for time-resolved diesel engine exhaust particle mass measurement'. Environmental Science and Technology 38:5704–5711. [Cited on p. 183]

Lehnert B.E. (1992) 'Pulmonary and thoracic macrophage subpopulations and clearance of particles from the lung'. Environmental Health Perspectives 97:17–46. [Cited on p. 366]

Leipertz A., Ossler F., Aldén M. (2002) 'Polycyclic aromatic hydrocarbons and soot diagnostics by optical techniques'. In *Applied Combustion Diagnostics*, Kohse-Höinghaus K., Jeffries J.B. (editors). Taylor and Francis, pp. 358–383. [Cited on pp. 188, 266]

Lejeune M., Lortet D., Benajes J., Riesco J.-M. (2004) 'Potential of premixed combustion with flash late injection on a heavy-duty diesel engine'. Society of Automotive Engineers 2004-01-1906. [Cited on p. 292]

Lemaire J., Khair M. (1994) 'Effect of cerium fuel additive on the emissions characteristics of a heavy-duty diesel engine'. Society of Automotive Engineers 942067. [Cited on p. 168]

Lenane D.L. (1990) 'Effect of a fuel additive on emission control systems'. Society of Automotive Engineers 902097. [Cited on p. 317]

Leonardi A., Burtscher H., Siegmann H.C. (1992) 'Size-dependent measurement of aerosol photoemission from particles in diesel exhaust'. Atmospheric Environment 26A:3287–3290. [Cited on p. 237]

Lepperhoff G., Houben M. (1990) 'Particulate emission and soot formation processes by diesel engines'. Institution of Mechanical Engineers C394/032. [Cited on p. 264]

Lepperhoff G., Houben M. (1993) 'Mechanisms of deposit formation in internal combustion engines and heat exchangers'. Society of Automotive Engineers 931032. [Cited on p. 274]

Lepperhoff G., Kroon G. (1985) 'Impact of particulate traps on the hydrocarbon fraction of diesel particles'. Society of Automotive Engineers 850013. [Cited on p. 304]

Lepperhoff G., Hüthwohl G., Lüers-Jongen B., Hammerle R.H. (1994) 'Methods to analyze non-regulated emissions from diesel engines'. Society of Automotive Engineers 941952. [Cited on pp. 169, 170, 175, 179, 230]

Lestz S.S., Geyer S.M., Jacobus M.J. (1984) 'Performance of alternative fuels in diesel engines'. FISITA 845078. [Cited on p. 260]

Lev-On M., Zielinska B. (2004) 'Compilation of PM characterization data for diesel fueled heavy-duty trucks and buses: Impact of tailpipe controls and fuel sulfur on the EC/OC ratio and the particles' chemical species content'. Society of Automotive Engineers 2004-01-1991. [Cited on p. 375]

Li F., Zhai J., Fu X., Sheng G. (2006) 'Characterization of fly ashes from circulating fluidized bed combustion (CFBC) boilers cofiring coal and petroleum coke'. Energy and Fuels 20:1411–1417. [Cited on p. 237]

Li L., Wang Z., Deng B., Xiao Z., Su Y., Wang H. (2004) 'Characteristics of particulate emissions fueled with LPG and gasoline in a small SI engine'. Society of Automotive Engineers 2004-01-2901. [Cited on p. 325]

Li N., Sioutas C., Cho A., Schmitz D., Misra C., Sempf J., Wang M., Oberley T., Froines J., Nel A. (2003) 'Ultrafine particulate pollutants induce oxidative stress and mitochondrial damage'. Environmental Health Perspectives 111:455–460. [Cited on p. 368]

Li W., Collins J.F., Norbeck J.M., Cocker D.R., Sawant A. (2006) 'Assessment of particulate matter emissions from a sample of in-use ULEV and SULEV vehicles'. Society of Automotive Engineers 2006-01-1076. [Cited on p. 392]

Li W., Collins J.F., Durbin T.D., Huai T., Ayala A., Full G., Mazzoleni C., Nussbaum N.J., Obrist D., Zhu D., Kuhns H.D., Moosmüller H. (2007) 'Detection of gasoline vehicles with gross PM emissions'. Society of Automotive Engineers 2007-01-1113. [Cited on pp. 120, 388]

Li X., Wallace J.S. (1995) 'A phenomenological model for soot formation and oxidation in direct injection diesel engines'. Society of Automotive Engineers 952428. [Cited on pp. 65, 80]

Li X., Chippior W.L., Gülder Ö. (1996) 'Effects of fuel properties on exhaust emissions of a single cylinder DI diesel engine'. Society of Automotive Engineers 962116. [Cited on pp. 256, 263, 264]

Li X.L., Wang J.S., Tu X.D., Liu W., Huang Z. (2007) 'Vertical variations of particle number concentration and size distribution in a street canyon in Shanghai, China'. Science of the Total Environment 378:306–316. [Cited on p. 361]

Liang C., Pankow J.F. (1996) 'Gas/particle partitioning of organic compounds to environmental tobacco smoke: Partition coefficient measurements by desorption and comparison to urban particulate material'. Environmental Science and Technology 30:2800–2805. [Cited on p. 117]

Liang F., Lu M., Birch M.E., Keener T.C., Liu Z. (2006) 'Determination of polycyclic aromatic sulfur heterocycles in diesel particulate matter and diesel fuel by gas chromatography with atomic emission detection'. Journal of Chromatography A 1114:145–153. [Cited on pp. 123, 233, 253]

Lim M.C.H., Ayoko G.A., Morawska L., Ristovski Z.D., Jayaratne E.R. (2007) 'The effects of fuel characteristics and engine operating conditions on the elemental composition of emissions from heavy duty diesel buses'. Fuel 86:1831–1839. [Cited on pp. 96, 283]

Lim Y.B., Ziemann P.J. (2005) 'Products and mechanism of secondary organic aerosol formation from reactions of n-alkanes with OH radicals in the presence of NO_X'. Environmental Science and Technology 39:9229–9236. [Cited on p. 37]

Limbach L.K., Li Y., Grass R.N., Brunner T.J., Hintermann M.A., Muller M., Gunther D., Stark W.J. (2005) 'Oxide nanoparticle uptake in human lung fibroblasts: Effects of particle size, agglomeration, and diffusion at low concentrations'. Environmental Science and Technology 39:9370–9376. [Cited on p. 367]

Lin C.-L., Wey M.-Y., Yu W.-J. (2005) 'Emission characteristics of organic and heavy metal pollutants in fluidized bed incineration during the agglomeration/defluidization process'. Combustion and Flame 143:139–149. [Cited on p. 98]

Lin C.Y., Pan J.-Y. (2001) 'The effects of sodium sulfate on the emissions characteristics of an emulsified marine diesel oil-fired furnace'. Ocean Engineering 28:347–360. [Cited on p. 98]

Lin Y.-C., Cheng M.T. (2007) 'Evaluation of formation rates of NO_2 to gaseous and particulate nitrate in the urban atmosphere'. Atmospheric Environment 41:1903–1910. [Cited on p. 46]

Lind T., Kauppinen E.I., Jokiniemi J.K., Maenhaut W. (1994) 'A field study on the trace metal Behaviour in atmospheric circulating fluidized-bed coal combustion'. Twenty-Fifth Symposium (International) on Combustion, The Combustion Institute, pp. 201–209. [Cited on p. 98]

Lindgren Å (1996) 'Asphalt wear and pollution transport'. Science of the Total Environment 189/190:281–286. [Cited on p. 344]

Linteris G.T., Katta V.R., Takahashi F. (2004) 'Experimental and numerical evaluation of metallic compounds for suppressing cup-burner flames'. Combustion and Flame 138:78–96. [Cited on p. 98]

Lipkea W.H., DeJoode A.D. (1994) 'Direct injection diesel soot modeling: Formulation and results'. Society of Automotive Engineers 940670. [Cited on pp. 63, 67]

Lipkea W.H., Johnson J.H., Vuk C.T. (1978) 'The physical and chemical character of diesel particulate emissions – measurement techniques and fundamental considerations'. Society of Automotive Engineers 780108. [Cited on pp. 17, 23, 26, 44, 213, 214]

Lippert A.M., Stanton D.W., Rutland C.J., Hallett W.L.H., Reitz R.D. (2000) 'Multidimensional simulation of diesel engine cold start with advanced physical submodels'. International Journal of Engine Research 1:1–27. [Cited on p. 121]

Lippmann M. (1998) 'The 1997 US EPA standards for particulate matter and ozone'. In *Air Pollution and Health*, Hester R.E., Harrison R.M. (editors), Issues in Environmental Science and Technology, Volume 10. Royal Society of Chemistry, pp. 75–99. [Cited on p. 371]

Lipsky E.M., Robinson A.L. (2006) 'Effects of dilution on fine particle mass and partitioning of semivolatile organics in diesel exhaust wood smoke'. Environmental Science and Technology 40:155–162. [Cited on p. 145]

Litzinger T., Stoner M., Hess H., Boehman A. (2000) 'Effects of oxygenated blending compounds on emissions from a turbo-charged direct injection diesel engine'. International Journal of Engine Research 1:57–70. [Cited on pp. 257, 258, 259]

Liu J., Zhao Z., Xu C., Duan A., Zhu L., Wang X. (2005) 'Diesel soot oxidation over supported vanadium oxide and K-promoted vanadium oxide catalysts'. Applied Catalysis B: Environmental 61:36–46. [Cited on p. 295]

Liu J., Zhao Z., Xu C.-m., Duan A.-j., Meng T., Bao X.-j. (2007) 'Simultaneous removal of NO_x and diesel soot particulates over nanometric $La_{2-x}K_xCuO_4$ complex oxide catalysts'. Catalysis Today 119:267–272. [Cited on p. 296]

Liu S., Obuchi A., Oi-Uchisawa J., Nanba T., Kushiyama S. (2001) 'Synergistic catalysis of carbon black oxidation by Pt with MoO_3 or V_2O_5'. Applied Catalysis B: Environmental 30:259–265. [Cited on p. 295]

Liu S., Obuchi A., Uchisawa J., Nanba T., Kushiyama S. (2002) 'An exploratory study of diesel soot oxidation with NO_2 and O_2 on supported metals oxide catalysts'. Applied Catalysis B: Environmental 37:309–319. [Cited on p. 125]

Liu X., van Espen P., Adams F., Cafmeyer J., Maenhaut W. (2000) 'Biomass burning in southern Africa: Individual particle characterization of atmospheric aerosols and savanna fire samples'. Journal of Atmospheric Chemistry 36:135–155. [Cited on p. 39]

Liu Y., Reitz R.D. (2005) 'Optimizing HSDI diesel combustion and emissions using multiple injection strategies'. Society of Automotive Engineers 2005-01-0212. [Cited on p. 277]

Liu Z., Lu M., Birch M.E., Keener T.C., Khang S.-J., Liang F. (2005) 'Variations of the particulate carbon distribution from a nonroad diesel generator'. Environmental Science and Technology 39:7840–7844. [Cited on p. 358]

Liu Z.G., Verdegan B.M., Badeau K.M.A., Sonsalla T.P. (2002) 'Measuring the fractional efficiency of diesel particulate filters'. Society of Automotive Engineers 2002-01-1007. [Cited on p. 304]

Liu Z.G., Skemp M.D., Lincoln J.C. (2003) 'Diesel particulate filters: Trends and implications of particle size distribution measurement'. Society of Automotive Engineers 2003-01-0046. [Cited on p. 17]

Liu Z.G., Chen D.-R., Perera N., Pingen G., Thurow E.M., Lincoln J.C. (2004) 'Transient analysis of engine nano-particles using a fast-scanning differential mobility particle analyzer'. Society of Automotive Engineers 2004-01-0971. [Cited on pp. 186, 190]

Lockwood F.C., van Niekerk J.E. (1995) 'Parametric study of a carbon black oil furnace'. Combustion and Flame 103:76–90. [Cited on p. 62]

Lockwood F.E., Zhang Z.G., Choi Yu W. (2001) 'Effect of soot loading on the thermal characteristics of diesel engine oils'. Society of Automotive Engineers 2001-01-1714. [Cited on pp. 286, 287]

Lohman K., Seigneur C. (2001) 'Atmospheric fate and transport of dioxins: Local impacts'. Chemosphere 45:161–171. [Cited on p. 233]

Lohmann R., Lammel G. (2004) 'Adsorptive and absorptive contributions to the gas–particle partitioning of polycyclic aromatic hydrocarbons: State of knowledge and recommended parametrization for modeling'. Environmental Science and Technology 38:3793–3803. [Cited on p. 117]

Lombaert K., le Moyne L., Amouroux J. (2002) 'Analysis of diesel particulate: Influence of air–fuel ratio and fuel composition on polycyclic aromatic hydrocarbon content'. International Journal of Engine Research 3:103–114. [Cited on pp. 109, 251]

Long T.C., Saleh N., Tilton R.D., Lowry G.V., Veronesi B. (2006) 'Titanium dioxide (P25) produces reactive oxygen species in immortalized brain microglia (BV2): Implications for nanoparticle neurotoxicity'. Environmental Science and Technology 40:4346–4352. [Cited on p. 393]

López-Fonseca R., Elizundia U., Landa I., Gutiérrez-Ortiz M.A., González-Velasco J.R. (2005) 'Kinetic analysis of non-catalytic and Mn-catalysed combustion of diesel soot surrogates'. Applied Catalysis B: Environmental 61:150–158. [Cited on p. 295]

López-Fonseca R., Landa I., Elizundia U., Gutiérrez-Ortiz M.A., González-Velasco J.R. (2006) 'Thermokinetic modeling of the combustion of carbonaceous particulate matter'. Combustion and Flame 144:398–406. [Cited on p. 168]

López Granados M., Larese C., Cabello Galisteo F., Mariscal R., Fierro J.L.G., Fernández- Ruíz R., Sanguino R., Luna M. (2005) 'Effect of mileage on the deactivation of vehicle-aged three-way catalysts'. Catalysis Today 107–108:77–85. [Cited on p. 296]

Lough G.C., Schauer J.J., Park J.-S., Shafer M.M., Deminter J.T., Weinstein J.P. (2005) 'Emission of metals associated with motor vehicle roadways'. Environmental Science and Technology 39:826–836. [Cited on pp. 342, 350]

Lu P.-H., Han J.-S., Lai M.-C., Henein N.A. (2001) 'Combustion visualization of DI diesel spray combustion inside a small-bore cylinder under different EGR and swirl ratios'. Society of Automotive Engineers 2001-01-2005. [Cited on p. 267, 277]

Lü X.-c., Yang J.-g., Zhang W.-g., Huang Z. (2005) 'Improving the combustion and emissions of direct injection conpression ignition engines using oxygenated fuel additives combined with a cetane number improver'. Energy and Fuels 19:1879–1888. [Cited on p. 259]

de Lucas A., Duràn A., Carmona M., Lapuerta M. (2001) 'Modeling diesel particulate emissions with neural networks'. Fuel 80:539–548. [Cited on p. 125]

Ludecke O.A., Bly K.B. (1984) 'Diesel exhaust particulate control by monolith trap and fuel additive regeneration'. Society of Automotive Engineers 840077. [Cited on p. 233]

Lüders H., Backes R., Hüthwohl G., Ketcher D.A., Horrocks R.W., Hurley R.G., Hammerle R.H. (1995) 'An urea lean NO$_X$ catalyst system for light duty diesel vehicles'. Society of Automotive Engineers 952493. [Cited on pp. 299]

Lüders H., Krüger M., Stommel P., Lüers B. (1998) 'The role of sampling conditions in particle size distribution measurements'. Society of Automotive Engineers 981374. [Cited on p. 185, 190]

Luo L., Pipho M.J., Ambs J.L., Kittelson D.B. (1989) 'Particle growth and oxidation in a direct-injection diesel engine'. Society of Automotive Engineers 890580. [Cited on p. 86]

Lynam D.R., Pfeifer G.D., Fort B.F., Ter Haar G.L., Hollrah D.P. (1994) 'Atmospheric exposure to manganese from use of methylcyclopentadienyl manganese tricarbonyl (MMT) performance additive'. Science of the Total Environment 146/147:103–109. [Cited on pp. 315, 317]

M

(*Note:* surnames beginning 'Mc' are listed under 'Mac'.)

McAughey J. (1999) 'Lung deposition of diesel particulates'. In *Particulate Matter: Properties and Effects upon Health*, Maynard R.L., Howard C.V. (editors). Bios Scientific Publishers, pp. 85–96. [Cited on p. 367]

McDonald J.D., Barr E.B., White R.K., Chow J.C., Schauer J.J., Zielinska B., Grosjean E. (2004) 'Generation and characterization of four dilutions of diesel engine exhaust for a subchronic inhalation study'. Environmental Science and Technology 38:2513–2522. [Cited on pp. 349, 377]

McDonald J.F., Purcell D.L., McClure B.T., Kittelson D.B. (1995) 'Emissions characteristics of soy methyl ester fuels in an IDI compression ignition engine'. Society of Automotive Engineers 950400. [Cited on p. 238]

MacDonald J.S. (1983) 'The effect of operating conditions on the effluent of a wall-flow monolith particulate trap'. Society of Automotive Engineers 831711. [Cited on p. 304]

MacDonald J.S., Barsic N.J., Gross G.P., Shahed S.P., Johnson J.H. (1984) 'Status of diesel particulate measurement methods'. Society of Automotive Engineers 840345. [Cited on pp. 116, 212]

McEnally C.S., Pfefferle L.D. (2007) 'Improved sooting tendency measurements for aromatic hydrocarbons and their implications for naphthalene formation pathways'. Combustion and Flame 148:210–222. [Cited on p. 65]

McEnally C.S., Pfefferle L.D., Atakan B., Kohse-Höinghaus K. (2006) 'Studies of aromatic hydrocarbon formation mechanism in flames: Progress toward closing the fuel gap'. Progress in Energy and Combustion Science 32:247–294. [Cited on p. 68]

McGeehan J.A. (2004) 'Diesel engines have a future and that future is clean'. Society of Automotive Engineers 2004-01-1956. [Cited on pp. 303, 396]

McGeehan J.A., Fontana B.J. (1980) 'Effect of soot on piston deposits and crankcase oils – infrared spectrometric technique for analyzing soot'. Society of Automotive Engineers 801368. [Cited on pp. 280, 287]

McGeehan J.A., Rynbrandt J.D., Hansel T.J. (1984) 'Effect of oil formulations on minimizing viscosity increase and sludge due to diesel engine soot'. Society of Automotive Engineers 841370. [Cited on pp. 287, 288, 289]

McGeehan J.A., Rutherford J.A., Couch M.C. (1991) 'Clean diesel exhaust but sooty engines: The importance of the crankcase oil'. Society of Automotive Engineers 912342. [Cited on p. 287]

McGeehan J.A., Shamah E., Couch M.C., Parker R.A. (1993) 'Selecting diesel crankcase oils to use with low-sulfur fuel'. Society of Automotive Engineers 932845. [Cited on p. 252]

McGeehan J.A., Alexander W., Ziemer J.N., Roby S.H., Graham J.P. (1999) 'The pivotal role of crankcase oil in preventing soot wear and extending filter life in low emission diesel engines'. Society of Automotive Engineers 1999-01-1525. [Cited on pp. 286, 289, 290]

McGeehan J.A., Yeh S., Couch M., Hinz A., Otterholm B., Walker A., Blakeman P. (2005) 'On the road to 2010 emissions: Field test results and analysis with DPF-SCR system and ultra low sulfur diesel fuel'. Society of Autmotive Engineers 2005-01-3716. [Cited on pp. 169, 233]

Machemer S.D. (2004) 'Characterization of airborne and bulk particulate from iron and steel manufacturing facilities'. Environmental Science and Technology 38:381–389. [Cited on p. 31]

Machta L. (1971) 'Water vapor pollution of the upper atmosphere by aircraft'. Society of Automotive Engineers 710323. [Cited on p. 41]

Macián V., Payri R., Tormos B., Montoro L. (2006) 'Applying analytical ferrography as a technique to detect failures in diesel engine fuel injection systems'. Wear 260:562–566. [Cited on p. 95]

McKinley T.L. (1997) 'Modeling sulfuric acid condensation in diesel engine EGR coolers'. Society of Automotive Engineers 970636. [Cited on p. 274]

McKinnon J.T., Meyer E., Howard J.B. (1996) 'Infrared analysis of flame-generated PAH samples'. Combustion and Flame 105:161–166. [Cited on p. 106]

McMahon M.A., Smolon W.J., DeSousa D.J. (1984) 'The effect of an alumina-coated metal mesh filter on the mutagenic activity of diesel particulate emissions'. Society of Automotive Engineers 840363. [Cited on p. 306]

McMillian M.H., Cui M., Gautam M., Keane M., Ong T.-M., Wallace W., Robey E. (2002) 'Mutagenic potential of particulate matter from diesel engine operation on Fischer-Tropsch fuel as a function of engine operating conditions and particle size'. Society of Automotive Engineers 2002-01-1699. [Cited on pp. 184, 238, 239, 251]

McMurry P.H. (2000a) 'The history of condensation nucleus counters'. Aerosol Science and Technology 33:297–322. [Cited on p. 190]

McMurry P.H. (2000b) 'A review of atmospheric aerosol measurements'. Atmospheric Environment 34:1959–1999. [Cited on p. 158]

McTaggart-Cowan G.P., Rogak S.N., Hill P.G., Bushe W.K., Munshi S.R. (2006) 'The effects of reingested particles on emissions from a heavy-duty direct injection natural gas engine'. Society of Automotive Engineers 2006-01-3411. [Cited on p. 274]

Madhavi Latha M., Highwood E.J. (2006) 'Studies on particulate matter (PM10) and its precursors over urban environment of Reading, UK'. Journal of Quantitative Spectroscopy and Radiative Transfer 101:367–379. [Cited on pp. 23, 202]

Magnussen B.F., Hjertager B.H. (1977) 'On mathematical modeling of turbulent combustion with special emphasis on soot formation and combustion'. 16th Symposium (International) on Combustion, The Combustion Institute, pp. 719–729. [Cited on p. 84]

Maheswaran R., Elliott P. (2003) 'Stroke mortality associated with living near main roads in England and Wales: A geographical study'. Stroke December 2003, pp. 2776–2780. [Cited on p. 373]

Majestic B.J., Schauer J.J., Shafer M.M., Turner J.R., Fine P.M., Singh M., Sioutas C. (2006) 'Development of a wet-chemical method for the speciation of iron in atmospheric aerosols'. Environmental Science and Technology 40:2346–2351. [Cited on p. 231]

Majewski W.A., Pietrasz E. (1992) 'On-vehicle exhaust gas cooling in a diesel emissions control system'. Society of Automotive Engineers 921676. [Cited on p. 274]

van Malderen H., van Grieken R., Bufetov N.V., Koutzenogii K.P. (1996a) 'Chemical characterization of individual aerosol particles in central Siberia'. Environmental Science and Technology 30:312–321. [Cited on p. 41]

van Malderen H., Hoornaert S., van Grieken R. (1996b) 'Identification of individual aerosol particles containing Cr, Pb, and Zn above the North Sea'. Environmental Science and Technology 30:489–498. [Cited on p. 315]

Malhotra V.M., Valimbe P.S., Wright M.A. (2002) 'Effects of fly ash and bottom ash on the frictional behavior of composites'. Fuel 81:235–244. [Cited on p. 35]

Maly R.R. (1994) 'State of the art and future needs in S.I. engine combustion'. Twenty-Fifth Symposium (International) on Combustion, The Combustion Institute, pp. 111–124. [Cited on p. 320]

Mandel B.M., Ineichen B. (1995) 'Investigations of soot particle growth in a diffusion flame considering the temperature dependent optical constant of the particles'. Society of Automotive Engineers 950849. [Cited on p. 182]

Manni M., Florio S., Gommellini C. (1997) 'An investigation on the reduction of lubricating oil impact on diesel exhaust emissions'. Society of Automotive Engineers 972956. [Cited on pp. 279, 282]

Manni M., Pedicillo A., Bazzano F. (2006) 'A study of lubricating oil impact on diesel particulate filters by means of accelerated engine tests'. Society of Automotive Engineers 2006-01-3416. [Cited on p. 233]

Mansouri S.H., Heywood J.B., Radhakrishnan K. (1982) 'Divided-chamber diesel engine, Part I: A cycle-simulation which predicts performance and emissions'. Society of Automotive Engineers 820273. [Cited on pp. 63, 82, 83]

Maricq M.M. (2004) 'Size and charge of soot particles in rich premixed ethylene flames'. Combustion and Flame 137:340–350. [Cited on p. 71]

Maricq M.M. (2005) 'The dynamics of electrically charged soot particles in a premixed ethylene flame'. Combustion and Flame 141:406–416. [Cited on p. 225]

Maricq M.M. (2006) 'A comparison of soot size and charge distributions from ethane, ethylene, acetylene, and benzene/ethylene premixed flames'. Combustion and Flame 144:730–743. [Cited on p. 71]

Maricq M.M., Xu N. (2004) 'The effective density and fractal dimension of soot particles from premixed flames and motor vehicle exhaust'. Journal of Aerosol Science 35:1251–1274. [Cited on pp. 218, 220]

Maricq M.M., Podsiadlik D.H., Chase R.W. (1999a) 'Examination of the size-resolved and transient nature of motor vehicle particulate emissions'. Environmental Science and Technology 33:1618–1626. [Cited on pp. 320, 323, 326, 333, 334]

Maricq M.M., Chase R.E., Podsiadlik D.H., Vogt R. (1999b) 'Vehicle exhaust particle size distributions: A comparison of tailpipe and dilution tunnel measurements'. Society of Automotive Engineers 1999-01-1461. [Cited on p. 147]

Maricq M.M., Podsiadlik D.H., Chase R.E. (1999c) 'Gasoline vehicle particle size distributions: Comparison of steady state, FTP, and US06 measurements'. Environmental Science and Technology 33:2007–2015. [Cited on pp. 320, 321]

Maricq M.M., Podsiadlik D.H., Brehob D.D., Haghgooie M. (1999d) 'Particulate emissions from a direct-injection spark-ignition (DISI) engine'. Society of Automotive Engineers 1999-01-1530. [Cited on pp. 164, 333]

Maricq M.M., Munoz R.H., Yang J., Anderson R.W. (2000a) 'Sooting tendencies in an air-forced direct injection spark-ignition (DISI) engine'. Society of Automotive Engineers 2000-01-0255. [Cited on pp. 333, 334]

Maricq M.M., Podsiadlik D.H., Chase R.E (2000b) 'Size distributions of motor vehicle exhaust PM: A comparison between ELPI and SMPS measurements'. Aerosol Science and Technology 33:239–260. [Cited on pp. 11, 125, 183, 185, 220]

Maricq M.M., Chase R.E., Xu N., Podsiadlik D.H. (2002a) 'The effects of the catalytic converter and fuel sulfur level on motor vehicle particulate matter emissions: Gasoline vehicles'. Environmental Science and Technology 36:276–282. [Cited on p. 326]

Maricq M.M., Chase R.E., Xu N., Laing P.M. (2002b) 'The effects of the catalytic converter and fuel sulfur level on motor vehicle particulate matter emissions: Light duty diesel vehicles'. Environmental Science and Technology 36:283–289. [Cited on pp. 144, 145, 298]

Marjamäki M., Virtanen A., Moisio M., Keskinen J. (1999) 'Modification of electrical low pressure impactor for particles below 30nm'. Journal of Aerosol Science 30:S393–S394. [Cited on p. 185]

Marjamäki M., Ntziachristos L., Virtanen A., Ristimäki J., Keskinen J., Moisio M., Palonen M., Lappi M. (2002) 'Electrical filter stage for the ELPI'. Society of Automotive Engineers 2002-01-0055. [Cited on p. 185]

Mark D. (1998) 'Atmospheric aerosol sampling'. In *Atmospheric Particles*, Harrison R.M., van Grieken R.E. (editors), IUPAC Series on Analytical and Physical Chemistry of Environmental Systems. John Wiley & Sons, Inc., pp. 29–94. [Cited on pp. 10, 26, 37, 39, 40, 165, 182, 183, 358, 375]

Markel V.A., Shalaev V.M. (1999) 'Absorption of light by soot particles in micro-droplets of water'. Journal of Quantitative Spectroscopy and Radiative Transfer 63:321–339. [Cited on p. 42]

Marr L.C., Grogan L.A., Wöhrnschimmel H., Molina L.T., Molina M.J., Smith T.J., Garschick E. (2004) 'Vehicle traffic as a source of particulate polycyclic aromatic hydrocarbon exposure in the Mexico City metropolitan area'. Environmental Science and Technology 38:2584–2592. [Cited on p. 359]

Marshall J.D., Behrentz E. (2005) 'Vehicle self-pollution intake fraction: Children's exposure to school bus emissions'. Environmental Science and Technology 39:2559–2563. [Cited on p. 360]

Marshall J.D., Behrentz E. (2006) 'Response to comment on 'vehicle self-pollution intake fraction: Children's exposure to school bus emissions'. Environmental Science and Technology 40:3124–3125. [Cited on p. 359]

Martinot S., Beard P., Roesler J., Garo A. (2001) 'Comparison and coupling of homogeneous reactor and flamelet library soot modeling approaches for diesel combustion'. Society of Automotive Engineers 2001-01-3684. [Cited on pp. 68, 72, 75]

Masjuki H.H., Kalam M.A., Maleque M.A. (2000) 'Combustion characteristics of biological fuel in diesel engine'. Society of Automotive Engineers 2000-01-0689. [Cited on p. 260]

Maskos Z., Khachatryan L., Dellinger B. (2005) 'Precursors of radicals in tobacco smoke and the roles of particulate matter in forming and stabilizing radicals'. Energy and Fuels 19:2466–2473. [Cited on p. 368]

Matheaus A.C., Rynan T.W., Daly D., Langer D.A., Musculus M.P.B. (2002) 'Effects of PuriNOx water – diesel fuel emulsions on emissions and fuel economy in a heavy-duty diesel engine'. Society of Automotive Engineers 2002-01-2891. [Cited on p. 251]

Mather D.K., Reitz R.D. (1998) 'Modeling the influence of fuel injection parameters on diesel engine emissions'. Society of Automotive Engineers 980789. [Cited on p. 91]

Mathis U., Mohr M., Zenobi R. (2004) 'Effect of organic compounds on nanoparticle formation in diluted diesel exhaust'. Atmospheric Chemistry and Physics 4:609–620. [Cited on p. 128]

Mathis U., Mohr M., Kaegi R., Bertola A., Boulouchos K. (2005) 'Influence of diesel engine combustion parameters on primary soot particle diameter'. Environmental Science and Technology 39:1887–1892. [Cited on pp. 215, 216, 251]

Matsui Y., Kamimoto T., Matsuoka S. (1982) 'Formation and oxidation processes of soot particulates in a D.I. diesel engine – an experimental study via the two-color method'. Society of Automotive Engineers 820464. [Cited on pp. 85, 203]

Matsumoto Y., Sakai S., Kato T., Nakajima T., Satoh H. (1998) 'Long-term trends of particulate mutagenic activity in the atmosphere of Sapporo. 1. Determination of mutagenic activity by the conventional tester strains TA98 and TA100 during an 18-year period (1974–1992)'. Environmental Science and Technology. 32:2665–2671. [Cited on p. 112]

Matsura Y., Nakazawa N., Kobayashi Y., Ogita H., Kawatani T. (1992) 'Effects of various methods for improving vehicle startability and transient response of turbocharged diesel trucks'. Society of Automotive Engineers 920044. [Cited on p. 276]

Matter D., Mohr M., Fendel W., Schmidt-Ott A., Burtscher H. (1995) 'Multiple wavelength aerosol photomission by excimer lamps'. Journal of Aerosol Science 26:1101–1115. [Cited on p. 194]

Matter U., Siegmann H.C., Burtscher H. (1999) 'Dynamic field measurements of submicron particles from diesel engines'. Environmental Science and Technology 33:1946–1952. [Cited on pp. 165, 194, 195]

Matthias-Maser S. (1998) 'Primary biological aerosol particles: Their significance, sources, sampling methods and size distribution in the atmosphere'. In *Atmospheric Particles*, Harrison R.M., van Grieken R.E. (editors), IUPAC Series on Analytical and Physical Chemistry of Environmental Systems. John Wiley & Sons, Inc., pp. 349–368. [Cited on pp. 27, 28, 29]

Mauderly J.L., Banas D.A., Griffith W.C., Hahn F.F., Henderson R.F., McClellan R.O. (1996) 'Diesel exhaust is not a pulmonary carcinogen in CD-1 mice exposed under conditions carcinogenic to F344 rats'. Fundamental and Applied Toxicology 30:233–242. [Cited on p. 378]

Mayer A., Buck A. (1992) 'Knitted ceramic fibers – a new concept for particulate traps'. Society of Automotive Engineers 920146. [Cited on p. 394]

Mayer A., Buck A., Bressler H. (1993) 'The knitted particulate trap: Field experience and development progress'. Society of Automotive Engineers 930362. [Cited on p. 237]

Mayer A., Czerwinski J., Scheidegger W., Kieser D., Bigga E., Wyser M. (1997) 'VERT – clean diesel engines for tunnel construction'. Society of Automotive Engineers 970478. [Cited on p. 95]

Mayer A., Matter U., Scheidegger G., Czerwinski J., Wyser M., Kieser D., Weidhofer J. (1998) 'VERT: Diesel nano-particulate emissions: Properties and reduction strategies'. Society of Automotive Engineers 980539. [Cited on pp. 250, 304, 305, 355, 391]

Mayer A., Matter U., Scheidegger G., Czerwinski J., Wyser M., Kieser D., Weidhofer J. (1999) 'Particulate traps for retro-fitting construction site engines VERT: Final measurements and implementation'. Society of Automotive Engineers 1999-01-0116. [Cited on p. 306]

Mayer A., Czerwinski J., Legerer F., Wyser M. (2002) 'VERT particulate trap verification'. Society of Automotive Engineers 2002-01-0435. [Cited on p. 301]

Mayer A., Heeb N., Czerwinski J., Wyser M. (2003) 'Secondary emissions from catalytic active particle filter systems'. Society of Automotive Engineers 2003-01-0291. [Cited on pp. 305, 306]

Mayer W.J., Lechman D.C., Hilden D.L. (1980) 'The contribution of engine oil to diesel exhaust particulate emissions'. Society of Automotive Engineers 800256. [Cited on pp. 282, 283]

Maynard R.L (1999) 'Introduction'. In *Particulate Matter: Properties and Effects upon Health*, Maynard R.L., Howard C.V. (editors). Bios Scientific Publishers, pp. 1–17. [Cited on p. 356]

Maynard R.L. (2000) 'New directions: Reducing the toxicity of vehicle exhaust'. Atmospheric Environment 34:2667–2668. [Cited on pp. 40, 375]

Maynard R.L. (2001) 'Particulate air pollution'. In *The Urban Atmosphere and Its Effects*, Brimblecombe P., Maynard R.L. (editors), Air Pollution Reviews, Vol. 1. Imperial College Press, pp. 163–194. [Cited on pp. 2, 3, 365, 372, 393]

Maynard R.L., Cameron K.M. (2001) 'Air pollution policy in the European Commission'. In *The Urban Atmosphere and Its Effects*, Brimblecombe P., Maynard R.L. (editors), Air Pollution Reviews, Vol. 1. Imperial College Press, pp. 273–300. [Cited on pp. 356, 396]

Medina S., Plasencia A., Ballester F., Mücke H.G., Schwartz J. (2004) 'Apheis: Public health impact of PM10 in 19 European cities'. Journal of Epidemiology and Community Health 58:831–836. [Cited on p. 375]

Mehta D., Alger T., Hall M., Matthews R., Ng H. (2001) 'Particulate characterization of a DISI research engine using a nephelometer and in-cylinder visualization'. Society of Automotive Engineers 2001-01-1976. [Cited on pp. 191, 328, 333]

Mehta P.S., Gupta A.K., Gupta C.P. (1988) 'Model for prediction of incylinder and exhaust soot emissions from direct injection diesel engines'. Society of Automotive Engineers 881251. [Cited on p. 67]

Meng Q.Y., Turpin B.J., Polidori A., Lee J.H., Weisel C., Morandi M., Colombe S., Stock T., Winer A., Zhang J. (2005) '$PM_{2.5}$ of ambient origin: Estimates and exposure errors relevant to PM epidemiology'. Environmental Science and Technology 39:5105–5112. [Cited on p. 371]

Mensink C., de Vlieger I. (1998) 'Regional-scale modelling of acidification associated with road transport and road transport systems'. International Journal of Vehicle Design 20:335–343. [Cited on p. 36]

Merola S.S., Vaglieco B.M., Tornatore C. (2005) 'Characterization of nanoparticles at the exhaust of a common rail diesel engine by optical techniques and conventional method'. Society of Automotive Engineers 2005-01-2155. [Cited on p. 191]

Merrion D.F. (2003) 'Heavy duty diesel emission regulations – past, present, and future'. Society of Automotive Engineers 2003-01-0040. [Cited on p. 6]

Merritt P., Huang Y., Khair M., Pan J. (2006) 'Unregulated exhaust emissions from alternate diesel combustion modes'. Society of Automotive Engineers 2006-01-3307. [Cited on p. 293]

Messerer A., Pöschl U., Niessner R., Rother D. (2006a) 'Soot particle deposition efficiency of diesel PM-catalyst structures – the influence of structure geometry and transient temperature inhomogeneities'. Society of Automotive Engineers 2006-01-3288. [Cited on p. 394]

Messerer A., Niessner R., Pöschl U. (2006b) 'Comprehensive kinetic characterization of the oxidation and gasification of model and real diesel soot by nitrogen oxides and oxygen under exhaust conditions: Measurement, Langmuir–Hinshelwood, and Arrhenius parameters'. Carbon 44:307–324. [Cited on pp. 168, 169]

Miguel A.G., Cass G.R., Glovsky M.M., Weiss J. (1999) 'Allergens in paved road dust and airborne particles'. Environmental Science and Technology 33:4159–4168. [Cited on p. 364]

Mihara K., Inoue H. (1995) 'Effect of piston top ring design on oil consumption'. Society of Automotive Engineers 950937. [Cited on p. 279]

Mikkanen P., Moisio M., Keskinen J., Ristimäki M., Marjamäki M. (2001) 'Sampling method for particle measurements of vehicle exhaust'. Society of Automotive Engineers 2001-01-0219. [Cited on pp. 164, 165]

Miller A., Ahlstrand G., Kittelson D., Zachariah M. (2007) 'The fate of metal (Fe) during diesel combustion: Morphology, chemistry, and formation pathways of nanoparticles'. Combustion and Flame 149:129–143. [Cited on pp. 101, 103, 104]

Mischler S.E., Volkwein J.C. (2005) 'Differential pressure as a measure of particulate matter emissions from diesel engines'. Environmental Science and Technology 39:2255–2261. [Cited on p. 212]

Mitchell D.L., Pinson J.A., Litzinger T.A. (1993) 'The effects of simulated EGR via intake air dilution on combustion in an optically accessible DI diesel engine'. Society of Automotive Engineers 932798. [Cited on p. 273]

Miyabara Y., Hashimoto S., Sagai M., Morita M. (1999) 'PCDDs and PCDFs in vehicle exhaust particles in Japan'. Chemosphere 39:143–150. [Cited on p. 231]

Miyahara M., Watanabe Y., Naitoh Y., Hosonuma K., Tamura K. (1991) 'Investigation into extending diesel engine oil drain interval (Part 1) – oil drain interval extension by increasing efficiency of filtering soot in lubricating oil'. Society of Automotive Engineers 912339. [Cited on p. 288]

Miyamoto N., Ogawa H., Goto N., Sasaki H. (1990) 'Analysis of diesel soot formation under varied ignition lag with a laser light extinction method'. Society of Automotive Engineers 900640. [Cited on p. 86]

Miyamoto N., Ogawa H., Nurun N.M., Obata K., Arima T. (1998) 'Smokeless, low NOx, high thermal efficiency, and low noise diesel combustion with oxygenated agents as main fuel'. Society of Automotive Engineers 980506. [Cited on p. 258]

Mogan J.P., Dainty E.D., Vergeer H.C., Lawson A., Westaway K.C., Weglo J.K., Thomas L.R. (1985) 'Investigation of the CTO emission control system applied to heavy-duty diesel engines used in underground mining equipment'. Society of Automotive Engineers 850151. [Cited on pp. 212, 238]

Mohr M., Burtscher H. (1997) 'Photoelectric aerosol charging at high particle concentrations'. Journal of Aerosol Science 28:613–621. [Cited on p. 193]

Mohr M., Forss A.-M., Steffen D. (2000) 'Particulate emissions of gasoline vehicles and influence of the sampling procedure'. Society of Automotive Engineers 2000-01-1137. [Cited on pp. 323, 324, 331]

Mohr M., Lehmann U., Margaria G. (2003a) 'ACEA program on the emissions of fine particulates from passenger cars (2) – Part 1: Particle characterisation of a wide range of engine technologies'. Society of Automotive Engineers 2003-01-1889. [Cited on pp. 243, 304, 331]

Mohr M., Lehmann U., Margaria G. (2003b) 'ACEA program on the emissions of fine particulates from passenger cars (2) – Part 2: Effect of sampling conditions and fuel sulphur content on the particle emission'. Society of Automotive Engineers 2003-01-1890. [Cited on pp. 283, 284, 286]

Mohr M., Lehmann U., Rütter J. (2005) 'Comparison of mass-based and non-mass-based particle measurement systems for ultra-low emissions from automotive sources'. Environmental Science and Technology 39:2229–2238. [Cited on p. 203]

Mohr M., Forss A.-M., Lehmann U. (2006) 'Particle emissions from diesel passenger cars equipped with a particle trap in comparison to other technologies'. Environmental Science and Technology 40:2375–2383. [Cited on p. 313]

Monn C. (2001) 'Exposure assessment of air pollutants: A review on spatial heterogeneity and indoor/outdoor/personal exposure to suspended particulate matter, nitrogen dioxide and ozone'. Atmospheric Environment 35:1–32. [Cited on pp. 358, 362, 371]

Monn C., Carabias V., Junker M. (1997) 'Small-scale spatial variability of particulate matter $\leq 10\mu$m (PM10) and nitrogen dioxide'. Atmospheric Environment 31:2243–2247. [Cited on p. 371]

Montajir R., Kusaka T., Kaori I., Kihara N., Asano I., Adachi M., Wei Q. (2007) 'Soot emission behavior from diverse vehicles and catalytic technologies measured by a solid particle counting system'. Society of Automotive Engineers 2007-01-0317. [Cited on pp. 301, 303, 389]

del Monte M., Rossi P. (2000) 'Calcium in the urban atmosphere'. In *Aerosol Chemical Processes in the Environment*, Spurny K.R., Hochrainer D. (editors). CRC Press LLC, pp. 347–364. [Cited on p. 32]

Montgomery D.T., Reitz R.D. (2001) 'Effects of multiple injections and flexible control of boost and EGR on emissions and fuel consumption of a heavy-duty diesel engine'. Society of Automotive Engineers 2001-01-0195. [Cited on p. 269]

Montgomery D.T., Chan M., Chang C.T., Farrell P.V., Reitz R.D. (1996) 'Effect of injector nozzle hole size and number on spray characteristics and the performance of a heavy duty D.I. diesel engine'. Society of Automotive Engineers 962002. [Cited on p. 266]

Montierth M.R. (1984) 'Fuel additive effect upon diesel particulate filters'. Society of Automotive Engineers 840072. [Cited on p. 233]

Moosmüller H., Arnott W.P., Rogers C.F., Bowen J.L., Gillies J.A., Pierson W.R., Collins J.F., Durbin W.R., Norbeck J.M. (2001) 'Time resolved characterization of diesel particulate emissions. 1. Instruments for particle mass measurements'. Environmental Science and Technology 35:781–787. [Cited on p. 203]

Moosmüller H., Mazzoleni C., Barber P.W., Kuhns H.D., Keislar R.E., Watson J.G. (2003) 'On-road measurement of automotive particle emissions by ultraviolet lidar and transmissometer: instrument'. Environmental Science and Technology 37:4971–4978. [Cited on p. 388]

Morawska L., Johnson G.R., He C., Ayoko G.A., Lim C.H., Swanson C., Ristovski Z.D., Moore M. (2006) 'Particle number emissions and source signatures of an industrial facility'. Environmental Science and Technology 40:803–814. [Cited on p. 357]

Morawska L., Ristovski Z.D., Johnson G.R., Jayaratne E.R., Mengersen K. (2007) 'Novel method for on-road emission factor measurements using a plume capture trailer'. Environmental Science and Technology 41:574–579. [Cited on p. 143]

Moreno T., Querol X., Alastuey A., Minguillón M.C., Pey J., Rodriguez S., Miró J.V., Felis C., Gibbons W. (2007) 'Recreational atmospheric pollution episodes: Inhalable metalliferous particles from firework displays'. Atmospheric Environment 41:913–922. [Cited on p. 31]

Morgan W.K.C., Reger R.B., Tucker D.M. (1997) 'Health effects of diesel emissions'. Annals of Occupational Hygiene 21:643–658. [Cited on p. 372]

Mori K., Kamikubo H., Kawatani T., Obara T., Fukano I., Sugawara K. (1990) 'Technology for meeting the 1991 U.S.A. exhaust emission regulations on heavy duty diesel engine'. Society of Automotive Engineers 902233. [Cited on pp. 279, 281]

Morimune T., Yamaguchi H., Yasukawa Y. (1998) 'Study of catalytic reduction of NO_x in exhaust gas from a diesel engine'. Experimental Thermal and Fluid Science 18:220–230. [Cited on p. 299]

Morin J.P., Leprieur E., Dionnet F., Robin L. (1999) 'The influence of a particulate trap on the in vitro lung toxicity response to continuous exposure to diesel exhaust emissions'. Society of Automotive Engineers 1999-01-2710. [Cited on p. 376]

Morin J.-P., Rumigny J.F., Bion A., Dionnet F. (2002) 'Isoflavones protect against diesel engine exhaust injury in organotypic culture of lung tissue'. Environmental Toxicology and Pharmacology 12:213–220. [Cited on pp. 376, 377]

Moriyoshi A., Takano S., Ono M., Ogasawara M., Tabata M., Miyamoto N., Ohta S. (2002) 'Analysis of contribution to SPM by organic matters using high-performance liquid chromatography (HPLC)'. Society of Automotive Engineers 2002-01-0653. [Cited on pp. 344, 347]

Mosleh M., Blau P.J., Dumitrescu D. (2004) 'Characteristics and morphology of wear particles from laboratory testing of disk brake materials'. Wear 256:1128–1134. [Cited on pp. 344, 345, 346]

Moss J.B., Stewart C.D., Young K.J. (1995) 'Modeling soot formation and burnout in a high temperature laminar diffusion flame burning under oxygen-enriched conditions'. Combustion and Flame 101:491–500. [Cited on p. 79]

Mueller C.J., Pitz W.J., Pickett L.M., Martin G.C., Sibers D.L., Westbrook C.K. (2003) 'Effects of oxygenates on soot processes in DI diesel engines: Experiments and numerical simulations'. Society of Automotive Engineers 2003-01-1791. [Cited on p. 84]

Mueller V., Christmann R., Muenz S., Gheorghiu V. (2005) 'System structure and controller concept for an advanced turbocharger/EGR system for a turbocharged passenger car diesel engine'. Society of Automotive Engineers 2005-01-3888. [Cited on pp. 274, 276]

Mukerjee S., Somerville M.C., Willis R.D., Fox D.L., Stevens R.K., Kellogg R.B., Stiles D.C., Lumpkin T.A., Shy C.M. (1996) 'Integrated assessment of reduced emission impacts from a biomedical waste incinerator. Atmospheric characterization and modeling applications on particulate matter and acid gases'. Environmental Science and Technology 30:1680–1686. [Cited on p. 35]

Mulawa P.A., Cadle S.H., Knapp K., Zweidinger R., Snow R., Lucas R., Goldbach J. (1997) 'Effect of ambient temperature and E-10 fuel on primary exhaust particulate matter emissions from light-duty vehicles'. Environmental Science and Technology 31:1302–1307. [Cited on p. 325]

Müller J.-O., Su D.S., Jenthoft R.E., Kröhnert J., Jenthoft F.C., Schlögl R. (2005) 'Morphology-controlled reactivity of carbonaceous materials towards oxidation'. Catalysis Today 102–103:259–265. [Cited on pp. 71, 215]

Müller J.-O., Su D.S., Jentoft R.E., Wild U., Schlögl R. (2006) 'Diesel engine exhaust emission: Oxidative behavior and microstructure of black smoke soot particulate'. Environmental Science and Technology 40:1231–1236. [Cited on pp. 214, 234]

Munro R. (1990) 'Emissions impossible – the piston & ring support system'. Society of Automotive Engineers 900590. [Cited on p. 279]

Muntean G.G. (1999) 'A theoretical model for the correlation of smoke number to dry particulate concentration in diesel exhaust'. Society of Automotive Engineers 1999-01-0515. [Cited on p. 203]

Murphy M.J., Hillenbrand L.J., Trayser D.A., Wasser J.H. (1981) 'Assessment of diesel particulate control – direct and catalytic oxidation'. Society of Automotive Engineers 810112. [Cited on pp. 80, 301]

Murtagh M.J., Sherwood D.L., Socha L.S. (1994) 'Development of a diesel particulate filter composition and its effect on thermal durability and filtration performance'. Society of Automotive Engineers 940235. [Cited on p. 150]

Musculus M.P., Dec J.E., Tree D.R. (2002a) 'Effects of fuel parameters and diffusion flame lift-off on soot formation in a heavy-duty DI diesel engine'. Society of Automotive Engineers 2002-01-0889. [Cited on pp. 84, 94, 258, 259]

Musculus M.P.B., Dec J.E., Tree D.R., Daly D., Langer D., Ryan T.W., Matheaus A.C. (2002b) 'Effects of water – fuel emulsions on spray and combustion processes in a heavy-duty DI diesel engine'. Society of Automotive Engineers 2002-01-2892. [Cited on pp. 84, 251]

Myers J.P. (1983) 'Timed sampling in the exhaust of a direct-injection diesel engine'. Society of Automotive Engineers 831710. [Cited on pp. 136, 266]

N

Nag P., Song K.-H., Litzinger T.A., Haworth D.C. (2001) 'A chemical kinetic modelling study of the mechanism of soot reduction by oxygenated additives in diesel engines'. International Journal of Engine Research 2:163–175. [Cited on p. 259]

Nagai I., Endo H., Nakamura H., Yano H. (1983) 'Soot and valve train wear in passenger car diesel engines'. Society of Automotive Engineers 831757. [Cited on pp. 290, 291]

Nagano S., Kawazoe H., Ohsawa K. (1991) 'Reduction of soot emission by air-jet turbulence in a DI diesel engine'. Society of Automotive Engineers 912353. [Cited on p. 91]

Nagase K., Funatsu K., Muramatsu Y., Kawakami M. (1985) 'An investigation of combustion in internal combustion engines by means of optical fibres'. Society of Automotive Engineers 851560. [Cited on p. 94]

Nagle J., Strickland-Constable R.F. (1962) 'Oxidation of carbon between 1000–2000°C'. Proceedings of the Fifth Conference on Carbon 1–2:154–164. Pergamon Press. [Cited on pp. 82, 83]

Nakakita K., Nagaoka M., Fujikawa T., Ohsawa K., Yamaguchi S. (1990) 'Photographic and three dimensional numerical studies of diesel soot formation process'. Society of Automotive Engineers 902081. [Cited on pp. 67, 84]

Nakane T., Ikeda M., Hori M., Bailey O., Mussman L. (2005) 'Investigation of the aging behavior of oxidation catalysts developed for active DPF regeneration systems'. Society of Automotive Engineers 2005-01-1759. [Cited on p. 296]

Nakayama S., Fukuma T, Matsunaga A., Miyake T., Wakimoto T. (2003) 'A new dynamic combustion control method based on charge oxygen concentration for diesel engines'. Society of Automotive Engineers 2003-01-3181. [Cited on p. 277]

Narusawa K., Hori S., Sato T., Abe T. (1995) 'The evaluation of oxidation catalysts for diesel trucks'. Society of Automotive Engineers 950157. [Cited on pp. 168, 173, 294]

Natarajan M., Frame E.A., Naegeli D.W., Asmus T., Clark W., Garbak J., Gonzalez M.A., Liney E., Piel W., Wallace J.P. (2001) 'Oxygenates for advanced petroleum-based diesel fuels: Part 1. Screening and selection methodology for the oxygenates'. Society of Automotive Engineers 2001-01-3631. [Cited on p. 257]

Nazha M.A.A., Rajakaruna H., Crookes R.J. (1998) 'Soot and gaseous species formation in a water-in-liquid fuel emulsion spray – a mathematical approach'. Energy Conversion and Management 39:1981–1989. [Cited on p. 251]

Neeft J.P.A., Makkee M., Moulijn J.A. (1996a) 'Diesel particulate emission control'. Fuel Processing Technology 47:1–69. [Cited on p. 295]

Neeft J.P.A., Hoornaert F., Makkee M., Moulijn J.A. (1996b) 'The effects of heat and mass transfer in thermogravimetrical analysis. A case study towards the catalytic oxidation of soot'. Thermochimica Acta 287:261–278. [Cited on p. 169]

Neer A., Koylu U.O. (2006) 'Effect of operating conditions on the size, morphology, and concentration of submicrometer particulates emitted from a diesel engine'. Combustion and Flame 146:142–154. [Cited on pp. 216, 218]

Nehmer D.A., Reitz R.D. (1994) 'Measurement of the effect of injection rate and split injections on diesel engine soot and NOx emissions'. Society of Automotive Engineers 940668. [Cited on p. 269]

Nelson A.W. (1974) 'Development of a reduced smoke combustor for the JT3D engine'. Society of Automotive Engineers 740484. [Cited on p. 4]

Nemmar A., Nemery B., Hoylaerts M.F., Vermylen J. (2002a) 'Air pollution and thrombosis: An experimental approach'. Pathophysiology of Haemostasis and Thrombosis 32:349–350. [Cited on p. 379]

Nemmar A., Hoet P.H.M., Vanquickenborne B., Dinsdale D., Thomeer M., Hoylaerts M.F., Vanbilloen H., Mortelmans L., Nemery B. (2002b) 'Passage of inhaled particles into the blood circulation in humans'. Circulation 105:411–414. [Cited on pp. 381]

Nemmar A., Nemery B., Hoet P.H.M., Vermylen J., Hoylaerts M.F. (2003a) 'Pulmonary inflammation and thrombogenicity caused by diesel particles in hamsters – role of histamine'. American Journal of Respiratory and Critical Care Medicine 168:1366–1372. [Cited on p. 379]

Nemmar A., Hoet P.H.M., Dinsdale D., Vermylen J., Hoylaerts M.F., Nemery B. (2003b) 'Diesel exhaust particles in lung acutely enhance experimental thrombosis'. Circulation 107:1202–1208. [Cited on p. 377]

Nemoto S., Kishi Y., Matsuura K., Miura M., Togawa S., Ishikawa T., Hashomoto T., Yamazaki T. (2004) 'Impact of oil-derived ash on continuous regeneration-type diesel particulate filter – JCAP Oil WG report'. Society of Automotive Engineers 2004-01-1887. [Cited on pp. 233, 283, 284]

Neukom H.-P., Grob K., Biedermann M., Noti A. (2002) 'Food contamination by C_{20}–C_{50} mineral paraffins from the atmosphere'. Atmospheric Environment 36:4839–4847. [Cited on pp. 40, 344]

Neyestanaki A.K., Klingstedt F., Salmi T., Murzin D.Y. (2004) 'Deactivation of postcombustion catalysts, a review'. Fuel 83:395–408. [Cited on pp. 95, 295, 296]

Ng H., Biruduganti M., Stork K. (2005) 'Comparing the performance of SunDiesel and conventional diesel in a light-duty vehicle and heavy-duty engine'. Society of Automotive Engineers 2005-01-3776. [Cited on p. 256]

Ni T., Gupta S.B., Santoro R.J. (1994) 'Suppression of soot formation in ethene laminar diffusion flames by chemical additives'. Twenty-Fifth Symposium (International) on Combustion The Combustion Institute, pp. 585–592. [Cited on p. 123]

Nichol J. (1997) 'Bioclimatic impacts of the 1994 smoke haze event in southeast Asia'. Atmospheric Environment 31:1209–1219. [Cited on p. 33]

Nielsen T. (1996) 'Traffic contribution of polycyclic aromatic hydrocarbons in the center of a large city'. Atmospheric Environment 30:3481–3490. [Cited on p. 35]

Nightingale J.A., Maggs R., Cullinan P., Donnelly L.E., Rogers D.F., Kinnersley R., Fan Chung K., Barnes P.J., Ashmore M., Newman-Taylor A. (2000) 'Airway inflammation after controlled exposure to diesel exhaust particulates'. American Journal of Respiratory Critical Care Medicine 162:161–166. [Cited on p. 382]

Nigge K.-M. (1998) 'A method for the site-dependent life cycle impact assessment of toxic air pollutants from traffic emissions'. Society of Automotive Engineers 982181. [Cited on p. 143]

Nikolic D., Wakimoto K., Takahashi S., Iida N. (2001) 'Effect of nozzle diameter and EGR ratio on the flame temperature and soot formation for various fuels'. Society of Automotive Engineers 2001-01-1939. [Cited on pp. 266, 278]

Nikula K.J., Avila K.J., Griffith W.C., Mauderly J.L. (1997) 'Lung tissue responses and sites of particle retention differ between rats and cynomolgus monkeys exposed chronically to diesel exhaust and coal dust'. Fundamental and Applied Technology 37:37–53. [Cited on p. 377]

Nikula K.J., Finch G.L., Westhouse R.A., Seagrave J., Mauderly J.L., Lawson D.R., Gurevich M. (1999) 'Progress in understanding the toxicity of gasoline and diesel engine exhaust emissions'. Society of Automotive Engineers 1999-01-2250. [Cited on p. 376]

Ning M., Yuan-Xian Z., Zhen-Huan S., Guo-dong H. (1991) 'Soot formation, oxidation and its mechanism in different combustion systems and smoke emission pattern in DI diesel engines'. Society of Automotive Engineers 910230. [Cited on pp. 85, 214]

Nishida K., Hiroyasu H. (1989) 'Simplified three-dimensional modelling of mixture formation and combustion in a D.I. diesel engine'. Society of Automotive Engineers 890269. [Cited on pp. 66, 81]

Nishiumi R., Yasuda A., Tsukasaki Y., Tanaka T. (2004) 'Effects of cetane number and distillation characteristics of paraffinic diesel fuels on PM emission from a DI diesel engine'. Society of Automotive Engineers 2004-01-2960. [Cited on p. 251]

Noble C.A., Prather K.A. (1999) 'Real-time measurement of correlated size and composition profiles of individual atmospheric aerosol particles'. Environmental Science and Technology 30:2667–2680. [Cited on p. 209]

Nogi T., Shiraishi T., Nakayama Y., Ohsuga M., Kurihara N. (1998) 'Stability improvement of direct fuel injection engine under lean combustion operation'. Society of Automotive Engineers 982703. [Cited on p. 333]

Nolte C.G., Schauer J.J., Cass G.R., Simoneit B.R.T. (1999) 'Highly polar organic compounds present in meat smoke'. Environmental Science and Technology 33:3313–3316. [Cited on p. 33]

Noppel M. (1998) 'Binary nucleation of water – sulfuric acid system: A reexamination of the classical hydrates interaction model'. Journal of Chemical Physics 109:9052–9056. [Cited on p. 126]

Nord A.G. (2000) 'Aerosol particles deposited on building stone'. In *Aerosol Chemical Processes in the Environment*, Spurny K.R., Hochrainer D. (editors) CRC Press LLC, pp. 297–308. [Cited on pp. 38, 344, 345]

Nord K., Haupt D., Ahlvik P., Egebäck K.-E. (2004) 'Particulate emissions from and ethanol fueled heavy-duty diesel engine equipped with EGR, catalyst and DPF'. Society of Automotive Engineers 2004-01-1987. [Cited on pp. 259, 260]

Nord K.E., Haupt D. (2005) 'Reducing the emission of particles from a diesel engine by adding an oxygenate to the fuel'. Environmental Science and Technology 39:6260–6265. [Cited on p. 260]

Norris-Jones S.R., Hollis T., Waterhouse C.N.F. (1984) 'A study of the formation of particulates in the cylinder of a direct injection diesel engine'. Society of Automotive Engineers 840419. [Cited on p. 85]

Ntziachristos L., Samaras Z. (2000) 'Sampling conditions effects on real-time particle measurements from a light duty vehicle'. Society of Automotive Engineers 2000-01-2049. [Cited on pp. 159, 219]

Ntziachristos L., Giechaskiel B., Pistikopoulos P., Samaras Z., Mathis U., Mohr M., Ristimäki J., Keskinen J., Mikkanen P., Casati R., Scheer V., Vogt R. (2004a) 'Performance evaluation of a novel sampling and measurement system for exhaust particle characterization'. Society of Automotive Engineers 2004-01-1439. [Cited on pp. 143, 144, 160, 164]

Ntziachristos L., Mamakos A., Samaras Z., Mathis U., Mohr M., Thompson R., Stradling R., Forti L., de Serves C. (2004b) 'Overview of the European "Particulates" project on the characterization of exhaust particulate emissions from road vehicles: Results for light-duty vehicles'. Society of Automotive Engineers 2004-01-1985. [Cited on pp. 222, 241, 243, 323, 331]

Ntziachristos L., Giechaskiel B., Pistikopoulos P., Samaras Z. (2005) 'Comparative assessment of two different sampling systems for particle emission type-approval measurements'. Society of Automotive Engineers 2005-01-0198. [Cited on p. 389]

Nussear D.L., Gautam M., Hong-Guang G., Clark N., Wallace W.E. (1992) 'Respirable particulate genotoxicant distribution in diesel exhaust and mine atmospheres'. Society of Automotive Engineers 921752. [Cited on p. 180]

Nyeki S., Colbeck I. (2000) 'The influence of morphological restructuring of carbonaceous aerosol on microphysical atmospheric processes'. In *Aerosol Chemical Processes in the Environment*, Spurny K.R., Hochrainer D. (editors), CRC Press LLC, pp. 505–523. [Cited on pp. 12, 33, 189]

O

Oakes D.J., Pollack J.K. (2000) 'The in vitro evaluation of the toxicities of three related herbicide formulations containing ester derivatives of 2,4,5-T and 2,4-D using sub-mitochondrial particles'. Toxicology 151:1–9. [Cited on p. 182]

Oberdorster G., Ferin J., Finkelstein J., Baggs R., Stavert D.M., Lehnert B.E. (1992) 'Potential health hazards from thermal degradation events: Particulate vs. gas phase effects'. Society of Automotive Engineers 921388. [Cited on p. 381]

Oberdörster G. (1995) 'Effects of ultrafine particles in the lung and potential relevance to environmental particles'. In *Proceedings of the 1995 Warsaw Particle Workshop*, Delft University Press. [Cited on p. 366]

Oberdörster G., Sharp Z., Atudorei V., Elder A., Gelein R., Kreyling W., Cox C. (2004) 'Translocation of inhaled ultrafine particles to the brain'. Inhalation Toxicology 16:437–445. [Cited on p. 368]

Obiols J., China P., Sibue L., Delvigne T., Deconninck B., Courtois O., Carlier P., Beziat J.-C., Da-Dilva M. (2005) 'An innovative on-line measurement method for studying the impact of lubricant formulations on poisoning and clogging of after-treatment devices'. Society of Automotive Engineers 2005-01-2178. [Cited on p. 303]

O'Byrne P.M., Postma D.S. (1999) 'The many faces of airway inflammation – asthma and chronic obstructive pulmonary disease'. American Journal of Respiratory and Critical Care Medicine 159:S41–S66. [Cited on p. 370]

Odaka M., Goto Y., Tsukamoto Y., Sekiya M., Yoshimura Y., Ikeda T. (2001) 'A new type partial flow dilution tunnel with geometrical partitioning for diesel particulate measurement'. Society of Automotive Engineers 2001-01-3579. [Cited on p. 160]

Ogawa T. (2005) 'Analytical conditions for field ionization mass spectrometry of diesel fuel'. Fuel 84:2015–2025. [Cited on p. 254]

Oguma M., Goto S. (2007) 'Evaluation of medium duty DME truck performance – field test results and PM characteristics'. Society of Automotive Engineers 2007-01-0032. [Cited on p. 260]

Oh K.C., Shin H.D. (2006) 'The effect of oxygen and carbon dioxide concentration on soot formation in non-premixed flames'. Fuel 85:615–624. [Cited on p. 187]

Oh K.C., Lee U.D., Shin H.D., Lee E.J. (2005) 'The evolution of incipient soot particles in an inverse diffusion flame of ethene'. Combustion and Flame 140:249–254. [Cited on pp. 106, 107]

Okada S., Kweon C.-B., Stetter J.C., Foster D.E., Shafer M.M., Christensen C.G., Schauer J.J., Schmidt A.M., Silverberg A.M., Gross D.S. (2003) 'Measurement of trace metal composition in diesel engine particulate and its potential for determining oil consumption: ICPMA (inductively coupled plasma mass spectrometer) and ATOFMS (aerosol time of flight mass spectrometer) measurements'. Society of Automotive Engineers 2003-01-0076. [Cited on p. 204]

Okamoto R.A., Kado N.Y., Kuzmicky P.A., Ayala A., Kobayashi R. (2006) 'Unregulated emissions from compressed natural gas (CNG) transit buses configured with and without oxidation catalyst'. Environmental Science and Technology 40:332–341. [Cited on p. 325]

Okona-Mensah K.B., Battershill J., Boobis A., Fielder R. (2005) 'An approach to investigating the importance of high potency polycyclic aromatic hydrocarbons (PAHs) in the induction of lung cancer by air pollution'. Food and Chemical Toxicology 43:1103–1116. [Cited on pp. 240, 369]

Okrent D.A. (1998) 'Optimization of a third generation TEOM monitor for measuring diesel particulate in real-time'. Society of Automotive Engineers 980409. [Cited on p. 199]

Öktem B., Tolocka M.P., Zhao B., Wang H., Johnston M.V. (2005) 'Chemical species associated with the early stage of soot growth in a laminar premixed ethylene – oxygen – argon flame'. Combustion and Flame 142:364–373. [Cited on pp. 73, 74, 106, 182]

Olivares G., Johansson C., Ström J., Hansson H.-C. (2007) 'The role of ambient temperature for particle number concentrations in a street canyon'. Atmospheric Environment 41:2145–2155. [Cited on p. 361]

Olson D.A., Burke J.M. (2006) 'Distributions of PM2.5 source strengths for cooking from the Research Triangle Park particulate matter panel study'. Environmental Science and Technology 40:163–169. [Cited on p. 359]

Opris C.N., Gratz L.D., Bagley S.T., Baumgard K.J., Leddy D.G., Johnson J.H. (1993) 'The effects of fuel sulfur concentration on regulated and unregulated heavy-duty diesel emissions'. Society of Automotive Engineers 930730. [Cited on pp. 114, 125, 167, 173, 238, 253]

Ormstad H. (2000) 'Suspended particulate matter in indoor air: Adjuvants and allergen carriers'. Toxicology 152:53–68. [Cited on p. 364]

Ormstad H., Gaarder P.I., Johansen B.V. (1997) 'Quantification and characterisation of suspended particulate matter in indoor air'. Science of the Total Environment 193:185–196. [Cited on p. 358]

Ormstad H., Gaarder P.I., Johansen B.V., Lvik M. (1998) 'Airborne house dust elicits a local lymph node reaction and has an adjuvant effect on specific IgE production in the mouse'. Toxicology 129:227–236. [Cited on p. 370]

Osada H., Okayama J., Ishida K., Saitoh O. (1982) 'Real-time measurement of diesel particulate emissions by the PAS method using a CO2 laser'. Society of Automotive Engineers 820461. [Cited on p. 192]

Osornio-Vargas Á.R., Bonner J.C., Alfaro-Moreno E., Martínez L., García-Cuellar C., Ponce-de-León Rosales S., Miranda J., Rosas I. (2003) 'Proinflammatory and cytotoxic effects of Mexico City air pollution particulate matter *in vitro* are dependent on particle size and composition'. Environmental Health Perpectives 111:1289–1293. [Cited on p. 28]

Ostermeyer G.P. (2003) 'On the dynamics of the friction coefficient'. Wear 254:852–858. [Cited on p. 344]

Osuka I., Nishimura M., Tanaka Y., Miyaki M. (1994) 'Benefits of new fuel injection system technology on cold startability of diesel engines – improvement of cold startability and white smoke reduction by means of multi injection with common rail fuel system (ECD-U2)'. Society of Automotive Engineers 940586. [Cited on p. 121]

Otani T., Shigemori M., Suzuki T., Shimoda M. (1988) 'Effects of fuel injection pressure and fuel properties on particulate emissions from H.D.D.I. diesel engine'. Society of Automotive Engineers 881255. [Cited on pp. 254, 256, 268]

Otto K., Sieg M.H., Zinbo M., Bartosiewicz L. (1980) 'The oxidation of soot deposits from diesel engines'. Society of Automotive Engineers 800336. [Cited on pp. 220, 223, 224, 230 231, 301]

den Ouden C.J.J., Clark R.H., Cowley L.T., Stradling R.J., Lange W.W., Maillard C. (1994) 'Fuel quality effects on particulate matter emissions from light- and heavy-duty diesel engines'. Society of Automotive Engineers 942022. [Cited on pp. 125, 255, 256, 263, 265]

Owega S., Khan B.-U.-Z., D'Souza R., Evans G.J., Fila M., Jervis R.E. (2004) 'Receptor modeling of Toronto PM$_{2.5}$ characterised by aerosol laser ablation mass spectrometry'. Environmental Science and Technology 38:5712–5720. [Cited on pp. 27, 342]

Owen K., Coley T. (1995) *Automotive Fuels Reference Book*. Society of Automotive Engineers. [Cited on pp. 95, 263, 264]

Owrang F., Olsson J., Pedersen J. (2004) 'Solid-state ^{1}H NMR spin–lattice relaxation in combustion chamber deposits from a gasoline direct injection engine'. Society of Automotive Engineers 2004-01-0042. [Cited on p. 330]

Oyama K., Kakegawa T. (2003) 'Evaluation of diesel exhaust emission of advanced emission control technologies using various diesel fuels, and sulfur effect on performance after mileage accumulation'. Society of Automotive Engineers 2003-01-1907. [Cited on p. 252]

Özkaynak H., Spengler J. (1996) 'The role of outdoor particulate matter in assessing total human exposure'. In *Particles in Our Air – Concentrations and Health Effects*, Wilson R., Spengler J. (editors). Harvard University Press, pp. 63–84. [Cited on pp. 358, 371]

P

Pachernegg S.J. (1975) 'Efficient and clean diesel combustion'. Society of Automotive Engineers 750787. [Cited on p. 120]

Pacyna J. (1998) 'Source inventories for atmospheric trace emissions'. In *Atmospheric Particles*, Harrison R.M., van Grieken R.E. (editors), IUPAC Series on Analytical and Physical Chemistry of Environmental Systems. John Wiley & Sons, Inc., pp. 385–423. [Cited on pp. 30, 31, 32, 35, 315]

Padovani S., Sada C., Mazzoldi P., Brunetti B., Borgia I., Sgamellotti A., Giulivi A., d'Acaptio F., Battaglin G. (2003) 'Copper glazes of Renaissance luster pottery: Nanoparticles, ions, and local environment'. Journal of Applied Physics 93:10058–10063. [Cited on p. 396]

Palke D.R., Tyo M.A. (1999) 'The impact of catalytic aftertreatment on particulate matter emissions from small motor cycles'. Society of Automotive Engineers 1999-01-3299. [Cited on p. 336]

Palmer H.B., Culliss C.F. (1965) In *Chemistry and Physics of Carbon*, Vol. 1, Walker P.L., Lu W. (editors), Marcel Dekker, pp. 265–325. [Cited on p. 62]

Pandis S.N., Pilinis C. (1995) 'In situ particle formation/reaction mechanisms'. In *Airborne Particulate Matter*, Volume 4, Part D, Kouimtzis T., Samara C. (editors). Springer, pp. 35–68. [Cited on pp. 124, 126]

Pandya R.J., Solomon G., Kinner A., Balmes J.R. (2002) 'Diesel exhaust and asthma: Hypotheses and molecular mechanisms of action'. Environmental Health Perspectives 110:103–112. [Cited on p. 370]

Pankow J.F. (1994) 'An absorption model of gas/particle partitioning of organic compounds in the atmosphere'. Atmospheric Environment 28:185–188. [Cited on p. 117]

Papaioannou E., Konstandopoulos A.G., Morin J.-P., Preterre D. (2006) 'A selective particle size sampler suitable for biological exposure studies of diesel particulate'. Society of Automotive Engineers 2006-01-1075. [Cited on p. 376]

Park C., Kook S., Bae C. (2004) 'Effects of multiple injections in a HSDI diesel engine equipped with common rail injection system'. Society of Automotive Engineers 2004-01-0127. [Cited on p. 269]

Park K., Cao F., Kittelson D.B., McMurry P.H. (2003) 'Relationship between particle mass and mobility for diesel exhaust particles'. Environmental Science and Technology 37:577–583. [Cited on pp. 219, 221]

Parker J. (1971) 'Air pollution at Heathrow Airport, London: April–September, 1970'. Society of Automotive Engineers 710324. [Cited on p. 4]

Parker T.E., Morncy J.R., Foutter R.R., Rawlins W.T. (1996) 'Infrared measurements of soot formation in diesel sprays'. Combustion and Flame 107:271–290. [Cited on p. 84]

Parkhurst G. (2004) 'Air quality and the environmental transport policy discourse in Oxford'. Transportation Research: Part D 9:419–436. [Cited on p. 357]

Parry M., George H., Edgar J. (2001) 'Understanding soot mediated oil thickening: Rotational rheology techniques to determine viscosity and soot structure in Peugeot XUD-11 BTE drain oils'. Society of Automotive Engineers 2001-01-1967. [Cited on p. 287]

Pataky G.M., Baumgard K.J., Gratz L.D., Bagley S.T., Leddy D.G., Johnson J.H. (1994) 'Effects of an oxidation catalytic converter on regulated and unregulated diesel emissions'. Society of Automotive Engineers 940243. [Cited on pp. 168, 236, 294, 297]

Pattas K.N., Michalopoulou C.C. (1992) 'Catalytic activity in the regeneration of the ceramic diesel particulate trap'. Society of Automotive Engineers 920362. [Cited on pp. 99, 233]

Pattas K., Samaras Z., Sherwood D., Umehara K., Cantiani C., Chariol O.A., Barthe P., Lemaire J. (1992) 'Cordierite filter durability with cerium fuel additive: 100,000 km of revenue service in Athens'. Society of Automotive Engineers 920363. [Cited on pp. 344, 350]

Pattas K., Kyriakis N., Samaras Z., Manikas T., Mihailidis A., Mustel W., Rouveirolles P. (1998) 'The behaviour of metal DPFs at low temperatures in conjunction with a cerium based additive'. Society of Automotive Engineers 980543. [Cited on p. 301]

Patterson J., Hassan M.G., Clarke A., Shama G., Hellgardt K., Chen R. (2006) 'Experimental study of DI diesel engine performance using three different biodiesel fuels'. Society of Automotive Engineers 2006-01-0234. [Cited on p. 257]

Payri F., Benajes J., Pastor J.V., Molina S. (2002) 'Influence of the post-injection pattern on performance, soot and NOx emissions in a HD diesel engine'. Society of Automotive Engineers 2002-01-0502. [Cited on p. 270]

Payri R., García J.M., Salvador F.J., Gimeno J. (2005) 'Using spray momentum flux measurements to understand the influence of diesel nozzle geometry on spray characteristics'. Fuel 84:551–561. [Cited on p. 265]

Pedersen D.U., Durant J.L., Penman B.W., Crespi C.L., Hemond H.F., Lafleur A.L., Cass G.R. (1999) 'Seasonal and spatial variations in human cell mutagenicity of respirable airborne particles in the northeastern United States'. Environmental Science and Technology 33:4407–4415. [Cited on p. 375]

Pentikäinen J., Rantanen L., Aakko P. (2004) 'The effect of heavy olefins and ethanol on gasoline emissions'. Society of Automotive Engineers 2004-01-2003. [Cited on p. 325]

Perez J.M., Lipari F., Seizinger D.E. (1984) 'Cooperative development of analytical methods for diesel emissions and particulates – solvent extractables, aldehydes and sulfate methods'. Society of Automotive Engineers 840413. [Cited on pp. 172, 173, 174]

Persiko-Karakash H., Sher E. (2006) 'Evaluation of various strategies for continuous regeneration of particulate filters'. International Journal of Vehicle Design 41:326–341. [Cited on p. 302]

Peters A., Döring A., Wichmann H.-E., Koenig W. (1997) 'Increased plasma viscosity during an air pollution episode: A link to mortality'. The Lancet 349:1582–1587. [Cited on p. 374]

Peters A., Dockery D.W., Muller J.E., Mittleman M.A. (2001) 'Increased particulate air pollution and the triggering of myocardial infarction'. Circulation 103:2810–2815. [Cited on p. 374]

Peterson R.C. (1987) 'The oxidation rate of diesel particulate which contains lead'. Society of Automotive Engineers 870628. [Cited on pp. 81, 223, 224, 230, 233]

Pethers P. (1998) 'The Business Perspective – Engineering Affordable Solutions to the HSDI Diesel EURO IV Challenge'. Institution of Mechanical Engineers: S490/001 (1998) [Cited on p. 6]

Petkar R.M., Kardile C.A., Deshpande P.V., Isenburg R., Soorajith R. (2004) 'Influence of increased diesel fuel spray velocities and improved spray penetration in DI engines'. Society of Automotive Engineers 2004-01-0031. [Cited on p. 265]

Petzold A., Niessner R. (1993) 'Photoacoustic sensor for carbon aerosols'. Sensors and Actuators B 13–14:640–641. [Cited on p. 192]

Phalen R.F. (2002) *The Particulate Air Pollution Controversy*. Kluwer Academic. [Cited on p. 372]

Phuleria H.C., Geller M.D., Fine P.M., Sioutas C. (2006) 'Size-resolved emissions of organic tracers from light- and heavy-duty vehicles measured in California'. Environmental Science and Technology 40:4109–4118. [Cited on p. 237]

Pickett L.M., Siebers D.L. (2004) 'Soot in diesel fuel jets: Effects of ambient temperature, ambient density, and injection pressure'. Combustion and Flame 138:114–135. [Cited on pp. 84, 89, 278, 279]

Pierson W.R., Brachaczek W.W. (1976) 'Particulate matter associated with vehicles on the road'. Society of Automotive Engineers 760039. [Cited on pp. 173, 315, 342, 347]

Pinson J.A., Ni T., Mitchell D.L., Santoro R.J., Litzinger T.A. (1993) 'Quantitative, planar soot measurements in a D.I. diesel engine using laser-induced incandescence and light scattering'. Society of Automotive Engineers 932650. [Cited on p. 91]

Pinson J.A., Ni T., Litzinger T.A. (1994) 'Quantitative imaging study of the effects of intake air temperature on soot evolution in an optically-accessible D.I. diesel engine'. Society of Automotive Engineers 942044. [Cited on pp. 86, 89, 91, 278]

Pipho M.J., Ambs J.L., Kittelson D.B. (1986) 'In-cylinder measurements of particulate formation in an indirect injection diesel engine'. Society of Automotive Engineers 860024. [Cited on pp. 85, 86]

Pirjola L., Parviainen H., Hämeri K., Hussein T. (2004) 'A novel mobile laboratory for "chasing". city traffic'. Society of Automotive Engineers 2004-01-1962. [Cited on pp. 142, 144]

Pitsch H., Barths H., Peters N. (1996) 'Three-dimensional modeling of NO_x and soot formation in DI-diesel engines using details chemistry based on the interactive flamelet approach'. Society of Automotive Engineers 962057. [Cited on pp. 272, 273]

Pitz M., Cyrys J., Karg E., Wiedensohler A., Wichmann H.-E., Heinrich J. (2003) 'Variability of apparent particle density of an urban aerosol'. Environmental Science and Technology 37:4336–4342. [Cited on p. 360]

Planer-Friedrich, Merkel B.J. (2006) 'Volatile metals and metalloids in hydrothermal gases'. Environmental Science and Technology 40:3181–3187. [Cited on p. 30]

Plee S.L., MacDonald, J.S. (1980) 'Some mechanisms affecting the mass of diesel exhaust particulate collected following a dilution process'. Society of Automotive Engineers 800186. [Cited on pp. 113, 114, 115, 116]

Podsiadlik D.H., Chase R.E., Lewis D., Spears M. (2003) 'Phase-based TEOM measurements compared with traditional filters for diesel PM'. Society of Automotive Engineers 2003-01-0783. [Cited on p. 200]

Poirier A., Gariépy C. (2005) 'Isotopic signature and impact of car catalysts on the anthropgenic osmium budget'. Environmental Science and Technology 39:4431–4434. [Cited on p. 351]

Pooley F.D., de Mille M.G. (1999) 'Microscopy and the characterization of particles'. In *Particulate Matter: Properties and Effects upon Health*, Maynard R.L., Howard C.V. (editors). Bios Scientific Publishers, pp. 19–38. [Cited on p. 32]

Poon W.S., Liu B.Y.H., Bugli N. (1997) 'Experimental measurement of clean fractional efficiency of engine air cleaning filters'. Society of Automotive Engineers 970675. [Cited on p. 344]

Pope A., Dockery D. (1996) 'Epidemiology of chronic health effects: Cross-sectional studies'. In *Particles in Our Air – Concentrations and Health Effects*, Wilson R., Spengler J. (editors). Harvard University Press, pp. 149–167. [Cited on pp. 373, 374]

Pope C.J., Howard J.B. (1994) 'Further testing of the fullerene formation mechanism with predictions of temperature and pressure trends'. Twenty-Fifth Symposium (International) on Combustion, The Combustion Institute, pp. 671–678. [Cited on p. 71]

Poster D.L., Baker J.E. (1996) 'Influence of submicron particles on hydrophobic organic contaminants in precipitation. 1. Concentrations and distributions of polycyclic aromatic hydrocarbons and polychlorinated biphenyls in Rainwater'. Environmental Science and Technology 30:341–348. [Cited on p. 117]

Postulka A., Lies K.-H. (1981) 'Chemical characterization of particulates from diesel-powered passenger cars'. Society of Automotive Engineers 810083. [Cited on pp. 168, 173]

Prado G., Lahaye J., Haynes B.S. (1983) 'Soot particle nucleation and agglomeration'. In *Soot in Combustion Systems and its Toxic Properties*, Lahaye J., Prado G. (editors). Plenum Press, pp. 145–161. [Cited on pp. 72, 73]

Pratt S.L., Grainger A.P., Todd J., Meena G.G., Rogers A.J., Davies B. (1997) 'Evaluation and control of employee exposure to diesel particulate at several Australian coal mines'. Applied Occupational and Environmental Hygiene 12:1032–1037. [Cited on p. 360]

Preining O. (1995) 'Fuchs award lecture 1994'. Journal of Aerosol Science 26:529–534. [Cited on p. 395]

Preining O. (1998) 'The physical nature of very, very small particles and its impact on their behaviour'. Journal of Aerosol Science 29:481–495. [Cited on p. 395]

Price P., Stone R., Collier T., Davies M. (2006) 'Particulate matter and hydrocarbon emissions measurements: Comparing first and second generation DISI with PFI in single cylinder optical engines'. Society of Automotive Engineers 2006-01-1263. [Cited on p. 333]

Price P., Stone R., Misztal J., Xu H., Wyszynski M., Wilson T., Qiao J. (2007) 'Particulate emissions from a gasoline homogeneous charge compression ignition engine'. Society of Automotive Engineers 2007-01-0209. [Cited on p. 293]

le Prieur E., Morin J.P., Vaz E., Bion A., Dionnet F. (2000) 'Toxicity of diesel engine exhausts: Induction of a pro-inflammatory response and apoptosis in an *in vitro* model of lung slices in biphasic organotypic culture'. Society of Automotive Engineers 2000-01-1928. [Cited on pp. 375, 376]

Puffel P.K., Thiel W., Frey R., Boesl U. (1998) 'A new method for the investigation of unburned oil emissions in the raw exhaust of SI engines'. Society of Automotive Engineers 982438. [Cited on p. 322]

Q

Qian S., Sakurai H., McMurry P.H. (2007) 'Characteristics of regional nucleation events in urban east St. Louis'. Atmospheric Environment 41:4119–4127. [Cited on p. 390]

Qu S.-X., Leigh J., Koelmeyer H., Stacey N.H. (1997) 'DNA adducts in coal miners: Association with exposures to diesel engine emissions'. Biomarkers 2:95–102. [Cited on p. 369]

Quader A.A. (1989) 'How injector, engine, and fuel variables impact smoke and hydrocarbon emissions with port fuel injection'. Society of Automotive Engineers 890623. [Cited on p. 319]

Quader A.A., Dasch C.J. (1992) 'Spark plug fouling: A quick engine test'. Society of Automotive Engineers 920006. [Cited on p. 319]

R

Raatz T., Mueller E. (2001) 'Examination of particle size distribution of homogeneous and conventional diesel combustion'. Society of Automotive Engineers 2001-01-3576. [Cited on pp. 268, 271, 293]

Raes F., van Dingenen R., Vignati E., Wilson J., Putaud J.-P., Seinfeld J.H., Adams P. (2000) 'Formation and cycling of aerosols in the global troposphere'. Atmospheric Environment 34:4215–4240. [Cited on pp. 23, 24]

Ragazzi R.A., Gallagher G.L., Barrett R.A. (1986) 'Inspection/maintenance for light-duty diesel vehicles'. Society of Automotive Engineers 860297. [Cited on p. 137]

Räisänen M., Kupiainen K., Tervahattu H. (2003) 'The effect of mineralogy, texture and mechanical properties of anti-skid and asphalt aggregates on urban dust'. Bulletin of Engineering Geology and the Environment 62:359–368. [Cited on p. 342]

Rakopoulos C.D., Hountalas D.T., Zannis T.C. (2004) 'Theoretical study concerning the effect of oxygenated fuels on DI diesel engine performance and emissions'. Society of Automotive Engineers 2004-01-1838. [Cited on pp. 258, 259]

Rakopoulos C.D., Antonopoulos K.A., Rakopoulos D.C. (2006) 'Multi-zone modeling of diesel engine fuel spray development with vegetable oil, bio-diesel or diesel fuels'. Energy Conversion and Management 47:1550–1573. [Cited on pp. 257, 258]

Rantanen L., Mikkonen S., Nylund L., Kociba P., Lappi M., Nylund N.-O. (1993) 'Effect of fuel on the regulated, unregulated and mutagenic emissions of DI diesel engines'. Society of Automotive Engineers 932686. [Cited on pp. 181, 256, 258, 260, 263, 265]

Rantanen L., Juva A., Niemi A., Mikkonen S., Aakko P., Lappi M. (1996) 'Effect of reformulated diesel fuel on unregulated emissions of light duty vehicles'. Society of Automotive Engineers 961970. [Cited on pp. 166, 240, 256]

Rapone M., Ragione L.D., Iaccio I., Meccariello G., Prati M.V. (2006) 'A multivariate statistical approach to evaluate the effect of after treatment device on bus particulate emissions by in-use testing'. Society of Automotive Engineers 2006-01-3396. [Cited on p. 200]

Rauch S., Morrison G.M., Moldovan M. (2002) 'Scanning laser ablation-ICP-MS tracking of platinum group elements in urban particles'. Science of the Total Environment 286:243–251. [Cited on pp. 350, 351]

Rauch S., Hemond H.F., Peuker-Ehrenbrink B. (2004) 'Recent changes in platinum group element concentrations and osmium isotopic composition in sediments from an urban lake'. Environmental Science and Technology 38:396–402. [Cited on pp. 350, 351]

Rauch S., Hemond H.F., Barbante C., Owari M., Morrison G.M., Peucker-Ehrenbrink B., Wass U. (2005a) 'Importance of automobile exhaust catalyst emissions for the deposition of platinum, palladium, and rhodium in the Northern Hemisphere'. Environmental Science and Technology 39:8156–8162. [Cited on pp. 351, 352]

Rauch S., Hemond H.F., Peucker-Enrenbrink B., Ek K.H., Morrison G.M. (2005b) 'Platinum group element concentrations and osmium isotopic composition in urban airborne particles from Boston, Massachusetts'. Environmental Science and Technology 39:9464–9470. [Cited on pp. 351, 352]

Rauterberg-Wulff A., Israel G.W., Pesch M., Schlums C. (1995) 'Bestimmung des Beitrags von Reifenabrieb zur Rußimmission an stark befahrenen Straßen', VDI Berichte Nr. 1228, pp. 81–92. [In German.] [Cited on p. 348]

Raux S., Forti L., Barbusse S., Plassat G., Pierron L., Monier R., Momique J.C., Pain C., Dionnet B., Zervas E., Rouveirolles P., Dorlhene P. (2005) 'French program on the impact of engine technology on particulate emissions, size distribution and composition heavy duty diesel study'. Society of Automotive Engineers 2005-01-0190. [Cited on p. 95]

Ray A., Wichman I.S. (1998) 'Influence of fuel-side heat loss on diffusion flame extinction'. International Journal of Heat and Mass Transfer 41:3075–3085. [Cited on p. 62]

Reading K., Roberts D.D., Evans T.M. (1991) 'The effects of fuel detergents on nozzle fouling and emissions in IDI diesel engines'. Society of Automotive Engineers 912328. [Cited on p. 261]

Reijnders L. (2006) 'Cleaner nanotechnology and hazard reduction of manufactured nanoparticles'. Journal of Cleaner Production 14:124–133. [Cited on pp. 368, 396]

Reinmann R., Saitzkoff A., Mauss F. (1997) 'Local air – fuel ratio measurements using the spark plug as an ionization sensor'. Society of Automotive Engineers 970856. [Cited on p. 225]

Rente T., Golovitchev V.I., Denbratt I. (2001) 'Effect of injection parameters on auto-ignition and soot formation in diesel sprays'. Society of Automotive Engineers 2001-01-3687. [Cited on pp. 69, 72]

Renwick L.C., Brown D., Clouter A., Donaldson K. (2004) 'Increased inflammation and altered macrophage chemotactic responses caused by two ultrafine particle types'. Occupational and Environmental Medicine 61:442–447. [Cited on p. 380]

Ressler T., Wong J., Roos J., Smith I.L. (2000) 'Quantitative speciation of Mn-bearing particulates emitted from autos burning (methylcyclopentadienyl) manganese tricarbonyl-added gasolines using XANES spectroscopy'. Environmental Science and Technology 34:950–958. [Cited on p. 317]

Reul-Chen C., Cabrera R., Ross C., Steele N.L.C. Winer A. (2004) 'Solid waste collection vehicle fleet maintenance in California'. Society of Automotive Engineers 2004-01-1717. [Cited on p. 388]

Rhead M.N., Trier C.J., Petch G.S. (1990) 'The development of a radiolabelling technique to unequivocally determine the products of combustion from specific components of diesel fuel'. In *Fuels for Automotive Diesel Engines*, Nov. 19–20. Institution of Mechanical Engineers. [cited on p. 68]

Richards P., Rogers T. (2002) 'Preliminary results from a six vehicle, heavy duty truck trial, using additive regenerated DPFs'. Society of Automotive Engineers 2002-01-0431. [Cited on p. 304]

Richards P., Terry B., Trivett A.S. (1998) 'Measurements of diesel particulate size distribution – using fuels with and without a smoke-reducing additive'. Institution of Mechanical Engineers S491/010/98. [Cited on p. 183]

Richards R.J., BéruBé K.A., Masek L., Symons D., Murphy S.A. (1999) 'Lung deposition of diesel particulates'. In *Particulate Matter: Properties and Effects upon Health*, Maynard R.L., Howard C.V. (editors). Bios Scientific Publishers, pp. 97–114. [Cited on p. 377]

Richardson D.E. (1996) 'Comparison of measured and theoretical inter-ring gas pressure on a diesel engine'. Society of Automotive Engineers 961909. [Cited on p. 279]

Richter H., Grieco W.J., Howard J.B. (1999) 'Formation mechanism of polycyclic aromatic hydrocarbons and fullerenes in premixed benzene flames'. Combustion and Flame 119:1–22. [Cited on p. 70]

Richter H., Howard J.B. (2000) 'Formation of polycyclic aromatic hydrocarbons and their growth to soot – a review of chemical reaction pathways'. Progress in Energy and Combustion Science 26:565–608. [Cited on pp. 63, 68]

Rickeard D.J., Bateman J.R., Kwon Y.K., McAughey J.J., Dickens C.J. (1996) 'Exhaust particulate size distribution: Vehicle and fuel influences in light duty vehicles'. Society of Automotive Engineers 961980. [Cited on p. 394]

Riddle S.G., Robert M.A., Jakober C.A., Hannigan M.P., Kleeman M.J. (2007) 'Size distribution of trace organic species emitted from heavy-duty diesel vehicles'. Environmental Science and Technology 41:1962–1969. [Cited on p. 176]

Riehle C., Wadenpohl C. (1996) 'Electrically stimulated agglomeration at an earthed surface'. Powder Technology 86:119–126. [Cited on p. 394]

Rietmeijer F.J.M., Janeczek J. (1997) 'An analytical electron microscope study of airborne industrial particles in Sosnowiec, Poland'. Atmospheric Environment 31:1941–1951. [Cited on p. 98]

Rijkboer R., Bremmers D., Samara Z., Ntiachristos L. (2005) 'Particulate matter regulation for two-stroke two wheelers: Necessity or haphazard legislation?'. Atmospheric Environment 39:2483–2490. [Cited on p. 335]

Rijkeboer R.C., van Beckhoven L.C. (1987) 'The lean-burn option – how does it compare on polynuclear aromatic hydrocarbons and mutagenicity'. Institution of Mechanical Engineers C328/87. [Cited on pp. 111, 240]

Rink K.K., Lefebvre A.H., Graves R.L. (1987) 'The influence of fuel composition and spray characteristics on particulate formation'. Society of Automotive Engineers 872035. [Cited on pp. 251, 256]

Ristovski Z.D., Morawska L., Bofinger N.D., Hitchins J. (1998) 'Submicrometer and supermicrometer particulate emission from spark ignition vehicles'. Environmental Science and Technology 32:3845–3852. [Cited on pp. 316, 325]

Ristovski Z.D., Jayaratne E.R., Lim M., Ayoko G.A., Morawska L. (2006) 'Influence of diesel fuel sulfur on nanoparticle emissions from city buses'. Environmental Science and Technology 40:1314–1320. [Cited on p. 253]

Ritrievi K.E., Longwell J.P., Sarofim A.F. (1987) 'The effects of ferrocene addition on soot particle inception and growth in premixed ethylene flames'. Combustion and Flame 70:17–31. [Cited on pp. 96, 99, 100, 101, 103]

Rivin D., Medalia A.I. (1983) 'A comparative study of soot and carbon black'. In *Soot in Combustion Systems and Its Toxic Properties*, Lahaye J., Prado G. (editors). Plenum Press, pp. 25–35. [Cited on p. 47]

Rodriguez S., Querol X., Alastuey A., Viana M.-M., Mantilla E. (2003) 'Events affecting levels and seasonal evolution of airborne particulate matter concentrations in the western Mediterranean'. Environmental Science and Technology 37:216–222. [Cited on p. 397]

Rodriguez-Navarro C., Sebastian E. (1996) 'Role of particulate matter from vehicle exhaust on porous building stones (limestone) sulfation'. Science of the Total Environment 187:79–91. [Cited on p. 39]

Rodríguez-Reinoso F. (1997) 'Activated carbon: structure, characterization, preparation and application'. In *Introduction to Carbon Technologies*, Marsh H., Heintz E.A., Rodríguez-Reinoso F. (editors). University of Alicante, pp. 35–101. [Cited on pp. 112, 223]

Rogers A., Davies B. (2005) 'Diesel particulates – recent progress on an old issue'. Annals of Occupational Hygiene 49:453–456. [Cited on p. 363]

Rogers B.J., Li G., Such C.H. (2002) 'The reduction of soot emissions by the use of a piston with micro-chambers on a medium duty diesel engine'. Society of Automotive Engineers 2002-01-1682. [Cited on p. 91]

Romero-Lopez A., Gutiérrez-Salinas R., Garcia-Moreno R. (1996) 'Soot combustion during regeneration of filter ceramic traps for diesel engines'. Society of Automotive Engineers 960469. [Cited on p. 80]

Romig C., Spataru A. (1996) 'Emissions and engine performance from blends of soya and canola methyl esters with ARB #2 diesel in a DDC 6V92TA MUI engine'. Bioresource Technology 56:25–34. [Cited on p. 258]

Rosinski J. (1996) Il Nuovo Cimento 19:217. [Cited on p. 24]

Rosinski J. (2000) 'On the role of aerosol particles in the phase transition in the atmosphere'. In *Aerosol Chemical Processes in the Environment*, Spurny K.R., Hochrainer D. (editors). CRC Press LLC, pp. 81–134. [Cited on pp. 23, 24, 25]

Rosner D.E. (2005) 'Flame synthesis of valuable nanoparticles: Recent progress/current needs in areas of rate laws, population dynamics, and characterization'. Industrial and Engineering Chemistry 44:6045–6055. [Cited on p. 98]

Roth B., Okada K. (1998) 'On the modification of sea-salt particles in the coastal atmosphere'. Atmospheric Environment 32:1555–1569. [Cited on p. 27]

Roth C.M., Goss K.-U., Schwarzenbach R.P. (2005) 'Sorption of a diverse set of organic vapors to diesel soot and road tunnel aerosols'. Environmental Science and Technology 39:6632–6637. [Cited on p. 117]

Roth P., Filippov A.V. (1996) 'In situ ultrafine particle sizing by a combination of pulsed laser heatup and particle thermal emission'. Journal of Aerosol Science 27:95–104. [Cited on p. 187]

Rothe D., Zuther F.I., Jacob E., Messerer A., Pöschl U., Niessner R., Knab C., Mangold M., Mangold C. (2004) 'New strategies for soot emission reduction of HD vehicles'. Society of Automotive Engineers 2004-01-3046. [Cited on p. 139]

Rothen-Rutishauser B.M., Schürch S., Haenni B., Kapp N., Gehr P. (2006) 'Interaction of fine particles and nanoparticles with red blood cells visualized with advanced techniques'. Environmental Science and Technology 40:4353–4359. [Cited on p. 368]

Rounds F.G. (1977) 'Carbon: Cause of diesel engine wear?'. Society of Automotive Engineers 770829. [Cited on p. 290]

Rounds F.G. (1981) 'Soots from used diesel engine oils – their effects on wear as measured in 4-ball wear tests'. Society of Automotive Engineers 810499. [Cited on pp. 214, 223, 290, 291]

Rudell B., Blomberg A., Helleday R., Ledin M.-C., Lundback B., Stjernberg N., Horstedt P., Sandstrom T. (1999) 'Bronchoalveolar inflammation after exposure to diesel exhaust: Comparison between unfiltered and particle trap filtered exhaust'. Occupational and Environmental Medicine 56:527–534. [Cited on p. 382]

Rule A.M., Chapin A.R., McCarthy S.A., Gibson K.E., Schwab K.J., Buckley T.J. (2005) 'Assessment of an aerosol treatment to improve air quality in a swine concentrated animal feeding operation (CAFO)'. Environmental Science and Technology 39:9649–9655. [Cited on pp. 29, 30]

Rumelhard M., Ramgolam K., Auger F., Dazy A.-C., Blanchet S., Marano F., Baeza- Squiban A. (2007) 'Effects of $PM_{2.5}$ components in the release of amphiregulin by human airway epithelial cells'. Toxicology Letters 168:155–164. [Cited on p. 376]

Rumminger M.D., Linteris G.T. (2000) 'The role of particles in the inhibition of premixed flames by iron pentacarbonyl'. Combustion and Flame 123:82–94. [Cited on p. 98]

Ruot C.D., Faure D., Blanc G. (2000) 'Adsorption of engine lubricant dispersants and polymers onto carbon black particles'. Society of Automotive Engineers 2000-01-1991. [Cited on p. 288]

Rushton L. (2001) 'Cancer and air pollution'. In *The Urban Atmosphere and Its Effects*, Brimblecombe P., Maynard R.L. (editors), Air Pollution Reviews, Vol. 1. Imperial College Press, pp. 129–162. [Cited on pp. 362, 371, 373]

Russell Jones R. (1987) 'The health effects of vehicle emissions'. Institution of Mechanical Engineers C355/87. [Cited on pp. 5, 315, 371]

Rutland C.J., Ayoub N., Han Z., Hampson G., Kong S.-C., Mather D., Montgomery D., Musculus M., Patterson M., Pierpont D., Ricart L., Stephenson P., Reitz R.D. (1995) 'Diesel engine model development and experiments'. Society of Automotive Engineers 951200. [Cited on pp. 66, 82, 83]

Ryan T.W., Owens E., Naegeli D., Doglio J., Bltyh G., van Dam W., Damin B., Olikara C., Villforth F. (1999) 'Effects of exhaust gas recirculation on the degradation rates of lubricating oil in a heavy-duty diesel engine'. Society of Automotive Engineers 1999-01-3574. [Cited on p. 289]

Ryason P.R., Hansen T.P. (1991) 'Voluminosity of soot aggregates: A means of characterizing soot-laden oils'. Society of Automotive Engineers 912343. [Cited on p. 287]

Ryason P.R., Hillyer M.J., Hansen T.P. (1994) 'Infrared absorptivities of several diesel engine soots: Application to the analysis of soot in used engine oils'. Society of Automotive Engineers 942030. [Cited on p. 286]

S

Sabbioni C. (2000) 'Aerosol and stone monuments'. In *Aerosol Chemical Processes in the Environment*, Spurny K.R., Hochrainer D. (editors). CRC Press LLC, pp. 327–345. [Cited on p. 38]

Sadezky A., Muckenhuber H., Grothe H., Niesner R., Pöschl U. (2005) 'Raman microspectroscopy of soot and related carbonaceous materials: Spectral analysis and structural information'. Carbon 43:1731–1742. [Cited on p. 214]

Sagai M., Saito H., Ichinose T., Kodama M., Mori Y. (1993) 'Biological effects of diesel exhaust particles. I. In vitro production of superoxide and in vivo toxicity in mouse'. Free Radical Biology and Medicine 14:37–47. [Cited on pp. 376, 378, 379]

Sagai M., Furuyama A., Ichinose T. (1996) 'Biological effects of diesel exhaust particles (DEP). III. Pathogenesis of asthma like symptoms in mice'. Free Radical Biology and Medicine 21:199–209. [Cited on p. 370]

Said R., Garo A., Borghi R. (1997) 'Soot formation modeling for turbulent flames'. Combustion and Flame 108:71–86. [Cited on pp. 80, 277]

Saito K. (1988) 'Development of diesel opacimeter for real-time measurement of low concentration smoke'. Society of Automotive Engineers 881319. [Cited on p. 203]

Saito K., Shinozaki O. (1990) 'The measurement of diesel particulate emissions with a tapered element oscillating microbalance and an opacimeter'. Society of Automotive Engineers 900644. [Cited on p. 200]

Saito M., Sato M., Sawada K. (1997) 'Variation of flame shape and soot emission by applying electric field'. Journal of Electrostatics 39:305–311. [Cited on p. 71]

Saito M., Arai T., Arai M. (1999) 'Control of soot emitted from acetylene diffusion flames by applying an electric field'. Combustion and Flame 119:356–366. [Cited on p. 71]

Saito S., Shinozaki R., Suzuki A., Jyoutaki H., Takeda Y. (2003) 'Development of urea-SCR system for commercial vehicle – basic characteristics and improvement of NOx conversion at low load operation'. Society of Automotive Engineers 2003-01-3248. [Cited on p. 299]

Saito T., Nabetani M. (1973) 'Surveying tests of diesel smoke suppression with fuel additives'. Society of Automotive Engineers 730170. [Cited on pp. 97, 261]

Saito T., Tokura N., Katoh T. (1982) 'Analysis of factors affecting the formation of major mutagenic substances in diesel particulate extracts'. Society of Automotive Engineers 821244. [Cited on pp. 111, 266]

Sakai T., Nakajima T., Yamazaki H. (1999) 'O-PM/emitted matters caused by two-stroke engine oil and its reduction'. Society of Automotive Engineers 1999-01-3260. [Cited on pp. 336, 338]

Sakamoto S., Saito J., Kishimoto T., Ishida K. (1997) 'Particulate characterization of automotive emissions by helium microwave-induced plasma atomic emission spectroscopy'. Society of Automotive Engineers 971017. [Cited on p. 230]

Sakurai H., Tobias H.J., Park K., Zarling D., Docherty K.S., Kittelson D.B., McMurry P.H., Ziemann P.J. (2003a) 'On-line measurements of diesel nanoparticle composition and volatility'. Atmospheric Environment 37:1199–1210. [Cited on pp. 178, 237, 392]

Sakurai H., Park K., McMurry P.H., Zaling D.D., Kittelson D.B., Ziemann P.J. (2003b) 'Size-dependent mixing characteristics of volatile and nonvolatile components in diesel exhaust aerosols'. Environmental Science and Technology 37:5487–5495. [Cited on p. 390]

Salatino P., Zimardi F., Paulicelli M. (1994) 'A transient kinetics study of the combustion reactivity of a coal char'. Twenty-Fifth Symposium (International) on Combustion The Combustion Institute, pp. 527–535. [Cited on p. 224]

Sallee G.P. (1970) 'Standard smoke measurement method'. Society of Automotive Engineers 700250. [Cited on p. 201]

Salma I., Maenhaut W., Zemplén-Papp É., Bobvos J. (1998) Microchemical Journal 58:291. [Cited on p. 24]

Salma I., Maenhaut W., Zemplén-Papp É., Bobvos J. (2000) 'Chemical characteristics and temporal variation of size-fractionated urban aerosols and trace gases in Budapest'. In *Aerosol Chemical Processes in the Environment*, Spurny K.R., Hochrainer D. (editors). CRC Press LLC, pp. 415–430. [Cited on pp. 23, 24]

Salvat O., Marez P., Belot G. (2000) 'Passenger car serial application of a particulate filter system on a common rail direct injection diesel engine'. Society of Automotive Engineers 2000-01-0473. [Cited on pp. 212, 303]

Salvi S., Holgate S.T. (1999) 'Mechanisms of particulate matter toxicity'. Clinical and Experimental Allergy 29:1187–1194. [Cited on p. 370]

Salvi S., Blomberg A., Rudell B., Kelly F., Sandstrom T., Holgate S.T., Frew A. (1999) 'Acute inflammatory responses in the airways and peripheral blood after short-term exposure to diesel exhaust in healthy human volunteers'. American Journal of Respiratory and Critical Care Medicine 159:702–709. [Cited on p. 382]

Samara C. (1995) 'Analysis of organic matter'. In *Airborne Particulate Matter*, Volume 4, Part D, Kouimtzis T., Samara C. (editors). Springer, pp. 233–252. [Cited on pp. 165, 175]

Samaras Z., Zierock K.-H. (1995) 'Off-road vehicles: A comparison of emissions with those from road transport'. Science of the Total Environment 169:249–255. [Cited on p. 358]

Samet J.M., Dominici F., Curriero F.C., Coursac I., Zegar S.L. (2000) 'Fine particulate air pollution and mortality in 20 US cities, 1987–1994'. New England Journal of Medicine 343:1742–1749. [Cited on p. 373]

Sams T., Tieber J. (1997) 'Emission behaviour of heavy-duty vehicles: A holistic calculation method'. International Journal of Vehicle Design 18:293–311. [Cited on p. 276]

Samson P.J. (1988) 'Atmospheric transport and dispersion of air pollutants associated with vehicular emissions'. In *Air Pollution, the Automobile, and Public Health*, Watson A.Y., Bates R.R., Kennedy D. (editors). National Academy Press, pp. 77–97. [Cited on pp. 143, 145, 357]

Sanders P.G., Xu N., Dalka T.M., Maricq M.M. (2003) 'Airborne brake wear debris: Size distributions, composition, and a comparison of dynamometer and vehicle tests'. Environmental Science and Technology 37:4060–4069. [Cited on pp. 159, 345, 346]

Sannigrahi P., Sullivan A.P., Weber R.J., Ingall E.D. (2006) 'Characterization of water-soluble organic carbon in urban atmospheric aerosols using solid-state 13C NMR spectroscopy'. Environmental Science and Technology 40:666–672. [Cited on p. 173]

Santamaría A., Mondragón F., Quiñónez W., Eddings E.G., Sarofim A.F. (2007) 'Average structural analysis of the extractabe material of young soot gathered in an ethylene inverse diffusion flame'. Fuel 86:1908–1917. [Cited on p. 73]

Santoro R.J., Shaddix C.R. (2002) 'Laser-induced incandescence'. In *Applied Combustion Diagnostics*, Kohse-Höinghaus K., Jeffries J.B. (editors). Taylor and Francis, pp. 252–286. [Cited on pp. 187, 188]

Sappok A.G., Wong V.W. (2007) 'Detailed chemical and physical characterization of ash species in diesel exhaust entering aftertreatment systems'. Society of Automotive Engineers 2007-01-0318. [Cited on pp. 232, 304]

Sarwar G., Corsi R. (2007) 'The effects of ozone/limonene reactions on indoor secondary organic aerosols'. Atmospheric Environment 41:959–973. [Cited on p. 359]

Sato H., Tokuoka N., Yamamoto H., Sasaki M. (1999) 'Study on wear mechanism by soot contaminated in engine oil (First report: Relation between characteristics of used oil and wear)'. Society of Automotive Engineers 1999-01-3573. [Cited on p. 290]

Sato H., Onose J.-i., Toyoda H., Toida T., Imanari T., Sagai M., Nishimura N., Aoki Y. (2001) 'Quantitative changes in glycosaminoglycans in the lungs of rats exposed to diesel exhaust'. Toxicology 166:119–128. [Cited on p. 378]

Schäfer J., Eckhardt J.-D., Berner Z.A., Stüben D. (1999) 'Time-dependent increase of traffic-emitted platinum-group elements (PGE) in different environmental compartments'. Environmental Science and Technology 33:3166–3170. [Cited on p. 351]

Schauer J.J., Rogge W.F., Hildemann L.M., Mazurek M.A., Cass G.R. (1996) 'Source apportionment of airborne particulate matter using organic compounds as tracers'. Atmospheric Environment 30:3837–3855. [Cited on p. 25]

Scherrer H.C., Kittelson D.B., Dolan D.F. (1981) 'Light absorption measurements of diesel particulate matter'. Society of Automotive Engineers 810181. [Cited on pp. 189, 200]

Schindler W., Nust M., Thaller W., Luxbacher T. (2001) 'Steady-state and transient measurement of low smoke values'. Mototechnische Zeitschrift 62:10. [Cited on pp. 203, 204]

Schindler W., Haisch C., Beck H.A., Niessner R., Jacob E., Rothe D. (2004) 'A photoacoustic sensor system for time resolved quantification of diesel soot emissions'. Society of Automotive Engineers 2004-01-0968. [Cited on pp. 192, 203]

Schlesinger R.B., Chen L.C., Driscoll K.E. (1984) 'Exposure – response relationship of bronchial mucociliary clearance in rabbits following acute inhalations of sulfuric acid'. Toxicology Letters 22:249. [Cited on p. 380]

Schlesinger R.B. (1988) 'Biological disposition of airborne particles: basic principles and application to vehicular emissions'. In *Air Pollution, the Automobile, and Public Health*, Watson A.Y., Bates R.R., Kennedy D. (editors). National Academy Press, pp. 239–298. [Cited on pp. 364, 368, 377]

Schmelzle P., Chandes K. (2004) 'The challenge facing AQUAZOLE: Compatibility with new engine and DPF technologies'. Society of Automotive Engineers 2004-01-1885. [Cited on p. 251]

Schmidt D., Hübsch U., Wurzer H., Heppt W., Aufderheide M. (1996) 'Development of an in vitro human nasal epithelium (HNE) cell model'. Toxicology Letters 88:75–69. [Cited on p. 375]

Schneider J., Hock N., Weimer S., Borrmann S., Kirchner U., Vogt R., Scheer V. (2005) 'Nucleation particles in diesel exhaust: Composition inferred from in situ mass spectrometric analysis'. Environmental Science and Technology 39:6153–6161. [Cited on pp. 128, 220, 221, 253, 298]

Schommers J., Duvinage F., Stotz M., Peters A., Ellwanger S., Koyanagi K., Gildein H. (2000) 'Potential of common rail injection system for passenger car DI diesel engines'. Society of Automotive Engineers 2000-01-0944. [Cited on p. 265]

Schraml S., Will S., Leipertz A. (1999) 'Simultaneous measurement of soot mass concentration and primary particle size in the exhaust of a DI diesel engine by time-resolved laser-induced incandescence'. Society of Automotive Engineers 1999-01-0146. [Cited on pp. 187, 188]

Schraml S., Will S., Leipertz A., Zens T., D'Alfonso N. (2000) 'Performance characteristics of TIRE-LII soot diagnostics in exhaust gases of diesel engines'. Society of Automotive Engineers 2000-01-2002. [Cited on p. 268]

Schug K.P., Guttmann H.-J., Preuss A.W., Schädlich K. (1990) 'Effects of ferrocene as a gasoline additive on exhaust emissions and fuel consumption of catalyst equipped vehicles'. Society of Automotive Engineers 900154. [Cited on pp. 98, 103]

Schulz H., de Melo G.B., Ousmanov F. (1999) 'Volatile organic compounds and particulates as components of diesel engine exhaust gas'. Combustion and Flame 118:179–190. [Cited on pp. 68, 169, 175, 263, 264]

Schwab S.D., Guinther G.H., Henly T.J., Miller K.T. (1999) 'The effects of 2-ethylhexyl nitrate and di-tertiary-butyl-peroxide in the exhaust emissions from a heavy-duty diesel engine'. Society of Automotive Engineers 1999-01-1478. [Cited on p. 264]

Schwartz J. (1994) 'What are people dying of on high air pollution days?'. Environmental Research 64:26–35. [Cited on p. 373]

Schwartz J. (2004) 'Air pollution and children's health'. Pediatrics 113:1037–1043. [Cited on p. 363]

Schweimer G.W. (1986) 'Ion probe in the exhaust manifold of diesel engines'. Society of Automotive Engineers 860012. [Cited on pp. 196, 225]

Schweizer T., Stein H.J. (2000) 'A new approach to particulate measurment on transient test cycles: Partial flow dilution as alternative to CVS full flow systems'. Society of Automotive Engineers 2000-01-1134. [Cited on p. 160]

Seaton A., MacNee W., Donaldson K., Godden D. (1995) 'Particulate air pollution and acute health effects'. The Lancet 345:176–178. [Cited on pp. 40, 367]

Seaton A. (1999) 'Airborne particles and their effects on health'. In *Particulate Matter: Properties and Effects upon Health*, Maynard R.L., Howard C.V. (editors). Bios Scientific Publishers, pp. 9–18. [Cited on pp. 367, 369]

Seinfeld J.H. (1986) *Atmospheric Chemistry and Physics of Air Pollution*. John Wiley & Sons, Inc. [Cited on pp. 13, 14, 25, 26, 36, 39, 124, 126, 143, 200, 364, 365]

Seizinger D.E., Eccleston B.H., Hurn R.W. (1979) 'Particulates and associated emissions from two medium-duty diesel engines'. Society of Automotive Engineers 790420. [Cited on p. 180]

Selby K. (1998) 'Rheology of soot thickened diesel engine oils'. Society of Automotive Engineers 981369. [Cited on pp. 287, 288]

Senior C.L., Lignell D.O., Sarofim A.F., Mehta A. (2006) 'Modeling arsenic partitioning in coal-fired power plants'. Combustion and Flame 147:209–221. [Cited on p. 98]

Sexton K., Ryan P.B. (1988) 'Assessment of human exposure to air pollution: Methods, measurements, and models'. In *Air Pollution, the Automobile, and Public Health*, Watson A.Y., Bates R.R., Kennedy D. (editors). National Academy Press, pp. 207–238. [Cited on pp. 356, 358]

Shaddix C.R., Palotás Á.B., Megaridis C.M., Choi M.Y., Yang N.Y.C. (2005) 'Soot graphitic order in laminar diffusion flames and a large-scale JP-8 pool fire'. International Journal of Heat and Mass Transfer 48:3604–3614. [Cited on p. 182]

Shafer M.M., Schauer J.J., Copan W.G., Peter-Hoblyn J.D., Sprague B.N. (2006) 'Investigation of platinum and cerium from use of a FBC'. Society of Automotive Engineers 2006-01-1517. [Cited on pp. 171, 350]

Shaffernocker W.M., Stanforth C.M. (1968) 'Smoke measurement techniques'. Society of Automotive Engineers 680346. [Cited on pp. 201, 202]

Shah S.D., Cocker D.R., Miller J.W., Norbeck J.M. (2004) 'Emission rates of particulate matter and elemental and organic carbon from in-use diesel engines'. Environmental Science and Technology 38:2544–2550. [Cited on p. 109]

Shah S.D., Ogunyoku T.A., Miller J.W., Cocker D.R. (2005) 'On-road emission rates of PAH and n-alkane compounds from heavy-duty vehicles'. Environmental Science and Technology 39:5276–5284. [Cited on p. 235]

Shamah E., Wagner T.O. (1973) 'Fuel quality or engine design: Which controls diesel emissions?'. Society of Automotive Engineers 730168. [Cited on p. 250]

Shayeson M.W. (1967) 'Reduction of jet engine exhaust smoke with fuel additives'. Society of Automotive Engineers 670866. [Cited on pp. 4, 98]

Sheng H.Z., Chen L., Zhang Z.P., Wu C.K., An C., Cheng C.Q. (1994) 'The droplet group microexplosions in water-in-oil emulsion sprays and their effects on diesel engine combustion'. Twenty-Fifth Symposium (International) on Combustion, The Combustion Institute, pp. 175–181. [Cited on p. 251]

Shenghua L., Hwang J.W., Park J.K., Kim M.H., Chae J.O. (1999) 'Multizone model for DI diesel engine combustion and emissions'. Society of Automotive Engineers 1999-01-2926. [Cited on p. 81]

Sherwood D.A., McKinnon D.L., von Hagn W.J. (1991) 'Evaluation of a wall flow diesel filter after >4000 hours of use on an underground mining vehicle'. Society of Automotive Engineers 912336. [Cited on p. 233]

Shi J.P., Harrison R.M. (1999) 'Investigation of ultrafine particle formation during diesel exhaust dilution'. Environmental Science and Technology 33:3730–3736. [Cited on pp. 127, 149]

Shimoda M., Suzuki T., Shigemori M. (1987) 'Observation of the particulate formation process in the cylinder of a direct injection diesel engine'. Society of Automotive Engineers 870268. [Cited on p. 109]

Shin Y., Cheng W.K. (1997) 'Engine-out "dry" particular matter emissions from SI engines'. Society of Automotive Engineers 972890. [Cited on pp. 321, 322]

Shine K.P., de F. Forster P.M. (1999) 'The effect of human activity on radiative forcing of climate change: A review of recent developments'. Global and Planetary Change 20:205–225. [Cited on p. 41]

Shirakawa T., Miura M., Itoyama H., Aiyoshizawa E., Kimura S. (2001) 'Study of model-based cooperative control of EGR and VGT for a low-temperature, premixed combustion diesel engine'. Society of Automotive Engineers 2001-01-2006. [Cited on p. 277]

Shore P.R. (1988) 'Advances in the use of tritium as a radiotracer for oil consumption measurement'. Society of Automotive Engineers 881583. [Cited on pp. 280, 281]

Literature Cited (Cross-referenced Against the Text) 467

Shore P.R., Tesh J.M., Bootman J. (1987) 'Application of short-term bioassays to the assessment of engine exhaust emissions'. Society of Automotive Engineers 870627. [Cited on pp. 323, 375, 376, 379]

Shrivastava M.K., Lipsky E.M., Stanier C.O., Robinson A.L. (2006) 'Modeling semivolatile organic aerosol mass emissions from combustion systems'. Environmental Science and Technology 40:2671–2677. [Cited on pp. 145, 161]

Sidhu S., Graham J., Striebich R. (2001) 'Semi-volatile and particulate emissions from combustion of alternative diesel fuels'. Chemosphere 42:681–690. [Cited on pp. 171, 257, 325, 358]

Siebers D., Higgins B. (2001) 'Flame lift-off on direct-injection diesel sprays under quiescent conditions'. Society of Automotive Engineers 2001-01-0530. [Cited on p. 84]

Siebers D., Higgins B., Pickett L. (2002) 'Flame lift-off on direct-injection diesel fuel jets: Oxygen concentration effects'. Society of Automotive Engineers 2002-01-0890. [Cited on p. 84]

Siegl W.O., Zinbo M., Korniski T.J., Richert J.J.O., Chladek E., Peck M.C.P., Weir J.E., Schuetzle D., Jensen T.E. (1994) 'Air toxics: A comparison of the gas- and particle-phase emissions from a high-emitter vehicle with those from a normal-emitter vehicle'. Society of Automotive Engineers 940581. [Cited on p. 175]

Siegmann K., Siegmann H.C. (2000) 'Fast and reliable "in situ" evaluation of particles and their surfaces with special reference to diesel exhaust'. Society of Automotive Engineers 2000-01-1995. [Cited on pp. 193, 194]

Siegmann K., Scherrer L., Siegmann H.C. (1999) 'Physical and chemical properties of airborne nanoscale particles and how to measure the impact on human health'. Journal of Molecular Structure (Theochem) 458:191–201. [Cited on p. 194]

Sienicki E.J., Jass R.E., Slodowske W.J., McCarthy C.I., Krodel A.L. (1990) 'Diesel fuel aromatic and cetane number effects on combustion and emissions from a prototype 1991 diesel engine'. Society of Automotive Engineers 902172. [Cited on pp. 256, 263, 264]

Signer M., Steinke R.E. (1987) 'Future trends in diesel engine design and their impact on lubricants'. Society of Automotive Engineers 871271. [Cited on p. 279]

Silva P.J., Prather K.A. (1997) 'On-line characterization of individual particles from automobile emissions'. Environmental Science and Technology 31:3074–3080. [Cited on p. 350]

Silva P.J., Liu D.-Y., Noble C.A., Prather K.A. (1999) 'Size and chemical characterization of individual particles resulting from biomass burning of local southern California species'. Environmental Science and Technology 33:3068–3076. [Cited on pp. 33, 35]

Silvis W.M., Marek G., Kreft N., Schindler W. (2002) 'Diesel particulate measurement with partial flow sampling systems: A new probe and tunnel design that correlates with full flow tunnels'. Society of Automotive Engineers 2002-01-0054. [Cited on pp. 146, 159, 160]

Simo R., Grimalt J.O., Albaiges J. (1997) 'Loss of unburned-fuel hydrocarbons from combusion aerosols during atmospheric transport'. Environmental Science and Technology 31:2697–2700. [Cited on pp. 106, 145]

Singh S.K., Agarwal A.K., Sharma M. (2006) 'Experimental investigations of heavy metal addition in lubricating oil and soot deposition in an EGR operated engine'. Applied Thermal Engineering 26:259–266. [Cited on pp. 274, 289]

Sioutas C., Koutrakis P. (1995) 'Methods for measuring atmospheric acidic particles and gases'. In *Airborne Particulate Matter*, Volume 4, Part D, Kouimtzis T., Samara C. (editors). Springer, pp. 200–232. [Cited on pp. 46, 124, 165]

Sirman M.B., Owens E.C., Whitney K.A. (2000) 'Emissions comparison of alternative fuels in an advanced automotive diesel engine'. Society of Automotive Engineers 2000-01-2048. [Cited on p. 241]

Sjöberg M. (2001) 'Correlation between flame pattern, heat-release and emissions for a DI diesel engine with rotating injector and variable swirl'. Society of Automotive Engineers 2001-01-2003. [Cited on pp. 92, 267]

Skippon S.M., Nattrass S.R., Kitching J.S., Hardiman L., Miller H. (1996) 'Effects of fuel composition on in-cylinder air/fuel ratio during fuelling transients in an SI engine, measured using differential infra-red absorption'. Society of Automotive Engineers 961204. [Cited on p. 320]

Slowik J.G., Stainken K., Davidovits P., Williams L.R., Jayne J.T., Kolb C.E., Worsnop D.R., Rudich Y., DeCarlo P.F., Jimenez J.L. (2004) 'Particle morphology and density characterization by combined mobility and aerodynamic diameter measurements. Part 2: Application to combustion-generated soot aerosols as a function of fuel equivalence ratio'. Aerosol Science and Technology 38:1206–1222. [Cited on p. 219]

Sluder C.S., Wagner R.M., Lewis S.A., Storey J.M.E. (2004) 'Exhaust chemistry of low-NO_X, low-PM diesel combustion'. Society of Automotive Engineers 2004-01-0114. [Cited on p. 293]

Smallwood G.J., Snelling D.R., Gülder Ö. L., Clavel D., Gareau D., Sawchuk R.A., Graham L. (2001) 'Transient particulate matter measurements from the exhaust of a direct injection spark ignition engine'. Society of Automotive Engineers 2001-01-3581. [Cited on pp. 188, 330, 331]

Smallwood G.J., Clavel D., Gareau D., Sawchuk R.A., Snelling D.R., Witze P.O., Axelsson B., Bachalo W.D., Gülder Ö.L. (2002) 'Concurrent quantitative laser-induced incandescence and SMPS measurements of EGR effects on particulate emissions from a TDI diesel engine'. Society of Automotive Engineers 2002-01-2715. [Cited on p. 274]

Smedley J.M., Williams A., Hainsworth D. (1995) 'Soot and carbon deposition mechanisms in ethane/air flames'. Fuel 74:1753–1761. [Cited on p. 109]

Smekens A., Godoi R.H.M., Berghmans P., van Grieken R. (2005) 'Characterisation of soot emitted by domestic heating, aircraft and cars using diesel or biodiesel'. Journal of Atmospheric Chemistry 52:45–62. [Cited on pp. 216, 223]

SMIC (1971) *Inadvertent Climate Modification: Report of the Study of Man's Impact on Climate*. MIT Press, Cambridge, MA. [Cited on p. 24]

Smith A.J., Tidmarsh D.H., Davies H., Willcock M. (1995) 'A review of the problems associated with diesel exhaust particles and the available methods for the extraction of the adsorbed hydrocarbon fraction'. Institution of Mechanical Engineers C498/28/249/96. [Cited on p. 171]

Smith D.J.T., Harrison R.M. (1998) 'Polycyclic aromatic hydrocarbons in atmospheric particles'. In *Atmospheric Particles*, Harrison R.M., van Grieken R.E. (editors), IUPAC Series on Analytical and Physical Chemistry of Environmental Systems. John Wiley & Sons, Inc., pp. 253–294. [Cited on pp. 35, 110, 117, 151, 235]

Smith D.M., Chughtai A.R (1995) 'The surface structure and reactivity of black carbon'. Colloids and Surfaces A: Physicochemical and Engineering Aspects 105:47–77. [Cited on p. 145]

Smith G.W. (1982) 'Kinetic aspects of diesel soot coagulation'. Society of Automotive Engineers 820466. [Cited on pp. 74, 76, 77, 80]

Smith J.R. (1971) 'Anatomy and physiology of the respiratory system'. Society of Automotive Engineers 710297. [Cited on p. 364]

Smith O.I. (1981) 'Fundamentals of soot formation in flames with application to diesel engine particulate emissions'. Progress in Energy and Combustion Science 7:275–291. [Cited on pp. 80, 123]

Smith R.G. (1971) 'Health aspects of atmospheric exposure to lead'. Society of Automotive Engineers 710302. [Cited on p. 315]

Smith S., Cheng Y.-S., Yeh H.C. (2001) 'Deposition of ultrafine particles in human tracheobronchial airways of adults and children'. Aerosol Science and Technology 35:697–709. [Cited on p. 365]

Smoluchowski M. von (1917) 'Versuch einer Mathematischen Theorie der Koagulationskinetik'. Zeitung für Physikalisches Chemie 92:129. [Cited on p. 21]

Smooke M.D., Long M.B., Connelly B.C., Colket M.B., Hall R.J. (2005) 'Soot formation in laminar diffusion flames'. Combustion and Flame 143:613–628. [Cited on pp. 62, 77]

Smyth K.C., Shaddix C.R. (1996) 'The elusive history of $m = 1.57 - 0.56i$ for the refractive index of soot'. Combustion and Flame 107:134–320. [Cited on p. 182]

Snelling D.R., Smallwood G.J., Sawchuk R.A., Neill W.S., Gareau D., Chippior W.L., Liu F., Gulder O.L., Bachalo W.D. (1999) 'Particulate matter measurements in a diesel engine exhaust by laser-induced incandescence and the standard gravimetric technique'. Society of Automotive Engineers 1999-01-3653. [Cited on p. 188]

Snelling D.R., Smallwood G.J., Sawchuk R.A., Neill W.S., Gareau D., Clavel D.J., Chippior W.L., Liu F., Gülder O.L., Bachalo W.D. (2000) 'In-situ real-time characterization of particulate emissions from a diesel engine exhaust by laser-induced incandescence'. Society of Automotive Engineers 2000-01-1994. [Cited on p. 188]

Sodeman D.A., Toner S.M., Prather K.A. (2005) 'Determination of single particle mass spectral signatures from light-duty vehicle emissions'. Environmental Science and Technology 39:4569–4580. [Cited on p. 119]

Song C. (2000) 'Introduction to chemistry of diesel fuels'. In *Chemistry of Diesel Fuels*, Song C., Hsu C.S., Mochida I. (editors). Taylor and Francis, pp. 1–60. [Cited on p. 123]

Song C., Na K., Cocker D.R. (2005) 'Impact of the hydrocarbon to NO_X ratio on secondary organic aerosol formation'. Environmental Science and Technology 39:3143–3149. [Cited on p. 37]

Song K.H., Lee Y-J., Litzinger T.A. (2000) 'Effects of emulsified fuels on soot evolution in an optically-accessible diesel engine'. Society of Automotive Engineers 2000-01-2794. [Cited on p. 251]

Song H., Ladommatos N., Zhao H. (2001) 'Diesel soot oxidation under controlled conditions'. Society of Automotive Engineers 2001-01-3673. [Cited on pp. 82, 84]

Song H., Ladommatos N., Zhao H. (2004) 'Morphology, size distribution, and oxidation of diesel soot'. Journal of the Energy Institute 77:26–36. [Cited on pp. 76, 80, 81, 82, 84, 223]

Song J., Alam M., Zello V., Boehman A.L., Bishop B., Walton F. (2002) 'Fuel sulfur effect on membrane coated diesel particulate filter'. Society of Automotive Engineers 2002-01-2788. [Cited on p. 123]

Song J., Lee K.O. (2007) 'Fuel property impacts on diesel particulate morphology, nanostructures, and NO_X emissions'. Society of Automotive Engineers 2007-01-0129. [Cited on p. 77]

Song J., Alam M., Boehman A.L., Kim U. (2006) 'Examination of the oxidation behavior of diesel soot'. Combustion and Flame 146:589–604. [Cited on pp. 81, 259]

Sorokin A., Arnold F. (2004) 'Electrically charged small soot particles in the exhaust of an aircraft gas-turbine engine combustor: Comparison of model and experiment'. Atmospheric Environment 38:2611–2618. [Cited on pp. 71, 123]

Soto K.F., Carrasco A., Powell T.G., Murr L.E., Garza K.M. (2006) 'Biological effects of nanoparticulate materials'. Materials Science and Engineering C 26:1421–1427. [Cited on p. 396]

Spengler J., Wilson R. (1996) 'Emissions, dispersion, and concentration of particles'. In *Particles in Our Air – Concentrations and Health Effects*, Wilson R., Spengler J. (editors). Harvard University Press, pp. 41–62. [Cited on pp. 36, 39, 40, 143]

Spicer P.T., Artelt C., Sanders S., Pratsinis S.E. (1998) 'Flame synthesis of composite carbon black – fumed silica nanostructured particles'. Journal of Aerosol Science 29:647–659. [Cited on p. 98]

Spindler G., Müller K., Brüggemann E., Gnauk T., Herrmann H. (2004) 'Long-term size-seggregated characterization of PM_{10}, $PM_{2.5}$, and PM_1 at the IfT research station Melpitz downwind of Leipzig (Germany) using high and low-volume filter samplers'. Atmospheric Environment 38:5333–5347. [Cited on p. 36]

Springer K.J., Baines T.M. (1977) 'Emissions from diesel versions of production passenger cars'. Society of Automotive Engineers 770818. [Cited on p. 125]

Springer K.J., Stahman R.C. (1977) 'Removal of exhaust particulate from a Mercedes 300D diesel car'. Society of Automotive Engineers 770716. [Cited on p. 394]

Spurny K.R. (2000a) 'Aerosol chemistry and its environmental effects'. In *Aerosol Chemical Processes in the Environment*, Spurny K.R., Hochrainer D. (editors). CRC Press LLC, pp. 3–21. [Cited on pp. 38, 397]

Spurny K.R. (2000b) 'Radioactive labeling in experimental aersosol research'. In *Aerosol Chemical Processes in the Environment*, Spurny K.R. and Hochrainer D. (editors). CRC Press LLC, pp. 213–246. [Cited on pp. 169, 381]

Spurny K.R. (2000c) 'Corrosion of asbestos cement building materials by the action of atmospheric acid aerosols and precipitations'. In *Aerosol Chemical Processes in the Environment*, Spurny K.R., Hochrainer D. (editors). CRC Press LLC, pp. 365–375. [Cited on p. 31]

Spurny K.R. (2000d) 'Atmospheric contamination by fibrous aerosols'. In *Aerosol Chemical Processes in the Environment*, Spurny K.R., Hochrainer D. (editors). CRC Press LLC, pp. 525–558. [Cited on pp. 31, 345, 355]

Spurny K.R. (2000e) 'Atmospheric contamination by agroaerosols'. In *Aerosol Chemical Processes in the Environment*, Spurny K.R., Hochrainer D. (editors). CRC Press LLC, pp. 559–575. [Cited on p. 30]

Spurny K.R. (2000f) 'Atmospheric condensation nuclei P.J.Coulier 1875 and J.Aitken 1880 (historical review)'. Aerosol Science and Technology 32:243–248. [Cited on pp. 10, 26]

Stanglmaier R.H., Roberts C.E., Moses C.A. (2002) 'Vaporization of individual fuel drops on a heated surface: A study of fuel – wall interactions within direct-injected gasoline (DIG) engines'. Society of Automotive Engineers 2002-01-0838. [Cited on p. 329]

Stanmore B., Brilhac J.-F., Gilot P. (1999) 'The ignition and combustion of cerium doped diesel soot'. Society of Automotive Engineers 1999-01-0115. [Cited on pp. 220, 223]

Stanmore B.R., Brilhac J.F., Gilot P. (2001) 'The oxidation of soot: A review of experiments, mechanisms and models'. Carbon 39:2247–2268. [Cited on pp. 168, 224]

di Stasio S., Konstandopoulos A.G., Kostoglou M. (2002) 'Cluster – cluster aggregation kinetics and primary particle growth of soot nanoarticles in flame by light scattering and numerical simulations'. Journal of Colloid and Interface Science 247:33–46. [Cited on pp. 76, 78]

Steenland K., Deddens J., Stayner L. (1998) 'Diesel exhaust and lung cancer in the trucking industry: Exposure – response analyses and risk'. American Journal of Industrial Medicine 34:220–228. [Cited on p. 373]

Stehouwer D.M., Shank G., Herzog S.N., Hyndman C.W., Kinker B.G., Simko R.P. (2002) 'Soot diesel engine oil pumpability studies as the basis of a new heavy duty diesel engine oil performance specification'. Society of Automotive Engineers 2002-01-1671. [Cited on p. 289]

Stein H.J., Herden T. (1998) 'Worldwide harmonization of exhaust emission test procedures for nonroad engines based on the international standard ISO 8178'. Society of Automotive Engineers 982043. [Cited on p. 160]

Steiner D., Burtscher H., Gross H. (1992) 'Structure and disposition of particles from a spark-ignition engine'. Atmospheric Environment 26A:997–1003. [Cited on p. 315]

Stemler E., Lawless P. (1997) 'The design and operation of a turbocharger test facility designed for transient simulation'. Society of Automotive Engineers 970344. [Cited on p. 276]

Sternbeck J., Sjödin Å. Andréasson K. (2002) 'Metal emissions from road traffic and the influence of resuspension – results from two tunnel studies'. Atmospheric Environment 36:4735–4744. [Cited on pp. 345, 347]

Stetter J.C., Foster D.E., Schauer J.J. (2005) 'Modern diesel particulate matter measurements and the application of lessons learned to 2007 levels and beyond'. Society of Automotive Engineers 2005-01-0194. [Cited on p. 278]

Stevens E., Steeper R. (2001) 'Piston wetting in an optical DISI engine: Fuel films, pool fires, and soot generation'. Society of Automotive Engineers 2001-01-1203. [Cited on p. 328]

Stevenson R. (1982) 'The morphology and crystallography of diesel particulate emissions'. Carbon 20:359–365. [Cited on p. 215]

Stevenson R. (1984) 'The effect of aluminium oxide additions to diesel engine inlet air on the morphology and crystallinity of particulate emissions'. Carbon 22:199–202. [Cited on p. 344]

Stöber W. (1987) 'On the health hazards of particulate diesel engine exhaust emissions'. Society of Automotive Engineers 871988. [Cited on pp. 372, 378]

Stöber W., Miller F.J., McClellan R.O. (1998) 'Requirements for a credible extrapolation model derived from health effects in rats exposed to particulate air pollution: A way to minimize the risks of human risk assessment?'. Applied Occupational and Environmental Hygiene 13:421–431. [Cited on p. 367]

Stokes R.H., Robinson R.A. (1949) 'Standard solutions for humidity control at 25°C. Industrial Engineering and Chemistry 41:2013. [Cited on p. 174]

Stolz A., Fleischer K., Knecht W., Nies J., Strähle R. (2001) 'Development of EGR coolers for truck and passenger car application'. Society of Automotive Engineers 2001-01-1748. [Cited on p. 275]

Stone V., Donaldson K. (1998) 'Small particles – big problem'. The Aerosol Society: Newsletter No. 3, September. [Cited on p. 366]

Stoner M., Litzinger T. (1999) 'Effects of structure and boiling point of oxygenated blending compounds in reducing diesel emissions'. Society of Automotive Engineers 1999-01-1475. [Cited on p. 259]

Storey J.M.E., Domingo N., Lewis S.A., Iric D.K. (1999) 'Analysis of semivolatile organic compounds in diesel exhaust using a novel sorption and extraction method'. Society of Automotive Engineers 1999-01-3534. [Cited on pp. 166, 172]

Storey J.M.E., Sluder C.S., Blom D.A., Higinbotham E. (2000) 'Particulate emissions from a pre-emissions control spark-ignition vehicle: A historical benchmark'. Society of Automotive Engineers 2000-01-2213. [Cited on p. 316]

Strommen M.R., Kamens R.M. (1999) 'Simulation of semivolatile organic compound microtransport at different time scales in airborne diesel soot particles'. Environmental Science and Technology 33:1738–1746. [Cited on p. 117]

Struwe F.J., Foster D.E. (2003) 'In-cylinder measurement of particulate radiant heat transfer in a direct injection diesel engine'. Society of Automotive Engineers 2003-01-0072. [Cited on p. 62]

Stunnenberg F., Kleijwegt P., de Vries Feyens A.W.L. (2001) 'Future heavy duty diesel lubricants for low emission engines'. Society of Automotive Engineers 2001-01-3768. [Cited on pp. 283, 284]

Su D.S., Müller J.-O., Jentoft R.E., Rothe D., Jacob E., Schlögl R. (2004) 'Fullerene-like soot from EuroIV diesel engine: Consequences for catalytic automotive pollution control'. Topics in Catalysis 30–31:241–245. [Cited on p. 71]

Su T.F., Farrell P.V., Nagarajan R.T. (1995) 'Nozzle effect on high pressure diesel injection'. Society of Automotive Engineers 950083. [Cited on p. 265]

Su Y., Lei Y.D., Wania F., Shoeib M., Harner T. (2006) 'Regressing gas/particle partitioning data for polycyclic aromatic hydrocarbons'. Environmental Science and Technology 40:3558–3564. [Cited on p. 145]

Sugiura K., Kagaya M. (1977) 'A study of visible smoke reduction from a small two-stroke engine using various engine lubricants'. Society of Automotive Engineers 770623. [Cited on p. 336]

Suhre B.R., Foster D.E. (1992) 'In-cylinder soot deposition rates due to thermophoresis in a direct injection diesel engine'. Society of Automotive Engineers 921629. [Cited on pp. 92, 94, 95]

Summers C.E. (1925) 'The physical and road characteristics of road and field dust'. Society of Automotive Engineers 250010. [Cited on p. 342]

Sun J.D., Bond J.A., Dahl A.R. (1988) 'Biological disposition of airborne emissions: Particle-associated organic constituents'. In *Air Pollution, the Automobile, and Public Health*, Watson A.Y., Bates R.R., Kennedy D. (editors). National Academy Press, pp. 299–322. [Cited on p. 396]

Sun Y., Frenkel A.I., Isserhoff R., Shonbrun C., Forman M., Shin K., Koga T., White H., Zhang L., Zhu Y., Rafailovich M.H., Sokolov J.C. (2006) 'Characterization of palladium nanoparticles by using X-ray reflectivity, EXAFS, and electron microscopy'. Langmuir 22:807–816. [Cited on p. 396]

Sunderland P.B., Faeth G.M. (1996) 'Soot formation in hydrocarbon/air laminar jet diffusion flames'. Combustion and Flame 105:132–146. [Cited on pp. 68, 73]

Sunderland P.B., Köylü Ü.Ö., Faeth G.M. (1995) 'Soot formation in weakly buoyant acetylene-fueled laminar jet diffusion flames burning in air'. Combustion and Flame 100:310–322. [Cited on p. 73]

Suppes G.J., Lula C.J., Burkhart M.L., Swearingen J.D. (1999) 'Type performance of Fischer – Tropsch liquids (FTL) in modified off-highway diesel engine test cycle'. Society of Automotive Engineers 1999-01-1474. [Cited on p. 257]

Suresh A., Johnson J.H. (2001) 'A study of the dilution effects on particle size measurement from a heavy-duty diesel engine with EGR'. Society of Automotive Engineers 2001-01-0220. [Cited on p. 148, 304]

Sutela C., Collings N., Hands T. (2000) 'Real time CO_2 measurement to determined transient intake gas composition under EGR conditions'. Society of Automotive Engineers 2000-01-2953. [Cited on p. 272]

Sutton M., Britton N., Otterholm B., Tengström P., Frennfelt C., Walker A., Murray I. (2004) 'Investigations into lubricant blocking of diesel particulate filters'. Society of Automotive Engineers 2004-01-3013. [Cited on p. 282]

Suzuki H., Toyooka T., Ibuki Y. (2007) 'Simple and easy method to evaluate uptake potential of nanoparticles in mammalian cells using a flow cytometric light scatter analysis'. Environmental Science and Technology 41:3018–3024. [Cited on p. 368]

Suzuki J., Yamazaki H., Yoshida Y., Hori M. (1985) 'Development of a dilution mini-tunnel and its availability for measuring diesel exhaust particulate matter'. Society of Automotive Engineers 851547. [Cited on p. 160]

Swartz E., Stockburger L., Gundel L.A. (2003) 'Recovery of semivolatile organic compounds during sample preparation: Implications for characterization of airborne particulate matter'. Environmental Science and Technology 37:597–605. [Cited on p. 178]

T

Takada T., Ikezawa H., Kotani Y. (2001) 'Determination of polyaromatic hydrocarbons in particulate matter with HPLC and 3D-detector'. Society of Automotive Engineers 2001-01-3318. [Cited on p. 177]

Takahashi S., Wakimoto K., Iida N., Nikolic D. (2001) 'Effects of aromatics content and 90 % distillation temperature of diesel fuels on flame temperature and soot formation'. Society of Automotive Engineers 2001-01-1940. [Cited on p. 256]

Takakura T., Ishikawa Y., Ito K. (2005) 'The wear mechanism of piston rings and cylinder liners under cooled-EGR condition and the development of surface treatment technology for effective wear reduction'. Society of Automotive Engineers 2005-01-1655. [Cited on pp. 252, 274, 289]

Takatsuka T., Inoue S.-i., Wada Y. (1997) 'Deep hydrodesulphurization process for diesel oil'. Catalysis Today 39:69–75. [Cited on p. 252]

Takeuchi Y., Hirano S., Kanauchi M., Ohkubo H., Nakazato M., Sutherland M., van Dam W. (2003) 'The impact of diesel engine lubricants on deposit formation in diesel particulate filters'. Society of Automotive Engineers 2003-01-1870. [Cited on pp. 104, 281]

Talib R.J., Muchtar A., Azhari C.H. (2003) 'Microstructual characteristics on the surface and subsurface of semimetallic automotive friction materials during braking processes'. Journal of Materials Processing Technology 149:694–699. [Cited on p. 345]

Tamanouchi M., Morihisa H., Yamada S., Iida J., Sasaki T., Sue H. (1997) 'Effects of fuel properties on exhaust emissions for diesel engines with and without oxidation catalyst and high pressure injection'. Society of Automotive Engineers 970758. [Cited on p. 263]

Tambour Y., Khosid S. (1995) 'On the stability of the process of formation of combustion-generated particles by coagulation and simultaneous shrinkage due to particle oxidation'. International Journal of Engineering Science 33:667–687. [Cited on pp. 76, 78]

Tammet H. (1995) 'Size and mobility of nanometer particles, clusters and ions'. Journal of Aerosol Science 26:459–475. [Cited on p. 395]

Tan P.Q., Hu Z.Y., Deng K.Y., Lu J.X., Lou D.M., Wan G. (2007) 'Particulate matter emission modelling based on soot and SOF from direct injection engines'. Energy Conversion and Management 48:510–518. [Cited on p. 109]

Tan P.V., Malpica O., Evans G.J., Owega S., Fila M.S. (2002) 'Chemically-assigned classification of aerosol mass spectra'. Journal of the American Society of Mass Spectrometry 13:826–838. [Cited on p. 209]

Tanaka S., Shimizu T. (1999) 'A study of composition and size distribution of particulate matter from DI diesel engine'. Society of Automotive Engineers 1999-01-3487. [Cited on pp. 263, 264]

Tanaka T., Yasunishi S., Watanabe N., Kobashi K. (1989) 'Development of a measuring meter and a control device for diesel white smoke'. Society of Automotive Engineers 892044. [Cited on pp. 122, 204]

Tang I.N. (2000) 'Phase transformation and growth of hygroscopic aerosols'. In *Aerosol Chemical Processes in the Environment*, Spurny K.R., Hochrainer D. (editors). CRC Press LLC, pp. 61–80. [Cited on p. 19]

Tang S., Johnson R., Lanni T., Webster W. (2001) 'Monitoring of PM-bound polycyclic aromatic hydrocarbons from diesel vehicles by photoelectric aerosol sensor'. Society of Automotive Engineers 2001-01-3578. [Cited on p. 194]

Tanin K.V., Wickman D.D., Montgomery D.T., Das S., Reitz R.D. (1999) 'The influence of boost pressure on emissions and fuel consumption of a heavy-duty single-cylinder D.I. diesel engine'. Society of Automotive Engineers 1999-01-0840. [Cited on p. 278]

Tanner R.L. (1998) 'Speciation techniques for fine atmospheric aerosols'. In *Atmospheric Particles*, Harrison R.M., van Grieken R.E. (editors), IUPAC Series on Analytical and Physical Chemistry of Environmental Systems. John Wiley & Sons, Inc., pp. 147–171. [Cited on pp. 19, 175]

Tao F., Chomiak J. (2002) 'Numerical investigation of reaction zone structure and flame liftoff of DI diesel sprays with complex chemistry'. Society of Automotive Engineers 2002-01-1114. [Cited on p. 69]

Tao F., Golovitchev V.I., Chomiak J. (2004) 'A phenomenological model for the prediction of soot formation in diesel spray combustion'. Combustion and Flame 136:270–282. [Cited on pp. 72, 74, 79]

Tashiro K., Ito S., Oba A., Yokomizo T. (1995) 'Development of oxidation catalyst for diesel passenger car'. JSAE Review 16:131–136. [Cited on p. 120]

Taskinen P. (2000) 'Modelling medium speed diesel engine combustion, soot and NOx-emission formations'. Society of Automotive Engineers 2000-01-1886. [Cited on pp. 67, 83]

Taskinen P., von Hollen P., Karvinen R., Liljenfeldt G., Salminen H.J. (1998) 'Simulation of combustion, soot and NO_X-emissions in a large medium speed diesel engine'. Society of Automotive Engineers 981449. [Cited on p. 67]

Taylor G.W.R. (2001) 'The effect of lubricating oil volatility on diesel emissions'. Society of Automotive Engineers 2001-01-1261. [Cited on p. 281]

Taylor R. (1997) 'Carbon blacks: Production, properties and applications'. In *Introduction to Carbon Technologies*, Marsh H., Heintz E.A., Rodríguez-Reinoso F. (editors). University of Alicante, pp. 167–210. [Cited on pp. 62, 229]

Terzi E., Samara C. (2004) 'Gas-particle partitioning of polycyclic aromatic hydrocarbons in urban, adjacent coastal, and continental background sites of western Greece'. Environmental Science and Technology 38:4973–4978. [Cited on p. 117]

Tesner P.A., Snegirova T.D., Knorre V.G. (1971) 'Kinetics of dispersed carbon formation'. Combustion and Flame 17:253–260. [Cited on p. 67]

Tessier L.P., Sullivan H.F., Bragg G.M., Hermance C.E. (1980) 'The development of a high efficiency diesel exhaust particulate filter'. Society of Automotive Engineers 800338. [Cited on p. 301]

Thijssen J.H., Toqan M.A., Beér J.M., Sarofim A.F. (1994) 'The formation and destruction of aromatic compounds in a turbulent flame'. Twenty-Fifth Symposium (International) on Combustion, The Combustion Institute, pp. 1215–1222. [Cited on p. 69]

Thimsen D.P., Baumgard K.J., Kotz T.J., Kittelson D.B. (1990) 'The performance of an electrostatic agglomerator as a diesel soot emission control device'. Society of Automotive Engineers 900330. [Cited on p. 394]

Thorsen W.A., Cope W.G., Shea D. (2004) 'Bioavailability of PAHs: Effects of soot carbon and PAH source'. Environmental Science and Technology 38:2029–2037. [Cited on p. 35]

Tian K., Liu F., Thomson K.A., Snelling D.R., Smallwood G.J., Wang D. (2004) 'Distribution of the number of primary particles of soot aggregates in a nonpremixed laminar flame'. Combustion and Flame 138:195–198. [Cited on p. 217]

Literature Cited (Cross-referenced Against the Text)

Tian K., Thomson K.A., Liu F., Snelling D.R., Smallwood G.J., Wang D. (2006) 'Determination of the morphology of soot aggregates using the relative optical density method for the analysis of TEM images'. Combustion and Flame 144:782–791. [Cited on p. 215]

Timoney D.J. (1985) 'A simple technique for predicting optimum fuel – air mixing conditions in a direct injection diesel engine with swirl'. Society of Automotive Engineers 851543. [Cited on p. 277]

Timoney D.J., Brophy B., Smith W.J. (1997) 'Heat release and emissions results from a D.I. diesel with special shrouded intake valves'. Society of Automotive Engineers 970900. [Cited on p. 277]

Tokura N., Terasaka K., Yasuhara S. (1982) 'Process through which soot intermixes into lubricating oil of a diesel engine with exhaust gas circulation'. Society of Automotive Engineers 820082. [Cited on p. 287]

Tomiyasu B., Owari M., Nihei Y. (2006) 'TOF-SIMS measurements of the exhaust particles emitted from gasoline and diesel engine vehicles'. Applied Surface Science 252:7026–7029. [Cited on p. 95]

Toner S.M., Sodeman D.A., Prather K.A. (2006) 'Single particle characterization of ultrafine and accumulation mode particles from heavy duty diesel vehicles using aerosol time-of-flight mass spectrometry'. Environmental Science and Technology 40:3912–3921. [Cited on p. 209]

Tosaka S., Fujiwara Y. (2000) 'The characteristics of chemical reaction of diesel fuel'. JSAE Review 21:463–468. [Cited on pp. 107, 108]

Tosaka S., Fujiwara Y., Murayama T. (1989) 'The effect of fuel properties on particulate formation (the effect of molecular structure and carbon number)'. Society of Automotive Engineers 891881. [Cited on p. 107]

Trapel E., Ifeacho P., Roth P. (2004) 'Injection of hydrogen peroxide into the combustion chamber of diesel engine: Effects on the exhaust gas behaviour'. Society of Automotive Engineers 2004-01-2925. [Cited on p. 252]

Trapel E., Mayer C., Schulz C., Roth P. (2005) 'Effects of bio diesel injection in a DI diesel engine on gaseous and particulate emission'. Society of Automotive Engineers 2005-01-2204. [Cited on p. 258]

Tree D.R., Dec J.E. (2001) 'Extinction measurements of in-cylinder soot deposition in a heavy-duty DI diesel engine'. Society of Automotive Engineers 2001-01-1296. [Cited on pp. 84, 92, 93, 94, 95]

Tree D.R., Svensson K.I. (2007) 'Soot processes in compression ignition engines'. Progress in Energy and Combustion Science (in press). [Cited on pp. 87, 89, 90, 257, 267]

Treuhaft M.B., Wisnewski J.P. (1977) 'Trapping of lead particulates in automotive exhaust'. Society of Automotive Engineers 770059. [Cited on pp. 315, 316]

Tritthart P., Chichocki R., Cartellieri W. (1993) 'Fuel effects on emissions in various test cycles in advanced passenger car diesel vehicles'. Society of Automotive Engineers 932684. [Cited on p. 255]

Truhan J.J., Covington C.B., Wood L. (1995) 'The classification of lubricating oil contaminants and their effect on wear in diesel engines as measured by surface layer activation'. Society of Automotive Engineers 952558. [Cited on p. 344]

Truhan J.J., Qu J., Blau P.J. (2005) 'The effect of lubricating oil condition on the friction and wear of piston ring and cylinder liner materials in a reciprocating bench test'. Wear 259:1048–1055. [Cited on p. 290]

Tsolakis A., Megaritis A. (2005) 'Partially premixed charge compression ignition engine with on-board H_2 production by exhaust gas fuel reforming of diesel and biodiesel'. International Journal of Hydrogen Energy 30:731–745. [Cited on pp. 252, 274]

Tsolakis A., Hernandez J.J., Megaritis A., Crampton M. (2005) 'Duel fuel diesel engine operation using H_2. Effect on particulate emissions'. Energy and Fuels 29:418–425. [Cited on p. 252]

Tsujimura T., Goto S., Matsubara M. (2007) 'A study of PM emission characteristics of diesel vehicle fueled with GTL'. Society of Automotive Engineers 2007-01-0028. [Cited on p. 256]

Tsunemoto H., Yamada T., Ishitani H. (1986) 'Behavior of adhering fuel on cold combustion chamber wall in direct injection diesel engines'. Society of Automotive Engineers 861235. [Cited on p. 120]

Tsunemoto H., Ishitani H., Yamazaki S., Nakajima S. (1999) 'New measuring method for blue and white smoke in diesel engines by a digital camera system'. Society of Automotive Engineers 1999-01-1503. [Cited on p. 204]

Tsunemoto H., Ishitani H., Kaimai T. (2001) 'Type effect of fuel properties on unburned HC and particulate matter emissions in a small DI diesel engine'. Society of Automotive Engineers 2001-01-3387. [Cited on pp. 254, 256]

Tuit C.B., Ravizza G.E., Bothner M.H. (2000) 'Anthropogenic platinum and palladium in the sediments of Boston Harbor'. Environmental Science and Technology 34:926–932. [Cited on p. 350]

Turpin B.J., Huntzicker J.J., Adams K.M. (1990) 'Intercomparison of photoacoustic and thermal-optical methods for the measurement of atmospheric elemental carbon'. Atmospheric Environment 24A:1831–1835. [Cited on pp. 169, 193]

Twomey S. (1977) *Atmospheric Aerosols*, Developments in Atmospheric Science, Vol. 7. Elsevier. [Cited on pp. 12, 23, 26, 27]

U

Uchida N., Daisho Y., Saito T. (1992) 'The control of diesel emissions by supercharging and varying fuel-injection parameters'. Society of Automotive Engineers 920117. [Cited on p. 277]

Ulfvarson U. (2000) 'Diesel-exhaust tests should be revised with respect to health-indicators'. Society of Automotive Engineers 2000-01-0235. [Cited on p. 250]

Ulfvarson U., Alexandersson R., Aringer L., Svensson E., Hedenstierna G., Hogstedt C., Holmberg B., Rosen G., Sorsa M. (1987) 'Effects of exposure to vehicle exhaust on health'. Scandanavian Journal of Work, Environment and Health 13:505–512. [Cited on p. 360]

Ullman T.L., Human D.M. (1991) 'Fuel and maladjustments effects on emissions from a diesel bus engine'. Society of Automotive Engineers 910735. [Cited on p. 263]

Ullman T.L., Hare C.T., Baines T.M. (1984) 'Preliminary trap tests on a 2-stroke diesel bus engine'. Society of Automotive Engineers 840079. [Cited on p. 304]

Uner D., Demirkol M.K., Dernaika B. (2005) 'A novel catalyst for diesel soot oxidation'. Applied Catalysis B: Environmental 61:2334–345. [Cited on p. 295]

Unsworth J.F., den Ouden C.J.J., Simm D.L., Wilson G.J. (1996) 'Fuel quality effects on oxidation exhaust catalysts in light-duty diesel motor vehicles'. Society of Automotive Engineers 961183. [Cited on p. 297]

Urabe M., Tomomatsu T., Ishiki K., Takiguchi M., Someya T. (1998) 'Variation of piston friction force and ring lubricating condition in a diesel engine with EGR'. Society of Automotive Engineers 982660. [Cited on pp. 279, 290]

Usui S., Ito K., Kato K. (2004) 'The effect of semi-circular micro riblets on the deposition of diesel exhaust particulate'. Society of Automotive Engineers 2004-01-0969. [Cited on pp. 274, 275]

Utell M., Samet J. (1996) 'Airborne particles and respiratory disease: Clinical and pathogenetic considerations'. In *Particles in Our Air – Concentrations and Health Effects*, Wilson R., Spengler J. (editors). Harvard University Press, pp. 169–188. [Cited on pp. 366, 371, 381]

Uyehara O.A. (1980) 'Diesel combustion temperature on soot'. Society of Automotive Engineers 800969. [Cited on p. 292]

V

Vaaraslahti K., Keskinen J., Giechaskiel B., Solla A., Murtonen T., Vesala H. (2005) 'Effect of lubricant on the formation of heavy-duty diesel exhaust nanoparticles'. Environmental Science and Technology 39:8497–8504. [Cited on pp. 112, 252, 283]

Valberg P.A., Long C.M. (2006) 'Comment on "Vehicle self-pollution intake fraction: Children's exposure to school bus emissions" '. Environmental Science and Technology 40:3123. [Cited on p. 359]

Valentine J.M., Peter-Hoblyn J.D., Acres G.K. (2000) 'Emissions reduction and improved fuel economy performance from a bimetallic platinum/cerium diesel fuel additive at ultra-low dose rates'. Society of Automotive Engineers 2000-01-1934. [Cited on p. 262]

Vander Wal R.L. (1998) 'Soot precursor carbonization: Visualization using LIF and LII and comparison using bright and dark field TEM'. Combustion and Flame 112:607–616. [Cited on p. 107]

Vander Wal R.L. (2002) 'Fe-catalyzed single-walled carbon nanotube synthesis within a flame environment'. Combustion and Flame 130:37–47. [Cited on p. 98]

Vander Wal R.L., Mueller C.J. (2006) 'Initial investigation of effects of fuel oxygenation on nanostructure of soot from a direct-injection diesel engine'. Energy and Fuels 20:2364–2369. [Cited on p. 259]

Vander Wal R.L., Tomasek A.J. (2004) 'Soot nanostructure: Dependence upon synthesis conditions'. Combustion and Flame 136:129–140. [Cited on p. 62]

Vander Wal R.L., Choi M.Y., Lee K.-O. (1995) 'The effects of rapid heating of soot: Implications when using laser-induced incandescence for soot diagnostics'. Combustion and Flame 102:200–204. [Cited on pp. 73, 81, 188]

Vander Wal R.L., Jensen K.A., Choi M.Y. (1997) 'Simultaneous laser-induced emission of soot and polycyclic aromatics hydrocarbons within a gas-jet diffusion flame'. Combustion and Flame 109:399–414. [Cited on p. 106]

Vander Wal R.L., Yezerets A., Currier N.W., Kim D.H., Wang C.M. (2007) 'HRTEM study of diesel soot collected from diesel particulate filter'. Carbon 45:70–77. [Cited on p. 301]

Vanrullen I., Chaumontet C., Pornet P., Véran F., Martel P. (2000) 'The oxidation catalytic converter reduces the inhibitory activity of soluble organic fractions of diesel particles on intercellular communication'. Environmental Science and Technology 34:1352–1358. [Cited on pp. 297, 375, 376]

Vehkamäki H., Kulmala M., Lehtinen K.E.J., Noppel K.E.J. (2003) 'Modelling binary homogeneous nucleation of water – sulfuric acid vapors: Parameterisation for high temperature emissions'. Environmental Science and Technology 37:3392–3398. [Cited on p. 126]

Venkatesan C.P., Abraham J. (2000) 'An investigation of the dependence of NO and soot emissions from a diesel engine on heat release rate characteristics'. Society of Automotive Engineers 2000-01-0509. [Cited on p. 66]

Verrrant J.A., Kittelson D.B. (1977) 'Sampling and physical characterization of diesel exhaust aerosols'. Society of Automotive Engineers 770720. [Cited on p. 143]

Vertin K.D., Ohi J.M., Naegeli D.W., Childress K.H., Hagen G.P., McCarthy C.I., Cheng A.S., Dibble R.W. (1999) 'Methyl and methyl-diesel blended fuels for use in compression-ignition engines'. Society of Automotive Engineers 1999-01-1508. [Cited on p. 257]

Vignati E., Berkowicz R., Palmgren F., Lyck E., Hummelshj P. (1999) 'Transformation of size distributions of emitted particles in streets'. Science of the Total Environment 235:37–49. [Cited on pp. 144, 145]

Villani K., Vermandel W., Smets K., Liang D., van Tendeloo G., Martens J.A. (2006) 'Platinum particle size and support effects in NO_X mediated carbon oxidation over platinum catalysts'. Environmental Science and Technology 40:2727–2733. [Cited on p. 295]

Villinger J., Federer W., Praun S., Zeiner W., Fürbacher R., Binder O., Kitzler H. (2002a) 'Comparative study of butadiene and B, T, X tailpipe emissions for gasoline of different octane levels'. Society of Automotive Engineers 2002-01-1643. [Cited on p. 330]

Villinger J., Federer W., Praun S. (2002b) 'Continuous pre- and post-catalyst hydrocarbon and nitrogen compounds-monitoring of various DeNOx reactions by twin chemical ionization spectrometry'. Society of Automotive Engineers 2002-01-1679. [Cited on p. 46]

Vincent M.W., Richards P.J. (2000) 'The long distance road trial of a combined diesel particulate filter and fuel additive'. Society of Automotive Engineers 2000-01-2849. [Cited on p. 285]

Vincent M.W., Richards P.J., Dementhon J.-B., Martin B. (1999) 'Improved diesel particulate filter regeneration performance using fuel soluble additives'. Society of Automotive Engineers 1999-01-3562. [Cited on p. 104]

Violi A., d'Anna A., d'Alessio A. (1999) 'Modeling of particulate formation in combustion and pyrolysis'. Chemical Engineering Science 54:3433–3442. [Cited on p. 69]

Virtanen A., Ristimäki J., Marjamäki M., Vaaraslahti K., Keskinen J., Lappi M. (2002) 'Effective density of diesel exhaust particles as a function of size'. Society of Automotive Engineers 2002-01-0056. [Cited on pp. 11, 219, 221, 222]

Virtanen A.K.K., Ristimäki J.M., Vaaraslahti K.M., Keskinen J. (2004) 'Effect of engine load on diesel soot particles'. Environmental Science and Technology 38:2551–2556. [Cited on pp. 77, 218]

Vogl G., Elstner E.F. (1989) 'Diesel soot particles catalyze the production of oxy-radicals'. Toxicology Letters 47:17–23. [Cited on p. 376]

Vogt R., Scheer V. (2002) 'Particles in diesel vehicle exhaust: A comparison of laboratory and chasing experiments'. Proceedings of the 11th International Symposium on Transport and Air Pollution, Graz, Austria, Pischinger R. (editor), June 19–21, pp. 79–84. [Cited on p. 144]

Vojtisek-Lom M., Allsop J.E. (2001) 'Development of heavy-duty diesel portable, on-board mass exhaust emissions monitoring system with NOX, CO2 and qualitative OM capabilities'. Society of Automotive Engineers 2001-01-3641. [Cited on p. 190]

Volckens J., Braddock J., Snow R.F., Crews W. (2007) 'Emissions profile from new and in-use handheld, 2-stroke engines'. Atmospheric Environment 41:640–649. [Cited on p. 336]

Vonarb R., Hachimi A., Jean E., Bianchi D. (2005) 'Catalytic oxidation of a diesel soot formed in the presence of a cerium additive. II. Temperature-programmed experiment on the surface-oxygenated complexes and kinetic modeling'. Energy and Fuels 19:35–48. [Cited on pp. 99, 231]

Voss K., Cioffi J., Gorel A., Norris M., Rotolico T., Fabel A. (1997) 'Zirconia based ceramic, in-cylinder coatings and aftertreatment oxidation catalysts for reduction of emissions from heavy duty diesel engines'. Society of Automotive Engineers 970469. [Cited on pp. 93, 294]

Vouitsis E., Ntziachristos L., Samaras Z. (2005) 'Modelling of diesel exhaust aerosol during laboratory sampling'. Atnospheric Environment 39:1335–1345. [Cited on p. 128]

Vouitsis E., Ntziachristos L., Samaras Z., Grigoratos T., Samara C., Miltsios G. (2007) 'Effect of a DPF and low sulfur lube oil on PM physicochemical characteristics from a Euro 4 light duty diesel vehicle'. Society of Automotive Engineers 2007-01-0314. [Cited on p. 283]

Vuk C.T., Jones M.A., Johnson J.H. (1976) 'The measurement and analysis of the physical character of diesel particulate emissions'. Society of Automotive Engineers 760131. [Cited on pp. 137, 184, 217, 219, 229, 231, 234, 387]

W

Wade W.R., Jones C.M. (1984) 'Current and future light duty diesel engines and their fuels'. Society of Automotive Engineers 840105. [Cited on p. 263]

Wade W.R., Hunter C.E., Trinker F.N., Hansen S.P. (1986) 'Future diesel engine combustion systems for low emissions and high fuel economy'. FISITA 865012. [Cited on p. 91]

Wadhwa A.R., Abraham J. (2000) 'An investigation of the dependence of NO and soot formation and oxidation in transient combusting jets on injection and chamber conditions'. Society of Automotive Engineers 2000-01-0507. [Cited on p. 84]

Wadhwa A.R., Gopalakrishnan V., Abraham J. (2001) 'A mixture fraction averaged approach to modeling NO and soot in diesel engines'. Society of Automotive Engineers 2001-01-1005. [Cited on p. 267]

Wagner R.M., Green J.B., Dam T.Q., Edwards K.D., Storey J.M. (2003) 'Simultaneous low engine-out NOx and particulate matter with highly diluted diesel combustion'. Society of Automotive Engineers 2003-01-0262. [Cited on p. 292]

Wagner U., Anca R., Velji A., Spicher U. (2003) 'An experimental study of homogeneous charge compression ignition (HCCI) with various compression ratios, intake air temperatures and fuels with port and direct fuel injection'. Society of Automotive Engineers 2003-01-2293. [Cited on p. 292]

Wåhlin P., Palmgren F., van Dingenen R., Raes F. (2001) 'Pronounced decrease of ambient particle number emissions from diesel traffic in Denmark after reduction of the sulphur content in diesel'. Atmospheric Environment 35:3549–3552. [Cited on p. 254]

Wakisaka Y., Azetsu A. (2002) 'Effects of fuel injection rate shaping on combustion and emission formation in intermittent spray'. Society of Automotive Engineers 2002-01-1159. [Cited on p. 267]

Waldenmaier D.A., Gratz L.D., Bagley S.T., Johnson J.H., Leddy D.G. (1990) 'The influence of sampling conditions on the repeatability of diesel particulate and vapor phase hydrocarbon and PAH measurements'. Society of Automotive Engineers 900642. [Cited on pp. 166, 176, 236]

Wall J.C., Hoekman S.K. (1984) 'Fuel composition effects on heavy-duty diesel particulate emissions'. Society of Automotive Engineers 841364. [Cited on pp. 125, 126, 152, 174, 179, 251, 252, 253, 256]

Wallace L.A., Emmerich S.J., Howard-Reed C. (2004) 'Source strengths of ultrafine and fine particles due to cooking with a gas stove'. Environmental Science and Technology 38:2304–2311. [Cited on p. 359]

Walsh M.P. (1983) 'The benefits and costs of light duty diesel particulate control'. Society of Automotive Engineers 830179. [Cited on pp. 358, 397]

Walsh M.P. (1985) 'The benefits and costs of diesel particulate control III – the urban bus'. Society of Automotive Engineers 850148. [Cited on p. 358]

Walsh M.P. (1987) 'Motor vehicle air pollution in Europe – a problem still not solved'. Institution of Mechanical Engineers C339/87. [Cited on p. 233]

Walsh M.P. (1993) 'Global trends in diesel particulate control, 1993 update'. Society of Automotive Engineers 930126. [Cited on p. 6]

Walsh M.P. (1995) 'Global trends in diesel particulate control – a 1995 update'. Society of Automotive Engineers 950149. [Cited on p. 371]

Walsh M.P. (2001) 'Global trends in diesel emissions regulation'. Society of Automotive Engineers 2001-01-0183. [Cited on pp. 358, 363, 373, 398]

Walsh M.P., Bradow R. (1991) 'Diesel particulate control around the world'. Society of Automotive Engineers 910130. [Cited on pp. 6, 362]

Walsh M.P., Branco G.M., Ryan J., Linke R.R.A., Romano J., Martins M.H.R.B. (2005) 'Clean diesels: The key to clean air in São Paulo'. Society of Automotive Engineers 2005-01-2215. [Cited on p. 360]

Walton O.R. (2004) 'Potential discrete element simulation applications ranging from airborne fines to pellet beds'. Society of Automotive Engineers 2004-01-2329. [Cited on p. 18]

Wan M.P., Chao C.Y.H., Ng Y.D., Sze To G.N., Yu W.C. (2007) 'Dispersion of expiratory droplets in a general hospital ward with ceiling mixing type mechanical ventilation system'. Aerosol Science and Technology 41:244–258. [Cited on p. 29]

Wang B., Lee S.C., Ho K.F., Kang Y.M. (2007) 'Characteristics of emissions of air pollutants from burning incense in temples, Hong Kong'. Science of the Total Environment 377:52–60. [Cited on p. 38]

Wang H., Frenklach M. (1997) 'A detailed kinetic modeling study of aromatics formation in laminar premixed acetylene and ethylene flames'. Combustion and Flame 110:173–221. [Cited on pp. 68, 69]

Wang Y.-F., Huang K.-L., Li C.-T., Mi H.-H., Luo J.-H., Tsai P.-J. (2003) 'Emissions of fuel metals content from a diesel vehicle engine'. Atmospheric Environment 37:4637–4643. [Cited on p. 96]

Warey A., Huang Y., Matthews R., Hall M., Ng H. (2002) 'Effects of piston wetting on size and mass of particulate matter emissions in a DISI engine'. Society of Automotive Engineers 2002-01-1140. [Cited on p. 328]

Warey A., Hendrix B., Hall M., Nevius T. (2004) 'A new sensor for on-board detection of particulate carbon mass emissions from engines'. Society of Autommotive Engineers 2004-01-2906. [Cited on p. 196]

Warner J.R., Huynh C., Janakiraman G., Johnson J.H., Bagley S.T. (2002) 'Oxidation catalytic converter and emulsified fuel effects on heavy-duty diesel engine particulate matter emissions'. Society of Automotive Engineers 2002-01-1278. [Cited on pp. 220, 251]

Warner J.R., Johnson J.H., Bagley S.T., Huynh C.T. (2003) 'Effects of a catalyzed particulate filter on emissions from a diesel engine: Chemical characterization data and particulate emissions measured with thermal optical and gravimetric methods'. Society of Automotive Engineers 2003-01-0049. [Cited on pp. 151, 169, 303, 304]

Wasil J.R., Montgomery D.T. (2003) 'A method to determine total PM emissions from marine outboard engines'. Society of Automotive Engineers 2003-32-0049. [Cited on p. 335]

Wasil J.R., Montogomery D.T., Strauss S., Bagley S.T. (2004) 'Life assessment of PM, gaseous emissions, and oil useage in modern marine outboard engines'. Society of Automotive Engineers 2004-32-0092. [Cited on pp. 336, 338]

Watabe Y., Irako K., Miyajima T., Yoshimoto T., Murakami Y. (1983) ' "Trapless" trap – a catalytic combustion system of diesel particulates using ceramic foam'. Society of Automotive Engineers 830082. [Cited on p. 295]

Watson H.C., Lu H. (1993) '25 years of emissions control: Its costs and benefits'. In *Worldwide Emissions Standards and How to Meet Them*, Institution of Mechanical Engineers. [Cited on p. 315]

Weaver C.S., Balam-Almanza M.V. (2001) 'Development of the "RAVEM" ride-along vehicle emission measurement system for gaseous and particulate emissions'. Society of Automotive Engineers 2001-01-3644. [Cited on p. 142]

Weaver C.S., Klausmeier R.F. (1986) 'Inspection and maintenance for heavy-duty diesel vehicles: Part I – evaluating the need'. Society of Automotive Engineers 861546. [Cited on p. 276]

Weaver C.S., Klausmeier R.J., Erickson L.M., Gallagher J., Hollman T. (1986) 'Feasibility of retrofit technologies for diesel emission control'. Society of Automotive Engineers 860296. [Cited on p. 368]

Weaver C.S., Turner S.H., Balam-Almanza M.V., Gable R. (2000) 'Comparison of in-use emissions from diesel and natural gas trucks and buses'. Society of Automotive Engineers 2000-01-3473. [Cited on p. 388]

Weber R.J., Stolzenburg M.R., Pandis S.P., McMurry P.H. (1998) 'Inversion of ultrafine condensation nucleus counter pulse height distributions to obtain nanoparticle ($\sim$3–10nm) size distributions'. Journal of Aerosol Science 29:601–615. [Cited on p. 190]

Weckwerth G. (2001) 'Verification of traffic emitted aerosol components in the ambient air of Cologne (Germany)'. Atmospheric Environment 35:5525–5536. [Cited on p. 31]

Wedekind B.G.A., Andersson J.D., Hall D., Stradling R., Barnes C., Wilson G. (2000) 'DETR/SMMT/CONCAWE particulate research program: Heavy-duty results'. Society of Automotive Engineers 2000-01-2851. [Cited on p. 391]

Wedlock D.J., Shuff P., Dare-Edwards M., Jia X., Williams R.A. (1999) 'Experimental and simulation approaches to understanding soot aggregation'. Society of Automotive Engineers 1999-01-1516. [Cited on p. 288]

Wei Q., Kittelson D.B., Watts W.F. (2001a) 'Single-stage dilution tunnel performance'. Society of Automotive Engineers 2001-01-0201. [Cited on pp. 146, 147, 149, 159, 160, 163, 164, 253, 254]

Wei Q., Kittelson D.B., Watts W.F. (2001b) 'Single-stage dilution tunnel design'. Society of Automotive Engineers 2001-01-0207. [Cited on pp. 159, 161, 162, 163]

Weilmünster P., Keller A., Homann K.-H. (1999) 'Large molecules, radicals, ions, and small soot particles in fuel-rich hydrocarbon flames Part I: Positive ions of polycyclic aromatic hydrocarbons (PAH) in low-pressure premixed flames of acetylene and oxygen'. Combustion and Flame 116:62–83. [Cited on p. 71]

Weinberg F.J. (1983) 'Electrical intervention in the sooting of flames'. In *Soot in Combustion Systems and its Toxic Properties*, Lahaye J., Prado G. (editors). Plenum Press, pp. 243–257. [Cited on p. 71]

Weingartner E., Keller C., Stahel W.A., Burtscher H., Baltensperger U. (1997) 'Aerosol emission in a road tunnel'. Atmospheric Environment 31:451–462. [Cited on p. 348]

Wellburn A. (1994) *Air Pollution and Climate Change – the Biological Impact*. Longman. [Cited on p. 36]

Wen J.Z., Thomson M.J., Lightstone M.F., Rogak S.N. (2006) 'Detailed kinetic modeling of carbonaceous nanoparticle inception and surface growth during pyrolysis of C_6H_6 behind shock waves'. Energy and Fuels 20:547–559. [Cited on p. 72]

Wendt J.O.L. (1994) 'Combustion science for incineration technology'. Twenty-Fifth Symposium (International) on Combustion, The Combustion Institute, pp. 277–289. [Cited on p. 98]

Westerholm R., Christensen A., de Serves C., Almén J. (1999) 'Determination of polycyclic aromatic hydrocarbons (PAH) in size fractionated diesel particles from a light duty vehicle'. Society of Automotive Engineers 1999-01-3533. [Cited on p. 237]

Westfield W.T. (1971) 'The current and future basis for aircraft air pollution control'. Society of Automotive Engineers 710339. [Cited on p. 201]

Wexler A.S., Potukuchi S. (1998) 'Kinetics and thermodynamics of tropospheric aerosols'. In *Atmospheric Particles*, Harrison R.M., van Grieken R.E. (editors), IUPAC Series on Analytical and Physical Chemistry of Environmental Systems. John Wiley & Sons, Inc., pp. 203–231. [Cited on p. 145]

Whitby R., Gibbs R., Johnson R., Hill B., Shimpi S., Jorgenson R. (1982) 'Real-time diesel particulate measurement using a tapered element oscillating microbalance'. Society of Automotive Engineers 820463. [Cited on pp. 198, 200]

Whitby R.W., Johnson R., Gibbs R. (1985) 'Second generation TEOM filters – diesel particulate mass comparisons between TEOM and conventional filtration techniques'. Society of Automotive Engineers 850403. [Cited on pp. 198, 200]

White J.J., Carroll J.N., Hare C.T. (1991) 'Emission factors for small utility engines'. Society of Automotive Engineers 910560. [Cited on p. 336]

Wickman D.D., Tanin K.V., Senecal P.K., Reitz R.D., Gebert K., Barkhimer R.L., Beck N.J. (2000) 'Methods and results from the development of a 2600 bar diesel fuel injection system'. Society of Automotive Engineers 2000-01-0947. [Cited on pp. 266, 269]

Wieser P., Wurster R. (1986) 'Application of laser-microprobe mass analysis to particle collections'. In *Physical and Chemical Characterization of Individual Airborne Particles*, Spurny K.R. (editor). Ellis Horwood, pp. 251–270. [Cited on p. 209]

Wijetunge R.S., Brace C.J., Hawley J.G., Vaughan N.D., Horrocks R.W., Bird G.L. (1999) 'Dynamic behaviour of a high speed direct injection diesel engine'. Society of Automotive Engineers 1999-01-0829. [Cited on p. 276]

William J., Dupont A., Bazile R., Marchal M. (2003) 'Study of geometrical parameter influence on air/EGR mixing'. Society of Automotive Engineers 2003-01-1796. [Cited on p. 272]

Williams A., McCormick R.L., Hayes R.R., Ireland J., Fang H.L. (2006) 'Effect of biodiesel blends on diesel particulate filter performance'. Society of Automotive Engineers 2006-01-3280. [Cited on p. 259]

Williams P.T., Andrews G.E., Bartle K.D. (1987) 'Diesel particulate emissions: The role of unburnt fuel in the organic fraction composition'. Society of Automotive Engineers 870554. [Cited on p. 123]

Williams R.L., Perez J.M., Griffing M.E. (1985) 'A review of sampling condition effects on polynuclear aromatic hydrocarbons (PNA) from heavy-duty diesel engines'. Society of Automotive Engineers 852081. [Cited on pp. 110, 151, 172, 178]

Literature Cited (Cross-referenced Against the Text)

Wilson M.R., Lightbody J.H., Donaldson K., Sales J., Stone V. (2002) 'Interactions between ultrafine particles and transition metals *in vivo* and *in vitro*'. Toxicology and Applied Pharmacology 184:172–179. [Cited on pp. 368, 376]

Wilson R. (1996) 'Introduction'. In *Particles in Our Air – Concentrations and Health Effects*, Wilson R., Spengler J. (editors). Harvard University Press, pp. 1–14. [Cited on pp. 1, 2, 3, 35, 363, 372, 385]

Wilson R., Spengler J. (1996) 'Policy implications: The national dilemma'. In *Particles in Our Air – Concentrations and Health Effects*, Wilson R., Spengler J. (editors). Harvard University Press, pp. 205–216. [Cited on pp. 363, 375]

Winkler M.F., Parker D.W. (1993) 'Ceramic thermal barrier coatings provide advanced diesel emissions control and improved management of combustion-exhaust system temperatures'. Society of Automotive Engineers 931106. [Cited on p. 93]

Wirojsakunchai E., Schroeder E., Kolodziej C., Foster D.E., Schmidt N., Root T., Kawai T., Suga T., Nevius T., Kusaka T. (2007) 'Detailed diesel exhaust particulate characterization and real-time filtration efficiency measurements during PM filling process'. Society of Automotive Engineers 2007-01-0320. [Cited on p. 301]

Wittmaack K., Strigl M. (2005) 'Novel approach to identifying supersaturated metastable ambient aerosol particles'. Environmental Science and Technology 39:8177–8184. [Cited on p. 215]

Witze P.O. (1989) 'Cycle-resolved multipoint ionization probe measurements in a spark ignition engine'. Society of Automotive Engineers 892099. [Cited on p. 195]

Witze P.O. (2002) 'Real-time measurement of the volatile fraction of diesel particulate matter using laser-induced desorption with elastic light scattering (LIDELS)'. Society of Automotive Engineers 2002-01-1685. [Cited on p. 191]

Witze P.O., Green R.M. (1997) 'LIF and flame-emission imaging of liquid fuel films and pool fires in an SI engine during a simulated cold start'. Society of Automotive Engineers 970866. [Cited on p. 319]

Witze P.O., Chase R.E., Maricq M.M., Podsiadlik D.H., Xu N. (2004) 'Time-resolved measurements of exhaust PM for FTP-75: Comparison of LII, ELPI, and TEOM techniques'. Society of Automotive Engineers 2004-01-0964. [Cited on p. 188]

Witze P.O., Gershenzon M., Michelson H.A. (2005) 'Dual-laser LIDELS: An optical diagnostic for time-resolved volatile fraction measurements of diesel particulate emissions'. Society of Automotive Engineers 2005-01-3791. [Cited on p. 191]

Wolff A., Boulouchos K., Mueller K. (1997) 'A computational investigation of unsteady heat flux through an I.C. engine wall including soot layer dynamics'. Society of Automotive Engineers 970063. [Cited on p. 92]

Wong C. (1988) 'Characterization of metal – soot systems by transmission electron microscopy'. Carbon 26:723–734. [Cited on p. 99]

Wong V.W., Hoult D.P. (1991) 'Experimental survey of lubricant-film characteristics and oil consumption in a small diesel engine'. Society of Automotive Engineers 910741. [Cited on p. 279]

Wong V.W., Yu M.L., Mogaka Z.N., Shahed S.M. (1984) 'Effects of catalytic wire-mesh traps on the level and measurement of heavy-duty diesel particulate emissions'. Society of Automotive Engineers 840172. [Cited on p. 173]

Wong W.Y., Midkiff K.C., Bell S.R. (1991) 'Performance and emissions of a natural gas dual-fueled, indirect injected diesel engine'. Society of Automotive Engineers 911766. [Cited on p. 251]

Wood A.D. (1975) 'Correlation between smoke measurements and the optical properties of jet engine smoke'. Society of Automotive Engineers 751119. [Cited on pp. 202, 203]

Wood K.V., Ciupek J.D., Cooks R.G., Ferguson C.R. (1982) 'Characterization of diesel particulates by mass spectrometry including MS-MS'. Society of Automotive Engineers 821217. [Cited on p. 175]

Woschni G., Huber K. (1991) 'The influence of soot deposits on combustion chamber walls on heat losses in diesel engines'. Society of Automotive Engineers 910297. [Cited on p. 92]

Wright J., Kukla P., Ball A., Gu F., Bann J. (2003) 'A novel electrostatic method of ultrafine PM control suitable for low exhaust temperature applications'. Society of Automotive Engineers 2003-01-0771. [Cited on p. 394]

Wu Y., Clark N.N., Carder D., Thompson G.J., Gautam M., Lyons D.W. (2007) 'Parametric study of 2007 standard heavy-duty diesel engine particulate matter sampling system'. Society of Automotive Engineers 2007-01-0060. [Cited on pp. 7, 160]

Wyatt M., Manning W.A., Roth S.A., D'Aniello M.J., Andersson E.S., Fredholm S.C.G. (1993) 'The design of flow-through diesel oxidation catalysts'. Society of Automotive Engineers 930130. [Cited on p. 295]

X

Xiao Y., Borgnakke C. (1991) 'A stochastic combustion model of direct injection diesel engines'. Society of Automotive Engineers 912354. [Cited on pp. 78, 79]

Xiaoguang X., Xiyan G., Xiancheng W., Chengbin L. (2001) 'After-treatment for reduction of diesel exhaust particulate'. Society of Automotive Engineers 2001-01-3204. [Cited on p. 301]

Xie X., Huang Z., Wang J., Xie Z. (2005) 'Thermal effects on vehicle emission dispersion in an urban street canyon'. Transportation Research: Part D 10:197–212. [Cited on p. 143]

Xu F., Faeth G.M. (2000) 'Structure of the soot growth region of laminar premixed methane/oxygen flames'. Combustion and Flame 121:640–650. [Cited on p. 69]

Xu F., Sunderland P.B., Faeth G.M. (1997) 'Soot formation in laminar premixed ethylene/air flames at atmospheric pressure'. Combustion and Flame 108:471–493. [Cited on pp. 75, 80, 106, 107, 216]

Xu F., Lin K.-C., Faeth G.M. (1998) 'Soot formation in laminar premixed methane/oxygen flames at atmospheric pressure'. Combustion and Flame 115:195–209. [Cited on p. 74]

Xu H., Myers P.S., Uyehara O.A. (1982) 'In-cylinder measurement of particulate number density and size'. Society of Automotive Engineers 820462. [Cited on pp. 86, 190]

Xu S., Clark N.N., Gautam M., Wayne W.S. (2005) 'Comparison of heavy-duty truck diesel particulate matter measurement: TEOM versus traditional filter'. Society of Automotive Engineers 2005-01-2153. [Cited on p. 200]

Y

Yakovleva E., Hopke P.K., Wallace L. (1999) 'Receptor modeling assessment of particle total exposure assessment methodology data'. Environmental Science and Technology 33:3645–3652. [Cited on pp. 358, 359]

Yamaguchi H., Tanabe H., Sato G.T. (1991) 'A study of particulate formation on the combustion chamber wall'. Society of Automotive Engineers 910488. [Cited on pp. 94, 109]

Yamaki Y., Mori K., Kamikubo H., Kohketsu S., Mori K., Kato T. (1994) 'Application of common rail fuel injection system to a heavy duty diesel engine'. Society of Automotive Engineers 942294. [Cited on p. 265]

Yamanaka C., Matsuda T., Ikeya M. (2005) 'Electron spin resonance of particulate soot samples from automobiles to help environmental studies'. Applied Radiation and Isotopes 62:307–311. [Cited on p. 301]

Yamane K., Chikahisa T., Murayama T., Miyamoto N. (1988) 'Measurement of particulate and unburnt hydrocarbon emissions from diesel engines'. Society of Automotive Engineers 880343. [Cited on p. 151]

Yan S., Eddings E.G., Palotas A.B., Pugmire R.J., Sarofim A.F. (2005a) 'Prediction of sooting tendency for hydrocarbon liquids in diffusion flames'. Energy and Fuels 19:2408–2415. [Cited on p. 65]

Yan S., Jiang Y.-J., Marsh N.D., Eddings E.G., Sarofim A.F., Pugmire R.J. (2005b) 'Study of the evolution of soot from various fuels'. Energy and Fuels 19:1804–1811. [Cited on pp. 68, 106, 214]

Yang B., Koylu U.O. (2005) 'Detailed soot field in a turbulent non-premixed ethylene/air flame from laser scattering and extinction measurements'. Combustion and Flame 141:55–65. [Cited on pp. 80, 106]

Yang G., Teague S., Pinketon K., Kennedy I.M. (2001) 'Synthesis of an ultrafine iron and soot aerosol for the evaluation of particle toxicity'. Aerosol Science and Technology 35:759–766. [Cited on p. 101]

Yang J., Kenney T. (2002) 'Some concepts of DISI engine for high fuel efficiency and low emissions'. Society of Automotive Engineers 2002-01-2747. [Cited on pp. 327, 333]

Yao X., Fang M., Chan C.K. (2003) 'The size dependence of chloride depletion in fine and coarse sea-salt particles'. Atmospheric Environment 37:743–751. [Cited on p. 27]

Yao X., Ling T.Y., Fang M., Chan C.K. (2007) 'Size dependence of in situ pH in submicron atmospheric particles in Hong Kong'. Atmospheric Environment 41:382–393. [Cited on p. 394]

Yashiro Y. (1987) 'Reduction of exhaust smoke and carbon deposit at exhaust port in two-stroke gasoline engines'. Society of Automotive Engineers 871216. [Cited on pp. 335, 336]

Yashiro Y., Takahashi K. (1991) 'Evaluation method of exhaust smoke of two-stroke engine oils'. Society of Automotive Engineers 911280. [Cited on p. 336]

Ye S.-H., Zhou W., Song J., Peng B.-C., Yuan D., Lu Y.-M., Qi P.-P. (1999) 'Toxicity and health effects of vehicle emissions in shanghai'. Atmospheric Environment 34:419–429. [Cited on p. 374]

Yeh L.I., Rickeard D.J., Duff J.L.C., Bateman J.R., Schlosberg R.H., Caers R.F. (2001) 'Oxygenates: An evaluation of their effects on diesel emissions'. Society of Automotive Engineers 2001-01-2019. [Cited on pp. 257, 258, 259]

Yezerets A., Currier N.W., Eadler H., Pupuri S., Suresh A. (2002) 'Quantitative flow-reactor study of diesel soot oxidation process'. Society of Automotive Engineers 2002-01-1684. [Cited on p. 169]

Yezerets A., Currier N.W., Kim D.H., Eadler H.A., Epling W.S., Peden C.H.F. (2005) 'Differential kinetic analysis of diesel particulate matter (soot) oxidation by oxygen using a step-response technique'. Applied Catalysis B: Environmental 61:120–129. [Cited on pp. 81, 168, 169, 224]

Yilmaz E., Thirouard B., Tian T., Wong V.W., Heywood J.B., Lee N. (2001) 'Analysis of oil consumption behavior during ramp transients in a production spark ignition engine'. Society of Automotive Engineers 2001-01-3544. [Cited on p. 280]

Yilmaz E., Tian T., Wong V.W., Heywood J.B. (2002) 'An experimental and theoretical study of the contribution of oil evaporation to oil consumption'. Society of Automotive Engineers 2002-01-2684. [Cited on p. 281]

Yoshida E., Nomura H., Sekimoto M. (1986) 'Fuel and engine effects on diesel exhaust emissions'. Society of Automotive Engineers 860619. [Cited on pp. 261, 264]

Yoshihara Y., Kazakov A., Wang H., Frenklach M. (1994) 'Reduced mechanism of soot formation – application to natural gas-fueled diesel combustion'. Twenty-Fifth Symposium (International) on Combustion The Combustion Institute, pp. 941–948. [Cited on pp. 72, 79]

You J.-H., Chiang P.-C., Chang S.-S., Wang-Wuu S. (1996) 'Polycyclic aromatic hydrocarbons (pahs) and mutagenicity in air emissions from the two-stage incineration of polystyrene with various metallic salt additives'. Journal of Hazardous Materials 48:68–82. [Cited on p. 36]

Young D.M., Hickman D.L., Bhatia G., Gunasekaran N. (2004) 'Ash storage concept for diesel particulate filters'. Society of Automotive Engineers 2004-01-0948. [Cited on p. 222]

Yu F. (2001) 'Chemiions and nanoparticle formation in diesel engine exhaust'. Geophysical Research Letters 28:4191–4194. [Cited on pp. 128, 129, 222]

Yu F. (2002) 'Chemiion evolution in motor vehicle exhaust: Further evidence of its role in nanoparticle formation'. Geophysical Research Letters 29:12-1 to 12-4. [Cited on p. 129]

Yu F., Turco R.P., Kärcher B. (1999) 'The possible role of organics in the formation and evolution of ultrafine aircraft particles'. Journal of Geophysical Research 104 D4:4079–4087. [Cited on pp. 128, 129]

Yu J.Z., Huang X.-F., Xu J., Hu M. (2005) 'When aerosol sulfate goes up, so does oxalate: Implications for the formation mechanisms of oxalate'. Environmental Science and Technology 39:128–133. [Cited on p. 36]

Yumlu S.V. (1988) 'Particulate traps: Some progress; some problems'. Society of Automotive Engineers 880347. [Cited on p. 233]

Z

Zabetta E.C., Hupa M., Niemi S. (2006) 'Bio-derived fuels may ease the regeneration of diesel particulate traps'. Fuel 85:2666–2670. [Cited on p. 259]

Zahdeh A.R., Henein N. (1992) 'Diesel engine cold starting: White smoke'. Society of Automotive Engineers 920032. [Cited on p. 120]

Zappia G. (2000) 'Effects of aerosol on modern and ancient building materials'. In *Aerosol Chemical Processes in the Environment*, Spurny K.R., Hochrainer D. (editors). CRC Press LLC, pp. 309–326. [Cited on p. 38]

Zayed J., Hong B., L'Espérance G.L. (1999) 'Characterization of manganese-containing particles collected from the exhaust emissions of automobiles running with MMT additive'. Environmental Science and Technology 33:3341–3346. [Cited on p. 317]

Zebel G. (1966) 'Coagulation of aerosols'. In *Aerosol Science*, Davies C.N. (editor). Academic Press, pp. 31–58. [Cited on p. 78]

Zelenka B., Hohenberg G., Thiel W., Ziegler P.M. (2004) 'Development and testing of a compact and mobile CVS system for passenger car particulate measurement'. Society of Automotive Engineers 2004-01-1444. [Cited on p. 160]

Zelenka P., Kriegler W., Herzog P.L., Cartellieri W.P. (1990) 'Ways toward the clean heavy-duty diesel'. Society of Automotive Engineers 900602. [Cited on pp. 152, 253]

Zelepouga S.A., Saveliev A.V., Kennedy L.A., Fridman A.A. (2000) 'Relative effect of acetylene and PAHs addition on soot formation in laminar diffusion flames of methane with oxygen and oxygen-enriched air'. Combustion and Flame 122:76–89. [Cited on p. 68]

Zellat M., Rolland T., Poplow F. (1990) 'Three dimensional modeling of combustion and soot formation in an indirect injection diesel engine'. Society of Automotive Engineers 900254. [Cited on p. 67]

Zerda T.W., Yuan X., Moore S.M., Leon y Leon C.A. (1999) 'Surface area, pore size distribution and microstructure of combustion engine deposits'. Carbon 37:1999–2009. [Cited on pp. 223, 322]

Zereini F., Alt F., Messerschmidt M., von Bohlen A., Liebl K., Püttman W. (2004) 'Concentration and distribution of platinum elements (Pt, Pd, Rh) in airborne particulate matter in Frankfurt am Main, Germany'. Environmental Science and Technology 38:1686–1692. [Cited on p. 350]

Zervas E., Dorlhène P., Forti L., Perrin C., Momique J.-C., Monier R., Ing H., Lopez B. (2005) 'Comparison between the exhaust particles mass determined by the European regulatory gravimetric method and the mass estimated by ELPI'. Society of Automotive Engineers 2005-01-2147. [Cited on p. 221]

Zhang H., Chen J., He Y., Xue X., Peng S. (1998) 'The preparation of carbon-coated iron nanocrystals produced from Fe_2O_3-containing composite anode in arc discharge'. Materials Chemistry and Physics 55:167–170. [Cited on p. 231]

Zhang J., Megaridis C.M. (1994) 'Iron/soot interaction in a laminar ethylene nonpremixed flame'. Twenty-Fifth Symposium (International) on Combustion, The Combustion Institute, pp. 593–600. [Cited on p. 101]

Zhang J., Megaridis C.M. (1994) 'Soot suppression by ferrocene in laminar ethylene/air nonpremixed flames'. Combustion and Flame 105:528–540. [Cited on p. 101]

Zhang K.M., Knipping E.M., Wexler A.S., Bhave P.V., Tonnesen G.S. (2005) 'Size distribution of sea-salt emissions as a function of relative humidity'. Atmospheric Environment 39:3373–3379. [Cited on p. 27]

Zhang L., Tsurushima T., Ueda T., Ishii Y., Itou T., Minami T., Yokota K. (1997) 'Measurement of liquid phase penetration of evaporating spray in a DI diesel engine'. Society of Automotive Engineers 971645. [Cited on pp. 266, 267, 278]

Zhang M., Yu J., Xu X. (2005) 'A new flame sheet model to reflect the influence of the oxidation of CO on the combustion of a carbon particle'. Combustion and Flame 143:150–158. [Cited on p. 79]

Zhang Q., Stanier C.O., Canagaratna M.R., Jayne J.T., Worsnop D.R., Pandis S.N., Jimenez J.L. (2004) 'Insights into the chemistry of new particle formation and growth events in Pittsburgh based on aerosol mass spectrometry'. Environmental Science and Technology 38:4797–4809. [Cited on pp. 128, 361, 362]

Zhang Y., Nishida K. (2003) 'Vapor/liquid behaviors in split-injection D.I. diesel sprays in a 2-D model combustion chamber'. Society of Automotive Engineers 2003-01-1837. [Cited on pp. 269, 270]

Zhao H., Ladommatos N. (1998) 'Optical diagnostics for soot and temperature measurement in diesel engines'. Progress in Energy and Combustion Science 24:221–255. [Cited on p. 84]

Zhao H., Ladommatos N. (2001) *Engine Combustion and Diagnostics*. Society of Automotive Engineers. [Cited on p. 84]

Zhao J., Turco R.P. (1995) 'Nucleation simulations in the wake of jet aircraft in stratospheric flight'. Journal of Aerosol Science 26:779–795. [Cited on p. 128]

Zhao X., Ren M., Liu Z. (2005) 'Critical solubility of dimethyl ether (DME) + diesel fuel and dimethyl carbonate (DMC) + diesel fuel'. Fuel 84:2380–2383. [Cited on p. 257]

Zheng H., Keith J.M. (2004) 'Ignition analysis of wall-flow monolith diesel particulate filters'. Catalysis Today 98:403–412. [Cited on p. 302]

Zhou Z.-Q., Ahmed T.U., Choi M.Y. (1998) 'Measurement of dimensionless soot extinction constant using a gravimetric sampling technique'. Experimental and Thermal Fluid Science 18:27–32. [Cited on p. 182]

Zhu X.L., Gore J.P. (2005) 'Radiation effects on combustion and pollutant emissions of high-pressure opposed flow methane/air diffusion flames'. Combustion and Flame 141:118–130. [Cited on p. 79]

Zhu Y., Hinds W.C. (2005) 'Predicting particle number concentrations near a highway based on vertical concentration profile'. Atmospheric Environment 39:1557–1566. [Cited on p. 143]

Zhu Y., Zhao H., Ladommatos N. (2003) 'Computational study of the effects of injection timing, EGR and swirl ratio on a HSDI multi-injection diesel engine emission and performance'. Society of Automotive Engineers 2003-01-0346. [Cited on p. 277]

Zhu Y., Hinds W.C., Shen S., Sioutas C. (2004) 'Seasonal trends of concentration and size distribution of ultrafine particles near major highways in Los Angeles'. Aerosol Science and Technology 38:5–13. [Cited on pp. 360, 361]

Zhu Y., Hinds W.C., Krudysz M., Kuhn T., Froines J., Sioutas C. (2005) 'Penetration of freeway ultrafine particles into indoor environments'. Journal of Aerosol Science 36:303–322. [Cited on p. 359]

Zhu Y., Kuhn T., Mayo P., Hinds W.C. (2006) 'Comparison of daytime and nighttime concentration profiles and size distributions of ultrafine particles near a major highway'. Environmental Science and Technology 40:2531–2536. [Cited on p. 361]

Zhu Y., Eiguren-Fernandez A., Hinds W.C., Miguel A.H. (2007) 'In-cabin commuter exposure to ultrafine particles in Los Angeles Freeways'. Environmental Science and Technology 41:2138–2145. [Cited on p. 359]

Ziejewski M., Goettler H.J., Cook L.W., Flicker J. (1991) 'Polycyclic aromatic hydrocarbons emissions from plant oil based alternative fuels'. Society of Automotive Engineers 911765. [Cited on p. 259]

Zielinska B., Sagebiel J., Arnott W.P., Rogers C.F., Kelly K.E., Wagner D.A., Lighty J.S., Sarofim A.F., Palmer G. (2004) 'Phase and size distribution of polycyclic aromatic hydrocarbons in diesel and gasoline vehicle emissions'. Environmental Science and Technology 38:2557–2567. [Cited on pp. 117, 237]

Zotin F.M.Z., Gomes O. d F. M., Oliveira C.H. d, Neto A.A., Cardoso M.J.B. (2005) 'Automotive catalyst deactivation: Case studies'. Catalysis Today 107–108:157–167. [Cited on p. 350]

Zufall M.J., Davidson C.I. (1998) 'Dry deposition of particles'. In *Atmospheric Particles*, Harrison R.M., van Grieken R.E. (editors), IUPAC Series on Analytical and Physical Chemistry of Environmental Systems. John Wiley & Sons, Inc., pp. 425–473. [Cited on pp. 22, 26]

Index

Absorption, light, *see* Light
Accumulation mode
   atmospheric aerosol   25
   brake wear   345
   definition   43–4
   in dilution tunnel   148
   effective density   220
   in exhaust   54
   survival in atmosphere   6
Acetylene
   soot inception   68
   surface growth   73
Acid dew point   125
Adsorption   20
   organic particulate   114–15
   photoelectric effect   194
Aftertreatment, *see under* Catalyst; Diesel particulate filter (DPF)
Agglomerates
   in atmospheric aerosol   33
   definition   44
   fractality   12, 77
   relation to spherule   216–17
Agglomeration   20
   charge effects   78, 228
   coalescent and noncoalescent   77
   in exhaust plume   144
   in lubricant   288
   making particles bigger   394
   Smoluchowski eqn.   21, 76
   turbulence   78

Aggregate, *see* Agglomerates
Agrochemical aerosols   30–1
Aircraft exhaust   128
Aitken nucleus   19, 26
Alkanes
   characterisations   235
   fuel formulation   256
   storage in exhaust   139
   survivability   108
Ames bioassay
   dose-response curve   180
   index   180
   indirect and direct mutagenicity   180
   NPAH and PAH   181
Ammonia
   deNO$_X$ catalysts   299
   motor vehicles   37
   natural sources   37
   nitrate   36, 46
   sulphate   36
Animal test (in vivo)
   epigenetic and genetic   378
   excessive exposures   378
   exposure to carbon black   379
   exposure to diesel particles   378–9
   exposure to sulphuric acid   379
   humans   381–2
   intracheal instillation   377
   upper respiratory tract   377
Arctic haze   41
Aromaticity   254, 256

Particulate Emissions from Vehicles   P. Eastwood
© 2008 John Wiley & Sons, Ltd

Asbestos 31
Ash particulate
  animal exposures 380
  chemical composition 104, 231
  chlorine 95
  coal 98
  definition 47
  density, bulk 220
  diesel particulate filter 303, 304
  flame suppressants 98
  fly ash 98
  fuel-borne elements 96, 261–2
  incombustible 232
  iron 101–4, 317
  leaded gasoline 315–16
  lubricant 283
  magnesium 95
  manganese 317
  nitration 112
  silicon 95
  storage in catalysts 296
  thermodynamics 97
  wear 51, 53
  zinc 95
Asteroid impact 41
Asthma 40
Attachment (to surfaces) 18

Bacteria 28, 29
Barium (fuel additive) 97, 98–9, 261
BET model 114, 222
Binary homogeneous nucleation (BHN) 126
Bioassays (in vitro)
  DNA damage 376
  oxidative stress 377–8
  types of 375
Biogenic aerosol 29
Biological aerosol 27–9
Bipolar charging 225
Blowdown 136
  charge sensing 196
Boltzmann charge distribution 227
Brake wear 344–6
  loss patterns 345
  particle composition 345
Bronchitis 40
Bubble bursting mechanism 27

Cancer 40
Candles 38
Carbon
  aciniform 47
  black carbon 47
  carbon black (manufacture) 62
  carbonate 32
  critical air-fuel ratio 65
  elemental carbon 32, 47
  organic 32
Carbon-hydrogen ratio
  carbonaceous fraction 229
  organic fraction 233
  thermogravimetry 170
  young spherules 107
Carbonaceous particulate
  air-fuel ratio 65, 320
  animal exposures 378–9
  catalytic formation 296
  catalytic oxidation 295
  climate 41
  damage to buildings 38
  definition 47
  density, bulk 219
  diesel particulate filter 302
  exhaust gas recirculation 272–3
  flame ionisation 197
  from lubricant 282–3
  fullerenes 71
  HACA mechanism 69
  heat release 66
  hydrophobicity 126
  ignition delay 66
  incandescence 62
  ions 71
  light absorption 192
  oxidants 79, 80
  oxygenates 257–8
  premixed burn 86
  pyrolysis 68–70
  soot-formation window 292
  soot-induced wear 289–90
  stabilomers 70
  thermodynamics 63
Cardiovascular illness 40–1
Catalyst
  ash storage 296
  attrition of 349–51
  carbonaceous storage 296
  deactivation 296
  formation of sulphuric acid 124
  hydrocarbon storage 294
  mutagenicity 297
  particle storage 140
  precious metal losses 350–1
  sulphate make 294–5, 317–18
  sulphate-organic trade-off 295

Index

sulphate storage 295
  three-way 326
Cenospheres 98
Cerium 103
Charring 169
Chase experiments 142
Chemiluminescence 87
Chemisorption, *see* Adsorption
Chromatography 175
  fuel-lubricant partitioning 178–9
Cigarette smoking 398
Climate (change) 41
Cloud formation 25
  cloud condensation nuclei 25, 26, 27, 41
  deposition 26
  ice formation nuclei 25
CNC 189–90
Coagulation, *see* Agglomeration
Coarse mode
  atmospheric aerosol 25, 40
  definition 42–3
  in exhaust 54
  formation 42
  isokinetic sampling 159
Combustion aerosols 33, 35–6
  biomass 33, 97
  climate 40–1
  coal 33, 98
  coke 43
  municipal, industrial, medical 35, 98
  open burning 39
Compensation culture 398
Condensation 18
  organic particulate 112–13
Condensation nuclei counter (CNC), *see* Light
Condensation particle counter (CPC)
Contact electrification 227
Contrails (condensation trails) 41
Cunningham slip correction 16
Cylinder dumping 84
Cylinder sampling 84, 109

Deliquescence 19
Demography
  alternative combustion systems 193
  catalysts 140–1, 298, 299
  diesel engines 241–4
  diesel particulate filter 242–3, 303–4
  dilution tunnel 149
  direct-injection gasoline 231
  EGR 274

exhaust plume 144
fuel additives 101–4, 261
injection pressure 268
leaded gasoline 316
lubricant 285
oxygenates 259–60
port-injection gasoline 323
sulphur 253–4
transfer line 146–7
two-stroke engines 337
Density
  bulk 220–1
  difficulties in measuring 221
  effective 221
  as a function of engine operation 221
  as a function of size 221–2
  incombustible ash 222
  surface area 223
Dew point 18, 52
Diameter, particle, *see* Size, particle
Diesel particulate filter (DPF)
  demography 241–3, 392–3
  evaluating efficiency of 151
  filtering mechanisms 301
  fuel additives 262
  ignition temperature 301–2
  regeneration methods 302
  similarity to dilution tunnel filter 150
  sulphate quantification 174
  toxicology 238, 305–6
  types of 300
Diffusion 16
  battery 158
  in transfer line 146
Dilution profile 163
Dilution ratio 51
  definition of 162
  in exhaust plume 141–2
  organic fraction 113
Dilution tunnel 51
  accumulation mode 148
  adiabaticity 161
  conditioning of air 161
  constant volume sampling (CVS) 160
  full- and partial-flow 160
  heat transfer characteristics 162
  interplay with saturation 148
  kinetic aspects 148–9
  laminar flow 164
  nucleation 149–50
  tasks of 160–1
  turbulent flow 164–5
Dimethyl sulphide (DMS) 27

Dioxins   36
Diseases
    allergens   364
    asthma   363, 370
    atherosclerosis   374
    blood clotting   369–70, 374
    bronchitis   370
    cancer   362–3, 369
    chronic obstructive pulmonary disease (COPD)   370, 374
    lung deposition   365
    lung function   363
    mortality   363
    myocardial infarctions   374
    PAH-DNA adducts   369
    respiratory, cardiovascular   363, 369–70
    rhinitis   364
Displacement   136
Drag force   16
Dry deposition   26
Dust   23

Efflorescence   19
Ejector diluter   164
Electrical charge
    charge cloud   195–6
    demography of   225–6
    indigenous   225–8
    mechanisms   225–6
    particle-sizing   185–6
    relation to smoke   197
Electrical low pressure impact (ELPI), *see* Impactor
Electrostatic attraction   16
ELPI   183–5
Epidemiology
    confounding factors   371–2
    lung cancer   371, 373
    occupational health   373
    other illnesses   374
    public health   373–4
    respiratory and cardiovascular   373
    risk   372
    threshold   371
Epithelium (lung)   39
Evaporation   19
    light scattering   191
    via LII   188
Exhaust gas recirculation (EGR)
    alternative combustion systems   292
    demography   274
    engine management   272
    engine wear   289

fouling of cooler and valve   274
gasoline engine   325
NO_X-soot trade-off   271–2
soot promoting mechanisms   272–4
Exhaust systems
    acid corrosion   349
    iron   349
    rust and scale   349
Extinction, light, *see* Light
Extraction, filter
    falsifications   171
    Soxhlet   171–2
    sulphates   171–3
    super- and subcritical   172

Ferrocene
    combustion   96
    gasoline   317
    nucleation of iron   103
    soot suppression   100–1
Filter, dilution tunnel
    acid-mediated falsifications   152
    blow-by   151, 165
    conditioning   167
    drawbacks   165
    extraction   171–4
    gas-particle partitioning   151
    inconstancy in characteristics   150
    nitration (PAH to NPAH)   152
    types   165
    vapour-phase trap   151, 165
    weighing   166–7
Fireworks   31
Fly ash   98
Fog   23, 38
Fractality   12
    agglomerate morphology   217–18
    spherule agglomeration   77–8
Fuel
    additives   261–2, 315–17
    aromaticity   254, 256
    ashing elements   96, 98
    desulphurisation   252
    distillation curves   263
    engine calibration   251, 265
    Fischer–Tropsch   251
    gasoline   325, 333
    partitioning from lubricant   178–9
    sulphur-bearing compounds   123
    volatility   263
    water-diesel emulsions   251
    white smoke   121

Index

Fuel injection
  atomisation 265–6, 268, 319
  demography 268
  direct, gasoline 326
  end of injection 266
  flow in nozzle 265
  'hang up' 320
  multiple injections 269
  port injection 318–19
  soot-suppression mechanisms 267–8, 269–70
Fullerenes 71
Fume 23

Gas-to-particle conversion
  ash in relation to soot 100–1
  diesel particulate filter 302, 305
  in exhaust 52–3
  hydrocarbons 118
  kinetic considerations 116
  leaded gasoline 315
  onto transfer line walls 146–7
  rise-and-fall characteristic 115
  soot (spherules) 71–2
Gasification 224
  see also Thermogravimetry
Gaussian plume dispersal 143
Geogenic aerosol 27–30
Global dimming 41–2
Gravitational settling 17, 25
Greenfield gap 26

HACA mechanism
  soot inception 69
  surface growth 74
Haze 38
Head vortex 88
Henry's law 19
Hydrocarbon
  sooting tendency 254
  storage 120
Hygroscopicity 19

Ignition delay 48
  correlation to smoke 65
  premixed burn 87
Impactor
  bounce 184
  plugging 184
  principle 184
  volatilisation 185
Incandescence 62, 87
  laser-induced 186–8
  in late burn 91

Induction
  air density 278
  airpath management 276
  charge temperature 278
  swirl 277
  'turbo-lag' 276
  turbulence 277
Industrial aerosols 31–2
Inertial impaction 17
  particle sizing 183–5
Interception 16
In-use compliance 389
Ion-mediated nucleation 128
Ionisation 193
Ions
  acquisition by particles 195, 225–6
  diffusion charging 195
  nucleation in exhaust 128
  soluble 173
  soot formation 70–1
Isokinetic sampling 17, 158

Kelvin diameter 19
Knudsen number 15
  in engine 76

Langmuir model 114
Laser ablation mass spectrometry 209
Laser imaging 84
Leaded gasoline 316–17
Leidenfrost effect 329
Lift-off length 88–9
Light
  absorption 191
  atmospheric aerosols 37
  fits to Mie theory 191–2
  Mie scattering, limitations 189
  nephelometers 190
  particle enumeration 190
  photoacoustic effect 191
  Rayleigh scattering 37
  refractive index 182
  smokemeters 200–4
  transmission (optical window) 94
  transmittance 202
Lubricant
  blowdown 136
  carbon-hydrogen ratio 233
  consumption versus combustion 280
  demography 285
  escape mode (vapour, liquid) 281
  escape routes 279, 281

Lubricant (*Continued*)
  fuel-oil interactions   282
  maintaining oil quality   286
  partitioning from fuel   178–9, 235–6
  quantification of soot   286
  reasons for soot entry   286
  reformulation   282
  rheology   287–8
  soot agglomeration   288
  soot-induced wear   289–90
  tracer studies   283–4
  transient escape   280

Methylcyclopentadienyl manganese tricarbonyl (MMT)   317
Mist   23
Morphology   12
  density   222
  in exhaust   137
  fractality   217–18
  gasoline particulate   322
  organic particulate   218–19
  radius of gyration   78, 217
  spherule agglomeration in engine   77
  surface area   223
Mucociliary escalator   39–40
Mutagenicity
  aftertreatment   238
  Ames bioassay   179–81
  catalysts   297
  diesel particulate filter   306
  fuel formulation   257
  gasoline engines   323
  index   180
  indirect and direct   180
  nitration   112
  oxygenates   260
  particle size   238
  subfractions   239–40
  synergisms   181–2
  two-stroke engines   337
  vehicle operation   238–9

Nagle and Strickland Constable (NSC)   82
Nanoparticles   26
  animal exposures   380–1
  chemical composition   362, 392
  electronic states   395–6
  flame synthesis   98
  following lung deposition   393
  lung deposition   365
  nucleation events   361, 393

public exposure   360–2
secondary versus primary   361–2, 390
tyre wear   347
very, very small particle   395
Nitrate particulate
  ammonia   37, 46
  atmospheric aerosol   36
  connection to sulphate particulate   125
  definition   46
Nitration (PAH to NPAH)   110–11, 238
  catalysts   297–8
  diesel particulate filter   305–6
  on dilution tunnel filter   152, 180
  metals   112
  mutagenicity   112
Nitro-PAH (NPAH)
  Ames bioassay   181
  characterisations   235
  quantification   177
  relation to PAH   110–11
Nuclear winter   41
Nucleation   19
  aircraft exhaust   128
  ash   100, 103, 104
  atmosphere   26
  charge   228
  direct-injection gasoline   331
  'events'   361
  in exhaust plume   143
  heterogeneous   19
  homogeneous   19
  implications for measurement   55
  meteorology   144
  oxygenates   260
  secondary nanoparticles   361
  soot inception   68
  sulphuric acid   126, 253
  in transfer line   147
Nucleation mode
  ash   261
  atmospheric aerosol   25
  catalysts   298
  definition   43
  diesel particulate filter   243–4, 303, 304
  in dilution tunnel   149–50
  effective density   220
  in exhaust   54

Occlusions   101, 230–1
Organic intermediates   105
Organic particulate
  adsorption   114–15
  animal exposures   379

Index   491

atmospheric aerosol   34–5
  boundary with HC   117–18, 175, 197–8
  boundary with soot   105
  catalytic oxidation   294
  condensation   112–13
  definition   45
  density, bulk   220
  diesel particulate filter   303, 304
  direct and indirect   50
  extraction   171–2
  flame ionisation   197
  fuel-lubricant partitioning   178, 235–6
  functional groups   233
  gasoline engine   321
  growth agents   128
  intermediates   105
  lubricant   50
  microtransport   116–17
  oxygenates   257–8
  particle size   237
  spiking experiments   178
  subfractions   175–7, 234
  survivability   108
  thermal transition   108
  vapour-particle partitioning   236
Outgassing   22
  in exhaust deposits   139
  from transfer line walls   147
Oxidation
  combusting plume   89
  late burn   90–1
  mixing of soot and oxygen   83, 91
  rate control   80
  rise-and-fall characteristic   85–6, 328
  soot suppressing additives   99, 100–1
  wall deposits   93
Oxidation catalyst, *see under* Catalyst
Oxygenates
  demography   259–60
  desirable properties   257
  organic-carbonaceous ratio   257
  PAH   259
  sequestration theory   258
  soot-suppressing mechanisms   259
  types of   256–7
Ozone   37, 38

Particulate
  ageing in atmosphere   145
  boundary problems   47–8
  boundary with HC   48, 117–18, 119
  definition of   8, 44

fractionation   167
fractions, amounts   47
lubricant contribution   50
wet and dry   44, 109
Petrogenic compounds   35, 70, 105
Photochemical smog, *see* Smog
Physisorption, *see* Adsorption
Pilot injection   270
Plant vitality   38
Plume stem   89
PM2.5 and PM10
  brake wear   345
  deposition curve (definition)   40
  public exposure   358
  relation to emissions legislation   397
  relation to vehicles   55
  road wear   343
Pneumoconiosis   40
Pollen   28
Polycyclic Aromatic Hydrocarbons (PAH)
  adducts   369, 375
  atmospheric aerosol   35
  atmospheric partitioning   117
  cancer   369
  catalytic oxidation   294
  cetane number   264
  characterisation   235
  chemiluminescence   87
  in food   35, 40
  fuel formulation   256
  gas-to-particle conversion   114, 117
  gasoline engines   322
  nitration (PAH to NPAH)   110–11, 261
  organic particulate   106
  oxygenates   259–60
  photoelectric effect   194
  proto-particles   106–7
  quantification   177
  soot formation   68–70
  source apportionment   35
  sulphur   253
  surface growth   73–4
  survivability   108
Polydispersity   12
Polyurethane foam (PUF)   165
Pool-fire combustion   328
Porosity   223
Porous tube diluter   164
Post injection   270
Premixed zone/flame   88–9
Public exposure
  gasoline versus diesel   358
  heavy-duty and light-duty diesels   358

Public exposure (*Continued*)
 indoors and outdoors   358, 359
 infiltration into buildings   359
 nanoparticles   361
 occupational   360
 particle ageing   360
 PM2.5 and PM10   359
 in school buses   359–60
 in vehicle compartments   359
Pyrogenic compounds   35, 70, 105
Pyrolysis
 coke   43
 in diesel engine   49, 63
 in exhaust   110
 soot formation   68–70
 on walls   109

Quartz crystal microbalance (QCM)   199

Re-entrainment   22
 duty cycle   137–8
 in engine   136
 in exhaust   137
Respiratory tract   39
 alveoli   366
 animals (various)   377
 antioxidant defence   368
 bioavailability   368
 defence mechanisms   365–6
 deposition curve   39–40, 364–5
 factors controlling deposition   365
 inflammation   367–8
 interstitialisation   368
 macrophages   366
 nanoparticles   368
 overload   367
 particle retention   366–7
 reactive oxygen species (ROS)   368
 translocation   368, 381
Road salt   95
Road wear   342
 particle composition   344
Rotating disc diluter   164

Saturation
 ash particulate   103
 definition   18
 in dilution tunnel   148
 implications for measurement   55
 long-term trends   391–2
 organic particulate   113
Scanning mobility particle sizer (SMPS)   185

Scattering, light, *see* Light
Sea salt   27
Secondary organic aerosol (SOA)   29, 33, 37
Signal-to-noise ratios   387
Silicosis   40
Size, particle
 biological aerosol   28
 density function   220–2
 deposition in lung   40
 distributions, formulae for   12–14
 instrumentation for   182
 Knudsen number   16
 spectrum   26
 spherules   43, 216
 transport   17
 typical distribution   54
Smog   23
 London   36
 Los Angeles   36
Smoke, colour   119–20, 336
Smoke detection (by charge)   227
Smoke point   65
Smokemeters
 correlations to other instruments   203
 opacimeter   202
 Ringelmann cards   201
 signal levels   203
 spotmeter   201–2
 white smoke   204
Smoluchowski eqn.   21
SMPS   185–6
Soluble organic fraction (SOF)   45, 168
Soot   47
 see also Carbonaceous particulate
Soot lamp   65
Source apportionment   25, 35, 194
Soxhlet extraction
 choice of solvent   172
 optimisation   171–2
 sulphates   172
Spherules
 assimilation of organics   106–7
 in atmospheric aerosol   33
 carbon-hydrogen ratio   107
 crystallites   214
 definition   43–4
 formation   72
 HACA mechanism   74
 internal oxidation   81
 nuclei   214
 oxygenates   259
 platelets   214
 relation to agglomerate   216–17

Index

size restriction 73–4
size via LII 187–8
sphericity 215–16
surface recession 80, 86
turbostratic structure 214
Stabilomers 70
Sticking probability 139
Sulphate particulate 36
  ammonia 37
  building damage 38–9
  characterisation 237–8
  climate 41
  connection to nitrate particulate 125
  definition 45–6
  density, bulk 220
  diesel particulate filter 303
  extraction 172–4
  gasoline engine 322
  lubricant 284
  manufacture by catalyst 123, 294–5, 317–18
  storage by catalysts 125–6, 297–8, 318
Sulphur
  carbonaceous fraction 229
  combustion chemistry 123, 252–3
  demography 253–4
  in fuel 252
  organic particulate 252, 253
Sulphur dioxide
  conversion to acid in atmosphere 36, 124
  conversion to acid in exhaust 53, 122–3
  inhalation 39
  oxidation on buildings 38
  volcanism 30
Sulphur trioxide
  acid dew point 125
  exhaust gas recirculation 271
Sulphuric acid
  animal exposures 379–80, 381
  decomposition 170
  dilution tunnel filter 152
  exhaust system corrosion 349
  formation in exhaust 53
  metal sulphation 97
  nucleation 126–8, 393
  quantification of water 174
  relation to sulphate particulate 45
  scrubbing effect 253
Surface area
  active surface 194
  BET method 222–3
  diesel particulate filter 302

diffusion charging 195
photoelectric effect 193–4
Surface tension 18, 219

Tapered element oscillating microbalance (TEOM) 198–200
Thermionic emission 225
Thermoelectric effect 228
Thermogravimetry 168
  falsifications 169
  protocols for 168, 169
Thermophoresis 16
  in catalysts 140
  in dilution tunnel 148
  in EGR circuit 274
  in engine 92
  in transfer line 146
Threshold sooting index 65
Two-stroke engines
  demography 337
  emission mechanisms 335
  mutagenicity 337
  oil-fuel ratio 335
  oil reformulation 336
  storage in exhaust 336
Tyre wear
  nanoparticles 347
  particle composition 347
  quantity lost 346

Ultrafine particles 26

van der Waals forces 18, 44
Vapour pressure 18
Viruses 27
Volatile organic fraction (VOF) 45, 168
Volcanism 29–30

Water
  atmospheric aerosol 23, 41
  deposition 26
  distinction from particulate 6, 44
  in exhaust 52
  nitrates 46
  sulphate extraction 274
  sulphuric acid 45, 174
Wet deposition 26
Wind-blown dust 30, 41
Wind tunnels 142
Work function 193

XAD-2 165